DENDRITIC CELLS IN FUNDAMENTAL AND CLINICAL IMMUNOLOGY

Volume 3

ADVANCES IN EXPERIMENTAL MEDICINE AND BIOLOGY

Recent Volumes in this Series

Volume 408
TOWARD ANTI-ADHESION THERAPY FOR MICROBIAL DISEASES
Edited by Itzhak Kahane and Itzhak Ofek

Volume 409
NEW HORIZONS IN ALLERGY IMMUNOTHERAPY
Edited by Alec Sehon, Kent T. HayGlass, and Dietrich Kraft

Volume 410
FRONTIERS IN ARTERIAL CHEMORECEPTION
Edited by Patricio Zapata, Carlos Eyzaguirre, and Robert W. Torrance

Volume 411
OXYGEN TRANSPORT TO TISSUE XVIII
Edited by Edwin M. Nemoto and Joseph C. LaManna

Volume 412
MECHANISMS IN THE PATHOGENESIS OF ENTERIC DISEASES
Edited by Prem S. Paul, David H. Francis, and David A. Benfield

Volume 413
OPTICAL IMAGING OF BRAIN FUNCTION AND METABOLISM 2: Physiological Basis and Comparison to Other Functional Neuroimaging Methods
Edited by Arno Villringer and Ulrich Dirnagl

Volume 414
ENZYMOLOGY AND MOLECULAR BIOLOGY OF CARBONYL METABOLISM 6
Edited by Henry Weiner, Ronald Lindahl, David W. Crabb, and T. Geoffrey Flynn

Volume 415
FOOD PROTEINS AND LIPIDS
Edited by Srinivasan Damodaran

Volume 416
PLATELET-ACTIVATING FACTOR AND RELATED LIPID MEDIATORS 2: Roles in Health and Disease
Edited by Santosh Nigam, Gert Kunkel, and Stephen M. Prescott

Volume 417
DENDRITIC CELLS IN FUNDAMENTAL AND CLINICAL IMMUNOLOGY, Volume 3
Edited by Paola Ricciardi-Castagnoli

A Continuation Order Plan is available for this series. A continuation order will bring delivery of each new volume immediately upon publication. Volumes are billed only upon actual shipment. For further information please contact the publisher.

DENDRITIC CELLS IN FUNDAMENTAL AND CLINICAL IMMUNOLOGY

Volume 3

Edited by

Paola Ricciardi-Castagnoli
University of Milan
Milan, Italy

SPRINGER SCIENCE+BUSINESS MEDIA, LLC

Library of Congress Cataloging-in-Publication Data

Dendritic cells in fundamental and clinical immunology, Volume 3 / edited by Paola Ricciardi-Castagnoli.
p. cm. -- (Advances in experimental medicine and biology ; v. 417)
"Proceedings of the 4th International Symposium on Dendritic Cells in Fundamental and Clinical Immunology, held October 5-10, 1996, in Venice, Italy"--T.p. verso.
Includes bibliographical references and index.
ISBN 978-1-4757-9968-2 ISBN 978-1-4757-9966-8 (eBook)
DOI 10.1007/978-1-4757-9966-8
1. Dendritic cells--Congresses. 2. Dendritic cells--Therapeutic use--Congresses. 3. Immunopathology--Congresses. I. Ricciardi-Castagnoli, Paola. II. International Symposium on Dendritic Cells in Fundamental and Clinical Immunology (4th : 1996 : Venice, Italy) III. Series.
[DNLM: 1. Dendritic Cells--congresses. 2. Dendritic Cells--immunology--congresses. W1 AD559 v.417 1997 / QW 568 D391 1997]
QR185.8.D45D46 1997
616.07'9--dc21
DNLM/DLC
for Library of Congress 97-8041
CIP

Proceedings of the 4th International Symposium on Dendritic Cells in Fundamental and Clinical Immunology, held October 5 – 10, 1996, in Venice, Italy

ISBN 978-1-4757-9968-2

Originally published by Plenum Press, New York in 1997
Softcover reprint of the hardcover 1st edition 1997

http://www.plenum.com

10 9 8 7 6 5 4 3 2 1

PREFACE

These proceedings contain selected contributions from the participants to the Fourth International Symposium on Dendritic cells that was held in Venice (Lido) Italy, from October 5 to 10, 1996. The symposium was attended by more than 500 scientists coming from 24 different countries.

Studies on dendritic cells (DC) have been greatly hampered by the difficulties in preparing sufficient cell numbers and in a reasonable pure form. At this meeting it has been shown that large quantities of DC can be generated from precursors in both mice and humans, and this possibility has enormously encouraged studies aimed to characterize DC physiology and DC-specific genes, and to employ DC therapeutically as adjuvants for immunization. The possibility of generating large numbers of autologous DC that can be used in the manipulation of the immune response against cancer and infectious diseases has tremendously boosted dendritic cell research and the role of DC in a number of medical areas has been heatedly discussed.

The availability of purified preparations of large numbers of DC is rapidly expanding our knowledge about these cells. The ontogeny of DC, their diversity and the fine mechanisms of antigen uptake and processing, and the identification of signal transduction pathways involved in DC activation are issues that are now being addressed more easily, and that will likely provide the information necessary for an optimal use of DC in immunotherapy. Several groups are also actively investigating the presence of DC-specific genes with the intent of identifying DC-specific markers. These studies may in perspective allow the identification of molecules suitable for a selective manipulation of DC functions.

One of the most important goals of DC research is the development of DC-based strategies for enhancing immune responses against tumors and infectious agents. Various studies in animal models have clearly shown that DCs pulsed with tumor antigens *in vitro* and then reinjected *in vivo* induce protective immune responses that block tumor growth. In addition, work in humans has confirmed the value of such approach and the final goal would be to use DC in the treatment of minimal residual disease after therapy of primary tumors. Two major challenges have been discussed in this respect: (i) the generation of sufficient numbers of DC to be reinfused into the patient, and (ii) how to properly load DC with antigens in order to promote a cytotoxic T cell response without inducing autoimmune responses.

In undamaged tissues DC are very capable in antigen capture and processing, but not efficient at activating T cells. When a signal of danger occurs in the tissue (bacterial and viral infections, necrosis, etc.), DCs rapidly initiate a maturation process that render them very potent APCs. This process is characterized by reduced antigen capture capacity and increased surface expression of MHC and co-stimulatory molecules, and can be reproduced *in vitro* by exposing immature DC to bacterial products and cytokines such as tumor necrosis factor (TNF-α) and interleukin 1β (IL-1β). The maturation of DC is completed upon interaction with T cells, and is characterized by loss of phagocytic capacity, further expression of co-stimulatory molecules and synthesis of cytokines, including TNF-α, IL-1β, and IL-12. This final maturation of DC is mediated by surface molecule interactions (e.g., CD40L/CD40) and by T cell-derived cytokines such as interferon-γ. The availability of long-term DC cultures has greatly enhanced our understanding of this phenomenon and will soon allow us to trace DC maturation at the molecular level.

An important characteristic of DCs is their motility and migratory capacity, which enables them to move from peripheral tissues to lymphoid organs where the pool of quiescent T cells recirculate. The selective immigration of DCs, their permanence in a given tissue as well as the migratory capacity are tightly regulated events. Chemotactic factors released by the target tissue and surface adhesins are involved in these processes. DC-activating stimuli (e.g., bacterial products, IL-β and TNF-α) induce rearrangements of the actin-based cytoskeleton, and are the main factors driving emigration of DC from peripheral tissues such as the skin and the intestine. Migration may be a relevant target for the manipulation of DC-driven immune responses: blocking the regression of DC from transplanted organs may prevent the initiation of immune responses against allografts and thus prolong graft survival.

The fifth meeting on Dendritic Cells will be held in Pittsburgh (USA) in 1998, organized by Dr. Michael Lotze from the Medical Centre of the University of Pittsburgh. There are no doubts that it will attract many immunologists, as well as scientists from other disciplines, including many molecular and cell biologists that already attended the Venice meeting.

The Organizing Committee

Dr. Paola Ricciardi-Castagnoli
CNR Center of Cytopharmacology

Dr. G. Girolomoni
IDI, IRCCS

Dr. A. Lanzavecchia
Basel Institute for Immunology

CONTENTS

Antigen Uptake, Processing, and Presentation

DC and Tolerance

DC in Cytokine Networks

DC and Viral Infections

Molecular Specializations

Follicular Dendritic Cells

Adjuvants for Immune Therapy

1

DENDRITIC CELL DEVELOPMENT AND MATURATION

Ralph M. Steinman,[1] Maggie Pack,[1] and Kayo Inaba[2]

[1]Laboratory of Cellular Physiology and Immunology
Rockefeller University
New York, New York 10021
[2]Zoology Department, Faculty of Science
Kyoto University
Kyoto 696, Japan

1. INTRODUCTION

This conference will open with a discussion of one of the most active and exciting areas in dendritic cell [DC] biology, their development from different populations of progenitors. The formation of potent, immunostimulatory DCs is an obvious control point for the onset of T-cell dependent immunity. In addition, the success in generating large numbers of DCs underlies two other topics that will recur throughout the congress: the use of DCs for active immunotherapy of human disease, and the dissection of DC function using cell biological and molecular methods.

There now appear to be multiple pathways for DC development, the functional consequences of which have yet to be established. These pathways will be detailed elsewhere by their discoverers, i.e., Banchereau, Bhardwaj, Caux, Liu, Ricciardi-Castagnoli, Schuler, Shortman, Young and others.

2. PROPERTIES OF MATURE DC

The term "mature DC" refers to the final and irreversible stage of development. This large, irregularly shaped cell has potent stimulatory activity for T cells, including naive cells in vivo. Take the allogeneic mixed leukocyte reaction as an example. A ratio of 1:100 - 1:1000 DC:T cells [stimulators:responders] results in responses that are greater than standard 1:1 mixes with bulk leukocytes.

Mature DCs, regardless of species [mice, rats, guinea pigs, sheep, monkeys, humans] and tissue source [epidermis, dermis, peripheral lymphoid organs, blood, afferent lymph, thymus medulla] have a set of common features [Table 1]. Two new markers help

Dendritic Cells in Fundamental and Clinical Immunology
edited by Ricciardi-Castagnoli, Plenum Press, New York, 1997

Table 1. Features of mature dendritic cells

Cell shape: Numerous processes [veils, dendrites]
Motility: Active process formation and movement
Antigen capture: Macrophage mannose receptor, DEC-205 receptor
Antigen presentation: High MHC class I and II, invariant chain
Abundance of molecules for T cell binding and costimulation, e.g. CD40, CD54/ICAM-1, CD58/LFA-3, CD80 and CD86/B7-1 and B7-2
Cytokines: Abundant IL-12 production; resistance to IL-10
DC-restricted molecules: p55/fascin, CD83, S100b
Absence of macrophage-restricted molecules and function: CD14, CD115/c-fms/M-CSF responsiveness, low CD68; Myeloperoxidase, lysozyme; Bulk endocytic activity [pinocytosis, phagocytosis]; Adherence to glass/plastic
Stability: No reversion/conversion to macrophages/lymphocytes

to distinguishing mature from less mature DCs. These are CD83, a member of the Ig superfamily, and p55 or fascin, a presumptive actin bundling protein. Both molecules were first identified in Epstein Barr Virus-infected cell lines. In Table 2, we list a number of new DC traits that are under study. These will be discussed in this congress and will no doubt contribute to the analysis of development and function.

3. IMMATURE DENDRITIC CELLS

The term "immature DC" was first used to describe the Langerhans cell, as it exists in the epidermis and shortly after isolation[1,2]. The features of epidermal DCs [Table 3] have in large part been encountered for other populations, especially a CD4+ cell in human blood, a CD8- cell in mouse spleen that is abundant in the periphery of the T cell area in the marginal zone, and the OX62+ DCs in rat airway epithelium and liver.

Table 2. New features of dendritic cells under study

Fas/fas-ligand
NF-κB/rel transcription factors
Antigen uptake, processing, and presentation MIIC's; the exogenous MHC I pathway
CD43
CD44 isoforms
DC-restricted promoters, e.g., CD11c and HIV-1 5'-LTR
Cytoskeleton e.g., p55/fascin and microtubules
FcγR and FcεR
Chemokines and chemokine receptors
Responses to cytokines like IL-10, TGF-β, flt-3 ligand
Markers for DC subsets, e.g., CD9, 13, 33, 36, factor XIII
Genetic modification of DCs

Table 3. Some features of immature dendritic cells

Presence of FcγR
More active endocytosis for certain particulates and proteins
Responsive to GM-CSF, but not M-CSF and G-CSF
Intracellular MHC II in the form of MIICs
Presence of CD1a
Low/absent adhesive and costimulatory molecules [CD40/54/58/86] and T cell sensitization in vitro
Low/absent CD83, p55, CD25, DEC-205, 2A1 antigen

Epidermal Langerhans cells require an exposure to cytokines like GM-CSF to mature. It has not been possible to induce these cells to develop along a macrophage pathway, e.g., with M-CSF.

The term "immature DC", if used in a literal sense, would denote a committed cell that can become DCs *but* lacks the potential to form phagocytes [macrophages, granulocytes] or lymphocytes.

Less mature progenitors in the DC developmental pathway would perhaps be termed "preDC" and "proDC". Examples of "preDC" might include the progenitors in human marrow and cord blood that have the potential to become either DCs or macrophages[3,4]. Another "preDC" or "proDC" would be the CD4+CD3- progenitor of Shortman that is postulated to give rise to both DCs and T cells[5,6].

Two papers from the Schuler and Bhardwaj labs[7,8] describe two stages in the development of DCs from human blood monocytes. In the first stage, a combination of GM-CSF and IL-4 induces the cells to develop some but not all features of mature DCs. Most important is that these GM-CSF and IL-4 treated cells will revert back to macrophages if the cytokines are removed. In the second stage, a monocyte conditioned medium completes the irreversible differentiation to DCs. The addition of the conditioned medium leads to increased expression of MHC II and CD86; lower CD1a, CD115, and CD14; and the induction of p55, CD83, and CD25. So the blood monocyte might be considered as a "preDC", i.e., a cell that is not committed but can be induced to differentiate to mature DCs with a complex series of cytokines.

4. CONTROL OF DENDRITIC CELL DEVELOPMENT

The stimuli that are being used to induce the development of DCs range from the ostensibly simple [CD40L, TNF, IL-1] to the more complex [cocktails of cytokines that are used to generate Shortman's lymphoid DCs from marrow, or Schuler's and Bhardwaj's monocyte-derived DCs from monocytes]. The former, more simple stimuli may in fact be more complex, i.e., CD40L and TNF may induce the formation of several cytokines from other cells in the culture, e.g., macrophages and fibroblasts.

Mature DCs [Table 1] express a large number of distinct features. These relate to low endocytic activity, strong costimulation, and active motility. Does the development of this phenotype require many different signalling pathways, or does development revolve entirely around the TNF-R/CD40 family and their associated TRAF proteins? Evidence is accumulating for the latter thesis. Even IL-1, long known for its capacity to enhance DC function[9-11], is now known to activate TRAF-6[12]. TRAFs lie upstream of NF-κB, and DCs are unusually rich in all members of the NF-κB/rel family[13]. The TNF-R also can couple via FAN, a nonTRAF molecule that activates neutral sphingomyelinase[14]. The latter in turn could catalyze the formation of ceramide that mediates the decrease in endocytic activity during DC maturation[15].

IL-4 [or IL-13] can promote DC development. The laboratories of H.Peters and F.Steinbach reported that IL-4 induces the blood monocyte to express certain features of DCs such as lower levels of CD14 and CD32, and higher levels of MHC II. Romani *et al.*, and Sallusto and Lanzavecchia, found that IL-4 in concert with GM-CSF induced many more features of DCs, although as mentioned above, the cells often are not irreversibly committed or fully mature. The initial rationale for the use of IL-4 by Romani et al was prior data that could block macrophage development in colony assays[16]. So IL-4 may function to direct "preDC" in human blood to a DC rather than macrophage pathway.

An important unknown is the role for IL-4 in marrow cultures in which DCs are developing from proliferating progenitors. Some investigators add IL-4 to mouse bone marrow cultures to enhance DC development in response to GM-CSF, but we have not noted this kind of effect. It is possibile that IL-4 or IL-13 is produced endogenously from other cells in these complex cultures, e.g., basophils, NK1.1 cells, stroma. In otherwords, IL-4 or some other macrophage-suppressive cytokine may not only be required to convert a nonproliferating monocyte to a DC but may also function to promote DC development from proliferating progenitors.

5. THREE PATHWAYS OF DENDRITIC CELL DEVELOPMENT

Since the last DC meeting in Annecy, France, several studies have been completed suggesting that there are three different pathways for the formation of mature DCs from CD34+ or other primitive progenitors [Table 4]. Each produces DCs with the properties of mature DCs [Table 1], but the three pathways differ in terms of progenitors and intermediate stages, cytokine requirements, some surface markers, and probably function. Our comments here are speculative. These pathways could be termed "sentinel, direct, or nonlymphoid", "migratory, bipotential, or myeloid", and "tolerogenic or lymphoid".

The sentinel, nonlymphoid pathway is illustrated by Caux's CD1a+CD14- cells that arise rapidly from CD34+ cord blood progenitors[4], and Young's committed CD34+ progenitor in human cord blood and adult marrow[17]. DCs develop very rapidly [5 days] and have the distinct markers of epidermal Langerhans cells. Other analogous "nonlymphoid"

Table 4. Three pathways of dendritic cell development

Function/Pathway	Locations	Features
Sentinel/nonlymphoid	Epidermis ? interstitial ? mucosal associated	Cadherin+
Migratory/myeloid	Spleen marginal zone ? afferent lymph monocyte/IL-4 airway epithelium	Bipotent precursors: CD11b/13/14/33+, c-fms+ precursor
Tolerogenic/lymphoid	T cell areas Thymic medulla fas-L+	CD11b/13/14/33-, CD4+CD11c- precursor

DCs may be those in the interstitial spaces of organs like heart and kidney, and beneath the follicular-associated, M-cell rich, epithelium of mucosal-associated lymphoid tissue. A distinguishing feature of the sentinel pathway may be the presence of cadherins that target progenitors to appropriate tissues, e.g., E-cadherin in epidermis, and liver/intestinal-cadherin for other DCs. This pathway may require TGF-β as discussed by Udey.

The lymphoid pathway was first described by Shortman's Lab[5,6]. The term "lymphoid" refers to several features. First the DC may share a precursor in common with T cells. Second, this pathway lacks several features of myeloid cells particularly CD11b, CD13, CD14, and CD33. In blood, the lymphoid precursor may be the CD4+CD11c- "plasma-like cell" that Liu will discuss in his chapter. Lymphoid DCs include those in the thymic medulla and many DCs in the T cell areas of all peripheral lymphoid organs. DCs in the T cell areas may include other DCs, e.g., sentinel and migratory DCs that are bringing antigens from the periphery. A distinguishing feature of lymphoid DC may be the capacity to express fas-L [and perhaps other death signals like membrane TNF and CD30L]. This would lead to T cell death and perhaps tolerance rather than immunity.

The migratory or myeloid pathway is distinguished by a developmental stage in which there is some expression of features that are shared with phagocytes. These include CD13, CD14, and CD33 and more active endocytic activity including for immune complexes. In fact a bipotential progenitor for macrophages and DC has been identified in marrow and cord blood[3,4]. The biopotential cells become macrophages in M-CSF and DCs in GM-CSF. The blood monocyte also is bipotential, but to date, the conversion of monocytes to DC initially requires the addition of IL-4 or IL-13, presumably to downregulate the propensity to form macrophages as discussed above.

These pathways of DC development may carry out three types of function. Nonlymphoid DCs, which may target to select tissues via cadherins, are resident sentinels that pick up antigens in the periphery but do not have a true macrophage potential. The myeloid/migratory DCs are short lived cells that are produced in large numbers to patrol the blood and afferent lymph, and to be mobilized as DCs or macrophages as needed e.g., into the airway epithelium. Through the expression of CD14, this pathway has the receptor that provides responsiveness to LPS and other bacterial ligands. The lymphoid pathway could have a major function in tolerance, both central tolerance in the thymus and peripheral tolerance in secondary lymphoid organs.

6. DENDRITIC CELLS IN THE T CELL AREAS [INTERDIGITATING CELLS]

One of the most prominent pools of DCs in vivo are those found in the T cell areas of peripheral lymphoid organs and thymic medulla. The first electron microscopic descriptions by Veldmann and von Ewijk in Holland drew attention to the complex processes, and to the paucity of cytoplasmic organelles but for mitochondria and a tubular smooth reticulum with an electron dense content. These "interdigitating cells" were considered to be mononuclear phagocytes, there being few other possible candidates. Subsequently, it became apparent that these cells showed little or no uptake of many different endocytic tracers, and there was absent or weak expression of many markers of phagocytes like Fc receptors and F4/80 antigen[18,19]. Markers of typical mature DCs were noted: high MHC II, invariant chain, CD86, CD40.

Until recently [Inaba *et al.*, in preparation], there has been relatively little progress in the isolation of T cell area DCs, particularly from the richest source, the peripheral lymph nodes. We have now accomplished this in the mouse.

Another issue concerns the developmental origin of T cell area DCs. All three pathways above can contribute, we think. The migration of LCs or airway DCs to the T cell areas is a key component of sentinel function in nonlymphoid organs. Myeloid DCs, such as those generated from mouse bone marrow, home efficiently to the T cell areas following subcutaneous injection. This must underlie the T cell priming that occurs when antigens are presented on the MHC class I or II molecules of these DCs. Finally, lymphoid DCs likely represent the bulk of the "resident" "interdigitating" DCs in the T cell areas.

REFERENCES

1. Schuler, G. & Steinman, R. M. *J. Exp. Med.* **161**, 526–546 (1985).
2. Romani, N., Koide, S., Crowley, M., et al. *J. Exp. Med.* **169**, 1169–1178 (1989).
3. Szabolcs, P., Avigan, D., Gezelter, S., et al. *Blood* **87**, 4520–4530 (1996).
4. Caux, C., Vanbervliet, B., Massacrier, C., et al. *J. Exp. Med.* **184**, 695–706 (1996).
5. Ardavin, C., Wu, L., Li, C. & Shortman, K. *Nature* **362**, 761–763 (1993).
6. Wu, L., Li, C.-L. & Shortman, K. *J. Exp. Med.* **184**, 903–911 (1996).
7. Romani, N., Reider, D., Heuer, M., et al. *J. Immunol. Methods* **196**, 137–151 (1996).
8. Bender, A., Sapp, M., Schuler, G., Steinman, R. M. & Bhardwaj, N. *J. Immunol. Methods* **196**, 121–135 (1996).
9. Koide, S. L., Inaba, K. & Steinman, R. M. *J. Exp. Med.* **165**, 515–530 (1987).
10. Inaba, K., Witmer-Pack, M. D., Inaba, M., Muramatsu, S. & Steinman, R. M. *J. Exp. Med.* **167**, 149–162 (1988).
11. Heufler, C., Koch, F. & Schuler, G. *J. Exp. Med.* **167**, 700–705 (1988).
12. Cao, Z., Xiong, J., Takeuchi, M., Kurama, T. & Goeddel, D. V. *Nature* **383**, 443–446 (1996).
13. Granelli-Piperno, A., Pope, M., Inaba, K. & Steinman, R. M. *Proc. Natl. Acad. Sci. USA* **92**, 10944–10948 (1995).
14. Adam-Klages, S., Adam, D., Wiegmann, K., et al. *Cell* **86**, 937–947 (1996).
15. Sallusto, F., Nicolo, C., De Maria, R., Corinti, S. & Testi, R. *J. Exp. Med.* **In Press**, (1996).
16. Jansen, J. H., Wientjens, G.-J. H. M., Fibbe, W. E., Willemze, R. & Kluin-Nelemans, H. C. *J. Exp. Med.* **170**, 577–582 (1989).
17. Young, J. W., Szabolcs, P. & Moore, M. A. S. *J. Exp. Med.* **182**, 1111–1120 (1995).
18. Witmer, M. D. & Steinman, R. M. *Am. J. Anat.* **170**, 465–481 (1984).
19. Inaba, K., Witmer-Pack, M., Inaba, M., et al. *J. Exp. Med.* **180**, 1849–1860 (1994).

2

GENERATION OF MATURE DENDRITIC CELLS FROM HUMAN BLOOD

An Improved Method with Special Regard to Clinical Applicability

Gerold Schuler[1] and Nikolaus Romani[2]

[1]Department of Dermatology
University of Erlangen-Nürnberg
Hartmannstrasse 14, D-92052
Erlangen, Germany
[2]Department of Dermatology
University of Innsbruck
Anichstrasse 35, A-6020
Innsbruck, Austria

1. INTRODUCTION

Efficient methods to generate large numbers of dendritic cells have been developed in the past five years. Caux *et al.*[1] have introduced the approach to grow dendritic cells from rare CD34+ progenitor cells in (cord) blood using GM-CSF and TNF-α as the critical cytokines. On the other hand, Sallusto et al.[2] and Romani et al.[3] have established procedures that make use of the more abundant monocytic CD34-negative and CD14+ precursors in peripheral blood. GM-CSF and IL-4 were the necessary cytokines. Both approaches have since been widely used, even up to the stage of clinical trials.

In the course of refining and improving the method developed by us[3] we made a critical observation. Dendritic cells grown from CD34-negative precursors in the presence of GM-CSF and IL-4 are not stable mature dendritic cells. Upon withdrawal of cytokines their morphology and adherence properties change: They become less irregular in shape and they adhere to the substrate to a substantial, though variable degree. They appear to revert to a monocytic stage. We present here a modification of the method that uses monocyte-conditioned media in order to achieve stably mature dendritic cells[4,5].

It is important to emphasize that the state of full maturation is defined by a number of criteria, all of which have to be fulfilled. Monitoring of only selected features (e.g. upregulation of CD83) may be misleading. In this article we will list and explain the criteria we consider critical.

2. MATERIALS AND METHODS

Lymphocyte-depleted human PBMC (i.e. mostly CD14+ cells) were plated into 6-well tissue culture plates and cultured for 7 days in the presence of 800 and 1000 U/ml of GM-CSF and IL-4, respectively ("priming culture"). From day 7 to day 10 cultures were supplemented with monocyte-conditioned media ("maturation culture"). Conditioned media were obtained from monocytes that had been stimulated via ligation of their Fc-receptors (i.e., adherence to Ig-coated dishes) for 24 hours. Cultures were performed with similar success in media containing either 10% fetal calf serum or 1% autologous plasma. This method has been published in detail recently[4,5].

3. RESULTS

Mature dendritic cells were consistently obtained by subjecting the cells to the "maturation culture" in the presence of monocyte-conditioned medium from day 7 to day 10. From a standard sample of 80 ml of blood we could obtain up to 8×10^6 and $2–4 \times 10^6$ mature dendritic cells in culture media containing 10% fetal calf serum and 1% autologous human plasma, respectively. These mature dendritic cells were compared with dendritic cells that were cultured from day 7 to day 10 in the absence of conditioned media or with dendritic cells that were harvested on day 7 following culture in GM-CSF plus IL-4 as described[3]. Both culture conditions yielded immature dendritic cells.

3.1. Morphology

Figure 1 shows clearly how different the morphology of immature and mature dendritic cells appears under the phase contrast microscope. Immature dendritic cells are non-adherent and irregular in shape with some cytoplasmic projections (Fig.1A). Mature dendritic cells, in contrast, display long and thin cytoplasmic processes. Only when these processes are elaborated to a degree as shown in Figure 1B can one speak of typical

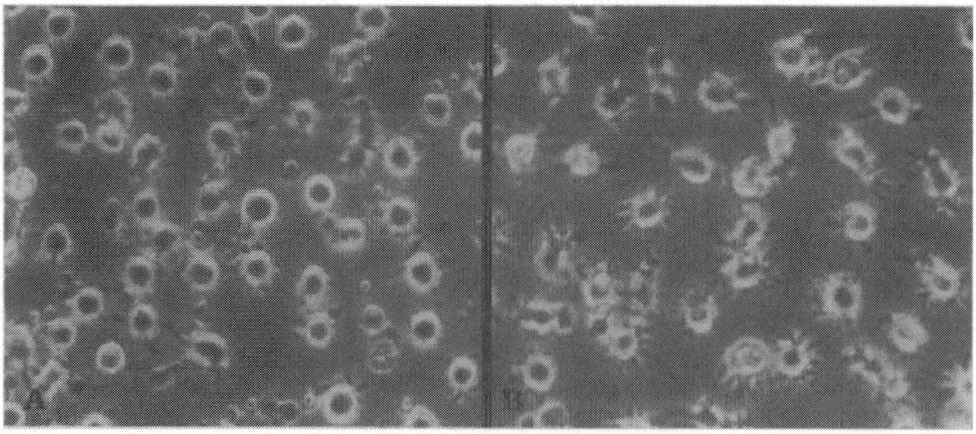

Figure 1. Phase contrast of immature (A) and mature (B) dendritic cells. Maturation of cells in panel B was induced by monocyte-conditioned media. Note the abundance of cytoplasmic veils displayed by the mature cells.

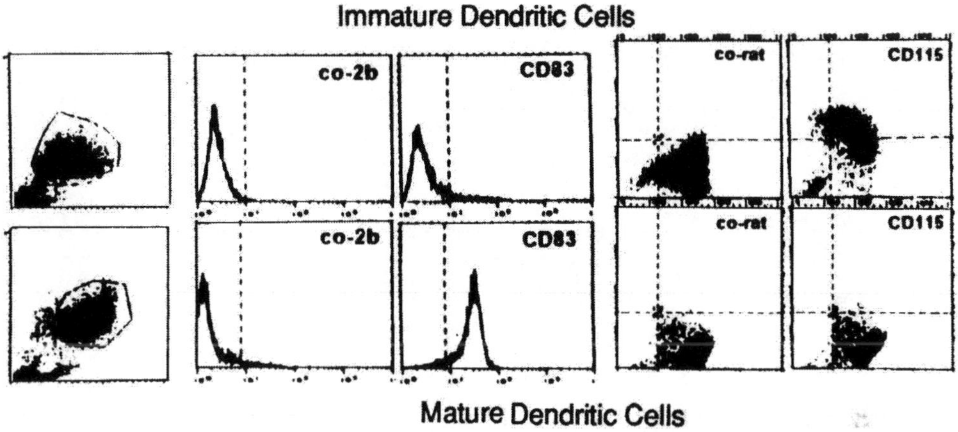

Figure 2. Comparative FACS analyses of dendritic cells matured in the presence of monocyte-conditioned medium (bottom panels) and of immature dendritic cells (top panels). Fluorescence of large cells (i.e. dendritic cells) as defined by light scatter properties (very left panels) is shown. The histograms on the left illustrate the upregulation of CD83 upon maturation and the dot plots on the right show the down-regulation of CD115 (y-axis; fluorescence on the x-axis represents HLA-DR staining).

"veils". Another important point that distinguishes genuine "veils" from other cytoplasmic structures is their high motility. This can be appreciated easily upon close (and patient) inspection under phase contrast or — better — by means of time lapse video recording.

3.2. Phenotype

The phenotype of immature dendritic cells has been described previously[3]. Two additional markers have turned out to be indicative of maturation: CD83[6] and CD115 (M-CSF receptor)(Figure 2). CD83 is not expressed on immature dendritic cells and is upregulated upon maturation in the presence of monocyte-conditioned media. CD115 expression is inverse: positive on immature and negative on mature dendritic cells. The costimulatory molecule CD86 (B7–2) is also markedly upregulated and CD68 expression changes from a non-characteristic cytoplasmic staining pattern in immature dendritic cells to a distinct perinuclear, spot-like pattern typical for mature dendritic cells (Figure 3). Langhoff et al. (this volume) describe an actin-bundling protein (p55) that is expressed in the same way as CD83.

3.3. Function

Maturation in monocyte-conditioned medium upregulates the capacity of dendritic cells to stimulate resting T cells in the allogeneic MLR. This also applies to naive, "virgin" T cells such as T cell populations from umbilical cord blood. Both CD4+ and CD8+ T cells can efficiently be sensitized by mature dendritic cells. Bhardwaj *et al.* (this volume) show that dendritic cells brought to full maturity by means of monocyte-conditioned media can induce antigen-specific cytotoxic T lymphocytes with much higher efficacy than immature dendritic cells. Conversely, the ability to process protein antigens like teta-

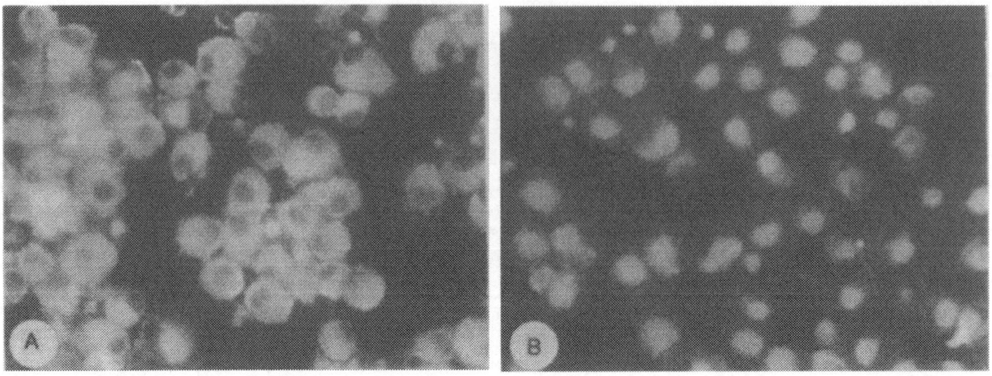

Figure 3. Expression of CD68 (red fluorescence) and HLA-DR (green fluorescence) on cytospins. Immature dendritic cells (A) are stained throughout the entire cytoplasm. In contrast, mature dendritic cells (B) display a pronounced spot-like staining pattern. An occasional mature dendritic cell can be identified by the spot-like staining within the population of immature dendritic cells (to the left of the A). Fluorescent rims around the CD68 spots in B represent HLA-DR surface staining.

nus toxoid is down-regulated upon maturation. One of the reasons for this is the greatly reduced activity of uptake mechanisms such as macropinocytosis[7]. In addition—as shown in other types of dendritic cells—mature dendritic cells have stopped the biosynthesis of MHC and invariant chain products[8–10] and have a diminished number of organelles where processing can take place[11,12].

3.4. Stability of the Mature, Immunostimulatory State

When immature dendritic cells obtained by culture for 7 days in GM-CSF and IL-4 are deprived of cytokines and cultured for another three to four days they become less irregular in shape and many of them adhere to and firmly spread on the tissue culture plastic. This is most pronounced under serum-free conditions in X-VIVO or AIM-V media or in media containing 10% human serum. CD14 re-expression has been shown for immature dendritic cells readhering on collagen substrates[13]. These data indicate that immature dendritic cells revert to a monocytic state. This phenomenon is never observed with fully mature dendritic cells such as dendritic cells obtained from skin by trypsinization of the epidermis and subsequent 3-day culture or by emigration from skin explants[14]. Importantly, reversion does also not occur with blood dendritic cells that have matured in the presence of monocyte conditioned media (Fig.4). CD83 expression also persists on such fully mature dendritic cells.

3.5. IL-12 production

Dendritic cells can synthesize and secrete bioactive IL-12[15,16]. Upon contact with T cells the release of IL-12 is markedly upregulated. Interactions between CD40 (on dendritic cells) and the CD40 ligand (on T cells) and MHC class II/peptide (on dendritic cells) and the T cell receptor are responsible for this boost in IL-12 production[17,18]. Dendritic cells are special in that the amounts of IL-12 released are much higher than in other cell types, e.g. in macrophages[19].

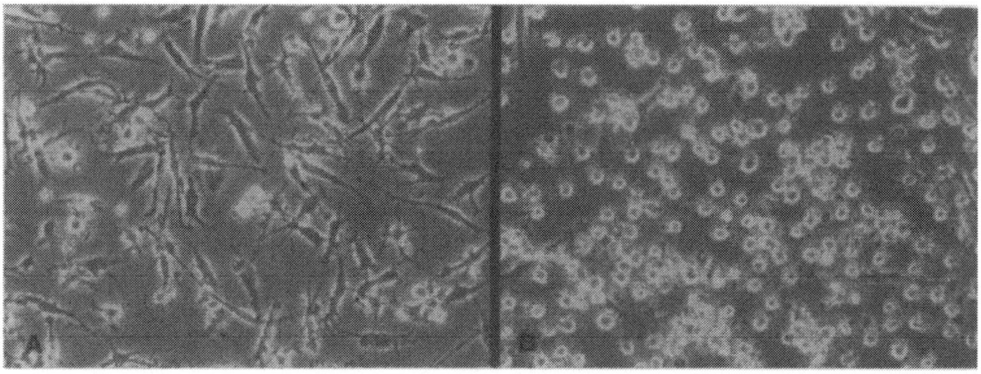

Figure 4. Immature (A) or mature (B) dendritic cells were deprived of cytokines and conditioned medium on day 10 of culture and were cultured in serum-free AIM-V medium for another 3 days. Immature dendritic cells become adherent (A) whereas dendritic cells that have matured in the presence of monocyte-conditioned medium remain non-adherent and still display many veils (that can be better appreciated with the help of magnifying glasses)(B).

3.6. Resistance to the Inibitory Effects of IL-10

When dendritic cells will be used for adoptive tumor therapy it is likely that, upon injection into the patient, they will encounter a milieu rich in IL-10[20]. IL-10 has been shown to interfere with the maturation process of dendritic cells[21,22]. This is also true for the culture system described here. Normal maturation does not occur in the presence of IL-10 and dendritic cells even convert into tolerogenic cells. However, once dendritic cells have fully matured in monocyte-conditioned medium they are not suceptible any longer to the effects of IL-10. In spite of the presence of this cytokine they remain CD83+ and immunostimulatory active (A. Enk, manuscript in preparation).

3.7. Migratory Capacity

Dendritic cells actively migrate from the tissues and organs where they reside in the body to the draining lymphoid organs[23,24]. This capacity is better developed in dendritic cells than in other types of antigen presenting cells. Dendritic cells that have matured in the presence of monocyte-conditioned media possess excellent migration properties in an experimental model using three-dimensional collagen lattices (Gunzer et al., this volume).

4. DISCUSSION

The ultimate goal of dendritic cell research is to find ways to use these cells for the successful treatment (and for prevention) of diseases, in particular of cancer. The first clinical trials using mature dendritic cells isolated directly from blood have proven successful and are expanded[25]. Different approaches to enhance the immunogenicity of dendritic cells are being chosen and compared to each other and dendritic cells generated in vitro will soon be explored. We feel that it is absolutely essential to also consider the maturational state of dendritic cells in these approaches. Only the profound knowledge of how dendritic cells mature and what their immunogenic properties (antigen processing, migration, T cell sensitization) are in the immature versus the mature state will ensure the successful clinical application of the dendritic cell concept.

Some of the maturational changes described here can apparently also be achieved when TNF-α is added to the cultures instead of monocyte-conditioned media. The upregulation of CD83[26] and the down-regulation of macropinocytosis and antigen processing[7] have been described. These experiments were all done in fetal calf serum, though. It remains to be determined whether TNF-α can substitute for *all the effects* of monocyte-conditioned medium, particularly in a clinically relevant setting with autologous plasma. Comparative analyses performed so far make this seem unlikely. Bhardwaj et al. (this volume) show that CD83 expression induced by TNF-α treatment does not stably remain on the surface of dendritic cells over a period of 2–3 days. In contrast, monocyte-conditioned medium-induced CD83 expression is stable. Further comparative analyses adressing the other criteria for maturity are being performed. Thus, it appears that the maturation induced by monocyte-conditioned media is of a different quality as maturity obtained by a single cytokine such as TNF-α. The exact role that TNF-α undoubtedly plays in the maturation process has still to be explored.

The method presented here[4,5] opens the way to adoptive immunotherapy. Sufficient numbers of dendritic cells can be procured. These dendritic cells can be harvested and used either in the immature (if processing is desired) or in the mature state (if processing is not necessary but costimulation is desired). We want to emphasize that our approach can be done entirely in an autologous system. Fetal calf serum is not essential. Culture media that are approved for clinical use (X-VIVO or AIM-V) can be used. We feel that this is a very important issue. In these days, when there appears to be a "rush" towards clinical trials with dendritic cells, one tends to be a bit more generous in interpreting the rules and guidelines for clinical trials. Yet, we are convinced that for ethical reasons one should rigorously comply with these rules that demand the use of approved reagents or reagents that are produced according to the principles of "good manufacturing practice" or at least "good laboratory practice". We are also convinced that dendritic cells used for adoptive immunotherapy must never have "seen" fetal calf serum. Furthermore we consider it essential that in vitro (IL-10 resistance), and even more importantly, in vivo controls (dendritic cells loaded with recall antigens) are incorporated into clinical trials. This is to assure that the dendritic cells administered to patients are not tolerogenic and thus harmful. The method reviewed here and described in detail elsewhere[4,5] should facilitate clinical trials employing dendritic cells.

REFERENCES

1. Caux, C., C. Dezutter-Dambuyant, D. Schmitt, and J. Banchereau. 1992. GM-CSF and TNF-α cooperate in the generation of dendritic Langerhans cells. *Nature* 360:258–261.
2. Sallusto, F. and A. Lanzavecchia. 1994. Efficient presentation of soluble antigen by cultured human dendritic cells is maintained by granulocyte/macrophage colony-stimulating factor plus interleukin 4 and downregulated by tumor necrosis factor a. *J. Exp. Med.* 179:1109–1118.
3. Romani, N., S. Gruner, D. Brang, E. Kämpgen, A. Lenz, B. Trockenbacher, G. Konwalinka, P. O. Fritsch, R. M. Steinman, and G. Schuler. 1994. Proliferating dendritic cell progenitors in human blood. *J. Exp. Med.* 180:83–93.
4. Romani, N., D. Reider, M. Heuer, S. Ebner, E. Kämpgen, B. Eibl, D. Niederwieser, and G. Schuler. 1996. Generation of mature dendritic cells from human blood: An improved method with special regard to clinical applicability. *J. Immunol. Methods* 196:137–151
5. Bender, A., M. Sapp, G. Schuler, R. M. Steinman, and N. Bhardwaj. 1996. Improved methods for the generation of dendritic cells from nonproliferating progenitors in human blood. *J. Immunol. Methods* 196:121–135

6. Zhou, L.-J. and T. F. Tedder. 1995. Human blood dendritic cells selectively express CD83, a member of the immunoglobulin superfamily. *J. Immunol.* 154:3821–3835.
7. Sallusto, F., M. Cella, C. Danieli, and A. Lanzavecchia. 1995. Dendritic cells use macropinocytosis and the mannose receptor to concentrate macromolecules in the major histocompatibility complex class II compartment: Downregulation by cytokines and bacterial products. *J. Exp. Med.* 182:389–400.
8. Kämpgen, E., N. Koch, F. Koch, P. Stöger, C. Heufler, G. Schuler, and N. Romani. 1991. Class II major histocompatibility complex molecules of murine dendritic cells: Synthesis, sialylation of invariant chain, and antigen processing capacity are down-regulated upon culture. *Proc. Natl. Acad. Sci. USA* 88:3014–3018.
9. Puré, E., K. Inaba, M. T. Crowley, L. Tardelli, M. D. Witmer-Pack, G. Ruberti, G. Fathman, and R. M. Steinman. 1990. Antigen processing by epidermal Langerhans cells correlates with the level of biosynthesis of major histocompatibility complex class II molecules and expression of invariant chain. *J. Exp. Med.* 172:1459–1469.
10. Becker, D., A. B. Reske-Kunz, J. Knop, and K. Reske. 1991. Biochemical properties of MHC class II molecules endogenously synthesized and expressed by mouse Langerhans cells. *Eur. J. Immunol.* 21:1213
11. Stössel, H., F. Koch, E. Kämpgen, P. Stöger, A. Lenz, C. Heufler, N. Romani, and G. Schuler. 1990. Disappearance of certain acidic organelles (endosomes and Langerhans cell granules) accompanies loss of antigen processing capacity upon culture of epidermal Langerhans cells. *J. Exp. Med.* 172:1471–1482.
12. Kleijmeer, M. J., V. M. J. Oorschot, and H. J. Geuze. 1994. Human resident Langerhans cells display a lysosomal compartment enriched in MHC class II. *J. Invest. Dermatol.* 103:516–523.
13. Radmayr, C., G. Böck, A. Hobisch, H. Klocker, G. Bartsch, and M. Thurnher. 1995. Dendritic antigen-presenting cells from the peripheral blood of renal-cell-carcinoma patients. *Int. J. Cancer* 63:627–632.
14. Pope, M., M. G. H. Betjes, H. Hirmand, L. Hoffman, and R. M. Steinman. 1995. Both dendritic cells and memory T lymphocytes emigrate from organ cultures of human skin and form distinctive dendritic-T-cell conjugates. *J. Invest. Dermatol.* 104:11–17.
15. Heufler, C., F. Koch, U. Stanzl, G. Topar, M. Wysocka, G. Trinchieri, A. Enk, R. M. Steinman, N. Romani, and G. Schuler. 1996. Interleukin-12 is produced by dendritic cells and mediates T helper 1 development as well as interferon-gamma production by T helper 1 cells. *Eur. J. Immunol.* 26:659–668.
16. Macatonia, S. E., N. A. Hosken, M. Litton, P. Vieira, C.-S. Hsieh, J. A. Culpepper, M. Wysocka, G. Trinchieri, K. M. Murphy, and A. O'Garra. 1995. Dendritic cells produce IL-12 and direct the development of Th1 cells from naive $CD4^+$ T cells. *J. Immunol.* 154:5071–5079.
17. Koch, F., U. Stanzl, P. Jennewein, K. Janke, C. Heufler, E. Kämpgen, N. Romani, and G. Schuler. 1996. High level IL-12 production by murine dendritic cells: Upregulation via MHC class II and CD40 molecules and downregulation by IL-4 and IL-10. *J. Exp. Med.* 184:741–746.
18. Cella, M., D. Scheidegger, K. Palmer-Lehmann, P. Lane, A. Lanzavecchia, and G. Alber. 1996. Ligation of CD40 on dendritic cells triggers production of high levels of interleukin-12 and enhances T cell stimulatory capacity: T-T help via APC activation. *J. Exp. Med.* 184:747–752.
19. Shu, U., M. Kiniwa, C. Y. Wu, C. Maliszewski, N. Vezzio, J. Hakimi, M. Gately, and G. Delespesse. 1995. Activated T cells induce interleukin-12 production by monocytes via CD40-CD40 ligand interaction. *Eur. J. Immunol.* 25:1125–1128.
20. Dummer, W., B. C. Bastian, N. Ernst, C. Schänzle, A. Schwaaf, and E. B. Bröcker. 1996. Interleukin-10 production in malignant melanoma: Preferential detection of IL-10-secreting tumor cells in metastatic lesions. *Int. J. Cancer* 66:607–610.
21. Buelens, C., F. Willems, A. Delvaux, G. Piérard, J. P. Delville, T. Velu, and M. Goldman. 1995. Interleukin-10 differentially regulates B7–1 (CD80) and B7–2 (CD86) expression on human peripheral blood dendritic cells. *Eur. J. Immunol.* 25:2668–2672.
22. Péguet-Navarro, J., C. Moulon, C. Caux, C. Dalbiez-Gauthier, J. Banchereau, and D. Schmitt. 1994. Interleukin-10 inhibits the primary allogeneic T cell response to human epidermal Langerhans cells. *Eur. J. Immunol.* 24:884–891.
23. Austyn, J. M. 1996. New insights into the mobilization and phagocytic activity of dendritic cells. *J. Exp. Med.* 183:1287–1292.
24. Lukas, M., H. Stössel, L. Hefel, S. Imamura, P. Fritsch, N. T. Sepp, G. Schuler, and N. Romani. 1996. Human cutaneous dendritic cells migrate through dermal lymphatic vessel culture model. *J. Invest. Dermatol.* 106:1293–1299.
25. Hsu, F. J., C. Benike, F. Fagnoni, T. M. Liles, D. Czerwinski, B. Taidi, E. G. Engleman, and R. Levy. 1996. Vaccination of patients with B-cell lymphoma using autologous antigen-pulsed dendritic cells. *Nature Med.* 2:52–58.
26. Zhou, L. J. and T. F. Tedder. 1996. $CD14^+$ blood monocytes can differentiate into functionally mature $CD83^+$ dendritic cells. *Proc. Natl. Acad. Sci. USA* 93:2588–2592.

3

GROWTH AND DIFFERENTIATION OF HUMAN DENDRITIC CELLS FROM CD34$^+$ PROGENITORS

Paul Szabolcs,[1,2] David H. Ciocon,[1] Malcolm A. S. Moore,[3] and James W. Young[1,4]

[1]Laboratory of Cellular Physiology and Immunology
The Rockefeller University
[2]Bone Marrow Transplantation Service
Department of Pediatrics
Memorial Sloan-Kettering Cancer Center
[3]James Ewing Laboratory of Developmental Hematopoiesis, Cell Biology and Genetics Program
Sloan-Kettering Institute for Cancer Research
[4]Allogeneic Bone Marrow Transplantation and Clinical Immunology Services
Division of Hematologic Oncology, Department of Medicine
Memorial Sloan-Kettering Cancer Center
New York, New York 10021

1. SUSPENSION CULTURES OF HUMAN CD34$^+$ BONE MARROW PROGENITORS REVEAL TWO PATHWAYS FOR DENDRITIC CELL DIFFERENTIATION

Several groups originally established the importance of GM-CSF and TNFα in the generation of dendritic cells from CD34$^+$ precursors (1–3), and most investigators have used cord blood as the starting source. Studies from our laboratories confirmed the essential role of these cytokines in supporting bone marrow CD34$^+$-derived dendritic cell growth and differentiation, and the addition of c-*kit*-ligand facilitated more detailed phenotypic and functional characterization of the dendritic cell progeny than had previously been possible (4).

Two new findings were realized. The first was that after two weeks' suspension culture in this cytokine-driven system, both CD14$^-$ HLA-DR$^{++/+++}$ dendritic cells and CD14$^+$ HLA-DR$^{+/++}$ macrophages shared expression of most accessory molecules. However, immunostimulatory activity segregated to the mature CD14$^-$ HLA-DR$^{++/+++}$ dendritic cells. Secondly, in contrast to the predominance of dendritic cells among the HLA-DR positive progeny grown from cord blood $CD34^+$ precursors under similar conditions (1,5), dendritic cells were less frequent than macrophages among the HLA-DR positive progeny of bone marrow $CD34^+$ cells. Dendritic cells constituted only 5–15% of the total myeloid population after 12–14 days' culture, the remainder being macrophages and intermediate granulocytic precursors.

When examined at earlier time points in the growth of bone marrow $CD34^+$ precursors, a discrete population of $CD14^-$ $HLA\text{-}DR^{++/+++}$ dendritic cells was present by day 6, and possibly as early as day 4 (6). In addition, a separate population of $CD14^+$ $HLA\text{-}DR^+$ cells was cytofluorographically identifiable, despite the removal of mature macrophages. This $CD14^+$ $HLA\text{-}DR^+$ population expressed phenotypic and functional properties that approximated but did not entirely coincide with those of either mature, terminally differentiated macrophages or dendritic cells. Approximately half were still in cell cycle, based on expression of Ki-67, but these cells had lost colony-forming capacity. This $CD14^+$ $HLA\text{-}DR^+$ population, however, proved to be bipotential in its further development, differentiating into mature dendritic cells in the presence of GM-CSF and TNFα, or into macrophages in the absence of exogenous cytokines. Macrophage differentiation of the $CD14^+$ $HLA\text{-}DR^+$ intermediates also occurred in the presence of M-CSF, which supported up to a two-fold expansion during reculture.

Thus monocytes/macrophages and dendritic cells can develop independently from bone marrow $CD34^+$ precursors during a first week of culture, as well as via a bipotential $CD14^+$ $HLA\text{-}DR^+$ intermediate after reculture *in vitro* for an additional week [(6); Figure 1]. This intermediate is not a monocyte/macrophage, despite its expression of CD14. Bone marrow $CD34^+$ -derived dendritic cell progeny by either route have not been distinguished using criteria evaluated to date. Caux and colleagues, however, have confirmed the existence of a $CD14^+$ non-monocyte/macrophage, bipotential intermediate from cord blood, which gives rise to dermal dendritic cells in the continued presence of GM-CSF and TNFα (7). These investigators have been able to distinguish these dendritic cells from

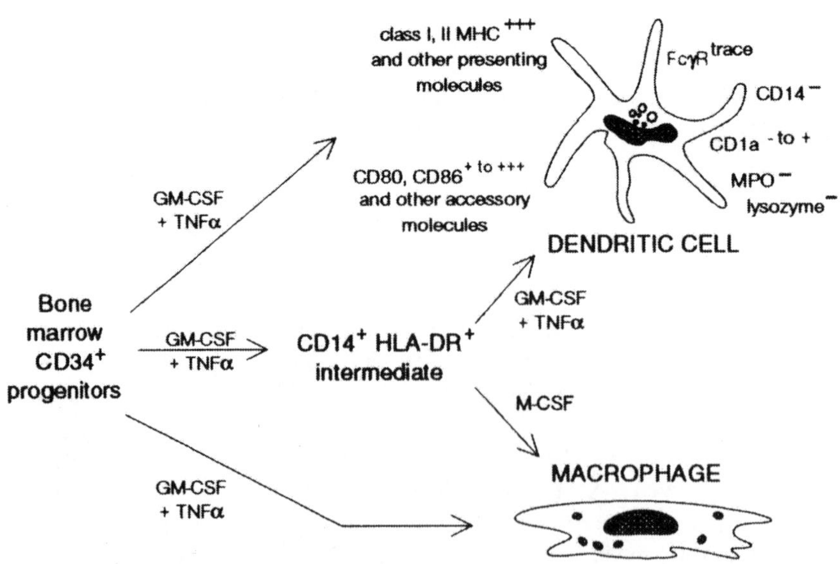

Figure 1. Alternative pathways for human dendritic cell differentiation from bone marrow CD34+ progenitors (see also ref. 6).

the epidermal dendritic or Langerhans cells arising directly from CD34⁺ progenitors in cord blood (7).

2. CFU-DC EXIST IN HUMAN BONE MARROW THAT GIVE RISE TO PURE DENDRITIC CELL COLONIES WHEN TNFα IS ADDED TO GM-CSF

Murine data from Inaba et al. (8) predicted that dendritic cells would constitute a trace presence among the myeloid progeny derived from CFU-GM, just as dendritic cells do in other sites in vivo. We found this to be the case when GM colonies were grown from human CD34⁺ bone marrow progenitors in the presence of GM-CSF and TNFα. However, pure dendritic cell colonies developed in addition to these GM colonies [(9); Figure 2]. The frequency of GM colonies was not significantly altered by the development of dendritic cell colonies, whether the former were grown in GM-CSF alone or in combination with TNFα. The addition of TNFα proved essential, however, to the presence of trace dendritic cells [~1%] in the GM colonies grown in GM-CSF and TNFα.

The addition of c-*kit*-ligand again enabled detailed characterization of the progeny of the respective colony types. This early acting growth factor, in combination with GM-CSF and TNFα, supported a two-fold increase in primary cloning efficiency of CFU-DC, and over two weeks, supported an almost 100-fold increase in CFU-DC based on secondary replating efficiency in clonogenic assays. The size of the colonies, and therefore the cell numbers when pooled, were also substantially greater. The purity of the dendritic cell colonies was confirmed by the uniform veiled morphology and motility when the colonies were plucked and pooled in suspension culture. Further, there was intense HLA-DR and absent CD14 expression by all clonogenic dendritic cell progeny. Variable expression of α-naphthyl butyrate esterase and CD1a was consistent with asynchronous terminal differentiation within individual colonies. CD83, a marker of mature myeloid dendritic cells

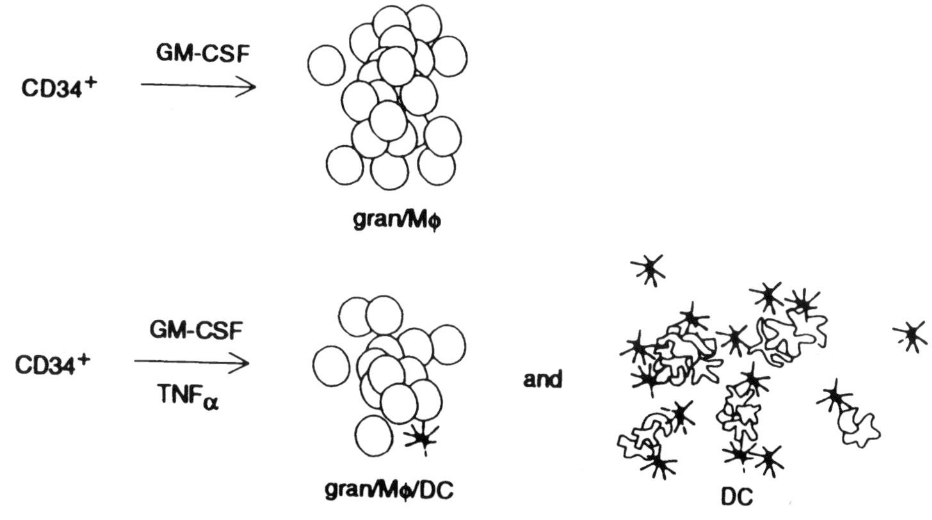

Figure 2. The addition of TNFα to GM-CSF supports the development of pure human dendritic cell colonies, in addition to typical GM colonies (see also ref. 9).

Table 1. Human myeloid colony growth [CFU-GM and CFU-DC] from cord blood CD34⁺

	CFU-GM/105 CD34+ cells plated ± SEM	CFU-DC/105 CD34+ cells plated ± SEM
Cytokine condition in clonogenic assay		
KL + GM-CSF	2,583 ± 328	17 ± 17
KL + GM-CSF + TNFα	433 ± 164	3,967 ± 109
Representative ^{3}HTdR incorporation [cpm ± SEM] by 5 × 10^4 allogeneic T cells, after stimulation by the respective colony type	373 ± 88	30,515 ± 8,749

(10), was not originally evaluated, but has since been found on a subset of dendritic cells from similar colonies and again likely reflects asyncronous maturation [not shown]. Finally, the progeny of CFU-DC were potent stimulators of resting allogeneic T cells, exceeding the activity of CFU-GM progeny grown under the same cytokine conditions by approximately 100-fold, consistent with the ~1% presence of dendritic cells in these GM colonies.

3. THE FREQUENCY OF HUMAN CFU-DC VARIES BETWEEN SOURCES OF CD34⁺ PROGENITORS

Because of differences in dendritic cell frequency among the myeloid progeny of CD34⁺ precursors isolated from cord blood (1,5,7) and bone marrow (2,4,6), we evaluated these two sources, as well as cytokine-elicited peripheral blood, at the clonogenic level. CD34⁺ cells were purified by positive selection and cultured in semisolid medium as described (9) or seeded to individual microwells. Several new findings have emerged. In contrast to bone marrow, even a few rare dendritic cell colonies can be identified in cultures of cord blood CD34⁺ progenitors supplemented with GM-CSF alone. This suggests endogenous production of TNFα, or a similarly acting cytokine. More notable, however, is the development of pure dendritic cell colonies at the expense of GM colony growth [Table 1], a phenomenon that was not observed in the bone marrow system (9). CFU-DC progeny seem to be morphologically and phenotypically similar, whether derived from bone marrow or cord blood CD34⁺ cells. CFU-DC also account for essentially all of the allostimulatory activity recovered from the clonogenic progeny in these cultures [Table 1]. Similar findings have recently been extended to CD34⁺ cells isolated from G-CSF-elicited leukopheresis products [not shown].

4. SUMMARY AND CONCLUSIONS

Human dendritic cells can be generated from bone marrow CD34⁺ progenitors in the presence of GM-CSF and TNFα. The addition of a factor like c-*kit*-ligand optimizes the expansion of dendritic cells, as well as the other myeloid progeny grown under the same conditions, and facilitates their identification and characterization. In contrast to cord blood, where dendritic cells account for the majority of the class II MHC positive myeloid progeny, bone marrow CD34⁺-derived dendritic cells are less frequent than macrophages. When mature macrophages are depleted from days 5–6 cultures, terminally differentiated

$CD14^-$ $HLA\text{-}DR^{+++}$ dendritic cells as well as non-monocyte/macrophage $CD14^+$ $HLA\text{-}DR^+$ cells can be distinguished. The latter are post-CFU, bipotential, intermediate precursors that can terminally differentiate into either dendritic cells or macrophages depending on subsequent cytokine exposure. Human $CD34^+$ progenitors isolated from bone marrow, as well as cord and peripheral blood, include CFU-DC that give rise to pure dendritic cell colonies in the combined presence of GM-CSF and TNFα. The different sources of $CD34^+$ progenitors are not equivalent, however, with respect to frequency of CFU-DC growth. Cord blood is relatively enriched for dendritic cell progenitors. The developmental relationship of CFU-DC and CFU-GM, to the early developing dendritic cells and the bipotential intermediates observed in suspension culture, is not yet established.

REFERENCES

1. Caux, C., Dezutter-Dambuyant, C., Schmitt, D., and Banchereau, J. 1992. GM-CSF and TNF-α cooperate in the generation of dendritic Langerhans cells. *Nature* 360:258–261.
2. Reid, C.D.L., Stackpoole, A., Meager, A., and Tikerpae, J. 1992. Interactions of tumor necrosis factor with granulocyte-macrophage colony-stimulating factor and other cytokines in the regulation of dendritic cell growth in vitro from early bipotent $CD34^+$ progenitors in human bone marrow. *J. Immunol.* 149:2681–2688.
3. Santiago-Schwarz, F., Belilos, E., Diamond, B., and Carsons, S.E. 1992. TNF in combination with GM-CSF enhances the differentiation of neonatal cord blood stem cells into dendritic cells and macrophages. *J. Leuk. Biol.* 52:274–281.
4. Szabolcs, P., Moore, M.A.S., and Young, J.W. 1995. Expansion of immunostimulatory dendritic cells among the myeloid progeny of human $CD34^+$ bone marrow precursors cultured with c-*kit* ligand, granulocyte-macrophage colony-stimulating factor, and TNF-α. *J. Immunol.* 154:5851–5861.
5. Caux, C., Vanbervliet, B., Massacrier, C., Durand, I., and Banchereau, J. 1996. Interleukin-3 cooperates with tumor necrosis factor alpha for the development of human dendritic/langerhans cells from cord blood CD34+ hematopietic progenitor cells. *Blood* 87:2376–2385.
6. Szabolcs, P., Avigan, D., Gezelter, S., Ciocon, D.H., Moore, M.A.S., Steinman, R.M., and Young, J.W. 1996. Dendritic cells and macrophages can mature independently from a human bone marrow-derived, post-CFU intermediate. *Blood* 87:4520–4530.
7. Caux, C., Vanbervliet, B., Massacrier, C., Dezutter-Dambuyant, C., de Saint-Vis, B., Jacquet, C., Yoneda, K., Imamura, S., Schmitt, D., and Banchereau, J. 1996. CD34+ hematopoietic progenitors from human cord blood differentiate along two independent dendritic cell pathways in response to GM-CSF+ TNF alpha. *J. Exp. Med.* 184:695–706.
8. Inaba, K., Inaba, M., Deguchi, M., Hagi, K., Yasumizu, R., Ikehara, S., Muramatsu, S., and Steinman, R.M. 1993. Granulocytes, macrophages, and dendritic cells arise from a common major histocompatibility complex class II-negative progenitor in mouse bone marrow. *Proc. Natl. Acad. Sci. USA* 90:3038–3042.
9. Young, J.W., Szabolcs, P., and Moore, M.A.S. 1995. Identification of dendritic cell colony-forming units among normal $CD34^+$ bone marrow progenitors that are expanded by c-*kit*-ligand and yield pure dendritic cell colonies in the presence of granulocyte/macrophage colony-stimulating factor and tumor necrosis factor α. *J. Exp. Med.* 182:1111–1120.
10. Zhou, L.-J. Tedder, T.F. 1995. Human blood dendritic cells selectively express CD83, a member of the immunoglobulin superfamily. *J. Immunol.* 154:3821–3835.

4

CD34$^+$ HEMATOPOIETIC PROGENITORS FROM HUMAN CORD BLOOD DIFFERENTIATE ALONG TWO INDEPENDENT DENDRITIC CELL PATHWAYS IN RESPONSE TO GM-CSF+TNFα

C. Caux,[1] C. Massacrier,[1] B. Vanbervliet,[1] B. Dubois,[1] B. de Saint-Vis,[1]
C. Dezutter-Dambuyant,[2] C. Jacquet,[2] D. Schmitt,[2] and J. Banchereau[1]

[1]Schering-Plough
Laboratory for Immunological Research
Dardilly, France
[2]INSERM U346
Hopital Edouard Herriot
Lyon, France

INTRODUCTION

DC are professional antigen presenting cells which are required for the initiation of immune responses. Many types of DC with subtle differences in phenotype have been described in peripheral blood, skin and lymphoid organs (1–8). Although each of these DC subsets display the ability to activate naive T cells, it is not clear whether they represent different stages of maturation of a unique DC lineage or whether they stem from different progenitors. In this study we demonstrate that in presence of GM-CSF+TNFα i) human myeloid progenitors can differentiate along two unrelated DC pathways: the Langerhans cells (LC) characterized by the expression of CD1a, Lag, Birbeck granules (BG) and E cadherin and a CD14-derived DC, related to monocyte-derived DC, characterized by the expression of CD1a, CD9, CD2 and factor XIIIa. ii) These two subsets display properties expected for DC, but in cocultures with CD40-activated naive B cells only the CD14$^+$ derived DC can induced the production of IgM in presence of IL-2. These results suggest that the CD14-derived DC type might be preferentially involved in development of humoral responses, while both populations can induce T cell priming.

MATERIALS AND METHODS

Isolation of cord blood CD34$^+$ progenitors was achieved using Minimacs separation columns (Miltenyi Biotec GmbH, Bergisch Gladbach, Germany). Cultures were estab-

Dendritic Cells in Fundamental and Clinical Immunology
edited by Ricciardi-Castagnoli, Plenum Press, New York, 1997

lished in the presence of SCF, GM-CSF and TNFα, as described (9–11). Cells were routinely collected after 5–6 days of culture for FACS-sorting according to CD1a and CD14 expression into $CD14^+CD1a^-$, $CD14^-CD1a^+$. Sorted cells were seeded in the presence of GM-CSF+TNFα for 6–7 additional days.

RESULTS

1. $CD34^+$ Cells Differentiate Along Two Dendritic Cell Pathways

Human DC can be generated in vitro in large numbers by culturing $CD34^+$ hematopoietic progenitors in presence of SCF+GM-CSF+TNFα during 12 days (9). At early time points (day 5–7) during the culture, two subsets of DC precursors identified by the exclusive expression of CD1a and CD14 emerge independently (11). Both precursor subsets mature at day 12–14 into DC with typical morphology and phenotype (expression of CD80, 83, 86, 58, high HLA class II). $CD1a^+$ precursors give rise to cells characterized by the expression of BG, the Lag antigen and E-cadherin, three markers specifically expressed on Langerhans cells in the epidermis. In contrast, the $CD14^+$ progenitors mature into $CD1a^+$ DC lacking Birbeck granules, E-cadherin and Lag antigen but expressing CD2, CD9, CD68 and the coagulation factor XIIIa described in dermal dendritic cells. The two mature DC were equally potent in stimulating allogeneic $CD45RA^+$ naive T cells. Interestingly, the $CD14^+$ precursors, but not the $CD1a^+$ precursors, represent bipotent cells that can

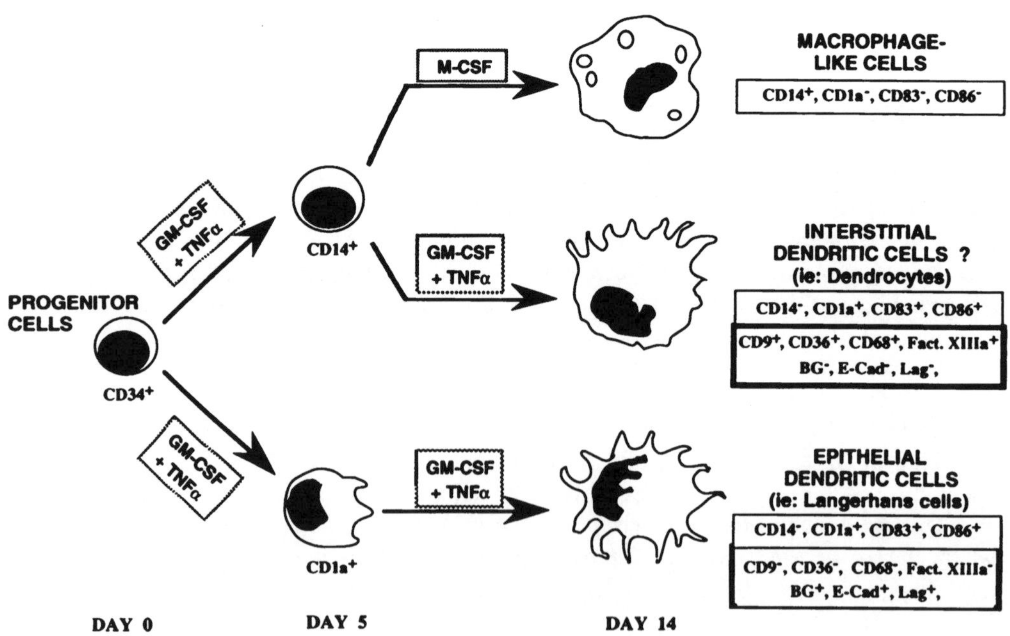

Figure 1. Different dendritic cell populations originate from different progenitor cells. $CD1a^+$ precursors represent DC specific precursors leading to epithelial type DC (ie: Langerhans cells). $CD14^+$ precursors represent precursors with dual differentiation potential leading to either interstitial type DC (ie: dermal DC) in response to GM-CSF+TNFα or to macrophage-like cells in response to M-CSF.

be induced to differentiate, in response to M-CSF, into macrophage-like cells, lacking accessory function for T cells.

Altogether, those results demonstrate that different pathways of DC development exist: the Langerhans cells and the CD14$^+$ derived DC related to dermal DC or circulating blood DC (Fig.1).

2. Relationships with Monocyte Derived Dendritic Cells (Mo-DC)

As reported by many groups (12–14), elutriated monocytes cultured with GM-CSF+IL-4 were found to differentiate into immature DC (CD1a$^+$, 14$^-$, 80$^-$, 86$^-$, 83$^-$, class II^{low}, moderate T cell priming). Following CD40 activation those cells develop into fully competent mature DC (CD1a^{low}, 14$^-$, 80$^+$, 86$^+$, 83$^+$, class II^{high}, strong T cell priming). We have investigated the effect of IL-4 on CD34$^+$ derived subsets. After facs-sorting cells were cultured in presence of GM-CSF+TNF with or without IL-4. As shown in Fig.2, IL-4 blocks maturation of CD14+ derived cells which develop a phenotype similar to Mo-DC (CD1a$^+$, 14$^-$, 80$^-$, 86$^-$, 83$^-$, class II^{low}, 11b$^+$, 23$^+$). In contrast, CD1a+ precursors were not affected by IL-4 in their maturation. Furthermore, CD14$^+$ derived cells cultured with IL-4 were 10 to 50 fold less efficient in activating naive T cells (not shown). The supperposable effects of IL-4 on monocytes and CD14$^+$ precursors suggest that Mo-DC and CD14$^+$ derived DC represent the same pathway of DC development.

3. Function of DC Subsets

CD14$^+$ derived DC demonstrate a potent antigen uptake activity (FITC-dextran or peroxidase) that is about 10 fold more efficient than that of CD1a$^+$ precursors. This uptake of CD14$^+$ derived DC was long lasting (from day 8 to day 13) while the weak activity of CD1a$^+$ cells was restricted to the immature stage (day 6) (Fig.3A). The capture of FITC-

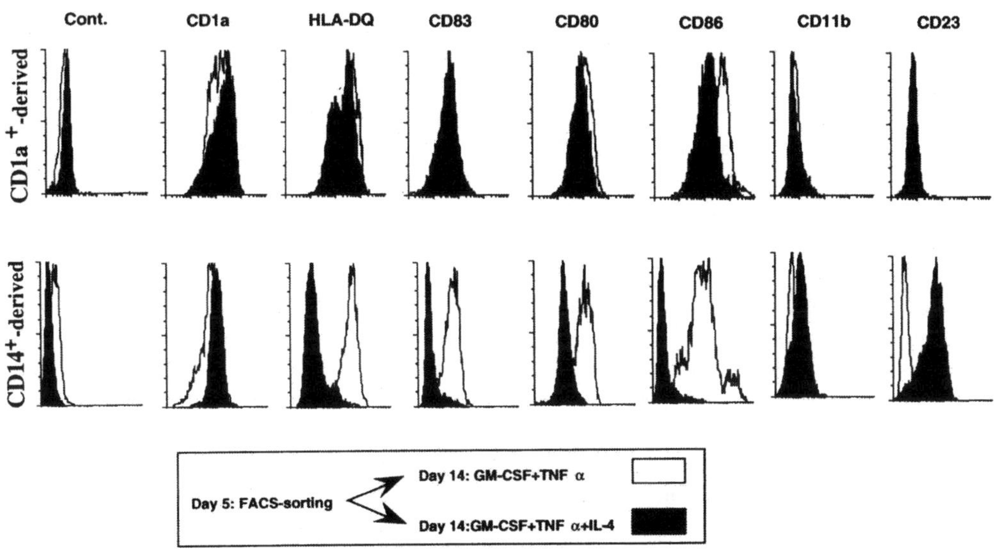

Figure 2. IL-4 blocks maturation of CD14 derived DC. After Facs-sorting, cells were cultured in presence of GM-CSF+TNFα with or without 50 U/ml IL-4. At day 12 cells were processed for cell surface phenotype.

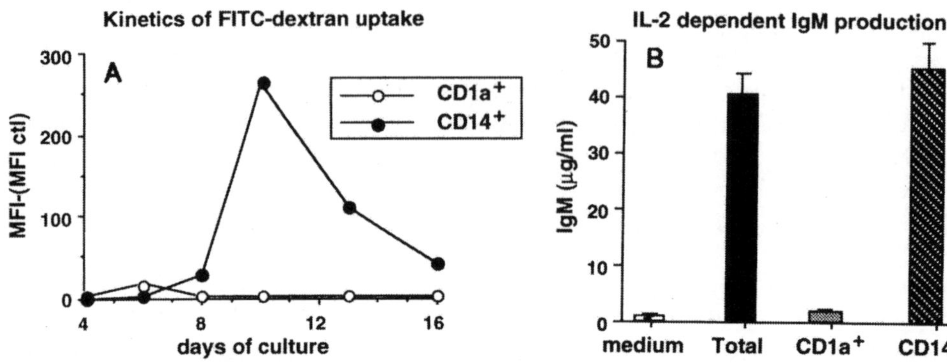

Figure 3. Differential functions of CD1a and CD14 derived DC. Panel A: At the indicated time points, cells were processed for FITC-dextran uptake (15). Cells were incubated in medium containing 0.1 mg/ml FITC-dextran at 37°C for 15 minutes, washed with cold medium, and analysed on a FACScan. MFI of histograms is shown. The background fluorescence (cells pulsed at 4°C) was subtracted. Panel B: 10^4 highly purified IgD^+ B cells were cultured over 2500 irradiated CD40-L transfected L cells in presence or absence of DC subsets (3000 cells per well), with or without 10 U/ml IL-2 (16). Supernatants were harvested after 15 days of culture and assayed for the presence of IgM.

dextran was in both cases exclusively mediated by receptors for mannose polymers. The high efficiency of antigen capture of $CD14^+$ derived cells is coregulated with the expression of non-specific esterase activity, a tracer of lysosomial compartment. In contrast the $CD1a^+$ population never express non-specific esterase activity (not shown). The most striking difference between the two DC populations is the unique capacity of $CD14^+$ derived DC to induce, in response to CD40 triggering and IL-2, naive B cells to proliferate and differentiate into IgM secreting cells (Fig.3B). Thus, while the two populations can allow T cell priming, only the CD14-derived DC may be involved in initiation of humoral responses.

DISCUSSION

The present studies lead us to propose that epithelial DC ($CD1a^+$ derived Langerhans cells) are preferentially involved in the cellular type immune responses, which is supported by the involvment of LC in delayed type hypersensitivity reactions. In contrast, the $CD14^+$ derived DC, potentially located in tissues such as dermis or blood, might after antigen capture migrate through the T cell rich area into the B cell follicules where they could be involved in the generation of primary B cell responses. This hypothesis, suggesting a preferential role of monocyte derived DC in humoral type immune responses, is in line with the, recently described, presence of typical DC of myeloid origin within human tonsil germinal centers (8). The present study demonstrate the existence of two independent pathways for the development of DC with discrete biological functions. The existence of different pathways of DC development is indeed supported by the description, in mice, of thymic DC originating from hematopoietic precursors with a lymphoid differentiation potential contrasting with myeloid precursors yielding splenic DC (9–11,17). This heterogeneity of DC requires further understanding, in particular in view of the development of strategies aimed at using DC as therapeutic entities to elicit immune responses specific for infectious agents as well as cancer.

REFERENCES

1. U. O'Doherty, M. Peng, S. Gezelter, W.J. Swiggard, M. Betjes, N. Bhardwaj, and R.M. Steinman. Human blood contains two subsets of dendritic cells, one immunologically mature and the other immature. *Immunology*. 82:487 (1994).
2. R. Thomas, and P.E. Lipsky. Human peripheral blood dendritic cell subsets. Isolation and characterization of precursor and mature antigen-presenting cells. *J. Immunol.* 153:4016 (1994).
3. D. Weissman, Y. Li, J. Ananworanich, L.-J. Zhou, J. Adelsberger, T.F. Tedder, M. Baseler, and A.S. Fauci. Three populations of cells with dendritic morphology exist in peripheral blood, only one of which is infectable with human immunodeficiency virus type 1. *Proc. Natl. Acad. Sci. USA*. 92:826 (1995).
4. A. Lenz, M. Heine, G. Schuler, and N. Romani. Human and murine dermis contain dendritic cells. *J. Clin. Invest.* 92:2587 (1993).
5. F.O. Nestle, X.-G. Zheng, C.B. Thompson, L.A. Turka, and B.J. Nickoloff. Characterization of dermal dendritic cells obtained from normal human skin reveals phenotypic and functionally distinctive subsets. *J. Immunol.* 151:6535 (1993).
6. R. Agger, M. Witmer-Pack, N. Romani, H. Stossel, W.J. Swiggard, J.P. Metlay, E. Storozynsky, P. Freimuth, and R.M. Steinman. Two populations of splenic dendritic cells detected with M342, a new monoclonal to an intracellular antigen of interdigitating dendritic cells and some B lymphocytes. *J. Leukocyte Biol.* 52:34 (1992).
7. B.L. Kelsall, and W. Strober. Distinct populations of dendritic cells are present in the subepithelial dome and T cell regions of the murine Peyer's patch. *J. Exp. Med.* 183:237 (1996).
8. G. Grouard, I. Durand, L. Filgueira, J. Banchereau, and Y.J. Liu. Dendritic cells capable of stimulating T cells in germinal centers. *Nature*. 384:364 (1996).
9. C. Caux, C. Dezutter-Dambuyant, D. Schmitt, and J. Banchereau. GM-CSF and TNF-a cooperate in the generation of dendritic Langerhans cells. *Nature*. 360:258 (1992).
10. P. Szabolcs, M.A.S. Moore, and J.W. Young. Expansion of immunostimulatory dendritic cells among the myeloid progeny of human CD34⁺ bone marrow precursors cultured with c-kit-ligand, GM-CSF, and TNFα. *J. Immunol.* 154:5851 (1995).
11. C. Caux, B. Vanbervliet, C. Massacrier, C. Dezutter-Dambuyant, B. de Saint-Vis, C. Jacquet, K. Yoneda, S. Imamura, D. Schmitt, J. Banchereau. CD34⁺ hematopoietic progenitors from human cord blood diffentiate along two independent dendritic cell pathways in response to GM-CSF+TNFα, *J. Exp. Med.* 184:695 (1996).
12. F. Sallusto, and A. Lanzavecchia. Efficient presentation of soluble antigen by cultured human dendritic cells is maintained by granulocyte/macrophage colony-stimulating factor plus interleukin 4 and downregulated by tumor necrosis factor alpha. *J. Exp. Med.* 179:1109 (1994).
13. N. Romani, S. Gruner, D. Brang, E. Kämpgen, A. Lenz, B. Trockenbacher, G. Konwalinka, P.O. Fritsch, R.M. Steinman, and G. Schuler. Proliferating dendritic cell progenitors in human blood. *J. Exp. Med.* 180:83 (1994).
14. L.-J. Zhou, and T.F. Tedder. CD14+ blood monocytes can differentiate into functionally mature CD83⁺ dendritic cells. Proc. Natl. Acad. Sci. USA. 93:2588 (1996).
15. F. Sallusto, M. Cella, C. Danieli, and A. Lanzavecchia. Dendritic cells use macropinocytosis and the mannose receptor to concentrate macromolecules in the major histocompatibility complex class II compartment: down-regulation by cytokines and bacterial products. *J. Exp. Med.* 182:389 (1995).
16. B. Dubois, B. Vanbervliet, J. Fayette, C. Massacrier, C. Van Kooten, F. Brière, J. Banchereau, C. Caux. Dendritic cells enhance growth and differentiation of CD40-activated B lymphocytes, *Submitted.*.
17. C. Ardavin, L. Wu, C.L. Li, and K. Shortman. Thymic dendritic cells and T cells develop simultaneously in the thymus from a common precursor population. *Nature*. 362:761 (1993).

5

MYELOID DCs DEDUCED FROM MONOCYTES

In-Vitro and in-Vivo Data Support a Monocytic Origin of DCs

Falko Steinbach,[1] Robert Gieseler,[2] Afsaneh Soruri,[2] Bianca Krause,[1,3] and J. Hinrich Peters[2]

[1]Institute for Virology, FU Berlin
Königin-Luise-Str. 49, D-14195 Berlin, Germany
[2]Department of Immunolgy
Georg-August-University
D-37075 Göttingen, Germany
[3]Department of Medicine III
Charité, HU Berlin
D-10117 Berlin, Germany

We recently presented cumulative evidence for the monocytic origin of myeloid dendritic cells (DC)[1]. Here we extend this concept and conclude that DC are to be regarded as members of the mononuclear phagocyte system (MPS).

1. LESSONS FROM IN-VIVO STUDIES

Features of monocytes (Mo) and macrophages (Mφ), earlier taken as steadfast, were, at the time, drawn upon to define the entire MPS. Therefore, antigen-presenting cells (APC) lacking such classical criteria were *per se* excluded from the MPS and hypothesized to constitute a separate lineage.

From this point of view, Takahashi and colleagues[2] excluded a monocytic origin of DC, because the generation of tissue-resident DC was not affected from the lack of M-CSF in osteopetrotic mice, otherwise coinciding with reduced numbers of osteoclasts and Mφ. However, DC development from a precursor common with Mφ may be less dependent on the presence of M-CSF. Indeed, DC development was achieved at only low concentrations of M-CSF[3], or without[4].

Moreover, early histological studies in neonatally thymectomized rats demonstrated promonocytic cells and monocytes in the splenic T-cell areas prior to the emergence of interdigitating reticulum cells (IDC), suggesting these to be part of the MPS[5]. Based on selected markers and electron-microscopic evidence, Hoefsmit et al.[6] concluded that Langerhans cells (LC), veiled cells and IDC may constitute discrete subpopulations of

Dendritic Cells in Fundamental and Clinical Immunology
edited by Ricciardi-Castagnoli, Plenum Press, New York, 1997

Mφ, which was also supported by Parwaresch et al.[7] An additionally report described the phenotypic transformation of Mφ to skin LC[8].

2. GENERATION OF DC FROM MONOCYTES

Monocytes isolated by adherence from the peripheral blood of humans or BALB/c mice, and cultured in media supplemented with GM-CSF plus IL-4 detach after 1–2 days and subsequently develop prominent veils as well as small protrusions (Fig.1).

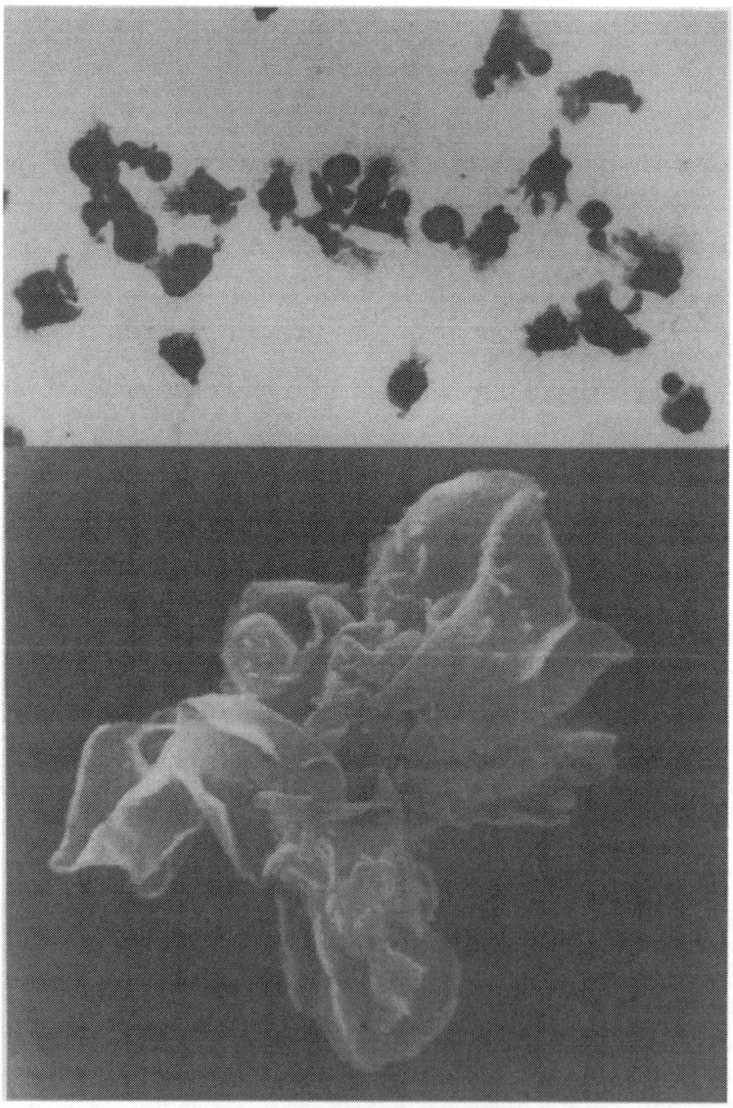

Figure 1. Human monocyte-derived dendritic cells (MoDC).

Table 1. Accessory activity of mouse MoDC

Stimulator cells	–	BMC	BMDC	BMDC	Mo	MoDC	MoDC
Stimulator cells / well	0	1×10^4	1×10^4	5×10^3	1×10^4	1×10^4	1×10^3
Responder cells / well	1×10^5	1×10^5	1×10^5	1×10^5	1×10^5	1×10^5	1×10^5
[^{3}H]-TdR (cpm)	107.4	460.33	12.1×10^3	9.6×10^3	6.3×10^3	11.5×10^3	5.5×10^3

BALB/c bone marrow-derived DC (BMDC) were generated using GM-CSF (10 ng/ml, Intergen). Monocytes were prepared from blood pooled after cardiac puncture. They were prepared by Ficoll-Metrizamide centrifugation and selective adherence on hydrophobic Teflon dishes (Petriperm, Bachofer, Reutlingen, FRG). They were differentiated by murine recombinant GM-CSF (5 ng/ml, Intergen), IL-4 (10 ng/ml, Intergen) and IFNγ (50 U/ml, Genzyme) in RPMI 1640 plus 10% FCS for 6 days, and thereafter used as stimulator cells. NMRI blood lymphocytes were used as responder cells. As controls, unmodified BM cells (BMC) and freshly prepared Mo were used. The resulting lymphocyte stimulation (cpm) is expressed as means of 5 parallel experiments.

They express high levels of class II antigen and are as potent in antigen presentation as are DC (for overview see Ref. 1). Such cells are now designated as monocyte-derived dendritic cells (MoDC), which present all phenotypic and functional key features of DC (T-cell stimulation both in allogeneic MLR and in antigen presentation assays)[4,9–12]. MoDC as the most easily accessible type of human DC bear an enormous impact for human therapeutic purposes e.g.[13] For developing an analogous animal model, we have successfully generated functionally potent murine MoDC (Table 1).

Beyond GM-CSF & IL-4, further factors may be added to differentiate human MoDC which are otherwise constitutively supplied by the monocytes themselves, such as IL-1 and IL-6[14], or TNFα[15]. Further factors, like IFNγ[14] derived from other cells (such as lymphocytes or platelets), have been detected in MoDC preparations. In other words, autocrine factors might also be present in "MoDC-conditioned media" and contribute to their differentiation[16].

3. THE INS AND OUTS OF BLOOD DC

Are blood DC a resident cell population or rather in transit from point A to B? Several subpopulations described may be attributed to migratory forms of DC. The first pathway to discuss will be the delivery of identifiable DC from the bone marrow, which then may appear as naive DC in the blood. Secondly and later in their life cycle, DC emigrating from the periphery to lymphoid organs may appear in the blood again.

O'Doherty et al. described two subsets of DC in the human blood[17], differing in their expression of the β_2-integrin (CD11c). The CD11c-negative cells were regarded as immature precursors. However, attempts to show the appearance of DC-lineage precursors in the periphery by intradermal GM-CSF injection failed. Moreover, no further or enhanced numbers of DC-precursors appeared in the blood[18]. The CD11c-positive DCs also expressed the activation antigen CD45R0 and were thus considered to be tissue-derived mature DC, possibly being *en route* to the spleen or lymph nodes. This notion correlates with results from transplantation studies: A direct release of allogeneic DC from the tissue into the blood was observed after allografting[19,20], and may as well occur in healthy, untreated individuals. This provides an explanation for the appearance of mature DC in the blood. Importantly, because of its lack in afferent lymphatics, the spleen recruits DC from the blood (Fig. 2).

DC may develop from monocytes as a result of prolonged purification. Accordingly, culture of monocytes leads to a transient peak of accessory activity and DC morphology after 24 hours[21]. Gradient medium containing Metrizamide has been described to induce differentiation of DC from monocytes[22], and many other protocols may exert similar ef-

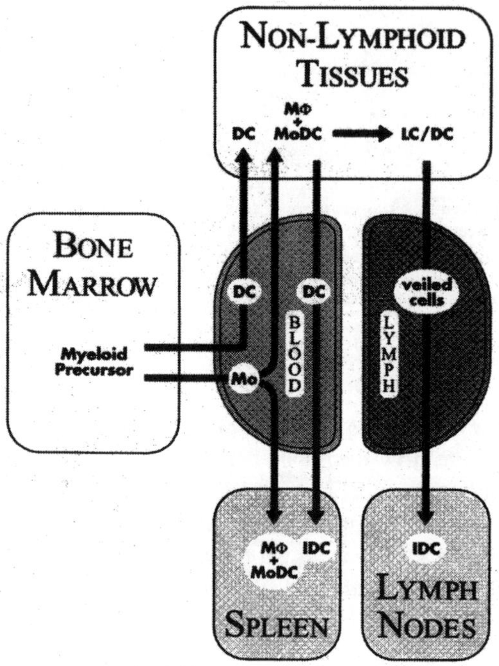

Figure 2. Traffic and homing of DC. DC may develop as early as in the bone marrow and be transported through the peripheral blood towards the non-lymphoid tissues. Alternatively, blood-borne monocytes may enter the peripheral organs to develop either into Mφ or MoDC. Langerhans cells (LC) have been mentioned representatively for various tissue-specific DC subtypes. When leaving the peripheral non-lymphoid organs, they migrate via peripheral blood or lymph into the spleen or lymph nodes, respectively.

fects. In early observations, and using alternative purification methods, some groups obtained many more veiled cells within blood-derived populations than others[22,23]. In retrospect, these observations are most likely explained by the plasticity of monocytes that were triggered towards DC differentiation. Likewise, when comparing peripheral blood DC with monocytes, a population of $CD14^{dim}$, $CD33^{+}$ cells was classified as DC precursors[24]. The assumption, however, that strong expression of CD14 is an irreversible marker of Mo/Mφ requires re-evaluation, as CD14 is subject to IL-4 regulation[10,25], which demonstrates that CD14 dim or negative cells can emerge from a post-monocytic stage[26].

4. DO DC BELONG TO THE MPS?

The case that DC, and Mφ alike, can be obtained homogenously from $CD34^{+}$ precursors, does not contradict their myelomonocytic origin: In rat bone marrow cultures, Gieseler et al.[3] were the first to describe a transient monocytic stage in developing DC and this was most recently confirmed[27]. Therefore, within the myeloid lineage, a pre-monocytic[28] as well as a post-monocytic origin of DC may be taken as equivalent sources of DC, which is in favour of a flexible, though perhaps heterogenous evolution of the immune system[29]. There is no better proof for the affiliation to the MPS than the existence of mixed DC-Mφ phenotypes, which can physiologically be encountered in mucosal tissues, and were first generated *in vitro* by Goordyal and Isaacson[30].

ACKNOWLEDGMENTS

This work was supported by the Deutsche Forschungsgemeinschaft, SFB 500, (C4), and grant Pe 192/6–3, as well as by grants from the state of Berlin. We thank Ms. S. Lenzner for artwork and Dr. P. Schwartz for REM.

REFERENCES

1. Peters, J.H., R. Gieseler, B. Thiele, and F. Steinbach. (1996). Dendritic cells – from ontogenetic orphans to myelo-monocytic descendants. *Immunol. Today, 17, 273–278*
2. Takahashi K, Naito M, Morioka Y, Schultz LD (1993) Immunophenotypic and ultrastructural differentiation and maturation of nonlymphoid dendritic cells in osteopetrotic (op) mice with the total absence of macrophage colony stimulating factor activity. Adv. Exp. Med. Biol. 329: pp 293–297
3. Gieseler, R.K., R.A. Röber, R. Kuhn, K. Weber, M. Osborn, and J.H. Peters. (1991). Dendritic accessory cells derived from rat bone marrow precursors under chemically defined conditions in vitro belong to the myeloid lineage. *Eur. J. Cell Biol.* 54, no. 1:171–181.
4. Xu, H., M. Krämer, H.-P. Spengler, and J.H. Peters. 1995. Dendritic cells differentiated from human monocytes through a combination of IL-4, GM-CSF and IFNγ exhibit phenotype and function of blood dendritic cells. *Adv. Exp. Med. Biol.378:75–78*
5. Veerman AJP (1974) On the interdigitating cells in the thymus-dependent area of the rat spleen: A relation between the mononuclear phagocyte system and T-lymphocytes. Cell Tissue Res. 148:247–257
6. Hoefsmit ECM, Duijvestijn AM, Kamperdijk EWA (1982) Relation between Langerhans cells, veiled cells, and interdigitating cells. Immunobiol. 161:255–265
7. Parwaresch MR, Radzun HJ, Feller AC, Peters K-P, Hansmann M-L (1983) Peroxidase-positive mononuclear leukocytes as possible precursors of human dendritic reticulum cells. J. Immunol. 131:2719–2725
8. Murphy GF, Messadi D, Forferko E, Hancock WW (1986) Phenotypic transformation of macrophages to Langerhans cells in the skin. Am. J. Pathol. 123: 401–406
9. Sallusto, F., and A. Lanzavecchia. 1994. Efficient presentation of soluble antigen by cultured human dendritic cells is maintained by GM-CSF plus IL-4 and downregulated by TNFα. *J. Exp. Med* 179:1109–1116.
10. Ruppert J, Schütt C, Ostermeier D, Peters JH (1993) Down-regulation and release of CD14 on human monocytes by IL-4 depends on the presence of serum or GM-CSF. Adv. Exp. Med. Biol. 329: 281–286
11. Steinbach, F., B. Krause, and B. Thiele. 1995. Monocyte derived Dendritic Cells (MoDC) present phenotype and functional activities of Langerhans Cells / Dendritic Cells. *Adv. Exp. Med. Biol.* 378, 151–153.
12. Gieseler R, Heise D, Soruri A, Schwartz P, Peters JH. In-vitro differentiation of mature-type dendritic cells from human blood monocytes. Develop. Immunol. (in press)
13. Mukherji B, Chakraborty NG, Yamasaki S, Okino T, Yamase H, Sporn JR, Kurtzman SK, Ergin MT, Ozols J, Meehan J, Mauri F (1995) Induction of antigen-specific cytolytic T cells *in situ* in human melanoma by immunization with synthetic peptide-pulsed autologous antigen-presenting cells. Proc. Natl. Acad. Sci. USA 92:8078–8082
14. Ruppert, J., and J.H. Peters. (1991). IL-6 and IL-1 enhance the accessory activity of human blood monocytes during differentiation to macrophages. *J. Immunol.* 146, 144–149.
15. Krause, B. Differenzierung Dendritischer Zellen aus Monozyten (MoDC): Vorbereitende Studie zur Differenzierung von MoDC bei SLE Patienten *MD thesis in prep.*, Charité der HU zu Berlin
16. Steinbach, F. et al., (in prep.), Endogenous factors play a central role in the differentiation of DC from monocytes (MoDC)
17. O'Doherty, U., M. Peng, S. Gezelter, W.J. Swiggard, M. Betjens, N. Bhardway, and R.M. Steinman. 1994. Human blood contains two subsets of dendritic cells, one immunologically mature and the other immature. *Immunology* 82:487–493.
18. Kaplan, G., G. Walsh, L.S. Giudo, P. Meyn, R.A. Burkhardt, R.M. Abalos, J. Barker, P.A. Frindt, T.T. Fajardo, R. Celona, Z.A. Cohn. 1992. Novel responses of human skin to intradermal recombinant granulocyte/macrophage-colony-stimulating factor: Langerhans cell recruitment, keratinocyte growth and enhanced wound healing. *J. Exp. Med.* 175:1717–1728.
19. Larsen CP, Morris PJ, Austyn JM (1990 a) Migration of dendritic leukocytes from cardiac allografts into host spleens: A novel pathway for initiation of rejection. J. Exp. Med. 171:307–314
20. Larsen CP, Steinman RM, Witmer-Pack M, Hankins DF, Morris PJ, and Austyn JM (1990 b) Migration and maturation of Langerhans cells in skin transplants and explants. J. Exp. Med. 172:1483–1493
21. Peters JH, Ruhl S, Friedrichs D (1987) Veiled accessory cells deduced from monocytes. Immunobiol. 176:154–166
22. Kabel PJ, de Haan-Meulman M, Voorbij HA, Kleingeld M, Knol EF, Drexhage HA (1989) Accessory cells with a morphology and marker pattern of dendritic cells can be obtained from elutriator-purified blood monocyte fractions. An enhancing effect of metrizamide in this differentiation. Immunobiol. 179: 395–341
23. Knight SC, Farrant J, Bryant A, Edwards AJ, Burman S, Lever A, Clarke J, Webster ADB (1986) Non-adherent, low-density cells from human peripheral blood contain dendritic cells and monocytes, both with veiled morphology. Immunology 57:595–603

24. Thomas, R., L.S. Davies, and P.E. Lipsky. 1993. Isolation and characterization of human peripheral blood dendritic cells. *J. Immunol.* 150:821–834.
25. Lauener RP, Goyert SM, Geha RS, Vercelli D (1990) Interleukin 4 down-regulates the expression of CD14 in normal human monocytes. Eur. J. Immunol. 20:2375–2381
26. Poulter LW, Campbell DA, Munro C, Janossy G (1986) Discrimination of human macrophages and dendritic cells by means of monoclonal antibodies. Scand. J. Immunol. 24:351–357
27. Chen-Woan M, Delaney CP, Fournier V, Wakizaka Y, Murase N, Fung J, Starzl TE, Demetris AJ (1996) In vitro characterization of rat bone marrow-derived dendritic cells and their precursors. J. Leuk. Biol. 59:196–207
28. Inaba K, Inaba M, Deguchi M, Hagi K, Yasumizu R, Ikehara S, Muramatsu S, Steinman RM (1993) Granulocytes, macrophages, and dendritic cells arise from a common major histocompatibility complex class II-negative progenitor in mouse bone marrow. Proc. Natl. Acad. Sci. USA 90:3038–3042
29. Leenen PJM, Campbell PA (1993) Heterogeneity of mononuclear phagocytes: An interpretive review. In: Horton MA (Ed.): Blood Cell Biochemistry, Vol. 5: Macrophages and Related Cells. Plenum Press New York pp.29–85
30. Goordyal P, Isaacson PG (1985) Immunocytochemical characterization of monocyte colonies of human bone marrow: a clue to the origin of Langerhans cells and interdigitating reticulum cells. J. Pathol. 146:189–195

6

DRAMATIC NUMERICAL INCREASE OF FUNCTIONALLY MATURE DENDRITIC CELLS IN FLT3 LIGAND-TREATED MICE

Eugene Maraskovsky,[1] Bali Pulendran,[1] Ken Brasel,[1] Mark Teepe,[1] Eileen R. Roux,[1] Ken Shortman,[3] Stewart D. Lyman,[2] and Hilary J. McKenna[1]

[1]Department of Immunobiology and
[2]Department of Molecular Genetics
Immunex Corporation
Seattle, Washington 98101
[3]The Walter and Eliza Hall Institute of Medical Research
Melbourne, Victoria 3050 Australia

1. INTRODUCTION

Dendritic cells (DC) are rare hematopoietic-derived cells that are predominantly found in the T-cell-dependent areas of lymphoid tissue, as well as other tissues of the body[1]. These cells express high levels of class I and class II major histocompatibility complex (MHC) proteins, CD11c[2], the mannose-receptor like protein DEC205[3,4] and adhesion and costimulatory molecules[1]. A substantial proportion of DC express CD8α as a homodimer[5]. DC specialize in processing and presenting foreign and self antigens to induce immunity[1] or tolerance[6,7]. The lineage derivation of DC remains controversial, but there is growing evidence that DC can be subdivided into myeloid-derived[8–17] and lymphoid-derived populations[17–19].

By virtue of their highly developed antigen-presenting capacity, the use of DC as cellular vectors for anti-tumour and infectious disease vaccines or as inducers of transplantation tolerance is a promising immunotherapy strategy[20–22]. However, the clinical feasibility of using DC as immunotherapy vectors is hampered because only limited DC numbers can be generated in vitro from bone marrow progenitors or peripheral blood mononuclear cells (PBMC) and only one lineage of DC (myeloid-derived) is generated due to the heavy reliance upon GM-CSF[8–17]. Interestingly, although DC can be generated in vitro using GM-CSF, the increased level of GM-CSF in GM-CSF transgenic mice does not increase the level of DC in lymphoid tissue suggesting that other growth factors are important for DC generation in vivo[23]. In search of the growth factor requirements for DC generation in vivo, we examined the effects of administering a recently identified hematopoietic growth factor, Flt3L[24]. Flt3L has been shown to stimulate the proliferation of hematopoietic stem and progenitor cells[25,26].

2. FLT3L TREATMENT OF MICE RESULTS IN A SIGNIFICANT INCREASE OF CELLS EXPRESSING DC MARKERS IN MULTIPLE ORGANS

DC are very rare in dissociated spleen cell suspensions. The majority of DC require release from the splenic stroma using enzymatic digestion, but even then they are an infrequent component of the suspension[27]. However, unlike spleen cell suspensions from mouse serum albumin (MSA)-treated control mice, mice treated daily for 9 days with sub-

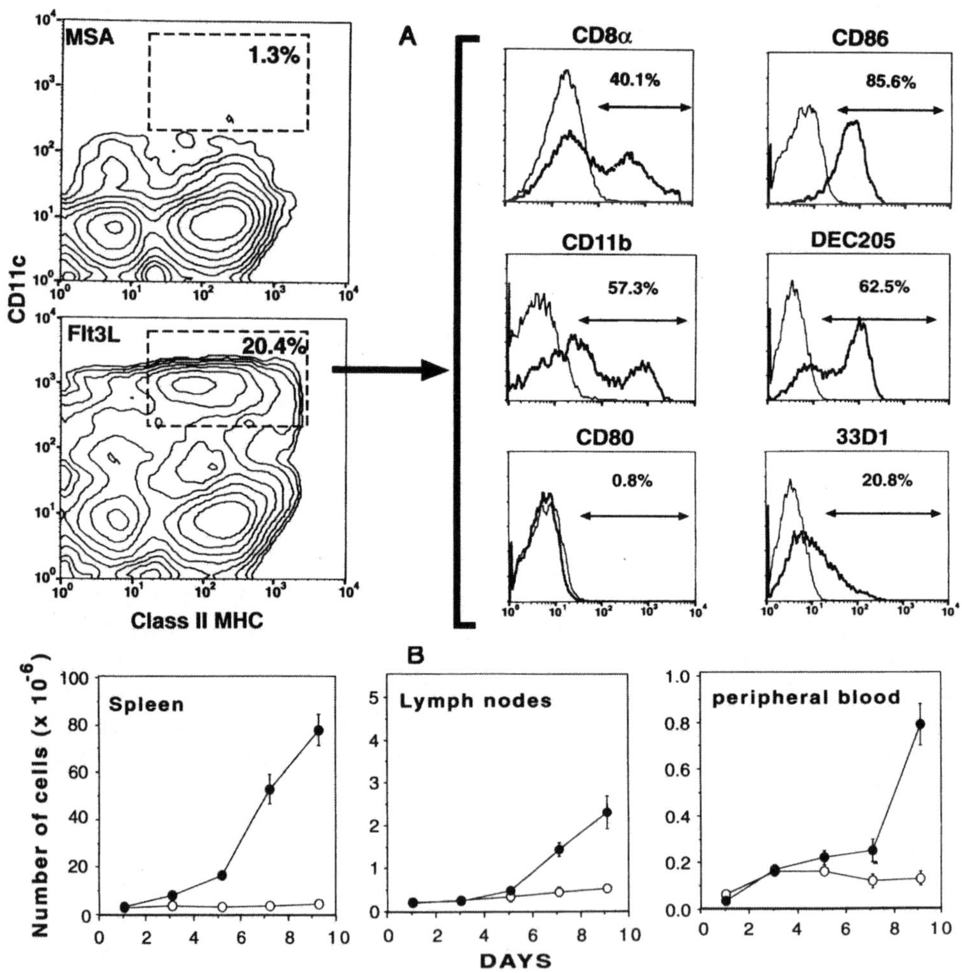

Figure 1. Flow cytometric analysis of spleen cells from mice treated with Flt3L. (A) The distribution of CD8α, CD11b, CD80, CD86, DEC205 or 33D1 on class II+ CD11c+ spleen cells. C57BL/6 mice (five per group) were injected once daily, subcutaneously, with MSA (1 μg) plus human Flt3L derived from Chinese hamster ovary cells (CHO) (10 μg) for 9 consecutive days. Flow cytometric analysis of spleen cells, LN (inguinal, axillary) and PB was performed on a FACStar Plus (Becton Dickinson, San Jose, CA). (B) The kinetics of generation of class II+ CD11c+ cells in spleen, LN and PB from mice treated daily for 9 days with either MSA (open circles) or with MSA plus Flt3L (closed circles). Values represent the mean ± SD of 5 mice. Figure 1B is reproduced from *The Journal of Experimental Medicine,* 180: p 53–62, by copyright permission of The Rockefeller University Press.

cutaneous injections of human Flt3L (derived from Chinese hamster ovary cells) resulted in 20% of spleen cells co-expressing class II MHC and the DC marker, CD11c (Figure 1A). Further phenotypic analysis revealed that class II$^+$ CD11c$^+$ cells were DEC205$^+$, CD80$^-$, CD86$^+$, heterogeneous for 33D1[28] and could be separated into three populations using CD11b expression (CD11b$^-$, CD11b^{dull} and CD11b^{bright}) (Figure 1A). Class II$^+$ CD11c$^+$ cells were CD3$^-$, CD19$^-$, Gr-1$^-$, NK1.1$^-$ and Ter119$^-$[29]. The absolute number of class II$^+$ CD11c$^+$ cells was increased by 17-fold in the spleen, 4-fold in the lymph nodes (LN) and 6-fold in the peripheral blood (PB) by day 9 (Figure 1B). Class II$^+$ CD11c$^+$ cells were detected as early as day 5 in the spleen and by day 7 in LN and PB (Figure 1B). Class II$^+$ CD11c$^+$ cells were also increased in number in the bone marrow, Peyer's patches, LN, liver, lung, thymus and peritoneal cavity[29].

3. FIVE DISTINCT POPULATIONS OF FUNTIONALLY MATURE DC CAN BE IDENTIFIED IN THE SPLEENS OF FLT3L-TREATED MICE

Lymphocyte-depleted spleen cells from Flt3L-treated mice were examined for the expression of CD11b and CD11c (Figure 2A). Five distinct populations were identified. Population A (CD11b^{bright} CD11c$^-$) was class II$^-$, and granulocyte marker Gr-1$^+$[29]; population B (CD11b^{bright} CD11c^{dull}) was heterogeneous for class II and Gr-1; whereas populations C (CD11b^{bright} CD11c$^+$) D (CD11b^{dull} CD11c$^+$) and E (CD11b$^-$ CD11c$^+$) were class II$^+$ and Gr-1$^-$[29]. Populations A, B and C were composed exclusively of CD8α$^-$ cells, whereas populations D, E and DC freshly isolated from the spleens of untreated mice (control DC)[5] could be separated into CD8α$^-$ and CD8α$^+$ populations (Figure 2A).

Wright-Giemsa staining of cytospins indicated that cells within populations A and B were highly enriched for immature myeloid and blast cells[29]. However, populations C, D and E were highly enriched for cells with veiled and dendritic processes that were indistinguishable from control DC[29]. Examination for in vitro antigen presenting capacity revealed that cells from populations C, D and E were as efficient as control DC at stimulating the proliferation of allogeneic DBA/2-derived T cells in a mixed lymphocyte reaction or of syngeneic keyhole limpet hemocyanin (KLH)-specific T cells in a KLH presentation assay (Figure 2B). In contrast, cells from populations A and B were 30-fold less efficient as antigen presenting cells compared with control DC (Figure 2B).

When class II$^+$ CD11c$^+$ spleen cells from Flt3L-treated mice were cultured in vitro with soluble KLH for 24 h and injected into the foot pads of naive mice, they were found to be as efficient as control DC at generating KLH-specific T cells in the draining LN (Figure 3)[30]. KLH-specific T cells were not detected in the contralateral LN draining the foot pads injected with either Flt3L-generated class II$^+$ CD11c$^+$ spleen cells or with control DC cultured in medium alone (-KLH) (Figure 3).

4. FLT3L TREATMENT, BUT NOT GM-CSF, GM-CSF PLUS IL-4, C-KIT LIGAND, OR G-CSF TREATMENT, CAN GENERATE LARGE NUMBERS OF DC

Class II$^+$ CD11c$^+$ cells were rare in the spleens of mice treated with either MSA alone or with muGM-CSF, or huG-CSF, or muGM-CSF plus muIL-4, or muciktL (Fig-

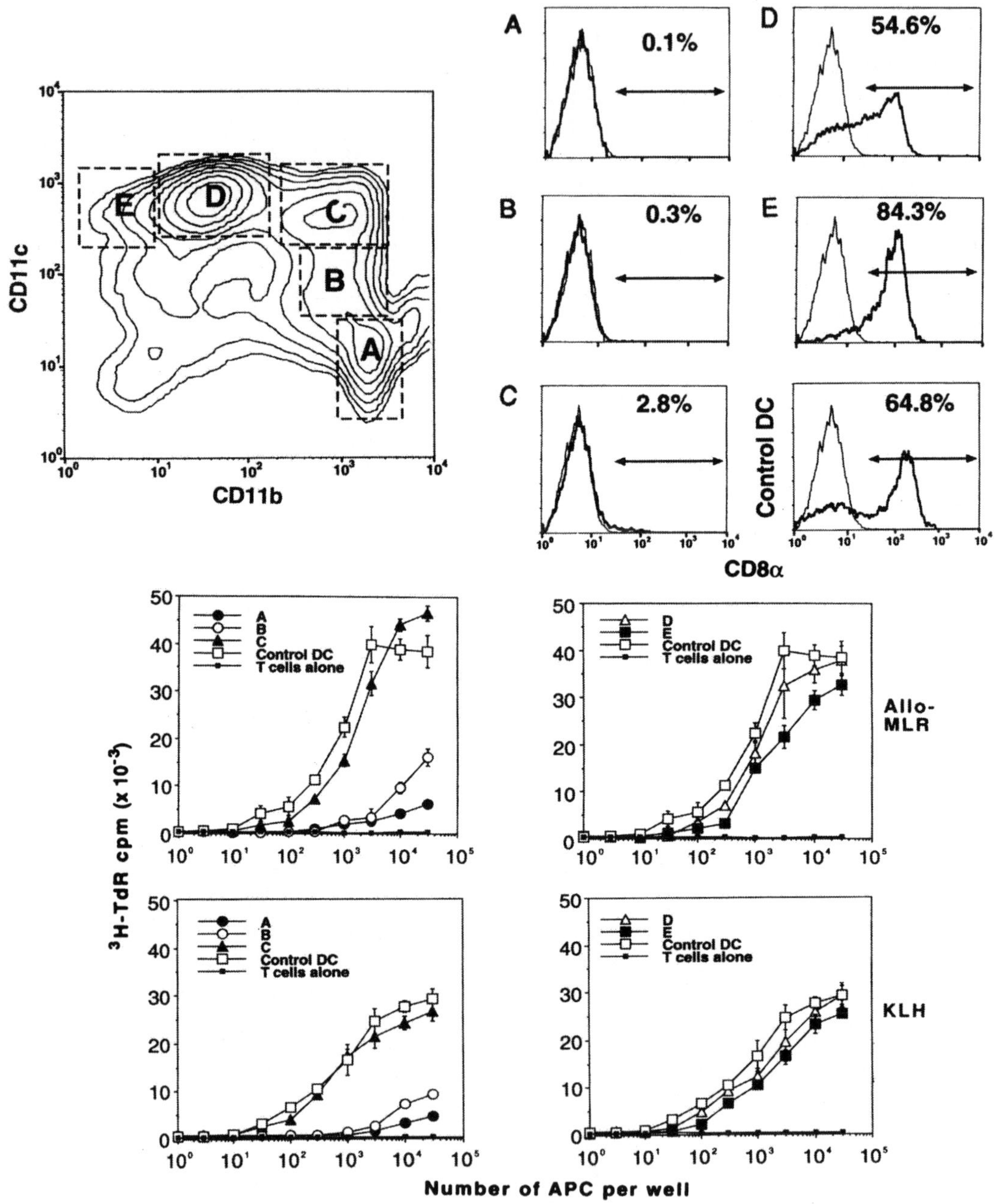

Figure 2. Detection and function of multiple spleen cell populations in mice after 11 days of treatment with Flt3L. (A) The distribution of CD11c and CD11b on lymphocyte-depleted spleen cells from Flt3L-treated mice showing the gates used for analysis of CD8α expression and for cell sorting. (B) Comparison of the antigen presenting capacity of the various spleen cell populations detected in Flt3L-treated mice to stimulate the proliferation of allogeneic or KLH-specific CD4+ T cells. Populations A (closed circles), B (open circles), C (closed triangles), D (open triangles), E (closed squares) were compared with control DC (open squares) isolated from the spleens of untreated mice. Background control cultures (closed star) contained either DBA/2 allogeneic T cells alone or C57Bl/6 syngeneic KLH-specific T cells plus KLH alone. The background counts were <100 cpm or 800 cpm respectively. Values represent the mean ± SEM of triplicate cultures. Figure 2 is reproduced from *The Journal of Experimental Medicine,* 184: p 53–62, by copyright permission of The Rockefeller University Press.

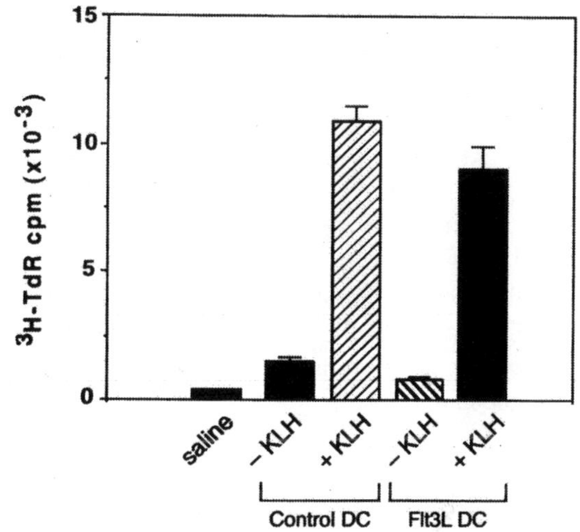

Figure 3. The induction of KLH-specific T cells in the draining LN of mice injected with DC pulsed with soluble KLH in vitro. Class II+ CD11c+ spleen cells from Flt3L-treated mice and control DC were pulsed with (+KLH) or without (-KLH) KLH antigen and injected into the foot pads of mice (4 per group). After 7 days, CD4+ T cells were isolated from the popliteal LN draining the foot pads and cultured with freshly isolated control DC in the presence or absence of KLH for 5 days and 3H-thymidine incorporation was measured. Values represent the mean ± SEM of triplicate cultures. Figure 3 is reproduced from *The Journal of Experimental Medicine*, 184: p53–62, by copyright permission of The Rockefeller University Press.

ure 4). However, treatment of mice with Flt3L alone or Flt3L plus GM-CSF or G-CSF significantly increased the relative numbers of class II$^+$ CD11c$^+$ cells in the spleen (Figure 4). Although all the single growth factor treatments or Flt3L-containing growth factor combinations resulted in an increase in total spleen cellularity over 11 days, only mice treated with Flt3L or Flt3L-containing combinations showed a 20- to 30-fold increase in class II$^+$ CD11c$^+$ cells in the spleen (Figure 4). Interestingly, the addition of either GM-CSF or G-CSF to Flt3L treatment increased the total number of class II$^+$ CD11c$^+$ cells only a further 1.3- and 1.5-fold respectively, indicating that for in vivo administration, Flt3L (but not

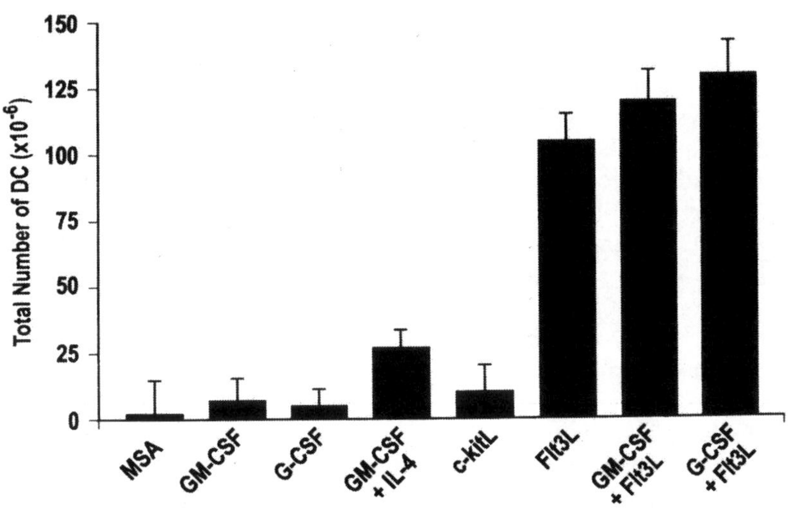

Figure 4. Detection of class II+ CD11c+ cells in growth factor treated mice. The generation of class II MHC+ and CD11c+ spleen DC from mice treated with MSA alone, or with either GM-CSF, G-CSF, GM-CSF + IL-4, c-kitL, Flt3L, GM-CSF + Flt3L, G-CSF + Flt3L (10 μg/mouse/day). Values represent the mean ± SD of 4 mice per group.

GM-CSF nor G-CSF) is the principal growth factor in the generation of class II$^+$ CD11c$^+$ cells. In addition, cessation of growth factor treatment at day 11 resulted in a reduction in the total number of class II$^+$ CD11c$^+$ cells by day 17, indicating that the generation of these cells was transient[29].

At least three phenotypically distinct DC subpopulations can be identified using CD11b and CD11c (populations C, D and E) and these appear to be functionally similar in the antigen-mediated T cell proliferation assays. Populations D and E, which are either dull or negative for CD11b expression, can be further separated using CD8α expression (a total of five DC subpopulations). This surface phenotype is similar to that of the lymphoid-derived DC populations[5,18]. Although the generation of these DC from a lymphoid precursor has only been established for thymic DC, a similar CD8α$^+$ population is found in spleen and LN[31].

Population C has not been previously described; however, the high level of CD11b expression suggests a relationship to the myeloid lineage and it may represent an immature stage of the myeloid-derived DC population[8–17]. Interestingly, cells within population C can be further subdivided on the basis of 33D1 expression[28]. 33D1 is not expressed on cells within populations D and E (Pulendran, B., manuscript in preparation). It is also of note that mature splenic monocytes and macrophages are not clearly discernible in these mice. It is possible that the in vivo effects of Flt3L on myelopoiesis may result in the terminal differentiation of developing monocytes and macrophages into DC analogous to human monocytes treated with GM-CSF and IL-4.[13,14] Recently, phenotypic, histological and functional studies suggest that the cells within population C are myeloid-derived DC (Pulendran, B., manuscript in preparation).

It is unlikely that Flt3L treatment simply results in the mobilization of existing mature DC from other sites into the lymphoid tissue. Firstly, Flt3 receptor is not detected on mature DC as assessed by flow cytometry and mature DC do not proliferate when cultured in Flt3L alone (data not shown). Secondly, elevated numbers of DC have been detected in multiple organs and tissues in Flt3L-treated mice, indicating that there is a generalized expansion of DC throughout these mice. Finally, Brasel et al. have recently shown that *in vivo* administration of Flt3L dramatically increases the numbers of hematopoietic progenitors in the bone marrow, peripheral blood and spleen, resulting in increased myelopoiesis and B lymphopoiesis[32,33]. These potent effects on hematopoiesis, as well as our observations on DC generation, suggest that in vivo administration of Flt3L facilitates the terminal development of a primitive, Flt3L-sensitive, rapidly expanding progenitor population into functionally mature DC. This hypothesis is further supported by studies showing that Flt3L can increase the absolute numbers of mature DC generated from cultured CD34+ human bone marrow progenitors.[34,35] In addition, the early detection of DC in the secondary lymphoid organs of mice after Flt3L-treatment may indicate that a more mature precursor population (such as a CFU-DC)[15,16] is also induced to differentiate into DC when exposed to Flt3L or Flt3L-inducible signals in vivo. We are currently investigating whether developmentally mature precursor cells are responsive to Flt3L in vitro.

These findings indicate that Flt3L is a potent inducer of DC when administered in vivo and may be an important growth factor for both the in vitro and in vivo generation of large numbers of functionally mature DC for use as adjuvants to immune based therapy. Furthermore, the identification of different DC subpopulations may ultimately lead to the identification of differing functions (e.g. stimulatory versus regulatory and control of T cell cytokine production). The identification of functionally distinct DC subpopulations represents the first step toward their appropriate use in immunotherapy.

5. REFERENCES

1. R.M. Steinman, The dendritic cell system and its role in immunogenicity, *Annu. Rev. Immunol.* 9:271 (1991).
2. J.P. Metlay, M.D. Witmer-Pack, R. Agger, M.T. Crowley, D. Lawless, and R.M. Steinman, The distinct leukocyte integrins of mouse spleen dendritic cells as identified with new hamster monoclonal antibodies, *J. Exp. Med.* 171:1753 (1990).
3. G. Kraal, M. Breel, M. Janse, and G. Bruin, Langerhans' cells, veiled cells, and interdigitating cells in the mouse recognized by a monoclonal antibody, *J. Exp. Med.* 163:981 (1986).
4. W. Jiang, W.J. Swiggard, C. Heufler, M. Peng, A. Mirza, R.M. Steinman, and M.C. Nussenzweig, The receptor DEC-205 expressed by dendritic cells and thymic epithelial cells is involved in antigen processing, *Nature* 375:151 (1995).
5. D. Vremec, M. Zorbas, R. Scollay, D.J. Saunders, C.F. Ardavin, L. Wu, and K. Shortman, The surface phenotype of dendritic cells purified from mouse thymus and spleen: investigation of the CD8 expression by a subpopulation of dendritic cells, *J. Exp. Med.* 176:47 (1992).
6. P. Matzinger and S. Guerder, Does T-cell tolerance require a dedicated antigen-presenting cell?, *Nature* 338:74 (1989).
7. O. Mazda, Y. Watanabe, J.-I. Gyotoku, and Y. Katsura, Requirement of dendritic cells and B cells in the clonal deletion of Mls-reactive T cells in the thymus, *J. Exp. Med.* 173:539 (1991).
8. M. Inaba, K. Inaba, M. Hosono, T. Kumamoto, T. Ishida, S. Muramatsu, T. Masuda, and S. Ikehara, Distinct mechanisms of neonatal tolerance induced by dendritic cells and thymic B cells, *J. Exp. Med.* 173:549 (1991).
9. K. Inaba, M. Inaba, N. Romani, H. Aya, M. Deguchi, S. Ikehara, S. Muramatsu, and R.M. Steinman, Generation of large numbers of dendritic cells from mouse bone marrow cultures supplemented with granulocyte/macrophage colony-stimulating factor, *J. Exp. Med.* 176:1693 (1992).
10. K. Inaba, R.M. Steinman, M.W. Pack, H. Aya, M. Inaba, T. Sudo, S. Wolpe, and G. Schuler, Identification of proliferating dendritic cell precursors in mouse blood, *J. Exp. Med.* 175:1157 (1992).
11. K. Inaba, M. Inaba, M. Deguchi, K. Hagi, R. Yasumizu, S. Ikehara, S. Muramatsu, and R.M. Steinman, Granulocytes, macrophages, and dendritic cells arise from a common major histocompatibility complex class II-negative progenitor in mouse bone marrow, *Proc. Natl. Acad. Sci. USA* 90:3038 (1993).
12. C. Caux, C. Dezutter-Dambuyant, D. Schmitt, and J. Banchereau, GM-CSF and TNF-α cooperate in the generation of dendritic Langerhans cells, *Nature* 360:258 (1992).
13. F. Sallusto and A. Lanzavecchia, Efficient presentation of soluble antigen by cultured human dendritic cells is maintained by granulocyte/macrophage colony-stimulating factor plus interleukin 4 and downregulated by tumor necrosis factor α, *J. Exp. Med.* 179:1109 (1994).
14. N. Romani, S. Gruner, D. Brang, E. Kämpgen, A. Lenz, B. Trockenbacher, G. Konwalinka, P.O. Fritsch, R.M. Steinman, and G. Schuler, Proliferating dendritic cell progenitors in human blood, *J. Exp. Med.* 180:83 (1994).
15. J.W. Young, P. Szabolcs, and M.A.S. Moore, Identification of dendritic cell colony-forming units among normal human CD34+ bone marrow progenitors that are expanded by c-*kit*-ligand and yield pure dendritic cell colonies in the presence of granulocyte/macrophage colony-stimulating factor and tumor necrosis factor α, *J. Exp. Med.* 182:1111 (1995).
16. P. Szabolcs, M.A.S. Moore, and J.W. Young, Expansion of immunostimulatory dendritic cells among the myeloid progeny of human CD34+ bone marrow precursors cultured with c-*kit* ligand, granulocyte-macrophage colony-stimulating factor, and TNF-α, *J. Immunol.* 154:5851 (1995).
17. C. Caux and J. Banchereau In vitro regulation of dendritic cell development and function, *in*: "Blood Cell Biochemistry, Vol.7: Hemopoietic growth factors and their receptors," A. Whetton and J. Gordon, eds., Plenum Press, London (1996, in press).
18. C. Ardavin, L. Wu, C.-L. Li, and K. Shortman, Thymic dendritic cells and T cells develop simultaneously in the thymus from a common precursor population, *Nature* 362:761 (1993).
19. A. Galy, M. Travis, D. Cen, and B. Chen, Human T, B, natural killer, and dendritic cells arise from a common bone marrow progenitor cell subset, *Immunity* 3:459 (1995).
20. J.I. Mayordomo, T. Zorina, W.J. Storkus, L. Zitvogel, C. Celluzzi, L.D. Falo, C.J. Melief, S.T. Ildstad, W.M. Kast, A.B. Deleo, and M.T. Lotze, Bone marrow-derived dendritic cells pulsed with synthetic tumour peptides elicit protective and therapeutic antitumour immunity, *Nat. Med.* 1:1297 (1995).
21. A.W. Thomson, L. Lu, N. Murase, A.J. Demetris, A.S. Rao, and T.E. Starzl, Microchimerism, dendritic cell progenitors and transplantation tolerance, *Stem Cells* 13:622 (1995).
22. J.W. Young and K. Inaba, Dendritic cells as adjuvants for class I major histocompatibility complex-restricted antitumor immunity, *J. Exp. Med.* 183:7 (1996).

23. D. Metcalf, K. Shortman, D. Vremec, S. Mifsud, and L. Di Rago, Effects of excess GM-CSF levels on hematopoiesis and leukemia development in GM-CSF/max 41 double transgenic mice, Leukemia 10:713 (1996).
24. S.D. Lyman, L. James, T. VandenBos, P. de Vries, K. Brasel, B. Gliniak, L.T. Hollingsworth, K.S. Picha, H.J. McKenna, R.R. Splett, F.A. Fletcher, E. Maraskovsky, T. Farrah, D. Foxworthe, D.E. Williams, and M.P. Beckmann, Molecular cloning of a ligand for the flt3/flk-2 tyrosine kinase receptor: A proliferative factor for primitive hematopoietic cells., *Cell* 75:1157 (1993).
25. S.D. Lyman, Biology of Flt3 ligand and receptor, *Int. J. Hematol.* 62:63 (1996).
26. H.J. McKenna and S.D. Lyman Biology of Flt3 ligand, a novel regulator of hematopoietic stem and progenitor cells, *in*: "Bone Marrow Transplantation: Basic and Clinical Studies," S. Ikehara, F. Takaku, and R.A. Good, eds., Springer, (1996).
27. M. Crowley, K. Inaba, M. Witmer-Pack, and R.M. Steinman, The cell surface of mouse dendritic cells: FACS analyses of dendritic cells from different tissues including thymus, *Cell. Immunol.* 118:108 (1989).
28. M.C. Nussenzweig, R.M. Steinman, M.D. Witmer, and B. Gutchinov, A monoclonal antibody specific for mouse dendritic cells, *Proc. Natl. Acad. Sci. USA* 79:161 (1982).
29. E. Maraskovsky, K. Brasel, M. Teepe, E.R. Roux, S.D. Lyman, K. Shortman, and H.J. McKenna, Dramatic increase in the numbers of functionally mature dendritic cells in Flt3 ligand-treated mice: multiple dendritic cell subpopulations identified, *J. Exp. Med., in press.* (1996).
30. K. Inaba, J.P. Metlay, M.T. Crowley, and R.M. Steinman, Dendritic cells pulsed with protein antigens in vitro can prime antigen-specific, MHC-restricted T cells in situ, *J. Exp. Med.* 172:631 (1990).
31. L. Wu, D. Vremec, C. Ardavin, K. Winkel, G. Süss, H. Georgiou, E. Maraskovsky, W. Cook, and K. Shortman, Mouse thymus dendritic cells: kinetics of development and changes in surface markers during maturation, *Eur. J. Immunol.* 25:418 (1995).
32. K. Brasel, H.J. McKenna, K. Charrier, P. Morrissey, D.E. Williams, and S.D. Lyman, Mobilization of peripheral blood progenitor cells with Flt3 ligand, *Blood* 86:463a (1995).
33. K. Brasel, H.J. McKenna, P.J. Morrissey, K. Charrier, A.E. Morris, C.C. Lee, D.E. Williams, and S.D. Lyman, Hematological effects of Flt3 ligand *in vivo* in mice, *Blood* 88:2004 (1996).
34. E. Maraskovsky, E. Roux, M. Tepee, H.J. McKenna, K. Brasel, S.D. Lyman, and D.E. Williams, The effect of Flt3 ligand and/or c-kit ligand on the generation of dendritic cells from human CD34+ bone marrow, *Blood* 86:420a (1995).
35. S. Siena, M. Di Nicola, M. Bregni, R. Mortarini, A. Anichini, L. Lombardi, F. Ravagnani, G. Parmiani, and A.M. Gianni, Massive ex vivo generation of functional dendritic cells from mobilized CD34+ blood progenitors for anticancer therapy, *Exp. Hematol.* 23:1463 (1995).

7

SELECTIVE EXPRESSION OF HUMAN FASCIN (p55) BY DENDRITIC LEUKOCYTES

Lene Sonderbye,[1] Roland Magerstadt,[1] Robert N. Blatman,[2]
Frederic I. Preffer,[3] and Erik Langhoff[1]*

[1]Renal Unit
Massachusetts General Hospital
Boston, Massachusetts
[2]Vincent Memorial Hospital
Massachusetts General Hospital
Boston, Massachusetts
[3]Department of Pathology
Massachusetts General Hospital, Massachusetts

1. INTRODUCTION

Dendritic cells are a heterogeneous group of antigen presenting leukocytes with distinctive cell morphology and function. The highly developed capacity of dendritic cells to present antigens and the way in which cells of variable maturity differ in their ability to take up, process, and present antigens have been well characterized (1–6). Dendritic cells are derived from cells in bone marrow in vivo and are released to peripheral blood and tissues (7,8), but dendritic cells can be generated in vitro from CD34 positive precursor cells (9–13). The position of dendritic cells in the hierarchy of hematopoietic cells remains to be established but the study of human blood dendritic and progenitor dendritic cells has been restricted by the lack of selective markers for this specialized subset of antigen presenting cells. Recently, an evolutionary conserved human actin-binding protein, fascin (p55), was demonstrated to be highly expressed by circulating blood dendritic cells (14). In the course of studying the development, migration and tissues distribution of dendritic cells we took advantage of a novel monoclonal antibody against p55 to examine the differential expression of p55 in immature, circulating, and tissue dendritic cells.

---

* Corresponding Author: Erik Langhoff, M.D., Ph.D., Renal Unit, Massachusetts General Hospital, Boston, Massachusetts 02114.

2. MATERIALS AND METHODS

2.1. Cell Fractionation and Cell Cultures

Progenitor and mature dendritic cells were isolated from peripheral or cord blood according to published procedures. Briefly, progenitor dendritic cells were isolated from umbilical cord blood or peripheral blood as CD34 positive cells using high gradient magnetic cell sorting (MiniMACS, Miltenyi). The isolated cells were cultured with GM-CSF (150 U/ml), IL-4 (150 U/ml), and TNFa (25U/ml) in Iscove's Medium (9,11,12). SCF (25 ng/ml) were added the first 7 days of culture (10). Techniques for isolation of mature dendritic cells have previously been published (14,15). Peripheral blood mononuclear cells were isolated by Ficoll-Hypaque (Pharmacia) from blood donor buffy coats or leukopaks and depleted of monocytes and T cells (16). After overnight culture dendritic cells were recovered as low density cells from a 14.5% metrizamide column.

The function of monocytes and dendritic cells was examined by comparing their antigen-presenting functions in the mixed leukocyte reaction (MLR) as previously described (16).

2.2. FACS Analysis and Immunohistochemistry

For immunohistochemistry, cells were cytocentrifuged and stored at -80°C until use. Blocks of tissue obtained at time of surgery or as biopsies were snap frozen in liquid nitrogen and stored at -80°C until use. Frozen cryostat-sectioned tissue or cytocentrifuged cells were fixed in acetone or acetone-methanol (1:1) and stained with monoclonal antibodies by the immunoperoxidase method (Vectastain, Vector). Cells or tissues were blocked with 1% serum-1% BSA before applying the primary antibody. Endogenous peroxidase activity was inhibited with 0.3% H_2O_2 before addition of the secondary biotinylated antibody. Binding of a complex of biotinylated peroxidase and avidin (ABC reagent, Vector) was followed by visualization with AEC (Sigma). Cells or sections were counterstained with hematoxylin with $LiCO_3$ before mounting with Glycergel (Dako).

A monoclonal antibody, K–2 (IgG1), to 55–kD protein was raised as previously described (14). Both hybridoma supernatant (undiluted) and ascites (diluted 1/500–1/1000) were used in the studies and gave consistent results.

Table 1. Phenotype (%) of cultured progenitor dendritic cells. CD1a, CD3, CD4, CD14, CD19, CD56, and HLA-DR analyzed by FACS. Expression of p55 determined by peroxidase stain of cytospin preparations

Culture days	p55	CD1a	CD34	HLA-DR	CD3	CD4	CD14	CD19	CD56
Day 0-3	6	1.8	82	86	2.5	<1	1.4	3.4	< 1
	n=3	n=23	n=33	n=26	n=28	n=5	n=33	n=25	n=2
Day 4-7	63	12	45	75	1.9	9	1.8	<1	< 1
	n=3	n=17	n=20	n=21	n=20	n=6	n=20	n=20	n=1
Day 8-11	82	13	4.9	66	11	6.0	12	<1	N.D.
	n=1	n=7	n=8	n=9	n=8	n=1	n=8	n=9	
Day 12-15	>90	24	1.0	63	3.8	20	1.6	1.2	N.D.
	n=1	n=12	n=9	n=12	n=11	n=3	n=11	n=11	
Day 16-21	>90	23	<1	43	1.1	32	<1	<1	N.D.
	n=1	n=5	n=5	n=5	n=5	n=2	n=5	n=5	

N.D. = not determined

Antibodies against leukocyte differentiation antigens were used according to the manufacturers instruction unless otherwise specified: anti–HLA–DR (9.3F10, IgG2a, ATCC or Anti–HLA–DR, IgG2a, Becton Dickinson (BD)), anti–CD3 (leu 4, IgG1, BD), anti–CD19/CD20 (leu 12, IgG1, BD/Leu 16, IgG1, BD), anti–CD14 IgG1,BD), anti–CD14 (3C10, IgG2a, American Type Culture Collection (ATCC), anti–CD56 (leu 19, IgG1, BD), anti–CD1a (OKT6, IgG1, BD).

Cytofluorography of cell isolates was performed as previously described (16). Cytospin centrifuge preparations were made of all isolated leukocyte subsets (20,000 cells per slide, Shandon, Southern Instruments). Procedures for cytospin immunofluorescense has previously been described (14,17).

3. RESULTS

Expression of p55 by Progenitor Dendritic Cells

Cultures of progenitor CD34 positive cells were established to monitor expression of p55 in parallel with a panel of defined leukocyte antigens. More than 98% purified $CD34^+$ cells were isolated and plated with GM-CSF, IL-4, and TNFα for 3 weeks to allow for differentiation into mature dendritic cells. Immediately after isolation (days 0–3) progenitor cells were positive for HLA-DR antigens. In contrast, CD1a, CD3, CD14, and CD19 were not detected by FACS analysis. At this time point cytospin preparations of the cultures showed only a few (6%) cells expressing p55 (Table 1).

In contrast, after two weeks of culture cytospin preparations of progenitor dendritic cells were > 90% positive for the p55 antigen. At this time point the progenitor cells had acquired the phenotypic veiled appearance of mature dendritic cells and were negative for CD34 with only a few contaminating T cells, B cells, and monocytes. A large proportion of the cells expressed the CD1a (24%) with a comparable fraction of cells expressing the CD4 antigen (20%) consistent with other reports (18). After two weeks of culture the progenitor dendritic cells have become highly potent MLR stimulatory cells in contrast to adherent monocytes and equally as potent as peripheral blood dendritic cells (Figure 1).

Expression of p55 by Circulating and Tissue Dendritic Cells

Dendritic cells isolated from peripheral blood demonstrated a strong cytoplasmic p55 expression. In contrast, T cells, B cells, and monocytes did not express the p55 antigen. In the thymus, the cortico-medullary junction stained positive for p55 antigen. In sections of lymph node and tonsil selective expression of the p55 antigen was present in the T cell dependent areas (not shown). The results also showed that the expression of p55 in lymph node and tonsil was not by follicular dendritic cells in the B cell follicles. In the skin preliminary studies showed no p55 staining of Langerhans cells in situ, but increased expression of the p55 antigen was found upon culture of isolated Langerhans cells (unpublished results).

4. CONCLUSION

P55 was recently demonstrated to be an actin binding protein (19). Actin binding proteins regulate rearrangements of the cytoskeleton or interaction between the cytoskele-

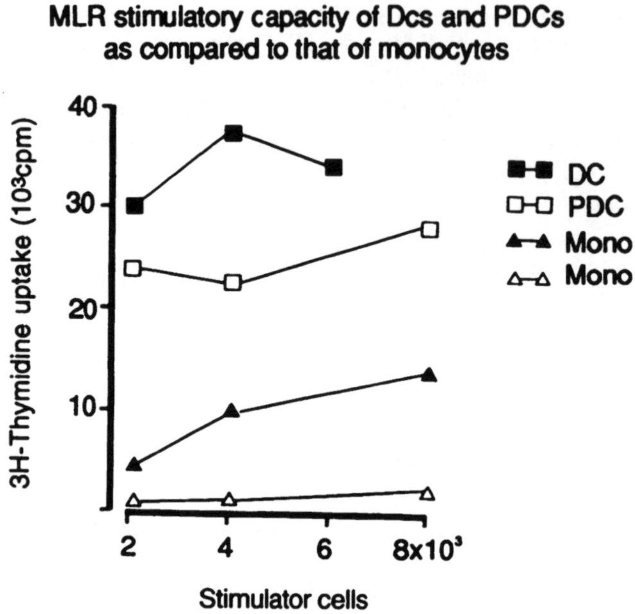

Figure 1. Stimulation of allogeneic T cells with : DCs (closed squares) , PDCs (open squares), and monocytes from 2 donors (open and closed triangles). 50,000 allogeneic T cells were used as responder cells and cultured for 5 days before overnight 3H-thymidine addition.

ton and the cell membrane in response to both intracellular and extra cellular signals, and direct the three dimensional polymerization of the cytoskeletal actin filaments at the migrating edge of cells. Related to cell motility is the ability to process and present antigens. Thus, the migratory nature and unique antigen presenting properties of dendritic cells suggest a key role of p55 in this specialized cell subset. The present results demonstrate that progenitor dendritic cells generated in vitro share the expression of the p55 antigen with mature circulating and tissue dendritic cells. The results also show that the expression of p55 in lymph node sections is restricted to cells resembling interdigitating dendritic cell and is not found expressed in the follicles. This is in agreement with the current theory that interdigitating dendritic cells are derived from a cell lineage different from follicular dendritic cells (20). Interdigitating dendritic cells in the lymph node are thought to be derived from circulating dendritic cells migrating to the lymphoid tissues (21,22). These results would suggest that expression of p55 is a shared phenotype in vitro and in vivo of mature dendritic cells.

A small number of CD1a positive dendritic cells was found in lymph node sections in the T cells zones (not shown). CD1a is an early maturation marker present on progenitor dendritic cells and Langerhans cells of the skin. The low frequency of CD1a positive dendritic cells could be explained by a small number of immature dendritic cells in the lymphoid tissues or a loss of the CD1a antigen expression by Langerhans cells upon migration from skin to lymph nodes (1,23).

Langerhans cells resident in the skin are derived from bone marrow cells (7,8). It is possible that the finding of a lack of p55 expression by Langerhans cells in situ but expression of p55 by cultured Langerhans cells reflects a developmental transition form resident to migrating dendritic cells.

Other markers for dendritic cells have been reported. Recent work has described expression of the CD83 antigen by a subset of dendritic cells. CD83 is, however, also expressed by follicular dendritic (FDC) cells and by activated T cells (24). The S–100 antigen has a wide tissue distribution and is expressed in many cell types of different somatic origin including dendritic cells (25). The combined use of several markers may provide new insights into the details of lineage differentiation of dendritic cells.

The ultimate potential of novel markers for dendritic cells is the possibility of demonstrating the envolvement of dendritic cells in disease processes. *In vivo* engagement of dendritic cells in pathological conditions is suggested from studies in rodents, but studies in humans have not been very revealing. Recent studies, however, now suggest that Reed-Sternberg cells of Hodgkin's Disease shares the expression of the p55. This observation adds additional support to the possible close relationship between Reed-Sternberg cells and dendritic cells (26).

ACKNOWLEDGMENTS

E.L. was supported by NIH grant AI28734. L.S.was supported by the Danish Medical Research Council. We thank D. Dombrowski for valuable technical assistance.

REFERENCES

1. Romani N, Lenz A, Glassel H, et al. Cultured human Langerhans cells resemble lymphoid dendritic cells in phenotype and function. Journal Investigative Dermatology 1989;93(5):600–609.
2. Schuler G, Steinman R. Murine epidermal Langerhans cells mature into potent immunostimulatory dendritic cells in vitro. Journal Experimental Medicine 1985;161:526–546.
3. Reis e Sousa C, Stahl PD, Austyn JM. Phagocytosis of antigens by Langerhans cells in vitro. Journal Experimental Medicine 1993;178:509–519.
4. Sallusto F, Lanzavecchia A. Efficient presentation of soluble antigen by cultured human dendritic cells is maintained by granulocyte/macrophage colony-stimulating factor plus interleukin-4 and downregulated by tumor necrosis factor alfa. Journal Experimental Medicine 1994;179:1109–1118.
5. Sallusto F, Cella M, Danieli C, Lanzavecchia A. Dendritic cells use macropinocytosis and the mannose receptor to concentrate macromolecules in the major histocompatibility complex class II compartment: Downregulation by cytokines and bacterial products. Journal Experimental Medicine 1995;182:389–400.
6. Levine TP, Chain BM. Endocytosis by antigen presenting cells: Dendritic cells are as endocytically active as other antigen presenting cells. Proceedings National Academy Sciences 1992;89:8342–8346.
7. Pelletier M, Perreault C, Landry D, David M, Monplasaisir S. Ontogeny of human epidermal Langerhans cells. Transplantation 1984;38:544–546.
8. Katz SI, Tamaki K, Sachs DH. Epidermal Langerhans cells are derived from cells originating in bone marrow. Nature 1979;282:324–326.
9. Caux C, Dezutter-Dambuyant C, Schmitt D, Banchereau J. GM-CSF and TNF-alpha cooperate in the generation of dendritic Langerhans cells. Nature 1992;360:258–261.
10. Szabolcs P, Moore MAS, Young JW. Expansion of immunostimulatory dendritic cells among the myloid progeny of human CD34+ bone marrow precursors cultured with c-kit ligand, granulocyte-macrophage colony-stimulating factor, and TNF-alpha. Journal Immunology 1995;154:5851–5861.
11. Romani N, Gruner S, Brang D, et al. Proliferating dendritic cell progenitors in human blood. Journal Experimental Medicine 1994;180:83–93.
12. Santiago-Schwarz F, Belilos E, Diamond B, Carsons SE. TNF in combination with GM-CSF enchances the differentiation of neonatal cord blood stem cells into dendritic cells and macrophages. Journal Leukocyte Biology 1992;52:274–281.
13. Inaba K, Steinman RM, Pack MW, et al. Identification of proliferating dendritic cell precursors in mouse blood. Journal Experimental Medicine 1992;175:1157–1167.
14. Mosialos G, Birkenbach M, Ayehunie S, Matsumura F, Pinkus GS,Kieff E, Langhoff E. Circulating human dendritic cells differentially express high levels of a p55-kD actin-bundling protein. American Journal Pathology 1996;148:593–600.
15. Langhoff E, Kalland K-H, Haseltine WA. Early molecular replication of Human Immunodeficiency Virus Type 1 in cultured blood-derived T helper dendritic cells. Journal Clinical Investigation 1993;91:2721–2726.

16. Langhoff E, Steinman RM. Clonal expansion of human T lymphocytes initiated by dendritic cells. Journal Experimental Medicine 1989;169:315–320.
17. Kalland K-H, Szilvay AM, Langhoff E, Haukenes G. Subcellular distributiion of human immunodeficiency virus Type 1 Rev and colocalization of Rev with RNA splicing factors in speckled pattern in the nucleoplasm. Journal Virology 1994;68(3):1475–1485.
18. Rosenzwajg M, Canque B, Gluckmaan JC. Human dendritic cell differentiation pathway for CD34+ hematopoietic precursor cells. Blood 1996;87(2):535–544.
19. Yamashiro-Matsumura S, Matsumura F. Purification and characterization of an F-actin-bundling 55-kilodalton protein form HeLa cells. Journal Biological Chemistry 1985;260:5087–5097.
20. Schreiver F, Nadler LM. The central role of follicular dendritic cells in lymphoid tissues. Advances Immunology 1992;51:243–283.
21. Larsen C, Steinman R, Witmer-Pack M, Hankins D, Morris P, Austyn J. Skin migration and maturation of Langerhans cells in skin transplants and explants. Journal Experimental Medicine 1990;172:1483–1493.
22. Macatonia SE, Knight SC, Edwards AJ, Griffiths S, Fryer P. Localization of antigen on lymph node dendritic cells after exposure to the contact sensitizer fluorescein isothiocyanate. Functional and morphological studies. Journal Experimental Medicine 1987;166:1654–1667.
23. Schuler G, Steinman RM. Murine epidermal Langerhans cells mature into potent immunostimulatory dendritic cells in vitro. Journal Experimental Medicine 1985;161:526–546.
24. Zhou L-J, Scharting R, Smith HM, Tedder TF. A novel cell-surface molecule expressed by human interdigitating reticulum cells, Langherhans cells and activated lymphocytes is a new member of the immunoglobullin superfamily. Journal Immunology 1992;149:735–742.
25. Vanstapel M-J, Gatter KC, de Wolfe-Peeters C, Manson DY, Desmet VD. New sites of human S-100 immunreactivity detected with monoclonal antibodies. American Journal Clinical Pathology 1986;85:160–168.
26. Pinkus GS, Pinkus JL, Langhoff E, et al. Fascin: A sensitive new marker for Reed-Sternberg cells of Hodgkin's disease - Evidence for a dendritic or B cell derivation? American Journal Pathology (in press).

8

DENDRITIC-LIKE CELLS FROM relB MUTANT MICE

Lisa DiMolfetto, Christina Reilly, Qun Wei, and David Lo

Department of Immunology IMM-25
The Scripps Research Institute
La Jolla, California

1. ABSTRACT

Mice deficient in the NF-κB transcription factor relB appear to have defects in the production of mature dendritic cells, as secondary lymphoid tissues are absent, and spleen cells show a significant loss of antigen presenting function. Moreover, the thymus appears to be impaired in negative selection, and immune responses in vivo are poor. Since dendritic cell precursors such as skin Langerhans cells appear to be normal, we sought information on the nature of the dendritic cell defect in these mice. Cultures of mutant bone marrow in the presence of GM-CSF revealed a delay in the accumulation of cells with dendritic cell features relative to controls; however, these cells were nearly as potent on a per cell basis as wild type cells in the stimulation of allogeneic mixed lymphocyte cultures. Similarly, skin Langerhans cells from mutant mice also showed significant ability to stimulate allogeneic T cells in culture. Since these findings cannot explain the defect in immune responses and the absence of secondary lymphoid tissues, we also looked at the ability of the relB mutant dendritic-like cells to form aggregates in vitro with naive syngeneic T cells. In this case, while wild type dendritic cells generated compact aggregates with T cells, relB mutant cells only formed irregular small aggregates. Thus, while relB mutant dendritic-like cells have some functions of mature dendritic cells, other functions are deficient. Understanding the role of relB in regulation of these functions should lead to a greater understanding of the molecular basis of dendritic cell development and function.

2. INTRODUCTION

Dendritic cells have a central role in several functions of the immune system. Aside from its well described role as the primary antigen presenting cell in the initiation of immune responses, the dendritic cell also is important in mediating negative selection of

Dendritic Cells in Fundamental and Clinical Immunology
edited by Ricciardi-Castagnoli, Plenum Press, New York, 1997

autoreactive T cells in the thymus. In addition, it appears to help in the organization of secondary lymphoid tissues. While these functions have been attributed to dendritic cells, many of the specific mechanisms have yet to be understood at a molecular level, due in part to the difficulty in generating large numbers of dendritic cells. Another problem stems from the limited information on the nature and function of dendritic cell developmental stages and functional subsets. Thus, while markers such as NLDC-145 and class II MHC identify many cells with dendritic morphology such as thymic dendritic cells and skin Langerhans cells, can it be said that both of these cells are of the same lineage and have the same function?

One approach to the molecular basis of dendritic cell development and function is the use of transgenic and knockout models with altered dendritic cell function. Here, we describe our studies on relB mutant cells with dendritic-like features. RelB, a transcription factor belonging to the NF-κB/Rel family of proteins, appears to be critical in some dendritic cell functions but not others. We believe that such studies will lead to a more detailed understanding of the genes regulating distinct dendritic cell developmental stages and functions.

3. RESULTS

3.1. Development of Dendritic-Like Cells in relB Mutant Mice

RelB was originally cloned as an activation-inducible gene from serum-activated fibroblasts.[1] Curiously however, relB was found to be expressed constitutively in dendritic cells and thymic medullary epithelium. When the relB gene was disrupted, the mutant mice developed an unusual syndrome of inflammation in several tissues, along with an absence of secondary lymphoid tissues and thymic medulla.[2–4]

Given the constitutive expression of relB in lymphoid dendritic cells, the relB deficient mouse phenotype could be explained as two related phenomena, both relating to a putative absence of mature dendritic cells. First, in the absence of dendritic cells, secondary lymphoid tissues might not be able to develop. Second, since granulocytes and dendritic cells are thought to come from a common stem cell, a feedback loop might be driving the stem cell to expand to compensate for the loss of dendritic cells. The inability to produce dendritic cells would instead result in an overproduction of granulocytes. These explanations remain to be proven, in part because our first assumption, the loss of dendritic cells, also requires a more direct demonstration.

Thus, to address the question of dendritic cell development in relB mutant mice, we looked to see which dendritic cell functions remained and which were deficient. Earlier studies suggested that spleen antigen presenting cell function was impaired in relB mutant mice. This was true for bulk cultures of spleen cell stimulators, but it was likely that the increased production of other cells such as macrophages and granulocytes were decreasing the actual numbers of dendritic cells. To determine whether production of dendritic cells was actually decreased in the relB mutant mice, we cultured bone marrow in the presence of GM-CSF to look at the rate of production of cells with dendritic cell characteristics (Figure 1).

We found that wild type bone marrow produced dendritic cells beginning by day three of culture using the criteria of dendritic morphology, high level expression of class II MHC antigens and expression of costimulatory molecules such as B7.2 (Figure 1). Significant numbers of these cells were generated by day seven of culture. In contrast, relB mutant bone marrow cultures began with an unusually high proportion of granulocytes (high side scatter characteristics). Dendritic-like cells were generated, but with delayed appearance, and a lower proportion even after seven days of culture. Thus, it appears that even outside the animal, bone marrow cells show a defect in the ability to generate dendritic-like cells, suggesting that the defect is intrinsic to the stem cells and not secondary to other effects in the mutant mice.

3.2. Antigen Presenting Function of relB Mutant Dendritic-Like Cells

Despite the lower production of dendritic-like cells in relB mutant bone marrow, the cells that were produced did express high levels of class II and B7, suggesting that they may have normal antigen presenting function. This was tested by using preparations of antigen presenting cells from spleen, skin, and bone marrow cultures as stimulators of both syngeneic and allogeneic lymph node T lymphocytes (Figure 2). Under these conditions, enriched dendritic cells actually stimulated higher responses from syngeneic responders as compared to allogeneic cells. As expected, moderately enriched spleen cells from relB mutant mice showed much less ability to stimulate T cell proliferation as compared to controls, in part because the enrichment probably did not eliminate the large numbers of granulocytes and macrophages (Figure 2a).

Another source of antigen presenting cells is the skin. As previously reported, relB mice have normal numbers of skin Langerhans cells. These cells are thought to be dendritic cell precursors, with maturation occurring during the activation and migration of these cells to draining lymph nodes. Consistent with this, normal Langerhans cells are negative for relB. It was possible that relB mutant mice may lack dendritic cells due to an inability of the cells to migrate from skin, but we found that floating cultures of ear skin from relB mutant mice yielded class II positive cells with a dendritic morphology migrating from the skin into the culture dish. These Langerhans cells isolated from the skin cultures showed similar stimulatory ability from both control and relB mutant mice, although the mutant cells were slightly less potent.

As discussed above, bone marrow cultured with GM-CSF generated cells with a dendritic cell phenotype even from relB mutant mice. Are these cells effective as antigen presenting cells? Bone marrow cultures enriched by sorting for class II^{hi} cells showed potent stimulation by both control and mutant cells, but again relB mutant cells showed slightly less stimulation on a per-cell basis (Figure 2). Thus, relB mutant stem cells appear to have two subtle defects: one defect in the rate of production of dendritic cells in culture and another minor decrease in their potency as antigen presenting cells.

3.3. Functions of Dendritic Cells in Vivo

Although relB mutant mice produced dendritic cells less efficiently, the few cells produced still had significant antigen presenting function. Yet how do we explain the observed immune system defects such as the absence of secondary lymphoid tissue, and the failure of the thymus to mediate normal negative selection? Given these disparities, it therefore seems possible that relB is necessary for dendritic cells to develop some functions (organization of lymphoid tissues, thymic negative selection) but not others (dendritic morphology, Langerhans cell development, some antigen presenting function).

One of the most difficult phenomena to reconcile here is the defect in thymic negative selection. We recently reported that in relB mutant thymuses, positive selection of T cells is normal, but deletion of superantigen/I-E reactive T cells expressing Vβ5 and Vβ11

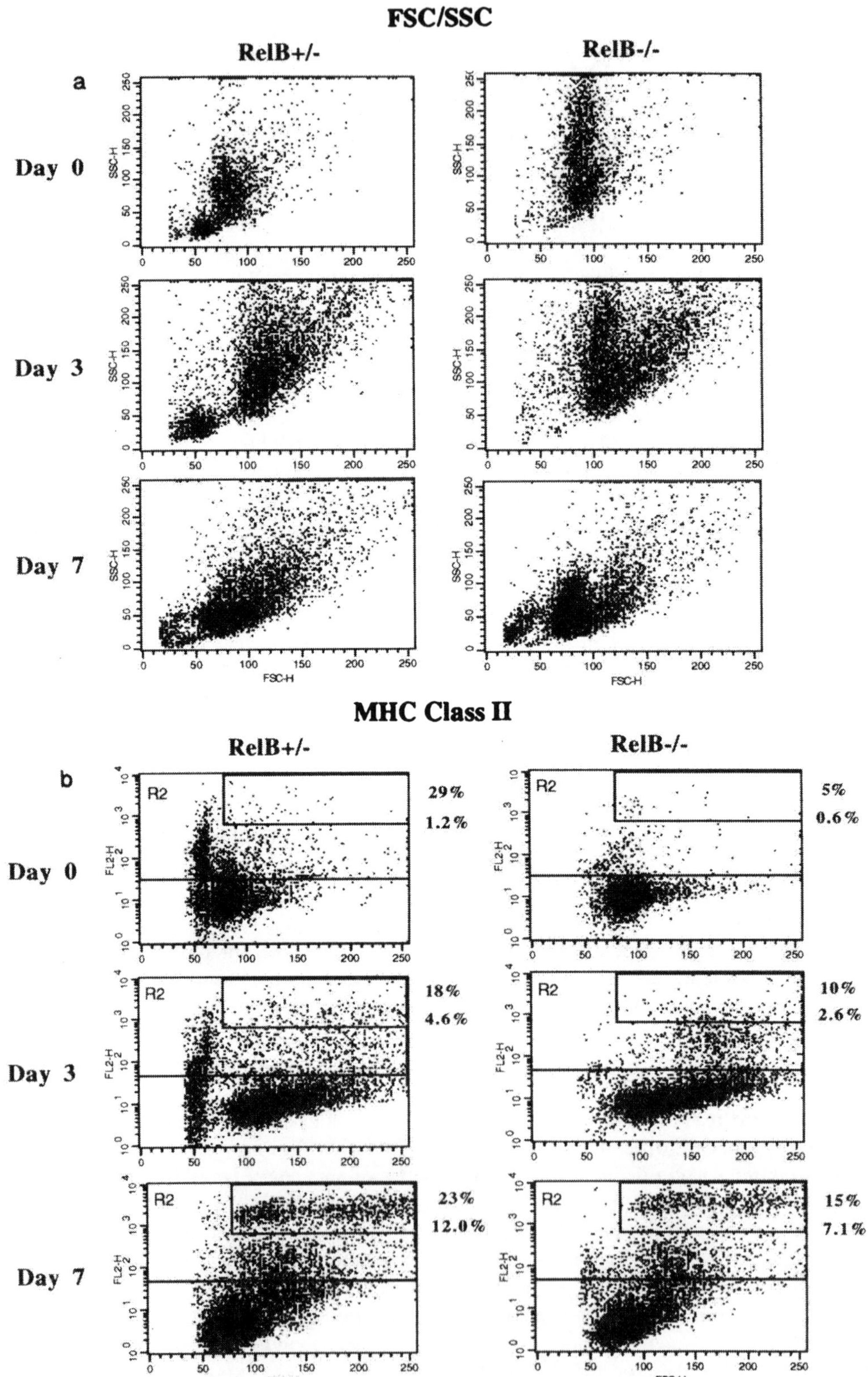
FSC/SSC
RelB+/-
RelB-/-
a
Day 0
Day 3
Day 7
SSC-H
FSC-H
MHC Class II
b
R2
FL2-H
29%
1.2%
5%
0.6%
18%
4.6%
10%
2.6%
23%
12.0%
15%
7.1%

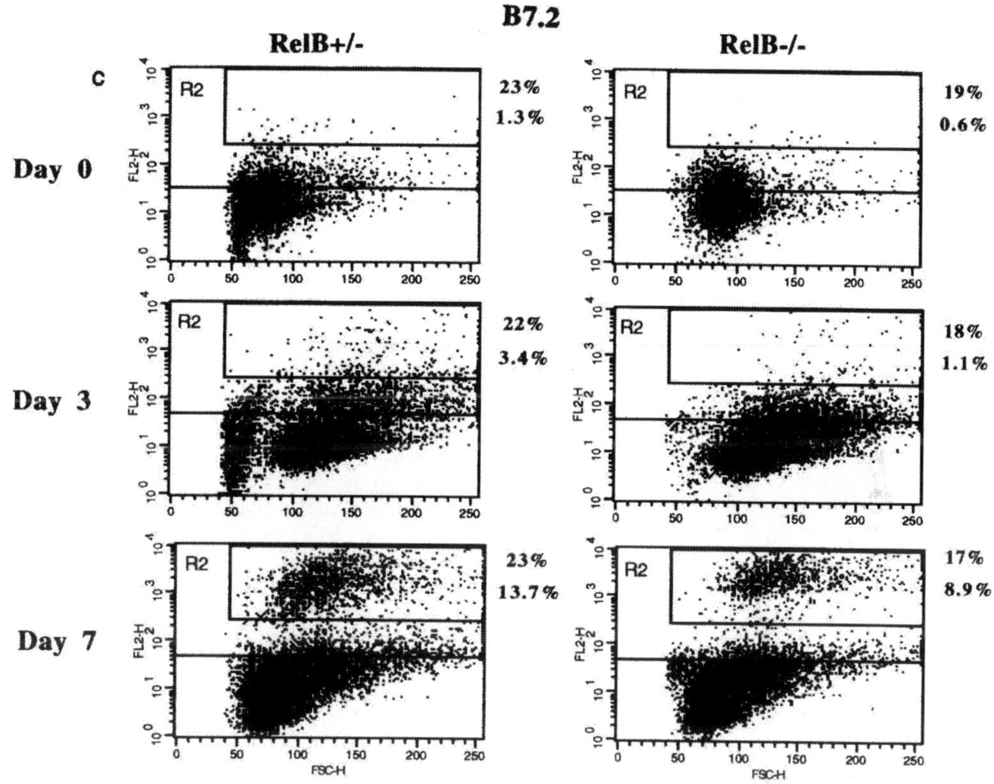

Figure 1. FACS analysis of bone cultures from heterozygote and relB mutant mice . Bone marrow cells were cultured in the presence of GM-CSF and analyzed on days 0, 3, and 7. (a), showing the forward and side scatter characteristics of the cultures showing the initial large numbers of granulocytes (high side scatter) in relB mutant bone marrow; (b), class II staining showing the later appearance and lower numbers of class II^{hi} cells from relB mutant cultures; (c), B7.2 staining showing that relB mutant cells also develop expression of costimulatory molecules. The numbers to the right of the dot plots reflect the percentage of cells above the marker (upper number), and the percentage of dendritic cells within the box (lower number).

is defective.[5] Moreover, T cells isolated from relB mutant mice show strong proliferative responses to normal syngeneic stimulators in vitro, in some cases nearly as high as responses to allogeneic stimulators.[6] It has been assumed that the ability of bone marrow derived antigen presenting cells to mediate negative selection in the thymus is related to their ability to present antigen and stimulate T cells. Unfortunately, there appears to be a significant disparity; is it possible that negative selection in the thymus involves distinct functions dependent on relB?

The absence of secondary lymphoid tissues is also complex. As described above, skin Langerhans cells are present in relB mutant mice, and they show normal ability to migrate from the skin in culture. Why do they fail to contribute to the development of lymph nodes - is it due to a failure to persist in draining lymph nodes, or a failure to help accumulate lymphocytes in the nodes?

As shown above relB mutant cells were capable of stimulating T cell proliferation. Yet many types of cells expressing MHC molecules and costimulatory signals are able to stimulate T cell proliferation in vitro, including various non-lymphoid (or non-mammal-

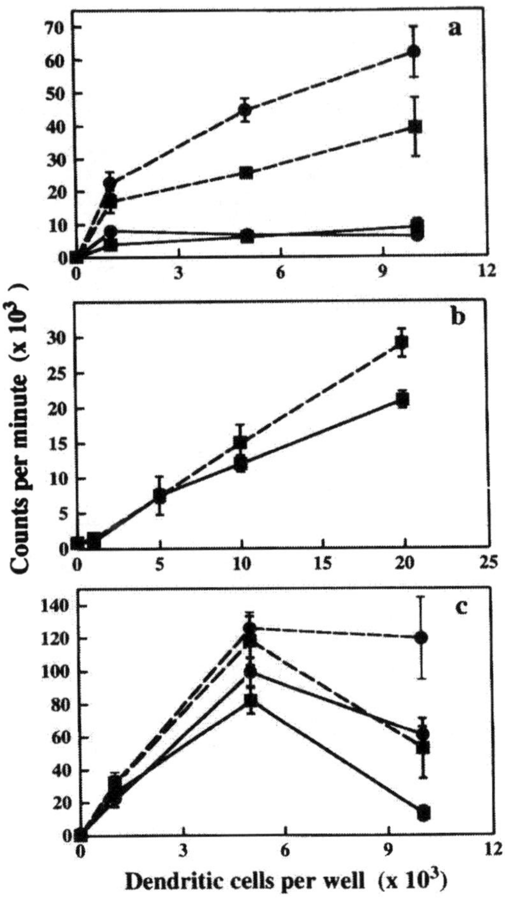

Figure 2. Antigen presenting function of relB mutant cells. (a), responses to spleen cells enriched by differential adherence to plastic, plus two day culture with GM-CSF. (b), responses to Langerhans cells collected from cultures of ear skin floating on media. (c), responses to dendritic cells enriched by culture of bone marrow plus GM-CSF for seven days, then sorting for class II^{hi} cells. Broken lines represent responses to heterozygote stimulators, solid lines represent responses to relB mutant stimulators. Circles indicate syngeneic responder cells, squares indicate allogeneic responders. METHODS: In (a,c), allogeneic (C57BL/6) and syngeneic (B10.D2) responder lymph node cells were added to enriched stimulators from relB heterozygote and mutant (knockout) mice on a B10.D2 background. In (b), allogeneic responders were from ko mice, which express a class II I-E transgene but no I-A, so the CD4 T cells will respond to any MHC class II haplotype. Proliferative responses were assessed after four days of culture.

ian!) cells transfected with the specific ligands. This in vitro capability therefore does not automatically make a bona fide dendritic cell, so we studied other functions of dendritic cells that might be absent in relB mutant cells which might explain the in vivo defects.

One known characteristic of dendritic cells is their ability to aggregate T cells. We tested this function by sorting for dendritic cells and naive syngeneic T cells, and culturing overnight to assess antigen-independent aggregate formation (Figure 3). While wild type dendritic cells produced compact aggregates consisting of large numbers of T cells, relB mutant dendritic-like cells only produced small irregular aggregates with fewer T cells per aggregate. Thus, it appears that relB mutant dendritic-like cells are less capable of holding resting T cells in aggregates, a defect which may account for the absence of secondary lymphoid tissues.

4. DISCUSSION

The relB deficient phenotype is clearly complex, potentially involving both lymphoid and non-lymphoid cells. With regard to antigen presenting function in the immune system and dendritic cell development, in vivo studies have suggested some interesting defects, but it was not established whether these were intrinsic to dendritic cell develop-

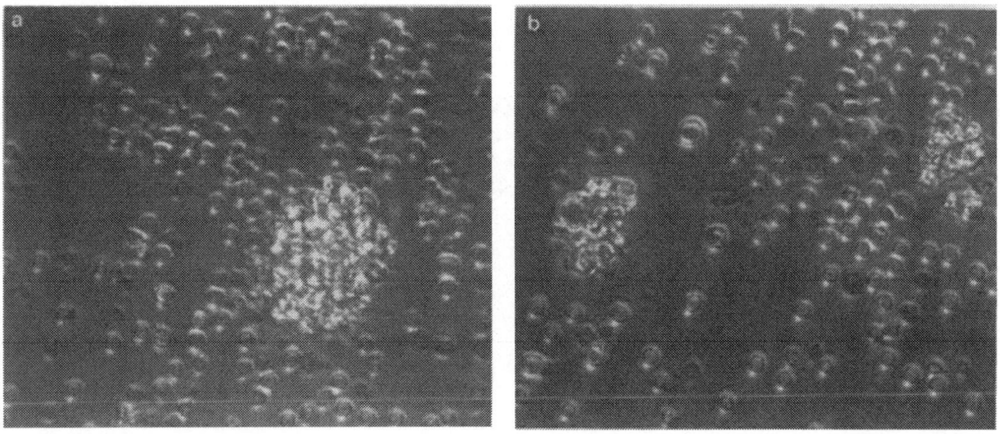

Figure 3. Aggregation of dendritic-like cells from normal and relB mutant spleen with sorted naive T cells. (a), showing that culture with wild type dendritic cells produced large rounded aggregates with T cells; (b), showing that relB mutant dendritic-like cells only produced small disorganized aggregates. METHODS: Spleen cells were enriched for dendritic cells by differential adherence and culture in the presence of GM-CSF for two days. Cultures were sorted for $B220^-$ $Thy1^-$ cells, then placed in overnight culture with sorted naive syngeneic wild type lymph node $Thy1^+$ cells.

ment, or secondary to other effects in the mutant mice. We have been able to use a number of in vitro studies to demonstrate that there are significant defects in dendritic cell development from bone marrow cultures. The dendritic-like cells generated in vitro from relB mutant cells show significant, albeit slightly diminished, antigen presenting cell function. However, these cells appear to have more significant functional defects that might at least account for the absence of secondary lymphoid tissues in vivo. Further studies should help explain the thymic defects in negative selection, and ultimately, it is hoped that these will lead to an understanding of how relB regulates the development of dendritic cells and the genes associated with specific dendritic cell functions.

5. ACKNOWLEDGMENTS

These studies were supported by grants AI29689, AI31583, and AI38375 to D.L. from the U.S. Public Health Service. This is manuscript number 10362-IMM from The Scripps Research Institute. Address correspondence to: David Lo, M.D., Ph.D., Department of Immunology IMM-25, The Scripps Research Institute, 10550 North Torrey Pines Road, La Jolla, CA 92037 USA. Phone: (619)784–8053; Fax: (619)784–9180; E-mail: davidlo@scripps.edu

6. REFERENCES

1. Ryseck, R.P., Bull, P., Takamiya, M., Bours, V., Siebenlist, U., Dobrzanski, P., and Bravo, R., RelB, a new Real family transcription activator that can interact with p50-NF-κB. Mol. Cell. Biol. 12:674–684, 1992.
2. Lo, D., Quill, H., Burkly, L., Scott, B., Palmiter, R.D., and Brinster, R.L., A recessive defect in lymphocyte and/or granulocyte function caused by an integrated transgene. Am. J. Pathol., 141:1237–1246, 1992.

3. Burkly, L., Hession, C., Ogata, L., Reilly, C., Marconi, L.A., Olson, D., Tizard, R., Cate, R., and Lo, D., Expression of relB is required for the development of thymic medulla and dendritic cells. Nature 373:531–536, 1995.
4. Weih, F., Carrasco, D., Durham, S.K., Barton, D.S., Rizzo, C.A., Ryseck, R.-P., Lira, S.A., and Bravo, R., Multiorgan inflammation and hematopoietic abnormalities in mice with a targeted disruption of relB, a member of the NF-KB/Rel family. Cell 80:331–340, 1995.
5. Laufer, T.M., DeKoning, J., Markowitz, J.S., Lo, D., Glimcher, L.H., Unopposed positive selection and autoreactivity in mice expressing class II MHC only on thymic cortex. Nature 382:81–85, 1996.
6. DeKoning, J., DiMolfetto, L., Reilly, C., Wei, Q., Havran, W., Lo, D., T cell development in relB deficient mice: Positive selection and acquisition of function do not require thymic medulla. Submitted, 1996.

9

DRIVING GENE EXPRESSION SPECIFICALLY IN DENDRITIC CELLS

Thomas Brocker, Mireille Riedinger, and Klaus Karjalainen

Basel Institute for Immunology
CH-4058 Basel
Grenzacherstr. 487, Switzerland

1. INTRODUCTION

Lymphoid dendritic cells (DC) play an important role in the immune system. They are potent inducers of primary T cell responses and play a crucial part as MHC class II$^+$ "interdigitating cells" in the thymus during thymocyte development. Most of our knowledge about DC was obtained using highly invasive and manipulatory experimental protocols. Since the functions of DC after these *in vitro* manipulations have been reported not to be identical to those of DC *in vivo* (1), we intended to establish a system that would allow us to investigate DC-function avoiding artificial interferences due to handling. Here we present a transgenic mouse system in which we targeted gene expression specifically to DC. Using the mouse CD11c promoter we expressed MHC class II I-E molecules specifically on DC of all tissues, but not on other cell types.

2. METHODS

A cDNA-pool from four human T cell clones, all expressing human CD11c, was used as template for PCR-amplification of a human CD11c gene-fragment that served as a probe to screen a mouse genomic library. We isolated two overlapping phages that were characterized by restriction mapping and partial sequencing of their inserts. DNA-sequence analysis of approximately 1000bp around the initiation codon showed 68.8% identity to the corresponding sequence of the human CD11c gene and were therefore accepted to represent mouse CD11c. We used a 5.5kb fragment that contained the 5′ region of mouse CD11c gene to drive the expression of the I-E_α^d -cDNA. This transgenic construct was injected into fertilized eggs of C57BL/6 mice, naturally lacking E_α^{d-}expression.

Expression of cell surface proteins was assayed by immunofluorescence analysis. Organs were teased through a mesh and cell suspensions of 1 × 10^5 viable cells were stained with 20 μg/ml mAb that was directly labeled. After washing, cells were analyzed using a FACScan.

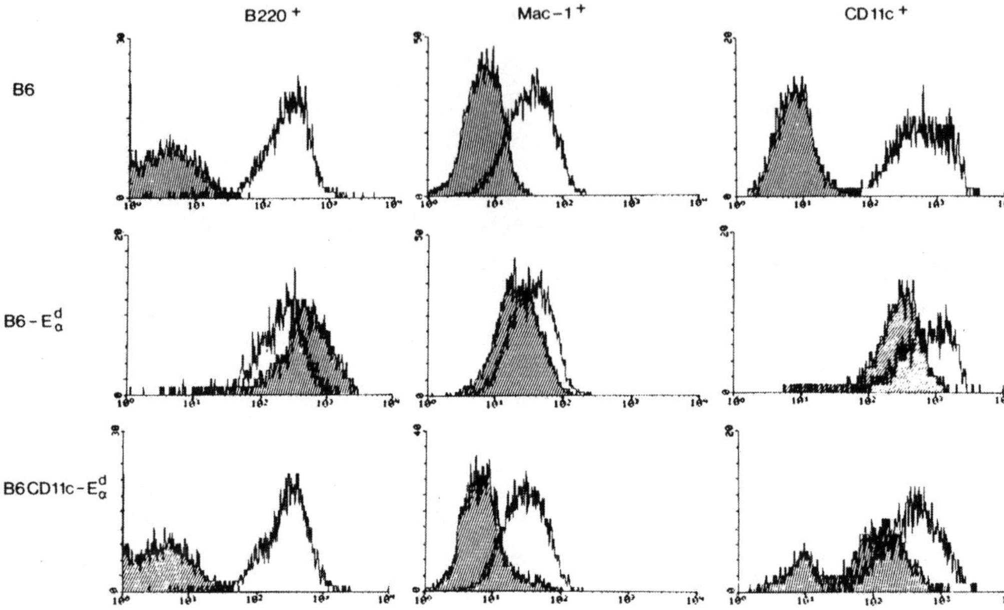

Figure 1. Cell suspensions of different origin were analyzed with mAbs specific for B220, Mac-1, CD11c, I-E, I-A . B cells from spleen were stained with anti-B220, -I-A and -I-E. Macrophages from peritoneum were stained with anti-Mac-1, -I-A and -I-E, while splenic DC from low density gradients were analyzed with anti-CD11c, -I-A and -I-E. Triple immunofluorescence was performed and the gates set on either B220+, Mac-1+ or CD11c+ cells respectively. Shown are the I-E stainings (grey histograms) and I-A stainings (white histograms) only.

3. RESULTS

Two different mouse strains were utilized as controls. An I-E transgenic line created by Widera et al. (2) was used as a "positive" control. This transgenic line 107.1 (here called B6-$E_\alpha{}^d$) expresses an I-$E_\alpha{}^d$ transgene under the control of a segment of the I-E MHC class II promoter. The expression pattern previously described for the B6-$E_\alpha{}^d$ line corresponded to wildtype I-E expression in I-E^+ strains including cortical and medullary thymic epithelial cells, as well as the BM-derived thymic fraction. Negative control mice were C57Bl/6-mice ("B6") that do not express the I-Eα genes, and therefore, no complementation with the β chain can occur, leading to absence of I-E MHC class II surface expression(3).

To analyze transgene expression in B lymphocytes we prepared cell suspensions from spleen and performed double immunofluorescence analysis with anti-B220 as a B cell marker and either anti-I-E or anti-I-A. As shown in Fig. 1, B cells from B6CD11c-$E_\alpha{}^d$ mice do not express I-E (grey histograms), while B cells from B6-$E_\alpha{}^d$ transgenic mice do express I-E. As expected, B cells from all three mice did express endogenous I-A (white histograms).

Furthermore we analyzed peritoneal macrophages 5 days after thiolglycollate injection into the three mouse types and after an additional 48h culture period in the presence of IFN-γ to upregulate class II expression. The FACS analysis in Fig. 1 shows that the Mac-1 positive cells from the three different mice all express similar levels of I-A (white histograms). But while macrophages from B6-$E_\alpha{}^d$ transgenic mice do express transgenic I-E, the majority of macrophages from B6CD11c-$E_\alpha{}^d$ transgenic mice does not express transgenic I-E (grey histograms). Only a few cells (usually 5–8%) from these preparations do express low levels of I-E (Fig. 1) and could very well be DC. When we isolated DC from spleen using a low buoyant density gradient, we find that usually 60–80% of the CD11c$^+$ cells do express the I-E transgene in B6CD11c-E_α^d transgenic mice (Fig. 1).

When the thymus of these animals was analyzed for I-E expression, we could not detect any I-E expression on cortical or medullary epithelial cells in the B6CD11c-E_α^d transgenic mice (data not shown); there the I-E expression was restricted the medulla and medullary-cortical junctions only (data not shown), the areas where thymic DC had been localized previously (4, 5).

4. DISCUSSION

Apparently the 5.5 kb fragment of the mouse CD11c 5' untranslated region is a valuable tool to explore function of DC *in vivo*, since it allows to express genes of interest exclusively in DC and not in other lymphoid or epithelial cells.We now do have a system to study more precisely the role of DC's in negative and positive selection of thymocytes as well as during immuneresponses and will investigate DC-function during immunisation, transplantation and memory.

5. ACKNOWLEDGMENTS

The authors would like to thank, M. Dessing for advice with FACS analysis, H. Stahlberger for graphic art work, U. Mueller for microinjections and Drs. C. Ruedl for help with DC isolations. The Basel Institute for Immunology was founded and is supported by Hoffmann-La Roche Ltd., Basel, Switzerland.

6. REFERENCES

1. Girolomoni, G., J.C. Simon, P.R. Bergstresser, and P. Cruz, Jr. 1990. Freshly isolated spleen dendritic cells and epidermal Langerhans cells undergo similar phenotypic and functional changes during short-term culture. *J Immunol* 145:2820–2826.
2. Widera, G., L.C. Burkly, C.A. Pinkert, E.C. Bottger, C. Cowing, R.D. Palmiter, R.L. Brinster, and R.A. Flavell. 1987. Transgenic mice selectively lacking MHC class II (I-E) antigen expression on B cells: an in vivo approach to investigate Ia gene function. *Cell* 51:175–187.
3. Mathis, D.B., C., Williams, V. Kanter, M., and McDevitt, H. 1983. Several mechanisms can account for defective Ea gene expression in different mouse haplotypes. *Proc. Natl. Acad. Sci.* 80:273–277.
4. Fairchild, P.J., and J.M. Austyn. 1990. Thymic dendritic cells: phenotype and function. *Int Rev Immunol* 6:187–196.
5. Barclay, A.N., and Mayrhofer, G. 1981. Bone marrow origin of Ia-positive Cells in the medulla of rat thymus. *J. Exp. Med* 153:1666–1671.

10

CHECKPOINTS AND FUNCTIONAL STAGES IN DC MATURATION

Claudia Winzler,[1] Patrizia Rovere,[2] Valerie S. Zimmermann,[2] Jean Davoust,[2] Maria Rescigno,[1] Stefania Citterio,[1] and Paola Ricciardi-Castagnoli[1*]

[1]CNR Centre of Cellular and Molecular Pharmacology
Milan 20129, Italy
[2]Centre d'Immunologie INSERM-CNRS de Marseille Luminy
Parc Scientifique de Luminy, Case 906
Marseille 13288, France

INTRODUCTION

A major breakthrough in our understanding of the regulation of immune responses has been the discovery of a number of integrated functions of the dendritic cells (DC) in the immune system. To better define the checkpoints and the functional stages in DC development one should be able to mimic and control DC maturation in vitro, from the initial irreversible commitment towards final stages of maturation.

In the last few years we have been able to grow in vitro immature DC long-term cultures, immortalized by the MIBψ2N11 retroviral vector[1] and their functional properties have been described[2,3,4]. Nevertheless these immortal cells, although functional, were unable to fully mature in vitro due to their continuous cell growth and our inability to stop them from proliferating. Thus, we investigated the possibility of a new approach aimed to maintain DC in long-term cultures using a cocktail of balanced growth-factors and differentiating cytokines. We succeeded to establish an in vitro system of growth-factor dependent cell line[5] and here we report the differentiation potential of immature spleen DC. Using this innovative approach we were able to reproduce and mimic *in vitro* the process of DC differentiation and to control and define the checkpoints of the so-called DC maturation.

* Correspondence to: Paola Ricciardi-Castagnoli, CNR (National Council Research) Centre of Cellular and Molecular Pharmacology, Via Vanvitelli 32, 20129 Milano, Italy (Phone: +39–2–70146283; Fax: +39–2–70146373; @ mail: Paola@farma8.csfic.mi.cnr.it).

Dendritic Cells in Fundamental and Clinical Immunology
edited by Ricciardi-Castagnoli, Plenum Press, New York, 1997

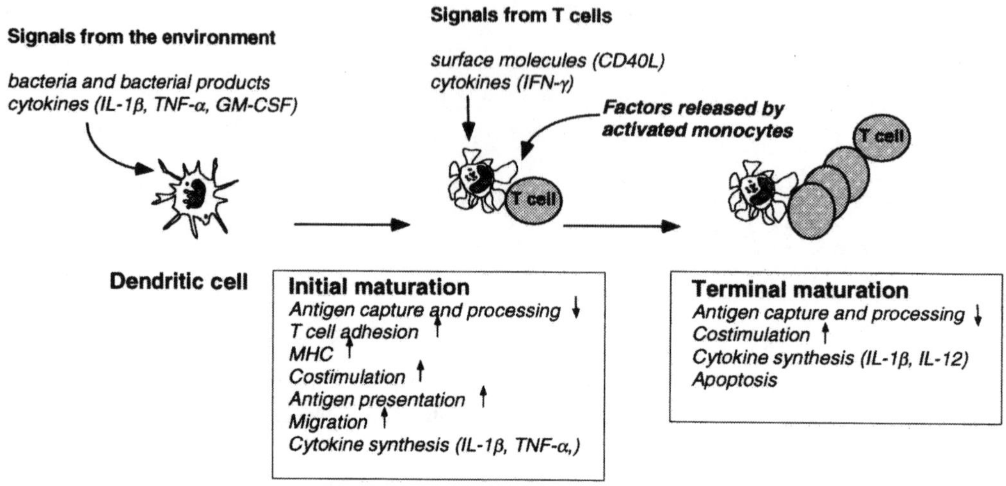

Figure 1. Model of DC maturation.

RESULTS

The immature DC population was driven to proliferate for as long as growth factors were provided. These long-term, growth-factor dependent immature DC cells (named D1) could be induced to mature *in vitro*, upon activation with a number of regulatory signals, into terminally-differentiated DC (Fig 1). Using this model system, we identified 3 sequential stages of the DC maturation and Fig. 2 summarizes the distinct features of each stage: DC population at stage 1 is characterized by high proliferation rate, low motility and highly organized cytoskeleton actin filaments and adhesion plaques; expression of MHC class II molecules is mostly restricted to cytoplasm and CD40 and B7.2 co-stimulatory molecules are expressed at low level. Furthermore, these

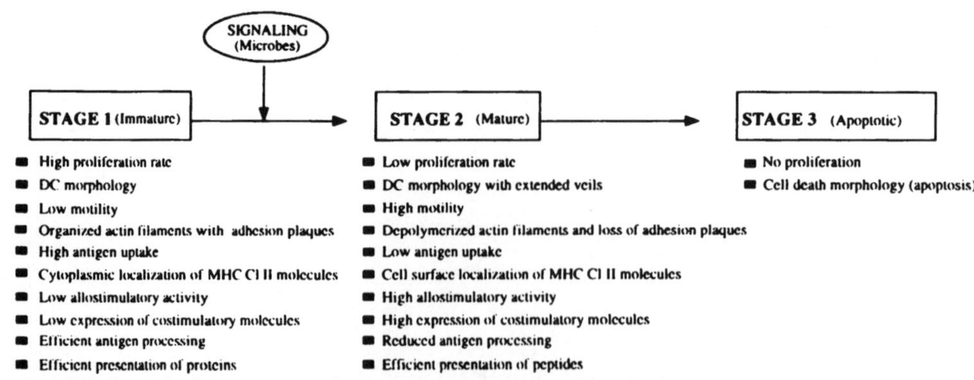

Figure 2. DC maturational stages in the proposed model system: correlation between phenotypical and functional maturation.

immature DC are inefficient in allostimulatory activity and are unable to secrete IL-12 but have a high capacity of antigen uptake and processing. Progression from stage 1 towards stage 2 requires signaling and this can be achieved using living bacteria, LPS or cytokines such as TNFα or IL-1ß but not IL-6. Stage 2, which is induced in few hours and is completed after one day, is characterized by low proliferative rate, high motility and complete depolarization of cytoskeleton actin filaments with loss of adhesion plaques. In addition, mature DC rearrange the cell surface expression of receptors and molecules: MHC class II, CD-40 and B7.2 molecules become highly expressed on the cell surface and this may correlate with the ability to produce IL-12. In contrast, mature DC disassemble actin-based cytoskeleton, lose adhesion plaques and become inefficient in protein antigen uptake and presentation. However, allostimulatory activity and peptide presentation remain very efficient.

Most of these cells will spontaneously progress towards stage 3 where they completely stop proliferation. In agreement with this hypothesis we also found that MHC class II ^{high} sorted D1 cells, if re-cultured after sorting, enter stage 3, characterized by cell growth arrest and final cell death by apoptosis. In contrast, sorting and re-culture of the MHC class II^{int} D1 subpopulation leads to the development of a population which resembles the initial bulk D1 cells (Fig 3). Thus, the D1 bulk population contains both immature and mature DC, and the former can be manipulated and controlled *in vitro*.

What stimuli induce full differentiation of mouse DC? We provided the first evidence that TNFα plays a crucial role in mouse DC maturation[5], as previously reported in human monocyte-derived[6] and bone marrow-derived DC[7], as well as in Langerhans cells[8]. We also showed that IL-1ß induces phenotypical DC maturation, most likely through a paracrine induction of TNFα expression, as shown in Langerhans cells[9]. Living bacteria and bacterial cell constituents are also strong inducers of full DC maturation, including upregulation of the costimulatory molecules B7.2 and CD40. It is likely that *in vivo*, the first signals inducing DC maturation are delivered by microbes since we previously showed a paracrine induction of TNFα expression in DC infected with Gram+ and Gram- bacteria[10]. Thus, TNFα is most likely a signal 2 for sequential DC maturation, whereas pathogens represent a signal 1.

In addition to its function in DC morphological maturation, TNFα also promotes DC emigration from a number of tissues[11]. The non-lymphoid DC need to move on to the regional lymph nodes in order to encounter naive T cells and elicit primary immune responses. Thus, the migratory properties are an essential component of the DC function. We also provided the first evidence that TNFα is able to induce a complete rearrangement of the actin-based cytoskeleton in DC (Fig.4). This actin disassembly could explain the decreased macropinocytosis ability of mature DC as proposed in other systems[12]. The disassembly of actin filaments upon TNFα treatment is correlated with a loss of endocytic capacity in DC. Macropinocytosis represents one of the most efficient antigen uptake mechanism in DC but requires the integrity of polymerized actin filaments; it is likely that DC maturation shuts off the micropynocytic pathway by inducing actin disassembly. The modification of the actin cytoskeleton that we observed after TNFα treatment (Fig 4) is also relevant in the acquisition of migration properties of mature DC. In fact, TNFα treatment induces cell detachment and formation of a organized centriolar array of microtubules which are likely to be highly dynamic. Membrane expansions of mature DC, which occur in response to the cytoskeleton remodelling, also appear highly dynamic: focal adhesion plaques containing vinculin, disappear and the cells acquire migratory properties. *In vivo*, motility should unable DC to encounter T cells and the final outcome should be T cell activation and Th1/Th2 polarization. It has been proposed that DC maturation continues until DC-T cell interaction occurs (Fig 1) but in the D1 model, mature DC were driven to terminal differentiation even in the absence of T cells. In contrast, interaction of DC with T cells was essential for IL-12 production. Interestingly, immature DC fail to produce IL-12 p75 either upon stimulation with heat treated S. aureus and IFNγ or after activation with antigen-specific T cell hybridoma[5]. In contrast, IL-12 p75 secretion is readily detectable in TNFα-treated DC following interaction with antigen specific T cells. This result indicates that the feedback between T cells and DC is dependent on TCR-peptide/class II interaction and this property is reached at stage 2 of functional maturation. These data also indicate that DC may play a major role in polarizing the immune response towards the Th1 pathway.

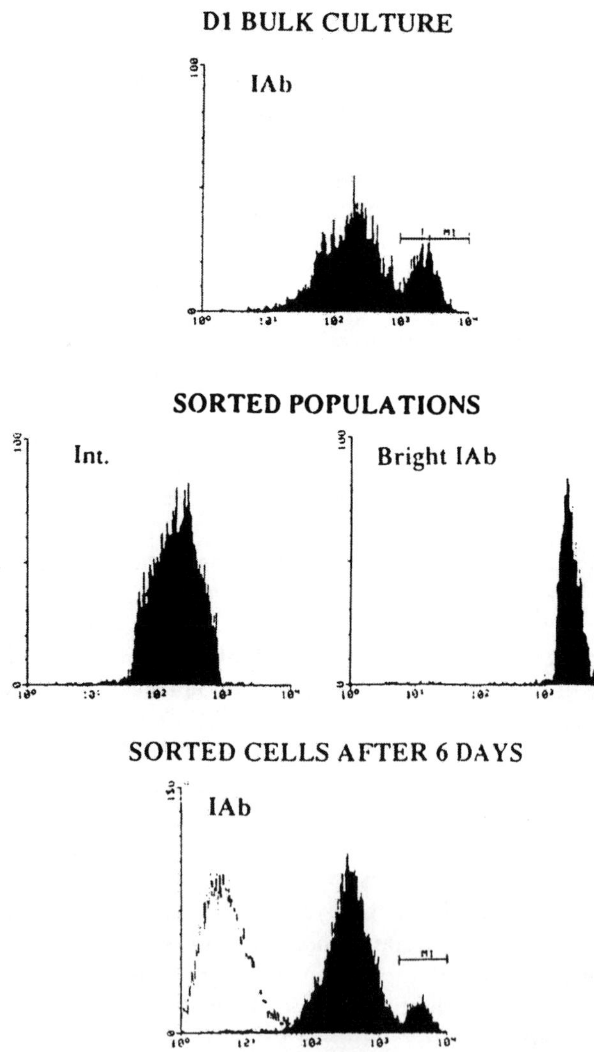

Figure 3. Cell sorting of Class II intermediate (int) and bright D1 cells. Sorted cells were replated in culture and re-analysed for class II expression after 6 days of culture.

ACKNOWLEDGMENTS

We wish to thank Giampiero Girolomoni for critical readings, Manfred Lutz for suggestions and Domenico Delia for cell sorting. This work was supported by the Italian Ministry of Public Health (Istituto Superiore di Sanità. AIDS project, grant 9403–92), by the National Council Research (CNR) and by the French Association against Cancer, grant 6058 (ARC).

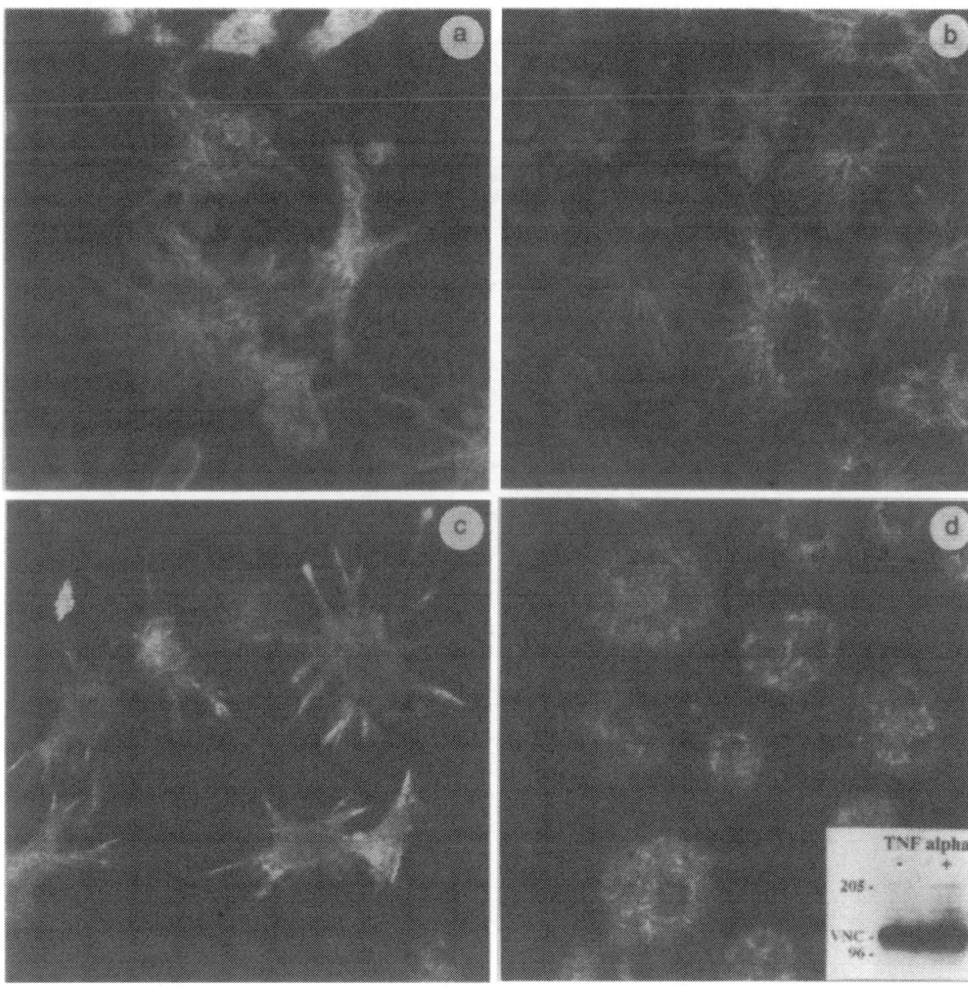

Figure 4. Cytoskeleton modifications of D1 cells after TNFα treatment. Confocal laser scanning microscopy of D1 cells stained with anti-vinculin (green colour, panels c and d) or anti-tubulin (green colour, panels a and b) mAbs in the presence of rhodamine-labelled phalloidin to visualize F-actin (red colour, panels a,b,c and d). Adhesion plaques can be detected as indicated by the simultaneous staining for F-actin and vinculin (panel c, yellow colour, due to superimposition of red-actin and green-vinculin). Treatment with TNFα (panels b and d) induces the loss of stress fibers and adhesion plaques are no longer detectable (panel d). In contrast to the disassembly of the actin filaments organization, microtubules are only partially affected by TNFα treatment (panel b, green colour). No differences in the amount of vinculin protein are detectable by Western blot analysis (panel d, insert).

REFERENCES

1. Sassano, M., F. Granucci, M. Seveso, G. Marconi, M. Foti, and P. Ricciardi-Castagnoli. 1994. Molecular cloning of a recombinant retrovirus carrying a mutated env^{AKR}-myc^{MH2} fusion gene immortalizing cells of the monocytic-macrophage lineage. *Oncogene*: 9: 1473–1477.
2. Paglia, P., G. Girolomoni, F. Robbiati, F. Granucci, and P. Ricciardi-Castagnoli. 1993. Immortalized Dendritic Cell Line Fully Competent in Antigen Presentation Initiates Primary T Cell Responses in Vivo. *J. Exp. Med.* 178: 1893–1901.
3. Girolomoni, G., M.B. Lutz, S. Pastore, C.U. Aßmann, A. Cavani, and P. Ricciardi-Castagnoli. 1995. Establishment of a cell line with features of early dendritic cell precursors from mouse fetal skin. *Eur. J. Immunol.* 25:2163–2169.
4. Lutz, M.B., C.U. Aßmann, G. Girolomoni, and P. Ricciardi-Castagnoli. 1996. Different cytokines regulate antigen uptake and presentation of a precursor dendritic cell line. *Eur. J. Immunol.* 26: 586–594.
5. Winzler, C., P. Rovere, M. Rescigno, F. Granucci, G. Penna, L. Adorini, V.S. Zimmermann, J. Davoust, and P. Ricciardi-Castagnoli. 1997. Maturation stages of mouse dendritic cells in growth-factor dependent long-term cultures. *J. Exp. Med.* 185: 317–328.
6. Sallusto, F., and A. Lanzavecchia. 1994. Efficient presentation of soluble antigen by cultured human dendritic cells is maintained by granulocyte/macrophage colony stimulating factor plus interleukin 4 and downregulated by tumor necrosis factor-α. *J. Exp. Med.* 179: 1109–1118.
7. Young, J., W. Szabolcs, and M.A.S. Moore. 1995. Identification of Dendritic Cell Colony-forming Units among Normal Human CD34+ bone marrow progenitors that are expanded by c-kit-ligand and yield pure dendtritic cell colonies in the presence of granulocyte/macrophage colony-stimulating Factor and Tumor Necrosis Factorα. *J. Exp. Med.* 182: 1111–1120.
8. Caux, C., D. Dezutter-Dambuyant, D. Schmitt, and J. Banchereau. 1992. GM-CSF and TNF-α cooperate in the generation of dendritic Langerhans cells. *Nature* 360: 258–261.
9. Heufler, C., F. Koch, and G. Schuler. 1988. Granulocyte/macrophage colony-stimulating factor and interleukin 1 mediate the maturation of epidermal Langerhans cells into potent immunostimulatory dendritic cells *J. Exp. Med.* 167: 700–712.
10. Koide,S.L., K. Inaba, and R.M Steinman. 1987. IL-1 enhances T-dependent immune responses by amplifying the function of dendritic cells. *J. Exp. Med.* 165: 515–1987.
11. Roake, J.A., A.S. Rao, P.J. Morris, C.P. Larsen, D.F. Hankins, and J.M. Austyn. 1995. Dendritic cell loss from nonlymphoid tissues after systemic administration of lipopolysaccharide, tumor necrosis factor, and interleukin 1. *J. Exp. Med.* 181: 2237–2247.
12. Pfeiffer J.R. J.C. Scagrve, B.H. Davis, G.G. Deanim and J.M. Oliver. 1985. Membrane and cytoskeleton changes associated with IgE-mediated serotonin release from rat basophlic leukemia. *J. Cell.Biol.* 101: 2145–2155.

11

T CELL-MEDIATED TERMINAL MATURATION OF DENDRITIC CELLS

Paul R. Bergstresser, Toshiyuki Kitajima, Shan Xu, Kiyoshi Ariizumi, and Akira Takashima

Department of Dermatology
University of Texas Southwestern Medical Center
Dallas, Texas

INTRODUCTION

Numerous investigators have supported, with compelling evidence, the concept that the activation of naive T cells occurs primarily through antigens (Ag) presented by a distinct class of leukocytes known as dendritic cells (DC). As skin biologists, our own efforts have focused on Langerhans cells (LC), a unique population of DC that resides normally within the epidermis of skin and whose identification even predates knowledge about DC by more than one century. Three years ago, we initiated a program to develop long-term DC lines from the epidermis of mice, largely to accommodate the cellular homogeneity that is required for molecular biologic techniques, but also to facilitate the economic use of laboratory animals. Two concepts guided these efforts: 1) cells derived during the perinatal period would possess a highest proliferative potential, and 2) GM-CSF, which had promoted the survival and maturation of LC cultured by other investigators, would be essential. Our efforts were ultimately rewarded by the development of stable DC lines, termed the "XS series", from specimens of epidermis in newborn BALB/c mice (1). These XS DC lines resemble resident LC in many respects, including: a) their dendritic morphology (1), b) Ag presenting capacity (1,2), and c) cytokine and cytokine receptor mRNA profiles (3–6). Importantly, our original XS cell cultures were "contaminated" by fibroblastoid cells, the "NS series," which could be removed through their selective sensitivity to treatment with trypsin. Once this was accomplished, however, proliferation by the XS cells diminished substantially. We have determined subsequently that XS52 cells require for their optimal growth not only GM-CSF, which was added continuously to XS cell cultures, but also CSF-1, which was secreted in relatively large amounts by the NS cells (4). These XS lines and similar DC lines developed since that time have proven to be useful tools in our efforts to characterize the function of DC.

Dendritic Cells in Fundamental and Clinical Immunology
edited by Ricciardi-Castagnoli, Plenum Press, New York, 1997

CYTOKINE-MEDIATED REGULATION OF THE MATURATIONAL STATE OF DENDRITIC CELLS

A critical issue in the biology of LC is their "maturation" when placed into *in vitro* culture. Maturational changes characteristic of cultured LC include: a) heightened expression of MHC class II molecules, CD80 (B7–1), CD86 (B7–2), and other adhesion molecules (7–10), b) increased ability to activate naive allogeneic T cells (11), c) a decreased ability to process complex protein antigens (12), d) increased mRNA expression of IL-1β (13) and an acquired capacity to secrete IL-6 (14), and e) a loss of adhesive capacity (8). Within this context, a unique aspect of XS cells was their retention of the "immature" features characteristic of resident LC, even after >3 years in culture in the continuous presence of added GM-CSF. Specifically, in the absence of stimulation, XS cells: a) express Ia, CD80, and CD86 molecules only at minimal levels (1); b) are relatively poor in their ability to stimulate allogeneic naive T cells (1); c) exhibit a potent capacity to process complex protein Ag (1,2); d) express relatively small amounts of IL-1β mRNA and no detectable IL-6 mRNA (3,5); and e) adhere progressively onto petri dishes and phagocytose latex particles (15).

Because XS cells had been maintained and expanded in the continuous presence of supernatant collected from NS cells (NS supernatant), in addition to GM-CSF, we considered the possibility that NS supernatant contained one or more soluble factors that suppressed the maturation of XS cells. In fact, when XS cells were cultured with GM-CSF in the absence of NS supernatant, they began to exhibit features characteristic of cultured LC. For example, they acquired elevated expression of Ia, CD80, and CD86 molecules, and they increased the ability to activate naive allogeneic T cells (2,3). Conversely, they even lost the surface expression of Ia molecules when cultured with NS supernatant alone (3). GM-CSF and NS supernatant also regulated, in a reciprocal manner, the phenotype and function of the XS-derived DC clones (16), indicating that these changes did not simply represent a selective expansion of different DC subpopulations. Rather, it was more likely that NS cells secreted soluble factors that prevented the ordinary maturation of XS DC. Studies are currently in progress to determine the identities of such DC maturation inhibitory factors. In sum, our observations demonstrate that the state of XS DC maturation can be regulated experimentally by altering the cytokine milieu in their environment.

T CELL-MEDIATED TERMINAL MATURATION OF DENDRITIC CELLS

During Ag presentation, DC are known to deliver activation signals to responding T cells, thereby altering their phenotype and function. We sought to determine whether responding T cells deliver signals back to DC. This possibility was assessed in studies in which the phenotypic and functional properties of XS cells were examined after Ag-specific interaction with T cells. When XS52 cells were incubated with the KLH-specific Th1 clone HDK-1 in the presence of Ag, they: a) increased the surface expression of CD86 (17); b) increased IL-1β mRNA expression and IL-1β converting enzyme (ICE) activity and began to secrete relatively large amounts (up to 1 ng/ml) of biologically active IL-1β (18); c) secreted IL-6 and TNFα as well (17), d) lost their adhesive and phagocytotic capacities (15), and e) lost their expression of CD115 (CSF-1 receptor) and at the same time their proliferative responsiveness to CSF-1 (19). None of these changes was observed

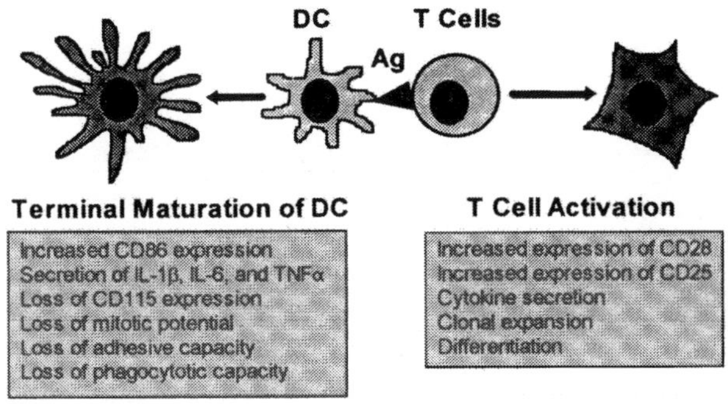

Figure 1

when XS52 cells were incubated with HDK-1 T cells or KLH alone, indicating the requirement for both T cells and Ag. Importantly, these changes, which occured in DC during Ag presentation, duplicated changes known to occur in LC during culture. [We have detected CD115 mRNA expression in freshly procured LC (4), whereas cultured LC have been reported to express no detectable CD115 (20), suggesting that lost CD115 expression is also a feature associated with LC maturation in culture.] Thus, we have postulated that during Ag presentation, DC undergo a critical transition from Ag presenting cells (APC) specialized for Ag uptake and processing into APC specialized for delivering T cell stimulatory signals, a process we have termed "T cell-mediated terminal maturation" of DC (Figure 1).

These observations also led to the conclusion that Ag-specific interaction between DC and T cells is not a unidirectional process, rather, that activation signals are also delivered "backward" from the responding T cells to the stimulating DC. With respect to the nature of DC-activating signals, the following observations indicate that interferon-γ (IFNγ), which is secreted by T cells upon DC-dependent stimulation, is responsible for at least some of the maturational changes. First, XS52 cells lost surface expression of CD115, proliferative responsiveness to CSF-1, and adhesive and phagocytotic capacities upon exposure to supernatant collected from complete cocultures of XS52 DC, HDK-1 T cells, and KLH, but not from incomplete cocultures that lacked one or more of the three components. Second, relatively large amounts of IFNγ were detected in the complete coculture supernatants, but not in incomplete coculture supernatants. Third, anti-IFNγ monoclonal antibody (mAb) neutralized all of the above biological activities of complete coculture supernatants. Finally, recombinant IFNγ alone was sufficient to diminish CD115 expression, proliferative responsiveness to CSF-1, and adhesive and phagocytotic capacities of XS52 cells (15,19). With respect to signals leading to IL-1β secretion, we have observed that anti-Ia mAb and CTLA4-Ig fusion protein: a) each inhibit T cell-mediated IL-1β secretion by XS52 cells when tested in soluble forms, and b) each trigger IL-1β secretion in the absence of T cells when tested in immobilized forms. Thus, we have concluded that ligation of Ia molecules and CD80/CD86 (on DC) with the T cell receptor/CD3 complex and CD28/CTLA4 (on T cells) delivers relevant signals leading to IL-1β secretion by DC (18). These observations demonstrate that during Ag presentation, responding T cells deliver different activation signals to trigger different maturational changes by DC. These observations also raise several questions. Because IFNγ is secreted primarily by the Th0/Th1 subsets of $CD4^+$ T cells, an obvious question concerns whether different T cell subsets (naive T cells versus primed T cells, and Th1 versus Th2) induce different forms of terminal maturation of DC. What is the *in vivo* relevance of our *in vitro* observations; do LC undergo terminal maturation in the draining lymph nodes upon Ag-specific interaction with T cells? Does pharmacologic interference with this transition lead to the suppression of T cell-mediated immune responses?

PHARMACOLOGICAL MODULATION OF DENDRITIC CELL TERMINAL MATURATION

Glucocorticoids (GCs) have been used for decades as immunosuppressive agents to treat inflammatory diseases, including diseases mediated by T cells. In terms of their pharmacologic mechanism of action, GCs are known to inhibit several activities of monocytes and macrophages, including a) secretion of proinflammatory cytokines and prostaglandins, b) expression of surface receptors, d) phagocytosis and pinocytosis, and e) their bactericidal and fungicidal activities. It is also known that GCs inhibit certain activities of T cells, including the secretion of IL-2 and cell division. Based on this knowledge, the immunosuppressive activities of GCs have been attributed primarily to their influence on monocytes/macrophages and on T cells.

Taking the advantage of long-term DC lines, we sought to examine the effect of GCs on DC. When dexamethasone (DEX) was added, at relatively low concentrations (10^{-9}–10^{-7}M), to the coculture of XS52 DC, HDK-1 T cells, and KLH, it inhibited substantially or completely T cell-induced maturational changes of XS52 cells, including: a) the secretion of IL-1β, IL-6, and TNFα, b) elevated expression of CD86, and c) loss of CD115 expression and proliferative responsiveness to CSF-1. By contrast, DEX at the same concentrations inhibited XS52 DC-stimulated IL-2 secretion by HDK-1 T cells, but not other changes that accompany T cell activation, including the secretion of IFNγ and TNFα and the elevated expression of CD25, CD28 and CD44 (17). Although the *in vivo* relevance of these observations remains to be determined, they provide a basis for the new concept that GCs suppress T cell-mediated immune responses by interfering with the terminal maturation of DC. This concept is mirrored in other reports. For instance, DEX has been shown to: a) inhibit APC-dependent activation of T cells, but not APC-independent activation, as is triggered by phorbol esters and calcium ionophores (21), b) to downregulate the costimulatory capacity of LC (22), and c) to downregulate the expression of CD80 and CD86 and to diminish the T cell stimulatory capacity of DC (23). We believe that this experimental system will serve as a useful method to study the impact of GCs and other immunosuppressive drugs on Ag-specific DC-T cell interaction and to search for and develop new compounds that inhibit DC maturation selectively.

REFERENCES

1. Xu S, Ariizumi K, Caceres-Dittmar G, Edelbaum D, Hashimoto K, Bergstresser PR, Takashima A: Successive generation of antigen-presenting, dendritic cell lines from murine epidermis. J Immunol 154:2697–2705, 1995.
2. Caceres-Dittmar G, Ariizumi K, Xu S, Tapia FJ, Bergstresser PR, Takashima A: Hydrogen peroxide mediates UVB-induced impairment of antigen-presentation and downregulation of IL-1β mRNA expression in a murine epidermal-derived dendritic cell line. Photochem Photobiol 62:176–183, 1995.
3. Xu S, Ariizumi K, Edelbaum D, Bergstresser PR, Takashima A: Cytokine-dependent regulation of growth and maturation in murine epidermal dendritic cell lines. Eur J Immunol 25:1018–1024, 1995.

4. Takashima A, Edelbaum D, Kitajima T, Shadduck RK, Gilmore GL, Xu S, Taylor RS, Bergstresser PR, Ariizumi K: Colony-stimulating factor-1 secreted by fibroblasts promotes the growth of dendritic cell lines (XS series) derived from murine epidermis. J Immunol 154:5128–5135, 1995.
5. Ariizumi K, Kitajima T, Bergstresser PR, Takashima A: Interleukin-1β converting enzyme in murine Langerhans cells and epidermal-derived dendritic cell lines. Eur J Immunol 25:2137–2141, 1995.
6. Mohamadzadeh M, Poltorak AN, Bergstresser PR, Beutler B, Takashima A: Dendritic cells produce macrophage inflammatory protein-1 gamma, a new member of the CC chemokine family. J Immunol 156:3102–3106, 1996.
7. Shimada S, Caughman SW, Sharrow SO, Stephany D, Katz SI: Enhanced antigen-presenting capacity of cultured Langerhans' cells is associated with markedly increased expression of Ia antigen. J Immunol 139:2551–2555, 1987.
8. Romani N, Lenz A, Glassel H, Stossel H, Stanzl U, Majdic O, Fritsch P, Schuler G: Cultured human Langerhans cells resemble lymphoid dendritic cells in phenotype and function. J Invest Dermatol 93:600–609, 1989.
9. Kawamura T, Furue M: Comparative analysis of B7–1 and B7–2 expression in Langerhans cells: differential regulation by T helper type 1 and T helper type 2 cytokines. Eur J Immunol 25:1913–1917, 1995.
10. Inaba K, Witmer-Pack M, Inaba M, Hathcock KS, Sakuta H, Azuma M, Yagita H, Okumura K, Linsley PS, Ikehara S, Muramatsu S, Hodes RJ, Stienman RM: The tissue distribution of the B7–2 costimulator in mice: Abundant expression on dendritic cells in situ and during maturation in vitro. J Exp Med 180:1849–1860, 1994.
11. Schuler G, Steinman RM: Murine epidermal Langerhans cells mature into potent immunostimulatory dendritic cells in vitro. J Exp Med 161:526–546, 1985.
12. Romani N, Koide S, Crowley M, Witmer-Pack M, Livingstone AM, Fathman CG, Inaba K, Steinman RM: Presentation of exogenous protein antigens by dendritic cells to T cell clones. J Exp Med 169:1169–1178, 1989.
13. Heufler C, Topar G, Koch F, Trockenbacher B, Kampgen E, Romani N, Schuler G: Cytokine gene expression in murine epidermal cell suspensions: interleukin 1 beta and macrophage inflammatory protein 1 alpha are selectively expressed in Langerhans cells but are differentially regulated in culture. J Exp Med 176:1221–1226, 1992.
14. Schreiber S, Kilgus O, Payer E, Kutil R, Elbe A, Mueller C, Stingl G: Cytokine pattern of Langerhans cells isolated from murine epidermal cell cultures. J Immunol 149:3525–3534, 1992.
15. Kitajima T, Caceres-Dittmar G, Tapia F, Jester J, Bergstresser P, Takashima A: T cell-mediated terminal maturation of dendritic cells: Loss of adhesive and phagocytotic capacities. J Immunol 157:2340–2347, 1996.
16. Xu S, Bergstresser PR, Takashima A: Phenotypic and functioinal heterogeneity among murine epidermal-derived dendritic cell clones. J Invest Dermatol 105:831–836, 1995.
17. Kitajima T, Ariizumi K, Bergstresser PR, Takashima A: A novel mechanism of glucocorticoid-induced immune suppression: The inhibition of T cell-mediated terminal maturation of a murine dendritic cell line. J Clin Invest 98:142–147, 1996.
18. Kitajima T, Ariizumi K, Mohamadzadeh M, Edelbaum D, Bergstresser PR, Takashima A: T cell-dependent secretion of IL-1β by a dendritic cell line (XS52) derived from murine epidermis. J Immunol 155:3794–3800, 1995.
19. Kitajima T, Ariizumi K, Bergstresser PR, Takashima A: T cell-dependent loss of proliferative responsiveness to colony-stimulating factor-1 by a murine epidermal-derived dendritic cell line, XS52. J Immunol 155:5190–5197, 1995.
20. Kampgen E, Koch F, Heufler C, Eggert A, Gill LL, Gillis S, Dower SK, Romani N, Schuler G: Understanding the dendritic cell lineage through a study of cytokine receptors. J Exp Med 179:1767–1776, 1994.
21. Furue M, Kawakami Y, Kawakami T, Katz SI: Differential inhibition of the T cell activation pathway by dexamethasone and cyclosporine. Transplantation 49:560–564, 1990.
22. Furue M, Katz SI: Direct effect of glucocorticosteroids on epidermal Langerhans cells. J Invest Dermatol 92:342–347, 1989.
23. Moser M, De Smedt T, Sornasse T, Tielemans F, Chentoufi AA, Muraille E, Van Mechelen M, Urbain J, Leo O: Glucocorticoids down-regulate dendritic cell function in vitro and in vivo. Eur J Immunol 25:2818–2824, 1995.

12

MIGRATION OF CULTURED CHIMPANZEE DENDRITIC CELLS FOLLOWING INTRAVENOUS AND SUBCUTANEOUS INJECTION

Simon M. Barratt-Boyes,[1] Simon C. Watkins,[2] and Olivera J. Finn[1,3]

[1]Department of Molecular Genetics and Biochemistry
[2]Department of Cell Biology and Physiology
University of Pittsburgh
Pittsburgh, Pennsylvania 15261
[3]Yerkes Regional Primate Research Center
Emory University
Atlanta, Georgia 30322

1. DENDRITIC CELLS AS ADJUVANTS FOR IMMUNOTHERAPY

In recent years much interest has been generated in using dendritic cells as vehicles for cancer immunotherapy[1]. This interest is based on the fact that dendritic cells function as specialized immunostimulatory cells in vivo, serving as initiators of T cell immune responses to antigen[2]. The dendritic cell-based approach to immunotherapy is feasible because of two advances: the identification and isolation of specific human tumor-associated antigens, and the development of techniques to propagate dendritic cells in vitro from precursor cells[3,4]. Hence, it is now possible to grow large numbers of dendritic cells from a patient's blood or bone marrow in vitro, treat these cells with tumor-associated antigens, and administer to the donor with the aim of inducing a strong and therapeutic immune response to the tumor. The therapeutic efficacy of the technique has been tested in murine tumor models[5], and clinical trials in cancer patients are being initiated[6].

The development of clinical trials using in vitro-differentiated dendritic cells has proceeded with little understanding of the characteristics of these cells in vivo. For example, a cardinal feature of the dendritic cell life cycle which is fundamental to its function in vivo is the ability to migrate[7]. Langerhans cells (immature dendritic cells) reside in the epidermis of the skin and specialize in antigen capture. Following antigen uptake Langerhans cells migrate via lymphatic vessels to the draining lymph node where they localize in the parafollicular cortex and paracortex. In this location the interdigitating dendritic cells, which have matured during the migration process into potent antigen-presenting cells, are ideally situated to interact with and stimulate T cells that passage through the lymphoid tissue[7]. This information on dendritic cell migration has been gleaned from studies using fluorescently-labeled antigens applied to the skin surface of mice (a model of contact hypersensitivity), and from studies using freshly isolated and labeled dendritic cells injected into mice[8–10]. No data currently exist documenting the migratory properties of dendritic cells differentiated in vitro, yet it is dendritic cells cultured from blood or bone marrow that are the focus of current efforts to develop immunotherapy protocols. It is important to determine if dendritic cells differentiated in vitro do migrate to lymph nodes and associate with T cells following administration to animals, if these cells are to be used to stimulate antigen-specific T cell responses.

2. A CHIMPANZEE MODEL OF DENDRITIC CELL-BASED IMMUNOTHERAPY

We have developed a chimpanzee model to test the preclinical application of dendritic cell-based immunotherapy for cancer[11]. Dendritic cells can be differentiated from peripheral blood mononuclear cells (PBMC) by culturing for 7 days with recombinant human granulocyte/macrophage colony-stimulating factor (GM-CSF) and interleukin 4 (IL-4). Approximately 10 million dendritic cells can be cultured and purified from a single 200 mL blood sample using this method[11]. These cells have all the morphologic, phenotypic and functional properties of dendritic cells cultured from human (and murine) blood[4]. The chimpanzee therefore is an ideal large animal model to study the in vivo migration of dendritic cells differentiated from precursor cells in vitro.

3. TRACKING THE IN VIVO MIGRATION OF DENDRITIC CELLS

Dendritic cells were grown for seven days from PBMC of three chimpanzees using GM-CSF and IL-4, purified and divided into two equal portions for fluorescent labeling. Cells were labeled either with 3,3′-dioctadecyloxacarbocyanine perchlorate [$DiOC_{18}(3)$; excitation/emission spectra = 484 nm/501 nm] or 1,1′-dioctadecyl-3,3,3′,3′-tetramethylindodicarbocyanine perchlorate [$DiIC_{18}(5)$; excitation/emission spectra = 644 nm/663 nm; both from Molecular Probes, Eugene, Oregon). In each animal four million dendritic cells labeled with $DiOC_{18}(3)$ were injected intravenously in the cephalic vein, and the same number of cells labeled with $DiIC_{18}(5)$ were injected subcutaneously in the inner thigh. Animals were anesthetized at either 24, 48 or 120 hours after injection, and skin and inguinal lymph nodes were harvested and examined for the presence of fluorescently-labeled cells. Frozen tissues were cryosectioned and stained with Hoescht 33342 dye (to label nuclei) with and without human-specific monoclonal antibodies and a Cy3-conjugated secondary antibody. Slides were examined using a Zeiss Axiovert 135 microscope fitted with appropriate filters.

3.1. Spontaneous Migration of Dendritic Cells to the Draining Lymph Node Following Subcutaneous Injection

Dendritic cells migrated rapidly from the site of subcutaneous injection, as fluorescent cells were identified sporadically in subcutaneous tissue at 24 hours after injection

Table 1. Dendritic cells differentiated in vitro from PBMC were injected intravenously and subcutaneously into three donor chimpanzees. Routes of administration were identified by labeling cells with fluorescent markers prior to injection. The region of skin encompassing the subcutaneous injection site, the inguinal lymph node draining the subcutaneous injection site, and the contralateral inguinal lymph node were examined for fluorescently-labeled dendritic cells at 24, 48 and 120 hours after injection in animals 1, 2 and 3, respectively

	Subcutaneous Route			Intravenous route		
Tissue Harvested	24 h	48 h	120 h	24 h	48 h	120 h
Skin/subcutaneous tissue	+	–	–	ND	ND	ND
Lymph node, draining	++	+++	++	–	–	–
Lymph node, contralateral	–	–	–	–	–	–

ND = not determined.
+ = rare fluorescent cells per high power field; ++ = several fluorescent cells per high power field; +++ = numerous fluorescent cells in several different high power fields.

but could not be identified after this time (Table 1). This is an important finding as it indicates efficient mobilization of injected dendritic cells from the site of injection. In contrast, murine studies have shown that the majority of freshly isolated dendritic cells remain in the subcutaneous tissue following injection[9]. Proinflammatory mediators such as tumor necrosis factor α are required for the efficient migration and maturation of Langerhans cells in these models[12]. A distinct difference between Langerhans cells and dendritic cells differentiated in vitro from precursor cells is the stage of maturation: the latter cells are very mature dendritic cells which express high levels of adhesion molecules known to be important in migration[13], whereas immature Langerhans cells have low expression of these molecules. This maturation state is likely to account for the spontaneous migration of injected dendritic cells.

Injected dendritic cells migrated from the subcutaneous tissue to the regional lymph node, as cells labeled with $DiIC_{18}(5)$ were identified in this node at 24, 48 and 120 hours after injection (Table 1). Injected cells were located in different areas of the lymph node at different time points, reflecting progressive migration from superficial to deep regions of the node (Table 2). In contrast, dendritic cells injected intravenously could not be found in inguinal lymph nodes at any time point, and no fluorescently labeled cells could be found in the inguinal lymph nodes contralateral to the subcutaneous injection site (Table 1).

Table 2. Location and cellular association of dendritic cells in draining lymph nodes harvested at 24, 48 and 120 hours after subcutaneous injection in animals 1, 2 and 3, respectively. Cellular association was determined by immunohistochemistry to detect B cells (CD19), T cells (CD3) and interdigitating dendritic cells (CD86)

Time after Injection	Principal Location in Lymph Node	Cell Association in Lymph Node
24 h	subcapsular	ND
48 h	parafollicular zone and paracortex	T lymphocytes, interdigitating DC
120 h	cortex and paracortex	ND

ND = not determined.

3.2. Localization of Dendritic Cells in T Cell Areas of the Lymph Node Following Subcutaneous Injection

Sections of the draining inguinal lymph node taken 48 hours after subcutaneous injection were analyzed by immunohistochemistry with cross-reactive monoclonal antibodies specific for human CD19, CD3 and CD86, to determine the location of injected dendritic cells with B cells, T cells and endogenous interdigitating dendritic cells, respectively. Injected cells were located in the parafollicular region of the cortex and the paracortical zone, and were associated with T lymphocytes in both regions (Table 2). Importantly, the injected dendritic cells had similar high level expression of MHC Class II molecules and the costimulatory molecules CD40 and CD86 as did endogenous interdigitating dendritic cells in the same location. These findings indicate that injected dendritic cells that localize in the draining lymph node retain the phenotype of activated antigen-presenting cells. Therefore we can predict that injected dendritic cells will serve as potent immunostimulatory cells in vivo, as they do in vitro.

4. WHAT ARE THE IMPLICATIONS OF THESE FINDINGS ON THE CLINICAL USE OF DENDRITIC CELLS IN IMMUNOTHERAPY?

These results indicate that dendritic cells, differentiated in vitro from peripheral blood precursor cells, have the ability to spontaneously migrate from a subcutaneous injection site and to localize in T cell areas of lymph nodes. The findings are very important for immunotherapy protocols as they suggest that dendritic cells grown and administered in this manner will be highly efficient mediators of T cell responses to specific tumor-associated antigens, loaded or genetically engineered[14] into dendritic cells prior to administration. Conversely, the data indicate that intravenously administered dendritic cells do not localize in the regional lymph nodes, although the precise location of these cells was not determined. This is in accordance with several murine studies using freshly isolated dendritic cells, which demonstrated that cells injected intravenously homed to the spleen, but were excluded from lymph nodes[10,15]. A further important finding of the murine studies was that splenic localization of dendritic cells induced a Th2, humoral immune response to soluble antigen presented by the injected cells, whereas homing to lymph nodes following subcutaneous administration of antigen-loaded dendritic cells resulted in a predominant Th1, cellular immune response[16]. Clearly, localization of antigen-treated dendritic cells to T cell areas of the lymph node and stimulation of a cellular response to antigen are desirable for the majority of immunotherapy protocols using tumor-associated antigens. Hence, we conclude that clinical studies using antigen-treated dendritic cells to elicit tumor-specific responses in patients should focus on the subcutaneous, and not intravenous, route of administration.

5. ACKNOWLEDGMENTS

The authors thank Walter Storkus and Thomas Tueting for reagents and technical advice; Henry Kao, William Konitsky, John McKolanis and Jenny Carnprobst for technical assistance; Brent Swenson and Elizabeth Strobert for veterinary services; and Michele

Saucier for generous use of laboratory facilities. This work was supported by NIH grants RR00119 to S.B.B. and CA57820 to O.J.F., NCI Cancer Center Support Grant 5P30CA47904 to the University of Pittsburgh Cancer Institute, and NIH grant RR00165 to Yerkes Regional Primate Research Center. Yerkes Regional Primate Research Center is fully accredited by the American Association for Accreditation of Laboratory Animal Care.

6. REFERENCES

1. Grabbe S, Biessert S, Schwarz T, Granstein RD (1995). Dendritic cells as initiators of tumor immune responses: a possible strategy for tumor immunotherapy? Immunol Today 16, 117–121.
2. Steinman RM (1991). The dendritic cell system and its role in immunogenicity. Annu Rev Immunol 9, 271–296.
3. Caux C, Dezutter-Dambuyant C, Schmitt D, Banchereau J (1992). GM-CSF and TNF-α cooperate in the generation of dendritic Langerhans cells. Nature 360, 258–261.
4. Sallusto F, Lanzavecchia A (1994). Efficient presentation of soluble antigen by cultured human dendritic cells is maintained by granulocyte/macrophage colony-stimulating factor plus interleukin 4 and downregulated by tumor necrosis factor α. J Exp Med 179, 1109–1118.
5. Mayordomo JI, Zorina T, Storkus WJ, Zitvogel L, Celluzzi C, Falo LD, Melief CJ, Ildstad ST, Kast WM, DeLeo AB, Lotze ML (1995). Bone marrow-derived dendritic cells pulsed with synthetic tumor peptides elicit protective and therapeutic antitumor immunity. Nature Med 1, 1297–1302.
6. Hsu FJ, Benike C, Fagnoni F, Liles TM, Czerwinski D, Taidi B, Engleman EG, Levy R (1996). Vaccination of patients with B-cell lymphoma using autologous antigen-pulsed dendritic cells. Nature Med 2, 52–58.
7. Austyn JM, Larsen CP (1990). Migration patterns of dendritic leukocytes. Transplantation 49, 1–7.
8. Macatonia SE, Edwards AJ, Knight SC (1986). Dendritic cells and contact sensitivity to fluorescein isothiocyanate. Immunology 59, 509–514.
9. Fossum S (1988). Lymph-borne dendritic leukocytes do not recirculate, but enter the lymph node paracortex to become interdigitating cells. Scand J Immunol 27, 97–105.
10. Kupiec-Weglinski JW, Austyn JM, Morris PJ (1988). Migration patterns of dendritic cells in the mouse: Traffic from the blood, and T cell-dependent and independent entry to lymphoid tissues. J Exp Med 167, 632–645.
11. Barratt-Boyes SM, Henderson RA, Finn OJ (1996). Chimpanzee dendritic cells with potent immunostimulatory function can be propagated from peripheral blood. Immunology 87, 528–534.
12. Roake JA, Rao AS, Morris PJ, Larsen CP, Hankins DF, Austyn JM (1995). Dendritic cell loss from nonlymphoid tissues after systemic administration of lipopolysaccharide, tumor necrosis factor, and interleukin 1. J Exp Med 181, 2237–2247.
13. Ma J, Wang J-H, Guo Y-J, Sy M-S, Bigby M (1994). In vivo treatment with anti-ICAM-1 and anti-LFA-1 antibodies inhibits contact sensitization-induced migration of epidermal Langerhans cells to regional lymph nodes. Cell Immunol 158, 389–399.
14. Henderson RA, Nimgoankar MT, Watkins SC, Robbins PD, Ball ED, Finn OJ (1996). Human dendritic cells genetically engineered to express high levels of the human epithelial tumor antigen mucin (MUC-1). Cancer Res 56, 3763–3770.
15. Austyn JM, Kupiec-Weglinski JW, Hankins DF, Morris PJ (1988). Migration patterns of dendritic cells in the mouse: Homing to T cell-dependent areas of spleen, and binding within marginal zone. J Exp Med 167, 646–651.
16. Morikawa Y, Furotani M, Kuribayashi K, Matsuura N, Kakudo K (1992). The role of antigen-presenting cells in the regulation of delayed-type hypersensitivity. I. Spleen dendritic cells. Immunology 77, 81–87.

13

THE LIVER SINUSOIDS AS A SPECIALIZED SITE FOR BLOOD-LYMPH TRANSLOCATION OF RAT DENDRITIC CELLS

Kenjiro Matsuno, Shunsuke Kudo, and Taichi Ezaki

Department of Anatomy II
Kumamoto University, School of Medicine
Kumamoto 860 Japan

1. INTRODUCTION

The fate as well as the significance of blood dendritic cells (DC) is not yet completely resolved. Some may migrate to the splenic white pulp[1] and other immature DC may reach the epithelial tissues to become resident DC. When particulates are injected intravenously, relatively immature particle-laden DC appear in the peripheral lymph draining the liver[2]. The results suggest that the rat DC lineage may be selectively recruited to the liver after intravenous injection of particulates and that they subsequently translocate from the liver vasculature to the hepatic lymph. The present study was performed to demonstrate the migration pathway of blood DC and the existence of specialized vasculature for DC, the liver sinusoids.

2. MATERIALS AND METHODS

2.1. Distribution of Particle-Laden DC after Intravenous Injection of Particulates

Inbred DA rats received an i.v. injection of latex particles, and cytosmears of liver perfusates and spleen cell suspensions were immunostained with the rat DC-specific mAb OX62[3]. Cryosections of the liver were triple immunostained with ED2, OX62 and anti-type IV collagen as described[2]. In the liver sections, the presence of $OX62^{+}$ latex-laden cells within the portal, sinusoidal and hepatic vein areas were counted. These 3 areas were clearly outlined by type IV collagen immunostaining and easily distinguished.

Dendritic Cells in Fundamental and Clinical Immunology
edited by Ricciardi-Castagnoli, Plenum Press, New York, 1997

2.2. Intravenous Transfer of DC to Allogeneic Hosts

Inbred DA ($RT1^a$) and Lewis ($RT1^l$) rats were used. The immature DC that had ingested paramagnetic latex (L0898, Sigma, St Louis, MO) in the blood were isolated from the hepatic lymph by magnetic attraction[2]. Lymph cells were treated with mitomycin C before isolation. Mature DC were also collected from the intestinal lymph and hepatic lymph without intravenous latex injection and were enriched on metrizamide gradients[4]. At various time intervals after cell transfer, host proliferating cells were labeled with 5-bromo-2′-deoxyuridine (BrdU) and host tissues were freshly frozen. Transferred DC was detected by immunostaining with mAb to donor type MHC class I antigen (anti-$RT1A^a$, MN_{9-41-6}). The host proliferative response, detected by BrdU immunostaining[5], was studied with respect to organ specificity, dose dependence on donor cells, time kinetics and phenotype of proliferating cells in order to reveal whether they respond to alloantigen presentation by the transferred DC or not. Control Lewis rats received an intravenous injection of unseparated cells from DA rats or of syngeneic latex-laden DC from Lewis rats.

2.3. In Situ Cell Binding Assay

This was performed after Austyn[6] to study binding of DC to frozen sections of various target tissues. Some sections were double immunostained with ED2 and $RT1A^a$ to detect Kupffer cells[2] and allogeneic DC, respectively.

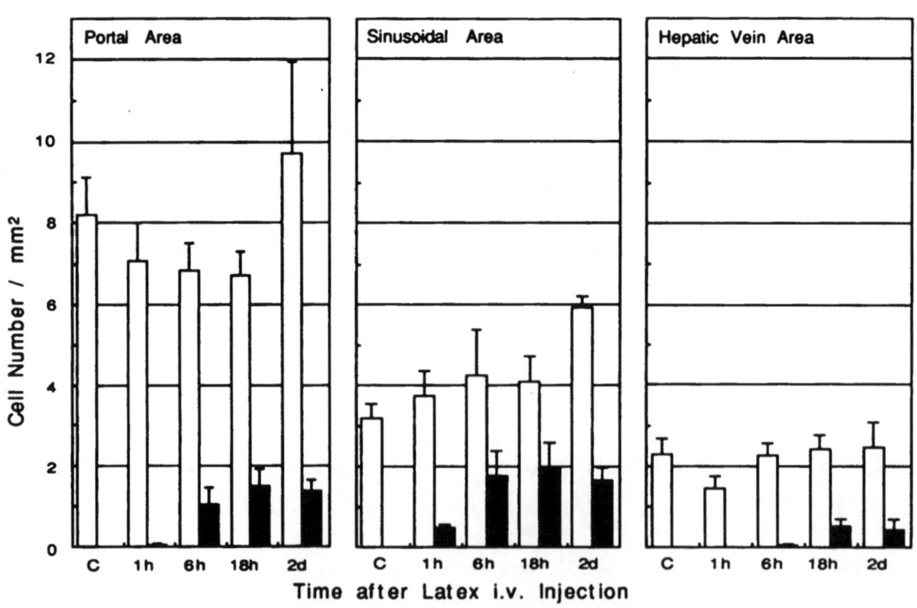

Figure 1. In vivo kinetics of total OX62+ cells and OX62+ latex-laden cells in the liver after i.v. injection of latex particles[2]. Cryosections were triple immunostained with ED2, OX62 and anti-type IV collagen and the proportion of total OX62+ cells and OX62+ latex-laden cells were estimated by thoroughly examining the portal, sinusoidal and hepatic vein areas. Data are expressed as mean ±SE. Bars represent SE. 4 to 6 rats per group were examined.

3. RESULTS

3.1. Time Kinetic Distribution of Particle-Laden DC

1 h after intravenous injection of latex particles, the particle-laden DC, defined as $OX62^+$ latex-laden cells, were first found in the sinusoidal area of the liver (Fig. 1), in the liver perfusate and in spleen cell suspensions (Fig. 2). At 6 h, particle-laden DC first appeared in the portal area. A close association between $OX62^+$ cells and Kupffer cells ($ED2^+$) was often observed in normal liver (Fig. 3) and liver following latex injection.

3.2. Selective Migration of Transferred DC to Regional Hepatic Nodes

Three types of donor DC ($RT1A^{a+}$, immature latex-laden and mature DC of hepatic lymph and mature DC of intestinal lymph) accumulated in regional lymph nodes of the host liver through lymph-borne pathway. In contrast, very few DC were found in the spleen and negligible in other lymph nodes. Syngeneic latex-laden DC, being traced by presence of latex particles, behaved similarly. Proliferative reaction of host cells in response to the alloantigen presentation by the donor DC were also restricted in the regional hepatic nodes (Fig. 4) and the stimulation index of experimental group were about 100 times higher than controls that received same number of mitomycin C-treated unseparated cells. While in the splenic periarterial lymphoid sheath, proliferative response was also observed but was not significantly higher than the allogeneic unseparated cells.

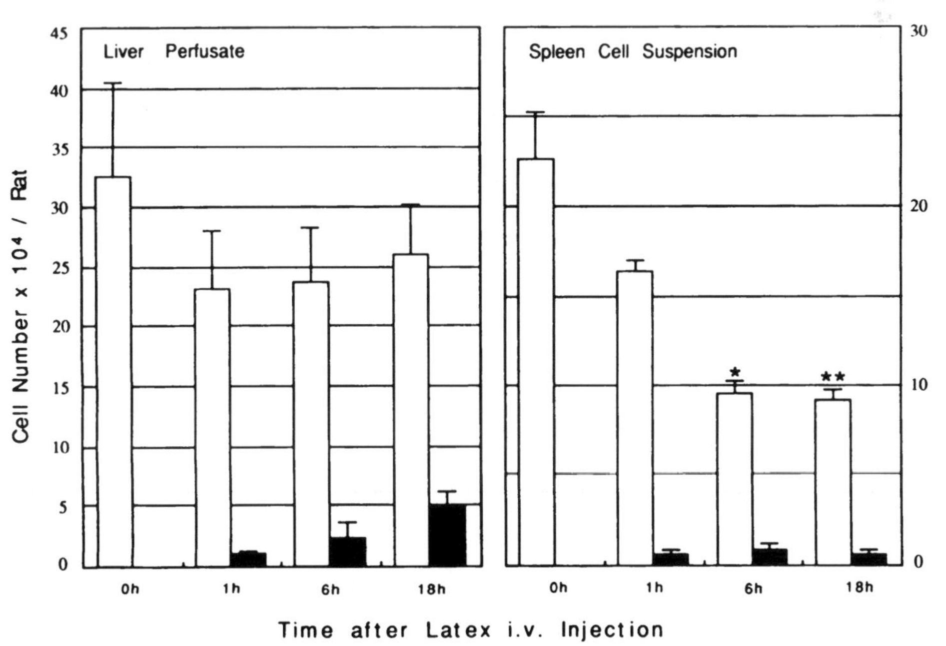

Figure 2. In vivo kinetics of total OX62+ and OX62+ latex-laden cell in the liver perfusate and spleen cell suspensions[2]. Data are expressed as mean ±SE. Bars represent SE. Three to 6 rats per group were examined.

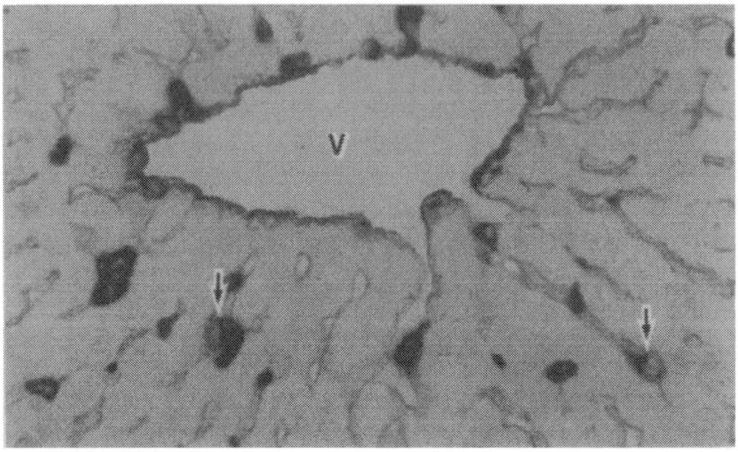

Figure 3. In vivo localization of OX62+ cells in the sinusoidal area of normal liver[2]. Triple immunostaining with ED2 (*black*), OX62 (*gray*) and anti-type IV collagen (*gray line*). Note that a close association between OX62+ cells and Kupffer cells (ED2+) is often observed (*arrows*). *V* central vein. (x460).

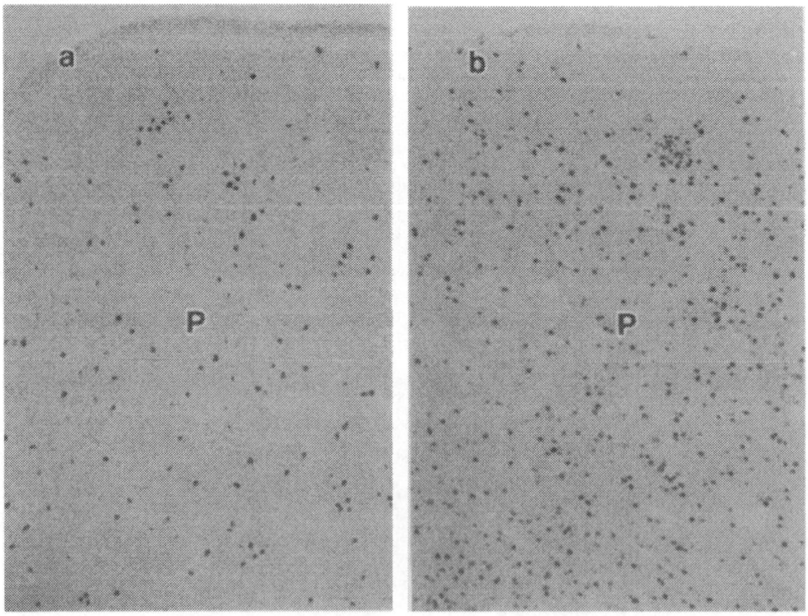

Figure 4. Proliferative response in the hepatic nodes 3 d after cell transfer detected by BrdU labeling (*black*). (*a*) 10^5 unseparated cells did not induce a significant increase in the number of $BrdU^+$ cells but (*b*) 10^5 latex-laden DC induced a significant proliferation in the paracortex (*P*). (x115).

3.3. In Situ Cell Binding Assay

Isolated lymph DC showed preferential binding to either allogeneic or syngeneic liver cryosections compared to other tissues such as spleen, lung, thymus and lymph nodes. The number of bound DC/mm^2 section in the liver was approximately 2–4 times more than those in the other tissues. The same concentration of unseparated cells showed less binding to the liver cryosections than DC. The ratio of $RT1A^{a+}$ cells associated or not, respectively, with $ED2^+$ cells was approximately 2.5, indicating a significant and selective binding of DC to Kupffer cells in the sections.

4. DISCUSSION

The time kinetic study of $OX62^+$ latex-laden cells indicated that the particle-laden DC migrated from the sinusoidal area to the portal area, then translocated to the hepatic lymph. The initial appearance of $OX62^+$ latex-laden cells in the sinusoidal area, in the liver perfusate and in spleen cell suspension implies that the particle-laden DC may be recruited to the liver from systemic circulation, especially the blood marginating pool[2].

The cell transfer study has demonstrated that not only immature syngeneic DC but also allogeneic DC in the blood efficiently undergo blood-lymph translocation through the liver sinusoids to the hepatic lymph. This result also confirms the speculation described above and suggests further that autologous particle-laden DC may be recruited as DC progenitors, possibly in response to the intravenous particulates[7], and then develop into a phagocytic stage prior to the translocation event. The transferred DC as well as autologous particle-laden DC may have passed through the space of Disse from the sinusoidal area to the connective tissue stroma of the portal area, then entered the initial lymphatic ducts located there[2].

It is demonstrated that blood DC perform the liver sinusoids-lymph translocation and accumulate very effectively in the regional hepatic nodes. The liver sinusoids might act as a biological concentrator of blood DC in the hepatic nodes. Kupffer cells may play as a trapping aid for blood DC for initiation of the translocation. The hepatic nodes should be regarded as an important lymphoid organ in not only infectious immunity but also transplantation immunity because graft-derived DC might also perform blood-lymph translocation into the nodes.

5. REFERENCES

1. Austyn, J. M., and C. P. Larsen. 1990. Overview. Migration patterns of dendritic leukocytes: implications for transplantation. Transplantation. 49:1–7.
2. Matsuno, K., T. Ezaki, S. Kudo, and Y. Uehara. 1996. A life stage of particle-laden rat dendritic cells in vivo: their terminal division, active phagocytosis, and translocation from the liver to the draining lymph. J. Exp. Med. 183:1865–1878.
3. Brenan, M., and M. Puklavec. 1992. The MRC OX-62 antigen: A useful marker in the purification of rat veiled cells with the biochemical properties of an integrin. J. Exp. Med. 175:1457–1465.
4. Matsuno, K., S. Kudo, T. Ezaki, and K. Miyakawa. 1995. Isolation of dendritic cells in the rat liver lymph. Transplantation. 60: 765–768.
5. Matsuno, K., T. Ezaki, and M. Kotani. 1989. Splenic outer periarterial lymphoid sheath (PALS): an immunoproliferative microenvironment constituted by antigen-laden marginal metallophils and ED2-positive macrophages in the rat. Cell Tissue Res. 257: 459–470.
6. Austyn, J. M., J. W. Kupiec-Weglinski, D. F. Hankins, and P. J. Morris. 1988. Migration patterns of dendritic cells in the mouse: homing to T cell-dependent areas of spleen, and binding within marginal zone. J. Exp. Med. 167:646–651.
7. Austyn, J. M. 1996. Commentary. New insights into the mobilization and phagocytic activity of dendritic cells. J. Exp. Med. 183:1287–1292.

14

REGULATION OF CD44 ISOFORM EXPRESSION AND CD44-MEDIATED SIGNALING IN HUMAN DENDRITIC CELLS

Hélène Haegel-Kronenberger,[1] Henri de la Salle,[1] Alain Bohbot,[2] Jérôme Galon,[3] Eric Tartour,[3] Jean-Pierre Cazenave,[4] and Daniel Hanau[1]

[1]INSERM CJF 94–03 and
[4]INSERM U.311, E.T.S.
Strasbourg, France
[2]Service d'Onco-Hématologie
Hôpital de Hautepierre, Strasbourg
[3]INSERM U.255
Institut Curie, Paris

ABSTRACT

Dendritic cells (DCs) express CD44, a cell surface receptor for the extracellular matrix ligand hyaluronate, involved in cell-cell interactions and cell migration. Besides the "standard" form of CD44, a variety of splice variants contain an additional extracellular region encoded by 10 "variable" exons termed v1 to v10. The standard form of CD44 as well as variants containing exon v6 (CD44v6) are known to play important roles in the immune system, yet largely unexplored in the DC lineage. In this study, we examined the regulation of CD44 isoforms in human DCs derived from monocytes cultivated in the presence of GM-CSF and IL-4. We found that v3, v6 and v9 variants are all up-regulated upon TNF-α stimulation of DCs. In addition, we show that stimulation of DCs using anti-CD44 mAbs can induce DC agregation, up-regulation of accessory molecule expression and secretion of cytokines. A mAb directed against CD44v6 variants was shown to mediate some of these effects.

1. INTRODUCTION

CD44 represents a family of cell surface and secreted glycoproteins encoded by a single gene[1]. A variable region composed of up to 10 variable exons can be inserted by alternative splicing into the membrane proximal domain of the smallest CD44 form (or standard CD44), resulting in a variety of CD44 variant isoforms. The expression of CD44 variants appears

Dendritic Cells in Fundamental and Clinical Immunology
edited by Ricciardi-Castagnoli, Plenum Press, New York, 1997

more restricted than that of the standard form and has been associated in particular with lymphocyte activation and tumor metastasis[2]. In addition to its function as main cell surface receptor for hyaluronate, an important component of the extracellular matrix[3], CD44 has been shown to play important roles in the immune system. Studies using mAbs have shown that CD44 is involved in hematopoïesis[4], homing to mucosal lymphoïd tissues[5], lymphocyte infiltration into cutaneous delayed type hypersensitivity sites in mice[6] and lymphocyte activation[7]. Langerhans cells as well as DCs from blood and lymph nodes strongly express CD44[8,9]. Human DCs differenciated from peripheral blood monocytes cultivated in the presence of GM-CSF and IL-4 also express CD44[10]. It has been shown previously that TNF-α, a known mediator of DC maturation, can induce the up-regulation of CD44 molecules expressed on the surface of monocyte-derived human DCs. In addition, the induction of a v9-containing CD44 isoform was reported on "mature" DCs[10].

In this study, we examined more closely the regulation of CD44 molecules expressed during DC differentiation from monocytes and activation by TNF-α. Furthermore we investigated the effects of DC culture in the presence of various anti-CD44 MAbs.

2. RESULTS

2.1. Up-Regulation of Variant CD44 Expression in TNF-α-Stimulated DCs

We asked whether the expression of different CD44 variants might be modulated upon TNF-α stimulation of DCs. We focused on different portions of the CD44 variable

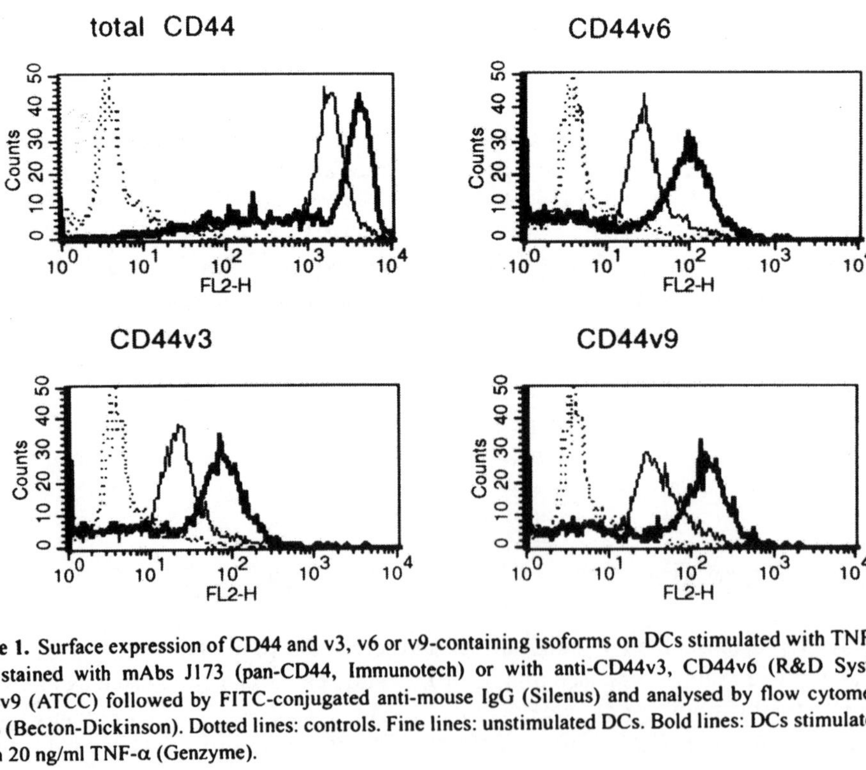

Figure 1. Surface expression of CD44 and v3, v6 or v9-containing isoforms on DCs stimulated with TNF-α. DCs were stained with mAbs J173 (pan-CD44, Immunotech) or with anti-CD44v3, CD44v6 (R&D Systems) or CD44v9 (ATCC) followed by FITC-conjugated anti-mouse IgG (Silenus) and analysed by flow cytometry on a FACS (Becton-Dickinson). Dotted lines: controls. Fine lines: unstimulated DCs. Bold lines: DCs stimulated for 24 h with 20 ng/ml TNF-α (Genzyme).

region: (i) v3, because it is present in glycosaminoglycan-modified CD44 isoforms known to bind growth factors[11]; (ii) v6, because it is contained in the CD44v6 form known to be induced in vivo in activated lymphocytes and macrophages[2, 12]; and (iii) v9, because it is present in the so-called "epithelial form" of CD44 containing the v8-v10 region[13].

Monocyte-derived DCs differentiated for 7 days in the presence of GM-CSF and IL-4 were treated with TNF-α for 24 hours. The variation in the expression of total CD44 as well as v3, v6, or v9-containing isoforms was examined by immunocytochemistry and flow cytometry. We found that all three variant isoforms were expressed by monocyte-derived DCs (Fig. 1). However, their level of expression was rather weak, with the mean fluorescence intensities about 100 fold lower than that of total CD44 stained with antibody J173. Upon TNF-α treatment, total CD44 expression was slightly but significantly up-regulated (1.5 to 2 fold). Proportionnally, the induction of the CD44 variants v3, v6 and v9 was more consequent (Fig. 1); the mean fluorescence intensities corresponding to v3, v6 and v9 isoforms increased up to 5 fold in TNF-α-stimulated DCs. Consequently, these variants represented approximately 1/20th (in the case of v9) to 1/40th (for v3 or v6) of the total CD44 expressed on stimulated cells, suggesting that they might play more important roles in "mature" than in immature DCs.

2.2. Modulation of CD44 Variant Transcripts Upon DC Differentiation and Maturation Induced by TNF-α

To look whether the regulation of CD44 isoforms on DC surfaces reflected a change in the level of transcripts, we examined the relative abundance of mRNAs encoding variable exons in unstimulated or 48 hours TNF-α-treated cells. We also compared the profile of CD44 variant transcripts in monocytes to that found in day-7 differentiated DCs. The entire variable region of CD44, spanning exons v1 to v10, was amplified using RT-PCR from mRNAs of fresh monocytes, day-7 DCs, or TNF-α treated DCs (Fig. 2). Some of the variable exons were barely detectable (v4, v5, v7 not shown), maybe due to a limitation in the size of PCR-amplified domains. The modulation of other exons revealed changes in their abundance. The relative amount of the v6-containing transcript and of two v3-containing mRNAs did not seem to vary significantly between monocytes and DCs. At least two CD44 variant transcripts containing v9 could be amplified from monocytes and DCs, giving bands around 380 and 520 bp. The 520 bp band, which is also found using probes for v8 (not shown) and v10 (Fig. 2) may well correspond to the "epithelial form" of CD44, CD44v8-v10. Both v9-containing transcripts were significantly decreased (around 4 fold) in DCs compared to monocytes. Hybridization to the v10 probe gave two doublets of 300–400 and 520–640 bp; the 520–640 bp band decreased (around 2 fold) in DCs compared to monocytes, probably reflecting the decrease in a CD44v8-v10 "epithelial type" transcript.

When DCs were induced to "mature" in the presence of TNF-α, the amount of all CD44 variant transcripts containing v3, v6, v9 or v10 increased. This increase was more striking in the case of the v6-encoding variant (more than 2 fold). This CD44 isoform appeared by its size to contain only the v6 variable exon, and might well correspond to the CD44v6 form known to be induced in activated lymphocytes and macrophages in the mouse[12]. The up-regulation of v3, v6 and v9-containing transcripts correlates with their increase in surface expression following TNF-α stimulation (Fig.1).

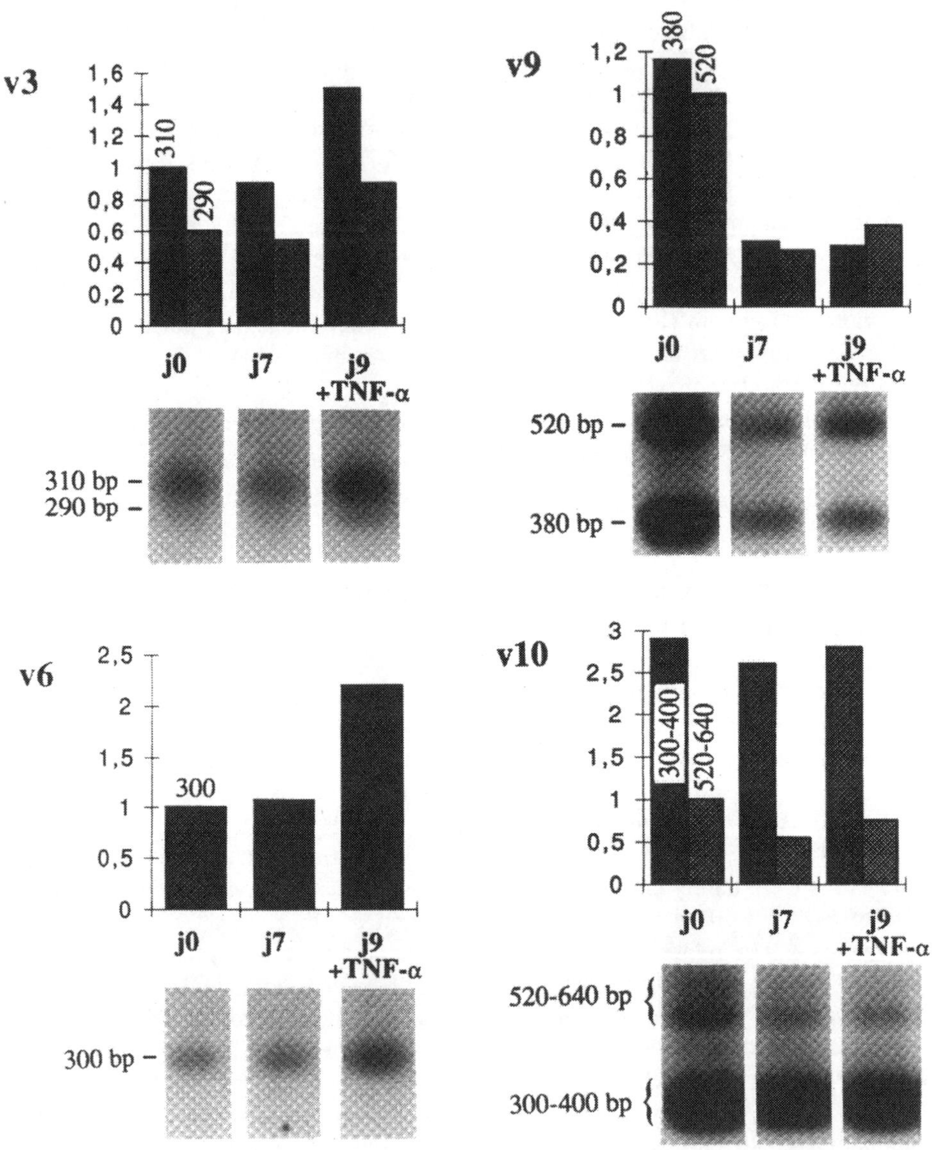

Figure 2. Regulation of CD44 variant transcripts in monocytes, day-7 differentiated DCs or TNF-α-treated DCs. RNA was extracted and the total CD44 variable region was PCR-amplified using oligonucleotides hybridizing 5' and 3' of the variable exons. Amplification products were separated on agarose gels and hybridized with oligonucleotides corresponding to each variable exon from v3 to v10. Autoradiograms were scanned and the band intensities quantified. Results were normalized by comparison with the intensity of a β-actin band amplified from the same cDNAs.

2.3. Stimulation by Anti-CD44 Antibodies Can Induce Phenotypic Maturation of DCs

2.3.1. Aggregation of DCs. To examine the effect of CD44 engagement on the cell surface, DCs were cultivated for 24 hours in the presence of immobilized antibodies directed either against all forms of CD44 (J173, 5F12) or against CD44v3 or v6 isoforms. Whereas the 5F12 or anti-CD44v3 antibodies did not seem to provoke any changes in the cultures as compared with a control IgG1, stimulation with the pan-CD44 antibody J173 or with the anti-CD44v6 antibody induced the formation of DC aggregates (not shown). The aggregates which formed upon stimulation with J173 were generally larger than those induced by anti-CD44v6 treatment, maybe due to the higher number of total CD44 epitopes on cell surfaces. However, it is interesting to note that anti-v6 but not anti-v3 antibodies induced cell aggregation although both variants showed comparable levels of surface expression (Fig. 1).

2.3.2. Cytokine Production by Anti-CD44-Stimulated DCs. The production of cytokines by DCs was investigated following culture on the immobilized anti-CD44 antibodies J173, 5F12, anti-v3 or anti-v6. Cytokine transcripts were PCR-amplified (not shown). Transcripts encoding TNF-α,IL-1β or IL-15 were amplified but their level did not seem to vary significantly between samples. In contrast, IL-6 and IL-8 mRNAs were strongly upregulated in DCs stimulated with the pan-CD44 antibody J173, whereas no effect was found using 5F12 or anti-variant CD44 antibodies. IL-6 and IL-8 production upon J173 stimulation were confirmed by ELISA testing of culture supernatants (Fig. 3).

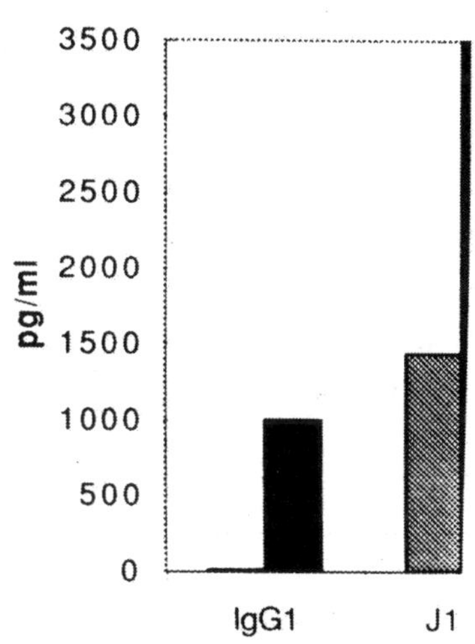

Figure 3. Induction of IL-6 and IL-8 secretion in DCs stimulated with anti-CD44 MAb J173. DCs were cultivated for 24 h in tissue-culture dishes coated with 10 μg/ml antibodies J173, 5F12 (pan-CD44), anti-CD44v3 or anti-CD44v6, or a control IgG1. IL-6 and IL-8 secretion were analysed by ELISA testing of culture supernatants. Controls (not shown) were performed to check that endotoxin contamination of J173 was not responsible for cytokine secretion.

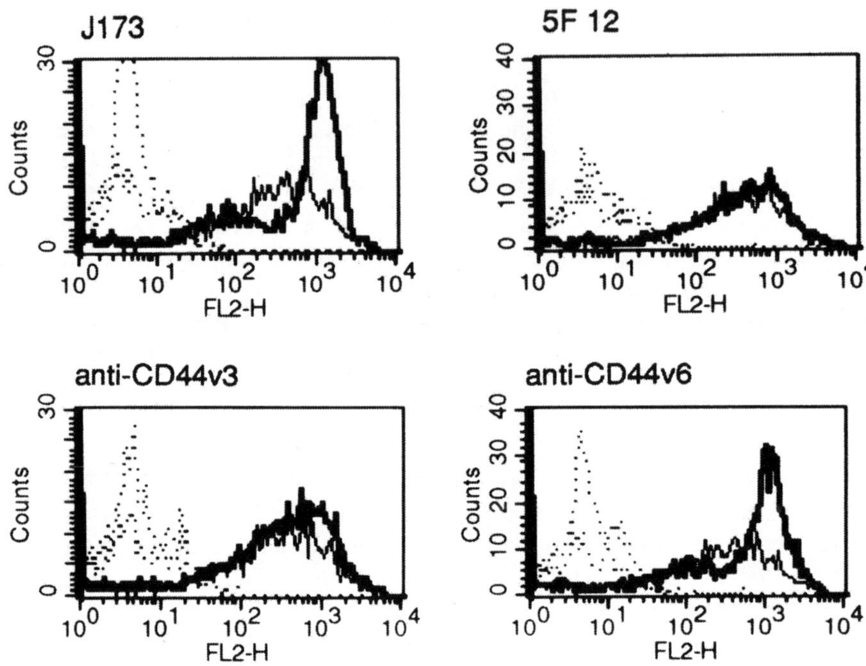

Figure 4. Up-regulation of MHC class II surface expression on DCs stimulated by J173 or anti-CD44v6 antibodies. DCs cultivated for 24 h on immobilized anti-CD44 mAbs were stained using Becton-Dickinson PE-conjugated anti-HLA-DR mAb (bold lines). Fine lines: DCs cultivated for 24 h on a control IgG1. Dotted lines: control staining with PE-conjugated IgG2a.

2.3.3. Anti-CD44 Antibodies Induce Phenotypic Maturation of DCs. Mature DCs are characterized notably by high levels of MHC class II surface expression[10]. In order to look whether anti-CD44 antibodies could induce DC maturation, cells were stimulated and subsequently analysed by flow cytometry. No significant variation in surface class II expression was found in 5F12- or anti-CD44v3-stimulated cells compared to control IgG1-treated cells. In contrast, DCs cultivated in the presence of the J173 or the anti-CD44v6 antibody showed enhanced class II surface expression (Fig. 4). In addition, surface expression of CD40 and B7–2 (CD86) were significantly up-regulated (not shown). These results suggest that anti-CD44 antibodies which induce DC aggregation also have an effect on MHC class II surface expression and phenotypic markers reflecting enhanced cell maturation.

3. DISCUSSION

This study shows that (i) the expression of CD44 and variant isoforms containing the v3, v6 or v9 exon undergoes regulation during DC differenciation from monocytes and subsequent DC stimulation by TNF-α; (ii) that mAb triggering of CD44 v6-containing isoforms induces DC aggregation and phenotypic maturation; and (iii) that stimulation with the J173 mAb, recognizing all CD44 forms, induces in addition IL-6 and IL-8 secretion.

CD44 variants containing exon v9 are down-regulated in DCs compared to monocytes (Fig.2). Variants containing v3, in contrast, are more strongly expressed by DCs than by monocytes. Activation of DCs with TNF-α induces the up-regulation of v3, v6 and v9-containing variants. Interestingly, v6-containing isoforms have been implicated both in tumor metastasis to lymph nodes[14,15] and in lymphocyte activation[12,16]. It has been suggested that retention of cells in the lymph nodes might be mediated through interaction of CD44v6 with an unknown ligand[2]. Such a function would be of interest in the case of DCs, known to migrate to draining lymph nodes following *in vivo* TNF-α stimulation[17].

To try to mimic the effect of ligand binding to CD44 molecules on the cell surface of DCs, a panel of mAbs was tested on cultured immature DCs. Stimulation with J173 may mimic ligation by the CD44 ligand hyaluronate, since this mAb has been reported to inhibit hyaluronate binding by T cells[18]. DC aggregation was induced not only upon treatment with J173 (anti-standard CD44) but also with an anti-CD44v6 mAb. Concommitantly, stimulation using J173 or anti-CD44v6 mAbs induced phenotypic maturation of DCs. Up-regulation of MHC class II antigen as well as B7–2 and CD40 expression is characteristic of DC maturation observed following TNF-α stimulation[10]. Triggering of CD44 molecules and particularly of v6-containing isoforms therefore appears to transduce an activation signal to DCs, resulting in enhanced cell-cell adhesiveness and up-regulation of accessory molecules involved in DC interaction with T cells. Stimulation with J173, but not with the anti-CD44v6 mAb, induced IL-6 and IL-8 release by DCs. This discrepancy can be explained either by the lower cell surface expression of CD44v6, whose triggering may not be sufficient to induce a signal for cytokine secretion, or by a distinct function of this CD44 isoform. Interestingly, monocytes stimulated with anti-standard CD44 mAbs have been found to secrete TNF-α, IL-1β and M-CSF[19,20]. Secretion of TNF-α or IL-1β is not induced upon J173-mediated triggering of CD44 on DCs (results not shown). This indicates that the pattern of cytokines secreted upon CD44 stimulation varies upon monocyte differentiation into DCs.

This study shows that stimulation of DCs via CD44 surface receptors (in particular v6-containing variants) can induce signals leading to phenotypic and functional changes characteristic of DC maturation, and suggests that CD44 binding on the cell surface may regulate important DC functions *in vivo*.

REFERENCES

1. Screaton G.R., Bell M.V., Jackson D.J., Cornelis F.B., Gerth U., Bell J.I. 1992. Proc. Natl. Acad. Sci. USA 88 : 7877–7881.
2. Herrlich P., Zöller M., Pals S., Ponta H. 1993. Immunol. Today 14 : 395–399.
3. Haynes B.F., Liao H.X., Patton K.L. 1991. Cancer Cells 3 : 347–350.
4. Miyake K., Medine K.L., Yayoshi S.I., Ono S., Hamaoka S., Kincade P. 1990. J. Exp. Med. 171 : 477–488.
5. Jalkanen S., Bargatze R.F., de los Toyos J., Butcher E.C. 1987. J. Cell. Biol. 105 : 983–990.
6. Camp R.L., Scheynius C., Johansson C., Puré E. 1993. J. Exp. Med. 178 : 497–507.
7. Huet S., Groux H., Caillou B., Valentin H., Prieur M., Bernard A. 1989. J. Immunol. 143 : 798–801.
8. Aiba S., Nakagawa S., Ozawa H., Miyake K., Yagita H., Tagami H. 1993. J. Invest. Dermatol. 100 : 143–147.
9. Hill S., Coates J.P., Kimber I., Knight S.C. 1994. Immunology 83 : 295–301.
10. Sallusto F., Lanzavecchia A. 1994. J. Exp. Med. 179 : 1109–1118.
11. Bennett K.L., Jackson D.G., Simon J.C., Tanczos E., Peach R., Modrell B., Stamenkovic I., Plowman G., Aruffo A. 1995. J. Cell Biol. 128 : 687–698.
12. Arch J.D., Wirth K., Hofman M., Ponta H., Matzku K., Herrlich P., Zöller M. 1992. Science 257 : 682–685.

13. Stamenkovic I., Amiot M., Pesando J.M., Seed B. 1991. EMBO J. 10 : 343–348.
14. Günthert U., Hofmann M., Rudy W., Reber S., Zöller M., Haussmann I., Matzku S., Wenzel A., Ponta H., Herrlich P. 1991. Cell 65: 13–24.
15. Seiter S., Arch R., Reber S., Komitowski D., Hofmann M., Ponta H., Herrlich P., Matsku S., Zöller M. 1993. J. Exp. Med. 177 : 443–455.
16. Moll J., Schmidt A., van der Putten H., Plug R., Ponta H., Herrlich P., Zöller M. 1996. J. Immunol. 156 : 2085–2094.
17. Cumberbatch M., Kimber I. 1992. Immunology 75 : 257–263.
18. Galandrini R., Galluzzo E., Albi N., Grossi C.E., Velardi A. 1994. J. Immunol. 153 : 21–31.
19. Webb D.S.A., Shimizu Y., van Sevanter G.A., Shaw S., Gerrard T.L. 1990. Science 249 : 1295–1297.
20. Gruber M.F., Webb D.S.A., Gerrard T. 1992. J. Immunol. 148 : 1113–1118.

15

MOUSE SPLEEN DENDRITIC CELLS

Phagocytic Activity and Expression of Macrophage Markers

Pieter J. M. Leenen,[1] Jane S. A. Voerman,[1] Katarina Radošević[1]
Nico van Rooijen,[2] and Willem van Ewijk[1]

[1]Department of Immunology
Erasmus University Rotterdam
[2]Department of Cell Biology and Immunology
Free University Amsterdam, The Netherlands

1. INTRODUCTION

Recent advances in the field have made increasingly clear that dendritic cells (DC) are heterogeneous in function, localization and developmental derivation[1]. In the spleen, two populations of DC have been demonstrated which show differences in location and immunophenotype[2,3]. Classic interdigitating cells (IDC) are located in the T cell areas of the splenic white pulp, whereas a less well characterized population of DC is situated in patches just outside the marginal zone. The developmental relationship between these two populations, however, is unclear. In the present study, we wished to explore the functional and phenotypic characteristics of splenic DC, and especially of the marginal DC population. The relevance of this aim was indicated by the observation that the majority of freshly isolated splenic DC is derived from this marginal population[3]. More specifically, we approached the question whether marginal DC are capable of endocytosing particulate antigens *in vivo*, since the spleen is important in the initiation of immune responses against circulating antigens. In immature developmental stages, DC may show uptake of solutes by means of macropinocytosis, but their phagocytic capacities are thought to be limited (discussed in ref. 4).

2. RESULTS AND DISCUSSION

To assess the *in vivo* phagocytic capacity of splenic DC, we intravenously injected-liposomes loaded with the drug clodronate (dichloromethylene-bisphosphonate, a gift from Boehringer Mannheim GmbH, Mannheim, Germany). Extensive characterization of this method has shown that clodronate, when encapsulated in liposomes and applied i.v., only affects the phagocytic cells that are in close contact to the circulation, i.e. macro-

Dendritic Cells in Fundamental and Clinical Immunology
edited by Ricciardi-Castagnoli, Plenum Press, New York, 1997

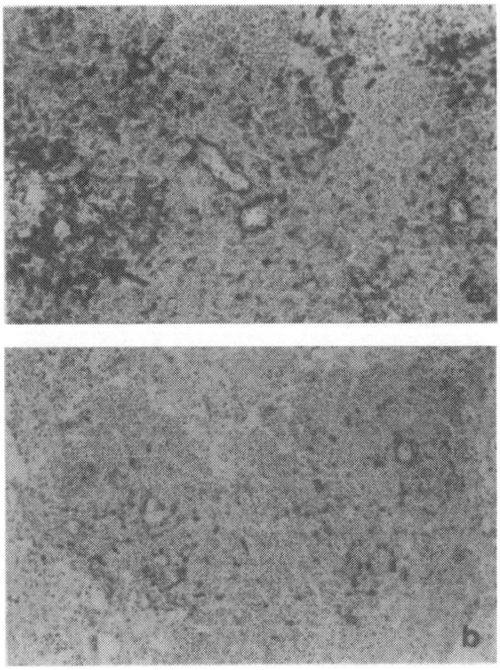

Figure 1. Splenic marginal DC, but not IDC are depleted by *in vivo* clodronate-liposome treatment as determined by CD13/ER-BMDM1 labeling. C57BL/6 mice were injected i.v. with 200 µl PBS (a) or 200 µl clodronate-liposomes (b), prepared as described[5]. Two days later, mice were sacrificed by CO_2 exposure and spleens were removed. Efficacy of clodronate-liposome treatment was determined by assessing the successful depletion of red pulp macrophages using F4/80 immunolabeling (not shown). Presence of DC was determined by ER-BMDM1 immunolabeling. The open arrow indicates interdigitating cells in the T cell area of the white pulp; the closed arrow indicates marginal dendritic cells at the border of the red pulp. Original magnification x 100.

phages in spleen and Kupffer cells in liver (reviewed in ref. 5). The presence of the different DC populations in clodronate- and PBS-treated spleens was detected using CD13 mAb ER-BMDM1[6]. In previous studies we have shown that CD13 is a useful marker for examining *in vivo* distribution of DC[6,7]. Under steady-state conditions, CD13 expression is relatively restricted to DC, although macrophages derived from culture or *in vivo* inflammation and macrophage cell lines may also express high levels of CD13. The same holds true, however, for other DC-markers such as NLDC-145 / DEC-205 and N418 / CD11c[8].

Figure 1a shows that the IDC population in the T cell area of the white pulp and the larger marginal DC population are readily detected in a control mouse by ER-BMDM1 labeling. Two days after clodronate-liposome treatment, however, the marginal DC population has vanished completely, whereas the IDC population is undisturbed (Fig. 1b).

To confirm that the disappearance of marginal DC by clodronate-liposomes was caused by actual phagocytosis of the liposomes, and not a mere stress-induced response, we injected i.v. liposomes labeled with the fluorochrome DiI and asked whether these were endocytosed by the DC.

Figure 2A shows that indeed the majority of splenic DC, identified as N418^{hi} cells, were labeled with DiI as indication of liposome phagocytosis during the overnight period after administration of the liposomes. Macrophages, identified as F4/80^{hi} cells, were similarly phagocytic, as expected (Fig. 2B). The white pulp IDC, which constitute about 30% of all isolated spleen DC, are not likely to contribute to the phagocytic DC population as they were not affected by the similarly applied clodronate-liposomes. Therefore, we conclude that most if not all marginal DC are phagocytic.

The typical macrophage characteristic of phagocytosis, expressed by marginal DC *in vivo*, raised the question whether these cells also express cell surface markers charac-

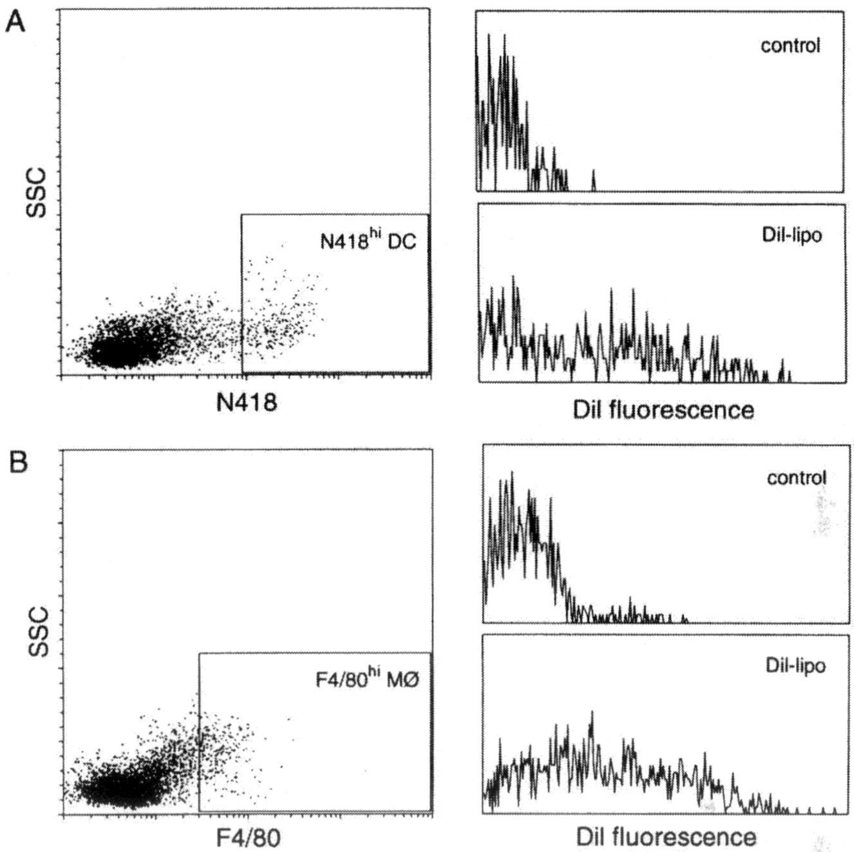

Figure 2. Splenic DC phagocytose *in vivo* applied DiI-liposomes. Spleens were obtained 18h after administration of DiI-liposomes (DiI-lipo) or PBS (control) and a single cell suspension was prepared over a 100 μm filter after collagenase / DNase treatment. Cells were stained for flowcytometry using N418 (CD11c) or F4/80 followed by the appropriate second step reagents to identify DC (A) and macrophages (B), respectively. Uptake of DiI liposomes was determined by gating on N418^{hi} or F4/80^{hi} cells as indicated.

teristic of macrophages. Therefore, we isolated total spleen cells from control mice and performed double labeling using N418, as a means to identify DC, and a panel of DC and macrophage markers.

Figure 3 shows that DC in freshly isolated spleen suspensions express high levels of MHC class II antigens and a low, but uniform level of ER-BMDM1 / CD13 (aminopeptidase N). Only a minority of these cells expresses NLDC-145 / DEC-205. This corresponds to the immunohistological findings on splenic DC, which indicate that only IDC in T cell areas express this marker. Thus, the phenotype of splenic DC reported here is in close agreement with data published before by others[2]. Interestingly, when we determined the expression levels of markers considered typical for tissue macrophages, we found fresh splenic DC to be uniformly positive for both F4/80 and Mac-1 / CD11b (Fig. 3). Therefore, these data indicate that freshly isolated splenic DC express markers specific for DC as well as others specific for macrophages.

In summary, in this study we confirm earlier reports that two subpopulations of splenic DC can be distinguished: interdigitating cells in the white pulp T cell area, and

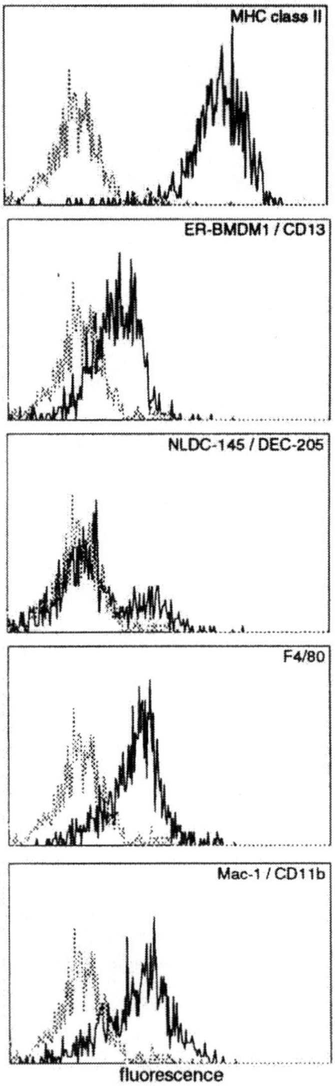

Figure 3. Splenic DC express both DC- and macrophage-specific markers. Fresh spleen cell suspensions were prepared from control mice as indicated earlier and labeled for flowcytometry using N418 (CD11c) combined with anti-MHC class II, ER-BMDM1 (CD13) and NLDC-145 (DEC-205) as DC markers and F4/80 and Mac-1 (CD11b) as macrophage markers. DC were identified in the samples by gating on N418^{hi} cells as indicated in Fig. 2A. Negative control fluorescence is indicated by the dotted line.

marginal dendritic cells located in patches in the red pulp, just outside the marginal zone. The vast majority of the latter population shows phagocytic activity *in vivo*, as indicated by the depletion by clodronate-liposome treatment and the uptake of i.v administered fluorescently labeled liposomes. This rather unexpected *in vivo* phagocytic activity by DC supports and extends recent findings by Matsuno et al.[9], who observed particle-laden DC in peripheral hepatic lymph after i.v. administration of ink or latex particles. The freshly isolated splenic DC display a mixed phenotype of DC and macrophage markers. Upon culture, however, the cells develop a more DC-restricted phenotype, as indicated by the increased expression of DC markers and the decrease of macrophage markers (data not shown). *In vitro* studies using dendritic cell lines or bone marrow cultures have suggested that immature DC pass through a similar phagocytic life stage, with concurrent expression of macrophage markers, before developing into the potent immunostimulatory, non-phagocytic DC (refs. 10,11, and Ricciardi-Castagnoli et al., personal communication).

Taken together, these data support the view that splenic marginal DC represent an immature life stage in DC development, which displays a mixed phenotype and function between macrophages and mature DC. Whether marginal DC develop into the more mature white pulp IDC remains to be established.

3. ACKNOWLEDGMENTS

This work was supported by grant 94.125 from the "Diabetes Fonds Nederland". We wish to thank Tar van Os for his expert (photo)graphical input in this paper.

4. REFERENCES

1. Stingl G, Bergstresser PR. Dendritic cells: a major story unfolds. Immunol.Today 16:330–333.
2. Steinman RM. 1991. The dendritic cell system and its role in immunogenicity. Annu.Rev.Immunol. 9:271–296.
3. Metlay JP, Witmer-Pack MD, Agger R, Crowley MT, Lawless D, Steinman RM. 1990. The distinct leukocyte integrins of mouse spleen dendritic cells as identified with new hamster monoclonal antibodies. J.Exp.Med. 171:1753–1771.
4. Steinman RM, Swanson J. 1995. The endocytic activity of dendritic cells. J.Exp. Med. 182:283–288.
5. van Rooijen N, Sanders A. 1994. Liposome mediated depletion of macrophages: mechanism of action, preparation of liposomes and applications. J.Immunol.Meth. 174:83–93.
6. Leenen PJM, Melis M, Kraal G, Hoogeveen AT, van Ewijk W. 1992. The monoclonal antibody ER-BMDM1 recognizes a macrophage and dendritic cell differentiation antigen with aminopeptidase activity. Eur.J.Immunol. 22:1567–1572.
7. Leenen PJM, de Bruijn MFTR, Voerman JSA, Campbell PA, van Ewijk W. 1994. Markers of mouse macrophage development detected by monoclonal antibodies. J.Immunol.Meth. 174:5–19.
8. Leenen PJM, Kraal G, Dijkstra CD. 1996. Markers of rodent myeloid cells. *In:* Immunology Methods Manual, Lefkovits I (ed.), section 34.2, Academic Press, New York, pp. 2467–2495.
9. Matsuno K, Ezaki T, Kudo S, Uehara Y.1996. A life stage of particle-laden rat dendritic cells *in vivo*: their terminal division, active phagocytosis, and translocation from the liver to the draining lymph. J.Exp.Med. 183:1865–1878.
10. Inaba K, Inaba M, Naito M, Steinman RM. 1993. Dendritic cell progenitors phagocytose particulates, including Bacillus Calmette-Guerin organisms, and sensitize mice to mycobacterial antigens *in vivo*. J.Exp.Med. 178:479–488.
11. Lutz MB, Aßmann CU, Girolomoni G, Ricciardi-Castagnoli P. 1996. Different cytokines regulate antigen uptake and presentation of a precursor dendritic cell line. Eur.J.Immunol. 26:586–594.

16

MIGRATION OF DENDRITIC CELLS IN 3D-COLLAGEN LATTICES

Visualisation of Dynamic Interactions with the Substratum and the Distribution of Surface Structures via a Novel Confocal Reflection Imaging Technique

Matthias Gunzer,[1] Eckhart Kämpgen,[2] Eva-B. Bröcker,[2] Kurt S. Zänker,[1] and Peter Friedl[1,2]

[1]Institute of Immunology
University of Witten/Herdecke
Witten, Germany
[2]Department of Dermatology
University of Würzburg
Würzburg, Germany

1. INTRODUCTION

Migration is an inherent and prominent quality of dendritic cells (DC)[1,2]. In the past years numerous studies have analysed this aspect of DC biology. Quantitative aspects of DC migration have been tested in whole animal assays using different methods such as lymph-cannulation[3] or fluorescence labelling of migrating cells[4,5]. Skin-explant emigration assays were applied for in vitro studies[6,7]. Single DC have been investigated migrating on plastic surfaces[8,9]. We describe here a novel method for the analysis of individual cells migrating within a 3D-collagen substratum. The technique applies and further extends a system previously described by Friedl[10] for the analysis of lymphocyte migration. Using viable or fixed cells, the simultaneous visualisation and 3D-reconstruction of dynamic cell-interactions with the surrounding collagen-environment as well as the distribution of structures on the cell surface allows a more detailed analysis of the cell biology underlying the process of DC migration.

2. MATERIALS AND METHODS

2.1. Cells

Murine epidermal Langerhans cells (LC) and bone marrow dendritic cells (bmDC) were prepared from specific pathogen-free BALB/c mice (Charles River Wiga GmbH,

Dendritic Cells in Fundamental and Clinical Immunology
edited by Ricciardi-Castagnoli, Plenum Press, New York, 1997

Sulzfeld) by methods described by Schuler and Koch[11] and Inaba, respectively[12]. Human DC were prepared from PBMC-enriched cultures as described by Romani[13]. All preparations contained DC at a purity of 75–95% as judged by morphology.

2.2. Preparation of Collagen-Matrices and Staining Procedures

Collagen matrices were prepared as described[10]. Briefly, purified cells were resuspended in a solution containing dermal bovine collagen (Collagen Corp., Palo Alto, Ca.) in minimal essential medium (Flow, McLean, Vi.). The final concentration for collagen was 1.7 mg/ml. This solution was filled into a self-constructed chamber, allowed to polymerise for 20 min (37°C, 5% CO_2) and subsequently overlaid with media (RPMI, 10% FCS (Seromed, Berlin, Germany), 1 mM L-Glutamine, 50 μM 2-ME, 50 μg/ml gentamicin sulphate, adjusted to pH 7.4). For the analysis of surface structures viable cells were immunostained before incorporation into the collagen lattice. Therefore, cells were incubated with the primary anti MHC II antibody (L243, ATCC HB 55 hybridoma supernatant; B21/2, ATCC TIB 229, hybridoma supernatant) for 20 min at 4°C, washed twice and then incubated with 10 μg/ml lissamine-rhodamine (LRSC)-conjugated secondary goat-anti mouse or goat-anti rat IgG-F(ab)′-fragment (Dianova, Hamburg, Germany) for 20 min at 4°C. After two additional washing steps cells were incorporated within collagen lattices.

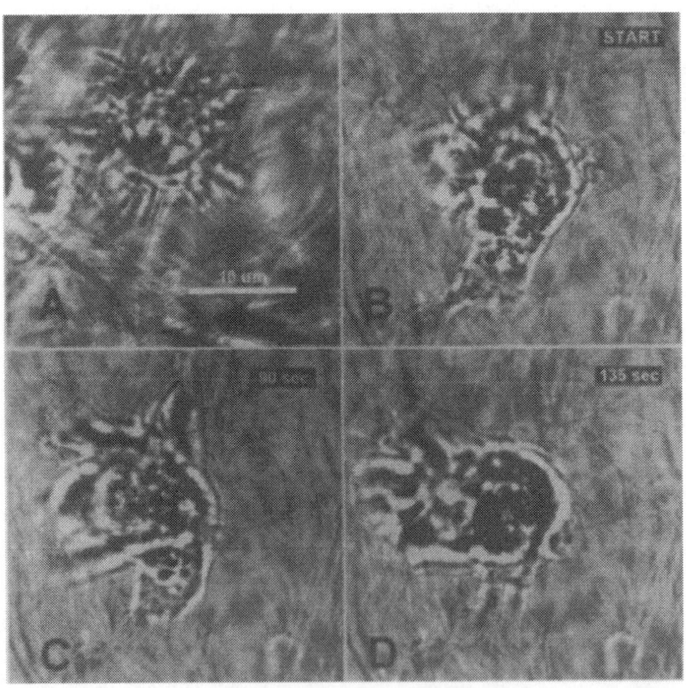

Figure 1. Morphology and migration of dendritic cells within a 3D-collagen matrix. Murine epidermal LC (A) or cultured human DC (B-D) were embedded within a 3D-collagen matrix (for details see materials and methods). The system was kept at 37°C and the shape of individual cells was recorded on a confocal-scanning microscope in the transmission light channel. (A) shows a single frame of a murine cultured LC optimised for the representation of dendritic processes. (B-D) show sequential states of a time-series analysis of a cultured human DC. The visual field was kept constant to elucidate the relative movement of the cell body. As indicated by the numbers, the cell made a 90° left turn within 135 seconds of observation.

Fixed samples were obtained by incubating a polymerised gel with a solution containing 4% para-formaldehyde in PBS (20 min., 37°C, 5% CO_2). The fixative was also prewarmed to 37°C to avoid a drop of the temperature during the reaction, which resulted in a loss of the typical dendritic appearance of the cells.

2.3. Analysis of Cell Morphology and Immunofluorescence by Confocal Microscopy

Cells embedded within collagen lattices were analysed on an inverted confocal-scanning microscope (Leica TCS 4D, Bensheim, Germany) as described (Friedl, manuscript submitted). For viable samples the system was kept at a temperature of 37°C. In all experiments a 63x oil immersion objective was used. Cell morphology was detected in the transmission light channel. For the analysis of cell-interactions with the surrounding substratum the reflection signals of both the cell body and the collagen fibres were recorded simultaneously. Laser light of 488 and 568 nm was introduced into the sample. The reflected light was passed through a trichroitic-Filter and short wavelengths were separated from the fluorescence using a 580 nm long-pass filter before detection in photomultiplier (PMT) 1. LRSC-fluorescence was simultaneously detected in PMT 2. Time series analyses of viable DC in the process of migration were recorded at a time interval of 15 sec/frame at fixed x-y-z-positions for a maximum period of 14 min. For the analysis of the 3D-structure the cell body was sectioned in steps of 1 μm in z-direction and the reflection signal in each x-y-plane was recorded. From these sections a topographical image was calculated using the TCS-4D software.

3. RESULTS

3.1. DC Embedded within 3D-Collagen Matrices Develop a Typical Morphology and Are Able to Migrate

The characteristical morphology of DC has been described for both cells in solution as well as in fixed histological samples[14]. Thus, the first question we addressed was, whether DC would be able to develop this phenotype again after the stressing procedure of separation and subsequent embedding within a 3-D collagen matrix. In collagen lattices DC formed the extended processes of the cell membrane which are characteristic for this cell type (Fig. 1). These pseudopods showed a permanent dynamic motility. Furthermore, cells were kept viable for several days (not shown) and exhibited considerable locomotion throughout (Fig. 1).

3.2. Upon Migration DC Attach to Collagen Fibres and Displace Them Elastically

As obvious from 3D-reconstructed z-sections of fixed samples (Fig. 2A), cell bodies were deeply embedded within a tight meshwork of collagen fibres. Considering this fact the question was, how the cells could accomplish the observed migration pattern mentioned in 3.1. Confocal reflection time series of viable cells demonstrated that upon migration DC interacted with individual collagen fibres and were able to bend and displace them (Figs. 2B-D). Additionally, cells could strongly reduce the cell diameter and squeeze

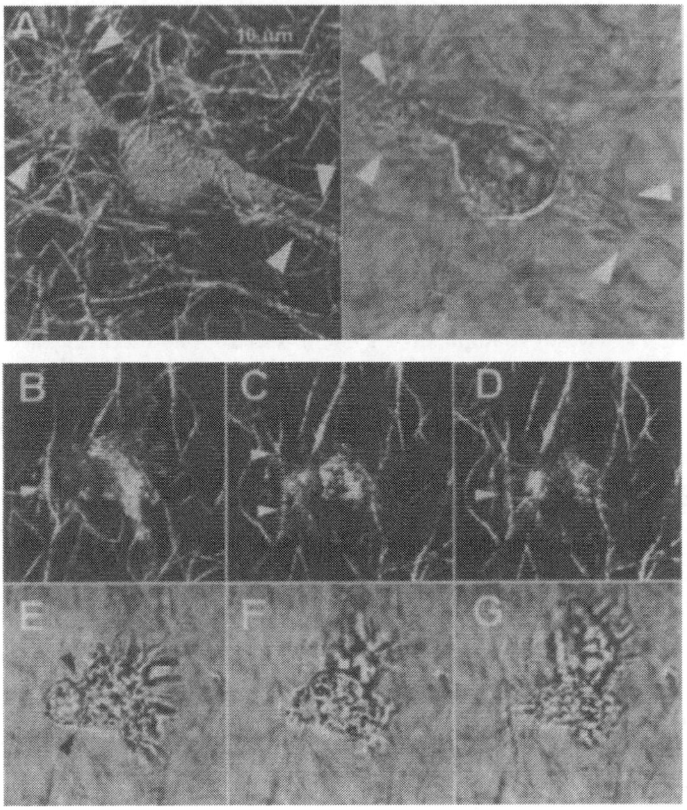

Figure 2. Confocal reflection analysis of human DC in 3D-collagen lattices. Human cultured DC were incorporated within a 3D-collagen lattice. The 3D-shape of a fixed cell in context with the surrounding collagen fibres reconstructed from sequential z-sections shows intensive interactions with the substratum on two opposite sides of the cell (A, arrowheads). The corresponding transmission light image shows the overall dimensions of the observed cell (A, right picture, arrowheads). Viable cells of the same type were recorded while migrating within a 3D-collagen matrix. The reflection signal of a cell is shown at the beginning of a movement (B) as well as 75 (C) and 105 seconds later (D). Note the pushing of a collagen fibre (B) followed by pulling (C). After its release (D) the fibre slid back into an intermediate position (B-D, arrowheads). The corresponding transmission light image of the same cell is shown in E-G. The trailing edge of the cell is extended and tied up (black arrowheads in E) during the crossing of a narrow pore and later retracted (F) while the cell orientates upwards (G).

themselves through narrow pores (Fig. 2 E-G). Thereby, the meshwork of collagen fibres imprinted deep constriction rings on the cell body (Fig. 2E, arrowheads). Also as a result of this process collagen fibres were delocalised up to several µm (Fig. 2B/C). This distortion of the collagen microenvironment was fully elastic and completely reversible. None of the changes did perpetually alter the collagen structure and no fibre destruction was observed.

3.3. MHC II Is Evenly Distributed on the Surface of Migrating DC

DC are known to express extremely high levels of MHC II[1]. Consequently, exploiting the ability of DC to transiently adhere to and spread on glass surfaces (Fig. 3A) we

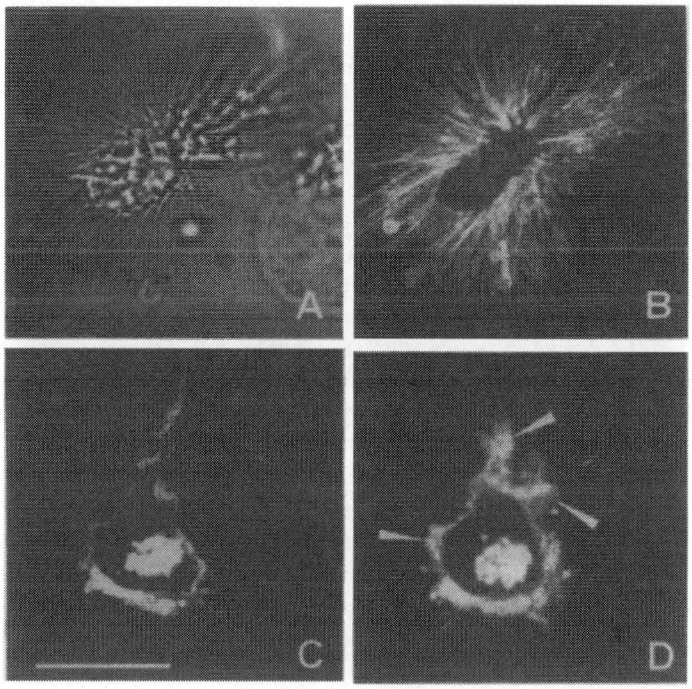

Figure 3. DC in liquid culture versus 3D-collagen lattice: comparison of the staining pattern for MHC II. Viable human cultured DC adhering on a glass surface (A transmission light / B fluorescence image) or murine bone marrow DC embedded within a 3D-collagen lattice (C/D) were stained for MHC II and the obtained fluorescent labelling subsequently analysed in a confocal-scanning microscope (B-D). (C) and (D) are two shapes of a motile cell as it appeared within 75 seconds of a movement. Note that a homogeneously strong staining of the whole membrane was achieved in media alone as well as in a collagen lattice. Upon migration long dendritic processes were retracted. A more intensive signal was detectable after involution of such processes (D, arrowheads). The strong intracellular fluorescence of the cell in (C) and (D), which is absent in the sample stained in media alone signals internalised MHC II molecules. Scalebar is 10 μm.

could use immunostaining of class II receptors to label even the finest dendritic processes of a DC in solution (Fig. 3B). We were interested, whether a comparable type of MHC II distribution would occur within a 3D-collagen matrix and whether the process of locomotion would affect this staining-pattern. In the 3D-collagen environment no patching or capping of MHC II was detectable (Fig. 3 C and D). Also upon migration MHC II remained evenly distributed on the cell surface. Interestingly, a major intracellular staining signal occurred, which represented internalised MHC II molecules. On areas of the cell surface, where long dendritic processes had been retracted into the mass of the cell body, an increased amount of staining was noticed (Fig. 3 D, arrowheads), suggesting higher concentrations of MHC II. However, further investigation is required whether this results from patching or clustering of MHC II on the cell surface or increased folding of the cell membrane below the limit of resolution of light microscopy.

4. DISCUSSION

Embedding purified DC within a 3D-lattice consisting of collagen offers the possibility to study the biology of individual DC in context with an ECM-like environment. We were able to show that within these collagen lattices DC develop a morphology characterised by many dendritic processes. Furthermore, time-series analyses revealed a high motility of these pseudopods which was accompanied by effective locomotion of the cell body (Fig. 1). Both observations, dendritic shape and migration, have always been described as hallmark of DC[1]. Confocal reflection analysis elucidates the mechanism by which DC may overcome the physical resistance of an ECM environment. Reflection time-series of migrating DC showed that the cells multifocally interacted with individual fibres. In addition, migrating DC were able to strongly decrease their diameter in order to cross narrow pores of the lattice (Fig. 2). In the process of migration DC were able to apply tensile strength to the fibres which resulted in their mechanical bending and dislocation, not, however, any mechanical or proteolytic cleavage. All processes affecting the microstructure of the collagen lattice were reversible and did not permanently alter the structure of the matrix. This is in sharp difference to observations made with migrating metastatic melanoma cells, which drastically and permanently reorganised the collagen lattice (Friedl et al, manuscript submitted).

Immunolabelling of DC with a class II monoclonal antibody and subsequent analysis of cell-dynamics within 3D-collagen lattices allowed further insight into membrane processes during migration. Comparable to results obtained in liquid culture it was possible to detect even the finest structures of the membrane of DC residing within a collagen environment (Fig. 3). Interestingly, upon migration, class II antigens remained homogeneously distributed on the surface of the DC. This might open a way to analyse interactions between MHC II on DC and CD4 on T-cells during priming reactions and their influence on the distribution pattern of these molecules in an ECM-like environment.

Furthermore, comparison between immature and mature states of DC, which are known to differ in their extent of MHC II-expression and -surface-stability[15,16] may lead to a better understanding of the cell biology underlying the maturation process[17]. Finally, using this technique for the analysis of surface receptors involved in migration (such as integrins and other adhesion molecules) could help identifying the relevant structures. Subsequently, approaches aiming towards the activation or inhibition of DC specific migration strategies[18], may be developed and tested with computer assisted cell-tracking in combination with confocal reflection analysis of migrating DC at the individual cell level in vitro.

Note

A video sequence of the cell shown in figure 3 (C/D) is available at the Internet-homepage of the Institute of Immunology, University of Witten/Herdecke under the following address: http://www.uwh.de/videos/videos.htm.

5. REFERENCES

1. Steinman, R. M. The dendritic cell system and its role in immunogenicity. *Annu. Rev. Immunol.* 9, 271–296 (1991).
2. Steinman, R., Hoffman, L. & Pope, M. Maturation and migration of cutaneous dendritic cells. *J. Invest. Dermatol.* 105, 2S-7S (1995).

3. Dandie, G. W., Ragg, S. J. & Muller, H. K. Migration of Langerhans cells from carcinogen-treated sheep skin. *J. Invest. Dermatol.* 99, 51S-53S (1992).
4. van Wilsem, E. J., Breve, J., Kleijmeer, M. & Kraal, G. Antigen-bearing Langerhans cells in skin draining lymph nodes: phenotype and kinetics of migration. *J. Invest. Dermatol.* 103, 217–220 (1994).
5. Wang, B., Kondo, S., Shivji, G. M., Fujisawa, H., Mak, T. W. & Sauder, D. N. Tumor necrosis factor receptor II (p75) signalling is required for the migration of Langerhans cells. *Immunology* 88, 284–288 (1996).
6. Larsen, C. P. & Austyn, J. M. Langerhans cells migrate out of skin grafts and cultured skin: a model in which to study the mediators of dendritic leukocyte migration. *Transplant. Proc.* 23, 117–119 (1991).
7. Larsen, C. P., Steinman, R. M., Witmer-Pack, M., Hankins, D. F., Morris, P. J. & Austyn, J. M. Migration and maturation of Langerhans cells in skin transplants and explants. *J. Exp. Med.* 172, 1483–1493 (1990).
8. Boehmelt, G., Madruga, J., Dörfler, P., et al. Dendritic cell progenitor is transformed by a conditional v-Rel estrogen receptor fusion protein v-RelER. *Cell* 80, 341–352 (1995).
9. Freudenthal, P. S. & Steinman, R. M. The distinct surface of human blood dendritic cells, as observed after an improved isolation method. *Proc. Natl. Acad. Sci. U. S. A.* 87, 7698–7702 (1990).
10. Friedl, P., Noble, P. B. & Zänker, K. S. Lymphocyte migration in three-dimensional collagen gels. Comparison of three quantitative methods for analysing cell trajectories. *J. Immunol. Meth.* 165, 157–165 (1993).
11. Schuler, G. & Koch, F. in *Epidermal Langerhans Cells* (ed Schuler, G.) 1st, 139–157 (CRC-Press, Boca Raton, Ann Arbor, Boston, 1991).
12. Inaba, K., Inaba, M., Romani, N., et al. Generation of large numbers of dendritic cells from mouse bone marrow cultures supplemented with granulocyte/macrophage colony-stimulating factor. *J. Exp. Med.* 176, 1693–1702 (1992).
13. Romani, N., Gruner, S., Brang, D., et al. Proliferating dendritic cell progenitors in human blood. *J. Exp. Med.* 180, 83–93 (1994).
14. Romani, N., Lenz, A., Glassl, H., et al. Cultured human Langerhans cells resemble lymphoid dendritic cells in phenotype and function. *J. Invest. Dermatol.* 93, 600–609 (1989).
15. Kämpgen, E., Koch, N., Koch, F., et al. Class II major histocompatibility complex molecules of murine dendritic cells: synthesis, sialylation of invariant chain, and antigen processing capacity are down-regulated upon culture. *Proc. Natl. Acad. Sci.* 88, 3014–3018 (1991).
16. Schuler, G., Koch, F., Heufler, C., Kämpgen, E., Topar, G. & Romani, N. Murine epidermal Langerhans cells as a model to study tissue dendritic cells. *Adv. Exp. Med. Biol.* 329, 243–249 (1993).
17. Schuler, G. & Steinman, R. M. Murine epidermal Langerhans cells mature into potent immunostimulatory dendritic cells in vitro. *J. Exp. Med.* 161, 526–546 (1985).
18. Roake, J. A. & Austyn, J. M. The role of dendritic cells and T cell activation in allograft rejection. *Exp. Nephrol.* 1, 90–101 (1993).

17

DENDRITIC CELLS OF THE MURINE PEYER'S PATCHES COLOCALIZE WITH *Salmonella typhimurium* AVIRULENT MUTANTS IN THE SUBEPITHELIAL DOME

S. A. Hopkins and J. -P. Kraehenbuhl

ISREC, Institute of Biochemistry
University of Lausanne
CH1066 Epalinges, Switzerland

1. INTRODUCTION

Salmonella typhimurium is an enteropathogenic bacterium that invades the murine Peyer's patches (PP), the mucosal lymphoid tissue of the small intestine (1), and later spreads systemically resulting in a disease similar to typhoid fever in mice (2).

Avirulent mutants of *Salmonella* have been generated which can be used as live vaccine vectors, carrying foreign antigens to the mucosal immune system without causing disease (3). It is not known which antigen presenting cells in the PP are responsible for the presentation of *Salmonella* vaccines to the mucosal immune system. The three possible candidates are macrophages, B cells or dendritic cells.

Our aim is to identify which mucosal antigen presenting cells in the PP associate with, process and present antigens derived from *Salmonella typhimurium* vaccine strains that express recombinant antigen.

2. MATERIALS AND METHODS

We used isogenic strains of a) wild type *Salmonella typhimurium*, b) *Salmonella* bearing the PhoP^c point mutation which attenuates bacterial survival within macrophages *in vitro* (4), and c) the Δ*cyacrp* double mutation which attenuates bacterial metabolism (5).

In order to visualize wild type and avirulent *Salmonella typhimurium* mutants after infection of murine PP, we transformed bacteria with a plasmid (pkk223.3, Pharmacia) containing the gene encoding green fluorescent protein (GFP), a jellyfish derived protein which absorbs light in the FITC range (6). Expression was verified by SDS PAGE analysis.

We then prepared frozen sections of PP derived from BALB/c mice immunized via the ligated loop technique (1), with either of the two mutants (PhoP^c,Δ*cyacrp*) or the wild type strain all of which expressed GFP. Specific markers (N418 for dendritic cells (7), B220 for B cells (8) and mac-1 (9) for macrophages and neutrophils) of different cell types were used in order to identify cells associated with bacteria.

Numbers of bacteria expressing the foreign antigen GFP that persisted in the PP after oral infection were quantified at different time points by homogenizing the PP and enumerating the plasmid containing bacteria on agar plates.

PP cell suspensions from mice immunized 3d previously were treated with gentamicin to kill extracellular bacteria, then washed and lysed to determine the numbers of persisting intracellular bacteria.

FACS anlaysis of PP cells 24h after oral immunization was performed to determine the percentage of cells that were associated with fluorescent bacteria.

3. RESULTS

The 65kD GFP was stably expressed *in vitro* by all strains of *Salmonella typhimurium*, as verified by SDS PAGE (data not shown), enabling them to absorb light in the FITC range and to emit green fluorescence.

All strains of fluorescent bacteria, were observed in the subepithelial dome (SED) region of PP 4h after ligated loop injection. These bacteria colocalized with the N418+ population of dendritic cells (data not shown).

There were also a large number of B220+ B cells in this region, while relatively few cells of the SED stained positively for the mac-1 marker of macrophages and neutrophils, and no colocalization with bacteria was observed for these cells (data not shown).

Avirulent strains of *Salmonella typhimurium* containing the GFP plasmid persisted for several weeks in PP of orally immunized mice, while infection with wild type *Salmonella typhimurium* resulted in lethality of mice after less than one week.

Incubation of PP cells with gentamicin killed extracellular bacteria and demonstrated that a proportion of the avirulent mutants were intracellular in the PP after 3 days.

FACS analysis 24h after oral immunization with the three strains expressing GFP showed only a very low percentage of PP cells associated with fluorescent bacteria in mice immunized with the two avirulent mutants. Mice infected with wild type *Salmonella typhimurium* expressing GFP had a 10% increase in the number of FITC+ cells compared to uninfected control PP cells, but this was not due to the GFP, as wild type bacteria alone also induced a 10% increase in autofluorescent cells.

Table 1. Persistence of different strains of *Salmonella typhimurium* expressing GFP in the PP of orally immunized mice (log number of colony forming units of bacteria/mouse)

Days post oral immunization	PhoP^c (GFP)	*cyacrp* (GFP)	Wild type (GFP)
1	4±0.7	2.2±0.3	2.4±0.5
3	3.4±0.05	3.8±0.01	3.7±0.3
10	3.1±0.2	3.4±0.2	no survivors

Table 2. FACS Analysis of PP cells from mice orally immunized 24h previously with avirulent and wild type strains expressing GFP, and also those infected with wild type bacteria alone

Salmonella strain	FITC + Peyer's patch cells
Phop^c (GFP)	0.5%
cya crp (GFP)	0.1%
Wild type (GFP)	10%
Wild type alone	10%

4. DISCUSSION

Although *in vitro* studies have concentrated on the uptake of *Salmonella typhimurium* by macrophages (4), we observed that both wild type and two attenuated mutants associated with N418+ dendritic cells 4h after the incubation of ligated loops of intestine with bacteria, and that there are in fact few macrophages in the SED of the PP, see figure 2.

PhoP^c mutants are attenuated in survival within macrophages *in vitro* (4), but nevertheless persisted for several weeks in PP after oral immunization and were shown to be intracellular. It is possible that they survived within the dendritic cells of the PP. The bacteria could be transported by migrating dendritic cells to the mesenteric lymph nodes and spleen. Mayrhofer et al.(10) showed that afferent intestinal lymph veiled cells from

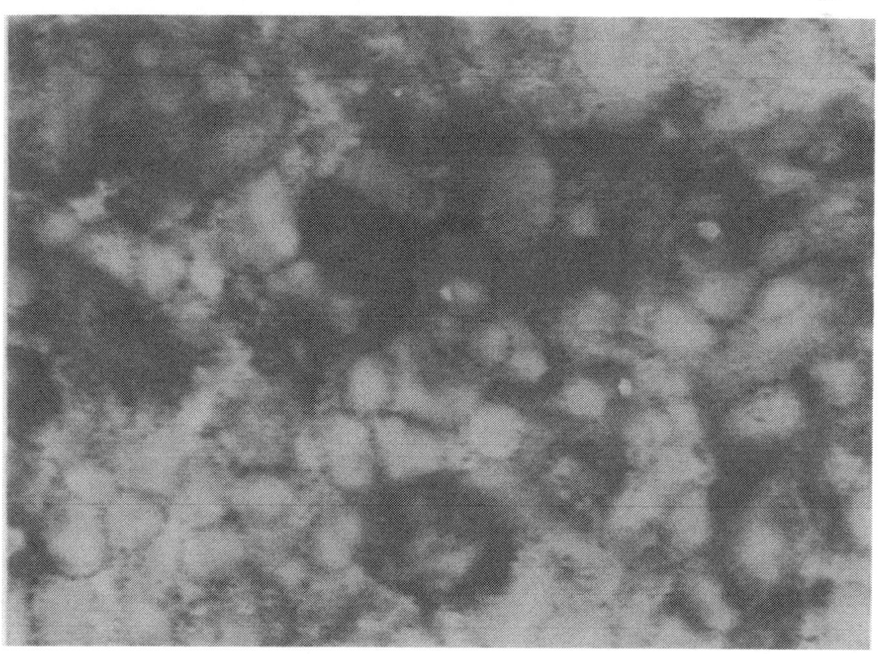

Figure 1. Gentamicin assay to identify intracellular bacteria in PP 3d after immunization with avirulent mutant *Salmonella typhimurium*.(log number of colony forming units of bacteria/mouse).

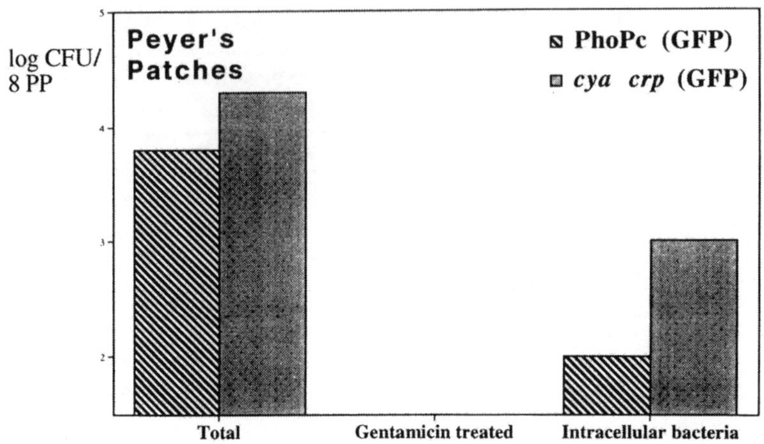

Figure 2. Schematic diagram of invasion of subepithelial dome by fluorescent bacteria and colocalization with dendritic cells.

rats infected with *Salmonella typhimurium* contained intracellular bacteria and could therefore be involved in carrying bacteria from the gut to the regional lymph nodes.

Wild type *Salmonella* but not the avirulent mutants induced an infiltration of autofluorescent mac1+ cells, probably neutrophils into the PP 24h after oral immunization. The characterization of these cells is underway.

As murine PP dendritic cells are highly potent antigen presenting cells (11&12), it is likely that they are responsible for presenting antigens derived from *Salmonella* attenuated vaccines to mucosal PP T cells, resulting in an antigen specific mucosal immune response at local and distant sites (13).

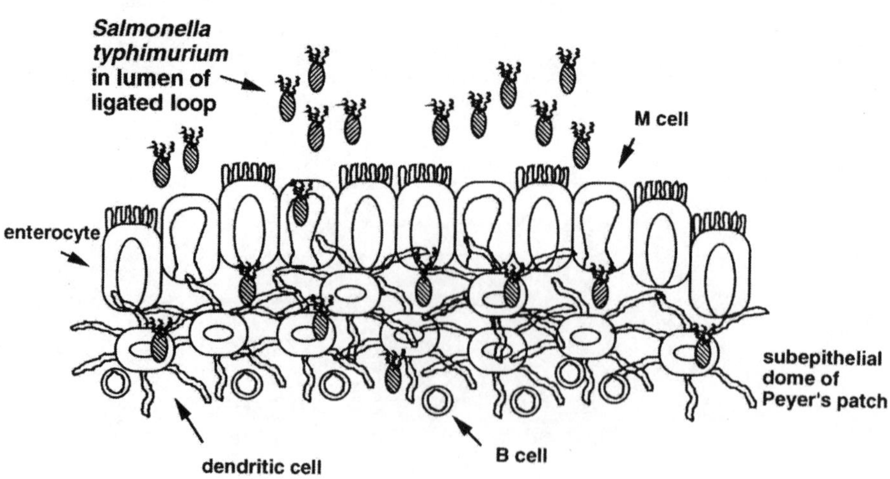

Diagram 1

The presentation of *Salmonella* vaccine strains by infected PP dendritic cells to primed T cells is currently under investigation, as is the precise cellular compartment occupied by the bacteria within the dendritic cell.

6. REFERENCES

1. Bradley, D. J., N. Ghori, Falkow, S. 1994. *Salmonella typhimurium* initiates murine infection by penetrating and destroying the specialized epithelial M cells of the Peyer's patches. J. Exp. Med. 180:15–23.
2. Hsu, H.S. 1989. Pathogenesis and immunity in murine salmonellosis. Microb. Rev. 53:390–409.
3. Curtiss, R. III 1990. Attenuated *Salmonella* strains as live vectors for the expression of foreign antigens, p347–383. In G. C. Woodrow and M.M. Levine (ed.), New generation vaccines. Marcel Dekker, Inc., New York.
4. Miller, S.I., Mekalanos, J.J. 1990. Constitutive expression of the *phoP* regulon attenuates *Salmonella* virulence and survival within macrophages. J. Bacteriol.172:2485–2489.
5. Curtiss, R., III, Kelly, S.M. 1987. *Salmonella typhimurium* deletion mutants lacking adenylate cyclase and cyclic AMP receptor protein are avirulent and immunogenic. Infect. Immun. 55:3035–3043.
6. Chalfie, M., Tu, Y., Euskirchen, G., Ward, W.W., Prasher, D.C.1994. Green fluorescent protein as a marker for gene expression. Science. 263:802–805
7. Metlay, J.P., Witmer-Pack, M.D., Agger, R., Crowley, M.T, Lawless, E., Steinman, R. M. 1990. The distinct leukocyte integrins of mouse spleen dendritic cells as identified with new hamster monoclonal antibodies. J. Exp. Med. 171:1753–1771.
8. Coffman, R.L., Weissman, I.L.1981. A monoclonal antibody that recognizes B cells and B cell precursors in mice. J. Exp. Med.153:269–279
9. Rosen, H., Gordon, S. 1987. Monoclonal antibody to the murine type 3 complement receptor inhibits adhesion of myelomonocytic cells *in vitro* and inflammatory cell recruitment *in vivo*. J. Exp. Med. 166:1685–1701.
10. Mayrhofer, G., Holt, P.G., Papadimitriou, J.M. 1986. Functional characteristics of the veiled cells in afferent lymph from the rat intestine. Immunol.58:379–387.
11. Kelsall, B. L. and W. Strober. 1996. Distinct populations of dendritic cells are present in the subepithelial dome and T cell regions of the murine Peyer's patch. J. Exp. Med. 183:237–247.
12. Ruedl, C., Rieser, C., Böck, G., Wick, G., Wolf, H. 1996. Phenotypic and functional characterization of CD11c+ dendritic cell population in mouse Peyer's patches. Eur. J. Immunol. 26:1801–1806
13. Hopkins, S., Kraehenbuhl, J.-P., Schödel, F. Potts, A., Peterson, D., De Grandi, P. Nardelli-Haefliger, D. 1995. A recombinant *Salmonella typhimurium* vaccine induces local immunity by four different routes of immunization. Infect.Immun. 63:3279–3286.

18

THE ROLE OF CD11c$^+$ CELLS AS POSSIBLE CANDIDATES FOR IMMATURE DENDRITIC CELLS IN THE MURINE PEYER'S PATCHES

C. Ruedl,[1] S. Hubele,[1] C. Rieser,[2] and H. Wolf[2]

[1]Basel Institute for Immunology
Basel, Switzerland
[2]Institute for General and Experimental Pathology
University of Innsbruck, Innsbruck, Austria

1. INTRODUCTION

The gastroinstestinal mucosa is continously exposed to food-derived antigens, resident microorganisms and invading pathogens. Correct antigen sampling and handling is essential for regulation and maintenance of mucosal immune responses against these factors However, antigen transport through the epithelium, uptake, processing and presentation by specialized cells in vivo are important steps in the initiation of a defined immune response (or tolerance) in the gut.

In the gastroinstestinal tract, antigen presenting cells, such as macrophages, dendritic cells (DCs) and B-cells, are mainly present in the lamina propria, Peyer's patches (PP´s) and draining lymph nodes. Recently, two different populations of dendritic cells could be observed in the PP's (1): a population of interdigitating dendritic cells (NLDC145$^+$,CD11c$^+$) in the interfollicular areas, whereas a dense layer of NLDC145$^-$ CD11c$^+$ cells could be observed just beneath the dome epithelium. This strategical location of PP´s CD11c$^+$ cells, points to a possible participation of this cell population in capturing and processing of luminal antigens crossing from the gut lumen into the lymphoid follicles via specialized M cells (Fig. 1). In this context, we are interested to investigate the possible role of PP's CD11c$^+$ cells as immature "sentinel" DC. We compared freshly isolated DC (f-DC) and in vitro matured DC (c-DC) to examine differences in morphology, surface marker expression, endocytic ability as well as antigen processing and presentation capacity.

2. MATERIALS AND METHODS

Fresh (day 0, f-DC) and cultured (day 1, c-DC) CD11c$^+$ from PP´s of BALB/c mice were prepared using a previously described method (2). In vitro maturation was induced by

Dendritic Cells in Fundamental and Clinical Immunology
edited by Ricciardi-Castagnoli, Plenum Press, New York, 1997

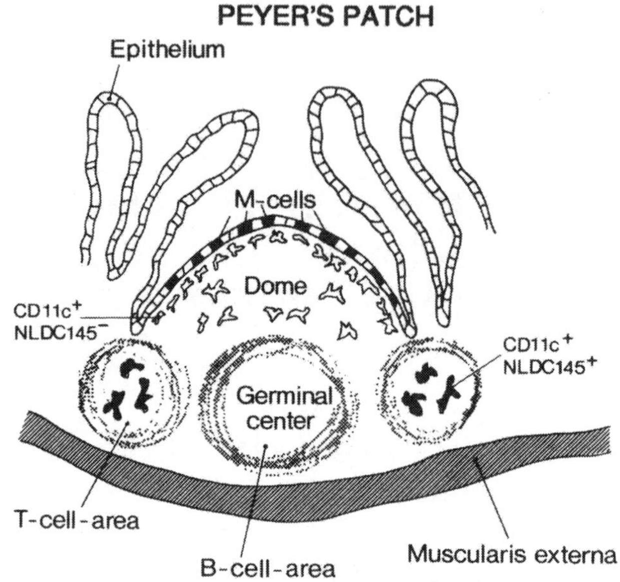

Figure 1. Two distinct populations of dendritic cells localized in the PP's. Interdigitating dendritic cells (NLDC145$^+$CD11c$^+$) in the interfollicular areas and a dense layer of NLDC145$^-$CD11c$^+$ dendritic cells just beneath the dome epithelium.

24 h culture in presence of GM-CSF (20 ng/ml) and TNF-α (50 ng/ml) or by binding the CD40 molecule. A panel of mAb (anti- I-A, anti-CD44, anti-ICAM-1, anti-NLDC145, anti-CD40, anti-B7.1, anti-B7.2) was tested on f-DC and c-DC, respectively, using a FACScan®.

For the FACS-analysis, DC were cocultured with irradiated J558 cells, expressing a membrane form of mouse CD40 ligand (mCD40L) (kindly provided by P. Lane, Basel Intitute for Immunology), for radioactive pulsing and antigen-uptake experiments the cells were stimulated with purified anti-CD40 antibody (kindly provided by T. Rolink, Basel Intitute for Immunology) because of the interference of the transfected CD40L/J558 cells.

T cell stimulation assays (MLR and oxidative mitogenesis) were performed as follows: DC from PP's (f-DC and c-DC) were irradiated with 2000 rad and mixed at graded doses with $2x10^5$ purified T cells in 96-well plates. For oxidative mitogenesis, we used syngeneic periodate-treated T cells (H-2^d), and for allogeneic MLR assays allogeneic T cells isolated from C3H/He mice (H-2^k). T cell proliferation was assessed by ^{3}H-thymidine (1μCi/well) uptake.

For metabolic labelling, freshly isolated and in vitro stimulated CD11c$^+$ cells (GM-CSF + TNF-α and anti-CD40) were incubated for 1 h with [^{35}S]methionine/cysteine mix, immunoprecipitated with MHC class II (N22), invariant chain (ln-1) antibodies and analyzed by SDS-PAGE.

Freshly sorted or in vitro stimulated cells were split into equal volumes and either kept on ice or incubated at 37°C in presence of FITC-labelled ovalbumin at a concentration of 1 mg/ml. The uptake of ovalbumin was monitored every 30 min for a period of 150 min. Three cell washings with cold PBS were performed and the fluorescence intensity was measured immediately on a FACScan®.

The processing and presentation activity of f-DC and c-DC (GM-CFS + TNF-α and anti-CD40) from PP's was measured by co-culturing with the ovalbumin-specific

DO11.10 T cells obtained from ovalbumin TRC transgenic mice. CD11c⁺ cells (1×10^4) were pulsed for 3 h with 1 mg/ml whole protein ovalbumin, with 10 μM 323–339 ovalbumin peptide and only medium without any antigen and cocultured with T cells (4×10^4). Proliferation was assessed by ^{3}H-thymidine (1μCi/well) incorporation.

3. RESULTS AND DISCUSSION

Previous findings have demostrated that populations of DC in certain tissues such as spleen (3), skin (4) and blood (5) are immunological immature because of reduced T cell stimulatory activity that rapidly develops upon in vitro stimulation via cytokines and CD40 crosslinking. We postulate that a similar population of immature DCs is localized in the dome region of the PP's capable to interact with antigens penetrating from the gut lumen into the PP's via the M-cells.

We developed a method for isolation of a highly purified population of CD11c⁺ cells involving enzymatic digestion, density centrifugation and cell sorting on the basis of CD11c epression. The obtained cells display a transient adherence capacity and when appropriately stimulated (anti-CD40 or GM-CSF and TNF-α, respectively) they undergo phenotypic and functional changes similar to those described for Langerhans cells. In vitro maturation is associated with upregulation of surface marker expression, such as MHC class II, adhesion molecules (ICAM-1, CD44), costimulatory molecules (B7.1 and B7.2) and, in the case of cytokine treatment, of CD40 expression. In addition, all stimulated CD11c⁺ cells expressed the NLDC145 marker restricted to interdigitating-DC known to be specific for mature DC. Furthermore, biosynthesis rates of MHC class II and invariant chain and pinocytosis of soluble antigens such as ovalbumin are markedely reduced or abolished during in vitro stimulation. On the other hand, freshly isolated PP's CD11c⁺ cells actively process exogenous antigens and synthetize at high rate MHC class II and invariant chain, essential requirements of an efficient antigen loading. However, downregulation of the capacity to process soluble native antigens recipocally correletes with the capacity to sentitize unprimed T cells. In fact, after 24 h stimulation, they differentiate into cells expressing the necessary accessory molecules and signals to stimulate non-sensitized naive T cells as shown in the illustrated MLR and oxidative mitogenesis assays.

These findings clearly demonstrate that freshly isolated murine PP's CD11c⁺ cells display characteristic features for immature DCs pointing to a crucial in vivo relevance of their counterparts localized in the dome region of the follicle. These observations further suggest that PP DC, like Langerhans cells of the skin, might exhist in two functionally distinct forms. The first, capable of antigen-processing ability, could be a sort of sentinel APC strategically located in the dome area were luminal antigen are penetreting, whereas the second, by upregulation of surface ligands and accessory signals, could be potent immunostimulatory dendritic cells able to stimulate naive or unprimed T cells. Whether this activation occurs in the T-cells areas of the PP´s itself or in distant lymphoid tissues will be topic of further investigations.

ACKNOWLEDGMENT

We thank A. Lanzavecchia and J. Pieters for critical reading of the manuscript. The Basel Institute was founded and is supported by Hoffman-LaRoche & Co. (Basel, Switzerland).

4. REFERENCES

1. B.L. Kelsall and W. Strober. Distinct population of dendritic cells are present in the subepithelial dome and T cell region of murine Peyer's patch. J. Exp. Med. 183: 237 (1996)
2. C. Ruedl, C. Rieser, G. BÖck, G. Wick and H. Wolf. Phenotypic and functional characterization of CD11c+ dendritic cell population in mouse Peyer's patches. Eur. J. Immunol 26: 1801–1806 (1996).
3. G. Girolomoni. J.C. Simon, P.R. Bergstresser and P.D. Cruz. Freshly isolated spleen dendritic cells and epidermal Langerhans cells undergo similar phenotypic and functional changes during short term culture. J. Immunol. 145: 2820 (1990)
4. N. Romani and G. Schuler. The immunologic properties of epidermal Langerhans cells as a part of the dendritic cell system. Springer Semin. Immunopathol. 13: 265 (1992).
5. U. O'Doherty, M. Peng, S Gezelter, W.J. Swiggard, M. Betjes, N. Bhardwaj and R.M. Steinman. Human blood contains two subsets of dendritic cells, one immunological mature and the other immature. Immunology 82: 487(1994).

19

THE INFLUENCE OF COLLAGEN, FIBRONECTIN, AND LAMININ ON THE MATURATION OF DENDRITIC CELL PROGENITORS PROPAGATED FROM NORMAL OR Flt3-LIGAND-TREATED MOUSE LIVER

Maureen L. Drakes,[1,2] Lina Lu,[1,2] Hilary J. McKenna,[4] and Angus W. Thomson[1,2,3]

[1]Thomas E. Starzl Transplantation Institute and
[2]Department of Surgery and
[3]Department Molecular Genetics and Biochemistry
University of Pittsburgh, Pennsylvania, 15213
[4]Immunex Research Development Corporation
Seattle, Washington 98101

1. INTRODUCTION

Dendritic cell (DC) progenitors can be propagated from normal mouse liver non-parenchymal cell suspensions in the presence of granulocyte/macrophage colony stimulating factor (GM-CSF). These cells strongly express CD45, CD11b, heat stable antigen and CD44 cell surface markers. They exhibit moderate to low levels of expression of mouse DC-restricted markers (DEC-205, 33D1 and N418 [CD11c]) as well as low levels of MHC class II and B7–2 molecules, features characteristic of "immature" mouse DC.[1,2,3] We have shown previously that these cells can be induced to express a mature DC phenotype and function (allostimulatory activity) following exposure to type-I collagen[1], a molecule that is spatially associated with DC in the portal triads of normal liver.[4–7] Here we report that these liver DC progenitors can also be induced to maturation in the presence of other extracellular matrix (ECM) proteins, namely fibronectin and laminin. Fibronectin is present along hepatic sinusoids of the liver, while laminin is localized on basement membranes along bile duct systems and blood vessels.[4]

Liver DC progenitors were propagated from normal mice or from animals treated with Flt3 ligand (FL). FL is a hematopoietic growth factor which markedly increases the number of progenitor/stem cells in vivo and strikingly augments DC numbers in mouse lymphoid and non-lymphoid tissues.[8,9] The DC progenitors propagated from normal of FL-treated mouse liver have a typical immature DC phenotype and weak allostimulatory capacity (Drakes et al., manuscript in preparation).

2. MATERIALS AND METHODS

2.1. Animals

Adult 8–12 week old male B10.BR (H-2^k, I-E^+), C57BL/10SJ (B10; H-2^b, I-A^b, I-E^-) and C3H/HeJ (C3H; H-2^k, I-A^k, I-E^k) were purchased from The Jackson Laboratory (Bar Harbor, ME).

2.2. Flt3 Ligand (FL) Treatment of Mice

CHO-cell derived human Flt3 ligand (FL) (0.025ng endotoxin/mg Flt3 L) (Immunex Corp., Seattle, WA) was administered intraperitoneally for 10 days to B10 mice at a dose of 10μg/mouse/day in Hanks' Balance Salts Solution (HBSS; Life Technologies, Gaithersburg, MD).

2.3. Isolation of Non-Parenchymal Cells (NPC) from Liver

Liver NPC were isolated from B10.BR or B10 mice as described elsewhere in detail.[1]

2.4. Culture of NPC with GM-CSF

2×10^6 liver NPC were cultured in 1ml volumes in each well of a 24-well plate in RPMI-1640 (Life Technologies) supplemented with 10% FCS and 4ng/ml GM-CSF (R&D Systems, Minneapolis, MN). Cultures were fed every other day by removing 50% supernatant from the top of the well and replenishing with an equivalent volume of GM-CSF containing medium. This process was designed to remove non-adherent granulocytes, without dislodging clusters of developing DC that attached loosely to firmly adherent macrophages. At day 7 for untreated mice, and day 4 for FL-treated mice, the non-adherent low buoyant-density cells were removed for study.

2.5. Culture of Liver DC Progenitors on Extracellular Matrix (ECM) Protein-Coated Plates

Each well of a 24-well plate was coated with either type-1 collagen purified from rat tail tendon (Sigma; 50μg/ml in 0.02 N acetic acid) or fibronectin from bovine plasma (Sigma, St. Louis, MO, 2μg/ml) or laminin from basement membrane of Engelbreth-Holm-Swarm mouse sarcoma (Sigma;2 μg/ml). Plates were left at 37°C for 1 hour, then allowed to air dry. They were washed twice in RPMI-1640 containing 10% heat inactivated FCS. Non-adherent cells from GM-CSF stimulated cultures were transferred to these plates or to uncoated or bovine serum albumin (BSA)-coated plates (1×10^6 cells/well) and grown in GM-CSF-containing medium as before for a further 3 days.

2.6. Cell Surface Immunophenotypic Analysis

The liver-derived cells were blocked with 10% goat serum for 20 min at 4°C and stained in HBSS with 1% BSA (Sigma). Monoclonal antibodies used included those directed against NLDC-145 (DEC-205; a kind gift from Dr. R.M. Steinman, the Rockefeller University, New York, N.Y.), CD40, CD32 (FcγRII), MHC class II (I-E^k or I-A^B), and

CD86 (B7–2), which were all obtained from PharMingen, San Diego, CA. After staining, cells were fixed in 1% paraformaldehyde and analyzed using a Coulter Elite ESP flow cytometer (Coulter Corp., Miami, FL).

2.7. Mixed Leukocyte Cultures

The allostimulatory activity of the liver-derived cells was measured in one-way mixed leukocyte cultures using spleen T cells as responders and γ-irradiated (20Gy) stimulators. Cultures were maintained in RPMI-1640 with 10% FCS for 72h in a 5% CO_2 humidified incubator, and pulsed with 1μCi [^{3}H] TdR per well for the final 18hr.

3. RESULTS

3.1. Upregulation of Cell-Surface Molecules on DC Progenitors Following Exposure to ECM Proteins

DC progenitors propagated from B10.BR liver NPC were grown in the presence of GM-CSF for 7 days and transferred to uncoated plates or plates coated with type-1 collagen or fibronectin and maintained for a further 3 days. Phenotypic analysis of the cells after exposure to the ECM proteins showed marked upregulation of DEC-205, MHC class II (I-E^k) and B7–2 molecules. There was a decrease in FcγRII expression and no dramatic change in CD40 expression (Table 1). Preliminary results for laminin also showed similar changes in the phenotype of the cultured cells (data not shown).

3.2. MLR Stimulatory Activity Following Exposure of Liver DC Progenitors to Fibronectin

DC progenitors propagated from livers of B10.BR mice were transferred to fibronectin-coated or uncoated plates for 3 days and their allostimulatory capacity assessed in mixed leukocyte reactions using varying numbers of stimulator cells and 2×10^5 B10 T cells per 96 well in round-bottomed plates as responders. DC removed from uncoated

Table 1. Influence of ECM proteins on the surface phenotype of normal liver-derived DC progenitors

Antigen	% Positive Cells Uncoated	Collagen	Fibronectin
DEC-205	15.6	47.5	35.7
CD40	30.2	28.4	18.8
FcγRII(CD32)	68.4	29.4	26.7
I-E^k (MHC class II)	5.0	13.1	28.4
B7-2 (CD86)	10.4	32.5	29.8

Phenotypic characterization of DC progenitors after exposure to uncoated, fibronectin- or collagen-coated plates. DC progenitors released from 7 day GM-CSF stimulated cultures of normal liver NPC from B10.BR mice were exposed to collagen or fibronectin in the continued presence of GM-CSF for a further 3 days. Percent positive cells was obtained by subtracting the value obtained with control isotype-matched antibody from that obtained with the antigen specific antibody.

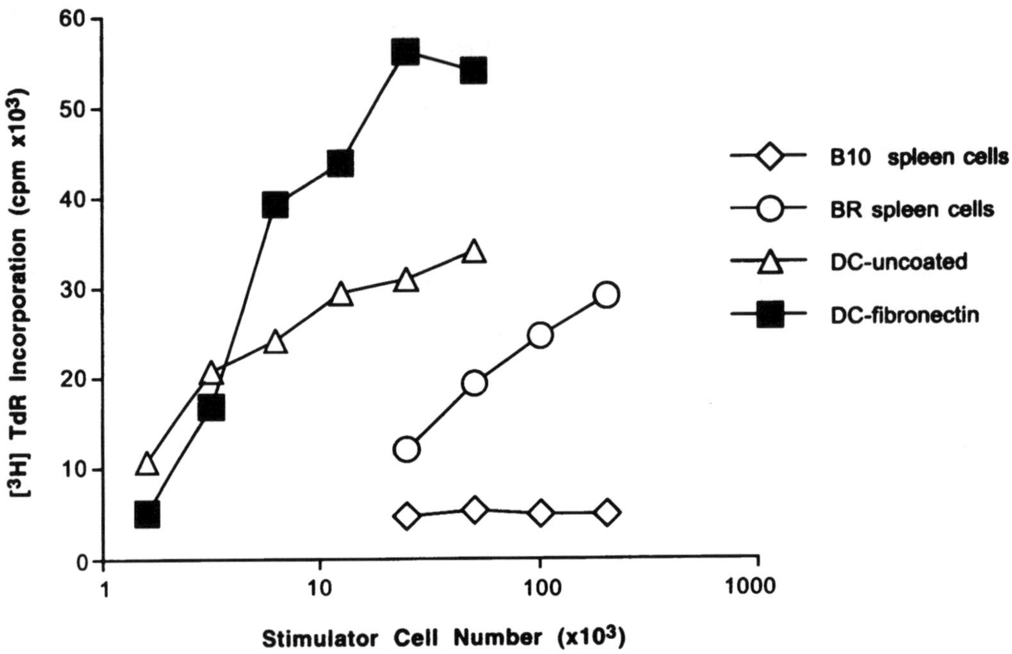

Figure 1. Allostimulatory activity of γ-irradiated, GM-CSF-stimulated B10.BR DC progenitors following exposure to fibronectin-coated or uncoated plates. Controls included allogeneic B10.BR spleen cells and syngeneic B10 spleen cells. Responder cells were 2 x 10^5 B10 T cells in each well of a 96-well round bottomed plate. Values represent mean cpm of [^{3}H]TdR for triplicate cultures. DC progenitors cultured on fibronectin-coated plates induced strong allogeneic T cell responses compared with controls.

plates were low stimulators of naive allogeneic T cells, while more potent stimulation of T cells was obtained with DC from fibronectin-coated plates (Figure 1). Similar investigations were undertaken with γ-irradiated DC progenitors from FL-treated B10 mice, after exposure to collagen, fibronectin or laminin as stimulators, and with C3H T cells as responders. Significant increases in allogeneic T cell stimulation were found in response to DC from ECM protein-coated plates as compared to those maintained on uncoated plates (Figure 2). This trend was similar to that observed in studies on DC progenitors from untreated mice.

4. DISCUSSION

GM-CSF stimulated DC progenitors propagated from liver NPC of normal B10.BR mice had low DEC-205, MHC class II (I-E^k) and B7–2 expression as is characteristic of "immature" liver DC. With exposure to the ECM proteins collagen, fibronectin or laminin, these parameters were upregulated, indicating phenotypic maturation of these cells. Changes in these parameters were also accompanied by an increase in allostimulatory capacity of the DC for T cells in MLR studies, as shown in the case of fibronectin in Figure 1. To further investigate the allostimulatory capacity of ECM protein treated DC, we used DC progenitors from livers of FL treated mice, since it has been shown recently that this molecule significantly increases the numbers of stem/progenitor cells in organs such as spleen, BM or LN[8,9] and in the

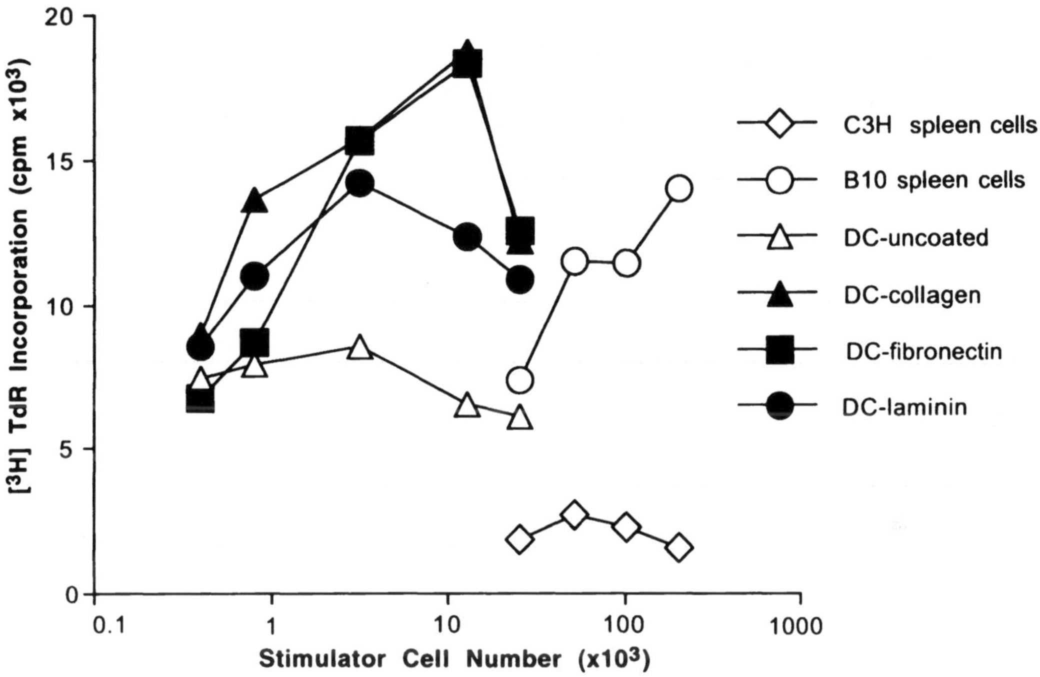

Figure 2. Allostimulatory activity of γ irradiated DC progenitors propagated from livers of FL treated B10 mice using C3H T cells as responders. DC progenitors were maintained on fibronectin-, collagen-, or laminin-coated, or uncoated plates for 3 days (day 4–7) of culture. Controls include syngeneic C3H spleen cells and allogeneic B10 spleen cells. Values represent mean cpm of [^{3}H]TdR for triplicate wells. DC grown on ECM protein-coated plates strongly stimulated allogeneic T cells.

liver (Drakes *et al.*, manuscript in preparation). This allowed us much greater numbers of progenitor cells for use in our investigations. The allostimulatory capacity of the DC progenitors was much enhanced after exposure to laminin, fibronectin or collagen-coated plates for 3 days, when compared with those maintained on uncoated plates (Fig. 2). It is noteworthy that DC progenitors placed on BSA-coated plates as a negative protein control resulted in comparative values for T cell stimulation to those obtained for DC on uncoated plates (data not shown). Collagen was used as a positive control since earlier studies in our laboratory showed increases in the parameters examined here when DC progenitors were subcultured on collagen-coated plates for 3 days.[1]

ECM proteins are widely distributed throughout the body and can support cell migration by providing attachment for many types of cells. Collagen for example, has been found in abundance in the portal triads of normal liver, fibronectin along hepatic sinusoids, and laminin along bile duct systems and blood vessels.[4,7] In addition, receptors for collagen and laminin have been identified on murine T lymphocytes.[10] Splenic and pulmonary DC have been shown to bind to fibronectin.[11]

The maturation of liver DC progenitors in the presence of these ECM proteins sheds light on possible microenvironmental factors which may influence the phenotype and function of DC following their migration to liver from bone marrow. Alterations in ECM deposition, as occurs in liver fibrosis[4,12], hepatic cancer and chronic liver disease[4] could possibly alter the function or phenotype of liver-infiltrating DC progenitors compared with those in normal liver.

Further studies to characterize the parameters that regulate liver DC maturation in the presence of ECM proteins and other molecules (in particular cytokines and liver cell growth factors) are under investigation. In this context FL is of value as an inducer of DC progenitors within the liver.

5. ACKNOWLEDGMENTS

We thank Ms. Francy G. Chambers for excellent technical assistance, Ms. Alison Logar for flow cytometric analysis and Ms. Shelly Lynn Conklin for secretarial assistance. This work was supported by National Institutes of Health grant DK 49745–01A1.

6. REFERENCES

1. L Lu, J Woo, AS Rao, Y Li, SC Watkins, S Qian, TE Starzl, AJ Demetris, AW Thomson. Propagatio of dendritic cell progenitors from normal mouse liver using granulocyte/macrophage colony-stimulating factor and their maturational development in the presence of type-1 collagen. J Exp Med 1994 179:1823.
2. L Lu, D McCaslin, TE Starzl, AW Thomson. Mouse bone marrow-derived dendritic cell progenitors (NLDC 145^{+}, MHC class II^{+}, B7–1^{dim}, B7–2^{-}) induce alloantigen-specific hyporesponsiveness in murine T lymphocytes. Transplantation 1995;60:1539.
3. AW Thomson, L Lu, S Qian, F Fu, JJ Fung, TE Starzl. Immature dendritic cells induce hyporesponsiveness to alloantigens *in vitro* and prolong mouse cardiac allograft survival. In: JW Alexander, and RA Good (eds) *Transplantation Tolerance Induction* RG Landes Co, Austin 1996;p.141.
4. T Takahashi, M Isemura, T Nakamura, S Matsui, Y Oyanagi, H Asakura. Immunolocalization of a fibronectin-binding proteoglycan (PG-P1) immunologically related to HSPG2/perlecan in normal and fibrotic human liver. J Hepatology 1994 21:500.
5. B Steiniger, J Klempnauer, K Wonigeit. Phenotype and histological distribution of interstitial dendritic cells in the pancreas, liver, heart and kidney. Transplantation 1984 38:169.
6. TCR Prickett, JL McKenzie, DNJ Hart. Characterization of interstitial dendritic cells in human liver. Transplantation 1988 46:754.
7. A Martinez-Hernandez. The hepatic extracellular matrix. I. Electron immunohistochemical studies in normal rat liver. Lab Invest 1984 51:57.
8. K Brasel, HJ McKenna, PJ Morrissey, K Charrier, AE Morris, CC Lee, DE Williams, SD Lyman. Hematological effects of FLT3 ligand in vivo in mice. Blood 1996 In press
9. E Maraskovsky, K Brasel, M Teepe, E Roux, K Shortman, SD Lyman, HJ McKenna. In vivo administration of flt3 ligand results in the generation of large numbers of dendritic cells in the lymphoid tissue of mice. J Exp Med 1996 In press.
10. S Miyake, T Sakurai, K Okumura, H Yagita. Identification of collagen and laminin receptor integrins on murine T lymphocytes. Eur J Immunol 1994;24:2000.
11. TC Tsaio, W Xia, GM Rodberg, CE Pinto, RL Kradin. Interferon-gamma and tumor necrosis factor-alpha promote the binding of dendritic cells to fibronectin. Pathobiology 1994;62:120.
12. MJ Arthur. Degradation of matrix proteins in liver fibrosis. Pathology, Research & Practice 1994;190:825.

20

STIMULATION OF LANGERHANS CELL MIGRATION IN MICE BY TUMOUR NECROSIS FACTOR α AND INTERLEUKIN 1β

M. Cumberbatch, R. J. Dearman, and I. Kimber

Zeneca Central Toxicology Laboratory
Alderley Park, Macclesfield
Cheshire, SK10 4TJ, United Kingdom

1. INTRODUCTION

Following contact sensitization or other forms of cutaneous trauma epidermal Langerhans cells (LC) are stimulated to migrate from the skin and to travel, via afferent lymphatics, to draining lymph nodes.[1] We have demonstrated previously that tumour necrosis factor α (TNF-α), a keratinocyte-derived epidermal cytokine, provides one stimulus for LC migration.[2] It has been shown that intradermal injection of mice with homologous recombinant TNF-α induces a rapid reduction in the frequency of epidermal LC local to the site of exposure and results, somewhat later, in the accumulation of dendritic cells (DC) in draining lymph nodes.[3,4] Systemic (intraperitoneal) treatment of mice with a neutralizing anti-TNF-α antibody prior to topical sensitization with oxazolone, a potent contact allergen, was found to inhibit almost completely the accumulation of DC in draining nodes normally provoked by exposure to this chemical. Such treatment was found also to suppress the development of contact sensitization.[5] In the present investigations we have examined whether another epidermal cytokine interleukin 1β (IL-1β), a product of LC, also plays a role in LC migration. Attention has focused on IL-1β for three reasons. First, it has been shown that the development of contact hypersensitivity is compromised in IL-1β gene deletion transgenics,[6] or in mice that have been treated with an anti-IL-1β antibody.[7] Second, there is evidence suggesting that IL-1β, a cytokine which is upregulated very rapidly following contact sensitization,[8] is able to provoke the increased expression of TNF-α by keratinocytes.[7] Third, it has been claimed on the basis of experiments performed with human skin explants that IL-1β may affect LC migration *in vitro*.[9] We here describe investigations in which the influence of IL-1β and TNF-α on LC migration and DC accumulation were examined.

Dendritic Cells in Fundamental and Clinical Immunology
edited by Ricciardi-Castagnoli, Plenum Press, New York, 1997

2. RESULTS

Groups of BALB/c strain mice (n=3 for measurement of LC frequency, or n=10 for evaluation of DC accumulation in draining lymph nodes) were treated, by intradermal injection (30μl) into both ear pinnae, either with 50ng of recombinant murine TNF-α (specific activity 2×10^8U/mg) or with the same amount of recombinant murine IL-1β (specific activity 1–2 × 10^8U/mg). Additional groups of mice were exposed in the same way to 30μl of 0.1% bovine serum albumin (BSA), the protein in which the cytokines were suspended. Further control animals were untreated.

The frequency of epidermal LC local to the site of exposure was measured 30 minutes and 2 hours following treatment. Epidermal sheets were prepared and LC identified as described previously using immunocytochemical assessment of MHC class II (Ia) antigen expression.[4] In untreated mice the frequency of LC in epidermal sheets remained relatively constant at approximately 900 LC/mm^2. LC density in mice exposed for either 30 minutes or 2 hours to BSA alone did not differ significantly from control values. Consistent with previous observations,[4] intradermal treatment of mice with homologous TNF-α induced a rapid reduction in the frequency of epidermal LC that was apparent within 30 minutes of exposure. In contrast, 30 minutes following treatment with IL-1β the density of LC in epidermal sheets was unchanged. Within 2 hours of treatment with either cytokine there was a clear and comparable reduction in LC numbers (Figure 1a).

In parallel experiments the ability of TNF-α and IL-1β to induce the accumulation of DC in draining lymph nodes was measured. Lymph nodes were excised 2 and 4 hrs after exposure, single cell suspensions prepared and DC enriched and enumerated as described previously.[3] In this series of experiments lymph nodes taken from untreated control mice were found to contain between 1,500 and 2,200 DC. Treatment of mice with BSA alone failed to influence DC numbers. It has been reported previously[3] that intradermal exposure of mice to homologous TNF-α induces within 2 hours an increase in the number of DC found within draining lymph nodes. In accord with those data such treatment was shown here to be associated with an approximate 2-fold increase in DC numbers by 2 hours and a further increase to over 5,000 DC/node when measurements were made at 4 hours. The frequency of DC in draining lymph nodes was increased by 4 hours following exposure to IL-1β, but not at 2 hours (Figure 1b).

3. DISCUSSION

The data presented here demonstrate that dermal exposure of mice to homologous TNF-α induces a rapid movement of a proportion of LC away from the epidermis and the accumulation, somewhat later, of DC in draining lymph nodes. As such these results are in agreement with those of previous investigations.[3,4] It is apparent, however, that IL-1β also has the ability to stimulate LC migration and DC accumulation, albeit with somewhat slower kinetics. One possibility is that IL-1β serves to induce or upregulate the production by keratinocytes of TNF-α that then acts on neighbouring LC to stimulate their migration. Certainly it has been shown that intradermal injection of mice with concentrations of IL-1β comparable to those used here caused a very substantial increase in the epidermal expression of mRNA for TNF-α.[7] The need for IL-1β to first induce the production of TNF-α would be consistent with the delayed kinetics of both LC migration and DC accumulation compared with those observed in mice exposed to TNF-α itself. Alternatively, or additionally, IL-1β may influence directly the activity of LC; this being a possibility since

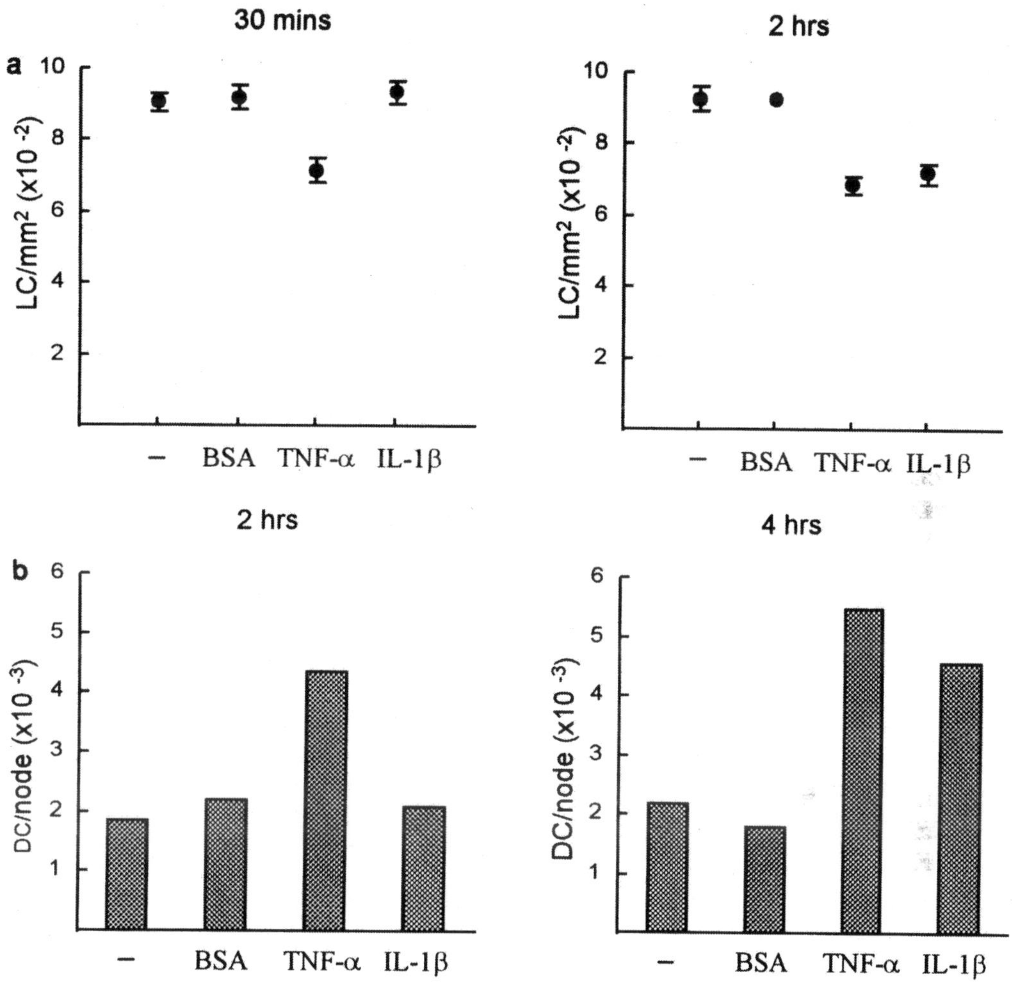

Figure 1. Influence of homologous TNF-α and IL-1β on epidermal LC frequency (a), and draining lymph node DC numbers (b). Groups of mice [(a) n=3; (b) n=10] received 30μl intradermal injections into both ear pinnae of equal amounts (50ng) of TNF-α or IL-1β in 0.1% bovine serum albumin (BSA). Control mice received an equivalent amount of BSA alone or were untreated (-). (a) Ears were removed 30 mins or 2 hrs following treatment, epidermal sheets prepared and the frequency of Ia^+ LC measured by indirect immunofluorescence. Results are recorded as mean $LC/mm^2 \pm$ SE from the examination of 10 fields/sample for each of 4 samples. (b) Draining lymph nodes were removed 2 or 4 hours following treatment. DC were prepared and numbers assessed by direct morphological examination using phase contrast microscopy. Results are expressed as number of DC/node.

freshly isolated LC have been found to express mRNA for both type I and type II receptors for IL-1.[10] Irrespective of the mode of action for IL-1β, the available evidence indicates that TNF-α interacts directly with epidermal LC and provides one of the proximate signals necessary for migration. Such evidence is based upon an understanding of receptors for TNF-α and the species-selectivity of epidermal responses to this cytokine. Two forms of the TNF-α receptor have been identified and are designated TNF-R1 and TNF-R2.[11] The former, a 55kDa receptor, displays most homology between species in the extracellular domain, whereas TNF-R2, a 75kDa receptor is most conserved in the intracellular region and exhibits less homology in the extracellular domain. For this reason TNF-R2, but not TNF-R1, is species specific with respect to the cytokine. Langerhans cells resident in the epidermis appear to express only TNF-R2[10,12] and such is compatible with evidence that, while homologous TNF-α will provoke in mice LC migration and DC accumulation, human TNF-α is without effect.[3,4] Human TNF-α will, however, induce responses by murine keratinocytes (increased expression of intercellular adhesion molecule-1)[4]; cells that are known to express the 55kDa TNF-R1.[13] If TNF-α were acting in mice via an indirect mechanism, involving perhaps an intermediary step requiring the activation of keratinocytes, then the human cytokine would be expected to elicit LC responses. The importance of signalling via TNF-R2 for the initiation of LC migration is supported also by the results of recent investigations in which it was found that in response to skin sensitization the accumulation of DC in draining lymph nodes is normal in TNF-R1 gene knockout mice. In such mice DC accumulation could, however, be inhibited if animals were treated with a neutralizing anti-TNF-α antibody.[14]

Taken together the data reported here indicate that, in addition to TNF-α, IL-1β is able to induce the migration of LC from the epidermis and the accumulation of DC in draining nodes. The mechanisms through which IL-1β is able to stimulate such responses are currently unknown. Despite the activity of IL-1β, the available evidence suggests that the initiation of LC migration requires as one stimulus the direct interaction of TNF-α with epidermal LC.

4. REFERENCES

1. I. Kimber and M.Cumberbatch, Dendritic cells and cutaneous immune responses to chemical allergens. Toxicol. Appl. Pharmacol. 117:137 (1992).
2. I. Kimber and M. Cumberbatch, Langerhans cell migration: initiation and regulation. In: The Immune Functions of Epidermal Langerhans Cells. H.Moll (ed.), R.G. Landes Co., Austin pp103 (1994).
3. M. Cumberbatch and I. Kimber, Dermal tumour necrosis factor-α induces dendritic cell migration to draining lymph nodes, and possibly provides one stimulus for Langerhans cell migration. Immunology 75:257 (1992).
4. M. Cumberbatch, I. Fielding and I. Kimber, Modulation of epidermal Langerhans cells frequency by tumour necrosis factor-α. Immunology 81:395 (1994).
5. M. Cumberbatch and I. Kimber, Tumour necrosis factor-α is required for accumulation of dendritic cells in draining lymph nodes and for optimal contact sensitization. Immunology 84:31 (1995).
6. L.P. Shornick, P. De Togni, S. Mariathasan, J. Goellner, J. Strauss-Schoenberger, R.W. Karr, T.A. Ferguson and D.D. Chaplin, Mice deficient in IL-1β manifest impaired contact hypersensitivity to trinitrochlorobenzene. J. Exp. Med. 183:1427 (1996).
7. A.H. Enk, V.L. Angeloni, M.C. Udey and S.I. Katz, An essential role for Langerhans cell-derived IL-1β in the initiation of primary immune responses in skin. J. Immunol. 150:3698 (1993).
8. A.H. Enk and S.I. Katz, Early molecular events in the induction phase of contact sensitivity. Proc. Natl. Acad. Sci. USA 89:1398 (1992).
9. A. Rambukkana, F.H.M. Pistoor, J.D. Bos, M.L. Kapsenberg and P.K. Das, Effects of contact allergens on human Langerhans cells in skin organ culture: migration, modulation of cell surface molecules, and early expression of interleukin 1β protein. Lab. Invest. 74:422 (1996).
10. A. Larregina, A. Morelli, E. Kolkowski and L. Fainboim, Flow cytometric analysis of cytokine receptors on human Langerhans cells. Changes observed after short-term culture. Immunology 87:317 (1996).
11. L.A. Tartaglia and D.V. Goeddel, Two TNF receptors. Immunol. Today 13:151 (1992).
12. B. Ryffel, M. Brockhaus, B. Greiner, M.J. Mihatsch and F. Gaudat, Tumor necrosis factor receptor distribution in human lymphoid tissue. Immunology 74:446 (1991).
13. U. Trefzer, M. Brockhaus, H. Loetscher, F. Parlow, A. Kapp, E. Schopf and J. Krutmann, 55-kDa tumor necrosis factor receptor is expressed by human keratinocytes and plays a pivotal role in regulation of human keratinocyte ICAM-1 expression. J. Invest. Dermatol. 97:911 (1991).
14. B. Wang, S. Kondo, C.M. Shivji, H. Fujisawa, T.W. Mak and D.N. Sauder, Tumour necrosis factor receptor II (p75) signalling is required for the migration of Langerhans cells. Immunology 88:284 (1996).

21

LANGERHANS CELLS REQUIRE SIGNALS FROM BOTH TUMOUR NECROSIS FACTOR α AND INTERLEUKIN 1β FOR MIGRATION

M. Cumberbatch, R. J. Dearman and I. Kimber

Zeneca Central Toxicology Laboratory
Alderley Park, Macclesfield
Cheshire, SK10 4TJ, United Kingdom

1. INTRODUCTION

Tumour necrosis factor α (TNF-α) is considered to play an important role in the initiation of epidermal Langerhans cell (LC) migration during the induction phase of cutaneous immune responses.[1] Intradermal injection of mice with homologous recombinant TNF-α stimulates both the migration away from the epidermis of a proportion of LC and the subsequent accumulation in draining lymph nodes of dendritic cells (DC).[2,3] Moreover, treatment of mice with a neutralizing anti-TNF-α antibody has been shown to inhibit markedly the increase in lymph node DC associated with exposure to skin sensitizing chemicals,[4,5] ultraviolet B light,[6] or the skin irritant sodium lauryl sulphate.[4]

There is evidence that another cytokine, interleukin 1β (IL-1β), a product exclusively of LC in murine epidermis,[7,8] is also important in this context. This cytokine has been shown to be required for the induction of contact sensitization.[9,10] In addition, IL-1β is able to stimulate the migration of LC *in vivo*[11] and possibly *in vitro*.[12]

In the present investigations we have examined, using the respective neutralizing anti-cytokine antibodies, the contributions made by TNF-α and IL-1β to LC migration and the accumulation of DC in draining lymph nodes. The influences of anti-TNF-α and anti-IL-1β on the accumulation of DC in draining lymph nodes induced by the contact allergen oxazolone have been measured, as have the effects of anti-TNF-α and anti-IL-1β on the stimulation of LC migration by, respectively, IL-1β and TNF-α.

2. RESULTS

In the first series of experiments the influence of anti-TNF-α and anti-IL-1β on the induction by oxazolone of DC accumulation in draining lymph nodes was investigated. Groups of BALB/c strain mice (n=10) were injected intraperitoneally with 100μl of either anti-TNF-α or anti-IL-1β, in both cases diluted 1:5 in sterile phosphate-buffered saline (PBS). Control mice received an equal volume of sterile normal rabbit serum (NRS) diluted with PBS to an equivalent extent. Two hours later all mice were exposed on the dorsum of both ears to 0.5% oxazolone in 4:1 acetone:olive oil. Draining auricular lymph nodes were excised 18 hours later, a single cell suspension prepared and DC enriched and enumerated as described previously.[2] Pretreatment of mice with anti-TNF-α or anti-IL-1β in each instance caused a very significant inhibition of oxazolone induced DC accumulation in draining lymph nodes (data not presented).

In a second series of experiments the influence of these same antibodies on the stimulation by cytokines of LC migration from the epidermis was investigated. Groups of BALB/c strain mice (n=3) were injected intraperitoneally with antibody or NRS as described above. Two hours later mice received, by intradermal injection (30μl) in both ear pinnae, 50ng of either homologous recombinant TNF-α (specific activity 2×10^8U/mg) or homologous recombinant IL-1β (specific activity $1\text{-}2 \times 10^8$U/mg) each suspended in 0.1% bovine serum albumin (BSA). Control mice were untreated. The frequency of epidermal LC local to the site of exposure was measured 30 minutes following the administration of TNF-α or 17 hours following similar treatment with IL-1β. Epidermal sheets were prepared and LC identified by expression of MHC class II (Ia) antigen using indirect immunofluorescence.[3] The results are summarized in Figure 1. Intradermal injection of mice with TNF-α (Figure 1b) or IL-1β (Figure 1c), in mice pretreated with NRS, caused a significant reduction in the numbers of LC identified in epidermal sheets. However, prior systemic exposure of mice to neutralizing anti-IL-1β inhibited significantly the induction of LC migration by dermal TNF-α (Figure 1b). Similarly, the stimulation of LC migration

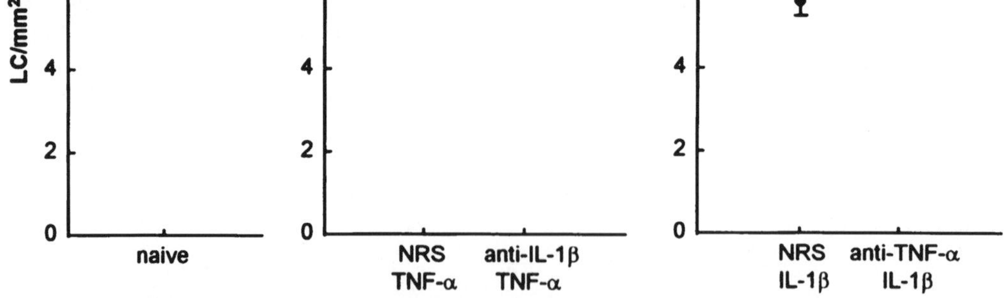

Figure 1. Influence of anti-TNF-α and anti-IL-1β on epidermal LC migration induced by cytokines. Groups of mice (n=3) received, by intraperitoneal injection, 100μl of anti-IL-1β (b) or anti-TNF-α (c). Two hours later mice were treated, by intradermal injection (30μl) into both ear pinnae, with 50ng of either TNF-α (b) or IL-1β (c) in 0.1% bovine serum albumin. Control mice were untreated (a). Ears were removed 30 minutes (b) or 17 hours (c) following dermal exposure to cytokine. Epidermal sheets were prepared and the frequency of Ia^+ LC measured by indirect immunofluorescence. Results are recorded as mean LC/mm^2 ± SE from the examination of 10 fields/sample for each of 4 samples.

by IL-1β was impaired markedly in mice that had been treated previously with neutralizing anti-TNF-α antibody (Figure 1c).

It was found also that treatment with neutralizing antibodies cross-inhibited cytokine-induced DC accumulation in draining lymph nodes. Thus, exposure of mice to anti-TNF-α prevented the stimulation by dermal IL-1β of DC accumulation and, in the same way, anti-IL-1β inhibited DC accumulation induced by TNF-α (data not presented).

3. DISCUSSION

The data presented here reveal that the effective migration of LC from the epidermis to draining lymph nodes is dependent upon the availability of both TNF-α and IL-1β. In part these results may be reconciled by the fact that IL-1β is able to induce keratinocyte TNF-α. Intradermal injection of mice with recombinant IL-1β was shown to cause a rapid and substantial increase in epidermal expression of mRNA for TNF-α.[9] One may speculate that the induced or increased expression of TNF-α following skin sensitization is effected exclusively by the paracrine action of LC-derived IL-1β on adjacent keratinocytes. Certainly such would be consistent with the observation that in response to topically applied contact allergens the expression by LC of mRNA for IL-1β is upregulated very quickly.[7] The assumption would be that this increased expression of IL-1β itself resulted from the interaction of sensitizing chemicals with epidermal LC, possibly via association with membrane Ia determinants.[13] Such a scheme, where LC are activated to synthesize and secrete IL-1β and where this cytokine is necessary for the elaboration by keratinocytes of the TNF-α required to stimulate LC migration, is an attractive one and would serve to explain why antibodies to both TNF-α and IL-1β inhibit oxazolone-induced DC accumulation and why anti-TNF-α is able to prevent the stimulation of LC migration by IL-1β. What can not be accomodated by this model, however, is the fact that the induction by TNF-α of LC migration and DC accumulation can be prevented or inhibited if mice are pretreated with a neutralizing anti-IL-1β antibody.

There is both direct and indirect evidence that one stimulus for LC migration results from the direct interaction of TNF-α with LC. First, LC migration in response to intradermal TNF-α is very rapid, occuring within 30 minutes of exposure;[3,11] a faster response than is seen when IL-1β is administered via the same route.[11] Second, epidermal LC express the species-specific (75kDa) form of the TNF-α receptor (TNF-R2)[14,15] and only homologous TNF-α is able to stimulate LC migration and DC accumulation in draining lymph nodes.[2,3] Third, in mice lacking a functional gene for TNF-R1, the species-unrestricted form of the receptor, allergen-induced DC accumulation in nodes is normal, but can be inhibited with an anti-TNF-α antibody; evidence indicative of signalling through LC TNF-R2 being required to initiate LC migration.[5] However, while the direct interaction of TNF-α with LC appears to be essential for migration, the evidence presented here indicates that, in the absence of IL-1β, signalling through LC TNF-R2 is either not sufficient for migration, or is not accomplished. With respect to the latter possibility it might be that IL-1β is required to maintain expression by LC of TNF-R2, although there is no evidence that this is the case. Alternatively, as LC are known to express receptors for this cytokine,[15] it is possible that IL-1β also induces in LC changes that are necessary for effective migration from the epidermis. Movement of LC through the skin and across the basement membrane to afferent lymphatics will undoubtedly require appropriate interactions with other cells and with extracellular matrices. It is known that the migration or activation of LC is associated with the altered expression of several adhesion molecules, (including intercellular adhesion molecule-1,[16] E-cadherin[17] and CD44[18]), that might facilitate their disengagement from keratinocytes in the epidermis or their movement through the skin and across the basement membrane. It remains a possibility that IL-1β, in addition to stimulating TNF-α production by keratinocytes, acts directly on LC to induce altered expression of adhesion molecules or other changes that are necessary for migration.

In conclusion, the results reported here demonstrate that both TNF-α and IL-1β contribute in important ways to the successful migration of LC from the epidermis. The changes induced by these cytokines have yet to be defined.

4. REFERENCES

1. I. Kimber and M. Cumberbatch, Langerhans cell migration: initiation and regulation. In: The Immune Functions of Epidermal Langerhans cells. H. Moll (ed.), R.G. Landes Co., Austin pp103 (1994).
2. M. Cumberbatch and I. Kimber, Dermal tumour necrosis factor-α induces dendritic cell migration and possibly provides one stimulus for Langerhans cell migration. Immunology 75:257 (1992).
3. M. Cumberbatch, I. Fielding and I. Kimber, Modulation of epidermal Langerhans cell frequency by tumour necrosis factor-α. Immunology 81:395 (1994).
4. M. Cumberbatch and I. Kimber, Tumour necrosis factor-α is required for accumulation of dendritic cells in draining lymph nodes and for optimal contact sensitization. Immunology 84:31 (1995).
5. B. Wang, S. Kondo, G.M. Shivji, H. Fujisawa, T.W. Mak and D.N. Sauder, Tumour necrosis factor receptor II (p75) signalling is required for the migration of Langerhans cells. Immunology 88:284 (1996).
6. A.M. Moodycliffe, I. Kimber and M. Norval, Role of tumour necrosis factor-α in ultraviolet-induced dendritic cell migration and suppression of contact hypersensitivity. Immunology 81:79 (1994).
7. A.H. Enk and S.I. Katz, Early molecular events in the induction phase of contact sensitivity. Proc. Natl. Acad. Sci. USA 89:1398 (1992).
8. C. Heufler, G. Topar, F. Koch, B. Trockenbacher, E. Kampgen, N. Romani and G. Schuler. Cytokine gene expression in murine epidermal cell suspensions. Interleukin 1β and macrophage inflammatory protein 1α are selectively expressed in Langerhans cells but are differentially regulated in culture. J. Exp. Med. 176:1221 (1992).
9. A.H. Enk, V.L. Angeloni, M.C. Udey and S.I. Katz, An essential role for Langerhans cell-derived IL-1β in the initiation of primary immune responses in skin. J. Immunol. 150:3698 (1993).
10. L.P. Shornick, P. De Togni, S. Mariathasan, J. Goellner, J. Strauss-Schoenberger, R.W. Karr, T.A. Ferguson and D.D. Chaplin, Mice deficient in IL-1β manifest impaired contact hypersensitivity to trinitrochlorobenzene. J. Exp. Med. 183:1427 (1996).
11. M. Cumberbatch, R.J. Dearman and I. Kimber, Stimulation of Langerhans cell migration in mice by tumour necrosis factor-α and interleukin 1β. In: Dendritic Cells in Fundamantal and Clinical Immunology. Vol.3, P. Ricciardi-Castagnoli (ed.), Plenum Press, New York pp (this volume).
12. A. Rambukkana, F.H.M. Pistoor, J.D. Bos, M.L. Kapsenberg and P.K. Das, Effects of contact allergens on human Langerhans cells in skin organ culture : migration, modulation of cell surface molecules, and early expression of interleukin 1β protein. Lab. Invest. 74:422 (1996).
13. N.S. Trede, R.S. Geha and T. Chatila, Transcriptional activation of IL-1β and tumour necrosis factor α genes by MHC class II ligands. J. Immunol. 146:2310 (1991).
14. B. Ryffel, M. Brockhaus, B. Greiner, M.J. Mihatsch and F. Gudat, Tumour necrosis factor receptor distribution in human lymphoid tissue. Immunology 74:446 (1991).
15. A. Larregina, A. Morelli, E. Kolkowski and L. Fainboim, Flow cytometric analysis of cytokine receptors on human Langerhans cells. Changes observed after short-term culture. Immunology, 87:317 (1996).
16. M. Cumberbatch, S.W. Peters, S.J. Gould and I. Kimber, Intercellular adhesion molecule-1 (ICAM-1) expression by lymph node dendritic cells : comparison with epidermal Langerhans cells. Immunol. Lett. 32:105 (1992).
17. K. Schwarzenberger and M.C. Udey, Contact allergens and epidermal proinflammatory cytokines modulate Langerhans cell E-cadherin expression in situ. J. Invest. Dermatol. 106:553 (1996).
18. D.I. Gabrilovich, G.M. Ward, S. Patterson, J.J. Harvey and S.C. Knight, Retrovirus-induced immunosuppression via blocking of dendritic cell migration and down-regulation of adhesion molecules. Immunology 82:82 (1994)

22

α6 INTEGRINS ARE REQUIRED FOR LANGERHANS CELL MIGRATION

A. A. Price,[1] M. Cumberbatch,[2] I. Kimber,[2] and A. Ager[1]

[1]National Institute for Medical Research
Mill Hill, London, NW7 1AA
[2]Zeneca Central Toxicology Laboratory
Alderley Park, Macclesfield
Cheshire, SK10 4TJ, United Kingdom

1. INTRODUCTION

Topical exposure of mice to contact allergens is associated with the migration of Langerhans cells (LC) from the epidermis and their accumulation as immunostimulatory dendritic cells (DC) in draining lymph nodes.[1] The movement of LC through the skin and across the basement membrane will undoubtedly be dependent upon their expression of appropriate adhesion molecules necessary for interactions with other cells and with extracellular matrices. Several adhesion molecules, the altered expression of which are associated with the activation or migration of LC, have been implicated in this process.[2] Thus, for instance, it has been found that the membrane expression of intercellular adhesion molecule-1 (ICAM-1, CD54) is upregulated markedly during LC migration such that the DC which accumulate in draining lymph nodes display a 40-fold increase in expression of this molecule compared with LC resident in the epidermis.[3] It has been shown that in mice treated with anti-ICAM-1 antibody the accumulation of DC in draining lymph nodes is significantly inhibited.[4] E-cadherin, a homotypic adhesion molecule expressed by both LC and keratinocytes may also be important. It has been proposed that this molecule serves to retain LC within the epidermis, the inference being that a prerequisite for migration is the reduced expression of E-cadherin by one or other cell type.[5,6] Certainly, it has been reported that, compared with LC in the epidermis, the DC found within lymph nodes express significantly less E-cadherin.[7]

Integrins are a family of widely expressed cell surface adhesion receptors and represent the main mechanism whereby cells interact with extracellular matrices. LC express β1 integrins and it has been suggested that very late antigen (VLA) proteins, VLA-4 (α4β1 integrins) and VLA-6 (α6β1 integrins) may, respectively, allow the association of LC with fibronectin and laminin.[8] It has been suggested also that (because α4 integrin expression is enhanced markedly during culture of LC) VLA-4 plays a major role in LC migration.[9] The present investigations were designed to examine the requirements for α4 and α6 integrins during LC migration and the accumulation of DC in draining lymph nodes.

2. RESULTS

In preliminary experiments the expression by epidermal LC and by lymph node DC of α4 and α6 integrins was measured in BALB/c strain mice using analytical flow cytometry. DC prepared from the draining lymph nodes of skin sensitized mice expressed high levels of α4, but little or no α6. In contrast, epidermal LC were found to display only low levels of α4, with a proportion of cells having high levels of α6 (data not presented).

The influence of antibodies to α4 and α6 integrins on the accumulation of DC in draining lymph nodes following topical exposure to contact allergens was investigated also. Groups of BALB/c mice (n=10) received by intraperitoneal injection (100μl) anti-α4 antibody, anti-α6 antibody or an isotype-matched control antibody. Two hours later mice were exposed topically on the dorsum of both ears to 25μl of 2% oxazolone in 4:1 acetone:olive oil. Draining auricular lymph nodes were excised 18 hours later, a single cell suspension prepared and DC enriched and enumerated as described previously.[10] It was found that pretreatment with anti-α6 antibody resulted in a dose-dependent inhibition of oxazolone-induced DC accumulation in lymph nodes. The same treatment with anti-α4 antibody was without effect (data not presented). In parallel experiments the influence of the same antibodies on the migration of LC from mouse skin explants was measured *in vitro*. When explants were floated on medium containing anti-α6 antibody there was a significant reduction in the number of cells migrating from explants at 70 hours compared with cultures containing no antibody or an isotype-matched control antibody. Consistent with the results of DC accumulation *in vivo*, anti-α4 antibody was without influence on the loss of LC from the skin explants (data not presented).

It has been demonstrated previously that intradermal injection of mice with homologous recombinant tumour necrosis factor α (TNF-α) stimulates the migration of LC away from the epidermis and the accumulation of DC in draining nodes.[10,11] In a final series of experiments the influence of anti-α6 antibody on the stimulation of LC migration by TNF-α was examined. Groups of BALB/c mice received anti-α6 or isotype control (IgG2a) antibody by intraperitoneal injection as described above. Two hours later mice were injected intradermally into both ear pinnae with 50ng of TNF-α (specific activity 2×10^8U/mg). Ears were removed 30 minutes following administration of TNF-α. Epidermal sheets were prepared and the frequency of MHC class II (Ia) antigen-positive LC measured by indirect immunofluorescence as described previously.[11] The results of a representative experiment are illustrated in Figure 1.

Consistent with previous observations[11] it was found that dermal exposure of mice to homologous TNF-α induced a rapid reduction in the frequency of Ia^+ epidermal LC. Prior exposure of mice to the isotype control (IgG2a) antibody was without effect. In contrast, systemic exposure of mice to anti-α6 antibody 2 hours prior to administration of TNF-α resulted in the failure of this cytokine to stimulate LC migration from the skin (Figure 1).

3. DISCUSSION

The data presented here demonstrate that the induced migration of LC from the epidermis and the subsequent accumulation of DC in draining lymph nodes is dependent

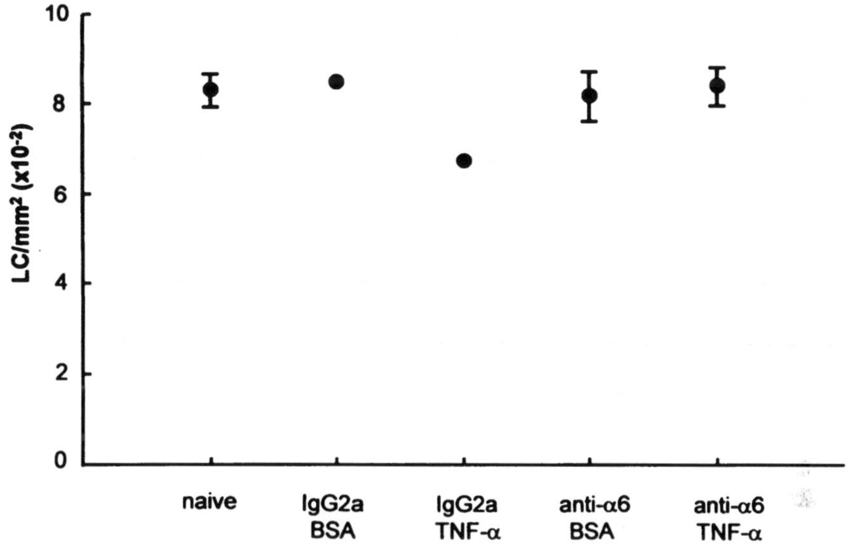

Figure 1. Influence of anti-α6 antibody on epidermal LC migration induced by TNF-α. Groups of mice (n=3) received, by intraperitoneal injection, 100μl of anti-α6 or isotype-matched control (IgG2a) antibodies. Two hours later mice were treated, by intradermal injection (30μl) into both ear pinnae, with 50ng of homologous TNF-α in 0.1% bovine serum albumin (BSA) or with BSA alone. Ears were removed 30 minutes later. Epidermal sheets were prepared and the frequency of Ia^+ LC measured by indirect immunofluorescence. Results are recorded as mean LC/mm^2 ± SE from the examination of 10 fields/sample for each of 4 samples. SE values are shown only when they exceed 0.2 ($x10^{-2}$) LC/mm^2.

upon α6 integrin expression. Neutralizing anti-α6 antibody was found to inhibit both the stimulation of LC migration induced by TNF-α and the accumulation of DC in draining lymph nodes that normally characterizes skin sensitization. The conclusion drawn is that α6 integrins are required for the passage of LC through the skin to afferent lymphatics via the basement membrane. The expression by LC of α6 is consistent with this.

It is known that LC express β1 integrins also.[8] VLA-6 comprises an α6β1 dimer that confers specificity for laminin and permits the association of cells with the basement membrane. The latter may represent a mandatory first step in the movement of LC across the basement membrane during their journey to draining lymph nodes.

Although these data reveal an important role for α6 integrin in the stimulation of LC migration and DC accumulation, it is important to emphasize that other adhesion molecules may also facilitate or be required for effective trafficking of LC.

4. REFERENCES

1. I. Kimber and M. Cumberbatch, Dendritic cells and cutaneous immune responses to chemical allergens. Toxicol. Appl. Pharmacol. 117:137 (1992).
2. I. Kimber and M. Cumberbatch, Langerhans cell migration: initiation and regulation. In: The Immune Functions of Epidermal Langerhans Cells. H. Moll (ed.), R.G. Landes Co., Austin (1994) pp103.
3. M. Cumberbatch, S.W. Peters, S.J. Gould and I. Kimber, Intercellular adhesion molecule-1 (ICAM-1) expression by lymph node dendritic cells: comparison with epidermal Langerhans cells. Immunol. Lett. 32:105 (1992).

4. J. Ma, J-H. Wang, Y-J. Guo, M-S. Sy and M. Bigby, In vivo treatment with anti-ICAM-1 and anti-LFA-1 antibodies inhibits contact sensitization-induced migration of epidermal Langerhans cells to regional lymph nodes. Cell. Immunol. 158:389 (1994).
5. A. Tang, M. Amagai, L.G. Granger, J.R. Stanley and M.C. Udey, Adhesion of epidermal Langerhans cells to keratinocytes mediated by E-cadherin. Nature 361:82 (1993).
6. A. Blauvelt, S.I. Katz and M.C. Udey, Human Langerhans cells express E-cadherin. J. Invest. Dermatol. 104:293 (1995).
7. T.A. Borkowski, B.J. Van Dyke, K. Schwarzenberger, V.W. McFarland, A.G. Farr and M.C. Udey, Expression of E-cadherin by murine dendritic cells: E-cadherin as a dendritic cell differentiation antigen characteristic of epidermal Langerhans cells and related cells. Eur. J. Immunol. 24:2767 (1994).
8. B. Le Varlet, M. J. Staquet, C. Dezutter-Dambuyant, P Delorme and D. Schmitt. In vitro adhesion of human epidermal Langerhans cells to laminin and fibronectin occurs through β1 integrin receptors. J. Leuk. Biol. 51:415 (1992).
9. S. Aiba, S. Nakagawa, H. Ozawa, K. Miyake, H. Yagita and H. Tagami, Up-regulation of α4 integrin on activated Langerhans cells: analysis of adhesion molecules on Langerhans cells relating to their migration from skin to draining lymph nodes. J. Invest. Dermatol. 100:143 (1993).
10. M. Cumberbatch and I. Kimber, Dermal tumour necrosis factor-α induces dendritic cell migration to draining lymph nodes and possibly provides one stimulus for Langerhans cell migration. Immunology 75:257 (1992).
11. M. Cumberbatch, I. Fielding and I. Kimber, Modulation of epidermal Langerhans cell frequency by tumour necrosis factor-α. Immunology 81:395 (1994).

23

FUNCTIONAL EXPRESSION AND MODULATION OF C5a RECEPTOR (CD88) ON SKIN DENDRITIC CELLS

A. Morelli, A. Larregina, E. Chuluyán, E. Kolkowski, and L. Fainboim

Laboratory of Immunogenetics
"Hospital de Clínicas"
School of Medicine
University of Buenos Aires (UBA)
Buenos Aires, Argentina

1. INTRODUCTION

The process of differentiation of resident epidermal Langerhans cells (LC) generates: i) changes in the level of expression and repertoire of cell surface molecules; ii) a decrease in the antigen processing capacity; iii) a significant increase in the ability to present peptides to T cells; and iv) the mobilisation of dendritic cells (DC) from the epidermis towards the draining lymph nodes[1]. There is little information about the factors involved in DC migration from non-lymphoid peripheral tissues as part of the process of skin DC differentiation. "In vivo" studies demonstrated that the administration of lipopolysaccharide (LPS), Tumour Necrosis Factor (TNF) α or Interleukin 1 (IL-1) can trigger the trafficking of DC from non-lymphoid tissues (i.e. skin, lung, gut, heart and kidney) towards secondary lymphoid organs[2–7]. A similar situation is observed in the inflammatory reaction that takes place during the first three days after skin transplantation; or after short term culture of skin organ explants[8]. In both cases it is likely that the skin injury by itself releases inflammatory mediators, sufficient to trigger both phenotypic and functional DC changes and migration of completely or partially mature skin DC.

The molecule C5a, produced as result of proteolitic cleavage of the complement factor C5 during inflammation, should be included in the list of putative mediators capable of inducing DC migration from non-lymphoid organs during the early phase of the inflammatory response. C5a has been reported as a chemoattractant for other myeloid cells such as neutrophils, eosinophils, basophils, monocytes and more recently for DC generated "in vitro" after treatment of peripheral blood mononuclear cells with Granulocyte macrophage colony stimulating factor (GM-CSF) plus IL-4[9,10]. These biological effects are elicited via binding to a C5a specific receptor (C5aR, clustered as CD88) which belongs to the family of seven transmembrane domain receptors that transduces a signal via guanosine triphosphate (GTP)-binding regulatory proteins[11].

Dendritic Cells in Fundamental and Clinical Immunology
edited by Ricciardi-Castagnoli, Plenum Press, New York, 1997

In this chapter, we will outline the studies performed by our laboratory concerning the expression of C5aR on skin DC; its modulation according to the state of cell differentiation; and its role as a chemotactic factor for fully differentiated skin DC[12].

2. RESULTS

2.1. Expression of C5aR on a of Epidermal DC

Epidermal cell suspensions prepared by tripsinization of split-thickness skin specimens and enriched in LC after Ficoll-Hypaque gradient ($\delta = 1.077$ g/cm^3) were analysed for the expression of C5aR by double labelling immunofluorescence flow cytometry. LC were identified by their positivity for CD1a. The percentage of CD1a positive cells was further enriched from 6–9% to 100% by electronic gating. A small subset (< 6%) of the epidermal population of CD1a positive cells were also positive for C5aR. Similar results were obtained using either the mAb S5/1 or the mAb W17/1, both specific for the extracellular domain of the C5aR[13]. Double immunostaining demonstrated that the subpopulation of LC positive for C5aR expressed higher levels of HLA-DR than the subpopulation of LC C5aR negative. This observation suggests that the stage of maturation of resident LC might be associated with the expression of receptors for chemotactic factors.

Double immunostaining of epidermal sheets microdissected from split-thickness skin after EDTA incubation (EDTA 20mM, pH 7.2–7.4, 2h, 37°C) confirmed the expression of C5aR on a low number of resident LC (< 3 cells/mm^2 of epidermis). Cells that carried C5aR, strongly expressed HLA-DR, were round-shaped and sometimes exhibited a cytoplasm tail. They were located at the same plane of focus as the basal and parabolas kenatinocytes, next to the basement membrane.

2.2. Epidermal LC Up-Regulate C5aR after Maturation "in Vitro"

To confirm the relationship between the stage of epidermal maturation of LC and the expression of C5aR, epidermal LC were induced to differentiate "in vitro" after 72 h culture in bulk epidermal cell suspensions supplemented with hrGM-CSF. In vitro differentiation of LC was corroborated by the increase in expression of MHC class I and II molecules; of the adhesion molecules ICAM-1 (CD54) and LFA3 (CD58); and the co-stimulatory signals B7.1 (CD80), B7.2 (CD86) and CD40. On the other hand, the expression of CD1a and CD32 antigens decreased their intensity but were still present. As a consequence of "in vitro" differentiation, LC up-regulated their expression of C5aR from the original 5–6% to approximately 70% of the whole population of DC.

Skin migratory DC obtained after culturing skin organ explants for 36–48 h and enriched up to 80–90% by Universal Separation Medium (Sigma) gradient (d=1.067 g/cm^3) were tested for their expression of C5aR. The dendritic lineage of migratory DC was confirmed by: (i) the high levels of MHC Class I and II molecules; (ii) the expression of CD1a and CD1c; (iii) the positivity for the myeloid markers CD13 and CD33 and the absence of the monocyte/macrophage markers CD14 and CD68, and the granulocyte markers CDw65, CD66acde and CD66b; and (iv) by ultrastructure. Based on the CD1a and C5aR positivity two subpopulations of skin migratory DC were distinguished: (i) a predominant population composed of CD1a^{low} C5aR positive DC; and (ii) a minority subset that comprised CD1a^{high} C5aR negative DC. A culture of dermal explants devoid of epidermis demonstrated that the CD1a^{low} C5aR positive DC subset was dermis derived. In contrast, the CD1a^{high} C5aR negative DC were only observed after culturing epidermal sheets (devoid of dermis) suggesting as putative origin of theses cells the epidermis, skin appendages or hair follicles.

2.3. Functional Studies of Chemotaxis Using Recombinant C5a and Skin Migratory DC

Chemotaxis of skin migratory DC was assayed "in vitro" by the micropore filter technique. Briefly, 8×10^4 migratory DC in 0.5 ml of RPMI 1640, 10 mM Hepes supplemented with 0.5% human serum albumin (HSA) were seeded in the upper compartment of the chemotactic chambers (Nunc, Denmark) placed in a 24 wells culture plate (Corning, USA). The upper and lower compartments were separated by a polycarbonate filter with a 8 μm pore size. The lower compartment was filled with 0.5 ml control medium (RPMI 1640, 10 mM Hepes supplemented with 0.5% HSA) or with different concentrations of recombinant human complement C5a (rC5a) (Sigma) diluted in the same medium. Chambers were incubated for 2 h at 37 °C in a 5% CO_2 humid atmosphere. At the end of the incubation, the upper chamber with the filter was removed and the number of DC present in the lower compartment was assessed by flow cytometry. DC which migrated to the lower compartment of chemotaxis chamber were collected, analysed and counted by flow cytometry. Optimum migration to rC5a was found at 10–8 M with a sigmoidal dose response curve, 2 hr after placing migratory DC in the upper chamber (fig 1).

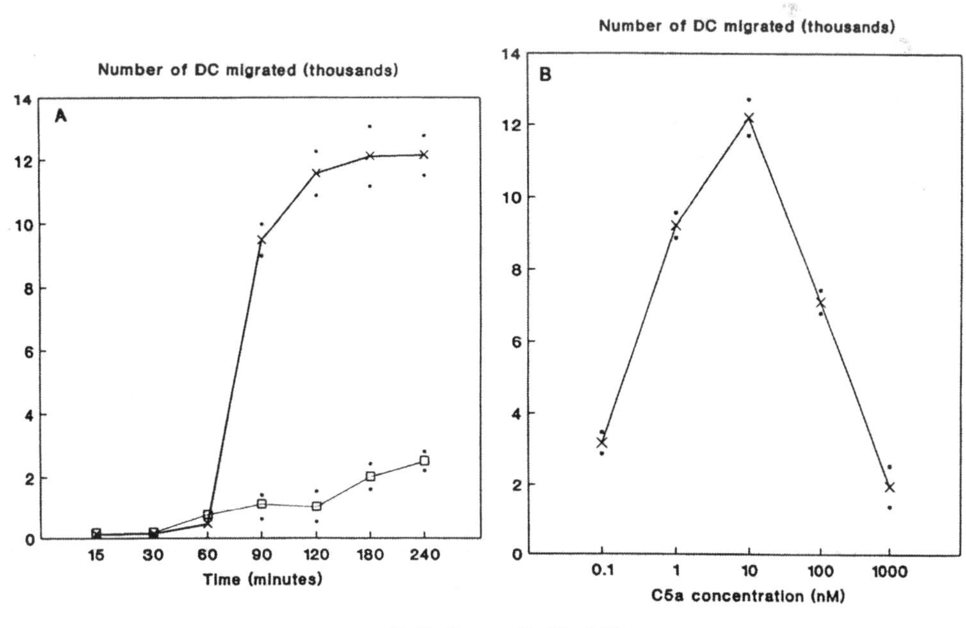

Figure 1. (A) Time course of migratory response of skin DC to rC5a. Migration of 8×10^4 DC seeded in the upper compartment of the chemotaxis chamber was assessed at different time intervals after exposure to a rC5a gradient (10^{-9} M) or diluent. (B) Dose-response curve of DC in response to different concentration of rC5a after 120 min of culture (Immunology, 89:126, 1996).

Table 1. Checkboard analysis to demonstrate the unidirectional migration of DC in response to rC5a (Immunology, 89:126, 1996)

Dose in lower compartment	Dose in upper compartment		
	0	10^{-9} M	10^{-8} M
0	1,485 ± 138	822 ± 12	1,038 ± 70
10^{-9} M	5,840 ± 215	1,654 ± 72	1,474 ± 145
10^{-8} M	9,355 ± 313	5,960 ± 267	1,703 ± 132

By means of a limited series of checkboard analysis, originally described by Zigmond and Hirsch, we investigate whether locomotion of migratory DC through the polycarbonate filter depended on the presence of a gradient of C5a[14]. Checkboard assay distinguishes chemotaxis (directional cell locomotion requiring a chemoattractant gradient) and chemokinesis (undirectional cell migration in the absence of a gradient) by varying the concentrations of chemoattractant above and below the filter in the migration chamber. Results obtained by means of checkboard analysis are shown in Table 1. The values presented along the diagonal, where there is no gradient, show only a modest augmentation of DC migration (chemokinetic effect) as compared to a baseline control, although it failed to reach the $p < 0.05$ level of significance. DC had a significant migratory response ($p < 0.01$) when the concentration of rC5a was higher in the lower compartment (positive gradient), demonstrating that DC migration was caused by chemotactic effect of rC5a.

3. CONCLUSIONS

Normal epidermis contains a minority of resident LC (5–15% of epidermal LC) that shows, as compared with the remaining LC, the following differences: (i) a higher level of HLA-DR and CD11c; (ii) positivity for CD40; (iii) a lower level of Fc gamma-RII (CD32); iv) a location in a lower position in the epidermis close to the basement membrane; (vi) a very strong immunostimulatory capacity in mixed leukocyte reactions; and vii) a higher level of cytosolic calcium. These features have led to the hypothesis that such a subset is composed of resident LC that have differentiated "in situ" and are about to leave the epithelium[15]. By flow cytometry and immunohistochemistry we were able to demonstrate that the subpopulation of epidermal LC positive for C5aR coexpress higher levels of HLA-DR than C5aR negative LC, and therefore, it can be included as part of the subset of mature epidermal rLC. The localisation of these C5aR positive cells in proximity to the epidermal/dermal basement membrane, and their round-shape morphology (sometimes with a small cytoplasm tail) support the hypothesis that they are migratory cell on their way out of the epidermis. The increase in the percentage of cells expressing C5aR along the differentiation pathway of DC was confirmed by: i) the up-regulation of C5aR after "in vitro" maturation of epidermal LC induced by GM-CSF; and ii) by the expression of C5aR by skin migratory DC and by Interdigitating DC (IDC) present in T cell areas of secondary lymphoid organs (data not shown).

Skin DC migrate unidirectionally in response to rC5a in a sigmoidal dose response curve, and as it was previously demonstrated for other cells, supersaturating concentrations of C5a exert an inhibitory effect on DC chemotaxis.

The expression of C5aR on DC as part of their process of cell differentiation represents one of the links between innate and adaptative immunity. Immature epidermal LC require the presence of the cytokines IL-1, TNF-alpha and GM-CSF to differentiate into fully mature antigen presenting cells. Epidermal LC express receptors for IL-1 and TNF. After generation of IL-1, TNF-alpha, and C5a during the early steps of an inflammatory response, resident LC might differentiate acquiring C5aR as part of this process. The up-regulation of the C5aR on LC might be one of the components that trigger LC migration and the subsequent adaptative immunity during an inflammatory process as recently postulated by Katz and co-workers[16].

ACKNOWLEDGMENTS

This work was supported in part by grants from the University of Buenos Aires (UBA), CONICET, and "Fundación Alberto J Roemmers", Argentina. We want to thank Dr Guillermo Vazquez for the skin specimens.

REFERENCES

1. Bos J. & Kapsemberg M. (1993) The skin immune system: progress in cutaneous biology. Immunol Today 14, 75.
2. MacPherson G.G. (1989) Properties of lymph-borne (veiled) dendritic cells in culture. I. Modulation of phenotype, survival and function: partial dependence on GM-CSF. Immunology 68, 102.
3. McWilliam A.S., Nelson D., Thomas J.A. & Holt P.G. (1994) Rapid dendritic cell recruitment is a hallmark of the acute inflammatory response at mucosal surfaces. J Exp Med 179, 1331.
4. MacPherson G.G., Jenkins C.D., Stein M.J. & Edwards C. (1995) Endotoxin mediated dendritic cell release from the intestine. Characterization of released dendritic cells and TNF dependence. J Immunol 154,1317.
5. Roake J.A., Rao A.S., Morris P.J., Larsen C.P., Hankins D.F., & Austyn J.M. (1995) Dendritic cell loss from nonlymphoid tissues after systemic administration of lipopolysaccharide, Tumor Necrosis Factor, and Interleukin 1. J Exp Med 181, 2237.
6. Lundqvist E. & Back O. (1990) Interleukin-1 decreases the number of Ia+ epidermal dendritic cells but increases their expression of Ia antigen. Acta Derm Venereol (Stockh) 70, 391.
7. Cumberbach M. & Kimber I. (1995) Tumour necrosis factor alpha is required for accumulation of dendritic cells in draining lymph nodes and for optimal contact sensitization. Immunology 84, 31.
8. Larsen C.P., Steinman R.M., Witmer-Pack M., Hankins D.F., Morris P.J. & Austyn J.M., (1990) Migration and maturation of Langerhans cells in skin transplants and explants. J Exp Med 172, 1483.
9. Frank M.M. & Fries L.F. (1991) The role of complement in inflammation and phagocytosis. Immunol Today 12, 322.
10. Sozzani S., Sallustro F., Luini W., Zhou D., Piemonti L., Allavena P., Van Damme J., Valitutti S., Lanzavecchia A. & Mantovani, A. (1995) Migration of dendritic cells in response to formyl peptides, C5a, and a distinct set of chemokines. J Immunol 155, 3292.
11. Gerard N.P. & Gerard C. (1991) The chemotactic receptor for human C5a anaphylatoxin. Nature 349, 614.

12. Morelli A., Larregina A., Chuluyan E., Kolkowski E., & L. Fainboim. (1996) Expression and modulation of C5a receptor (CD88) on skin dendritic cells. Chemmotactic effect of C5a on skin migratory dendritic cells. Immunology 89,126.
13. Oppermann, M. (1995) CD88 (C5a receptor) cluster workshop report. In: Leucocyte Typing V. White cell differentiation antigen (ed. S.F Schlossman, et al.), Vol I, p. 953. Oxford University Press Inc, New York.
14. Zigmond S.H. & Hirsch J.G. (1973) Leukocyte locomotion and chemotaxis. New methods for evaluation, and demonstration of a cell-derived chemotactic factor. J Exp Med 137, 387.
15. Shibaki A., Meunier L., Ra C., Shimada S., Ohkawara A. & Cooper K.D. (1995) Differential responsiveness of Langerhans cell subsets of varying phenotypic states in normal human epidermis. J Invest Dermatol 104, 42.
16. Ibrahim M.A., Chain B.M. & Katz D.R. (1995) The injured cell: the role of the dendritic cell system as a sentinel receptor pathway. Immunol Today 16,181.

24

UNILINEAGE DENDRITIC CELL CULTURES GENERATED BY PURIFIED HUMAN HEMATOPOIETIC PROGENITOR CELLS

E. Montesoro, U. Testa, M. Gabbianelli, and C. Peschle

Department of Hematology-Oncology
Istituto Superiore di Sanità
Rome, Italy
Thomas Jefferson University
Philadelphia, Pennsylvania

1. INTRODUCTION

It has been previously reported that normal human hematopoietic progenitor cells (HPCs) can be channelled into various differentiation pathways depending on exposure to appropriate growth factor (GF) stimuli.

Unilineage differentiation of HPCs has been hampered by two main factors: 1) the lack of purity of the starting population and 2) the lack of optimal combination of GFs selective for a specific lineage.

We have described a methodology to obtain highly purified (≥90%) human HPCs from normal human peripheral blood (PB)[1], and subsequently identified the culture conditions allowing the selective differentiation into a single hematopoietic lineage ("unilineage HPC cultures")[2–6]. In this study, our initial purpose was to identify the GFs necessary to support cultures highly enriched in dendritic cells (DCs) from human PB HPCs.The optimal cytokine combination was composed of GM-CSF, c-kit ligand (KL), flt3 receptor ligand (FL), IL-4, IL-6 and IL-7. Cultures of purified PB HPCs in the presence of these cytokines give rise to a cell progeny composed by 90% DCs with an overall expansion of about 50 fold of the initial cell number.

Clonogenetic assays confirmed the selective effect of this GF cocktail on DC HPCs (CFU-DC): 90% of the colonies developed in semisolid cultures were composed of DCs. The remaining were very small granulocyte colonies (<100 cells / colony).

We then tested the clonogenetic capacity of HPCs seeded in single cell liquid suspension cultures supplemented as bulk cultures. 80% of the resulting colonies were constituted exclusively by DCs, thus confirming the above data.

Dendritic Cells in Fundamental and Clinical Immunology
edited by Ricciardi-Castagnoli, Plenum Press, New York, 1997

2. MATERIALS AND METHODS

HPCs were purified from normal human PB according to a three-step procedure[3–6]:

1. separation over a Ficoll-Hipaque density gradient followed by three cycles of plastic adherence;
2. centrifugation on a discontinuous Percoll four-step gradient (d, 1.052, 1.056, 1.060, 1.065);
3. incubation of the two lowest density cell fractions with a panel of murine monoclonal antibodies directed against differentiated cells (B, T and NK lymphocytes, granulocytes, monocytes) and then with immunomagnetic beads coated with goat anti-mouse Ig, followed by separation with a magnet. Recovered cells were ≥90% purified $CD34^+$ clonogenetic HPCs.

Morphological characterization was performed on cells cytocentrifuged onto glass slides and stained with May-Grünwald Giemsa.

Monoclonal antibodies conjugated with fluorescein isothiocyanate or phycoerythrin were used for double-fluorescence analysis of membrane phenotype by FACScan.

PB HPCs were cultured in IMDM containing 30% heat-inactivated FCS and different combinations of growth factors, i.e., GM-CSF, KL, FL, IL-4, IL-6, IL-7. Cells were incubated in a fully humidified 5% $C0_2$, 5% O_2, 90% N_2 atmosphere. The initial density was 5×10^4 cells/ml. Every two days the cell concentration was adjusted at 1×10^5 cells/ml and fresh HGFs were added.

Clonogenetic efficiency of purified HPCs was assessed by seeding 500 cells/ml/dish in IMDM containing 0.9% methylcellulose, 30% FCS and the GFs listed above.

Single cell liquid suspension cultures were performed seeding one HPC/flat-bottomed 96 microwell plate containing IMDM supplemented as for bulk liquid cultures.

The capacity of DCs to induce proliferation of resting allogeneic T cells was assessed by seeding 10^5 T cells/well of microtest tissue-culture plates with graded numbers of DCs (50, 200, 500, 1000). After 6 days of culture in IMDM with 10% human AB serum, cells were pulsed with 1 μCi ^{3}H-thymidine overnight, collected and analyzed for radioactivity.

3. RESULTS

In a first series of experiments we evaluated the effect of different HGF combinations on HPC proliferation and DC differentiation.

Cells were grown in liquid cultures containing KL, FL, IL-4, +/- GM-CSF, +/- IL-7, +/- IL-6.

The combination of these GFs resulted in optimal cell number amplification (30 - 50 times) coupled with the most consistent differentiation to DCs (70 - 90%). In particular, IL-7 was not necessary for proliferation, but its presence increased the percentage of DC from 76% to 90% (mean values).

We also evaluated the effect of different concentrations of GM-CSF, KL and FL. KL concentrations ranging between 0.1 ng/ml and 1 ng/ml did not influence proliferation or DC differentiation. Reduced FL concentrations from 1 ng/ml to 0.1 ng/ml caused a lower amplification of cell number (45 vs 31 times) and a reduced DC yield (90% vs 69%). These two parameters were also influenced by the concentration of GM-CSF, which was more effective at very low dosages (0.001 - 0.0001 ng/ml).

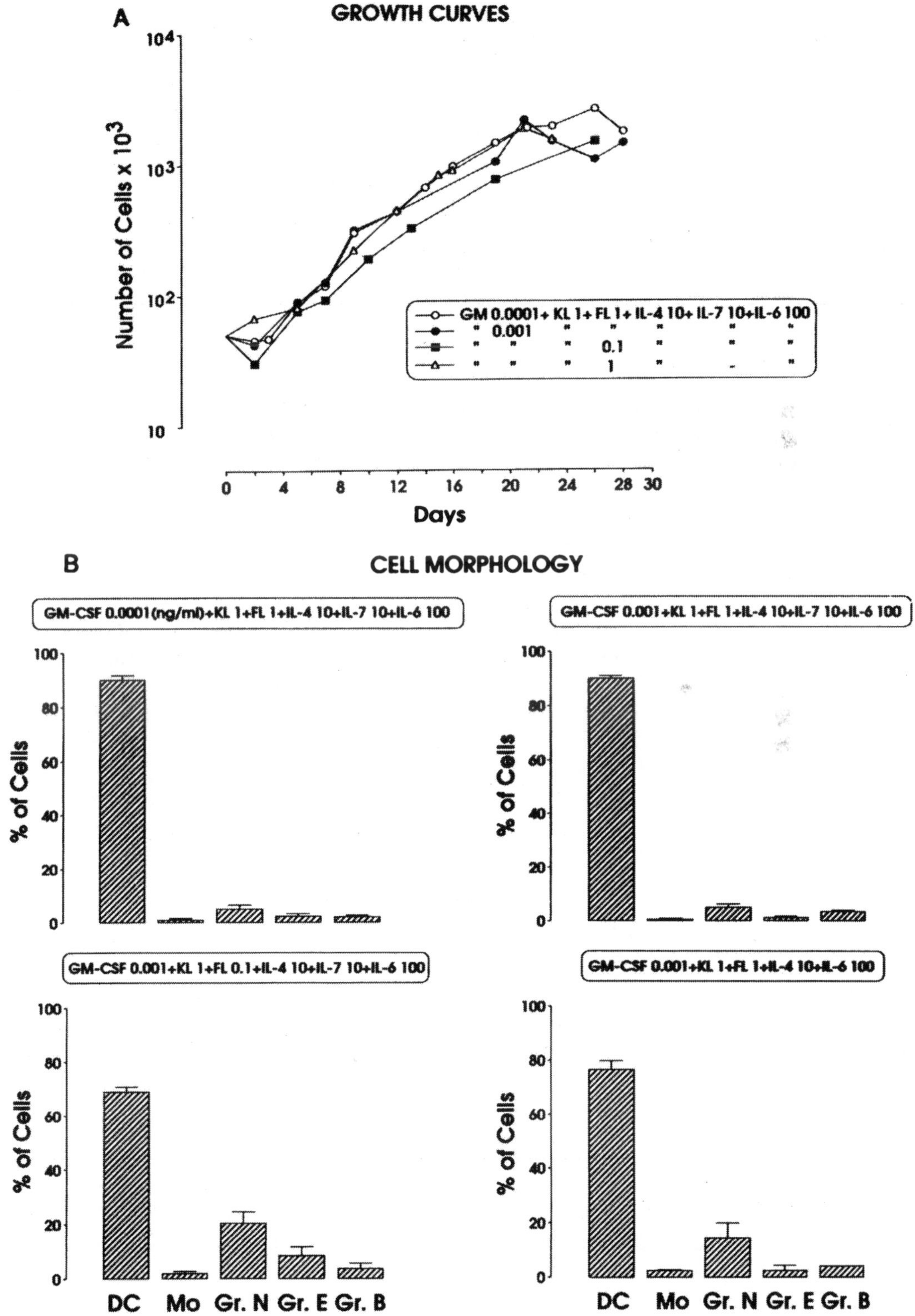

Figure 1. Effects of different GF combinations on (A) HPC growth and (B) differentiation.

On the basis of these observations we have identified an optimal cocktail to sustain the growth of DCs: GM-CSF 0.0001 ng/ml, KL 1 ng/ml, FL 1 ng/ml, IL-4 10 ng/ml, IL-7 10 ng/ml, IL-6 100 ng/ml. Morphological and phenotypical analyses showed that these culture conditions gave rise to a cell population which was amplified more than 50 times at day 26 and highly enriched (90%, mean value) in DCs.

Starting from day 7 of culture, cells with typical dendritic features were scored , although about 50% were not completely differentiated. After 14 days of culture all DCs showed characteristic mature morphology with contaminating cells being predominantly granulocytes (≤ 9%) and a very low number of monocytes/ macrophages (<1%).

Membrane phenotype analysis was consistent with the morphological data: specific dendritic surface markers appeared after 5–7 days of culture. In particular at day 7, 10% of the cells were $CD1a^+$, $CD80^+$, $CD86^+$ and increased up to 60% at the end of the culture. Conversely, CD34 antigen, present on virtually all cells (>90%) at the beginning of the culture, disappeared rapidly, being markedly reduced at day 7 (27%) and completely absent at day 14, a finding in line with a gradual differentiation of PB progenitors into DCs. HLA-DR, initially expressed on the large majority of the cells (90%), slightly decreased in the early days of culture but returned to original levels at the end of the culture, when the cells were also highly positive for CD13 and CD33 markers. $CD14^+$ cells remained less than 5% during the entire culture period (more than three weeks).

To further assess the capacity of HPCs to differentiate towards the dendritic lineage we seeded 500 cells/ml in semisolid medium containing the same GF cocktail used for bulk liquid cultures. In these conditions HPCs gave rise almost exclusively to DC colonies generated by CFU-DC (cloning efficiency ≥ 3%) with a negligible presence of G colonies (0.3%) which probably derived from late HPCs/early precursors, as suggested by their very small size (<100 cells/colony). A more accurate analysis of CFU-DC colonies revealed that their average size was about 500 cells/colony and that they were composed exclusively by dendritic cells, as assessed by scoring of cytospin preparations derived from single colonies stained with May - Grünwald Giemsa.

In parallel, we have developed liquid suspension cultures derived from single HPCs plated at 0.5 cell/well and grown as bulk cultures. Using this assay we showed that 3.1% of HPCs formed colonies entirely composed by DCs.

Finally, DCs cultivated for 14 days in liquid suspension were tested for their capacity to induce proliferation of resting allogeneic T lymphocytes. A significant stimulation (9 fold over the background) was observed at a DC/T cell ratio of 1/2000 and increased in parallel with DC concentration reaching, at the highest DC/T cell ratio tested (1/100), the level of 35 times the background.

4. DISCUSSION

We describe a liquid culture system which supports selective HPC differentiation towards the dendritic lineage. During the first week of culture HPCs differentiate into dendritic precursors as shown by the gradual decrease of $CD34^+$ cells coupled with the appearance of DC surface markers (CD1a, CD80, CD86)[7–10]. At day 14 $CD34^+$ HPCs are no longer detectable and the percentage of cells with dendritic phenotype continues to increase reaching a considerable level (up to 90%) at the end of the culture. The almost complete absence of CD14 marker is supported by morphological analysis, which revealed very low monocytic contamination[11,12], and only a minority of granulocytes. This finding may be explained by the blocking effect exerted by IL-4 on monocyte development:[13] this cytokine may also play an important role in promoting growth and differentiation of CFU-DC[14]. These observations were further supported by clonogenetic assays showing that purified HPCs grown in the presence of the optimal GF cocktail generate 90% "pure" CFU-DC colonies with the remaining 10% being small CFU-G colonies, and by liquid suspension cultures originated from single HPCs, where dendritic and granulocytic colonies represent 80% and 20% respectively of the cell progeny. Both culture systems indicate that CFU-DC represent ~ 3% of our purified HPCs and give rise to colonies of ~ 500 cells.

To gain more insight into the development of the dendritic lineage we will perform cell sibling analysis of HPC single cell cultures: this may allow to evaluate the very early phases of DC differentiation to determine whether DCs represent an independent lineage or are related to other series (particularly monocytes).[15–18]

5. REFERENCES

1. Gabbianelli M., Sargiacomo M., Pelosi E., Testa U., Isacchi G., Peschle C. Science 1990, 249: 1561.
2. Sposi N.M., Zou L.I., Carè A., Valtieri M., Testa U., Gabbianelli M., Mariani G., Bottero L., Mather C., Orkin S.H., Peschle C. Proc. Natl. Acad. Sci. USA 1992, 89: 6353.
3. Condorelli G.L., Testa U., Valtieri M., Vitelli L., De Luca A., Barberi T., Montesoro E., Campisi S., Giordano A., Peschle C. Proc. Natl. Acad. Sci. USA 1995, 92: 4808.
4. Labbaye C., Valtieri M., Barberi T., Meccia E., Masella B., Pelosi E., Condorelli G.L., Testa U., Peschle C. J. Clin. Invest. 1995, 95: 2346.
5. Gabbianelli M., Pelosi E., Montesoro E., Valtieri M., Luchetti L., Samoggia P., Vitelli L., Barberi T., Testa U., Lyman S., Peschle C. Blood 1995, 86: 1661.
6. Guerriero R., Testa U., Gabbianelli M., Mattia G., Montesoro E., Macioce G., Pace A., Ziegler B., Hassan H.J., Peschle C. Blood 1995, 86: 3725.
7. Steinman R.M. Annu. Rev. Immunol. 1991, 9: 271.
8. Caux C., Liu Y-J., Banchereau J. Immunol. Today 1995, 16: 2.
9. Young J.W., Koulova L., Soergel A., Clark E., Steinman R.M., Dupont B. J. Clin. Invest. 1992, 90: 229.
10. Caux C., Vanbervliet B., Massacrier C., Azuma M., Okumura K., Lewis L., Lanier L.L., Banchereau J. J. Exp. Med. 1994, 180: 1841.
11. Goyert S.M., Ferrero E., Rettig W.J., Yenamandra A.K., Obata F., LeBeau M.M. Science 1988, 239: 497.
12. Ziegler-Heitbrock H.W.L., Ulevitch R.J. Immunol. Today 1993, 14: 121.
13. Jansen J.H., Wientjens G.-J.H.M., Fibbe W.E., Willemze R., Kluin-Nelemans H.C. J. Exp. Med. 1989, 170: 577.
14. Sallusto F., Lanzavecchia A. J. Exp. Med. 1994, 179: 1109.
15. Inaba K., Inaba M., Deguchi M., Hagi K., Yasumizu R., Ikemara S., Muramatsu S., Steinman R.M. Proc. Natl. Acad. Sci. USA 1993, 90: 3038.
16. Szabolcs P., Moore M.A.S., Young J.W. J. Immunol. 1995, 154: 5851.
17. Res P., Martinez-Càceres E., Jaleco A.C., Staal F., Noteboom E., Weijer K., Spits H. Blood 1996, 87: 5196.
18. Szabolcs P., Avigan D., Gezelter S., Ciocon D.H., Moore M.A.S., Steinman R.M., Young J.W. Blood 1996, 87: 4520.

25

CD14 IS EXPRESSED BY SUBSETS OF MURINE DENDRITIC CELLS AND UPREGULATED BY LIPOPOLYSACCHARIDE

Karsten Mahnke,[1] Eva Becher,[1] Paola Ricciardi-Castagnoli,[2] Thomas A. Luger,[1] Thomas Schwarz,[1] and Stephan Grabbe[1*]

[1]Ludwig Boltzmann Institute for Cell Biology and Immunobiology of the Skin
Department of Dermatology
University of Münster, Germany
[2]CNR Center of Cytopharmacology
University of Milan
Milan, Italy

SUMMARY

The CD14 surface molecule is predominantly expressed by cells of myeloid origin and regarded as a specific marker for macrophages (Mϕ). Thus, in human mononuclear cell preparations, CD14 expression is a widely used parameter to distinguish Mϕ from dendritic cells (DC). Since a murine homologue of CD14 was recently identified, this study investigated expression of CD14 by murine Mϕ and DC. Flow cytometry with a monoclonal antibody directed against murine CD14 revealed that bone marrow-derived DC express CD14 to various extents during differentiation. Functionally, $CD14^{high}$ and $CD14^{low}$ DC did not differ significantly in their capacity to present alloantigen, protein antigen or immunogenic peptide. Furthermore, surface expression of CD14 could be modulated by interleukin (IL)-4 and LPS. Incubation of bone marrow-derived DC with IL-4 (100 U/ml) resulted in downregulation of CD14 surface expression, whereas exposure of BmDC to LPS (1 μg/ml) led to upregulation of CD14. After blockage of CD14 molecules by incubation of DC with anti-CD14 antibodies, downregulation of LPS triggered IL-1 release could be detected. In addition, other Mϕ markers such as CD11b, F4/80, BM8, and ER-TR9, are also expressed on DC. Therefore, we conclude that CD14, like other Mϕ markers, is expressed on murine DC during maturation. Thus, Mϕ and DC cannot be dis-

* Correspondence address: Dr. Stephan Grabbe, Department of Dermatology, University of Münster, Von-Esmarch-Str. 56, D-48149 Münster, Germany. Phone: +49–251–83 56582; Fax: +49–251–83 58579; e-mail: grabbe@uni-muenster.de.

tinguished by flow cytometry using these markers. Moreover, CD14 may be involved in mediating LPS-induced activation of murine DC.

1. INTRODUCTION

The CD 14 molecule is a GPI-anchored surface molecule, whose expression is restricted to the myeloid cell lineage [1]. Although granulocytes also express low levels of CD14 [2–4], it is widely used as a specific marker for human monocytes (MO**)/macrophages (Mϕ). CD14 may function as a receptor for lipopolysaccharide (LPS), since binding of LPS to CD14 results in release of proinflammatory cytokines such as tumor-necrosis-factor (TNF)-α and interleukin (IL)-1 by human MO. Moreover, in inflammatory situations where high amounts of LPS are present, CD14 molecules were shown to scavenge LPS complexes, contributing to elimination of LPS from the blood [4–6]. In addition to these findings in the human system, a murine protein with high homology to human CD14 could recently be identified [7]. It is expressed by different murine cell lines of myelo-monocytic origin and upregulated by LPS [6], suggesting that murine CD14 may also serve as a receptor for LPS.

Dendritic cells are highly specialized antigen-presenting cells (APC) which arise from a specific progenitor or differentiate from monocytes. They are characterized by their dendritic morphology as well as by a number of functional parameters, one of which is their uniquely potent capacity to prime naive T cells [for review, see ref. 8]. The ontogenetic and functional relationship between DC and Mϕ is still unresolved, and evidence exist that DC and Mϕ represent alternative differentiation forms of a common progenitor [9] as well as that both cell types belong to entirely seperate lineages [for review, see ref 10]. Although Caux et al. have already demonstrated that a subset of human DC transiently expresses CD14 during maturation [11], CD14 expression is widely used as a marker for Mϕ, and has often been applied to distinguish Mϕ from DC. Thus, we were interested whether murine dendritic cells (DC) also express CD14 molecules. By using DC lines as well as bone marrow-derived DC (BmDC) we demonstrate that DC are capable of expressing CD14 both at the mRNA and at the protein level. CD14 surface expression is modulated by IL-4 and LPS, and blockage of CD14 molecules by anti-CD14 antibodies (Ab) results in marked downregulation of LPS-induced IL-1 release by DC. We therefore conclude that CD14 functions as a receptor for LPS on murine DC, and that CD14 expression is not unique to Mϕ but can also be observed on DC.

2. MATERIAL AND METHODS

2.1. Cell Lines and Media

The cell lines CB-1, D2SC-1 and MT-2 were generated by P. Ricciardi-Castagnoli (CNR Center of Cytopharmacology, University of Milan, Milan, Italy), and maintained in Iscove´s medium containing 5 % FCS, 1 % non essential amino acids (NEAA), 1% HEPES, 1% glutamine, 50 μM 2-mercapto-ethanol, 1% penicillin/streptomycin (all tissue culture chemicals were obtained from PAA, Linz, Austria) as described elsewhere [12,13]. The skin-derived DC line, XS52, was kindly provided by A. Takashima (Dept. of Dermatology, Univ. of Texas, Dallas) [14]. P388D1 and A20 cells were obtained from ATCC (Rockville, MD, USA) and maintained in complete RPMI medium (CM) containing 10% FCS, 1% NEAA, 1% HEPES, 1% glutamine, 50 μM 2-mercaptoethanol, 1% penicillin/streptomycin.

2.2. Generation and Culture of DC

DC were generated by culture of bone marrow cells in the presence of GM-CSF as described previously [15,16]. Briefly, bone marrow was collected from tibias of female DBA/2N mice and suspended by vigorous pipetting. Erythrocytes were lysed and the remaining cells were passed through a nylon mesh to remove small pieces of bone and debris. The cells were washed twice with cold PBS, resuspended in CM containing 10% FCS and cultured in petri dishes (Becton-Dickinson, Heidelberg, Germany) for 4 h. Non adherent cells were collected, and aliquots of 1×10^6 cells were placed in 24-well plates (Becton-Dickinson, Heidelberg, Germany) in 1 ml CM medium, (5% FCS) supplemented with 200 U/ml GM-CSF (Laboserv, Giessen, Germany). After 4 days of incubation, 650 µl medium was removed and the same volume of fresh CM (5% FCS), containing 200 U/ml GM-CSF was added. After 3 days, non adherent cells were harvested, and DC aggregates were subcultured in 6-well plates in medium containing 200 U/ml GM-CSF. After overnight cultivation, most of the non adherent cells in culture had acquired typical dendritic morphology. These cells were further cultured with cytokines (IL-1ß, IL-4, TNF-α; R&D Systems, Wiesbaden, Germany) or LPS (serotype 055:B5; Sigma, Deisenhofen, Germany) and subjected to flow cytometric analysis.

2.3. RT-PCR

Total RNA was isolated from cells using 4 M guanidine isothiocyanate and chloroform extraction according to standard protocols. After treatment of RNA with RNAse-free DNAse I (Boehringer, Mannheim, Germany), 2 µg of total RNA were reverse transcribed using myeloblastosis virus reverse transcriptase (Promega, Heidelberg, Germany). For PCR amplification of the CD14 RNA the following primers were used: 3′ TGTTTGGGGGCGGCAGATGTG 5′ and 3′ GGTGGAGAGGGCAGGGAAGAC 5′ (Gibco, Eggenstein, Germany). The PCR samples were prepared according to the manufacturers protocol (Promega) and after thirty PCR cycles the reaction products were analysed on 1% agarose gels.

2.4. Flow Cytometry of DC

Cells were harvested on ice, washed with ice-cold PBS and aliquots of 2×10^5 cells were incubated with monoclonal Abs against murine CD14 (clone rmC5–3, Pharmingen, Hamburg, Germany), I-A^{b,d,q}, I-E^{d,k} (clone M5/114; Boehringer, Mannheim, Germany), F4/80 (Dianova, Hamburg, FRG), BM8 (kindly provided by Dr. C. Sorg, Institute for Experimental Dermatology, University of Münster), CD11b (Mac-1, clone M1/70, ATCC, 10% culture supernatant), B7–1 (clone 1G10, Pharmingen, San Diego, CA), B7–2 (clone Gl-1, Pharmingen), CD40 (clone 3/23, Pharmingen), CD11c (clone N418, Endogen, Cambridge, MA), or rat IgG as isotype control (Dianova, Hamburg, Germany) for 45 min on ice (1µg/ml diluted in PBS/0,5% BSA (v/w)). Thereafter, cells were washed twice with PBS/0.1% BSA (v/w) and incubated with FITC-conjugated goat anti-rat IgG (Dianova, diluted 1:50 in PBS/1% mouse serum (v/v)) for 45 min. on ice. Finally, propidium iodide (100 µM; Sigma) was added, cells were washed twice and subsequently analysed in an EPICS-XL flow cytometer (Coulter, Krefeld, Germany). Dead cells, determined by their propidium iodide fluorescence, were gated out. For double staining experiments, a FITC-coupled Ab directed against IA^d (clone AMS-32.1, Pharmingen) was used and the primary Ab specific for CD14 was detected by a phycoerythrin (PE) coupled secondary Ab (goat anti-rat IgG, Dianova).

2.5. Effect of Anti-CD14 Ab on LPS Induced Cytokine Release

DC were incubated for 1 h with anti-CD14 Ab (1 μg/ml) on ice. After removing unbound Ab by three washes with medium, the cells were cultured with or without LPS (1 μg/ml) for 8 h and the amounts of IL-1ß and TNF-α, respectively, released were determined by ELISA (Cytoscreen™, Laboserv, Giessen, Germany).

2.6. Mixed Lymphocyte Reaction

For primary mixed lymphoycte reactions (MLR), T cells were prepared by passing RBC-depleted spleen cells over a nylon wool column, followed by removal of remaining contaminants using the monoclonal antibodies M5/114, Mac-1 and B220 and immunomagnetic microbeads (MiniMACS, Miltenyi Biotech, Bergisch Gladbach, FRG). The resulting cell preparation contained less than 0.1% IA^+ cells. In the MLR's, freshly prepared DC from female DBA/2N mice (H-2^d) which had been separated into $CD14^+$ and $CD14^-$ populations by cell sorting with anti CD14-coated immunomagnetic microbeads (MiniMACS, Miltenyi Biotech) were cocultured with purified T cells from C3H/HeN (H-2^k) or C57/BL6 (H-2^b) mice in complete medium without essential amino acids and FCS, but supplemented with 1.5% mouse serum and 5 μg/ml indomethacin ('MELR medium'). Serial dilutions of triplicate samples of DC were mixed with a constant amount (2×10^5) of allogeneic T cells in 96 well culture dishes. Cells were cultured for 6 days, and pulsed with 1 mCi/well of [^{3}H]thymidine for 18 hr.

2.7. Presentation of Soluble Protein or Peptide Antigen

To assess the ability of DC to present soluble protein antigen or immunogenic peptides to primed T cells, the ovalbumin (OVA) specific T-T-hybridoma D011.1 was used, which produces IL-2 in response to OVA presented by I-A^d. The OVA-derived peptide, OVA323–339, contains the specific immunogenic epitope recognized by D011.1 and does not require further processing. BmDC were generated and sorted into $CD14^+$ and $CD14^-$ cell populations as described above. Graded numbers of BmDC were then incubated together with 1×10^5 D011.1 cells in the presence or absence of 1 mg/ml ovalbumin or 1 μg/ml OVA323–339 peptide for 24 hr at 37°C in a total volume of 200 μl. One hundred μl of culture supernatant was removed and assayed for IL-2 content using the IL-2 responsive cell line, CTLL-2, as described [17].

3. RESULTS

3.1. Expression of CD14 mRNA in Murine DC

To investigate whether murine DC are capable of expressing CD14, total RNA was isolated from BmDC, as well as from the dendritic cell lines CB-1 and D2SC-1 [13], and from the macrophage lines MT-2 and P388. RNA was treated with DNAse and subjected to RT-PCR. Using CD14 specific primers, a PCR product of the expected size could be amplified from BmDC and both DC lines (Fig. 1). As expected, a PCR product of identical size was also detectable in the monocyte/macrophage cell lines MT-2 and P388. To exclude the possibility that these PCR products were amplified from genomic DNA con-

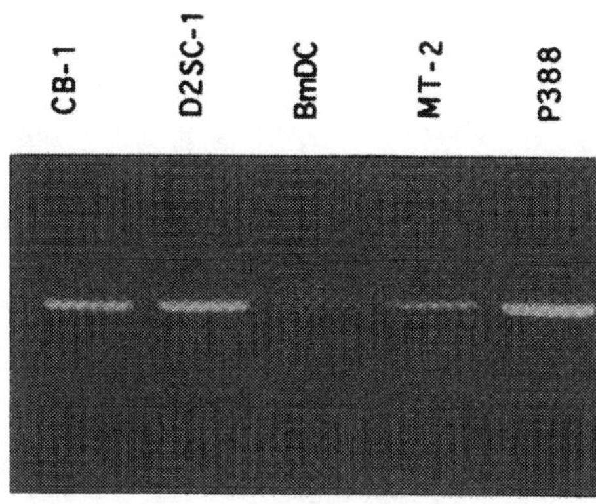

Figure 1. Expression of CD14 mRNA by DC. Total RNA of BmDC, CB-1, D2SC-1, MT-2, and P388 cells was isolated and subjected to RT-PCR, using CD14 specific primers. Samples were electrophoresed on 1% agarose gel.

taminations, PCR was performed with corresponding RNA preparations without reverse transcription, which did not reveal any products (data not shown).

3.2. Surface-Expression of CD14 by DC

To further investigate whether DC express CD14 on their cell surface, flow cytometric analysis was performed using a monoclonal Ab directed against murine CD14. As shown in Fig. 2, BmDC, the skin-derived DC line XS52, the splenic DC lines CB-1 and D2SC-1, as well as the monocytic cell line MT-2 all exhibited high surface expression of CD14, when compared to isotype control samples. Interestingly, frehly prepared or GM-CSF-cultured epidermal Langerhans cells did not express significant amounts of CD14 surface molecules (Fig. 2 and data not shown). As expected, no CD14 expression could be detected in A20 B-cells, confirming the specificity of the CD14 Ab used. In BmDC cultures, most experiments revealed a minor population of $CD14^-$ and a major population of $CD14^+$ cells, whereas by light microscopy, >80% of all cells had typical dendritic morhology (not shown), and by flow cytometry, >70% of the cells expressed high amounts of Ia, CD11c, CD40 and B7–2 (Fig.3). Since high surface expression of MHC class II molecules is characteristic for DC, double staining experiments were performed with anti-CD14 Ab vs. anti-I-A Ab. As shown in Fig. 4, the majority of BmDC, which co-express high levels of IA, also expresses CD14 (75%). However, a subpopulation of $I\text{-}A^+$ BmDC lacks expression of CD14 (10%). In addition, BmDC also express high amounts of CD11b (Mac-1) as well as the surface molecules recognized by antibodies BM8, F4/80 and ER-TR9 (Fig. 3), which are also regarded as specific markers for Mϕ [1–3]. Moreover, upon two colour staining for CD14 vs. F4/80 and BM8, respectively, most cells were found to be double positive for both markers (data not shown). Thus, these data indicate that at least a subpopulation of murine BmDC express surface CD14.

3.3. Effects of Cytokines and LPS on Surface Expression of CD14 by DC

In a next series of experiments, we tested whether the cytokines IL-1ß, IL-4, and TNF-α, which are known to exert effects on DC maturation and differentiation, are involved in regulation of surface expression of CD14 by BmDC. Accordingly, BmDC as

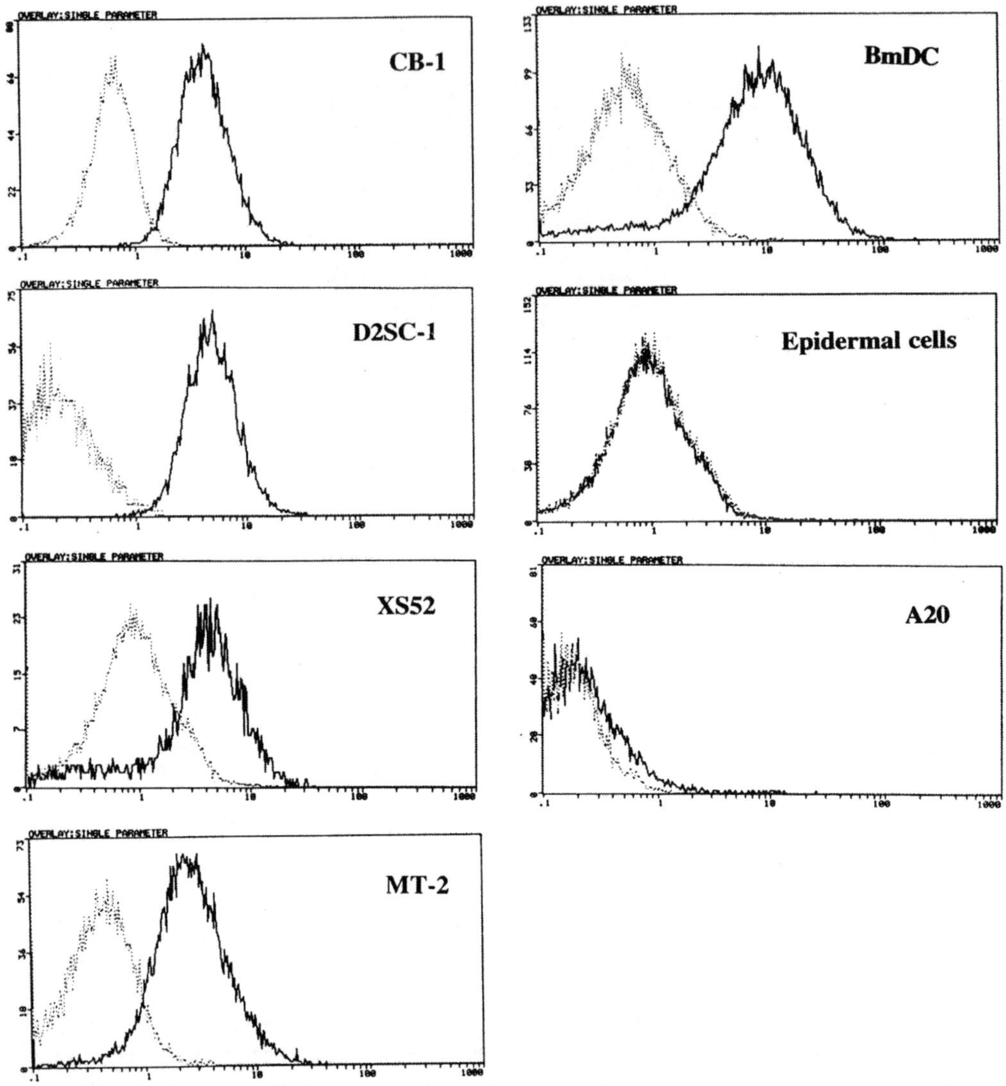

Figure 2. CD14 is expressed on the surface of DC. Dendritic cells (CB-1, D2SC-1, XS52), BmDC, freshly prepared epidermal Langerhans cells, monocytic cells (MT-2) and B cells (A20) were stained for cell surface expression of CD14 molecules (solid lines) by using the monoclonal antibody rmC5–3 and analysed by flow cytometry. Staining with IgG served as control for unspecific binding (dotted lines).

well as CB-1 and D2SC-1 cells, respectively, were incubated with these cytokines for 48 h. Among the cytokines tested, only IL-4 exerted a significant downregulatory effect on CD14 expression by BmDC (Tab. 1). However, CD14 expression on DC lines CB-1 and D2SC-1 was not affected by any of the cytokines tested in these experiments.

Since human CD14 is known to serve as a receptor for LPS which in turn upregulates CD14 expression on MO/Mϕ, we further investigated whether CD14 expression on murine DC is affected by LPS. For this purpose, DC lines and BmDC were incubated with 5 μg/ml LPS for different periods of time and analysed for surface expression of CD14

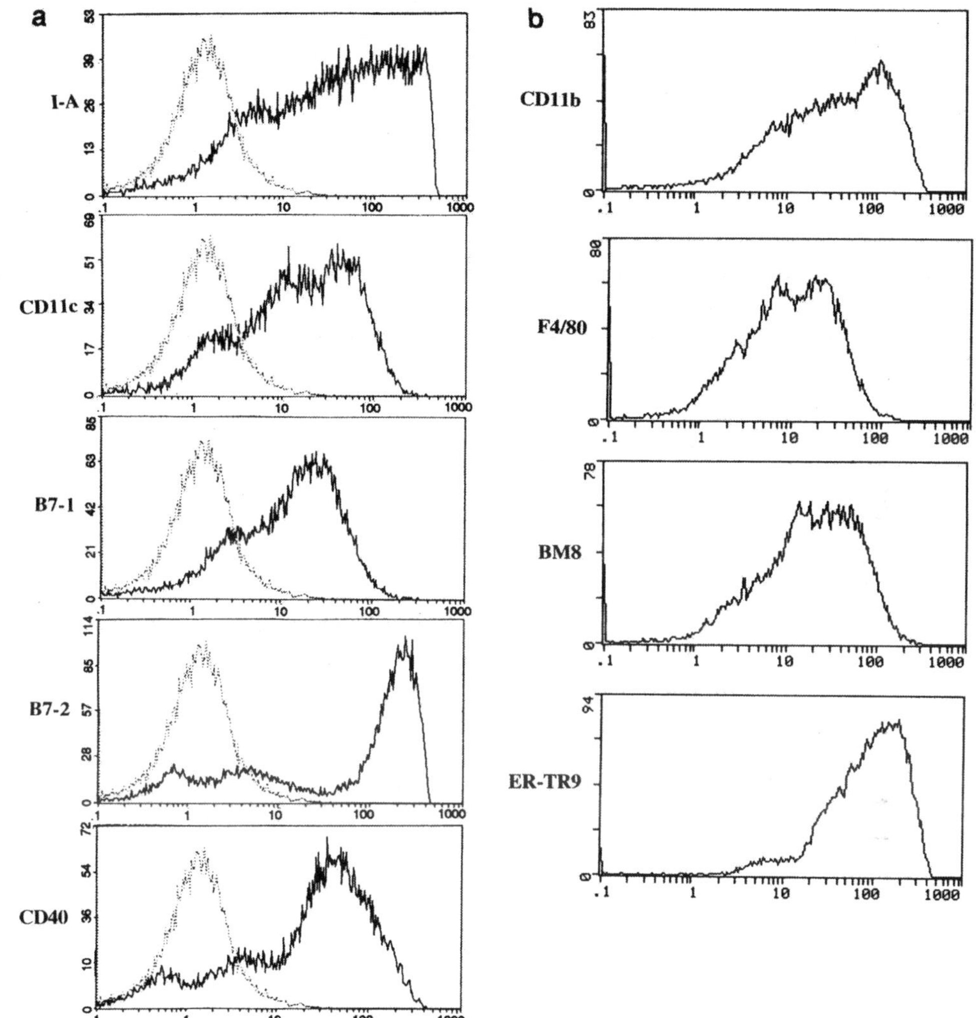

Figure 3. Surface phenotype of BmDC. BmDC were stained for cell surface expression of the DC surface molecules CD11c, CD40, CD80, CD86 and Ia (**a**), as well as the macrophage surface molecules CD11b (Mac-1), F4/80, BM8 and ER-TR9 (**b**) as described above.

molecules using flow cytometry. In these experiments, we show (Fig. 5) that cell surface expression of CD14 on BmDC is rapidly upregulated by LPS as compared to samples cultured without LPS for the same period of time. In the first 6 hours after addition of LPS, CD14 surface expression increased constantly, reaching a maximum with up to 70% of BmDC expressing CD14. Thereafter, the increase of CD14 expression declined and even after an additional overnight incubation period (14 h) with LPS the CD14 expression did not change significantly as compared to a 6 h stimulation period. In these experiments surface expression of MHC class II molecules (IA) by BmDC was analysed in parallel (Fig. 5, upper panel). A constantly high expression of MHC class II molecules (IA) could be detected which was not affected by LPS, thus verifying that the BmDC investigated here indeed belong to the DC lineage and that contaminating MO/Mϕ did not contribute to

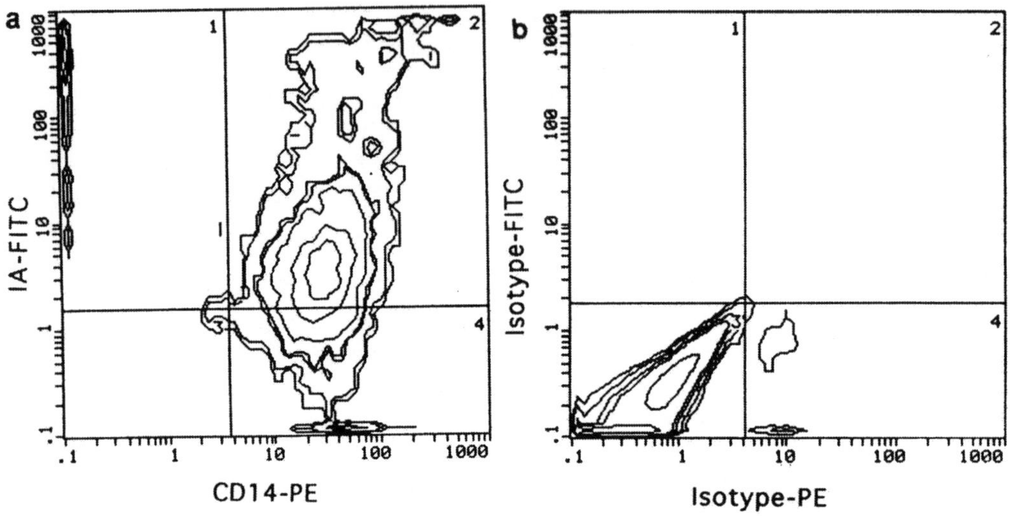

Figure 4. Coexpression of MHC class II (IA) molecules and CD14 by BmDC. **(a)** BmDC were double stained with anti-CD14 (rmC5–3, detected by phycoerythrin-conjugated secondary antibodies) and a FITC-labeled monoclonal antibody directed against MHC class II (IA) molecules, followed by flow cytometric analysis. Percentage of cells in each quadrant: 1=10%; 2=75%; 3=1%; 4=14%. **(b)** Samples stained with isotype matched antibodies. Percentage of cells in each quadrant: 1=1%; 2=1%; 3=94%; 4=4%.

the observed upregulation of CD14. However, when the DC lines CB-1 (Fig. 5) and D2SC-1 (data not shown) were exposed to LPS, no significant effects on CD14 surface expression could be recorded after 6 h, but after an ovenight incubation period (14 h), a downregulation rather than upregulation of CD14 by LPS could be observed, indicating that LPS might regulate surface expression of CD14 differentially in DC of different origin.

3.4. LPS Induced Release of IL-1 by DC Is Suppressed by Anti-CD14 Abs

As LPS is known to induce the release of IL-1ß by DC [18], we further tested whether CD14 expressed on the surface of murine DC may be involved in mediating these LPS induced effects. Accordingly, the DC lines D2SC-1 and CB-1 cells were incubated

Table 1. Effect of different cytokines on CD14 surface expression by DC

	BmDC	CB-1	D2SC-1
GM-CSF	82 ±5	95 ±4	98 ±7
GM-CSF + TNF-α	78 ±4	96 ±6	95 ±6
GM-CSF + IL-1ß	79 ±4	96 ±5	97 ±3
GM-CSF + IL-4	69 ±5[#]	98 ±3	97 ±4

BmDC and DC lines CB-1 and D2SC-1 were incubated with GM-CSF, TNF-α, IL-1ß, IL-4 (100 U/ml) for 48 h, thereafter CD14 surface expression was determined by flow cytometry. The percentage (mean ± SEM) of $CD14^+$ cells is given. [#]Significantly ($p<0.05$) different compared with GM-CSF treated BmDC.

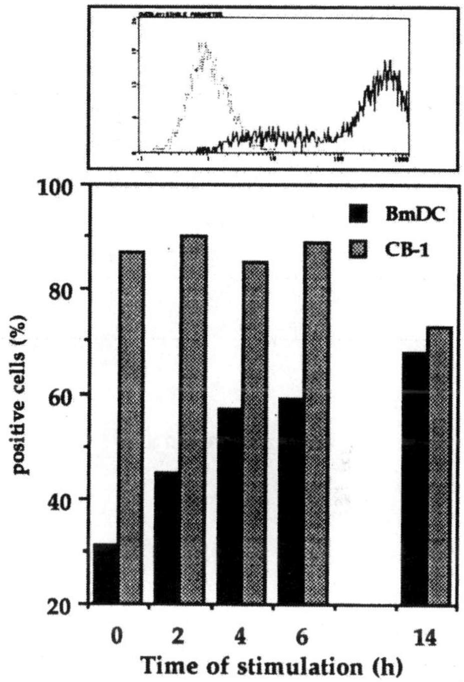

Figure 5. LPS upregulates CD14 expression on murine DC. BmDC and the DC line CB-1 were cultured in the presence of LPS for different time intervals. Thereafter, cells were harvested, stained and analysed by flow cytometry for presence of CD14 and IA molecules. The *lower panel* shows the percentage of CD14$^+$ cells of a representative experiment. Less than 1% of the cells exhibited non-specific binding as determined by IgG stained control. In the *upper panel*, the high expression of MHC class II molecules (solid lines) by BmDC is shown in comparison to isotype stained samples (dotted lines), verifying that the cells investigated here belong to the DC lineage.

with anti-CD14 Ab, followed by stimulation with LPS. Thereafter, supernatants were harvested and tested for the presence of IL-1ß and TNF-α using ELISA. LPS induced release of IL-1ß was significantly reduced in CB-1 cells and completely abolished in D2SC-1 cells by CD14 blockage (Fig. 6). In contrast, LPS induced TNF-α release was not affected by anti-CD14 treatment in each cell line (data not shown), indicating that different pathways are involved in IL-1ß and TNF-α release by LPS.

3.5. Allostimulatory Capacity of CD14-Sorted BmDC

To further investigate the influence of CD14-expression on DC function, cells were seperated by immunomagnetic beads into CD14$^+$ or CD14$^-$ cells and tested for their allostimulatory capacity. The proliferation of allogeneic responder T cells was determined by incorporation of [^{3}H]thymidine. As shown in Fig. 7, the two populations did not differ significantly in their capacity to present alloantigen in the MLR.

3.6. Presentation of Ovalbumin or OVA-Peptide 323–339 by CD14-Sorted BmDC

To assess the ability of CD14$^+$ and CD14$^-$ BmDCs to present soluble protein antigen to primed T cells, the ovalbumin specific T-T hybridoma D011.1 was used. The CD14$^+$ BmDC population induced a slightly weaker T cell response then the CD14$^-$ BmDC population (Fig. 8a). A similiar effect was seen using native ovalbumin protein as antigen (Fig. 8b). Thus, both cell populations were found to be able to present soluble antigen to responder T cells, with the CD14$^-$ population being approximately five-fold less efficient than the CD14$^+$ population.

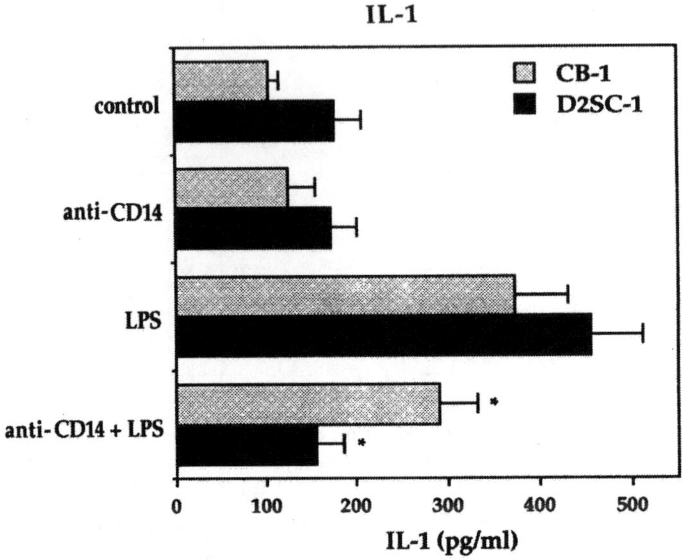

Figure 6. Incubation of DC with anti-CD14 antibodies reduces LPS induced release of IL-1ß. CB-1 and D2SC-1 cells were incubated with anti-CD14 antibodies for 1 h, washed and incubated for 3 h with or without LPS. Control cells were not treated with antibodies prior to their cultivation. Thereafter, IL-1ß content in supernatants was determined using an ELISA. Mean values (± SEM) of three experiments are shown, * indicates significant ($p < 0.01$) difference compared to controls.

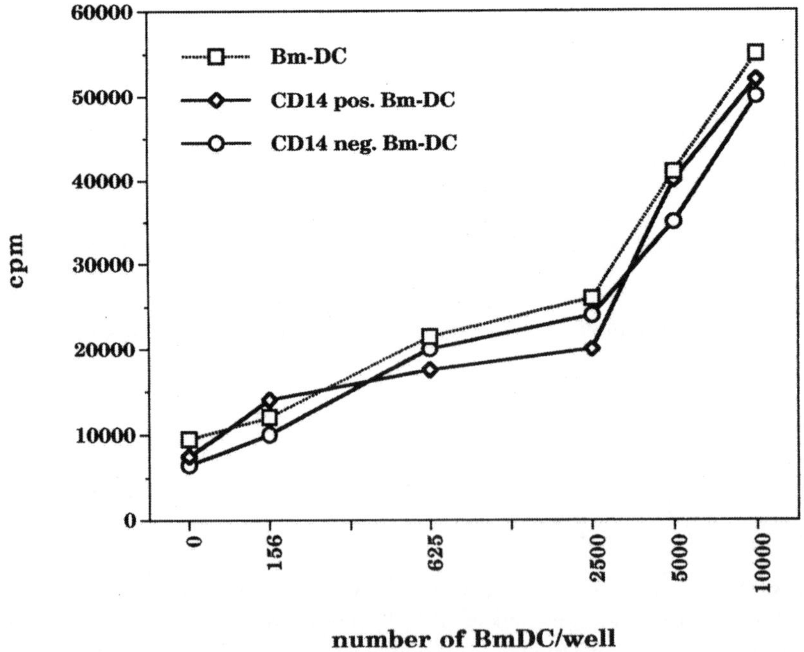

Figure 7. Allogeneic mixed lymphocyte reaction of CD14+ and CD14- BmDC. Graded numbers of BALB/c-derived (H-2^d) $CD14^+$ and $CD14^-$ BmDC were coincubated with 2×10^5 allogeneic purified T cells from C57BL/6 mice (H-2^b) in an allogeneic MLR and incubated for 5 days at 37°C, followed by overnight pulse with ^{3}H-thymidine. Background proliferation of BmDC or T cells alone was always below 10 000 cpm.

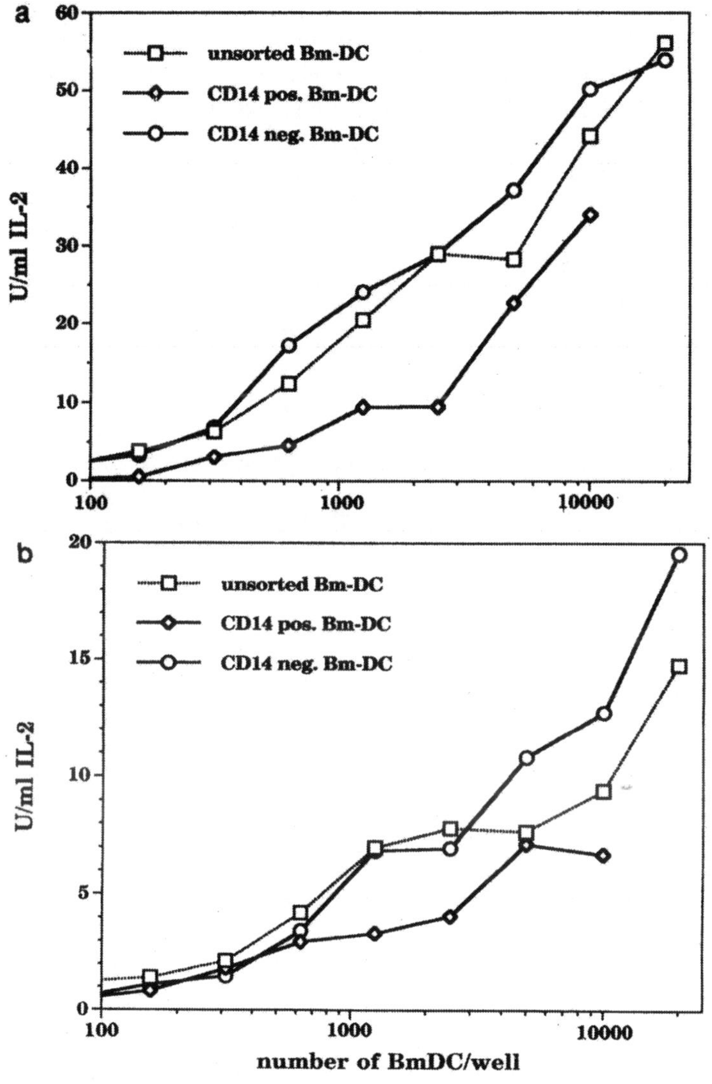

Figure 8. Protein and peptide antigen presentation by CD14+ and CD14- BmDC. **(a)** Presentation of OVA323–339 peptide and **(b)** native OVA by CD14^{+} and CD14^{-} BmDC. The graphs show IL-2 production of the OVA323–339 specific T cell hybridoma, DO11.1, in response to BmDC plus 1 mg/ml OVA and 1 μg/ml OVA323–339, respectively.

4. DISCUSSION

Traditionally, CD14 expression has been a hallmark of Mϕ, since almost all subtypes of human Mϕ express CD14, whereas only a few other human cell types were found to also express CD14 [3]. Likewise, in vitro generated DC derived from CD34^{+} cord blood progenitors, as well as human Langerhans cells did not express CD14 on their surface [19,20]. Consequently, absence of CD14 has been accepted as one indicator for DC [21]. However, recent investigations showed that CD14 can be expressed on human DC as well.

In lymphoid tissues of the nose, Langerhans cells were found to co-express CD14 as well as CD1a [22], and CD14⁺ precursors of DC could also be detected in human peripheral blood [23–25]. In addition, Caux et al. demonstrated that two distinct pathways may exist for differentiation of human DC from precursor cells, one involving CD14-expressing progenitor cells that differentiate into DC preferentially in the presence of GM-CSF and IL-4, and the other involving precursor cells which never express CD14 during GM-CSF and TNFα-dependent differentiation into mature DC [11]. Similar data have been generated by others, also suggesting that human DC derive either from a distinct DC-precursor cell, or from a $CD14^+$ cell type which has already acquired a monocytic phenotype [26]. Moreover, an interconversion from $CD14^+$ Mϕ to DC seems to be possible, as in the human system *in vitro* incubation of $CD14^+$ Mϕ with high doses of GM-CSF and IL-4 results in differentiation into mature DC [20]. During this process, Mϕ downregulate surface expression of CD14 and acquire phenotypical as well as functional features of DC. Taken together, these data show that CD14 is not an irreversible marker of terminally differentiated Mϕ in the human system. Our data now show that this also holds true for the murine system.

Our double staining experiments using BmDC indicate that at least two different populations of DC might exist in murine GM-CSF supplemented bone marrow cultures. One DC population is characterised by its co-expression of high levels of MHC class II and CD14 molecules on their surface. In parallel, another DC population is detectable, expressing only high levels of MHC class II molecules. Both populations appear to belong to the DC lineage due to their high expression of MHC class II molecules, their weak adherence to plastic surfaces and their dendritic morphology, but both populations clearly differ in their expression of CD14 molecules. However, we observed that in most experiments, CD14 expression of the $CD14^+$ DC subset was downregulated and sometimes lost during terminal maturation of BmDC (S.G., unpublished observation). Moreover, peptide and protein antigens were presented more efficiently by $CD14^-$ DC, suggesting that $CD14^-$ cells represent more mature DC. Nevertheless, CD14 was clearly still expressed on many BmDC at a differentiation state when these cells already exhibited the functional capacity to stimulate naive T cells, distinct dendritic morphology, and the immunophenotype of DC. We therefore suggest that $CD14^+$ DC represent a population of DC with a weaker antigen presenting capacity, which may not yet be fully differentiated. Thus, murine bone marrow-derived $CD14^+$ DC may represent DC which are still in the process of differentiation from a pluripotent precursor, and depending on cytokine signals differentiation into either Mϕ or into DC occurs. Since these cells clearly exhibit at least some functional capacities of DC, such as priming of naive alloreactive T cells in vitro, it appears to us a rather academic question whether to call these cells 'DC', 'immature DC' or 'activated Mϕ'.

Addition of IL-4 to bone marrow cultures results in a downregulation of surface expression of CD14, even when IL-4 is added at a timepoint when most cells have already differentiated into functionally active DC. However, generation of BmDC in the continous presence of GM-CSF and IL-4 leads to BmDC which express little or no CD14 at day 7–9 of culture, while BmDC generated by culture of bone marrow in the presence of GM-CSF alone express abundant CD14 (data not shown). It is still unclear whether addition of IL-4 to the bone marrow cultures merely suppresses CD14 expression or alters the differentiation of DC, potentially by selectively stimulating separate DC precursors which do not express CD14 during maturation into DC. However, sequential immunophenotyping of bone marrow cultures suggest that in GM-CSF plus IL-4 treated cultures, almost all cells in the cultures at least transiently express CD14 (S.G., unpublished observation).

Little is known about the functional significance of CD14 expression on DC and MO/Mϕ in mice. In Kupffer cells, LPS has been shown to upregulate CD14 expression both at the mRNA and the protein level [6], which is in analogy to our own results using murine BmDC. The time course of this upregulation is comparable to results obtained with Mϕ and Kupffer cells [6], showing a rapid upregulation of CD14 after an one h incubation period with LPS. Moreover, after an 18 h incubation period with LPS a downregulation of CD14 expression by CB-1 cells was detectable which confirms previous results obtained in the human system, showing that high doses of LPS downregulate the expression of CD14 by means of shedding CD14 molecules from the surface of Mϕ [27]. In contrast to these results obtained with BmDC, the surface expression of CD14 on the splenic DC-lines CB-1 and D2SC-1 did not alter significantly after treatment with LPS for 8 h. Therefore these data may indicate, that LPS differentially modulates CD14 expression in various types of murine DC. In vivo, LPS as well as TNFα and IL-1 induce DC emigration from nonlymphoid organs [28]. As LPS is known to induce leukocyte production of IL-1ß and TNF-α [13], we further investigated whether CD14 might be involved in mediating these LPS induced effects. Occupation of CD14 molecules by anti-CD14 Ab diminished the LPS induced release of IL-1, indicating that CD14 is involved in transmitting LPS mediated effects to DC and may function as a receptor for LPS on murine DC. However, since TNF-α release of DC could not be blocked by Ab directed against CD14, other structures expressed by DC may also serve as a receptor for LPS. This observation is further supported by a report on human MO, which demonstrates the presence of additional, as yet uncharacterized surface receptor(s) for LPS [29]. Thus, CD14 may not be the only surface molecule which is capable to trigger cytokine release from LPS stimulated MO.

Taken together, we demonstrate that murine DC lines as well as BmDC, at least at certain differentiation stages, express CD14, and that expression of CD14 as well as of other Mϕ-associated surface molecules is not a valid criteria to differentiate between Mϕ and DC in the murine system. Moreover, we demonstrate that LPS affects BmDC functionally, which can be blocked by administration of anti-CD14 antibodies. Although our data provide only preliminary evidence that CD14 may function as LPS receptor on murine DC, they demonstrate that binding sites for LPS are present on DC. Thus, LPS-like bacterial products may affect DC function, which could be of relevance for antibacterial immune responses in vivo.

ABBREVIATIONS USED IN THIS PAPER

Ab: antibody; **BmDC:** bone marrow derived dendritic cells; **CM**: complete medium; **DC**: dendritic cells; **FCS**: fetal calf serum; **FITC**: fluorescein isothiocyanate; **GM-CSF**: granulocyte/monocyte-colony stimulating factor; **IL-1**: interleukin 1; **LPS**: lipopolysaccharide; **MHC**: major histocompatibility complex; **Mϕ**: macrophages; **MO**: monocytes; **OVA**: ovalbumin; **PBS**: phosphate-buffered saline; **PCR:** polymerase chain reaction; **TNF-α**: tumor necrosis factor α.

5. ACKNOWLEDGMENTS

The authors wish to thank Meike Steinert for her excellent technical assistance. This work was supported by grants from the German Research Foundation (GR1022/3–1, SFB293/96-B1).

6. REFERENCES

1. Todd, R.F., Nadler, L.M., Schlossman, S.F. (1981) Antigens on human monocytes identified by monoclonal antibodies. *J. Immunol.* **126**, 1435–1442.
2. Ball, E.D., Graziano, R.F., Shen, L., Fanger, M.W. (1982) Monoclonal antibodies to novel myeloid antigens reveal human neutrophil heterogeneity. *Proc. Natl. Acad. Sci. USA* **79**, 5374–5377.
3. Kielian, T.L., Blecha, F. (1995) CD14 and other recognition molecules for lipopolysaccharide: a review. *Immunopharmacology* **29**, 187–205.
4. Wright, S.D. (1991) CD14 and immune response to lipopolysaccharide. *Science (Wash. DC)*. **252**, 1321–1325.
5. Wright, S.D., Ramos, R.A., Tobias, P.S., Ulevitch, R.J., Mathison, J.C. (1990) CD14, a receptor for complexes of lipopolysaccharide (LPS) and LPS binding protein. *Science (Wash. DC)* **249**, 1431–1434.
6. Matsura, K., Ishida, T., Setoguchi, M., Higuchi, Y., Akizuki, S., Yamamoto, S. (1994) Upregulation of mouse CD14 expression in Kupffer cells by lipopolysaccharide. *J. Exp. Med.* **179**, 1671–1676.
7. Ferrero, E., Hsieh, C.L., Francke, U., Goyert, S.M. (1990) CD14 is a member of the family of leucine-rich proteins and is encoded by a gene syntenic with multiple receptor genes. *J. Immunol.* **145**, 331–336.
8. Steinman, R.M., Witmer-Pack, M., Inaba, K. (1993) Dendritic cells: antigen presentation, accessory function and clinical relevance. *Adv. Exp. Med. Biol.* **329**, 1–9.
9. Inaba, K., Inaba, M., Deguchi, M., Hagi, K., Yasumizu, R., Ikehara, S., Muramatsu, S., Steinman, R.M. (1993) Granulocytes, macrophages, and dendritic cells arise from a common major histocompatibility complex class II-negative progenitor in mouse bone marrow. *Proc Natl Acad Sci U S A* **90**, 3038–42.
10. Peters, J.H., Gieseler, R., Thiele, B., Steinbach, F. (1996) Dendritic cells: from ontogenetic orphans to myelomonocytic descendants. *Immunol. Today* **17**, 273–278.
11. Caux, C., Vanbervliet, B., Massacrier, C., Dezutter Dambuyant, C., de Saint Vis, B., Jacquet, C., Yoneda, K., Imamura, S., Schmitt, D., Banchereau, J. (1996) $CD34^+$ hematopoietic progenitors from human cord blood differentiate along two independent dendritic cell pathways in response to GM-CSF plus TNFα. *J. Exp. Med* **184**, 695–706.
12. Paglia, P., Girolomoni, G., Robbiati, F., Granucci, F., Ricciardi-Castagnoli, P. (1993) Immortalized dendritic cell line fully competent in antigen presentation initiates primary T cell responses in vivo. *J. Exp. Med.* **178**, 1903–10.
13. Granucci, F., Girolomoni, G., Lutz, M.B., Foti, M., Marconi, G., Gnocchi, P., Nolli, L., Ricciardi Castagnoli, P. (1994) Modulation of cytokine expression in mouse dendritic cell clones. Eur. J. Immunol 24, 2522–6.
14. Xu, S., Ariizumi, K., Caceres, D.G., Edelbaum, D., Hashimoto, K., Bergstresser, P.R., Takashima, A. (1995) Successive generation of antigen-presenting, dendritic cell lines from murine epidermis. *J Immunol* **154**, 2697–705.
15. Scheicher, C., Mehlig, M., Zecher, R., Reske, K. (1992) Dendritic cells from bone marrow: in vitro differentiation using low doses of recombinant granulocyte-macrophage colony stimulating factor. *J. Immunol. Meth.* **154**, 153–264.
16. Inaba, K., Inaba, M., Romani, N., Aya, H., Deguchi, M., Ikehara, S., Muramatsu, S., Steinman, R. (1992) Generation of large numbers of dendritic cells from mouse bone marrow cultures supplemented with granulocyte/macrophage colony stimulating factor. *J. Exp. Med.* **176**, 1693–1702.
17. S. Grabbe, K. Steinbrink, M. Steinert, T.A. Luger, T. Schwarz. 1995. Removal of the majority of epidermal Langerhans cells by topical or systemic steroid application enhances the effector phase of murine contact hypersensitivity. *J. Immunol.* **155**:4207–4217.
18. Ruppert, J., Peters, J.H. (1991) Accessory cell function during monocyte/macrophage differentiation: relation to interleukin-1 (IL-1 beta) production and release. *Eur J Cell Biol* **55**, 352–61.
19. Romani, N., Lenz, A., Glassel, H., Stossel, H., Stanzl, U., Majdic, O., Fritsch, P., Schuler, G. (1989) Cultured human Langerhans cells resemble lymphoid dendritic cells in phenotype and function. *J. Invest. Dermatol.* **93**, 600–609.
20. Sallusto, F., Lanzavecchia, A. (1994) Efficient presentation of soluble antigen by cultured human dendritic cells is maintained by granulocyte / macrophage colony-stimulating factor plus interleukin 4 and downregulated by tumor necrosis factor alpha. *J. Exp. Med.* **179**, 1109–18.
21. Steinman, R.M. (1991) The dendritic cell system and its role in immunogenicity. *Annu. Rev. Immunol.* **9**, 271–296.
22. Greame-Cook, F., Bhan, A.K., Harris, N.L. (1993) Immunohistochemical characterization of intraepithelial and subepithelial mononuclear cells of the upper airways. *Am. J. Pathol.* **143**, 1416–1422.

23. O'Doherty, U., M. Peng, S. Gezelter, W. J. Swiggard, M. Betjes, N. Bhardwaj, and R. M. Steinman. (1994) Human blood contains two subsets of dendritic cells, one immunologically mature and the other immature. *Immunology* **82**, 487–493.
24. Thomas, R., and P. E. Lipsky (1994) Human peripheral blood dendritic cell subsets. Isolation and characterisation of precursor and mature antigen- presenting cells. *J. Immunol.* **153**, 4016–4027.
25. Zhou, L.J., Tedder, T.F. (1996) $CD14^+$ blood monocytes can differentiate into functionally mature $CD83^+$ dendritic cells. *Proc. Natl. Acad. Sci. USA.* **93**, 2588–92
26. Ruppert, J., Friedrichs, D., Xu, H., Peters, J.H. (1991) IL-4 decreases the expression of monocyte differentiation marker CD14, paralleled by an increasing accessory potency. *Immunobiology* **182**, 449–464.
27. Bazil, V., Strominger, J.L. (1991) Shedding as a mechanism of down- modulation of CD14 on stimulated human monocytes. *J. Immunol.* **147**, 1567–1571.
28. Roake, J.A., Rao, A.S., Morris, P.J., Larsen, C.P., Hankins, D.F., Austyn, J.M. (1995) Dendritic cell loss from nonlymphoid tissues after systemic administration of lipopolysaccharide, tumor necrosis factor, and interleukin 1. *J. Exp. Med.* **181**, 2237–2247.
29. Couturier, C., Jahns, G., Kazatchine, M., Haefner-Cavaillon, N. (1992) Membrane molecules which trigger the production of interleukin-1 and tumor necrosis factor-α by lipopolysaccharide-stimulated human monocytes. *Eur. J. Immunol.* **22**, 1461–1466.

26

TGF-β1 DEPENDENT GENERATION OF LAG$^+$ DENDRITIC CELLS FROM CD34$^+$ PROGENITORS IN SERUM-FREE MEDIUM

Herbert Strobl,[1] Elisabeth Riedl,[1] Clemens Scheinecker,[1]
Concha Bello-Fernandez,[1] Winfried F. Pickl,[2] Otto Majdic,[2] and
Walter Knapp[2]

[1]Institute of Immunology
Vienna International Research Cooperation Center (VIRCC) at SFI
University of Vienna
Brunnerstraβe 59, A-1235 Vienna, Austria
[2]Institute of Immunology
University of Vienna
Borschkegasse 8A, A-1090 Vienna, Austria

1. INTRODUCTION

Several studies have shown that substantial numbers of functional dendritic cells (DC) can be generated *in vitro* from human CD34$^+$ progenitor cells upon culture with the two cytokines GM-CSF and TNFα.[1,2,3] More recent data suggest that at least two *in vitro* differentiation pathways for the development of DC seem to exist.[4,5,6] One pathway gives rise to Langerhans (LC) type DC. Phenotypically, this pathway follows the route CD34$^+$ to CD1a$^+$ to CD1a$^+$/Lag$^+$ and does not involve a CD14$^+$ monocytoid intermediate cell stage.[5]

In vitro development of DC (including LC type DC) from hemopoietic progenitor cells critically depends on serum/plasma supplementation of the growth medium. Serum-free cultures containing GM-CSF plus TNFα were reported to be highly inefficient in promoting DC development.[1]

We recently established a serum-free culture system which efficiently and selectively promotes the development of LC-type DC from hemopoietic progenitor cells.[7] These LC type DC develop along a recently described differentiation pathway.[5] Simultaneous generation of CD14$^+$ monocytoid cells is however minimal. This selective outgrowth of LC type DC (Lag$^+$) in serum free medium could be induced upon addition of the cytokine transforming growth factor-beta 1 (TGF-β1) to the culture medium.

Dendritic Cells in Fundamental and Clinical Immunology
edited by Ricciardi-Castagnoli, Plenum Press, New York, 1997

2. MATERIALS AND METHODS

CD34⁺ cells were isolated from umbilical cord blood mononuclear cells using the MACS CD34 Progenitor Cell Isolation Kit (Miltenyi Biotec, Bergisch Gladbach, Germany), according to the instructions of the manufacturer. The purity of the $CD34^+$ population ranged from 87% to 98% (mean = 94%). Culture of purified $CD34^+$ cells was performed in 24 well plates (Costar, Cambridge, MA) ($1–3x10^4$ cells in 1 ml/ well) at 37°C in a humidified atmosphere and in the presence of 5% CO_2. The serum-free medium X-VIVO 15 (Bio Whittaker, Walkersville, MD) contained L-glutamine (2.5 mM), penicillin (125 IE/ml) and streptomycin (125 μg/ml). Cultures were supplemented with optimized concentrations of the following human cytokines: transforming growth factor-beta 1 (TGF-β1; 0,5 ng/ml; purified from platelets; British Biotechnology, Abington, UK), tumor necrosis factor alpha (rhTNFα; 50 U/ml; Bender, Vienna, Austria), granulocyte-macrophage colony-stimulating factor (rhGM-CSF; 100 ng/ml; Sandoz, Basel, Switzerland) and stem cell factor (rhSCF; 20 ng/ml; Amgen, Thousand Oaks, CA). When indicated, serum-free cultures were supplemented with 10% pooled human umbilical cord blood plasma (CBPl) obtained from at least three individuals as described previously.[8]

For indirect suspension stainings of the intracellular Lag antigen (mAb clone Lag, IgG1, kindly provided by Dr. Imamura, Kyoto, Japan), we used the commercially available reagent combination Fix&Perm from An der Grub (A-2572 Kaumberg, Austria) according to the manufacturer's procedure. Cells were analyzed by flow cytometry (FACScan, Becton Dickinson Immunocytometry Systems, San Jose, CA).

3. RESULTS AND DISCUSSION

Our initial experiments showed that the cytokine combination GM-CSF plus TNFα even when further supplemented with SCF (c-kit ligand) is inefficient in promoting DC development in serum-free medium. The same growth factor combination, however, effectively promoted generation of both $CD1a^+/CD14^-$ dendritic cell and $CD1a^-/CD14^+$ monocytoid cell populations when added to serum/plasma supplemented medium.[7]

In our search for constituents of plasma or serum potentially involved with the observed DC growth promoting effect, we evaluated the cytokine transforming growth factor-beta 1 (TGF-β1). This highly pleiotropic 25-kDa homodimeric polypeptide[9] has been shown before to be present in serum and plasma samples[10] and to promote the in vitro growth of committed hemopoietic progenitors and monocytoid cells.[11,12,13]

Addition of TGF-β1 at low concentrations (0.5ng/ml) to GM-CSF plus TNFα and SCF supplemented serum-free culture medium indeed significantly ($p<0.05$) enhanced both total cell growth and increased percentages of $CD1a^+$ cells. At day 10, 35 ± 4% of all cultured cells in the presence of TGF-β1 expressed CD1a and cultures initiated with 1×10^4 $CD34^+$ cells yielded on average $8 \pm 3\times10^4$ $CD1a^+$ cells. In contrast, $CD14^+/CD1a^-$ monocytic cells remained infrequent in the presence of TGF-β1 (≤8%, n=6). Parallel cultures set up in the presence of 10% cord blood plasma but not of exogenous TGF-β1 gave rise to significantly higher percentages of $CD14^+/CD1a^-$ monocytoic cells (25 ± 3%, n=6, $p<0.05$) but significantly lower percentages of $CD1a^+$ cells (18 ± 3%, n=6, $p<0.05$).

The morphological appearances of cells grown for 10 days in either culture medium alone, plasma-supplemented culture medium or TGF-β1-supplemented culture medium, each in the presence of GM-CSF plus TNFα and SCF, were strikingly different (Figure 1). Cells grown in plasma- and TGF-β1-free culture medium were round shaped, devoid of cells with

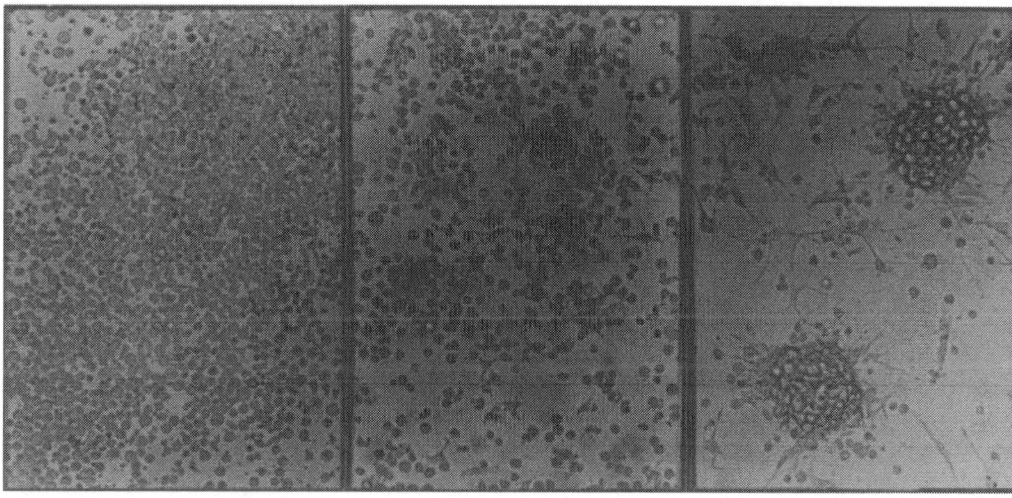

Figure 1. Microscopic appearance of 10-day cultures. $CD34^+$ CB cells were cultured in GM-CSF+TNFα+SCF supplemented plasma-free medium (left), or with additional supplementation with 10% cord blood plasma (center) or 0.5 ng/ml TGF-β1 (right).

dendritic morphology and contained very few small aggregates. Cultures with 10% cord blood plasma-supplemented medium were composed mainly of small dendritic cell clusters and of adherent macrophage-type cells. On the other hand, TGF-β1-supplemented cultures, contained large aggregates consisting of most of the cells present in the culture. Loosely adherent to these aggregates, cells with highly dendritic morphology predominated.

The acquisition of highly dendritic morphology of cells after culture for 10 days in TGF-β1 supplemented serum-free medium was correlated with acquisition of Lag. Lag antigen represents a 40-kDa glycoprotein which is associated with Langerhans cells/Birbeck granules.[14] As shown in Figure 1, a substantial subset of cells generated in the presence of TGF-β1 was found to intensely stain positive for Lag. This was indeed indicative for the presence of numerous typical cytoplasmic racket-shaped and rod-like Langerhans cell/Birbeck granules as confirmed by electron microscopy.[7] Strikingly, Lag^+ cells were only observed in TGF-β1 supplemented serum-free cultures and could neither be detected in parallel cultures without TGF-β1-supplementation nor with plasma-supplementation (Figure 2).

Thus we demonstrated that TGF-β1 represents an important co-stimulatory cytokine for the in vitro development of Langerhans cells in serum free medium. Recent data showed that TGF-β1 null mice (-/-) lack epidermal Langerhans cells (M. Udey, this issue), suggesting that also *in vivo* TGF-β1 is required for the development and/or migration of this particular cell type.

The differentiation pathway by which Lag^+ DC develop in our serum-free culture system certainly does not involve a $CD14^+/CD1a^-$ monocytic intermediate cell stage since $CD14^+/CD1a^-$ cells remained highly infrequent throughout the 10 day culture period. This clearly fits to recent observations by Caux et al.[5] who showed that Lag^+ DC unlike simulataneously developing Lag^- DC do not develop from an intermediate $CD14^+/CD1a^-$ monocytoid cell stage.

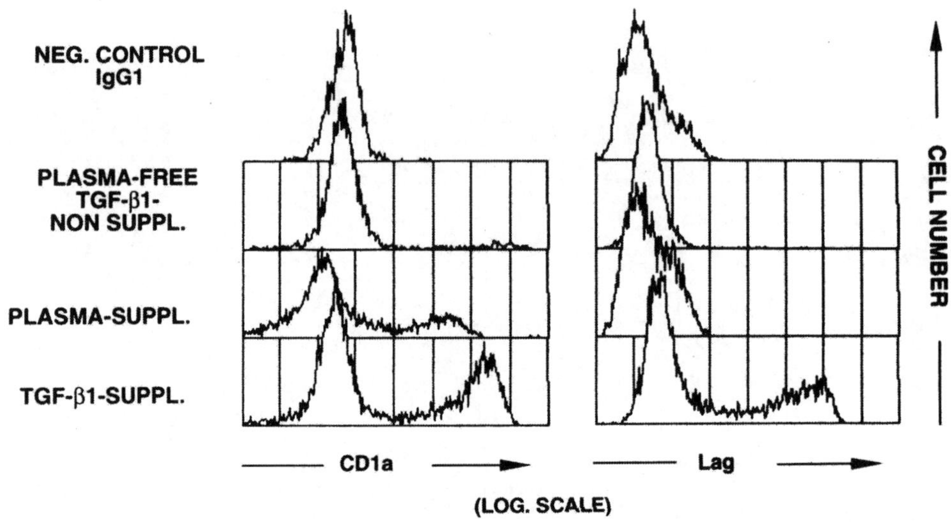

Figure 2. Analysis of Lag expression by cultured cells. Cells generated were stained for CD1a or intracellular Lag expression as described in Materials and Methods. Representative histograms show cells analyzed in parallel for Lag or CD1a expression.

Strikingly in plasma-supplemented medium substantial proportions of cells acquire CD1a expression but do not show Lag reactivity. One explanation could therefore be that these $Lag^-/CD1a^+$ DC derive from a distinct progenitor cell subset and/or differentiation pathway.[5] Alternatively, simultaneously developing monocytes/macrophages may inhibit further DC differentiation, as recently described.[4]

REFERENCES

1. Reid, C. D. L., Stackpoole, A., Meager, A. and Tikerpace, J. 1992. Interactions of tumor necrosis factor with granulocyte-macrophage colony-stimulating factor and other cytokines in the regulation of dendritic cell growth in vitro from early bipotent $CD34^+$ progenitors in human bone marrow. *J Immunol* 149:2681.
2. Caux, C., Dezutter-Dambuyant, C., Schmitt, D. and Banchereau, J. 1992. GM-CSF and TNF-α cooperate in the generation of dendritic Langerhans cells. *Nature* 360:258.
3. Santiago-Schwarz, F., Belilos, E., Diamond, B. and Carsons, S. E. 1992. TNF in combination with GM-CSF enhances the differentiation of neonatal cord blood stem cells into dendritic cells and macrophages. *J Leukoc Biol* 52:274.
4. Szabolcs, P., Avigan, D., Gezelter, S., Ciocon, D. H., Moore, M. A. S., Steinman, R. M. and Young, J. W. 1996. Dendritic cells and macrophages can mature independently from a human bone marrow-derived, post-colony-forming unit intermediate. *Blood* 87:4520.
5. Caux, C., Vanbervliet, B., Massacrier, C., Dezutter-Dambuyant, C., de Saint-Vis, B., Jacquet, C., Yoneda, K., Imamura, S., Schmitt, D. and Banchereau, J. 1996. $CD34^+$ hematopoietic progenitors from human cord blood differentiate along two independent dendritic cell pathways in response to GM-CSF+TNFα. *J Exp Med* 184:695.
6. Galy, A., Travis, M., Cen, D. and Chen, B. 1995. Human T, B, natural killer cells, and dendritic cells arise from a common bone marrow progenitor cell subset. *Immunity* 3:459.
7. Strobl, H., Riedl, E., Scheinecker, C., Bello-Fernandez, C., Pickl, W. F., Rappersberger, K., Majdic, O. and Knapp, W. 1996. TGF-β1 promotes in vitro development of dendritic cells from $CD34^+$ hemopoietic progenitors. *J Immunol* 157:1499.

8. Scheinecker, C., Strobl, H., Fritsch, G., Csmarits, B., Krieger, O., Majdic, O. and Knapp, W. 1995. Granulomonocyte-associated lysosomal protein expression during in vitro expansion and differentiation of CD34⁺ hemopoietic progenitor cells. *Blood* 86:4115.
9. Massagué, J., Attisano, L. and Wrana, J. L. 1994. The TGF-β family and its composite receptors. *Trends in Cell Biol.* 4:172.
10. Childs, C. B., Proper, J. A., Tucker, R. F. and Moses, H. L. 1982. Serum contains a platelet-derived transforming growth factor. *Proc Natl Acad Sci USA* 79:5312.
11. Keller, J. R., Jacobsen, S. E. W., Sill, K. T., Ellingsworth, L. R. and Ruscetti, F. W. 1991. Stimulation of granulopoiesis by transforming growth factor β: Synergy with granulocyte/macrophage-colony-stimulating factor. *Proc Natl Acad Sci USA* 88:7190.
12. Piabicello, W., Ferrero, D., Sanavio, F., Badoni, R., Stacchini, A., Severino, A. and Aglietta, M. 1991. Responsiveness of highly enriched CFU-GM subpopulations from bone marrow, peripheral blood, and cord blood to hemopoietic growth inhibitors. *Exp Hematol* 19:1084.
13. Bursuker, I., Neddermann, K. M., Petty, B. A., Schacter, B., Spitalny, G. L., Tepper, M. A. and Pasternak, R. D. 1992. In vivo regulation of hemopoiesis by transforming growth factor beta 1: Stimulation of GM-CSF- and M-CSF-dependent murine bone marrow precursors. *Exp Hematol* 20:431.
14. Kashihara, M., Ueda, M., Horiguchi, Y., Furukawa, F., Hanaoka, M. and Imamura, S. 1986. A monoclonal antibody specifically reactive to human Langerhans cells. *J Invest Dermatol* 87:602.

27

A NEWLY IDENTIFIED ANTIGEN RETENTION COMPARTMENT IN THE FSDC PRECURSOR DENDRITIC CELL LINE

Manfred B. Lutz,[1] Patrizia Rovere,[2] Monique J. Kleijmeer,[3] Caroline U. Aßmann,[1] Viola M. J. Oorschot,[3] Maria Rescigno,[1] Hans. J. Geuze,[3] Jean Davoust,[2] and Paola Ricciardi-Castagnoli[1]

[1]CNR Cellular and Molecular Pharmacology Center
Milan, Italy
[2]Centre d'Immunologie INSERM-CNRS de Marseille-Luminy
France
[3]Department of Cell Biology, School of Medicine
Institute of Biomembranes
Utrecht University, The Netherlands

INTRODUCTION

Antigen uptake and presentation by dendritic cells (DC) occur at different stages of their maturation and are directed by certain cytokines (1–3). Phagocytosis (4) as well as mannose receptor-mediated endocytosis through clathrin-coated pits and fluid-phase macropinocytosis have been described for human and murine DC at the immature developmental stage (5, 6). Antigen presentation capacity of those cells is weak. After antigen contact with an atnigen or proinflammatory cytokines they mature under loss of the antigen uptake capabilities, but upregulate surface MHC class II and costimulatory molecules for antigen presentation. In vivo, resting DC of non-lymphoid organs internalize antigens and transport them to the draining lymph nodes. In the T cell areas of the lymph node the now fully mature interdigitating DC is able to initiate primary T cell responses. During migration the DC undergoes maturation, but antigen processing should be delayed until it reaches the lymph node. In the precursor DC line FSDC (6, 7), we have now identified a specialized antigen retention compartment, and propose a mechanism of how antigens within these vesicles are prevented from degradation by lysosomal enzymes.

MATERIALS AND METHODS

Lucifer Yellow, Fluorescein-conjugated ovalbumin (FITC-OVA), dextran (FITC-DX, MW 40000), and transferrin were purchased from Molecular Probes (Eugene, OR) and fluorescence detected by FACS (6), confocal microscopy (8) or detection with immunogold-labeled secondary antibodies for electron microscopy (9). Antigen pulses of 1 × 10^6 FSDC with OVA and DX were for 30–60 min at 1 mg/ml in a 37°C CO_2-in-air incubator (6). To follow antigen retention, the cells were washed and chased in tracer-free medium for 1–24 h before analysis.

To detect acidic vesicles the weak base DAMP (N-3-2,4-dinitrophenylaminopropyl-N-(3-aminopropyl)-methylamine dihydrochloride, Molecular Probes) was added through the last 60 min of chase. Cells were fixed in 2% paraformaldehyde and 0.2% glutaraldehyde in PBS before DAMP was detected with an anti-DNP antibody and FITC-labeled tracers with an anti-FITC antibody. Then the cells were processed for electron microscopy as described (10).

RESULTS AND DISCUSSION

The FSDC show a constitutively high rate of both macropinocytosis of soluble tracers, such as FITC-OVA, and endocytosis of FITC-DX via the mannose receptor (6). In contrast to other cell types (B cells or monocytes/macrophages) these tracers remain inside the FSDC for a long time. Kinetic studies analysed by FACS, revealed that FITC-OVA, FITC-DX and the small molecule Lucifer Yellow could still be detected inside the cells 24 h after antigen pulse. FITC-coupled transferrin, which is known to recycle after binding to its receptor very fast, could not be retained.

Confocal and immunoelectron microscopy studies clearly showed the two separate ways of internalization for FITC-OVA, preferentially into large macropinosomes, and for FITC-DX, mainly via clathrin-coated pits, within the first hour of antigen pulse. After 6 h of chase in tracer-free medium, both pathways most likely share a common retention com-

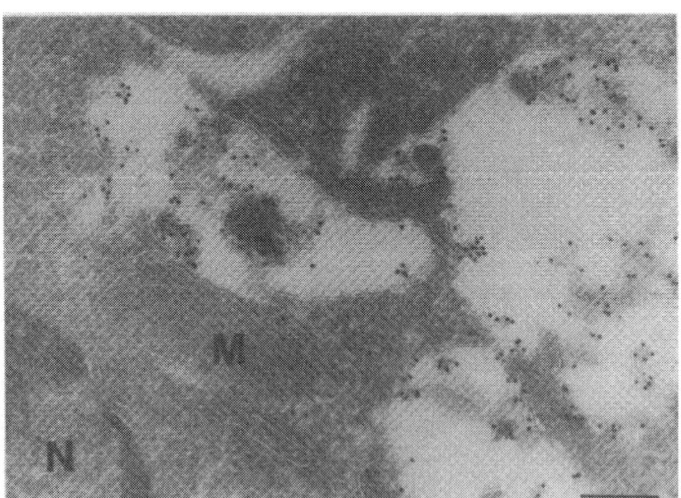

Figure 1. Large macropinosomal compartments in the FSDC cell line retain antigens at mildly acidic/neutral pH. FSDC cells were pulsed with FITC-DX for 1 h, washed and chased for 6 h. During the last hour of chase DAMP was added. After fixation the cells were processed for immunogold labeling and electron microscopy (see Materials and Methods). The weak base DAMP accumulates in acidic organelles, such as early and late endosomes and lysososmes (not shown). Only little DAMP (5 nm gold particles) can be detected in compartments containing abundant labeling for FITC-DX (10 nm gold particles) in the FSDC. Bar 200 nm. N nucleus; M mitochondria.

partment (Fig. 1). Those vesicles resemble large electron-lucent macropinosomes, which have acquired the lysosomal membrane protein Lamp-1 and the lysososmal protease cathepsin D. Staining for the small GTPase rab7, to identify endosomal compartments, did not colocalize with the retained tracers.

During the endosomal pathway of protein degradation, increasing vesicle acidification is observed (11). An acidic pH is necessary to enable the enzymatic activity of cathepsin D (12, 13). Intracellular vesicles with acidic pH can be visualized by selective accumulation of the weak base DAMP (10). Within the retention compartments however, DAMP was rarely detected (Fig. 1), in contrast to small endosomes (not shown). This might explain why OVA and cathepsin D can coexist without degradation of OVA. Interestingly, a similar pH dependent mechanism of intracellular persistence has been described for Mycobacteria (14, 15).

After longer chase times of 8–24 h increasing acidification of large retention compartments with loss of tracer content can be observed. This might represent the start of antigen processing and loading of MHC class II molecules for antigen presentation as observed by DC in lymphoid organs. Further studies have to elucidate the relation of the retention compartments described with the MHC molecule loading compartments (MIIC) described in more mature DC (16).

REFERENCES

1. Steinman, R.M. 1991. The dendritic cell system and its role in immunogenicity. *Annu. Rev. Immunol.* 9:271.
2. Steinman, R.M., and J. Swanson. 1995. The endocytic activity of dendritic cells. *J. Exp. Med.* 182:283.
3. Lutz, M.B., G., Girolomoni, and P. Ricciardi-Castagnoli. 1996. The role of cytokines in functional regulation and differentiation of dendritic cells. *Immunobiol.* 195:431.
4. Reis e Souza, C., P.D. Stahl, and J.M. Austyn. 1993. Phagocytosis of antigens by Langerhans cells in vitro. *J. Exp. Med.* 178:509.
5. Sallusto, F., M. Cella, C. Danieli, and A. Lanzavecchia. 1995. dendritic cells use macropinocytosis and the mannose receptor to concentrate macromolecules in the major histocompatibility complex class II compartment: down regulation by cytokines and bacterial products. *J. Exp. Med.* 182:389.
6. Lutz, M.B., C.U. Aßmann, G. Girolomoni, and P. Ricciardi-Castagnoli. 1996. Different cytokines regulate antigen uptake and presentation of a precursor dendritic cell line. *Eur. J. Immunol.* 26:586.
7. Girolomoni, G., Lutz, M.B., S. Pastore, C.U. Aßmann, and P. Ricciardi-Castagnoli. 1995. Establishment of a cell line with features of early dendritic cell precursors from mouse fetal skin. *Eur. J. Immunol.* 25:2163.
8. Demandolx, D. and J. Davoust. 1995. Multicolor analysis in confocal immunofluorescence microscopy. *J. Trace Microprobe Tech.* 13:217.
9. Liou, W., H.J. Geuze, and A.J. Weerkamp. 1985. Localization of macromolecular components by application of the immunogold technique on cryosected bacteria. *Methods Microbiol.* 21:211.
10. Anderson, R.G.W., J.R. Falk, J.L. Goldstein, and M.S. Brown1984. Visualization of acidic organelles in intact cells by electron microscopy. Proc. Natl. Acad. Sci USA 81:4838.
11. Mellman, I., R. Fuchs, and A. Helenius. 1986. Acidification of the endocytic and exocytic pathways. *Annu. Rev. immunol.* 55:663.
12. Rodriguez, G.M., and S. Diment. 1995. Destructive proteolysis by cysteine protease in antigen presentation of ovalbumin. *Eur. J. Immunol.* 25:1823.
13. Bohley, P., and P.O. Seglen. 1992. Protease and proteolysis in the lysosome. *Experientia* 48:151.
14. Crowle, A.J., R. Dahl, E. Ross, and M.H. May. 1991. Evidence that vesicles containing living, virulent Mycobacterium tuberculosis or Mycobacterium avium in cultured human macrophages are not acidic. *Infect Immun.* 59:1823.
15. Pancholi, P., A. Mirza, N. Bhardwaj, and R.M. Steinman. 1993. Sequestration from immune $CD4^+$ T cells of Mycobacteria growing in human macrophages. *Science* 260:984.
16. Kleijmeer, M.J., M.A. Ossevoort, C.J.H. van Veen, J.J. van Hellmond, J.J Neefjes, W.M. Kast, C.J.M. Melief, and H.J. Geuze. 1995. MHC class II compartment and the kinetics of antigen presentation in mouse spleen dendritic cells. *J. Immunol.* 154:5715.

28

MANNOSE RECEPTOR MEDIATED UPTAKE OF ANTIGENS STRONGLY ENHANCES HLA-CLASS II RESTRICTED ANTIGEN PRESENTATION BY CULTURED DENDRITIC CELLS

M. C. Agnes A. Tan,[1] A. Mieke Mommaas,[2,3] Jan Wouter Drijfhout,[1] Reina Jordens,[1] Jos J. M. Onderwater,[2] Desiree Verwoerd,[4] Aat A. Mulder,[3] Annette N. van der Heiden,[1] Tom H. M. Ottenhoff,[1] Marina Cella,[5] Abraham Tulp,[4] Jacques J. Neefjes,[4] and Frits Koning[1]

[1]Department of Immunohematology and Blood Bank
[2]Laboratory for Electron Microscopy
[3]Department of Dermatology of the Leiden University Hospital in Leiden
The Netherlands
[4]Division of Cellular Biochemistry of The Netherlands Cancer Institute
Amsterdam, The Netherlands
[5]Basel Institute for Immunology
Basel, Switserland

1. SUMMARY

Dendritic cells (DCs) use macropinocytosis and mannose receptor mediated endocytosis for the uptake of exogenous antigens. Here we show that the endocytosis of the mannose receptor and mannosylated antigen is distinct from that of a non-mannosylated antigen. Shortly after internalization, however, both mannosylated and non-mannosylated antigen are found in an MIIC like compartment. The mannose receptor itself does not reach this compartment, and probably releases its ligand in an earlier endosomal structure. Finally, we found that mannosylation of peptides strongly enhanced their potency to stimulate HLA class II-restricted peptide-specific T cell clones. Our results indicate that mannosylation of antigen leads to selective targeting and subsequent superior presentation by DCs which may be useful for vaccine design.

2. INTRODUCTION

As professional antigen presenting cells, dendritic cells (DC) are able to capture, process and present antigens. The efficiency of antigen uptake critically affects peptide loading and presentation by MHC class II molecules (1,2). DCs internalize soluble proteins via fluid phase macropinocytosis and mannose receptor-mediated endocytosis (3). In both cases, antigens are delivered to endocytic compartments where they can be processed to antigenic peptides that can bind to MHC class II molecules and be transported to the cell surface (4,5). The mannose receptor has been shown to select ligands carrying terminal sugars such as mannose, fucose, and N-acetylglucosamine (6,7), and to have the capacity to recycle (8), allowing continuous uptake of ligands.

An important question that remained sofar unresolved, is whether the mannose receptor-ligand pathway intersects with that of the MHC class II molecules. We have therefore studied the intracellular transport and subcellular distribution of the mannose receptor, mannosylated and non-mannosylated proteins, HLA-DM, and MHC class II molecules using density gradient electrophoresis (DGE) in combination with an immunoelectronmicroscopical analysis. In addition, we determined the effect of mannosylation of peptide antigens on MHC class II restricted antigen presentation by DCs.

3. EXPERIMENTAL PROCEDURES

3.1. Antibodies

The following antibodies were used: anti-human class I mAb HC10, anti-human class II α-chain mAb 1B5, anti-human class II ß-chain mAb anti-ß, anti-human DMA mAB 5C1, mouse anti-human class II mAb TU36, anti-human class II mAb PdV 5.2, anti-biotin, anti-CD63 mAb CLB/Gran12, anti-human calnexin, anti-human mannose receptor D547, anti-human cathepsin D, anti-human rab 5, anti-human lamp-1 mAb BB6, and anti-ß-hexosaminidase.

3.2. Isolation of Human DCs from Peripheral Blood

DCs were isolated from PBMC according to (9), and cultured in RPMI-10% FCS supplemented with 800 U/ml GM-CSF and 1000 U/ml IL-4. DCs were used for experiments between day 6–8.

3.3. Subcellular Fractionation

Subcellular vesicles were fractionated according to their cell surface charge by density gradient electrophoresis as described in (10).

3.4. Immunoelectron Microscopy

For immunoelectron microscopy DCs were processed as described in (11). Ultrathin cryosections were incubated with antibodies followed by immunogold staining.

3.5. Proliferation Assay

1.0×10^4 T cells were cultured in 96 flat-bottom microplates with 5×10^3 irradiated (3000 rad) HLA-typed DCs or 5×10^4 irradiated (3000 rad) HLA-typed PBMCs in the absence or presence of various concentrations of peptides. [^{3}H]Thymidine incorporation was measured at day 3 after a 20 h pulse with [^{3}H]thymidine (1 Ci/well).

4. RESULTS AND DISCUSSION

4.1. Subcellular Distribution of the Mannose Receptor and Proteins Involved in Antigen Presentation

In order to label endocytic vesicles, cultured DCs were pulsed with horse radish peroxidase (HRP), a mannose receptor ligand (12), and chased. Subsequently subcellular vesicles were fractionated according to their cell surface chargeand analysed. Already after a 3 min pulse HRP appeared to be internalized, as it was located at a position distinct from surface bound HRP. The HRP traveled through endocytic compartments to lysosomes within 25 min. Subcellular distribution of the mannose receptor and proteins associated with antigen presentation showed that, as expected, mature HLA-DM was present in lysosomes, immature HLA-DM in the ER and cis-Golgi, MHC class I molecules on the plasma membrane and early endosomes, and MHC class II molecules on the cell surface and intracellularly. The mannose receptor, however, had a distribution that was distinct from the other proteins analyzed.

4.2. Internalization of the Ligand-Mannose Receptor Complex and Fluid Phase Endocytosis

To study internalization of the mannose receptor-ligand complex and fluid phase endocytosis, DCs were pulsed with bovine serum albumin (BSA) or mannosylated-BSA and processed for immunoelectronmicroscopical analysis. We observed that uptake of mannose-receptor-antigen complexes took place by a coated pit and coated vesicle pathway, while non-mannosylated antigens were endocytosed in larger vesicles, most likely macropinosomes. Shortly after internalization, however, these two pathways converged and within 25 minutes, both mannosylated and non-mannosylated antigens co-localised in an MIIC/lysosomal like compartment. In contrast, the mannose receptor itself did not reach this compartments, suggesting that it releases its ligand in an earlier endosomal structure. Thus, antigens internalized via receptor-mediated and fluid phase endocytosis enter MIIC compartments and lysosomes, but the mannose receptor appears to release its ligand in an earlier endocytic compartment. Together these new findings indicate that after endocytosis the receptor releases its ligand, allowing recycling of the receptor and transport of the mannosylated ligand to the lysosomal system including the MIIC.

4.3. Antigen Presentation of Mannosylated Antigens

To examine the effect of mannosylation on antigen presentation, we used three different known T cell peptide epitopes. Using DCs as antigen presenting cells we observed, that approximately 300 – 10,000 fold less of the mannosylated peptide was required for the induction of half-maximal proliferation compared to the non-mannosylated analogue.

In summary, our results indicate that DCs are extremely efficient in the uptake, transport and presentation of mannosylated antigens. This may have important implications for the improvement of immunisation protocols.

5. ACKNOWLEDGMENTS

We thank Dr. J. Trowsdale for the generous gift of the anti-human HLA-DMA antibody, Dr. P. Chavrier for the generous gift of the anti-human rab 5 antibody. This work was financially supported by the Netherlands Organisation for Scientificesearch Grant 030–93–001.

6. REFERENCES

1. Lanzavecchia, A. (1990). Annu. Rev. Immunol. 8, 773–793.
2. West, M.A., Lucocq, J.M., and Watts, C. (1994). Nature 369, 147–151.
3. Sallusto, F., Cella, M., Danieli, C., and Lanzavecchia, A. (1995). J. Exp. Med. 182, 389–400.
4. Cresswell, P. (1994). Curr. Biol. 4, 541–543.
5. Wolf, P.R. and Ploegh, H.L. (1995). Annu. Rev. Cell Dev. Biol. 11, 267–306.
6. Stahl, P.D. (1992). Curr. Opin. Immunol. 4, 49–52.
7. Avrameas, A., McIlroy, D., Hosmalin, A., Autran, B, Debre, P., Monsigny, M., Roche, A.C., and Midoux, P. (1996). 26, 394–400.
8. Stahl, P.D., Schesinger, P.H., Sigardson, E., Rodman, J.S., and Lee, Y.C. (1980). Cell 19, 207–215.
9. Sallusto, F. and Lanzavecchia, A. (1994). J. Exp. Med. 179, 1109–1116.
10. Tulp., A., Verwoerd, D., Fernandez-Borja, M., Neefjes, J.J., Hart, A.A.A. (1996). Electrophoresis 17, 173–178.
11. Mommaas, A.M., Wijsman, M.C., Mulder A.A., Van Praag, M.C.G., Vermeer, B.J., and Koning, F. (1992). Human Immunol. 34, 99–106.
12. Sung, S.-S., Nelson, R.S., and Silverstein, S.C. (1983). J. Cell. Physiol. 116, 21–25.

29

MECHANISMS OF FcεRI-IgE-FACILITATED ALLERGEN PRESENTATION BY DENDRITIC CELLS

Dieter Maurer,[1*] Christof Ebner,[3] Bärbel Reininger,[1] Peter Petzelbauer,[2] Edda Fiebiger,[1] and Georg Stingl[1]

[1]Division of Immunology
Allergy and Infectious Diseases (DIAID)
[2]Division of General Dermatology
Department of Dermatology, and
[3]Institute of General and Experimental Pathology
Division of Immunopathology
University of Vienna Medical School
Vienna, Austria

INTRODUCTION

Dendritic cells (DC) are of key importance for the induction of antigen-specific T cell immunity. In their immature state, DC capture antigen at peripheral sites and, thereafter, migrate to secondary lymphoid organs where they, as matured cells, efficiently present antigen-derived peptides to naive T cells (1). It is obvious that the antigen recognition, capture and internalization functions of immature DC are of major biological relevance as they are the prerequisite for the occurrence, quantity and, perhaps, quality of antigen-specific T cell responses. For a long period, it was generally believed that antigen uptake by DC is mainly accomplished by micropinocytosis. In addition to this non-selective process, three other antigen uptake mechanisms of DC were recently discovered (2). The first is macropinocytosis, an actin-dependent type of fluid phase endocytosis that is mediated by membrane ruffling and characterized by the appearance of large cytoplasmic vesicles in which macrosolutes become concentrated (3). The two other uptake mechanisms are antigen/IgG complex- or glycosylation pattern-specific and rely on the presence of cell surface receptors. DC express FcγRII- and the mannose receptor that enable efficient capture of IgG complexed antigens (4, 5) and mannosylated/fucosylated antigens (3, 6), respec-

* Address correspondence to: Dieter Maurer, M.D., Division of Immunology, Allergy and Infectious Diseases (DIAID), Department of Dermatology, University of Vienna Medical School, Währinger Gürtel 18–20, A-1090 Vienna, Austria. Telephone: +43 1 4036933; Fax: +43 1 4031900.

tively. In rodents, DC express a further C-type lectin receptor, DEC-205, which recognizes specific glycosylation patterns on proteins and, thus, may be involved in the selective uptake and presentation of certain pathogens (7).

Recently, we reported that FcεRI can function as an allergen-focusing structure which, in the case of monocyte-enriched mononuclear cells of atopic individuals, amplifies allergen presentation in an IgE-dependent manner (8). In this chapter we will describe the characteristics of a newly discovered FcεRI-expressing DC population and the consequences of IgE-FcεRI interactions for the antigen presentation capacity of these cells.

VARIOUS TYPES OF DENDRITIC CELLS EXPRESS FcεRI

In humans, the high affinity receptor for IgE (FcεRI) is expressed on Langerhans cells (9, 10), dermal DC (9, 11), and monocytes (12). To better characterize the spectrum of FcεRI-expressing antigen-presenting cells (APC), we recently screened peripheral blood for HLA-DR$^+$ (i.e., presumably antigen-presenting) cells that can bind monomeric IgE via FcεRI but fail to express monocyte-specific markers (e.g., CD14). We found that, in addition to basophils and monocytes, a third type of FcεRI-expressing cells circulates in the peripheral blood of both atopic and healthy individuals (13). On the surface of these cells, FcεRI occurs as a multimeric structure containing FcεRIα and FcεRIγ chains but, unlike its counterpart on basophils, lacking FcεRIβ. Further experiments revealed that these FcεRIαγ-expressing cells are peripheral blood DC (PB-DC) based on their immunophenotype (HLA-DR^{high}, HLA-DQ^{high}; CD4$^+$, CD11a$^+$, CD32$^+$, CD33$^+$, B7/2 [CD86]$^+$; CD11b^{low}, CD14^{low}, CD40^{low}, CD54^{low}, CD64^{low}), typical cell morphology, and the capacity to induce primary alloreactive T cell responses (13).

FcεRI-IgE-FACILITATED ANTIGEN UPTAKE AND PRESENTATION BY DENDRITIC CELLS

To see whether targeting of allergen to FcεRI on DC is followed by efficient antigen presentation, we used an allergen focusing system that allows the binding of haptenized (i.e., nitrophenacetyl [NP]-conjugated) recombinant major birch pollen allergen (r*Bet v* 1) to Fc-IgE receptors via NP-specific monomeric IgE and measured the proliferative response of allergen-specific, peptide-mapped T cell clones (TCC, ref. 8). In these experiments, the addition of hapten-specific IgE to DC cultures results in an approximately 10-fold amplified proliferative TCC response at low concentrations of the haptenized allergen (Fig. 1). Furthermore, the presence of hapten-specific IgE reduced the amount of haptenized allergen required to yield a significant TCC proliferation by a factor of 100. These experiments identified FcεRI as# the principal IgE receptor on DC mediating IgE-dependent allergen uptake and presentation since the addition of an anti-FcεRIα mAb - but not of an isotype-matched control mAb - reduced the DC-induced TCC proliferation to levels seen with allergen in the absence of hapten-specific IgE (Fig. 1).

In recent experiments we started to explore the molecular mechanisms involved in FcεRI/IgE-mediated antigen uptake by DC. We found that multimerization but not monovalent ligation of FcεRI-bound IgE leads (i) to the efficient internalization of triggered receptors and, subsequently, (ii) to the focusing of FcεRI/IgE complexes into few distinctive cytoplasmic DC compartments which contain MHC class II antigens, HLA-DM and lysosome-associated proteins (Maurer et al., manuscript in preparation). Crosslinking-induced activation of FcεRI rather than constitutive membrane turnover is responsible for

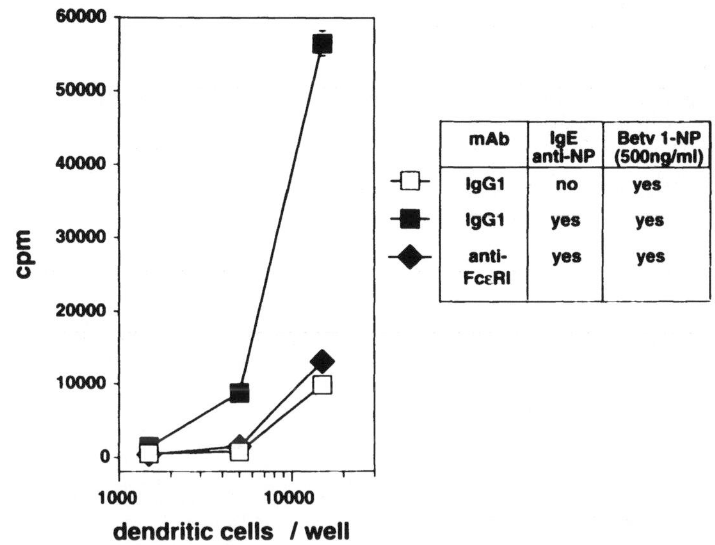

Figure 1. DC use FcεRI to present allergen to HLA-DR-restricted T cells in an IgE-dependent manner. Indicated numbers of peripheral blood DC were reacted with IgG1 (open squares), IgG1 plus NP-specific cIgE (closed squares), or mAb 15–1 (anti-FcεRIα) plus NP-specific cIgE (diamonds) followed by the addition of 500 ng/ml NP-conjugated r*Bet v* 1. T cells of TCC WD24 were added as responders and [^{3}H]-thymidine uptake was measured (cpm; ordinate).

receptor internalization since compounds interfering with protein tyrosine phosphorylation but not agents that perturb actin and/or microtubule organization/remodeling can inhibit this process (Maurer et al., manuscript in preparation). Tyrosine-based sequences have been identified in the cytoplasmic tail of certain other receptors and mediate their internalization (14), interaction with clathrin-associated protein complexes (15) and, in some cases, delivery of internalized and/or newly synthesized receptors into peptide-loading compartments (16, 17). It is thus tempting to speculate that the crosslinking-induced tyrosine phosphorylation of some of the YXXL sequences contained within the ITAM motifs of the FcεRI γ-chains critically controls receptor internalization and subcellular targeting.

IMPLICATIONS OF FcεRI-IgE-DEPENDENT ALLERGEN UPTAKE FOR ATOPIC ALLERGY

Dendritic cells, as opposed to semi-professional APC, have the indigenous ability to stimulate primary T cell responses (1). FcεRI-expressing DC may acquire the additional property to activate and prime naive T cells to hitherto unrecognized epitopes of IgE-reactive antigens/allergens since the IgE-dependent allergen uptake may influence the repertoire of allergen-derived peptides being generated and presented. As previously described for membrane Ig-dependent antigen presentation by B cells (18), epitopes located outside the IgE recognition site on the allergen may be preferentially presented whereas the HLA-class II-peptide assembly of those located within the binding site of IgE may be suppressed. In summary, FcεRI/IgE-dependent allergen uptake by DC may both quantitatively and qualitatively modulate allergen presentation in vitro and may have profound implications on the magnitude and diversification of allergen-specific T cell responses in human disease.

ACKNOWLEDGMENTS

This work was supported, in part, by grants from the Austrian Science Foundation (S06702-MED and S06704-MED) and from the Sandoz Research Institute, Vienna, Austria.

REFERENCES

1. R.M. Steinman, The dendritic cell system and its role in immunogenicity, *Annu. Rev. Immunol.* 9:271 (1991).
2. R.M. Steinman and J. Swanson, The endocytotic activity of dendritic cells, *J. Exp. Med.* 182: 283 (1995).
3. F. Sallusto, M. Cella, C. Danielli, and A. Lanzavecchia, Dendritic cells use macropinocytosis and the mannose receptor to concentrate antigen in the MHC II compartment. Downregulation by cytokines and bacterial products, *J. Exp. Med.* 182:389 (1995).
4. F. Manca, D. Fenoglio, G.L. Pira, A. Kunkel, and F. Celada, Effect of antigen/antibody ratio on macrophage uptake, processing, and presentation to T cells of antigen complexed with polyclonal antibodies, *J. Exp. Med.* 173:37 (1991).
5. F. Sallusto and A. Lanzavecchia, Efficient presentation of soluble antigen by cultured human dendritic cells is maintained by granulocyte/macrophage colony-stimulating factor plus interleukin 4 and downregulated by tumor necrosis factor α. *J. Exp. Med.* 179:1109 (1994).
6. P.D. Stahl, The mannose receptor and other macrophage lectins, *Curr. Opin. Immunol.* 4: 49 (1992).
7. W. Jiang, W.J. Swiggard, C. Heufler, M. Peng, A. Mirza, R.M. Steinman, and M.C. Nussenzweig, DEC-205, a receptor expressed by dendritic cells and thymic epithelial cells, has ten C-type lectin domains and is involved in antigen presentation, *Nature* 305:151 (1995).
8. D. Maurer, C. Ebner, B. Reininger, E. Fiebiger, D. Kraft, J.-P. Kinet, and G. Stingl, The high affinity immunoglobulin E receptor (FcεRI) mediates IgE-dependent allergen presentation, *J. Immunol.* 154: 6285 (1995).
9. B. Wang, A. Rieger, O. Kilgus, K. Ochiai, D. Maurer, D. Födinger, J.-P. Kinet, and G. Stingl, Epidermal Langerhans cells from normal human skin bind monomeric IgE via FcεRI, *J. Exp. Med.* 175:1353 (1992).
10. T. Bieber, H. de la Salle, A. Wollenberg, J. Hakimi, R. Chizzonite, J. Ring, D. Hanau, and C. de la Salle, Human epidermal Langerhans cells express the high affinity receptor for immunoglobulin E (FcεRI), *J. Exp. Med.* 175:1285 (1992).
11. B. Osterhoff, C. Rappersberger, B. Wang, F. Koszik, K. Ochiai, J.-P. Kinet, and G. Stingl, Immunomorphological characterization of FcεRI-bearing cells within the human dermis, *J. Invest. Dermatol.* 102:315 (1994).
12. D. Maurer, E. Fiebiger, B. Reininger, B. Wolff-Winiski, M.-H. Jouvin, O. Kilgus, J.-P. Kinet, and G. Stingl, Expression of functional high affinity IgE receptors (FcεRI) on monocytes of atopic individuals, *J. Exp. Med.* 179:745 (1994).
13. D. Maurer, E. Fiebiger, C. Ebner, B. Reininger, G.F. Fischer, S. Wichlas, M.-H. Jouvin, M. Schmitt-Egenolf, D. Kraft, J.-P. Kinet, and G. Stingl, Peripheral blood dendritic cells express FcεRI as a complex composed of FcεRIα- and FcεRIγ-chains and can use this receptor for IgE-mediated allergen presentation, *J. Immunol.* 157: 607 (1996).
14. I.S. Townbridge and J.F. Collawn, Signal-dependent membrane protein trafficking in the endocytotic pathway, *Annu. Rev. Cell Biol.* 9: 129 (1993).
15. H. Ohno, J. Stewart, M.-C. Fournier, H. Bosshart, I. Rhee, S. Miyatake, T. Saito, A. Gallusser, T. Kirchhausen, and J.S. Bonifacio, Interaction of tyrosine-based sorting signals with clathrin-associated proteins, *Science* 269:1872 (1995).
16. M. Sugita, R.M. Jackman, E. van Donselaar, S.M. Behar, R.A. Rogers, P.J. Peters, M.B. Brenner, and S.A. Porcelli, Cytoplasmic tail-dependent localization of CD1b antigen presenting molecules to MIICs, *Science* 273: 349 (1996).
17. R. Lindstedt, M. Liljedahl, A. Péléraux, P.A. Peterson, and L. Karlsson, The MHC class II molecule H2-M is targeted to an endosomal compartment by a tyrosine-based targeting motif, *Immunity* 3: 561 (1995).
18. C. Watts and A. Lanzavecchia, Suppressive effect of antibody on processing of T cell epitopes, *J. Exp. Med.* 187:1459 (1993).

30

LOCALIZATION AND INTRACELLULAR TRANSPORT OF MHC CLASS II MOLECULES IN BONE MARROW-DERIVED DENDRITIC CELLS

P. Pierre,[1] S. J. Turley,[1] J. Meltzer,[1] A. Mirza,[1] R. Steinman,[2] and I. Mellman[1]

[1]Department of Cell Biology
Yale School of Medicine
New Haven, Connecticut
[2]Laboratory of Cellular Physiology and Immunology
The Rockefeller University
New York, New York

INTRODUCTION

Expression of class II molecules is restricted to certain cell types, including B lymphocytes, dendritic cells (DC) from various tissues (e.g. Langerhans cells from skin) and macrophages. Dendritic cells are a system of potent antigen presenting cells (APC) that are characterized by their strong capacity to stimulate immunologically naive T cells[1,2]. A key function of DC is thought to be the acquisition of antigens in peripheral tissues and their transport to draining lymph nodes for presentation of the processed peptides to the T lymphocytes. Until recently, the major limitation to study the cell biology of dendritic cells has been the absence of long term cell lines and clones. However, the finding that granulocyte-macrophage colony stimulating factor (GM-CSF)[3] promotes growth and maturation of large quantities of DC issued from bone marrow progenitors[4] provided us with a tool to study the MHC class II transport and distribution in these important APCs. Antigens must be converted in short peptides and loaded on the major histocompatibility complex (MHC) molecules before they can trigger an immune response[5]. Class II associated peptides are derived from extracellular proteins or endogenous proteins that have access to the endocytic pathway. Class II molecules associate with the Invariant chain (Ii) in the endoplasmic reticulum (ER)[6] and are targeted to the endocytic pathway. Class II molecules accumulate transiently in endocytic compartments designated MHC class II compartment (MIIC)[7,8,9] or class II vesicles (CIIV)[10,11]. CIIV are physically and biochemically distinct from endosomes and lysosomes but do, however, contain early endocytic markers such as transferrin receptor and surface immunoglobulins. MIIC, on the other hand, are depleted of recycling receptors and are enriched in late endocytic markers such as Lamp 1. Bone marrow derived DC were cultivated in presence of GM-CSF and a cell biological analysis was performed at different periods of their development. we have shown that major changes in the distribution of class II molecules can be observed during the maturation of DC. Additionaly, we have characterized a new population of DC, representing an intermediate stage of maturation in which CIIV and MIIC can coexist. This developmental switch is coordinated with changes in the distribution of the invariant chain, as well as a dramatic acceleration of class II transport to the plasma membrane.

RESULTS

Mouse bone marrow progenitor cells act as a major source of dendritic cells when cultivated with GM-CSF[4]. Non adherent mature dendritic cells are generated in one week from loosely adherent proliferation clusters upon treatment with GM-CSF. We have produced large amounts of dendritic cells *in vitro* to study MHC class II distribution and transport during maturation. Immunofluorescence microscopy studies allowed us to identify and characterize 3 major developmental stages of DC throughout the culture (Fig. 1). Early in development (day 3–4), DCs are mainly concentrated in proliferating clusters which are loosely attach to a more firmly adherent stroma[4]. These clustered DCs have been shown to give rise to mature dendritic cells when purified and left in culture with GM-CSF[4]. Most of the cells present in clusters show little class II molecules on their surface and have their internal class II molecules concentrated in scattered MIIC/lysosomes[12]. These observations were confirmed by percoll gradient fractionation of early DC, in which MHC class II molecules are found co-distributing with heavy density lysosomes. Later during maturation (day 5–6), an intermediate population of DC could be detected. These non adherent cells display a large pool of class II-positive/LGP-negative intracellu-

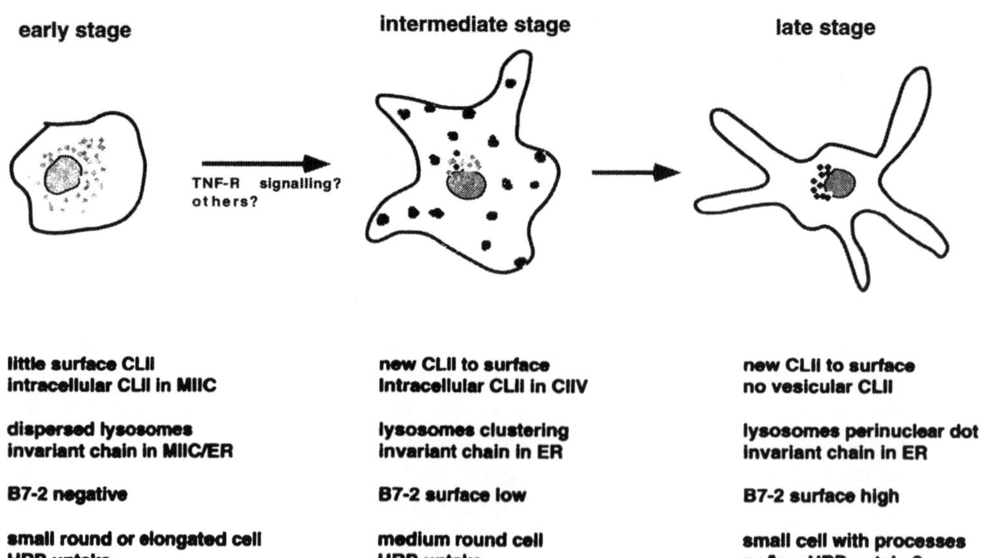

Figure 1. Developmental cell biology of mouse dendritic cells.

lar vesicles localized at the periphery of the cell and likely to be CIIV. At the final stages of maturation, the DC population (day 7–8) reaches a stage of development described by most investigators as "mature" dendritic cells[2]. These cells display a characteristic dendritic morphology with most of their class II molecules on the cell surface and their intracellular compartments strongly depleted of class II molecules (in particular MIIC/lysosomes). These observations were also confirmed by percoll gradient fractionation of late DC in which MHC class II molecules do not co-distribute with heavy density lysosomes. Interestingly, in mature cells, the MHC class II depleted MIIC/Lysosomes are still strongly positive for H2-M[13].

Class II dimers and Ii are strongly interconnected, Ii serves as a chaperone and provides the targeting signal necessary for the class II molecules delivery to the endocytic pathway[14]. In the early cells, Invariant chain can be detected as diffuse pattern likely to be ER and in sharper vesicular structures positive for class II and LGP. Such a localization was confirmed by percoll gradient fractionation where Ii and MHC class II molecules are found co-distributing with heavy density lysosomes. Interestingly, only the CLIP containing p10 proteolytic fragment of Ii[15] could be detected in the heavy density fractions. The p10 fragment is normally generated in early endosomes and is very short lived, unless the cells are incubated with a protease inhibitor such as leupeptin[16]. In contrast, invariant chain could only be found within the ER of intermediate and late DC, where it colocalizes with Calnexin[17]. The distribution of invariant chain in late DC was confirmed by percoll fractionation where it could only be detected in light density fraction. Interestingly P10 can also be found in light density fractions (both in early and late DC), confirming that the Ii does not need to reach the MIIC/lysosomes to get degraded[18]. P10, still able to influence the targeting of the class II dimers, could induce the class II molecules transport to MIIC/lysosomes, in a similar fashion as an artificial treatment with leupeptin[19,16]. In summary, the changes in class II distribution are probably dependent on invariant chain degradation rate in the endocytic pathway.

A series of pulse chase experiments, followed by surface biotinylation were performed to monitor the rate of class II transport to the plasma membrane. In early DC, MHC class II molecules show a rather poor rate of arrival at the plasma membrane when compared with late DC or B cells. Class II molecules begin to appear at the surface of intermediate/late DC after one hour of chase reaching a maximum after four hours as already described for B cells[10]. In the case of early DC, little class II molecules have reached the plasma membrane after 4 hours, as suggested by the low amount of class II observed on their cell surface by immunofluorescence.

CONCLUSION

Dendritic cells have been shown to have dramatic changes in their ability to internalize and to present antigens, while treated with different cytokines such as GM-CSF or TNF-α. These changes were generally coordinated with down regulation of several molecules (e. g. Fc receptor) or upregulation of adhesion or costimulatory molecules (e.g. B7–2, DEC-205) (Fig. 1). We demonstrated, that in the mouse bone marrow system, GM-CSF while influencing the differentiation of progenitor cells in mature DC, induces dramatic changes in the distribution and the transport of class II molecules in the maturing cells. These changes in class II molecules distribution and trafficking define 3 major stages of DC maturation (Fig.1). At an early maturation stage, the reduced transport of class II molecules to the plasma membrane is concomitant with their targeting to

MIIC/lysosomes and a slow degradation of invariant chain in this compartment. Later stages of development are charaterized by a kinetic of transport of the class II molecules to the cell surface comparable to the one observed for B cells and the absence of class II molecules and invariant chain in the MIIC/lysosomes. This developmental switch is also characterized by the presence of Class II molecules into an LGP-negative compartment likely to be the equivalent of CIIV. In the bone marrow system a slight increase (25%) of the class II synthesis rate could be observed. This difference in synthesis is unlikely to be responsible for the dramatic changes in the transport and the targeting of the class II molecules during maturation. In conclusion, these observations suggest that class II molecules transport in DCs treated with GM-CSF is developmentally regulated. The rate of Ii degradation seems to affect directly the class II trafficking, the amount and the type of specific proteases available (cathepsin D, L, S) could play a direct role in the developmental regulation of the transport. The nature of this cytokine dependent regulatory mechanism still remain unclear and the culture of bone marrow derived DC will be of great help to elucidate this process.

REFERENCES

1. Fossum, S. *Curr. Top. Pathol.* **15**, 101–124 (1989).
2. Steinman, R.M. *Annual Review of Immunology* **9**, 271–96 (1991).
3. Rasko, J.E. & Gough, N.M. in *The Cytokine Handbook, 2nd ed* 344–369 (Academic Press Limited, 1994).
4. Inaba, K., *et al. Journal of Experimental Medicine* **176**, 1693–702 (1992).
5. Mellman, I., Pierre, P. & Amigorena, S. *Curr. opin. Cell Biol.* **7**, 564–572 (1995).
6. Lamb, C.A. & Cresswell, P. *J. Immunol.* **148**, 3478–82 (1992).
7. Peters, P.J., Neefjes, J.J., Oorschot, V., Ploegh, H.L. & Geuze, H.J. *Nature* **349**, 669–676 (1991).
8. Tulp, A., Verwoerd, D., Dobberstein, B., Ploegh, H.L. & Pieters, J. *Nature* **369**, 120–126 (1994).
9. Qiu, Y., Xu, X., Wandinger-Ness, A., Dalke, D.P. & Pierce, S.K. *J. Cell Biol.* **125**, 595–605 (1994).
10. Amigorena, S., Drake, J.R., Webster, P. & Mellman, I. *Nature* **369**, 113- 120 (1994).
11. West, M.A., Lucocq, J.M. & Watts, C. *Nature* **369**, 147–151 (1994).
12. Kleijmeer, M.J., *et al. J. Immunol.* **154**, 5715–5724 (1995).
13. Sanderson, F., *et al. Science* **266**, 1566–1569 (1994).
14. Cresswell, P. *Cell* **84**, 505–507 (1996).
15. Romagnoli, P. & Germain, R.N. *J. Exp. Med.* **180**, 1107–1113 (1994).
16. Amigorena, S., *et al. J. Exp. Med.* **181**, 1729–1741 (1995).
17. Hammond, C. & Helenius, A. *J. Cell Biol.* **126**, 41–52 (1994).
18. Romagnoli, P., Layet, C., Yewdell, J., Bakke, O. & Germain, R.N. *J. Exp. Med.* **177**, 583–96 (1993).
19. Neefjes, J.J. & Ploegh, H.L. *EMBO Journal* **11**, 411–6 (1992).

31

MANNOSE RECEPTOR MEDIATED ANTIGEN UPTAKE AND PRESENTATION IN HUMAN DENDRITIC CELLS

Anneke J. Engering,[1,2] Marina Cella,[1] Donna M. Fluitsma,[2]
Elisabeth C. M. Hoefsmit,[2] Antonio Lanzavecchia,[1] and Jean Pieters[1]

[1]Basel Institute for Immunology
Basel, Switzerland
[2]Department Cell Biology and Immunology
Vrije Universiteit
Amsterdam, The Netherlands

1. ABSTRACT

In an immature state, dendritic cells (DC) can capture antigen via at least two mechanisms. First, DC use macropinocytosis for continuous uptake of large amounts of soluble antigens. Second, they express high levels of mannose receptor that can mediate internalization of glycosylated ligands. We found that dendritic cells can present mannosylated antigen 100 – 1000 fold more efficiently than non-mannosylated antigen. Immunocytochemistry as well as subcellular fractionation demonstrated that the mannose receptor and MHC class II molecules were located in distinct subcellular compartments. These results demonstrate that the mannose receptor endows DC with a high capacity to present glycosylated antigens at very low concentrations.

2. INTRODUCTION

Antigen presenting cells can internalize antigens via fluid phase or receptor-mediated uptake.[1] After uptake, antigens are processed in the endosomal/lysosomal pathway and peptides are loaded onto class II molecules in specialized compartments, MHC class II compartments.[2–6] From these compartments, class II/peptide complexes are transported to the plasma membrane for presentation to T cells.

Dendritic cells (DC) are highly efficient antigen presenting cells capable of inducing a naive T cell response (reviewed in 7). Features that contribute to efficient antigen presentation include high expression of class II molecules and a capacity to endocytose large amounts of material.[8,9] Two distinct mechanisms of uptake have been identified in DC;

Dendritic Cells in Fundamental and Clinical Immunology
edited by Ricciardi-Castagnoli, Plenum Press, New York, 1997

first, macropinocytosis, that allows high levels of fluid phase uptake and second, receptor-mediated uptake via the mannose receptor.[9,10] Here, we analyzed presentation of antigens internalized via the second route. Mannosylated antigen was presented at very low concentrations compared to non-mannosylated antigen. In addition, the intracellular compartment containing the mannose receptor was found to be distinct from MHC class II compartments. These results suggest that the mannose receptor efficiently concentrates antigens for transport to processing and peptide loading compartments.

3. MATERIALS AND METHODS

3.1. Cells

Human dendritic cells were obtained by *in vitro* culture of elutriated peripheral blood monocytes in RPMI-1640 containing 10% fetal calf serum supplemented with 50 ng/ml GM-CSF and 1,000 U/ml IL-4, as described.[8]

3.2. Immuno-Electron Microscopy

Cells were fixed at room temperature in 2% paraformaldehyde and 0.5% glutaraldehyde in phosphate buffered saline (PBS); after 1 hour, the cells were pelleted, resuspended in 2% paraformaldehyde and transferred to 4°C. After infusion with 2.3 M sucrose, samples were quickly frozen in liquid nitrogen.[11] Ultrathin cryosections were prepared as described[12] and labeled with antibodies, followed by colloidal gold particles coupled to protein A.[13]

3.3. Subcellular Fractionation

Subcellular fractionation was performed as described.[5] Markers for lysosomes, as well as the total amount of protein in the different fractions were assayed as described.[14,15] The migration of plasma membrane was analyzed by cell surface iodonation.[5] The distribution of MHC class II molecules and mannose receptor was determined by immunoblotting using anti-MHC class II and anti-mannose receptor antibodies followed by incubation with peroxidase-coupled secondary antibodies and visualization by chemoluminescence.

4. RESULTS AND CONCLUSIONS

To localize the mannose receptor intracellularly, dendritic cells were fixed and ultrathin cryosections were prepared. Immunolabeling using anti-mannose receptor antibodies revealed the presence of mannose receptor in coated pits and small vesicles (Fig. 1, 10 nm gold). Double labeling with anti-class II antibodies showed the majority of MHC class II molecules to be present in multivesicular and multilaminar compartments (Fig. 1, 15 nm gold). No colocalization of the mannose receptor and MHC class II molecules was found.

To physically isolate MHC class II and mannose receptor positive compartments, DC were homogenized and the organelles separated by electrophoresis (see methods). Fractions were collected, and the distribution of various markers for the different subcellular organelles was analyzed. As shown in Figure 2A, most organelles migrated around fraction 34, as analyzed by total protein content. Lysosomal compartments, identified by

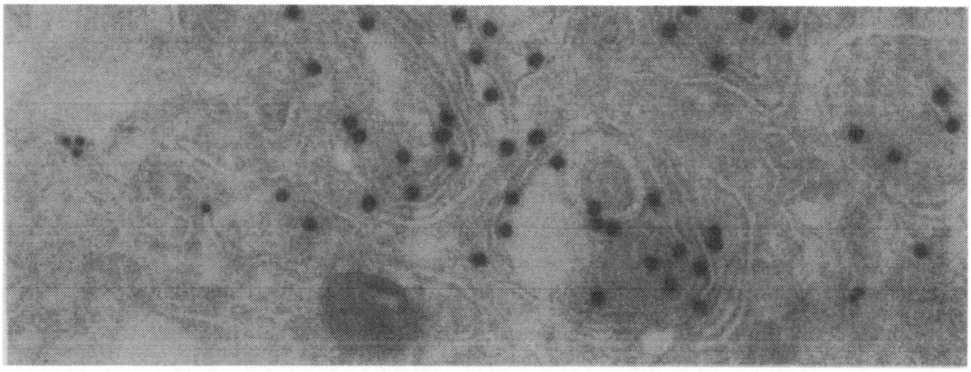

Figure 1. Immunocytochemical analysis of mannose receptor and MHC class II containing compartments. Ultrathin sections from fixed DC were double labeled with antibodies against the mannose receptor (3.29; IgG1; 10 nm gold) and MHC class II molecules (polyclonal serum against the α-chain; gift from H.L. Ploegh; 15 nm gold) as described in Methods. Magnification: 130,000x.

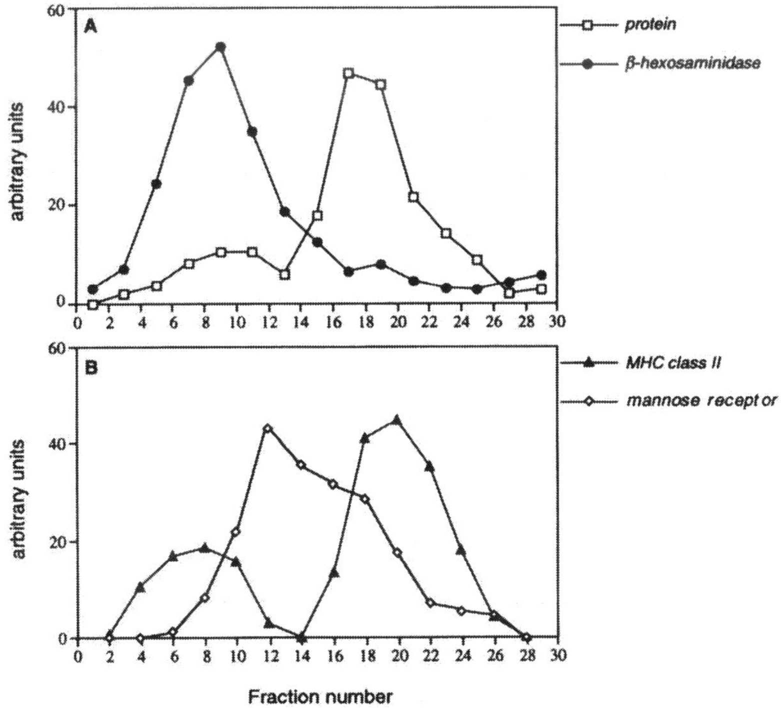

Figure 2. Isolation of distinct subcellular compartments containing mannose receptor or MHC class II molecules. DC were homogenized and the subcellular organelles were separated by electrophoresis. In each fractions the amount of protein (squares) and β-hexosaminidase activity (circles) was determined (A) as well as the amount of MHC class molecules (triangles) and of the mannose receptor (diamonds) (B).

the presence of β-hexosaminidase activity in the different fractions, migrated towards the anode (peak at fraction 24) and were well separated from most of organelles (Figure 2A).

The presence of MHC class II and mannose receptor in the different fractions was analyzed by immunoblotting; results are shown in Figure 2B. MHC class II molecules were found to be present in two distinct populations; the majority of MHC class II containing membranes migrated in fractions (34–38) containing plasma membrane (Fig. 2B and data not shown). A second population of MHC class II containing organelles migrated in anodically shifted fractions (fractions 22–26), representing MHC class II compartments. In contrast, mannose receptor containing compartments migrated at a distinct position (Fig 2B), confirming the immunocytochemical analysis. Taken together these results show that MHC class II and mannose receptor molecules are present in distinct compartments.

To analyze presentation of antigens internalized via the mannose receptor, DC were pulsed either with glycosylated bovine serum albumin (BSA) or with non-glycosylated BSA and their capacity to induce proliferation of a BSA-specific class II restricted T cell clone was analyzed. Mannosylated as well as fucosylated BSA was in presented 100–1000 fold more efficient than non-glycosylated BSA (Fig 3 and data not shown). T cell proliferation was still detectable at very low (<1 ng/ml) concentrations of mannosylated BSA (data not shown).

In conclusion, internalization of antigens via the mannose receptor results in highly efficient antigen presentation. The mannose receptor containing compartment involved in antigen uptake is distinct from MHC class II positive organelles, thus suggesting a sequential transit of mannosylated antigens from this compartment to processing and peptide loading compartments.

It is interesting to note that a variety of infectious microorganisms, as opposed to higher eukaryotes, express terminally mannosylated proteins.[16] The efficient presentation of mannosylated antigens together with the ability to generate naive T cell responses may render human dendritic cells especially effective during the early phases of infection.

5. ACKNOWLEDGMENTS

We thank Doris Scheidegger for expert technical assistance and Marco Colonna and Christiane Ruedl for critical reading of the manuscript. The Basel Institute for Immunology was founded and is supported by F. Hoffmann-La Roche & Co. Ltd. (Basel, Switzerland).

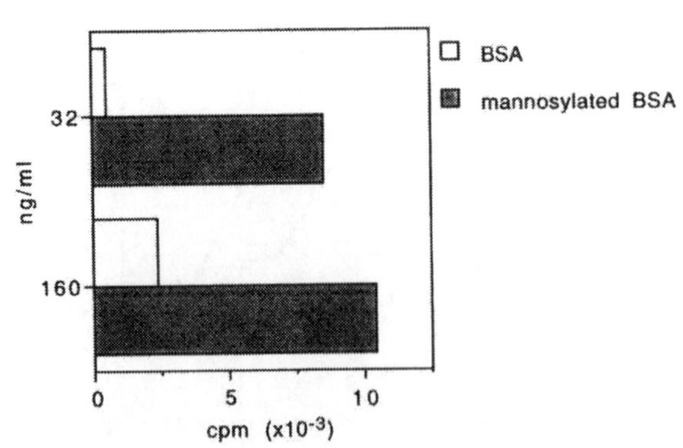

Figure 3. Enhanced antigen presentation of mannosylated BSA. DCs were incubated with mannosylated BSA or BSA and their ability to stimulate T cell proliferation was analyzed using a proliferation assay.[8]

REFERENCES

1. Lanzavecchia, A. *Current Opinion in Immunol.* **8**, 348–354 (1996).
2. Peters, P.J., Neefjes, J.J., Oorschot, V., Ploegh, H.L. & Geuze, H.J. *Nature* **349**, 669–676 (1991).
3. Pieters, J., Horstmann, H., Bakke, O., Griffiths, G. & Lipp, J. *J. Cell Biol.* **115**, 1213–1223 (1991).
4. Amigorena, S., Drake, J.R., Webster, P. & Mellman, I. *Nature* **369**, 113- 120 (1994).
5. Tulp, A., Verwoerd, D., Dobberstein, B., Ploegh, H.L. & Pieters, J. *Nature* **369**, 120–126 (1994).
6. West, M.A., Lucocq, J.M. & Watts, C. *Nature* **369**, 147–151 (1994).
7. Williams, L.A., Egner, W. & Hart, D.N.J. *Int. Rev. Cytol.* **153**, 41–103 (1994).
8. Sallusto, F. & Lanzavecchia, A. *J. Exp. Med.* **179**, 1109–1118 (1994).
9. Sallusto, F., Cella, M., Danieli, C. & Lanzavecchia, A. *J. Exp. Med.* **182**, 389–400 (1995).
10. Stahl, P.D. *Current Biology* **4**, 49–52 (1992).
11. Geuze, H.J. & Slot, J.W. *Eur. J. Cell. Biology* **21**, 93–100 (1980).
12. Liou, W., Geuze, H.J. & Slot, J.W. *Histochem. Cell. Biol.* **106**, 41–58 (1996).
13. Geuze, H.J., Slot, J.W., van der Ley, P. & Scheffer, R.C.T. *J. Cell. Biol.* **89**, 653–659 (1981).
14. Bradford, M. *Analyt. Biochem.* **72**, 248–254 (1976).
15. Koldovsky, O. & Palmieri, M. *Biochem. J.* **125**, 697–701 (1971).
16. Stahl, P.D. *Am. J. Respir. Cell Mol. Biol.* **2**, 317–318 (1990).

32

ENHANCED ANTIGEN PRESENTING CELL FUNCTION FOLLOWING *IN VIVO* PRIMING

Geneviève De Becker, Philippe Mockel, Jacques Urbain, Oberdan Leo, and Muriel Moser

Département de Biologie Moléculaire
Université Libre de Bruxelles
B-1640 Rhode-Saint-Genèse, Belgium

INTRODUCTION

The population of cells able to present antigen to class II MHC-restricted, $CD4^+$ T helper cells appears heterogenous and, in the mouse, includes B lymphocytes, macrophages and DC. R. Steinman and collaborators[1] have shown that an antigen injected intravenously at high dose in a naive mouse was present in an immunogenic form on DC only, a finding that correlates with the unique capacity of the DC in stimulating antigen-specific, naive T cells. The cellular interactions leading to a specific immune response following the encounter with an antigen may be different in a primary and a secondary response. In particular, B lymphocytes which bind proteins to surface Ig receptors have been shown to present these proteins at very low concentrations *in vitro* and *in vivo*[2]. Since antigen-specific B cells undergo clonal expansion, the B cells in a primed mouse represent a significant proportion among all antigen-presenting cells. Furthermore, circulating antigen-specific antibodies may form complexes with the antigen which can be taken up by FcR^+ cells[3]. In this paper, we tested whether B cells and macrophages could play a role in antigen presentation in an anamnestic response *in vivo*, and compared the antigen presentation during a secondary versus a primary response.

RESULTS

Splenic B Cells from Preimmunized But Not from Naive Mice Express Peptide/MHC Complexes

To define the role of splenic APC in antigen presentation *in vivo*, we injected naive and BALB/c mice preimmunized with antigen in complete Freund's adjuvant (thereafter referred to as immune animals) with 4 mg of soluble myoglobin, according to a protocol

Dendritic Cells in Fundamental and Clinical Immunology
edited by Ricciardi-Castagnoli, Plenum Press, New York, 1997

described by Crowley et al.[1], and harvested their spleens 2 h later. The myoglobin-specific, T cell hybridoma 13–26–8 was cultured with varying numbers of unseparated spleen cells. The use of an hybridoma diminishes the contribution of accessory molecules to T cell stimulation and provides a rather direct method for assessing peptide-MHC complex expression. The data in Figure 1A show that unseparated spleen cells from naive mice, pulsed *in vivo* with soluble antigen, were poor stimulators of IL-2 production, whereas spleen cells from immune mice induced a detectable response in the same conditions. Addition of antigen in medium resulted in strong activation of T cell hybridoma in all cultures (Figure 1B). Spleen cells from mice, naive or immune, which were not injected with 4mg soluble myoglobin ("unpulsed") did not bear immunogenic fragments of the antigen (Figure 1A). Depletion of B lymphocytes from spleen cells from immune mice (Figure 1C) resulted in strong reduction of IL-2 production, as compared to control spleen cells, thereby showing that B cells from immune mice have processed the antigen, following injection of 4 mg of myoglobin *in vivo*.

Dendritic Cells Bear Immunogenic Fragments of Myoglobin in Naive and Primed Mice

A similar experiment was performed with DC, isolated from *in vivo* pulsed naive or immune animals. T cell hybridoma 13–26–8 was cultured with varying numbers of purified DC. The results in Figure 2A show that DC from naive mice bear immunogenic frag-

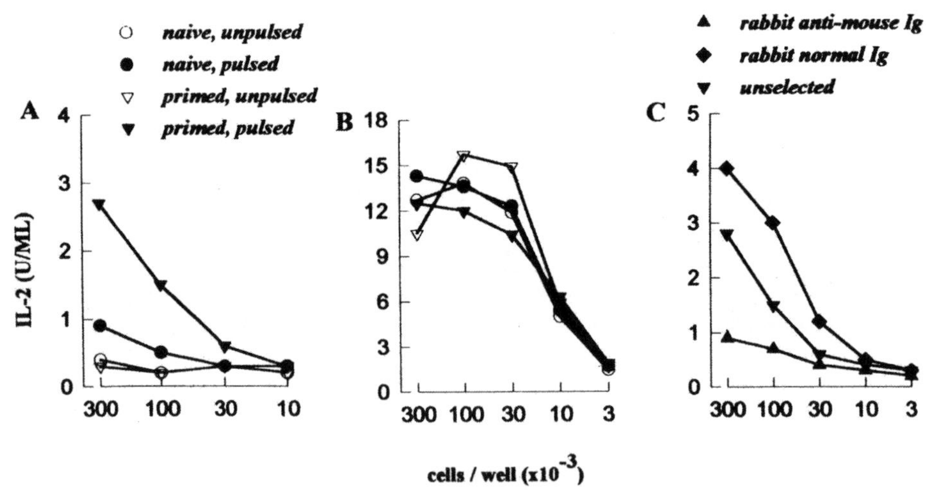

Figure 1. Splenic B cells from immune but not from naive mice retain the antigen in an immunogenic form following *in vivo* antigen pulse. (A, B) Four BALB/c mice were either untreated or injected with 100 μg of myoglobin emulsified in CFA. Half of the mice (closed symbols) were injected 14 days later with 4 mg of myoglobin intravenously (*in vivo* pulsing). 2 h later, the spleens were harvested and unseparated spleen cells were counted and cultured with 3×10^4 T hybridoma cells with (B) or without (A) addition of 20 μg/ml of myoglobin. (C) Spleen cells from *in vivo* pulsed, primed mice (triangle down) were incubated with biotinylated rabbit anti-mouse Ig antibodies (triangle up) or with normal rabbit Ig (diamonds), passed over a MACS column and cultured with the I-E^d-restricted, myoglobin-specific hybridoma 13–26–8[4]. 24 h culture supernatants were assayed for IL-2 production. Each point represents the mean of triplicates. Similar data were obtained in 5 independent experiments.

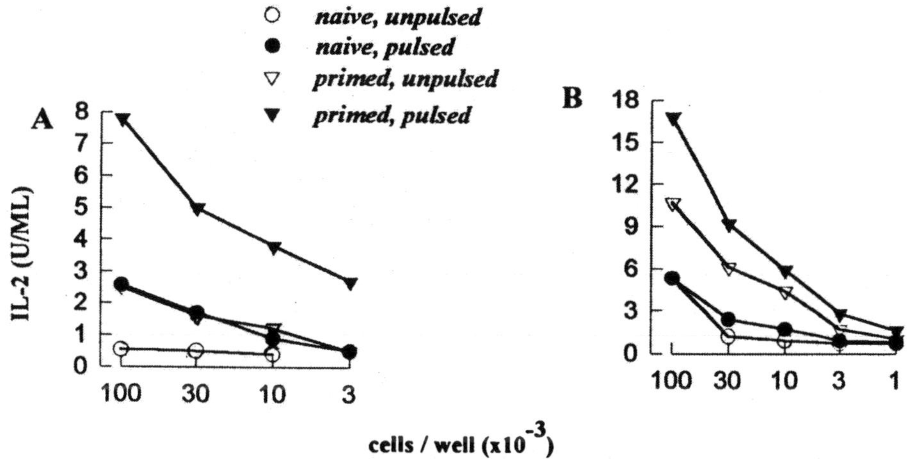

Figure 2. DC isolated from mice pulsed *in vivo* with myoglobin present immunogenic fragments of the antigen. Ten BALB/c mice were either untreated or injected with 100 μg of myoglobin emulsified in CFA. Half of the mice (closed symbols) were injected 14 days later with 4 mg of myoglobin intravenously (*in vivo* pulsing). After 2 h, the spleens were harvested and the DC were purified[5]. At the end of the purification steps (after overnight culture), the various DC preparations were counted and cultured with 3×10^4 T hybridoma cells with (B) and without (A) addition of 20 μg/well of myoglobin. 24 h culture supernatants were assayed for IL-2 production, as described in Material and Methods. Each point represents the mean of triplicates. Similar data were obtained in 6 independent experiments.

ments of the antigen following *in vivo* pulsing. DC from immune animals induced secretion of higher levels of IL-2 by T cell hybridoma than DC from naive mice. The potency of the uptake, processing and presentation of protein antigen by DC was reflected in the very low numbers required to activate antigen-specific T cell hybridoma. It should be noted that, in all experiments performed, DC isolated from primed mice, which have not been pulsed *in vivo*, significantly stimulated 13–26–8 hybridoma. This observation may suggest that levels of processed antigen, sufficient to trigger T cell hybridoma, are still expressed on DC 2 weeks after immunization. The addition of antigen in culture (Figure 2B) only partially increased activation of T cell hybridoma, a finding which correlates with the downregulation of the processing capacity of DC *in vitro*[6,7].

Peritoneal Adherent Cells Are a Source of Antigen/MHC Complexes in Immune Mice, Not in Naive Mice

Peritoneal adherent cells (comprising mainly macrophages) were purified from antigen-pulsed naive and immune mice, and cultured with T cell hybridoma. In three out of five experiments (Figure 3 represents one positive experiment), adherent cells from immune mice expressed higher levels of MHC/peptide complexes than adherent cells isolated from antigen-pulsed, naive mice, as assessed by IL-2 secretion by 13–26–8 T cells. However, addition of antigen only slightly enhanced activation by adherent cells from naive mice, which remained much weaker than the activation induced by adherent cells from primed mice in the same conditions. These data suggest that adherent cells from naive mice express very few complexes of antigen/MHC, not only because they do not efficiently capture the antigen *in vivo*, but also because they express low levels of class II antigens.

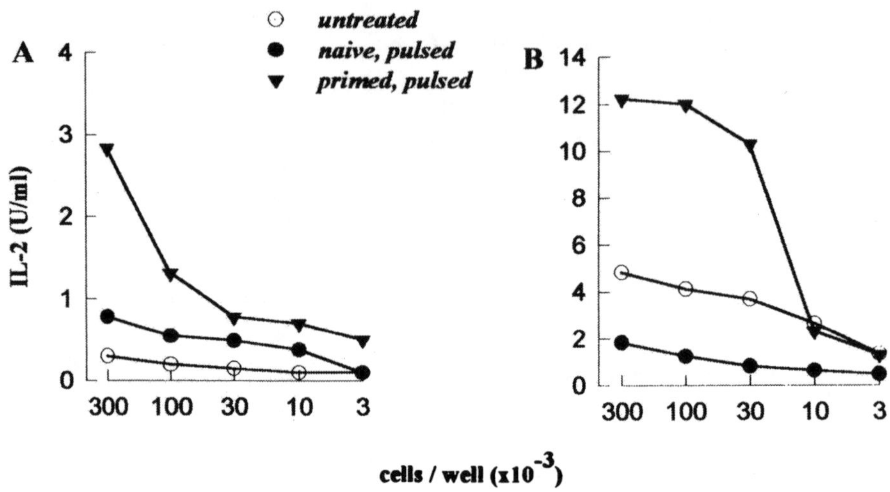

Figure 3. Peritoneal macrophages from immune but not from naive mice, pulsed *in vivo*, display immunogenic fragments of the antigen. BALB/c mice were either untreated or injected with 100 μg myoglobin emulsified in CFA. Half of the mice (closed symbols) were pulsed 14 days later with 4 mg myoglobin intravenously. Two hours later, the mice were killed and the cells harvested by extensive washing of the peritoneal cavity[5]. Adherent cells were cultured overnight, collected using a rubber policeman, and cultured with 3×10^4 T hybridoma cells with (B) or without (A) addition of 20 μg/ml of antigen. 24 h culture supernatants were assayed for IL-2 production.

DISCUSSION

The major observation of this study is that the population of cells bearing immunogenic forms of antigen is quantitatively and qualitatively different in primed versus naive mice. In naive mice injected with high doses of spermwhale myoglobin, the uptake of myoglobin was restricted to DC, with no apparent contribution of B cells or macrophages. By contrast, in primed animals, splenic B lymphocytes, peritoneal macrophages and DC pulsed *in vivo* bear immunogenic fragments of myoglobin and can efficiently present the antigen to specific T cell hybridoma.

The increased antigen-presenting capacity (on a per cell number basis), in mice injected with myoglobin emulsified in CFA but not in animals injected with CFA alone (data not shown), suggests that the capture of the antigen is specific. In particular, it has been shown that B lymphocytes which bind proteins to surface Ig receptors present these proteins at very low concentrations, whereas effective presentation by non specific B cells requires much higher antigen concentrations[2]. Since the number of antigen-specific B lymphocytes is increased in primed versus naive mice, our data are consistent with the capture and presentation of the antigen by specific B cells. Moreover, the capture of the antigen by peritoneal macrophages and B cells in immunized mice could be mediated by myoglobin specific antibodies which bind to Fc receptors present at the surface of B cells and macrophages[3]. The same antibody-mediated capture of the antigen may explain why DC (which express low albeit detectable levels of FcR) from primed animals were consistently better stimulators than DC from naive mice.

Numerous studies have clearly shown that optimal activation of naive T cells requires presentation of antigen on DC *in vitro* and *in vivo*[6–9]. By contrast, there is some evidence that antigen-experienced T cells have less stringent conditions of activation. In particular, Inaba and Steinman[8] have shown, in a mixed leukocyte reaction model, that responses of naive helper T cells are initiated by antigen plus DC, whereas sensitized T cells could be stimulated by allogeneic macrophages or B cells. These findings have been confirmed by more recent studies showing that Ag-experienced cells secrete cytokines in response to many APC types, whereas naive cells only react to DC and to a lesser extent to activated B cells *in vitro*[8–14] and *in vivo*[9].

Our data suggest that, during the course of an immune response, antigen is first presented to naive T cells via DC, and that only subsequently primed T cells can be stimulated by antigen presented by B cells or macrophages. Therefore, the enhanced secondary T cell response is most probably due to a combination of the increased frequency of antigen-specific cells and their ability to react with antigen presented on a wider range of potential APC types.

ACKNOWLEDGMENTS

We thank Dr A. Livingstone for helpful discussions and for providing 13–26–8 T cell hybridoma.

REFERENCES

1. Crowley M, Inaba K, Steinman R. J Exp Med 1990;172:383–86.
2. Abbas AK, Haber S, Rock KL. J Immunol 1985;135:1661–7.
3. Lanzavecchia A. Annu Rev Immunol 1990;8:773–93.
4. Rubberti G, Sellins KS, Hill CM, Germain RN, Fathman CG, Linvingstone A. J Exp Med 1992;175:157–162.
5. De Becker G, Sornasse T, Nabavi N, *et al.* Eur J Immunol 1994;24:1523–1528.
6. Inaba K, Metlay JP, Crowley MT, Steinman RM. J Exp Med 1990;172:631–40.
7. Sornasse T, Flamand V, De Becker G, *et al.* J Exp Med 1992;175:15–21.
8. Inaba K, Steinman RM. J Exp Med 1984;160:1717–35.
9. Ronchese F, Hausmann B. J Exp Med 1993;177:679–90.
10. Luqman M, Bottomly K. J Immunol 1992;149:2300–6.
11. Constant S, Zain M, West J, Pasqualini T, Ranney P, Bottomly K. Eur J Immunol 1994;24:1073–9.
12. Croft M, Duncan DD, Swain SL. J Exp Med 1992;176:1431–7.
13. Croft M, Bradley LM, Swai SL. J Immunol 1994;152:2675–85.
14. Cassel DJ, Schwartz RH. J Exp Med 1994;180:1829–40.

33

DENDRITIC CELLS FROM MICE LACKING THE INVARIANT CHAIN EXPRESS HIGH LEVELS OF MEMBRANE MHC CLASS II MOLECULES *IN VIVO*

Patrizia Rovere,[1] Frédérique Forquet,[1] Valérie S. Zimmermann,[1] Jeannine Trucy,[1] Paola Ricciardi-Castagnoli,[2] and Jean Davoust[1]

[1]Centre d'Immunologie INSERM-CNRS de Marseille-Luminy 13288 France
[2]CNR Cellular and Molecular Pharmacology Center 20129, Milano, Italy

1. SUMMARY

We investigated in H-2^k mice bearing a genetically disrupted invariant chain (Ii) gene[1], the MHC class II expression and antigen presentation ability of dendritic cells (DC) freshly purified from the spleen (SpDC) or derived from bone marrow precursors (BMDC) upon treatment with GM-CSF. In the absence of Ii, class II α/β heterodimers are expressed on the DC membranes to a similar extent than in control mice, in contrast to splenic B cells. Class II molecules immunoprecipitated from the plasma membrane of Ii deficient DC are compact indicating that the dimers are stabilized by antigenic peptides[2]. Furthermore DC from Ii mutant mice are able to present to $CD4^+$ T lymphocytes, epitopes derived from the processing of the hen egg lysozyme (HEL) that normally require expression of the Ii molecule for presentation by B cells. All together, our results show that the antigen processing machinary of DC provides peptides that can reach class II molecules and stabilize their conformation in the absence of Ii mediated targeting of class II complexes.

2. INTRODUCTION

Presentation of exogenous antigens to $CD4^+$ T lymphocytes requires antigen internalization and processing in endocytic compartments[3]. Newly synthetized α and β chains of class II heterodimers associate in the endoplasmic reticulum (ER) with Ii as a third partner. This complex is then transported to a specialized compartment along the endocytic route[4–8] where loading with antigenic peptides derived from exogenous proteins occurs. The intracellular route followed by class II-Ii complexes from the trans-Golgi network to the endocytic pathway remains unclear in DC. Among the different models an initial targeting to early endosomes, possibly via the plasma membrane, has been proposed[9,10].

In contrast to dendritic cells, B cells from mice bearing a genetically disrupted Ii gene[1,11,12] show a striking alteration in the maturation of class II molecules: *i)* most complexes are retained into the ER; *ii)* few complexes reach the plasma membrane; *iii)* the few complexes expressed at the cell surface do not have the "compact" conformation indicative of tight peptide binding. Furthermore Ii is required for the efficient assembly of class II dimers in allelic variants only: B cells from mice with an $H\text{-}2^b$ background mostly accumulate in the ER free α and β chains in the absence of Ii, while class II dimers from $H\text{-}2^k$ and $H\text{-}2^d$ background are efficiently assembled even in the absence of Ii[13]. In the Ii knock out (ko) mice $CD4^+$ T cells are partially positively selected (~15–30% of the normal)[1,11]. However, the general health status and growth rate of these mice are not different from wild type littermates[1] and although primary IgM responses against nominal antigens were impaired, recall vaccinations induced the production of antigen-specific IgG at levels comparable with those of wild type mice[1]. The production of IgG upon *in vivo* challenge with antigen requires efficient antigen presentation to class II restricted $CD4^+$ T cells. However, B cells from Ii deficient mice do not meet the full requirements for antigen presentation to T cells[1]. Furthermore, data from several laboratories point out to an initial role of DC in the induction of T cell responses in experimental animals[14–17]. In this study we investigated in SpDC and BMDC the Ii dependence of class II association with antigen peptides and their presentation to class II restricted $CD4^+$ T lymphocytes.

3. MATERIALS AND METHODS

3.1. Mice

C57BL6 and B10.BR inbred mice bearing a genetic disruption of the Ii gene were a generous gift of D. Mathis and C. Benoist (Strasbourg, France)[1]. The absence of the gene was tested in all mice by Southern blots of EcoR1-digested tail DNA as described[1].

3.2. Antibodies (Abs)

Anti-IA^k mAbs (10.2.16, ATCC) were purified by protein A affinity chromatography from the B cell hybridoma culture supernatant. Anti-cytoplasmic domain Ii and anti-H2M β chain Abs were kindly provided by Nicolas Barois (Marseille, France). Anti-CD45R and anti-CD3 Abs were purchased from Pharmingen. NLDC-145 and N-418 mAbs were a gift of Luciano Adorini (Milano, Italy). All fluorescent conjugated anti-mouse, anti-rat, anti rabbit and anti-hamster Abs were from Jackson Immunoresearch Laboratories Inc.

3.3. Cells

Spleen cells were derived from physical dissociation of the organs. Spleen derived B cells and SpDC were isolated as described in[18]. Briefly, after collagenase digestion, an adhesion step was performed at 37 °C for 2 hours. B cells were retrieved from the non adherent cell population after T cell depletion by Ab-mediated complement lysis. Adherent spleen cells were incubated overnight at 37°C. SpDC were obtained from the de-adhered cell population, after centrifugation over a 50% discontinuous Percol gradient. BMDC were derived as described in[19] modified. $CD34^+$ precursors were grown for 15 days in RPMI additioned with 20 ng/ml of recombinant murine GM-CSF (Granzyme). The culture medium was changed every three days. All the $I\text{-}A^K$ restricted T cell hybridomas utilized in this study (3A9, IC5.1, 2B6, 3B11) were generously provided by Luciano Adorini (Milano, Italy). IL-2 dependent CTLL-2 cells were purchased from ATCC.

3.4. Flow Cytometry, Immunohistology, and Confocal Microscopy

Staining was performed as described in[20]. Samples were analyzed in a FACScan apparatus analysis (Becton Dickinson, San Jose, CA). Staining of cryosections was performed as described in[1] and mounted as described for confocal microscopy. Confocal immunofluorescence was performed as in[21] and results visualized by a Leica TCS 4D (Leica Lasertechnik, Heidelberg, Germany).

3.5. SDS-PAGE of Surface Class II Complexes

15×10^6 spleen derived B cells or BMDC from mutant and control mice were labelled by lacto peroxidase-catalyzed iodination. Briefly, after extensive washing in ice-cold PBS, cells were incubated for four minutes with ^{125}I (1mCi), lactoperoxidase and H_2O_2 (10 µl of a 1/30000 dilution in PBS). H_2O_2 addition was repeated after the first two minutes of incubation. After repeated washings, cells were lysed using a buffer containing 1% NP40 and proteases inhibitors, and $I\text{-}A^k$ immunoprecipitated with the 10.2.16 mAb. Before electrophoresis on a SDS-polyacrylamide (12.5 %) gel, the immunoprecipitated class II was divided in two aliquots: the first was fully denatured at 95 °C for 5 min; the second aliquot was incubated 1 hour at room temperature in SDS sample buffer containing 5% β-mercaptoethanol to preserve the peptide loaded, compact, MHC class II heterodimers[22].

3.6. Antigen Presentation

Antigen presentation was performed incubating in 96 wells flat bottom plates: T hybridoma cells (3A9, IC5.1, 2B6, 3B11) (5×10^4 /well), in the presence or in the absence of 10-fold serial dilution of the relevant antigens with different APC populations (BMDC, 5×10^4/well; spleen derived B cells, 10^5/well, or SpDC pulsed overnight with different antigens and further purified over a Percol gradient as above, 5×10^5/well. After 24 hours, supernatants were collected and tested for the presence of IL-2 using the IL-2 dependent CTLL-2 cells.

4. RESULTS AND DISCUSSION

4.1. DC but Not B Cells from Ii Mutant $H\text{-}2^k$ Mice Express High Levels of Surface MHC Class II Molecules

All mice were tested at three weeks of age for Ii gene disruption by southern blot (Fig.1, panel A). In vitro derived cells were tested for Ii expression by western blot (Fig.1, panel B). The class II expression of spleen derived B cells and SpDC was analyzed by flow cytometry with the 10.2.16 mAb, immediately after *in vitro* purification. Figure 2, panel A shows that

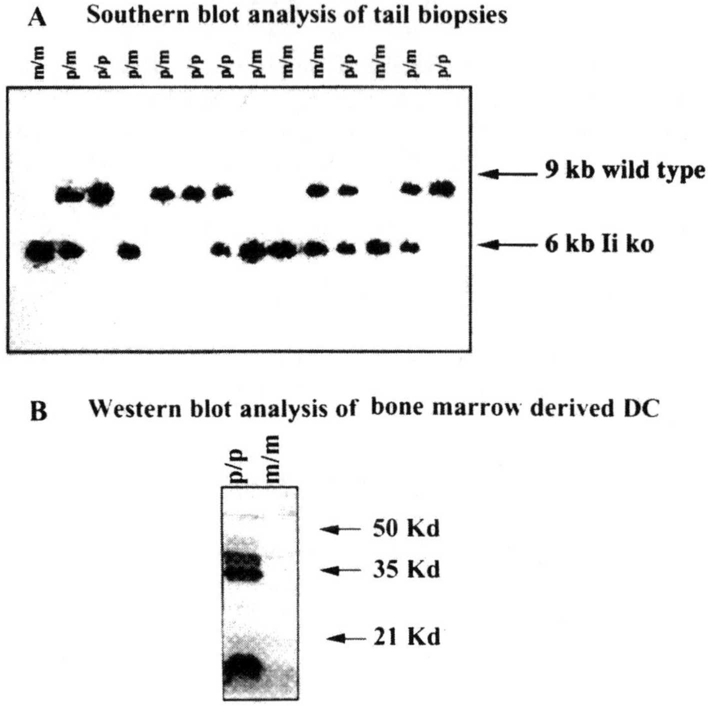

Figure 1. Analysis of Ii ko mice. All mice were tested at three weeks of age by southern blot analysis of genomic DNA (A). The expression of Ii molecule was also verified in bone marrow-derived DC by western blot (B).

class II surface expression is highly reduced in Ii mutant mice spleen derived B cells (left) and only partially compromised in SpDC (right). Class II membrane expression was also preserved in other DC populations, as BMDC (Fig.2, panel B). These results are not due to the cell purification procedures or to *in vitro* culture since DC but not B cells express *in vivo* amounts of membrane class II molecules comparable to control mice, as demonstrated by histological sections and class II staining of different organs (not shown).

4.2. MHC Class II Molecules at the DC Plasma Membrane Are Compact or Associated with Unusual Long Peptides

The 10.2.16 mAb is specific for the β chain of class II dimers and immunoprecipitate both "compact" class II dimers, associated with antigenic peptides, and empty or Ii associated unstable dimers. To better define the characteristics of class II molecules at the cell surface of Ii mutant mice-derived DC, BMDC membranes were iodinated and immunoprecipitated, and class II molecules analyzed by SDS-polyacrylamide electrophoresis. Table 1 shows the densitometric analysis of iodinated bands: unstable dimers, that almost represented the totality of immunoprecipitated class II from Ii mutant spleen derived B cells (not shown), were only the 27.8% of the total immunoprecipitated class II in DC. "Compact", peptide-associated dimers from mutant mice DC represented the 52% (90% in normal mice). SDS stable complexes of higher molecular weight were also detectable (19.2%), possibly deriving from the association of the class II dimers, in the absence of Ii, with long polypeptides, as described in other systems[23].

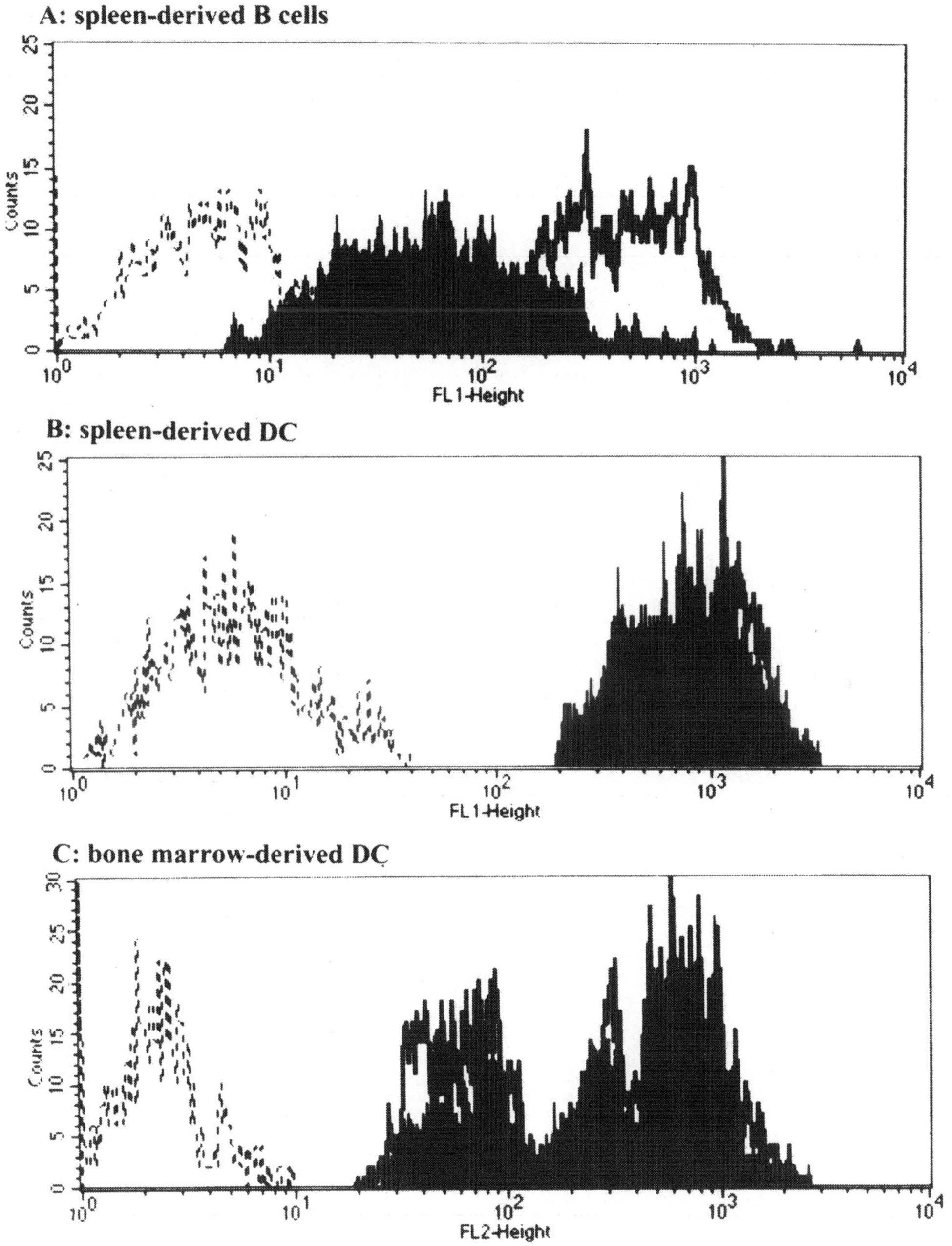

Figure 2. FACS analysis of class II expression. Spleen-derived B cells (A), spleen-derived DC (B) and bone-marrow-derived DC (C) were analyzed for class II expression using the 10.2.16 mAg. Ii positive cells (empty fill), Ii ko cells (full fill), negative control (second step reagent alone) (hatched line).

Table 1. Relative surface expression of compact and unstable MHC class II dimers in DC and B cells

	DC				B Cells			
	Ii +/+		Ii -/-		Ii+/+		Ii -/-	
	25°C	95°C	25°C	95°C	25°C	95°C	25°C	95°C
HMW	–	–	19.2	–	–	–	–	–
CF	90	–	52	–	96	–	–	–
Unstable	10	100	27.8	100	4	100	100	100

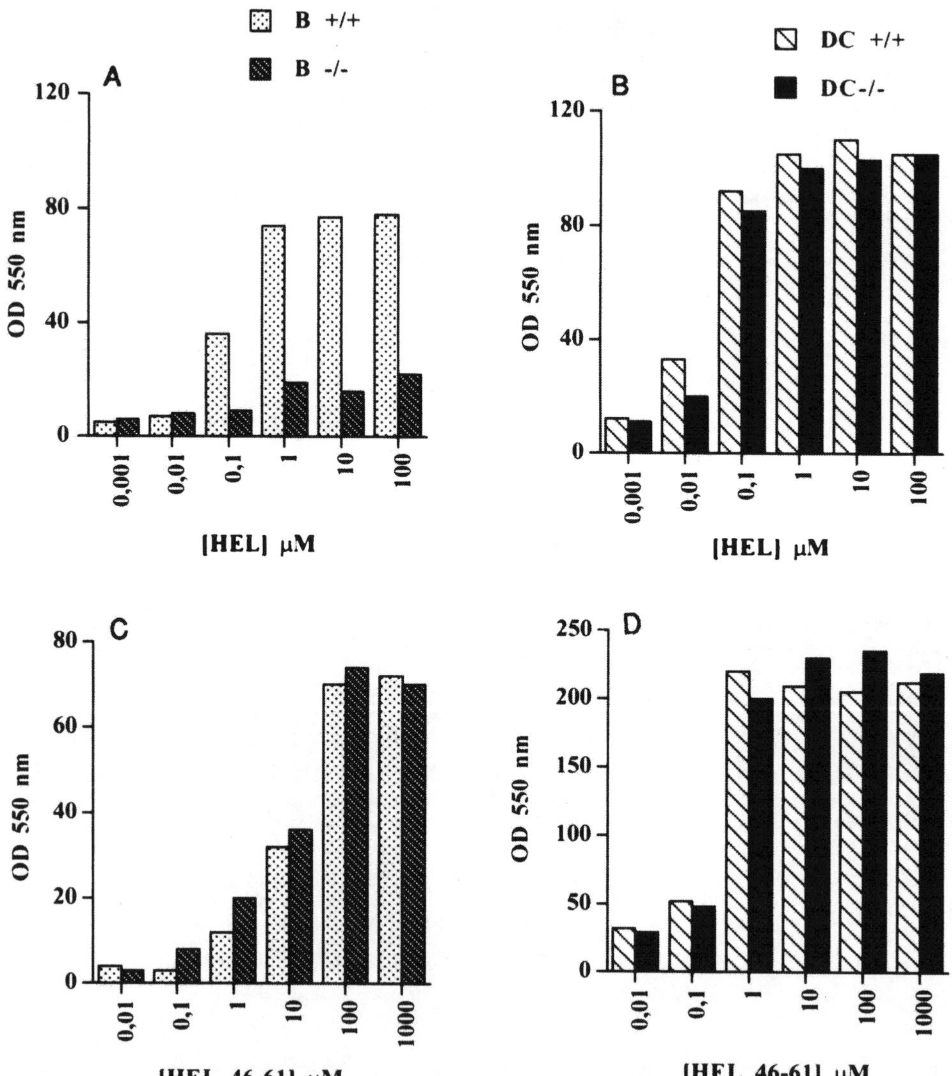

Figure 3. HEL antigen presentation. The ability of Ii positive B cells (lightly dotted bars), Ii ko B cells (), Ii positive DC () and Ii ko Dc (heavily dotted bars) to present to the 3A9 hybridoma (specific for the HEL-derived 46–61 peptide) HEL protein (panel A) was tested. As control (panel B) the presentation of the in vitro syntetized 46–61 peptide was also tested.

4.3. DC from Ii Ko Mice Present with High Efficiency Ii-Dependent Epitopes to Class II Restricted T Cells

In the absence of Ii resting IgM positive B cells, although able to efficiently present peptide antigens to $CD4^+$ T cells, are unable to present internal epitopes derived from the processing of internalized HEL protein[1]. However this does not apply to Ii deficient DC (Fig.3), which present different antigens derived from HEL protein, in contrast to Ii deficient B cells, with almost the same efficiency than DC from control mice. Different hypothesis have been formulated to explain the difference between Ii-dependent and -independent epitopes: in particular, it has been proposed that different epitopes may require different pH and intracellular conditions to be generated[24]. Ii could indeed be required to bring class II molecules in the same intracellular compartment where these particular peptide epitopes are generated, while the α/β chain targeting motives alone could be sufficient to reach the compartments where other Ii-independent peptides are generated. The ability of DC to present also in the absence of Ii both classes of epitopes could reflect either a difference in the intracellular sites where peptides are generated, or a difference in class II trafficking in DC compared to B cells. In DC class II dimers may reach all sites of epitope generation allowing peptide loading to occur through an Ii independent pathway. This characteristic may reflect their unique properties in antigen uptake, storage and presentation to $CD4^+$ T cells.[25,26]

5. REFERENCES

1. Viville et al., 1993 *Cell* 72:635
2. Sadegh-Nasseri and Germain, 1991 *Nature* 353:167
3. Germain, 1994 *Cell* 76:287
4. Amigorena et al., 1994 *Nature* 369:113
5. Tulp et al., 1994 *Nature* 369:120
6. West et al., 1994 *Nature* 369:147
7. Rudensky et al., 1994 *Immunity* 1:585
8. Nijman et al., 1995 *J.Exp.Med.* 182:163
9. Roche et al., 1993 *Proc.Natl.Acad.Sci. USA* 90:8581
10. Castellino and Germain, 1995 *Immunity* 2:73
11. Bikoff et al., 1993 *J.Exp.Med.* 177:1699
12. Elliott et al, 1994 *J.Exp.Med.* 179:681
13. Bikoff et al., 1995 *Immunity* 2:301
14. Inaba et al., 1990 *J.Exp.Med.* 172:631
15. Ronchese and Hausmann, 1993 *J.Exp.Med.* 177:679
16. Ebstein et al., 1995 *J.Exp.Med.* 182:915
17. Guéry et al., 1996 *J.Exp.Med.* 183:751
18. Guéry and Adorini, 1995 *J.Immunol.* 154:536
19. Inaba et al., 1992 *I.Exp.Med.* 176:1693
20. Rovere et al., 1996 *J.Immunol.* 156:2273
21. Demandolx and Davoust, 1996 *J. Microscopy* in press
22. Germain and Hendrix, 1991 *Nature* 353:134
23. Busch et al., 1996 *EMBO J.* 15:418
24. Germain, et al., 1996 *Immunological Rev.* 151:5
25. Schuler and Stenmann, 1985 *J. Exp. Med.* 181:2237
26. Steinman, 1991 *Annual Rev. Immunol.* 9:1693

34

T LYMPHOCYTE MEDIATED REGULATION OF COSTIMULATOR MOLECULE EXPRESSION ON HUMAN DENDRITIC CELLS

A. D. McLellan,[1] R. V. Sorg,[1] D. B. Fearnley,[1] B. D. Hock,[1] R. J. Tiedemann,[2] J. D. Fraser,[2] and D. N. J. Hart[1]

[1]Haematology/Immunology Research Group
Christchurch Hospital
Christchurch, New Zealand
[2]Department of Molecular Medicine
University of Auckland
Auckland, New Zealand

The potent T lymphocyte stimulatory capacity of DC and other antigen presenting cell types is not constitutive and requires an activation/maturation event that can be provided by *in vitro* culture or enhanced by both membrane bound and soluble factors.[1–4] Culture or cytokine induced increases in DC stimulatory activity correlate with the upregulation of the CD40 and CD86 costimulatory molecules, which are essential for optimal T lymphocyte activation. Although the allostimulatory activity of fresh peripheral blood DC (fPBDC) is inhibited by antibodies or soluble recombinant constructs which block the CD86:CD28 or CD40:CD40L interaction,[1,5,6] paraformaldehyde fixed, freshly isolated DC are incapable of initiating T lymphocyte proliferation unless they are subjected to a period of culture prior to fixation.[3,7] This implies that costimulatory molecules are upregulated on fresh DC during DC:T lymphocyte coculture. Studies of the kinetics of costimulation indicate that DC must provide this "second signal" in the very early stages of DC:T lymphocyte contact for full T lymphocyte activation to occur.[5,8,9]

DC costimulator expression *in vitro* may be enhanced by monocyte conditioned media,[2] however fresh DC populations purified by depletion of lineage marker positive cells are still capable of costimulator upregulation in the absence of conditioned media[1,6] and comparable levels of costimulator expression also occurs in serum free conditions (data not shown). Costimulator expression on purified DC *in vitro* may be due to solid phase stimulus (plastic contact), autocrine effects from DC:DC cytokine or membrane interactions or factors released by necrotic cells within the cultures. Alternatively, contaminating lineage marker negative non-DC cells may contribute soluble or membrane molecules that lead to *in vitro* DC activation/maturation. As RT-PCR analysis of these lineage marker negative populations has failed to demonstrate high level transcription of GM-CSF, IL-3

Dendritic Cells in Fundamental and Clinical Immunology
edited by Ricciardi-Castagnoli, Plenum Press, New York, 1997

or TNF-α (Sorg RV & Hart DNJ unpublished data), all of which have been shown to augment the culture induced levels of CD80, CD86 and CD40 on DC and increase DC costimulatory capacity,[1,10,11] other factors may be involved.

A potent activation maturation signal for fresh DC or LC can be provided via membrane bound or soluble trimeric CD40L.[1,10,11] DC freshly isolated from peripheral blood express only low levels of CD40 antigen, although significant CD40 expression is attained rapidly after brief periods of *in vitro* culture.[1,2] Freshly isolated LC express moderate levels of CD40, which is also increased upon *in vitro* culture.[11,12] In addition, the functional relevance of DC CD40 expression has been underscored by studies showing that optimal antigen presentation by DC to T lymphocytes requires an intact CD40:CD40L interaction.[1,11,13] Ligation of DC CD40 to CD40L during DC:T lymphocyte interactions leads to increased levels of CD80 and CD86 expression and the functional significance of increased CD86 expression on LC as a result of CD40 cross linking has been documented.[11] In addition to providing a potent signal for CD28 ligand upregulation on DC, the CD40:CD40L interaction can provide a costimulus for T lymphocyte proliferation as well as IL-2[1,14] and IL-4[15] secretion by $CD4^+$ T lymphocytes. This interaction also appears to act in synergy with TCR/CD3 and CD28 ligation for T lymphocyte proliferation and IL-4 secretion.[15] CD40 signalling has also been shown to enhance the *in vitro* differentiation of DC from $CD34^+$ or PBMC precursors leading to increased levels of DC CD80 and CD86 and a more potent T lymphocyte activating phenotype.[16,17]. CD40 signalling can also drive the maturation of CMRF-44^+ DC from fresh blood.[18] CD40L expression on T lymphocytes is induced within a few hours of TCR engagement, but CD40L is rapidly downregulated by transcriptional shut off, proteolytic cleavage from the T lymphocyte membrane and endocytosis.[19–21] Given the potent effects of CD40 engagement on DC costimulatory molecule expression, the transient expression of CD40L on TCR activated T lymphocytes probably serves to limit DC activation to levels sufficient for cognate T lymphocyte activation and avoids inappropriate bystander activation. It is also tempting to speculate that *in vivo*, withdrawal of the CD40L stimulus following antigen presentation leads to DC apoptosis as has been shown for migratory LC.[10]

We have further investigated the interaction of T lymphocytes and fPBDC with regard to the kinetics of costimulatory molecule expression on immature DC, seeking evidence for effects attributable to cognate MHC class II:TCR interactions. A co-culture system was devised where fPBDC were incubated with autologous or allogeneic $CD4^+$ T lymphocytes. After a 16–24h period of *in vitro* culture the DC: T lymphocyte culture was separated into clustered and non-clustered fractions. Our analyses of single cells liberated from these cultures showed a several fold increase in the MFI of CD40, CD80 and CD86 on fPBDC isolated from $CD4^+$ T lymphocyte clusters (range 12–27% of total fPBDC form clusters in these co-cultures). These observed increases were antigen dependent in that allogeneic DC:T lymphocyte clusters exhibit a higher level of costimulator induction on DC than autologous DC:T lymphocyte clusters. SEA was also added at the initiation of culture using autologous T lymphocytes as responders. The highest levels of costimulator induction on fPBDC were observed in these autologous clusters pulsed with SEA. These data suggest that fPBDC which are in intimate contact with T lymphocytes exhibit high levels of costimulator expression as a result of signals derived from antigen triggered T lymphocytes.

In direct contrast, fPBDC that were not associated with T lymphocyte clusters at the time of analysis (range 73–88% of total fPBDC) showed depressed levels of costimulator expression below that of control cultures (DC cultured in the absence of T lymphocytes). This suppressive effect was dependent on the interaction of live T lymphocytes with

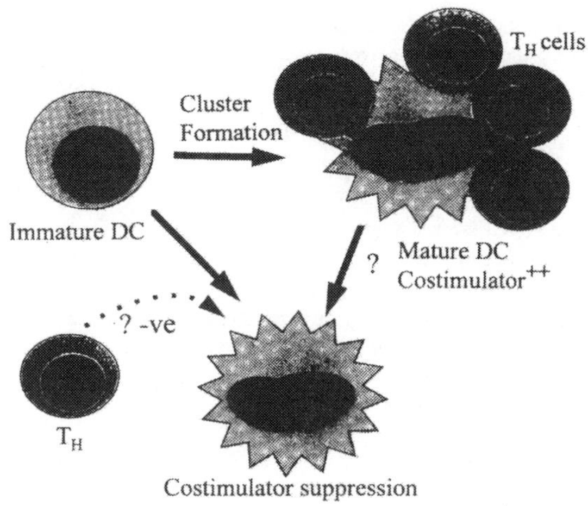

Figure 1. Schema of T lymphocyte modulation of costimulator expression on fPBDC in *in vitro* coculture (see text).

fPBDC, as paraformaldehyde fixation of the T lymphocytes prior to initiation of DC: T lymphocyte cultures abrogated the suppression of fPBDC costimulator expression. Our preliminary data suggests that prior culture of fPBDC renders the DC resistant to T lymphocyte suppression of costimulator expression. Other studies using a similar co-culture system[13,22] but with DC cultured prior to isolation and T lymphocyte contact, have found that CD80 and CD86 were upregulated following 1–2 day culture with allogeneic T lymphocytes.

A possible series of events which may contribute to the outcome observed in these co-culture systems is shown below in Fig. 1.

Suppression of costimulator expression on fPBDC not in intimate association with T lymphocytes at the end of the culture period may be the result of a prior clustering event leading to triggering and subsequent downregulation of CD40, CD80 and CD86. Alternatively, the unclustered fPBDC may have interacted with suppressive soluble or membrane factors produced during the culture period by the T lymphocytes and these suppressive effects were not offset by stimulatory signals from antigen dependent T lymphocyte clustering or the other antigen independent stimuli which influence DC cultured in the absence of T lymphocytes. It is probable that the small percentage of fPBDC capable of clustering with T lymphocytes in these co-cultures represent an *in vivo* activated population of DC that is somehow "immunocompetent" at the time of isolation or is selected in some other way by antigen specific or nonspecific T lymphocyte interaction. We are currently investigating the effects of MHC class II signalling on DC and nature of the suppressive factor/s involved in costimulator suppression on fPBDC populations by T lymphocytes.

REFERENCES

1. McLellan AD, Sorg RV, Williams LA, Hart DNJ: Human dendritic cells activate T lymphocytes via a CD40:CD40 ligand-dependent pathway. Eur J Immunol 26:1204, 1996
2. O'Doherty U, Steinman RM, Peng M, Cameron PU, Gezetter S: Dendritic cells freshly isolated from human blood express CD4 and mature into typical immunostimulatory dendritic cells after culture in monocyte-conditioned medium. J Exp Med 179:1067, 1993

3. Peguet-Navarro J, Dalbiez-Gauthier C, Schmidt D: Accessory function of human Langerhans cells in the primary allogeneic T-cell response. J Invest Dermatol 99:87s, 1992
4. Morrissey PJ, Bressler L, Park LS, Alpert A, Gillis S: Granulocyte-macrophage colony stimulating factor augments the primary antibody response by enhancing the function of antigen presenting cells. J Immunol 139:1113, 1987
5. Thomas R, Davis LS, Lipskey PE: Comparative accessory cell function of human peripheral blood dendritic cells and monocytes. J Immunol 151:6840, 1993
6. McLellan AD, Starling GC, Williams LA, Hock B, Hart DNJ: Activation of human peripheral blood dendritic cells induces the CD86 costimulatory molecule. Eur J Immunol 25:2064, 1995
7. Mohamadzadeh M, Pavlidou A, Enk A, Knop J, Rude J, Gradehandt G: Freshly isolated mouse 4F7+ splenic dendritic cells process and present exogenous antigens to T cells. Eur J Immunol 24:3170, 1994
8. Prickett TCR, McKenzie JL, Hart DNJ: Adhesion molecules on human tonsil dendritic cells. Transplantation 53:483, 1992
9. Ding L, Shevach EM: Activation of $CD4^+$ cells by delivery of the B7 costimulatory signal on bystander antigen-presenting cells (trans-costimulation). Eur J Immunol 24:859, 1994
10. Ludewig B, Graf D, Gelderblom HR, Becker Y, Kroczek RA, Pauli G: Spontaneous apoptosis of dendritic cells is efficiently inhibited by TRAP (CD40 ligand) and TNF-α, but strongly enhanced by interleukin-10. Eur J Immunol 25:1943, 1995
11. Peguet-Navarro J, Dalbiez-Gauthier C, Rattis F, van Kooten C, Banchereau J, Schmitt D: Functional expression of CD40 antigen on human epidermal Langerhans cells. J Immunol 155:4241, 1995
12. Romani N, Lenz A, Glassel H, Stossel H, Stanzl U, Majdic O, Fritsch P, Schuler G: Cultured human Langerhans cells resemble lymphoid dendritic cells in phenotype and function. J Invest Dermatol 93:600, 1989
13. Zhou L, Tedder TF: Human blood dendritic cells selectively express CD83, a member of the immunoglobulin superfamily. J Immunol 154:3821, 1995
14. Cayabyab M, Phillips JH, Lanier LL: CD40 preferentially costimulates activation of $CD4^+$ T lymphocytes. J Immunol 152:1523, 1994
15. Blotta MH, Marshall JD, DeKruyff RH, Umetsu DT: Cross linking of the CD40 ligand on human $CD4^+$ T lymphocytes generates a costimulatory signal that up-regulates IL-4 synthesis. J Immunol 156:3133, 1996
16. Sallusto F, Lanzavecchia A: Efficient presentation of soluble antigen by cultured human dendritic cells is maintained by granulocyte/macrophage colony-stimulating factor plus interleukin 4 and downregulated by tumour necrosis factorα. J Exp Med 179:1109, 1994
17. Caux C, Massacrier C, Vanbervliet B, Dubois B, Van Cooten C, Durand I, Banchereau J: Activation of human dendritic cells through CD40 cross-linking. J Exp Med 180:1263, 1994
18. Fearnley DB, McLellan AD, Mannering SI, Hock BD, Hart DNJ: Isolation of human blood dendritic cells using the CMRF-44 monoclonal antibody: implications for studies on antigen presenting cell function and immunotherapy. Blood In press:1996
19. Graf D, Muller S, Korthauer U, van Kooten C, Weise C, Kroczek RA: A soluble form of TRAP (CD40 ligand) is rapidly released after T cell activation. Eur J Immunol 25:1749, 1995
20. van Kooten C, Gaillard C, Gallizi J, Herman P, Fossiez F, Banchereau J, Blanchard D: B cells regulate the expression of CD40 ligand on activated T cells by lowering the mRNA level and through the release of soluble CD40. Eur J Immunol 24:787, 1994
21. Yellin MJ, Sippel K, Inghirami G, Covey LR, Lee JJ, Sinning J, Clark EA, Chess L, Lederman S: CD40 molecules induce down-modulation and endocytosis of T cell surface T cell-B cell activating molecule/CD40-L. J Immunol 152:598, 1994
22. Young JW, Koulova L, Soergel SA, Clarke EA, Steinman RM, Dupont B: The B7/BB-1 antigen provides one of several co-stimulatory signals for the activation of $CD4^+$ T lymphocytes by human blood DC in-vitro. J Clin Invest 90:229, 1992

35

ACTIVATION OF IMMATURE DENDRITIC CELLS VIA MEMBRANE SIALOPHORIN (CD43)

Emanuele Fanales-Belasio,[1] Giovanna Zambruno,[2] Andrea Cavani,[1] and Giampiero Girolomoni[1]

[1]Laboratory of Immunology and
[2]Laboratory of Molecular and Cellular Biology
Istituto Dermopatico dell'Immacolata
IRCCS, 00167 Roma, Italy

1. INTRODUCTION

Sialophorin (leukosialin, CD43) is a leukocyte membrane glycoprotein very rich in sialic acid residues and with an extended, mucine-like structure[1]. Various studies have suggested that CD43 is involved in the regulation of cell-cell interactions as an anti-adhesion molecule and that CD43 can mediate an independent pathway of cytoplasmic signaling and activation in monocytes, T and B lymphocytes[2–4]. Dendritic cells (DC) are the antigen presenting cells (APC) most efficient in the activation of naive T cells and in the induction of primary T cell-mediated immune responses. In unperturbed tissue, DC exhibit elevated antigen capture capacity but low APC activity (immature DC). Upon activation, DC increase expression of surface MHC and accessory molecules, and become very potent APC. The signals that drive DC maturation are bacterial products and cytokines such as TNF-α and IL-1β. Maturation of DC, however, is completed only during antigen presentation by interactions between cell surface molecules expressed by T cells and DC (e.g., CD40L/CD40) and cytokines released by T cells[5]. Unraveling the factors and the mechanisms involved in DC activation is a very important issue, not only for a better understanding of DC physiology, but also for developing more effective DC-based strategies in the immunotherapy of tumors and infections. In this study, we have investigated the expression and function of CD43 on epidermal Langerhans cells (LC) and in DC generated from peripheral blood monocytes incubated with GM-CSF and IL-4 for 6–10 days (Mo-DC).[6,7]

2. DC EXPRESS MEMBRANE CD43

Sialophorin is widely expressed among hematopoietic cells and, in keeping with this notion, we observed that cutaneous DC (i.e. epidermal LC and dermal DC) in situ were

Dendritic Cells in Fundamental and Clinical Immunology
edited by Ricciardi-Castagnoli, Plenum Press, New York, 1997

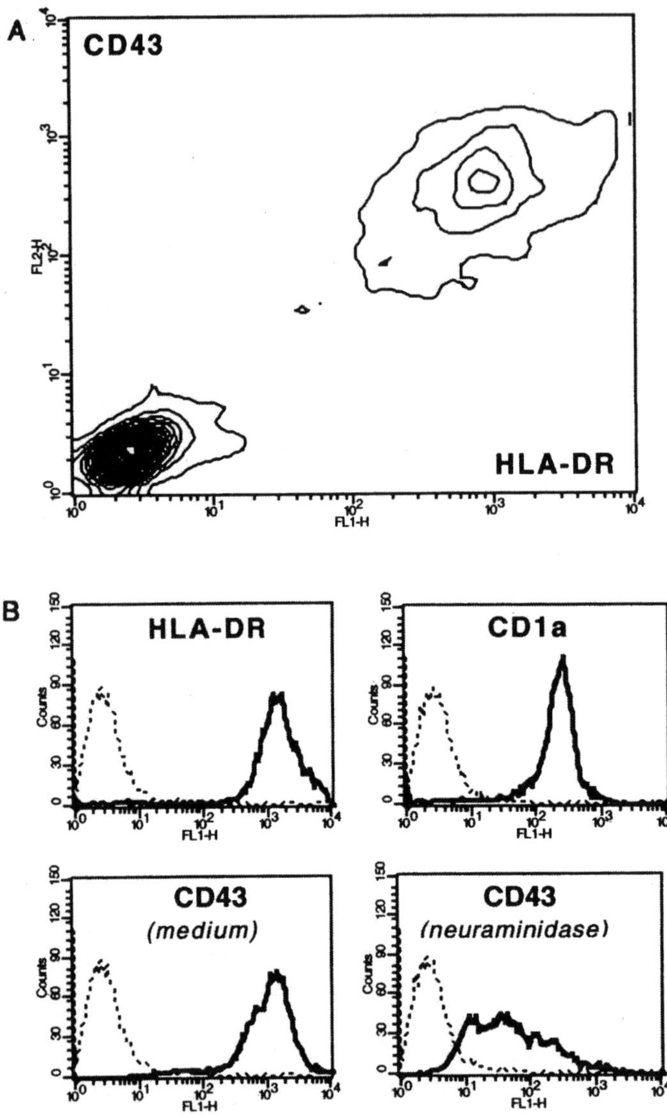

Figure 1. Epidermal LC in suspension and Mo-DC express CD43. (A) Freshly prepared epidermal cell suspensions were enriched for LC and double stained with control IgG1 or anti-CD43 mAb (DFT-1) followed by PE-conjugated goat anti-mouse Ig and FITC-conjugated anti-HLA-DR. About 30% of epidermal cells co-express HLA-DR and CD43. (B) Mo-DC were strongly positive for CD1a, CD1b, HLA-DR and CD43 (MEM-59) and negative for CD14, CD16, CD19 and CD3. Incubation with neuraminidase (from *Vibrio cholerae* and *Arthobacter ureafaciens*) for 40 min at 37°C resulted in a marked down-regulation of CD43 expression.

stained by anti-CD43 mAbs. High expression of CD43 was confirmed on freshly isolated epidermal LC (fLC) and on Mo-DC (Fig. 1). In contrast to other surface molecules that undergo modulation during DC activation, CD43 levels did not change in LC after culture or in Mo-DC treated with TNF-α. CD43 expression on both LC and Mo-DC was drastically reduced by neuraminidase.

3. ANTIBODIES AGAINST CD43 INCREASE THE CAPACITY OF IMMATURE DC TO AGGREGATE AND ACTIVATE T CELLS

To investigate the functional role of CD43 on DC, we incubated cells with anti-CD43 mAb (MEM-59[4], IgG1; 1–50 μg/ml) or with control mouse IgG1 in complete RPMI with 5% human serum for 25 min on ice and then for 20 min at 37°C. Cells were then ex-

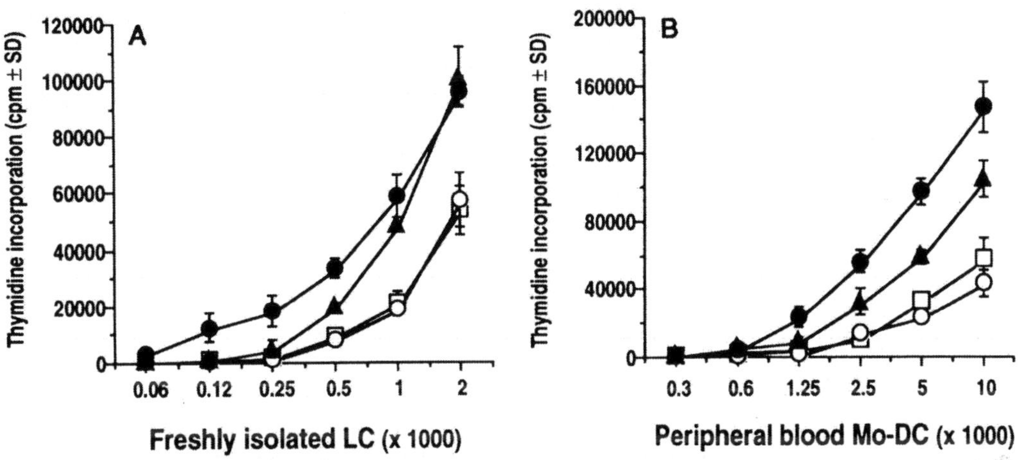

Figure 2. Antibodies to CD43 or treatment with neuraminidase augment the capacity of immature DC to activate naive allogeneic T cells. fLC (A) or Mo-DC (B) were incubated with control IgG1 (50 μg/ml) (O) or with MEM-59 (50 μg/ml, ●) for 25 min on ice and then for 20 min at 37°C, extensively washed and then co-cultured with 2 x 10^5/well naive allogeneic T cells. Alternatively, APC were incubated with neuraminidase (from *Vibrio cholerae* and *Arthobacter ureafaciens*) for 40 min at 37°C, washed and then co-cultured with T cells (▲). Proliferation of T cells alone was less than 1,000 cpm.

tensively washed in order to remove unbound Abs completely, and co-cultured in 96-well microculture plates, together with T lymphocytes. Mo-DC and fLC when co-cultured for 2–3 days with allogeneic T lymphocytes formed small aggregates. In contrast, fLC or Mo-DC transiently treated with anti-CD43 could form more numerous and larger clusters with T lymphocytes as compared to cells treated with control IgG1. In addition, fLC and Mo-DC temporarily incubated with anti-CD43 mAb displayed a markedly enhanced capacity to induce proliferation of naive allogeneic T lymphocytes (Fig. 2). As expected, cultured LC (cLC) were more efficient in clustering T cells compared to fLC and Mo-DC. When cLC were treated with anti-CD43 mAb, they increased slightly the T cell aggregating capacity and did not augment significantly their already elevated T cell activating function. Similarly, T cell response to Mo-DC treated with TNF-α did not increase further when they were incubated with anti-CD43. Results similar to those observed with anti-CD43 mAb were obtained using neuraminidase, which removes sialic acid from the cell surface. Incubation of fLC and Mo-DC with neuraminidase was able to augment APC-T cell cluster formation as well as T cell proliferation, but had only little or no effect on the same properties of cLC and TNF-α-treated Mo-DC. Finally, Mo-DC that had been transiently treated with anti-CD43 were also more efficient at presenting tetanus toxoid to primed T cells.

4. MECHANISMS OF DC ACTIVATION VIA CD43 TRIGGERING

Two mechanisms have been proposed to explain the effects of anti-CD43 mAbs on DC. The first mechanism is based on the physico-chemical properties of sialophorin, which being highly negatively charged and extending 45 nm in length with an unfolded configuration, provides a repulsive barrier that inhibits cell-cell contacts[2,3]. Using immunogold electron microscopy on LC-enriched epidermal cell suspensions maintained at 4°C we could observe CD43 expression on the cell surface, especially on the cell projections. Upon warming to 37°C, gold granules were rapidly internalized by receptor-mediated endocytosis, indicating that anti-CD43 mAb can remove the molecule from the cell membrane and thus allow closer contacts between DC and responding T cells. A similar mechanism can be postulated for neuraminidase, which removes sialic acid residues from the cell surface. The second mechanism relies on the capacity of CD43 to act as a receptor that mediates intracellular signaling[8,9]. To test the hypothesis that anti-CD43 mAb can induce DC maturation, we evaluated the expression of accessory molecules on DC 24 h after incubation with anti-CD43 mAb. As shown in Fig. 3, Mo-DC that were exposed to anti-CD43 mAb expressed higher levels of CD54, CD40, CD80 and CD86. In addition,

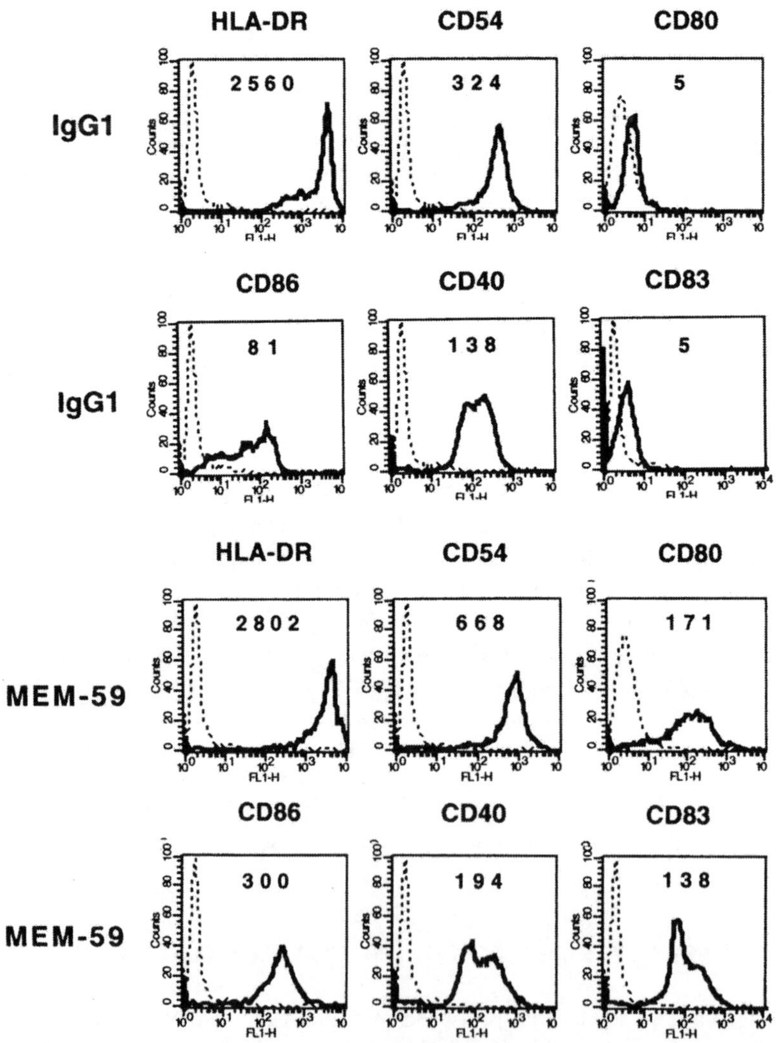

Figure 3. Transient incubation with the anti-CD43 mAb MEM-59 up-regulates HLA-DR, ICAM-1, B7–1, B7–2, CD40 and CD83 on Mo-DC. Cells were incubated with control IgG1 or MEM-59 (both at 50 μg/ml) as described in Fig. 2, and were analyzed after 24 h.

DC expressed larger amounts of CD83, a marker of DC maturation. These results suggest that anti-CD43 mAb (or the natural CD43 ligand) triggers cytoplasmic signal pathways that eventually promote up-regulation of molecules critical for antigen presentation.

5. CONCLUSIONS

Sialophorin belongs to a growing family of mucins sharing similar structural and functional properties and that are implicated in cell adhesion and migration, as well as in cell activation[10]. Epidermal LC, peripheral blood DC[11,12] as well as Mo-DC express CD43. The natural ligand(s) of CD43 has not yet been defined, but it is possible that CD43 binds to selectins, perhaps expressed on T lymphocytes. The function of CD43 may be to limit (nonspecific) contacts with T cells, but once interactions become strong enough, it can deliver activation signals in DC that reinforce adhesion with T cells and enhance antigen presenting capacity. Targeting DC with anti-CD43 mAb may favor initial antigen-independent contact with T cells[13] and subsequently promote DC maturation. Thus, membrane CD43 may provide a novel tool for potentiating the adjuvant functions of ex vivo generated DC to be used in the immunotherapy of cancer[14–16].

6. ACKNOWLEDGMENTS

The authors would like to thank Dr. V. Horejsi (Institute of Molecular Genetics, Academy of Sciences of the Czech Republic, Praha) for the generous gift of MEM-59. This work was supported by Istituto Superiore di Sanità (progetto AIDS, grant 9304.5) and by the Italian Ministry of Health.

7. REFERENCES

1. Remold-O'Donnel, E., D.M. Kenney, R. Parkman, L. Cairns, B. Savage, and F.S. Rosen. 1984. Characterization of a human lymphocyte sialoglycoprotein that is defective in Wiskott-Aldrich syndrome. *J. Exp. Med.* 159:1705–1723.
2. Nong, Y.H., E. Remold-O'Donnell, T.W. LeBien, and H.G. Remold. 1989. A monoclonal antibody to sialophorin (CD43) induces homotypic adhesion and activation of human monocytes. *J. Exp. Med.* 170:259–267.
3. De Smet, W., H. Walter, and L. Van Hove. 1993. A new CD43 monoclonal antibody induces homotypic aggregation of human leukocytes though a CD11a/CD18-dependent and -independent mechanism. *Immunology* 79:46–54.
4. Alvarado M., Klassen C., Cerny J., Horejsi V., and Schmidt R.E. 1995. MEM-59 monoclonal antibody detects a CD43 epitope involved in lymphocyte activation. *Eur. J. Immunol.* 25:1051–1055.
5. Lutz, M.B., G. Girolomoni, and P. Ricciardi-Castagnoli. 1996. The role of cytokines in the functional regulation and differentiation of dendritic cells. *Immunobiology* 195:431–455, 1996.
6. Sallusto, F., and A. Lanzavecchia. 1994. Efficient presentation of soluble antigen by cultured human dendritic cells is maintained by granulocyte/macrophage colony-stimulating factor plus interleukin 4 and downregulated by tumour necrosis factor-α. *J. Exp. Med.* 179:1109–1118.
7. Romani, N., S. Gruner, D. Brang, E. Kämpgen, A. Lenz, B. Trockenbacher, G. Konwalinka, P.O. Fritsch, R.M. Steinman, and G. Schuler. 1994. Proliferating dendritic cell progenitors in human blood. *J. Exp. Med.* 180:83–93.
8. Wong, R.C.K., E. Remold-O'Donnell, D. Vercelli, J. Sancho, C. Terhorst, F. Rosen, R. Geha, and T. Chatila. 1990. Signal trasduction via leukocyte antigen CD43 (sialophorin). Feedback regulation by protein kinase C. *J. Immunol.* 144:1455–1460.
9. Manjunath, N., and B. Ardman. 1995. CD43 regulates tyrosine phosphorilation of a 93-kD protein in T lymphocytes. *Blood* 86:4194–4198.

10. Varki, A. 1994. Selectin ligands. *Proc. Natl. Acad. Sci. USA* 91:7390–7397.
11. O'Doherty, U., R.M. Steinman, M. Peng, C.U. Cameron, S. Gezelter, I. Kopeloff, W.J. Swiggard, M. Pope, and N. Bhardwaj. 1993. Dendritic cells freshly isolated from human blood express CD4 and mature into typical immunostimulatory dendritic cells after culture in monocyte-conditioned medium. *J. Exp. Med.* 178:1067–1078.
12. Zhou, L.-J., and T.F. Tedder. 1995. Human blood dendritic cells selectively express CD83, a member of the immunoglobulin superfamily. *J. Immunol.* 154:3821–3835.
13. Inaba, K., N. Romani, and R.M. Steinman. 1989. An antigen-independent contact mechanism as an early step in T cell-proliferative response to dendritic cells. *J. Exp. Med.* 170:527–542.
14. Strunk, D., K. Rappersberger, C. Egger, H. Strobl, E. Krömer, A. Elbe, D. Maurer, and G. Stingl. 1996. Generation of human dendritic cells/Langerhans cells from circulating $CD34^+$ hematopoietic progenitor cells. *Blood* 87:1292–1302.
15. Young J.W., and K. Inaba. 1996. Dendritic cells as adjuvants for class I major histocompatibility complex-restricted antitumor immunity. *J. Exp. Med.* 183:7–11.
16. Hsu, F.J., C. Benike, F. Fagnoni, T.M. Liles, D. Czerwinski, B. Taidi, E.G. Engleman, and R. Levy. 1996. Vaccination of patients with B-cell lymphoma using autologous antigen-pulsed dendritic cells. *Nat. Med.* 2:52–58.

36

PROCESSING OF EXOGENOUS PROTEIN ANTIGEN BY MURINE DENDRITIC CELLS FOR PRESENTATION TO CYTOTOXIC T LYMPHOCYTES

Marloes L. H. De Bruijn, Danita H. Schuurhuis, Hans Vermeulen, Karin A. J. de Cock, and Cornelis J. M. Melief

Department of Immunohematology and Bloodbank
University Hospital Leiden
Building 1, E3-Q, P.O. Box 9600
2300 RC Leiden, The Netherlands

Generation of cytotoxic T lymphocytes (CTL) from quiescent CD8+ precursors can be achieved by stimulation with professional antigen-presenting cells (APC), dendritic cells (DC). With the elucidation of CTL epitopes that can be manufactured synthetically, in vitro CTL responses specific for these peptides can be established with DC (1–3). In these studies, DC of supposedly mature phenotype were used, i.e. derived from murine spleens.

Langerhans cells directly isolated from the skin have been reported to exemplify immature progenitors of DC (4–6). In addition, immature DC can be cultured from cord blood, peripheral blood and bone marrow in the presence of different sets of cytokines, of which GM-CSF is essential (7–12).

Immature DC have been reported to be well-versed in antigen uptake and processing, whereas mature DC appear to have lost this ability but are specialized in optimally stimulating T lymphocytes (5, 6, 13–15). In contrast to this concept are reports describing the processing abilities of mature DC, e.g. of cultured Langerhans cells and splenic DC (16, 17). These studies involved activation of class II-restricted T lymphocytes.

Although exogenous antigens are presented by MHC class II molecules as a general rule, class I-restricted presentation of exogenous antigens is possible. This may take place when exogenous antigen gains access to the cytosol (e.g. influenza virus presented by dendritic cells, ref. 18) or when it is administered in a particulate form (to macrophages: ref. 19–22). Because Langerhans cells and progenitors of DC are able to phagocytose (23–25), particulate exogenous antigen may well be presented by class I molecules of DC.

Recently, DC became the centre of attention in the field of tumor immunotherapy. As was shown with SC DC (26), immunizations with antigen-pulsed BM DC can protect

Dendritic Cells in Fundamental and Clinical Immunology
edited by Ricciardi-Castagnoli, Plenum Press, New York, 1997

mice against a subsequent tumor challenge (27, 28), but the novelty lies with therapeutical applications of DC immunizations (29–32). These therapy studies were performed with IL4 + GM-CSF -cultured BM DC ('immature') in mice, with the exception of the study by Hsu et al (32), that involved vaccination of patients with B-cell lymphoma using autologous antigen-pulsed dendritic cells, isolated from the blood by density gradient and adherence ('mature'). Immature and mature DC have not been evaluated in a comparative study.

From our experience with spleen-derived DC (SC DC), we were anxious to compare them with IL-4 plus GM-CSF -cultured DC derived from bone marrow (BM DC). We wondered whether these DC preparations, 'mature' versus 'immature', were functionally similar or if one was superior to the other in CTL response induction and antitumor activity.

Phenotypic differences between SC DC and BM DC were apparent (Figure 1). SC DC preparations contained 80–100% class II- and B7- positive cells, that also expressed 33D1 and NLDC-145 antigens. BM DC preparations contained 30–40% class II- and B7- positive cells, (almost) no 33D1, and variable levels of NLDC-145 molecules. By comparison, the level of MHC class II expression on SC DC is higher than on BM DC. Similarly, B7, P150,95 and ICAM-1 molecules are more abundantly expressed on SC DC than on BM DC. (Note difference in background staining of secondary antibody only.) SC DC and BM DC did not express F4/80.

BM DC were generated by culturing bone marrow cells that were depleted of erythrocytes (with ammonium chloride) and lymphocytes (with Abs against CD4, CD8, B220 and Ia^b and magnetic beads) overnight at 37°C in 24 well plates in medium only. Next day, the nonadherent cells were collected and cultured in the presence of IL-4 (10^3 U/ml) and GM-CSF (10^3 U/ml) for three days. On day 4, the plates were gently swirled and 75% of the medium with floating cells was removed. New medium containing IL4- and GM-CSF was added. On day 8, BM DC were harvested by collecting the nonadherent cells. FACScan analysis of a SC DC preparation (no gate, 100% of the cells) and a BM DC preparation (gated on cells with a high side scatter, 35% of the cells). Staining was performed with M5/114 (anti-Ia^b), 33D1 (DC marker), NLDC-145 (DEC-205, DC marker), F4/80 (macrophage marker), CTLA4Ig (ligand of B7.1 and B7.2), N418 (P150,95) and 3E2 (ICAM-1). The dotted lines represent staining with the secondary Ab only.

CTL response induction experiments in vitro and in vivo using CTL epitopes as antigen also revealed differences between SC DC and BM DC (Table 1). Based on the same number of class II- and B7- positive cells in these preparations, we observed that 5- to 25-times more BM DC than SC DC were required to obtain similar peptide-specific CTL responses in vitro. In vivo immunization with 2–4 times as many peptide-loaded BM DC as SC DC resulted in CTL responses of about half the lytic capacity as observed with SC DC. Thus, SC DC were superior to BM DC in CTL response induction against peptide antigens.

Albeit phenotypic differences and differences in capacity to induce CTL responses against peptides antigens, we were interested in comparing the presentation abilities of protein antigen by SC DC and BM DC to T lymphocytes in view of the difference in maturation state. Both SC DC and BM DC pulsed with ovalbumin were specifically recognized by class II-restricted T cells (data not shown). The data obtained with SC DC are in concordance with data on protein-pulsed DC that were capable of priming antigen-specific T cells in vivo (33, 34). Also, we have previously demonstrated class II-restricted presentation of SC DC pulsed with hen egg-white lysozyme protein (17), showing that SC DC have processing abilities at least for class II-restricted presentation. When investigating processing abilities for class I-restricted presentation, we found that both SC DC and BM DC were specifically recognized by CTL after pulsing with ovalbumin protein (Table 2). Pulsing either early or late in the cultures of SC DC and BM DC gave similar results. As an indication for intracellular processing might serve the observation that the supernatants of protein-pulsed SC DC and BM DC did not contain detectable peptide fragments. Class I-restricted presentation of another protein antigen, i.e. E7 of human papilloma virus type 16 (HPV16), by BM DC was also observed (Table 2). Similar experiments with SC DC are in progress. In addition, we would like to investigate by what mechanism this presentation takes place in SC DC and BM DC (by interfering with class I- and class II- processing pathways).

Because CTL recognition of protein-pulsed SC DC and BM DC was successful, we investigated whether CTL responses could be induced by them. We found that the CTL bulks that were generated in vitro by one stimulation event with protein-loaded DC, either SC DC or BM DC, lysed target cells in an antigen-independent fashion (Figure 2). However, when we restimulated these bulks for at least two times with a cell line endogenously expressing the antigen, lysis by the remaining CTL was antigen-specific (Figure 2). When protein-pulsed SC DC or BM DC were administered in vivo, subsequent in vitro restimulation of in vivo primed T cells yielded antigen-specific CTL bulks (Figure 2). Thus, antigen-specific CTL can be generated by in vitro and in vivo administration of SC DC or BM DC that were pulsed with protein antigen when followed by subsequent restimulation events.

In different tumor models (HPV16-containing and adenovirus type 5 E1-containing tumors), we were able to provide complete protection against a tumor challenge by intraveneous administration of SC DC and BM DC that were pulsed with peptides antigens presented by the tumor (De Bruijn and Toes, unpublished data). Yet, we noticed that tu-

Table 1. Peptide-loaded SC DC are more efficient than BM DC at CTL response induction in vitro and in vivo. Listed are the number of unprimed T cells (spleen cell population of naive C57 BL/6 mice depleted for B cells and adherent cells), class II- and B7-positive SC DC and BM DC of C57/BL6 origin, that are required in a 5-day in vitro culture in 24-well plates to yield comparable CTL responses in vitro, i.e. 60–80% peptide-specific CTL lysis at E/T ratio 50 (data obtained from five independent experiments). In vivo immunizations were performed with 2- to 4- fold more peptide-loaded BM DC than SC DC. Splenic T cells of immunized mice were restimulated in vitro with cells endogenously expressing the antigen. By immunization with BM DC lower CTL responses were obtained than with SC DC, depicted as percentage antigen-specific lysis at E/T ratio 50 (data obtained from three independent experiments). The following peptides were used: K^b-binding OVA8, SIINFEKL; K^b-binding VSV8, RGYVYQGL; D^b-binding HPV9, RAHYNIVTF

in vitro requirements			
Responders	unprimed T	5×10^6	/ well
Stimulators	SC DC	1×10^4 - 5×10^4	/ well
	BM DC	25×10^4	/ well

		One i.v. immunization with
in vivo requirements		
10^5	SC DC	60 - 80 % Ag-spec. lysis at E/T ratio 50
$2 - 4 \times 10^5$	BM DC	20 - 40 % Ag-spec. lysis at E/T ratio 50

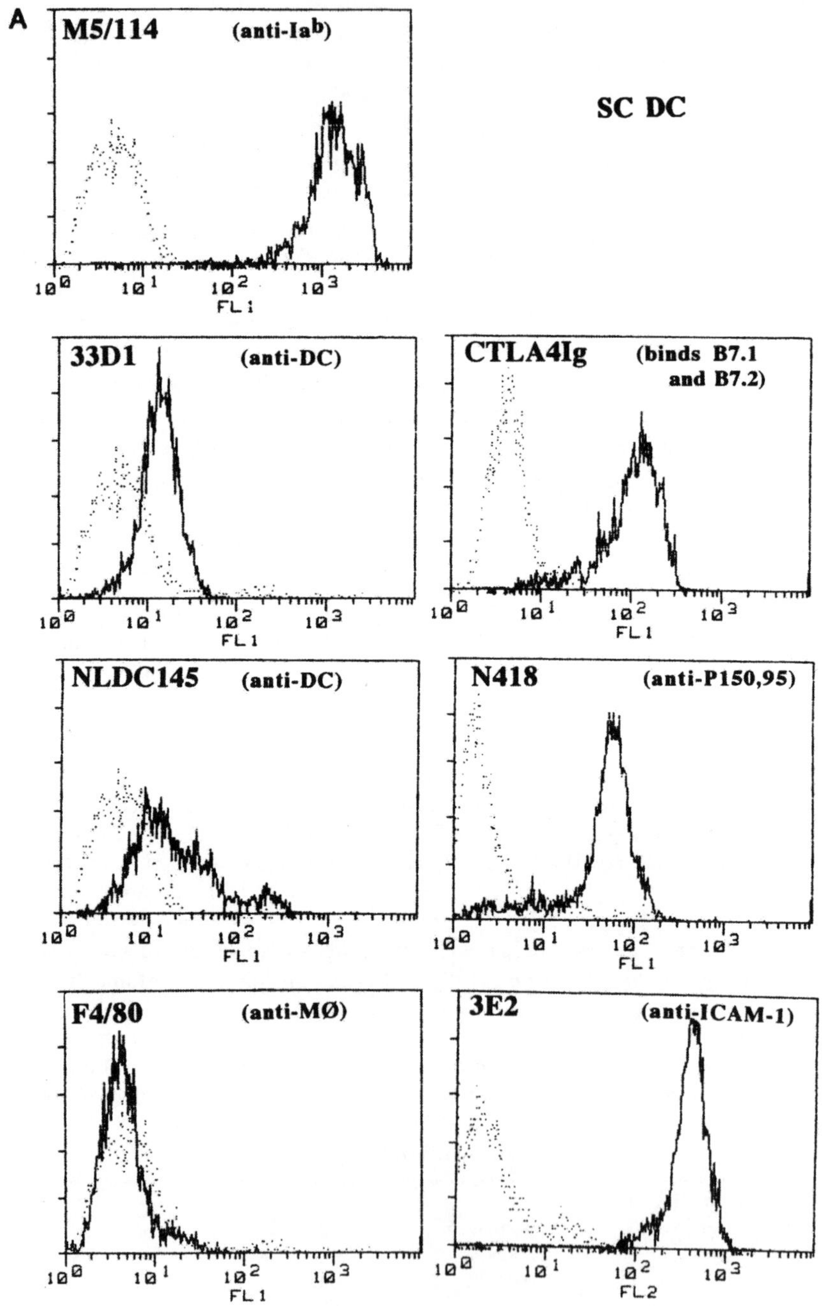
A
M5/114 (anti-Iab)
SC DC
33D1 (anti-DC)
CTLA4Ig (binds B7.1 and B7.2)
NLDC145 (anti-DC)
N418 (anti-P150,95)
F4/80 (anti-MØ)
3E2 (anti-ICAM-1)
10⁰ 10¹ 10² 10³
FL1
FL2

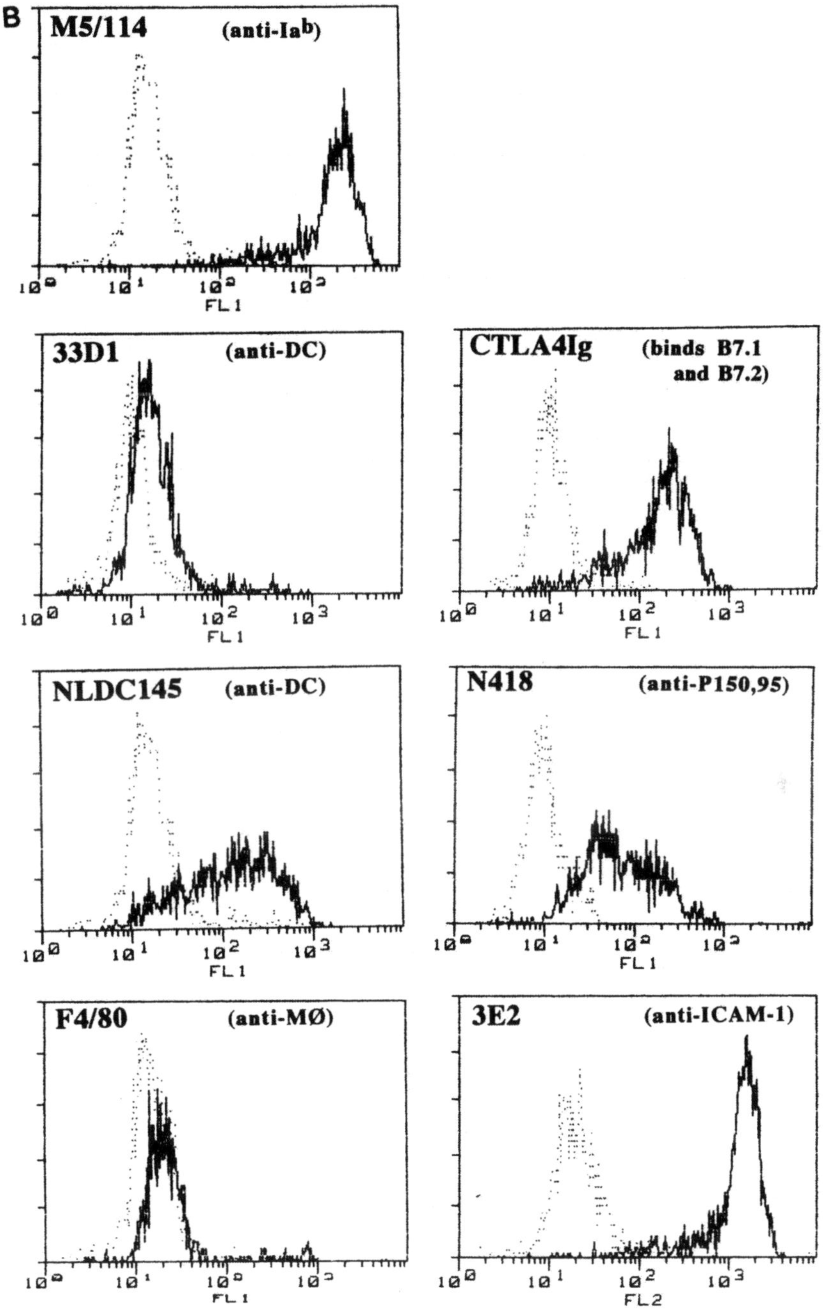

Figure 1. Phenotypic differences between SC DC and BM DC. SC DC were isolated by treating spleens with collagenase and passing spleen fragments through a nylon mesh filter. This cell suspension was spun on a discontinuous BSA gradiënt of 10, 28 and 35 % BSA at 4°C for 30 min at 10,000 g. The low density cells at the interphase between 10 and 28% BSA were collected and cultured for 90 min at 37°C in glass petri dishes. Nonadherent cells were discarded and medium refreshed. After 18 hours culture at 37°C, SC DC were harvested by collecting the nonadherent cells.

Table 2. Protein-pulsed BM DC and also protein-pulsed SC DC are specifically recognized by class I-restricted T cells. SC DC and BM DC were pulsed overnight with 10 μM protein (from day 0 to day 1 and from day 7 to day 8, respectively). Class II- and B7- positive DC previously pulsed with protein antigen and CTL were co-cultured for 18 hours. The results of three independent experiments are shown as percentage response, i.e. in a TNFalfa -release assay as a percentage of maximal WEHI cell death (exp. I and II) and in a blue cell assay as a percentage of maximal triggering of a CTL clone that carries an inducible Lac-Z gene fused to the IL-2 enhancer (exp. III). Similar results were obtained when SC DC were pulsed for 4 hours on day 1 of the isolation procedure, and when BM DC were pulsed earlier in the culture period, namely on day 2 or day 5 (data not shown). Supernatants of protein-pulsed SC DC or BM DC did not contain peptide fragments that could be detectected in a CTL assay (data not shown)

	OVA8 + K^b CTL			HPV9 + D^b CTL	
Experiment nr.	I	II	III	I	II
SC DC			18		
SC DC + OVA8			**95**		
SC DC + HPV9					
OVA- pulsed SC DC			**48**		
E7 - pulsed SC DC					
BM DC	20	20	12	20	16
BM DC + OVA8		**81**	**65**		21
BM DC + HPV9		23			**60**
OVA- pulsed BM DC	**85**	**77**	**42**	28	21
E7 - pulsed BM DC	32	16		**61**	**78**

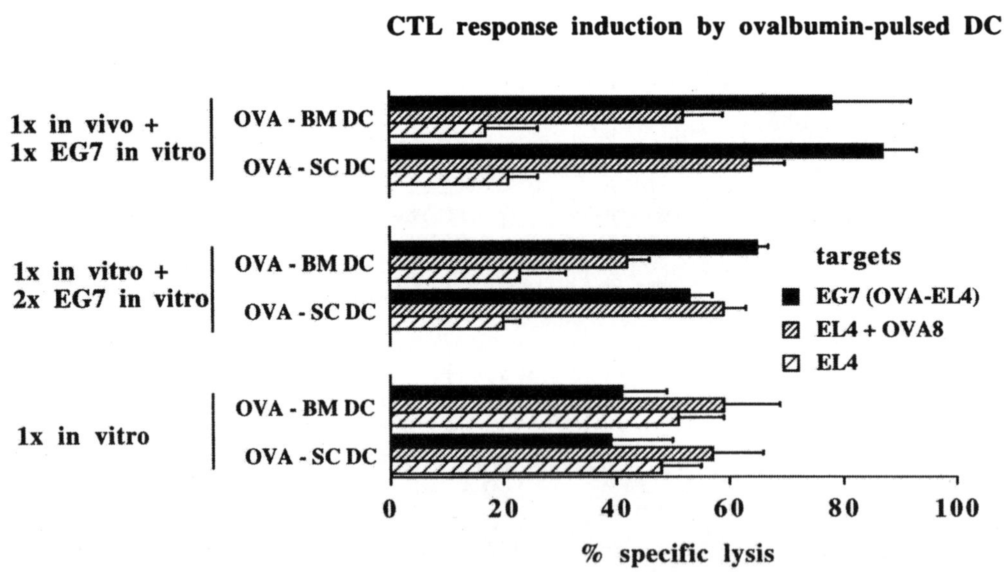

Figure 2. Antigen-specific CTL generated by protein-pulsed SC DC and BM DC were selected by subsequent restimulation. Ovalbumin-pulsed syngeneic SC DC and BM DC were cultured for 5 days with unprimed T cells of C57BL/6 mice. The CTL bulks were tested for lysis of EL4 cells with or without OVA8 and of EG7, the ovalbumin-transfected EL4 cell line. Similar experiments were performed with CTL bulks induced by ovalbumin-pulsed SC DC and BM DC, but these were restimulated twice in vitro with EG7 before the cytotoxicity assay was performed. In vivo administration of ovalbumin-pulsed SC DC and BM DC was followed by in vitro restimulation with EG7 cells. Depicted in the graph are the percentages specific lysis at E/T ratio 20.

mor protection was not always antigen-specific. Thus, SC DC and BM DC without tumor antigen can also protect against a tumor challenge. Protection experiments with protein-pulsed SC DC and BM DC are in progress.

In summary, SC DC and BM DC either loaded with peptides comprising CTL epitopes or pulsed with whole protein antigen, can induce CTL responses. Presenting peptide antigens, SC DC were more efficient than BM DC. With protein antigens, they both were equally effective. From our CTL data and tumor protection experiments involving BM DC and SC DC we also conclude that antigen-independent effector mechanisms can be anticipated.

ACKNOWLEDGMENTS

This work was supported by grants from the Dutch Cancer Society (RUL 94–881) and the Netherlands Organization for Scientific Research (NWO 901–09–241).

REFERENCES

1. Inaba, K., Young, J.W. and Steinman, R.M., 1987, Direct activation of CD8+ cytotoxic T lymphocytes by dendritic cells, J. Exp. Med. 166: 182.
2. Macatonia, S.E., Taylor, P.M., Knight, S.C. and Askonas, B.A., 1989, Primary stimulation by dendritic cells induces antiviral proliferative and cytotoxic T cell responses in vitro, J. Exp. Med. 169:1255.
3. De Bruijn, M.L.H., Nieland, J.D., Schumacher, T.N.M., Ploegh, H.L., Kast, W. Martin and Melief, C.J.M., 1992, Mechanisms of induction of primary virus-specific cytotoxic T lymphocyte responses, Eur. J. Immunol. 22: 3013.
4. Schuler, G. and Steinman, R.M., 1985, Murine epidermal Langerhans cells mature into potent immunostimulatory dendritic cells in vitro. J. Exp. Med. 172: 383.
5. Romani, N., Koide, S., Crowley, M., Witmer-Pack, M., Livingstone, A.M., Fathman, C.G., Inaba, K. and Steinman, R.M., 1989, Presentation of exogenous protein antigen by dendritic cells to T cell clones. Intact protein is presented best by immature epidermal Langerhans cells, J. Exp. Med. 169: 1169.
6. Streilein, J.W. and Grammer, S.F., 1989, In vitro evidence that Langerhans cells can adopt two functionally distinct forms of antigen presentation to T lymphocytes, J Immunol. 143: 3925.
7. Inaba, K., Inaba, M., Romani, N., Aya, H., Deguchi, M., Ikehara, S., Muramatsu, S., and Steinman, R.M., 1992, Generation of large numbers of dendritic cells from mouse bone marrow cultures supplemented with granulocyte/macrophage colony-stimulating factor, J. Exp. Med. 176:1693.
8. Inaba, K., Steinman, R.M., Witmer Pack, M., Aya, H., Inaba, M., Sudo, T., Wolpe, S. and Schuler, G., 1992, Identification of proliferating dendritic cell precursors in mouse blood, J. Exp. Med. 175: 1157.
9. Scheicher, C., Mehlig, M., Zecher, R. and Reske, K., 1992, Dendritic cells from mouse bone marrow: in vitro differentiation using low doses of recombinant granulocyte-stimulating factor, J. Immunol. Methods 154: 253.
10. Reid, C.D., Stackpoole, A., Meager, A. and Tikerpae, J., 1992, Interactions of tumor necrosis factor with granulocyte-macrophage colony-stimulating factor and other cytokines in the regulation of dendritic cell growth in vitro from early bipotent CD34+ progenitors in human bone marrow, J. Immunol. 149: 2681.
11. Romani, N., Gruner, S., Brang, D., Kämpgen, E., Lenz, A., Trockenbacher, B., Konwalinka, G., Fritsch P.O., Steinman, R.M. and Schuler, G., 1994, Proliferating dendritic cell progenitors in human blood, J. Exp. Med. 180: 83.
12. Sallusto, F. and Lanzavecchia, A., 1994, Efficient presentation of soluble antigen by cultured human dendritic cells is maintained by granulocyte/macrophage colony-stimulating factor plus interleukin 4 and downregulated by tumor necrosis factor alfa, J. Exp. Med. 179: 1109.
13. Girolomoni, G., Simon, J.C., Bergstresser, P.R. and Cruz, P.D. Jr., 1990, Freshly isolated spleen dendritic cells and epidermal Langerhans cells undergo similar phenotypic and functional changes during short term culture, J. Immunol. 145: 2820.
14. Austyn, J.M., 1992, Antigen uptake and presentation by dendritic leukocytes, Semin. Immunol. 4:227.
15. Nijman, H.W., Kleijmeer, M.J., Ossevoort, M.A., Oorschot V.M.J., Vierboom, M.P.M., van de Keur, M., Kenemans, P., Kast, W.M., Geuze, H.J. and Melief, C.J.M., 1995, Antigen capture and MHC class II compartments of freshly isolated and cultured human blood dendritic cells, J. Exp. Med. 182: 163.

16. Aiba, S. and Katz, S.I., 1991, The ability of cultured Langerhans cell to process and present protein antigens is MHC-dependent, J. Immunol. 146: 2479.
17. De Bruijn, M.L.H., Nieland, J.D., Harding, C.V. and Melief, C.J.M., 1992, Processing and presentation of intact hen egg-white lysozyme by dendritic cells, Eur. J. Immunol. 22: 2347.
18. Bender, A., Kim Bui, L., Feldman, M.A.V, Larsson, M. and Bhardwaj, N., 1995, Inactivated Influenza virus, when presented on dendritic cells, elicits human CD8+ cytolytic T cell responses, J. Exp. Med. 182: 1663.
19. Pfeifer, J.D., Wick, M.J., Roberts, R.L., Findlay, K., Normark, S.J. and Harding, C.V., 1993, Phagocytic processing of bacterial antigens for class I MHC presentation to T cells, Nature 361:359.
20. Kovacsovics-Bankowski, M., Clark, K., Benacerraf, B. and Rock, K.L., 1993, Efficient major histocompatibility complex class I presentation of exogenous antigen upon phagocytosis by macrophages, Proc. Natl. Acad. Sci. USA 90: 4942.
21. Harding, C.V. and Song, R., 1994, Phagocytic processing of exogenous particulate antigens by macrophages for presentation by class I MHC molecules, J. Immunol. 153: 4925.
22. De Bruijn, M.L.H., Jackson, M.R. and Peterson, P.A., 1995, Phagocyte-induced antigen-specific activation of unprimed CD8+ T cells in vitro, Eur. J. Immunol. 25: 1274.
23. Reis e Sousa, C., Stahl, P.D. and Austyn, J.M., 1993, Phagocytosis of antigens by Langerhans cells in vitro, J. Exp. Med. 178: 509.
24. Inaba K., Inaba., M., Saito, M. and Steinman, R.M., 1993, Dendritic cells progenitors phagocytose particulates, including bacillus Calmette-Guerin organisms, and sensitize mice to mycobacterial antigens in vivo, J. Exp. Med. 178: 479.
25. Austyn., J., 1996. New insights into mobilization and phagocytic activity of dendritic cells, J. Exp. Med. 183: 1287.
26. Ossevoort, M.A., Feltkamp, M.C., van Veen, K.J., Melief, C.J.M. and Kast, W.M., 1995, Dendritic cells as carriers for a CTL epitope-based peptide vaccin in protection against a HPV16-induced tumor, J. Immunotherapy 18: 86.
27. Celluzzi, C.M., Mayordomo, J.I., Storkus, W.J., Lotze, M.T. and Falo, L.D. Jr., 1996, Peptide-pulsed dendritic cells induce antigen-specific, CTL-mediated protective tumor immunity, J. Exp. Med. 183: 283.
28. Porgador, A., Snyder, D., and Gilboa, E., 1996, Induction of antitumor immunity using bone marrow-generated dendritic cells, J. Immunol. 156: 2918.
29. Mayordomo, J.I., Zorina, T., Storkus, W.J., Zitvogel, L., Celluzi, C., Falo, L.D., Melief, C.J.M, Ildstad, S.T., Kast, W.M., DeLeo, A.B. and Lotze, M.T., 1995, Bone marrow-derived dendritic cells pulsed with synthetic tumour peptides elicit protective and therapeutic antitumour immunity, Nature Medicine 1: 1297.
30. Mayordomo, J.I., Loftus, D.J., Sakamoto, H., De Cesare, C.M., Appasamy, P.M., Lotze, M.T., Storkus, W.J., Appella, E. and DeLeo A.B., 1996, Therapy of murine tumors with p53 wild-type and mutant sequence peptide-based vaccines, J. Exp. Med. 183:1357.
31. Zitvogel L., Mayordomo, J.I., Tjandrawan, T., DeLeo, A.B., Clarke, M.R., Lotze, M.T. and Storkus, W.J., 1996, Therapy of murine tumors with tumor peptide pulsed dendritic cells: dependence on T cells, B7 costimulation, and Th1-associated cytokines, J.Exp. Med. 183: 87.
32. Hsu, F.J., Benike, C., Fagnoni, F., Liles, T.M., Czerwinski D., Taidi, B., Engelman, E.G. and Levy, R., 1996, Vaccination of patients with B-cell lymphoma using autologous antigen-pulsed dendritic cells, Nature Medicine 2: 52
33. Inaba, K., Metlay, J.P., Crowley, M.T. and Steinman, R.M., 1990, Dendritic cells pulsed with protein antigens in vitro can prime antigen-specific, MHC-restricted T cells in situ, J. Exp. Med. 172: 631.
34. Paglia, P., Chiodoni, C., Rodolfo, M.. and Colombo, M.P., 1996. Murine dendritic cells loaded in vitro with soluble protein prime CTL against tumor antigen in vivo, J. Exp. Med. 183: 317.

37

THE TH1 LYMPHOKINE INTERFERON-γ IS A POTENT UPREGULATOR OF DENDRITIC CELLS WITH PHAGOCYTIC CAPACITY IN GM-CSF SUPPLEMENTED BONE MARROW CULTURES

Christoph Scheicher, Heike Corban, Vera Hof, Michael Robbers, and Konrad Reske

Institut für Immunologie
Johannes Gutenberg Universität Mainz
D-55101 Mainz, Germany

1. INTRODUCTION

Myeloid dendritic cells (DC), macrophages and granulocytes are descendants of a hematopoietic progenitor cell that originates in the bone marrow[1]. Thus, bone marrow derived cells distributed in tissue culture in the presence of GM-CSF give rise to the three leukocyte populations which under various in vitro culture conditions proceed in differentiation and phenotypic maturation[2–7].

The efficacy of antigen presentation by in vitro grown DC correlates with the mode of acquisition of the antigen. Internalization of antigen by phagocytosis is superior to uptake of antigen by pinocytosis or endocytosis and results in unsurpassed levels of T cell stimulation[8–11]. Notably, engulfment of antigen in particulate form triggered a number of gene activation events in the DC. Apart from an increase in synthesis of class II elements, augmentation of the level of IL-1α and the de novo synthesis of IL-12 were detected[11]. The production of the latter lymphokine suggested that DC appropriately sensitized might alter the balance between TH1 and TH2 cells and thus might influence the outcome of an antigen specific immune response.[12]

In this work we investigated the possibility that lymphokines derived from an ongoing TH1 response might on the other hand influence recruitment and functional capacity of DC. We found that administration of IFN-γ to a GM-CSF supplemented in vitro culture of mouse bone marrow cells had pronounced effects on the uptake of microparticles by the DC. Considering that mature DC are phagocytosis-negative whereas progenitor DC are prone to engulf microparticles our findings suggest that in the presence of IFN-γ the number of ProDC with phagocytic capacity increases significantly. This might be accomplished by a more efficient generation of ProDC from committed DC precursors or by an IFN-γ mediated modulation of the ProDC's differentiation.

2. MATERIALS AND METHODS

2.1. Animals

Bone marrow of 8 to 10 weeks old BALB/c/ANN mice of either sex was used.

2.2. Antibodies

The mAb MK-D6 (anti-I-A^d, ATCC HB-3)[13] was used as tissue culture supernatant. The secondary reagent was phycoerythrin-conjugated goat anti-mouse IgG2a (Medac/Hamburg) used at a 1:1000 dilution. For neutralization of mouse IFN-γ the rat anti-mouse mAb XMG1.2 (IgG2a) was used with a neutralizing capacity of 10 U IFN-γ/5μl mAb preparation as determined by ELISA.

2.3. Cell Culture

Bone marrow cells were cultivated as described[3]. Cells were cultured for 7 days at 37°C with 10% CO_2 in humidified air.

2.4. Recombinant Mouse Interferon-γ

IFN-γ was used as a supernatant of transfected BHK cells (kindly provided by S. Jin, Institut für Immunologie, Mainz) with a concentration of 50.000 U IFN-γ/ml, as determined by ELISA, or as a purified preparation from the BHK supernatant.

2.5. Administration of Stimuli

Bone marrow cells were cultured under standard conditions (see 2.3). Stimuli comprised: i) recombinant mouse IFN-γ (added to the cells every 24h), ii) mAb XMG1.2 (5μl/ml) (added to the cells every 24h), iii) FITC-labeled microparticles (given one pulse 2 days after culture onset).

2.6. FCM-analysis

Total cultured BMC were obtained by detachment of the adherent cell fraction for 10 min at 4°C with Versen buffer containing 1mM EDTA and pooling with the nonadherent cell fractions. After fixation of the cells in 1% paraformaldehyde in PBS at ambient temperature all further washings were done in PBS/1%FCS supplemented with 0.1% NaN_3. 10^6 cells were incubated with anti-MHC class II mAb MK-D6 for 1h followed by two cycles of washing and 1h incubation with the phycoerythrin-conjugated secondary reagent.

3. RESULTS AND CONCLUSION

We previously described the uptake of FITC-labeled polystyrol particles by developing dendritic precursor cells in GM-CSF supplemented bone marrow cultures[11]. To reveal the pro-

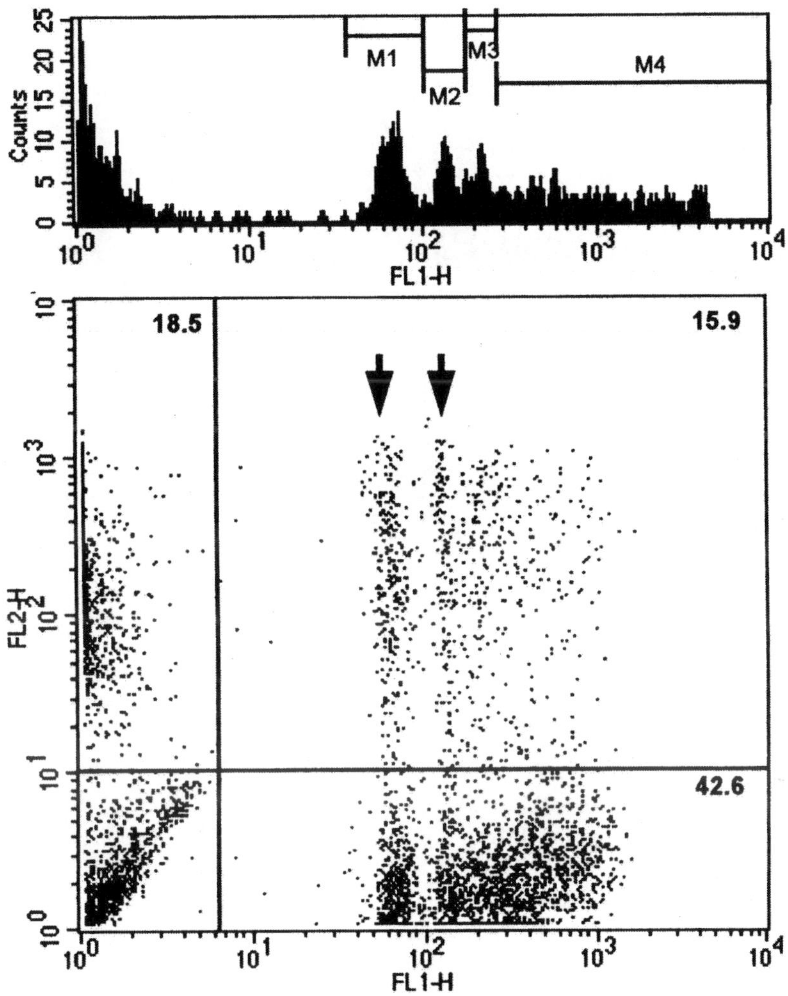

Figure 1. DC precursor cells ingest few fluorescent beads unlike BM-macrophages which are heavily bead-loaded. FL1: FITC fluorescence; FL2: PE fluorescence.

portion of DC phagocytosis$^+$ by FACS analysis standard BM cultures were set up and were given a single dose of microparticles on day 2. On day 7 cells were PE-stained for class II expression and subjected to double fluorescent FACS-analysis. The histogram in Fig. 1 reveals a distinct set of class II$^+$ (upper left) and class II$^+$ as well as phagocytosis$^+$ (upper right) DC. The stepwise increase in FITC-fluorescence intensity (FL1) is correlated with the uptake of one (M1, 14.9%), two (M2, 8.7%), three (M3, 5.6%) and more than three (M4, 20.0%) microparticles per cell as depicted in the upper panel. The level of particle load achieved by the DC is rather low and only a small number of cells is found within the high fluorescence intensity range. In contrast a large number of class II$^-$ macrophages showing extensive engulfment of particles can be distinguished (lower right in Fig. 1).

To assess the influence of IFN-γ on the development of DC and on the DC's potential to engulf microparticles, four parallel GM-CSF supplemented bone marrow cultures were set up. To maintain a certain level of IFN-γ throughout the entire culture period 10

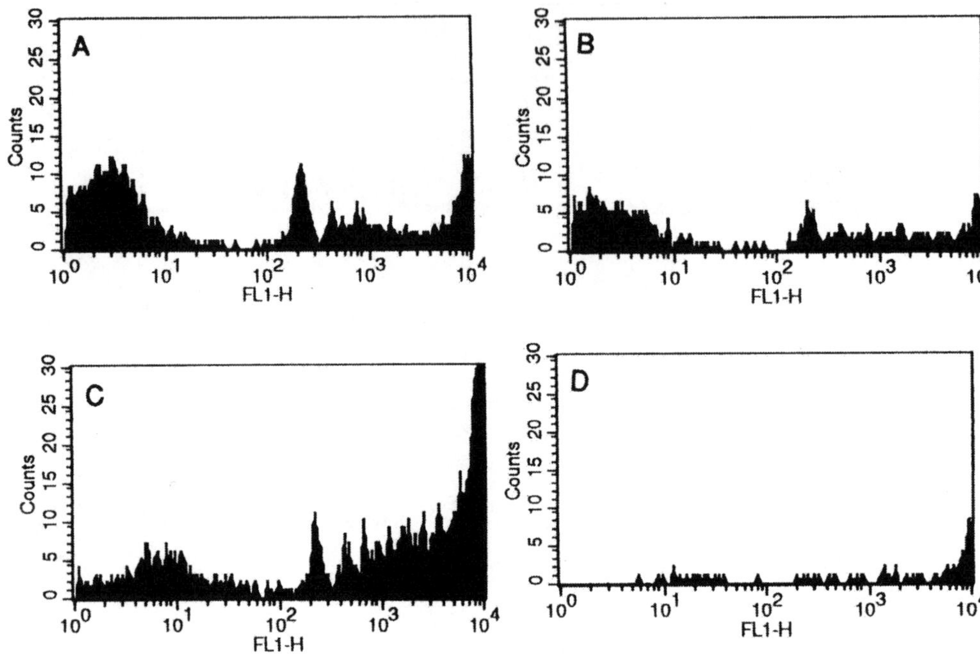

Figure 2. Development of DC in BM cultures in the presence of IFN-γ results in higher quantities of phagocytozing progenitor DC showing augmented particle uptake.

units of IFN-γ were added every 24h to the test culture (Fig. 2B). The second test culture received a daily dose of 5 μl/ml of anti-IFN-γ mAb XMG 1.2 instead of IFN-γ (Fig. 2C). Two control cultures were included, one GM-CSF supplemented bone marrow culture which did not receive IFN-γ (Fig. 2A) and one culture for the isotype class II staining control (Fig.2D). Particles were added to the four sets of cultures once on day 2. The cultures were harvested on day 7 and class II expression as well as bead fluorescence were measured by FACS double staining.

In accord with prior observations addition of beads to control culture 2A resulted in roughly 20% of double positive cells that expressed class II and had engulfed FITC-labeled microparticles. In the presence of a low dose of IFN-γ a shift of the PE^+ cells towards the double positive (PE^+ and $FITC^+$) cell population and towards higher particle numbers became apparent (Fig. 2B) suggesting that the number of class II^+ DC and their phagocytic capacity was augmented in response to the lymphokine. If on the other hand uptake of particles by the DC was responsive to IFN-γ then blocking of endogenous quantities of this lymphokine should depress the DC's phagocytic capacity. Therefore the effect of the anti-IFN-γ monoclonal antibody on the developing DC culture was estimated. As shown in Fig. 2C a clearcut decline in the double positive population was noticed, although phagocytosis was not abrogated.

In summary, IFN-γ, a lymphokine secreted by TH1 cells, has a distinct impact on developing DC in GM-CSF supplemented, microparticle containing mouse bone marrow cultures. In response to IFN-γ the proportion of progenitor DC that take up microparticles increases significantly, and also the number of particles phagocytozed by the DC is elevated.

REFERENCES

1. K. Inaba, M. Inaba, M. Deguchi, K. Hagi, R. Yasumizu, S. Ikehara, S. Muramatsu, and R. M. Steinman, Granulocytes, makrophages, and dendritic cells arise from a common major histocompatibility complex class II-negative progenitor in mouse bone marrow. Proc. Natl. Acad. Sci. USA 90: 3038–3042 (1993).
2. K. Inaba, R. M. Steinman, M. Witmer Pack, H. Aya, M. Inaba, T. Sudo, S. Wolpe, and G. Schuler, Identification of proliferating dendritic cell precursors in mouse blood. J. Exp.Med. 175: 1157–1167 (1992).
3. C. Scheicher, M. Mehlig, R. Zecher and K. Reske, Dendritic cells from mouse bone marrow: in vitro differentiation using low doses of recombinant granulocyte-macrophage colony-stimulating factor. J. Immunol. Meth. 154: 253–264 (1992).
4. K. Inaba, M. Inaba, N. Romani, H. Aya, M. Deguchi, S. Ikehara, S. Muramatsu, and R. M. Steinman, Generation of large numbers of dendritic cells from mouse bone marrow cultures supplemented with granulocyte/macrophage colony-stimulating factor. J. Exp. Med. 176: 1693–1702 (1992).
5. C. Caux, C. Dezutter-Dambuyant, D. Schmitt and J. Banchereau, GM-CSF and TNF-α cooperate in the generation of dendritic Langerhans cells. Nature 360: 258–261 (1992).
6. C. Reid, A. Stackpoole, A. Meager, and J. Tikerpae, Interactions of tumor necrosis factor with granulocyte-macrophage colony-stimulating factor and other cytokines in the regulation of dendritic cell growth in vitro from early bipotent $CD34^+$ progenitors in human bone marrow. J. Immunol. 149: 2681–2688 (1992).
7. F. Sallusto and A. Lanzavecchia, Efficient presentation of soluble antigen by cultured human dendritic cells is maintained by granulocyte/macrophage colony-stimulating factor plus interleukin 4 and downregulated by tumor necrosis factor α. J. Exp. Med. 179: 1109–1118 (1994).
8. K. Inaba, M. Inaba, M. Naito, and R. M. Steinman, Dendritic cell progenitors phagocytose particulates, including Bacillus Calmette-Guerin organisms, and sensitize mice to mycobacterial antigens in vivo. J. Exp. Med. 178: 479–488 (1993).
9. C. Reis e Sousa, P. D. Stahl, and J. M. Austyn, Phagocytosis of antigens by Langerhans cells in vitro. J. Exp. Med. 178: 509–519 (1993).
10. C. Blank, H. Fuchs, K. Rappersberger, M. Röllinghoff, and H. Moll, Parasitism of epidermal Langerhans cells in experimental cutaneous leishmaniasis with Leishmania major. J. Infect. Dis. 167: 418–425 (1993).
11. C. Scheicher, M. Mehlig, H.-P. Dienes and K. Reske, Uptake of microparticle-adsorbed protein antigen by bone marrow-derived dendritic cells results in up-regulation of interleukin-1α and interleukin-12 p40/p35 and triggers prolonged, efficient antigen presentation. Eur. J. Immunol. 25: 1566–1572 (1995).
12. S. E. Macatonia, C.-S. Hsieh, K. M. Murphy, and A. O'Garra, Dendritic cells and macrophages are required for Th1 development of $CD4^+$ T cells from αβ TCR transgenic mice: IL-12 substitution for macrophages to stimulate IFN-γ production is IFN-γ-dependent. Internat. Immunol. 5: 1119–1128 (1993).
13. J. W. Kappler, B. Skidmore, J. White, P. Marrack, Antigen-inducible, interleukin-2-producing T cell hybridomas. Lack of independent antigen and H-2 recognition. J. Exp. Med. 153:1198–1214 (1981).

38

FUNCTIONAL ROLE OF CD101 ON SKIN DENDRITIC CELLS

Martine Bagot,[1] Isabelle Martinel,[1] Dominique Charue,[1] Marie-Laure Boulland,[2] Janine Wechsler,[2] Armand Bensussan,[1] and Laurence Boumsell[1]

[1]INSERM U448, IM3
Paris XII University
Hôpital Henri Mondor, Créteil, France
[2]Department of Pathology
Paris XII University
Hôpital Henri Mondor, Créteil, France

1. CD101: A NEW LEUKOCYTE SURFACE ANTIGEN INVOLVED IN T CELL ACTIVATION

1.1. Generation of Anti-CD101 Monoclonal Antibodies (mAbs)

CD101 is a cell surface protein with a molecular weight of about 140 kDa and seven Ig-like loops in its extra-cellular domain, that was first described in our laboratory, using two different monoclonal antibodies (mAb), BB27 and BA27 (1,2). The two anti-CD101 mAbs BB27 and BA27 were generated by immunization of BALB/c mice with the CD4+CD8+ thymic clone B12 (3). These mAbs recognize CD101, a disulfide bonded-homodimer, on subsets of circulating T lymphocytes.

1.2. Cell Distribution and Function

CD101 is detected at low density on 25–30% of resting T lymphocytes in normal individuals and mainly expressed by granulocytes and cells of the monocytes/macrophages lineage. After polyclonal activation, the majority of T cells highly express CD101. CD101 participates in T lymphocyte activation and is probably involved in intracellular signalling. CD101 molecule seems to be required for anti-CD28 mAb and PMA stimulation of PBL (1,2). Recently, CD101 has been described as a costimulatory molecule preferentially expressed on mucosal T lymphocytes (4).

Dendritic Cells in Fundamental and Clinical Immunology
edited by Ricciardi-Castagnoli, Plenum Press, New York, 1997

1.3. Cloning of the CD101 Gene

The gene coding for the V7 antigen, recognized by the V7.1 mAb, has been independently cloned (5,6), and next found to be almost identical to the gene coding for the CD101 molecule (Boumsell L, Freeman GJ. & Bensussan A., CD101 Workshop Panel Report, 6th International Workshop and Conference, Kobe, Japan , manuscript in press). V7 is a recently described leukocyte cell surface glycoprotein expressed on activated T cells, monocytes and granulocytes. The mAb V7.1, that recognizes the V7 antigen, was generated by repeated immunization of female BALB/c mice with the long-term alloactivated CD3+CD4-CD8+ noncytolytic T cell clone CS1 (5). V7.1 immuno-precipitates a single polypeptide chain of approximately 135 kDa under reducing conditions from alloactivated T cells or monocytes, which migrated to approximately 110 kDa after treatment with N-glycanase. The gene coding V7 has been mapped to chromosome 1p13 (6). The V7 glycoprotein is a type I trans-membrane protein with seven disulfide-linked Ig-like loops in its extracellular domain. It is a novel member of the Ig superfamily with an unknown function.

2. EXPRESSION OF CD101 BY SKIN DENDRITIC CELLS

2.1. *In Situ* Expression of CD101

As CD101 appeared to be expressed mainly on cells of the myeloid/monocytic lineage, we studied the expression and the function of this molecule in the skin. Incubation of skin sections with BB27 mAb in APAAP technique resulted in the marked labeling of cells with dendritic morphology. These labeled cells were located mainly in the dermis, whereas only few DC were stained in the epidermis. In addition, double staining of skin sections of Langerhans cell histiocytosis showed that histiocytosis cells infiltrating both epidermis and dermis strongly expressed both CD1a and CD101.

2.2. Expression of CD101 by Skin Dendritic Cells (DC) Migrating Out of Skin Explants

To define surface proteins that are involved in skin dendritic cells localization or function, we looked for the expression of CD101 on skin dendritic cells migrating from human skin explants. It has been shown that skin DC can be enriched, using a new technique based on the migratory properties of these cells, and requiring no use of proteolytic enzyme (7–12). These skin DC were studied by double immunostaining and flow cytometric analysis. The majority of the DC population migrating from skin explants had a phenotype of LC-like mature DC, i.e. they expressed HLA-DR, CD1a, CD1c, CD11a, CD11c, CD40, CD50, CD54, CD58, CD80, CD83, and CD86. We found that CD101 was expressed by a major subset of these HLA-DR+CD1a+CD1c+ LC-like skin DC.

3. FUNCTIONAL ROLE OF CD101 ON SKIN DC

3.1. BB27 and V7.1 Recognize Different Epitopes of the CD101 Molecule

As BB27 and V7.1 are both anti-CD101 mAb, we addressed the question of whether they recognized the same or distinct epitopes of the molecule. Cross-block experiments

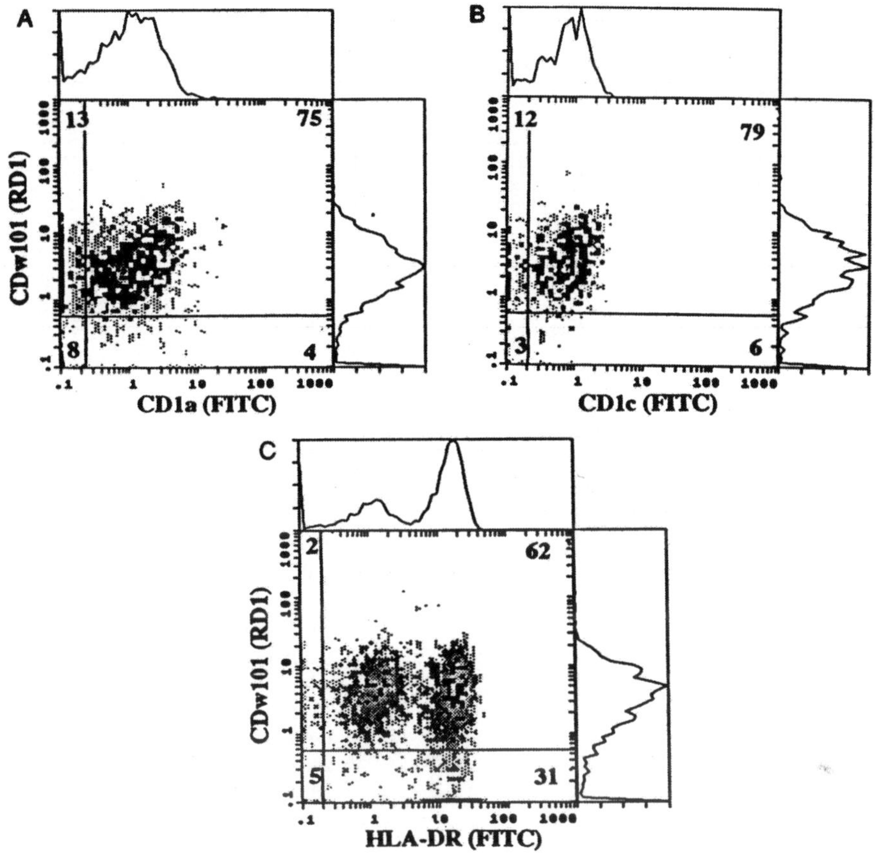

Figure 1. Double immunostaining flow cytometric analysis of gated enriched skin DC suspensions. DC suspensions were subjected to immunofluorescence double-labeling using RD1-conjugated anti-CD101 BB27, and FITC-conjugated anti-CD1a WM35 (A), FITC-conjugated anti-CD1c L161 (B), or FITC-conjugated anti-HLA DR B8.12.2 (C). CD101 was expressed by a major subset of CD1a+, CD1c+, and HLA-DR+ skin DC.

using the YT2C2 natural killer cell line (13), that constitutively expresses high levels of CD101, demonstrated that BB27 and V7.1 recognize distinct epitopes of the CD101 molecule.

3.2. Anti-CD101 mAbs Inhibit Allogeneic and Soluble Antigen-Specific Mixed Skin DC-Lymphocyte Reactions

We next studied the effect of anti-CD101 mAbs on primary allogeneic and on soluble antigen-specific mixed skin DC-lymphocyte reactions. Enriched skin DC migrating from skin explants induced strong proliferations of lymphoid cells and purified T cells. The two different mAbs, BB27 and V7.1, induced a dose-dependent inhibition of the T lymphocytes proliferative responses. This inhibitory effect could be overcome by high doses of exogenous IL-2. Preincubation assays showed that the inhibitory effects of anti CD101 mAbs were observed when either peripheral blood lymphoid cells or skin DC were preincubated with the mAb.

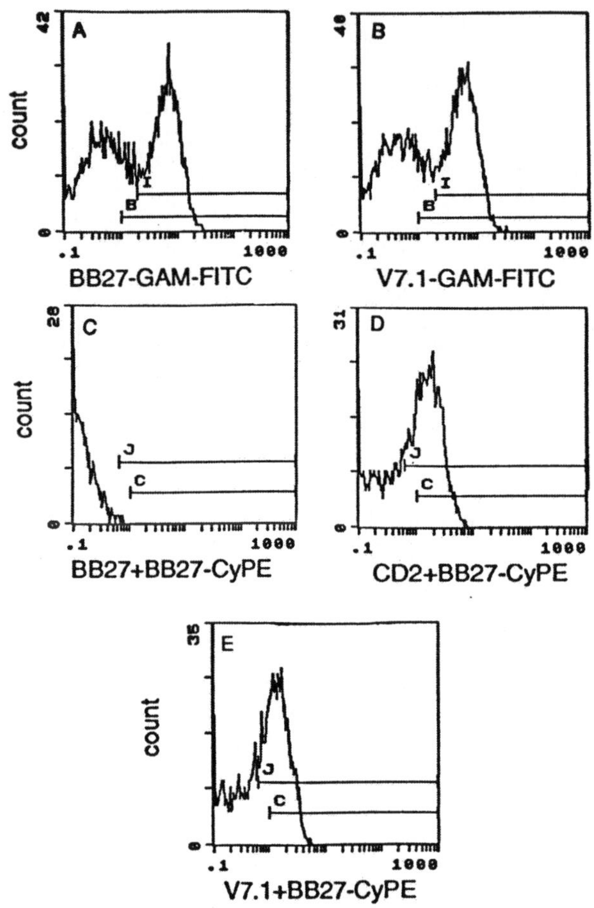

Figure 2. Cross-block experiments. The YT2C2 natural killer cell line constitutively expresses CD101, as shown by staining with FITC-conjugated BB27 (A) or V7.1 (B) mAbs. The YT2C2 cells were incubated with the unlabeled anti-CD101 mAb (BB27 or V7.1) or an isotype-matched anti-CD2 mAb (CD2X11) during 30 min, washed, and stained with a labeled anti-CD101 mAb (BB27-Cy5-PE) (C,D,E). Flow cytometric analysis was then performed. Results show that V7.1 and BB27 recognize distinct epitopes of the CD101 molecule.

3.3. CD101: A Co-Stimulatory Molecule?

V7.1 mAb was reported to inhibit the proliferative response of T cells to allogeneic cells or immobilized anti-CD3 mAb, but not to mitogenic lectins (5). It must be noted that anti-CD101 mAbs were also reported to amplify mucosal T lymphocyte (4). This contradictory results could be due to the fact that most intestinal mucosal lymphocytes expressed higher level of CD101, whereas only a minor subset of PBL expressed lower level of CD101. Therefore, signaling through CD101 might be different in both cell populations. Taking together, these results suggest that CD101 plays a role in TCR/CD3-mediated T cell activation. The expression of CD101 during activation seems to be required as a CD28 costimulatory signal (1). Co-stimulatory signals play a crucial role for the activity of antigen-presenting cells, since abrogation of these accessory signals may lead to clonal anergy. Cultured epidermal LC express several costimulatory molecules, including CD80, CD86, CD40, CD44, CD50, and CD54 (14–19). The activation of T lymphocytes by enriched LC is inhibited by anti-CD86 and anti-CD50 mAbs and at a lesser extent by anti-CD80 mAbs (15,18,20). Anti-CD101 mAbs strongly inhibited the activation of T lymphocytes by skin DC in allogeneic and soluble antigen-specific proliferations. These results demonstrate that CD101 plays a major role for the activation of T lymphocytes by skin DC. Further, our results showed that the inhibition induced by anti-CD101 mAbs

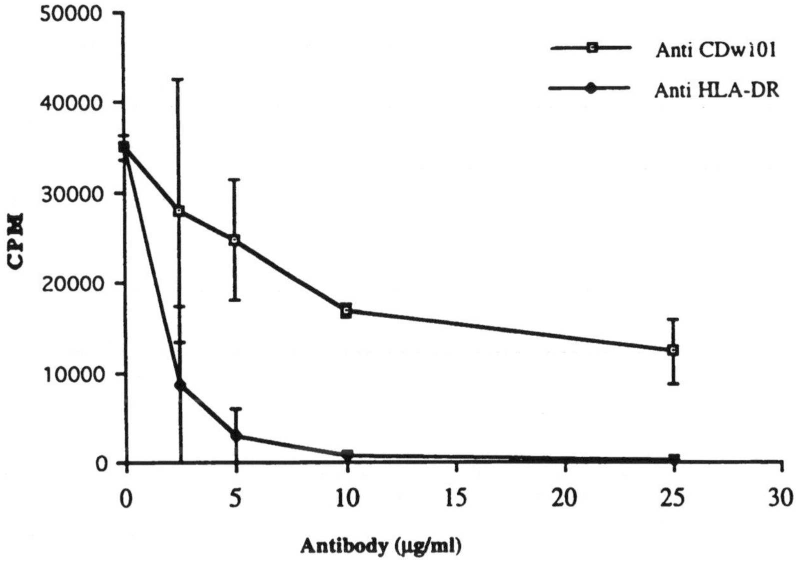

Figure 3. The concentration-dependent effects of anti-CD101 antibodies on enriched skin DC-induced T-cell alloreaction. 5×10^4 T lymphocytes were seeded in 96-well round-bottomed culture plates in the presence of 5×10^3 allogeneic enriched DC migrating from skin explants after 3 d. mAbs against CD101 (BB27), or HLA-DR (B8.12.2) were added at the beginning of the culture. Proliferation was measured by tritiated thymidine incorporation after 6 d of culture. Results are representative of six experiments. Values are expressed as mean cpm for triplicate. Error bars, SD.

could be overcome by the addition of high doses of exogenous IL-2, which suggests that the inhibition is related to a defect in the interaction of T lymphocytes and DC involving the production of cytokines. CD101 could thus function as a co-stimulatory molecule, promoting interactions between CD101 positive cells and other cells expressing a ligand for CD101 that remains to be determined.

REFERENCES

1. Gouttefangeas, C., S. Jacquot, E. Meffre, M. Schmid, L. Boumsell, and A. Bensussan. 1993. Differential proliferative responses in subsets of human CD28+ cells delineated by BB27 mAb. *Int. Immunol. 6:423–430.*
2. Gouttefangeas, C., M. Schmid, L. Boumsell, and A. Bensussan. 1995. Delineation of human CD28 functional subsets by a novel 140-kDa homodimeric surface lymphocyte molecule defining CDw101. In: Schlossman SF, Boumsell L, Gilks W, Harlan JM, Kishimoto T, Morimoto C, Ritz J, Shaw S, Silverstein R, Springer T, Tedder TF, Todd RF (eds). *Leukocyte Typing V. White Cell Differentiation Antigens.* Oxford University Press, New York, pp 293–295.
3. Boumsell, L., M. Schmid, H. Dastot, C. Gouttefangeas, D. Mathieu-Mahul, and A. Bensussan. 1990. In vitro differentiation from a pluripotent human CD4+CD8+ thymic cloned cell into four phenotypically distinct subsets. *J. Immunol. 145:2797–2802.*
4. Russell, G.J., C.M. Parker, A. Sood, E. Mizoguchi, E.C. Ebert, A.K. Bhan, and M.B. Brenner. 1996. p126 (CDw101): a costimulatory molecule preferentially expressed on mucosal T lymphocytes. *J. Immunol 157:3366–74.*
5. Rivas, A., C.L. Ruegg, J. Zeitung, R. Laus, R. Warnke, C. Benike, and E.G. Engleman. 1995. V7, a novel leukocyte surface protein that participates in T cell activation. I. Tissue distribution and functional studies. *J. Immunol. 154:4423–33.*

6. Ruegg, C.L., A. Rivas, N.D. Madani, J. Zeitung, R. Laus, and E.G. Engleman. 1995. V7, a novel leukocyte surface protein that participates in T cell activation. II. Molecular cloning and characterization of the V7 gene. *J. Immunol. 154:4434–4443.*
7. Nestle, F.O., X.G. Zheng, C.B. Thompson, L.A. Turka, and B.J. Nickoloff. 1993. Characterization of dermal dendritic cells obtained from normal human skin reveals phenotypic and functionally distinctive subsets. *J. Immunol. 151:6535–6545.*
8. Larsen, C.P., R.M. Steinman, M.D. Witmer-Pack, D.F. Hankins, P.J. Morris, and J.M. Austyn. 1990. Migration and maturation of Langerhans cells in skin transplants and explants. *J. Exp. Med.172:1483–1493.*
9. Lenz, A., M. Heine, G. Schuler, and N. Romani. 1993. Human and murine dermis contain dendritic cells. Isolation by means of a novel method and phenotypical and functional characterization. *J. Clin. Invest. 92:2587–2596.*
10. Richters, C.D., M.J. Hoekstra, J. Van Baare, J.S. Du Pont, E.C.M. Hoefsmit, and E.W.A. Kamperdijk. 1994. Isolation and characterization of migratory human skin dendritic cells. *Clin. Exp. Immunol. 98:330–336.*
11. Pope, M., G.H. Betjes, H. Hirmand, L. Hoffman, and R.M. Steinman. 1995. Both dendritic cells and memory T lymphocytes emigrate from organ cultures of human skin and form distinctive dendritic-T-cell conjugates. *J. Invest. Dermatol. 104:11–17.*
12. Lukas, M., H. Stössel, L. Hefel, S. Imamura, P. Fritsch, N.T. Sepp, G. Schuler, and N. Romani. 1996. Human cutaneous dendritic cells migrate through dermal lymphatic vessels in askin organ culture model. *J. Invest. Dermatol. 106:1293–99.*
13. Teshigawara, K.,.H.M. Wang, K. Kato, and K.A. Smith. 1987. Interleukin 2 high-affinity receptor expression requires two distinct binding proteins. *J. Exp. Med. 165:223–38.*
14. Symington, F.W., W. Brady, and P.S. Linsley. 1993. Expression and function of B7 on human epidermal Langerhans cells. *J. Immunol. 150:1286–1295.*
15. Yokozeki, H., I. Katayama, O. Ohki, T. Matsunaga, K. Watanabe, T. Satoh, M. Azuma, K. Okumura, and K. Nishioka. 1996. Functional CD86 (B7–2/B70) on cultured human Langerhans cells. *J. Invest. Dermatol. 106:147–153.*
16. Peguet-Navarro, J., C. Dalbiez-Gauthier, F.M. Rattis, C. Van Kooten, J. Banchereau, and D. Schmitt. 1995. Functional expression of CD40 antigen on human epidermal Langerhans cells. *J. Immunol. 155:4241–4247.*
17. Aiba, S., S. Nakagawa, H. Ozawa, K. Miyake, H. Yagita, and H. Tagami. 1993. Up-regulation of α4 integrin on activated Langerhans cells: analysis of adhesion molecules on Langerhans cells relating to their migration from skin to draining lymph nodes. *J. Invest. Dermatol. 100:143–147.*
18. Griffiths, C.E.M., D. Railan, W.M. Gallatin, and K.D. Cooper. 1995. The ICAM-3/LFA-1 interaction is critical for epidermal Langerhans cell alloantigen presentation to CD4+ T cells. *Br. J. Dermatol. 133:823–829.*
19. De Panfilis, G., G.M. Manara, C. Ferrari, and C. Torresani. 1990. Adhesion molecules on the plasma membrane of epidermal cells. II. The intercellular adhesion molecule-1 is constitutively present on the cell surface of human resting Langerhans cells. *J. Invest. Dermatol. 94:317–21.*
20. Teunissen, M.B., C.W. Koomen, and J.D. Bos. 1995. Intercellular adhesion molecule-3 (CD50) on human epidermal Langerhans cells participates in T-cell activation. *J. Invest. Dermatol. 104:995–998.*

39

PRESENTATION OF TETANUS TOXOID TO AUTOLOGOUS T CELLS BY DENDRITIC CELLS GENERATED FROM HUMAN BLOOD

Improved Specificity with Dendritic Cells Generated Without Fetal Calf Serum

Susanne Büchele and Reinhard Höpfl

Department of Dermatology
University of Innsbruck
A-6020 Innsbruck, Austria

1. SUMMARY

Dendritic cells (DC) are highly specialised to initiate primary immune responses and may therefore serve as natural adjuvant in future strategies for specific immunotherapy, e. g. with tumor antigens. The originally developed culture system to generate DC from peripheral human blood with GM-CSF and IL-4 was dependent on the use of fetal calf serum. We employed such DC as antigen presenting cells in a modified lymphocyte proliferation assay to measure the response of autologous T cells to tetanus toxoid. However, a substantial proliferative response of T cells was also observed in control wells without antigen, i.e. in the setting of a syngeneic mixed leukocyte reaction. This makes it difficult, if not impossible, to monitor antigen-specific responses in vitro. In a recently developed improved method fetal calf serum was replaced by 1% autologous human plasma. Using such DC in our lymphocyte proliferation assay background proliferation was markedly reduced. T cell responses to tetanus toxoid were strongest when the antigen was added to DC three days before cocultivation with T cells. We conclude that DC cultured in FCS-free autologous systems, suitable for clinical use, can process and present tetanus protein to autologous T cells. Using such DC in a lymphocyte proliferation assay may facilitate the measurement of antigen-specific T cell responses.

2. MATERIAL AND METHODS

Dendritic cells (DC) were generated from peripheral human blood mononuclear cells (PBMC) depleted of lymphocytes by E-rosetting. Those cells were cultured for 7

Dendritic Cells in Fundamental and Clinical Immunology
edited by Ricciardi-Castagnoli, Plenum Press, New York, 1997

Table 1. Time course of DC-culture

<u>**day 0-7:**</u> * Generation of DC with GM-CSF and IL-4 from PBMC depleted of lymphocytes by E- rosetting

in

A)	B)
RPMI-1640 medium + 10% FCS	RPMI-1640 medium + 1% autologous human plasma

<u>**day 7:**</u> **immature DC** **! PROCESSING !**

+ conditioned medium
+ antigen (tetanus toxoid)
+ GM-CSF and IL-4

<u>**day 7-10:**</u> Maturation of DC

days in RPMI-1640 medium supplemented with 1% autologous human plasma instead of 10% fetal calf serum (FCS) and with GM-CSF and IL-4 [1,2]. Maturation of DC in FCS-free medium was then achieved during an additional three day culture period with monocyte-conditioned medium as described [3,4]. DC generated by both methods (see table 1) were employed and compared in a modified lymphocyte proliferation assay to measure proliferative T-cell responses to tetanus toxoid (TT) in a vaccinated individual. A detailed description of this assay is given in table 2.

3. RESULTS

Only DC generated without FCS (see table 1, method B) were useful in our proliferation assay. In contrast DC cultured in RPMI-1640 medium enriched with 10% FCS (see table 1, method A) induced high background levels of thymidine uptake in autologous T-cells despite the absence of antigen. Figure 1 shows a comparison of DC generated by method A) or B) in the setting of a syngeneic mixed leukocyte reaction.

Table 2. Time course of the modified lymphocyte proliferation assay

<u>**day 10:**</u> **mature DC** **! PRESENTATION ! of antigen**, processed between days 7-10

* wash DC
* irradiate DC with 30 gray

<u>**day 10-17:**</u>

* co-culture of DC and autologous T-cells in triplicates

in RPMI-1640 medium + 10% autologous human serum or pooled human serum

<u>**day 16:**</u> pulse cells with ^{3}H- thymidine
<u>**day 17:**</u> measure incorporation of labeled thymidine

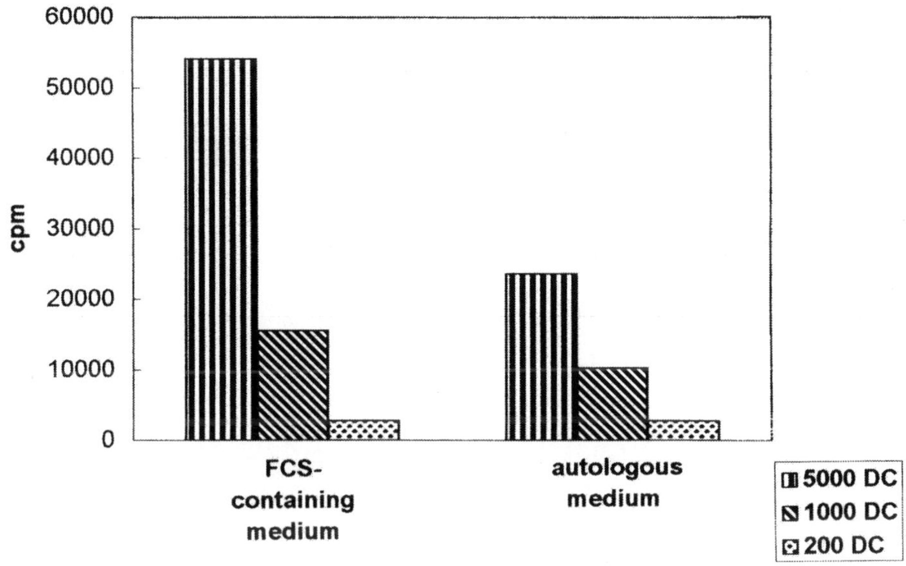

Figure 1. Response [incorporation of ^{3}H-thymidine measured in counts per minute (cpm)] of a syngeneic mixed leukocyte reaction (background proliferation). DC were investigated under FCS- and under autologous conditiones. Note that the syngeneic mixed leukocyte reaction is substantially lower in medium containing autologous human plasma.

In the following experiments DC generated in the autologous culture system and such DC which had matured in FCS containing medium were used to present tetanus toxoid (TT) to autologous T cells in a modified lymphocyte proliferation assay. Stimulation indices (i. e., ratio of proliferation in the presence of antigen to proliferation in the absence of antigen) were higher with DC generated in FCS-free medium. Figure 2 shows that proliferative responses to TT remained positive with antigen dilutions up to 1 : 2 000 000.

Background proliferation was further reduced by using autologous human serum instead of pooled human serum for the proliferation assay (figure 3), thus making the assay more sensitive.

4. CONCLUSIONS

A precondition for the application of dendritic cells (DC) for immunotherapy is to avoid the use of fetal calf serum and to prepare the starting population of lymphocyte-depleted PBMC by procedures and with reagents approved for clinical use.

It will be important to control T cell responses in vitro. In some cases such immune responses will be weak and therefore conventional lymphocyte transformation assays will be insufficient for monitoring them. Consistent with this hypothesis is the observation that DC generated from progenitors in peripheral human blood are more potent stimulators of lymphocyte reactions than PBMC. So DC can be used as antigen presenting cells in lymphocyte proliferation assays and may facilitate the measurement of weak T cell responses. Our data indicate that DC may be particularly helpful when entirely autologous culture systems are used. Those requirements are essential for clinical application due to ethical considerations.

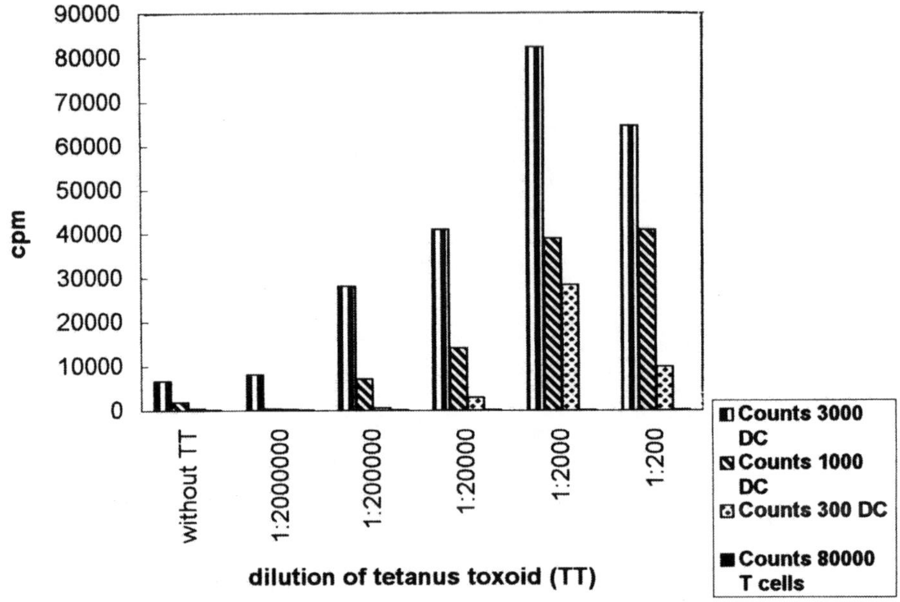

Figure 2. Proliferative immune responses to tetanus toxoid (TT) [incorporation of ^{3}H-thymidine measured in cpm] using different dilutions of antigen. TT is presented by DC generated in autologous medium to autologous T-cells. The proliferation was done in medium containing 10% pooled human serum.

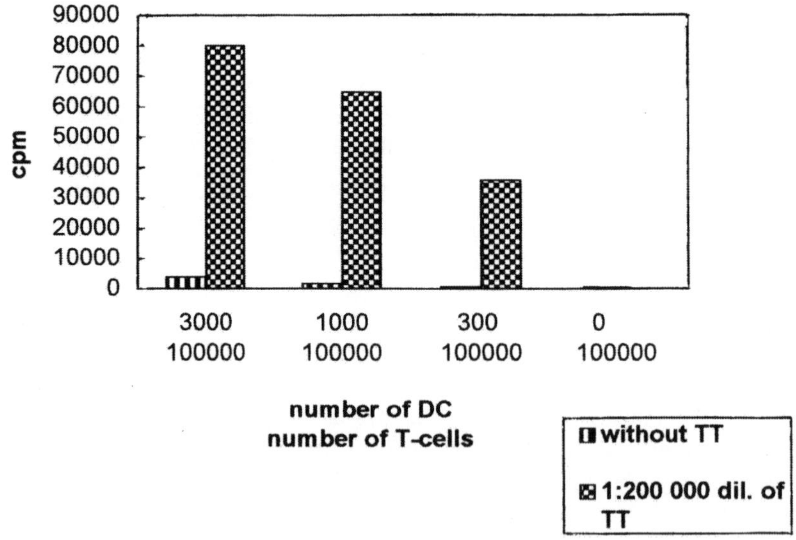

Figure 3. Specific immune response to tetanus toxoid (TT) by 100 000 autologous T-cells [incorporation of ^{3}H-thymidine measured in cpm]. Antigen was presented by DC generated in autologous medium. Also the proliferation assay was performed in medium containing autologous human serum. Note that background proliferation in the absence of antigen is very low even when high doses of DC were used.

5. ACKNOWLEDGMENTS

This research was supported by a grant (P-10896-MED) from the Austrian Science Foundation.

We thank Niki Romani and his "laboratory crew" for their advice and technical assistance.

6. REFERENCES

1. Sallusto, F. and A. Lanzavecchia. 1994. Efficient presentation of soluble antigen by cultured human dendritic cells is maintained by granulocyte/macrophage colony-stimulating factor plus interleukin 4 and downregulated by tumor necrosis factor a. *J. Exp. Med.* 179:1109–1118.
2. Romani, N., S. Gruner, D. Brang, E. Kämpgen, A. Lenz, B. Trockenbacher, G. Konwalinka, P. O. Fritsch, R. M. Steinman, and G. Schuler. 1994. Proliferating dendritic cell progenitors in human blood. *J. Exp. Med.* 180:83–93.
3. Romani, N., D. Reider, M. Heuer, S. Ebner, E. Kämpgen, B. Eibl, D. Niederwieser, and G. Schuler. 1996. Generation of mature dendritic cells from human blood: An improved method with special regard to clinical applicability. *J. Immunol. Methods* 196:137–151
4. Bender, A., M. Sapp, G. Schuler, R. M. Steinman, and N. Bhardwaj. 1996. Improved methods for the generation of dendritic cells from nonproliferating progenitors in human blood. *J. Immunol. Methods* 196:121–135

40

THE REGULATION OF T CELL RESPONSES BY A SUBPOPULATION OF $CD8^{+}DEC205^{+}$ MURINE DENDRITIC CELLS

Vadim Kronin, Gabriele Süss, Ken Winkel, and Ken Shortman

The Walter and Eliza Hall Institute of Medical Research
Melbourne, Victoria 3050, Australia

1. INTRODUCTION

Dendritic cells (DC) are migratory bone-marrow-derived cells of sparse but broad tissue distribution[1,2]. They are not homogeneous and even within a particular lymphoid organ distinct subpopulations can be separated[3–5]. If procedures are used which extract all DC, both $CD8\alpha^{+}$ and $CD8\alpha^{-}$ DC can be isolated from mouse spleen[5]. Under these conditions the DC isolated from thymus are predominantly $CD8^{+}$, whereas those isolated from lymph nodes (LN) are predominantly $CD8^{-}$. This heterogeneity may be of functional importance, since there is evidence that the $CD8^{+}$ DC of the thymus are of different developmental origin from the classical $CD8^{-}$ DC typically found in LN[6,7]. Since conventional isolation procedures yielded mainly the $CD8^{-}$ DC population the functional capacity of $CD8^{+}$ DC had not previously been explored. To determine whether $CD8^{+}$ and $CD8^{-}$ splenic DC differed in function, we tested their capacity to stimulate purified allogeneic CD4 and CD8 T cells in a primary mixed leucocyte reaction.

2. REGULATION OF CD4 T CELL RESPONSES VIA FAS/FAS-LIGAND-INDUCED APOPTOSIS

The $CD8^{+}$ and $CD8^{-}$ subpopulations of splenic DC were tested for their capacity to stimulate proliferation of purified allogeneic mature naive CD4 T cells in culture. The proliferative responses obtained from these pure T cells after stimulation with pure DC were measured by ^{3}H-TdR uptake. The DC were isolated by collagenase digestion, then EDTA release from spleen fragments, selection of light density cells, immunomagnetic bead depletion of other cell lineages, then as a final step sorting for $CD11c^{high}$ DC which were either positive or negative for $CD8\alpha$[5,7,8]. CD4 T cells from CBA mice proliferated in a dose-dependent manner to both DC populations from C57BL/6 (B6) mice. However, as

Dendritic Cells in Fundamental and Clinical Immunology
edited by Ricciardi-Castagnoli, Plenum Press, New York, 1997

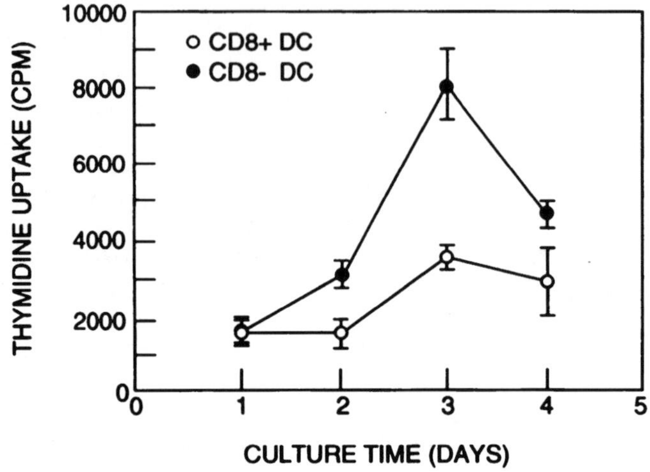

Figure 1. The kinetics of proliferative response of CD4 T cells stimulated with allogeneic CD8$^+$ and CD8$^-$ DC. DC were purified from C57BL/6 mouse spleen; 1000 DC were used per culture. CD4 T cells were purified from CBA mouse LN and used at 20,000 cells per culture. Proliferation of T cells was assessed by ^{3}H-TdR incorporation after a 12h pulse. Full details are given elsewhere[20]. This figure is reprinted with permission from.[20]

shown in Fig. 1, there was a marked difference in the extent of the response, with CD8$^-$ DC inducing much greater proliferation[8]. Similar results were obtained using other allogeneic combinations as well as using a defined peptide antigen and CD4 T cells from a hemagglutinin specific class II MHC-restricted T-cell receptor (TCR) transgenic mouse.[8] The differential T-cell response was not due to insufficient production of cytokines such as interleukin (IL)-2 in the CD8$^+$ DC stimulated cultures. As shown in Fig. 2 (top), addition of exogenous IL-2 maintained the marked difference between the proliferative responses to CD8$^+$ and CD8$^-$ DC[8].

The major reason for the reduced CD4 T cell proliferation response was a high rate of T-cell death in cultures stimulated by CD8$^+$ DC. This T-cell death was revealed to be apoptotic both by propidium iodide (PI) staining to detect hypodiphoid cells and by the TUNEL labelling technique to detect DNA strand breaks[8]. Since the signal for the apoptosis of mature T cells is normally delivered through the Fas molecule (CD95/APO-1) expressed on the surface of activated T cells[9,10] we used Fas-deficient C3H/HeJ*lpr* CD4 T cells[11,12] stimulated with B6 CD8$^+$ or CD8$^-$ DC to assess the role of Fas in the apoptotic T-cell death. As shown in Fig. 2, Fas-deficient *lpr* CD4 T cells responded at least as well to CD8$^+$ as to CD8$^-$ DC whereas the control C3H/HeJ CD4 T cells responded much less well to CD8$^+$ DC[8]. This demonstrates that lack of inherent stimulatory ability by CD8$^+$ DC was not the reason for the reduced response of normal T cells, which agrees with the observations of D. Vremec (submitted for publication) demonstrating that CD8$^+$ and CD8$^-$ DC express similar levels of class II MHC, of B7–1 and of B7–2. The decreased proliferation appeared to be due to Fas-mediated killing of the CD4 T cells, following their activation by CD8$^+$ DC.

The ligand for Fas (FasL) has been demonstrated on the surface of T cells after activation[13,14], leading to a T cell-T cell killing[15–17]. In our cultures, however, the main source of the FasL appeared to be the antigen-presenting DC, since the difference in proliferation in response to CD8$^+$ versus CD8$^-$ DC persisted when CD4 T cells isolated from FasL mutant *gld* mouse strain[18,19] were used as responders. To test directly for the presence of

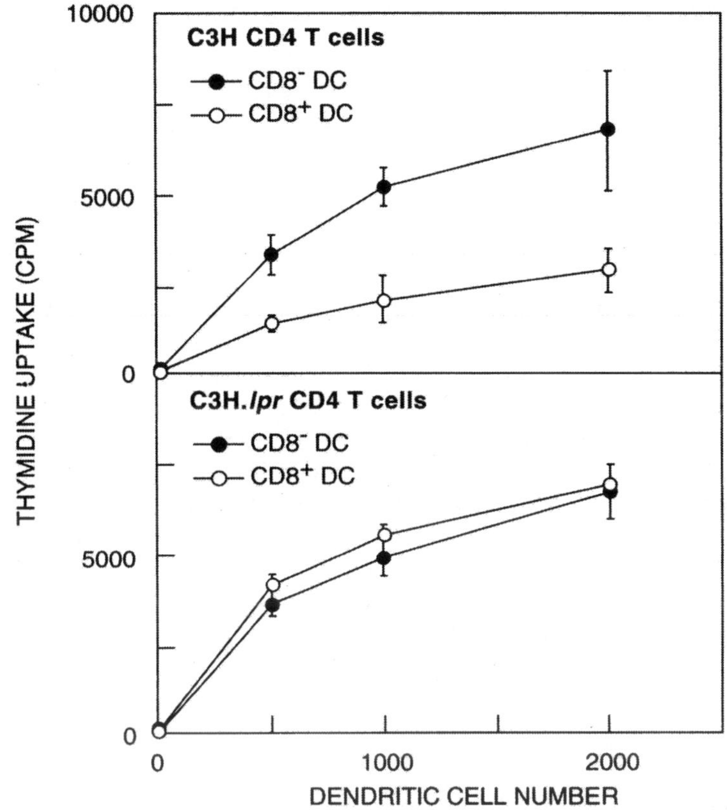

Figure 2. The proliferative response of normal and Fas-deficient mice to allogeneic $CD8^-$ and $CD8^+$ DC. Top, The response of the control C3H/HeJ CD4 T cells to C57BL/6 $CD8^+$ or $CD8^+$ DC. Bottom, The response of the Fas-deficient C3H/HeJ*lpr* CD4 T cells to C57BL/6 $CD8^+$ or $CD8^-$ DC. Recombinant mouse IL-2 was added to the cultures at 100 U/ml; these results were confirmed in experiments where no IL-2 was added to the cultures. Proliferation was assessed after day 3 of the culture. Full details are given elsewhere.[8] This figure is reprinted with permission from.[8]

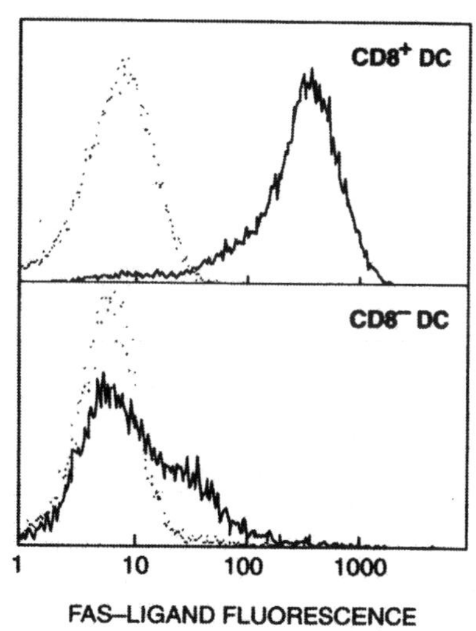

Figure 3. Expression of FasL on the surface of $CD8^+$ and $CD8^-$ DC. Immunofluorescence profiles show the binding of a Fas-Fc fusion protein to C57BL/6 splenic DC. The broken line shows background staining with purified human IgG. Full details are given elsewhere[8]. This figure is reprinted with permission from.[8]

FasL, we stained CD8⁺ and CD8⁻ DC with the Fas-Fc fusion protein obtained from D. Lynch[14]. As shown in Fig. 3, the CD8⁺ DC all stained exceptionally strongly with this reagent, whereas the CD8⁻ DC showed low or negative staining[8]. This result leads to the conclusion that CD8⁺ DC, as they exist in the spleen, express high levels of FasL and are pre-armed for killing of activated, Fas-expressing CD4 T cells[8].

3. REGULATION OF CD8 T CELL RESPONSES BY LIMITING THEIR IL-2 PRODUCTION

To determine whether CD8⁺ DC have the same "regulatory" effect on CD8 T cells as on CD4 T cells, we studied the proliferative responses of CD8 T cells purified from CBA mice to CD8⁻ and CD8⁺ DC isolated from spleens of B6 mice. As shown in Fig.4, CD8⁻ DC induced much higher levels of proliferation by CD8 T cells than that obtained with CD8⁺ DC if assayed late after culture initiation. However, a key difference in the way CD8 and CD4 T cells respond to CD8⁺ DC was apparent from the kinetics of the proliferative response. The reduced response of CD4 T cells to CD8⁺ DC was apparent from the earliest times of culture (Fig. 1). In contrast, up to day 2.5 of culture CD8 T cells responded to CD8⁺ DC at least as well as to CD8⁻ DC. After this time, proliferation in response to CD8⁺ DC dropped rapidly, whereas the response to CD8⁻ DC continued to increase. Similar results were obtained using other allogeneic mouse strain combinations and also using CD8 T cells purified from two TCR transgenic mouse stains, namely H-2K^b-specific αβ-TCR transgenic mice and HY-specific, H-2^b restricted, αβ-TCR transgenic mice.[20]

To test if this reduced CD8 T-cell proliferative response late in culture was also due to Fas-mediated negative signals from the CD8⁺ DC, CD8 T cells were purified from Fas-deficient C3H.*lpr* mice, and compared in their responses to CD8 T cells purified from normal C3H mice. However, the proliferative responses of normal and mutant CD8 T cells to the allogeneic CD8⁻ and CD8⁺ DC were identical.[20] Both the normal and the Fas-deficient

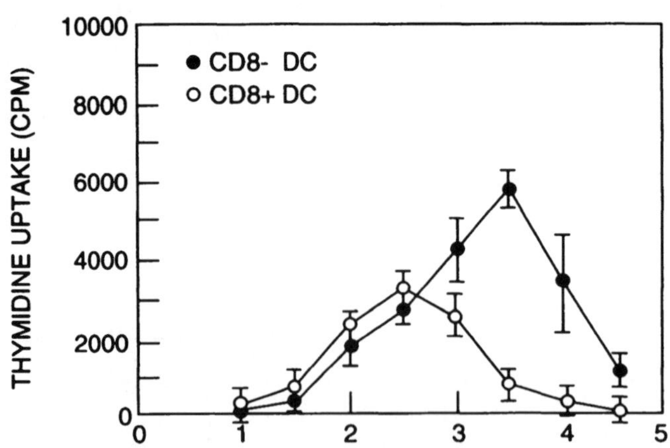

Figure 4. The kinetics of proliferative response of CD8 T cells stimulated with allogeneic CD8⁺ and CD8⁻ DC. DC were purified from C57BL/6 mouse spleen; 500 DC were used per culture. CD8 T cells were purified from CBA mouse LN and used at 20,000 cells per culture. Proliferation of T cells was assessed after a 12h pulse. Full details are given elsewhere[20]. This figure is reprinted with permission from.[20]

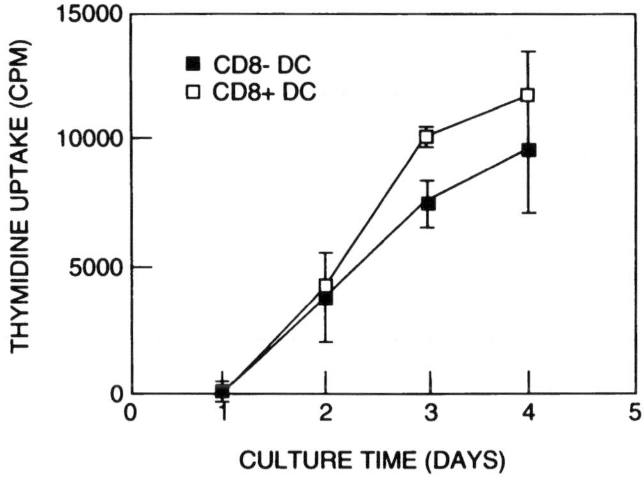

Figure 5. The effect of exogenous IL-2 on the kinetics of the CD8 T cells response to $CD8^+$ and $CD8^-$ DC. CD8 T cells were isolated from LN of CBA mice and stimulated with allogeneic DC from C57BL/6 mice. 100 U/ml of IL-2 was added to all cultures. Full details are given elsewhere.[20] This figure is reprinted with permission from.[20]

CD8 T cells gave, after day 3 of culture, a reduced response to $CD8^+$ DC, compared to $CD8^-$ DC (data not shown). In contrast to the results with CD4 T cells[8], Fas-mediated apoptosis of T cells was not the reason for the differences in the response of CD8 T cells to these splenic DC subpopulations[20].

The explanation for the restricted response of CD8 T cells to $CD8^+$ DC was a limitation in endogenous cytokine, and in particular IL-2, production. This was in total contrast to the previous results with CD4 T cells. The results of Fig. 4 were obtained when no exogenous cytokines were added to the cultures. When 100 U/ml of IL-2 was added to the culture medium and the proliferative response of CD8 T cells from CBA mice to $CD8^+$ and $CD8^-$ DC from B6 mice measured, proliferation was extended in both cultures but the stimulation by $CD8^+$ DC was enhanced to a greater extent (Fig. 5). The result was that the $CD8^+$ DC were at least as effective as $CD8^-$ DC at stimulating CD8 T-cell proliferation[20]. This result demonstrates that $CD8^+$ DC were fully equipped to stimulate CD8 T cells into division, provided adequate IL-2 was present to maintain proliferation.

The restoration by IL-2 of the CD8 T-cell responses to $CD8^+$ DC pointed to a deficit in endogenous IL-2 production. To check this, the supernatants from cultures of CD8 T cells from CBA mice stimulated by $CD8^+$ and $CD8^-$ DC from B6 mice were collected and analyzed for IL-2. In the $CD8^+$ DC stimulated cultures the levels of free supernatant IL-2 in the $CD8^-$ DC stimulated cultures were much higher than in $CD8^+$ DC stimulated cultures (Fig. 6). In the $CD8^+$ DC stimulated cultures the levels of IL-2 in the supernatants were at all times below the threshold of sensitivity. Overall the results indicate that the rate of IL-2 production by CD8 T cells was much lower when stimulated by $CD8^+$ DC than by $CD8^-$ DC[20].

4. THE ROLE OF CD8 ON DENDRITIC CELLS IN THE REGULATION OF CD4 AND CD8 T CELL RESPONSES

There is evidence that CD8, if expressed on an antigen presenting cell, can deliver negative signals which either inhibit T-cell responses or kill interacting T cells[21–23]. To test

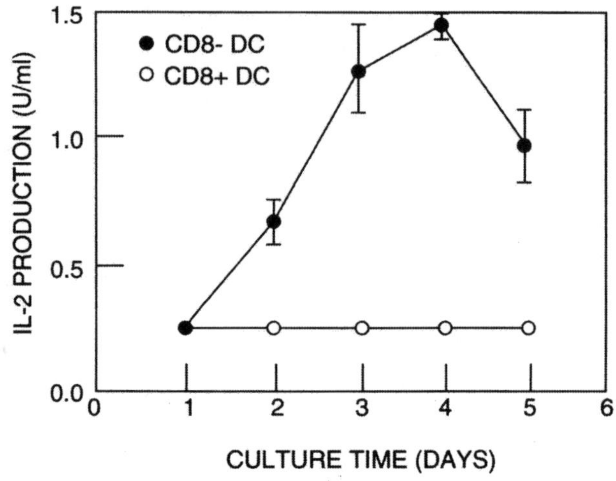

Figure 6. The IL-2 production in cultures of CD8 T cells stimulated by allogeneic CD8$^+$ or CD8$^-$ DC. The conditions were as in Figs. 4, except that C57BL/6 DC were used at 1000 cells per culture. The levels of IL-2 were measured using an IL-2 dependent cell line. Full details are given elsewhere[20]. This figure is reprinted with permission from.[20]

the possible role of CD8 on DC in the regulation of the responses of both CD8 and CD4 T cells we employed mutant CD8$^{-/-}$ "knockout" mice[24] on a B6 background. However, before using these mice as a model to investigate the function of the CD8 molecule, we had to establish that the same populations of DC were present, and we had to devise a strategy for isolation the equivalents of the CD8$^+$ and CD8$^-$ DC in the absence of the CD8 marker.

We established that the absence of the CD8 molecule in the CD8$^{-/-}$ mice did not influence the numbers of DC in the spleens we used as a DC source. The spleens and LN of the CD8 "knockout" mice had normal numbers of DC which, apart from CD8α, expressed the normal spectrum of surface markers. DC development thus appeared to proceed normally in the absence of CD8α. We also established a correlation in the normal mouse spleen between the expression of CD8 and another DC surface molecule, the interdigitating DC marker DEC-205[25], in order to use the latter as a substitute marker. This correlation is shown in Fig. 7. Accordingly, DEC-205 was employed as a substitute marker for CD8.

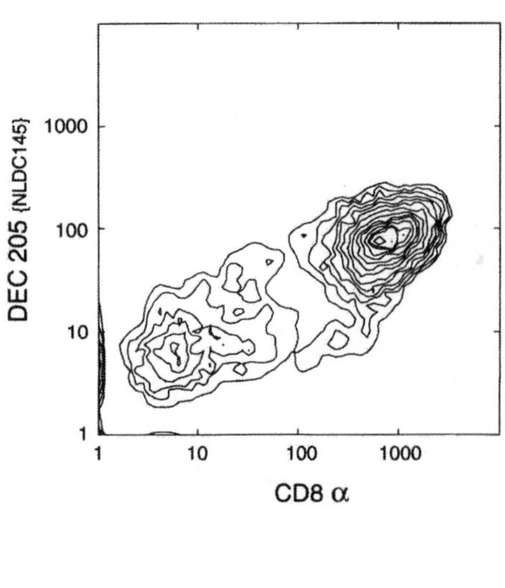

Figure 7. Expression of CD8α and DEC-205 on the splenic DC surface. DC were isolated from the spleens of normal C57BL/6 mice. The enriched DC were stained with: allophycocyanin-conjugated anti-MHC class II; fluorescein-conjugated anti-CD8α; biotinylated anti-DEC-205, followed by phycoerythrin-conjugated streptavidin as second stage; PI to stain dead cells. The figure shows the distribution of DEC-205 and CD8α fluorescence on the DC population, gated as class II MHC highly positive cells with high forward and side scatter, and with PI-positive cells excluded.

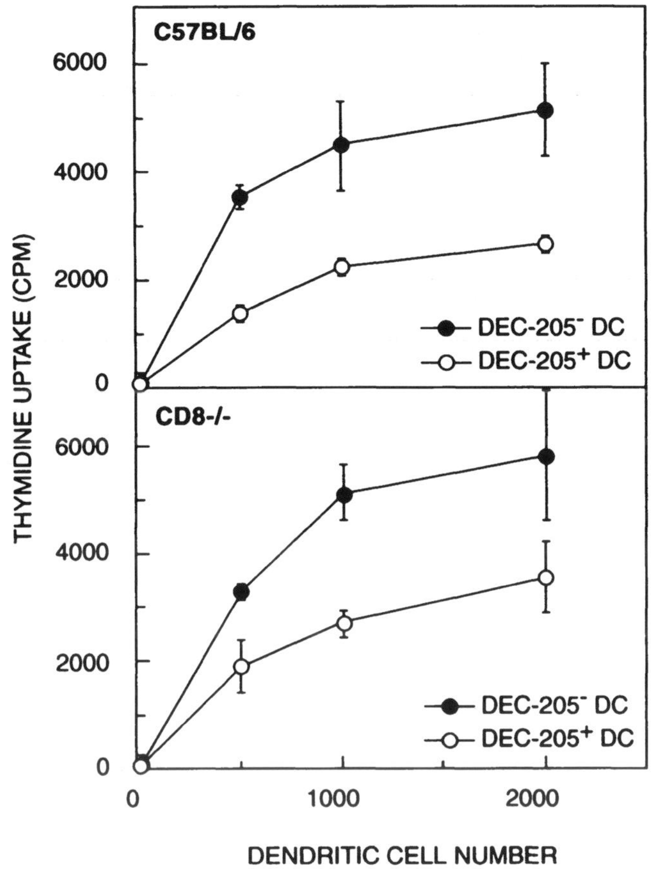

Figure 8. Comparison of the proliferation of CD4 T cells stimulated by allogeneic DEC-205⁺ and DEC-205- splenic DC, isolated from normal or CD8⁻/⁻ "knockout" mice. DC were enriched from normal C57BL/6 or CD8⁻/⁻ mice, then stained with: fluorescein-conjugated anti-CD11c; biotinylated anti-DEC-205, followed by phycoerythrin-conjugated streptavidin as second stage. The DC were then sorted into CD11c⁺DEC-205⁺ and CD11c⁺DEC-205⁻ populations. CD4 T cells were purified from LN of CBA mice and used at 20,000 cells per culture. The proliferative response in DC-T cell cultures were determined after day 3 of culture using a 4h pulse.

The two populations of DC were isolated from normal B6 mice, using N418 monoclonal antibodies (mAb) to stain CD11c and to define DC, together with NLDC145[26] mAb to stain DEC-205 and to divide the DC into two subpopulations. These pure, separated DC subpopulations were tested for capacity to stimulate the proliferation of allogeneic CD4 T cells or CD8 T cells. The results were equivalent to those obtained using separation based on CD8 expression. The DEC-205⁻ DC gave much greater proliferation of both CD4 and CD8 T cells at day 3, than did the DEC-205⁺ DC (Figs.8 and 9). This correspondence between DEC-205⁻ and CD8⁻ DC, and between DEC-205⁺ and CD8⁺ DC applied throughout the whole time-course of the response.

Since DEC-205 appeared to be a satisfactory substitute marker for CD8, we were able to test whether the CD8 molecule was responsible for any of the regulatory effects. We compared the T-cell stimulatory abilities of DEC-205⁺ and DEC-205⁻ DC isolated from CD8⁻/⁻ mice, with DEC-205⁺ and DEC-205⁻ DC from the CD8⁺/⁺ B6 background

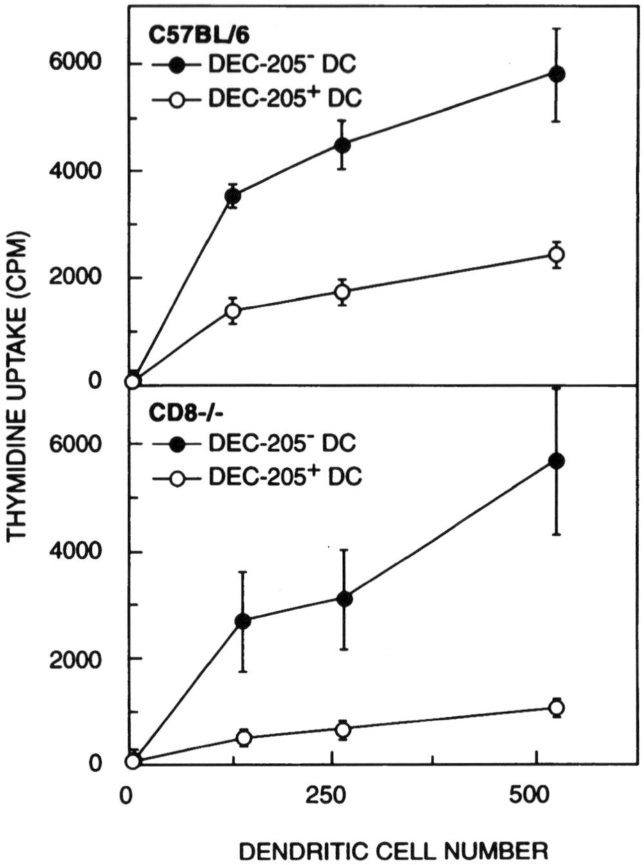

Figure 9. Comparison of the proliferation of CD8 T cells stimulated with allogeneic DEC-205$^+$ and DEC-205$^-$ splenic DC, isolated from normal or CD8$^{-/-}$ "knockout" mice. The procedure was similar to that described in Fig. 8, except that CD8$^+$ T cells rather than CD4$^+$ T cells were isolated from CBA mice, and that the ^{3}H-TdR pulse was for 12h.

mice, using CD4 T cells and CD8 T cells purified from the CBA mice. The "regulatory" effects were the same regardless of whether CD8α was present or not (Figs. 8 and 9). In both cases the DEC-205$^-$ DC produced much higher levels of responding T cell proliferation on day 3 than did the DEC-205$^+$ DC. The CD8 molecule is apparently not required for the regulatory function of DEC-205$^+$ DC.

5. CONCLUSIONS

These studies indicate that the DC subclasses defined by surface markers are functionally specialized. The CD8$^-$ DC of spleen have the very high T-cell stimulatory capability ascribed to classical antigen-presenting DC[1,2]. The CD8$^+$ DC of spleen while also able to activate T cells, are equipped with additional mechanisms to regulate or limit both CD4 and CD8 T cell responses. Although CD8α marks this regulatory DC population, it does not itself play a direct role in the regulatory process. For interaction with CD4 T cells,

FasL on the CD8⁺ DC is the molecule crucial for the regulatory function. For interaction with CD8 T cells we have not identified the key signalling molecule, although DEC-205 must now be tested for possible involvement.

The developmental relationship between these two functionally distinct classes of DC is clearly of interest. We have no direct evidence on this issue for splenic DC. However, it should be noted that thymic DC, which are predominantly CD8⁺ and whose interaction with developing T cells is also regulatory rather than stimulatory, appear to represent a lymphoid-related developmental lineage differing in origin from the classical, myeloid derived DC[6,7]. It is possible that the differences in function we have observed reflect the existence of distinct developmental lineages or sublineages of splenic DC.

ACKNOWLEDGMENTS

This work was supported by the Cooperative Research Centre for Vaccine Technology, by the National Health and Medical Research Council, Australia and by the Human Frontier Science Program Grant Organization.

REFERENCES

1. R. M. Steinman, The dendritic cell system and its role in immunogenicity, Annu. Rev. Immunol. 9:271 (1991)
2. S. C. Knight and A. J. Stagg, Antigen-presenting cell types, Curr. Opin. Immunol. 5:374 (1993)
3. M. Crowley, K. Inaba, M. Witmer-Pack and R. M. Steinman, The cell surface of mouse dendritic cells: FACS analysis of dendritic cells from different tissues including thymus., Cell. Immunol. 118:108 (1989)
4. M. T. Crowley, K. Inaba, M. D. Witmer-Pack, S. Gezelter and R. M. Steinman, Use of the fluorescence activated cell sorter to enrich dendritic cells from mouse spleen., J. Immunol. Methods. 133:55 (1990)
5. D. Vremec, M. Zorbas, R. Scollay, D. J. Saunders, C. F. Ardavin, L. Wu and K. Shortman, The surface phenotype of dendritic cells purified from mouse thymus and spleen: Investigation of the CD8 expression by a subpopulation of dendritic cells., J. Exp. Med. 176:47 (1992)
6. C. Ardavin, L. Wu, C. Li and K. Shortman, Thymic dendritic cells and T cells develop simultaneously within the thymus from a common precursor population., Nature. 362:761 (1993)
7. L. Wu, D. Vremec, C. Ardavin, K. Winkel, G. Suss, H. Georgiou, E. Maraskovsky, W. Cook and K. Shortman, Mouse thymus dendritic cells: kinetics of development and changes in surface markers during maturation, Eur. J. Immunol. 25:418 (1995)
8. G. Süss and K. Shortman, A subclass of dendritic cells kills CD4 T cells via Fas/Fas-ligand-induced apoptosis, J. Exp. Med. 183:1789 (1996)
9. P. H. Krammer, I. Behrmann, P. Daniel, J. Dhein and K.-M. Debatin, Regulation of apoptosis in the immune system, Curr. Opin. Immunol. 6:279 (1994)
10. G. G. Singer and A. K. Abbas, The Fas Antigen is Involved in Peripheral but not Thymic Deletion of T Lymphocytes in T Cell Receptor Transgenic Mice, Immunity. 1:365 (1994)
11. R. Watanabe-Fukunaga, C.I. Brannan, N.G. Copeland, N.A. Jenkins, S. Nagata, Lymphoproliferation disorder in mice explained by defects in Fas antigen that mediates apoptosis, Nature. 356:314 (1992)
12. J. H. Russell, B. Rush, C. Weaver and R. Wang, Mature T cells of autoimmune *lpr/lpr* mice have a defect in antigen-stimulated suicide, Proc. Natl. Acad. Sci. USA. 90:4409 (1993)
13. T. Suda, T. Takahashi, P. Golstein and S. Nagata, Molecular cloning and expression of the Fas ligand, a novel member of the tumor necrosis factor family, Cell. 75:1169 (1993)
14. D. H. Lynch, M. L. Watson, M. R. Alderson, P. R. /. Baum, R. E. Miller, T. Tough, M. Gibson, T. Davis-Smith, C. A. Smith, K. Hunter, D. Bhat, W. Din, R. G. Goodwin and M. F. Seldin, The mouse Fas-ligand gene is mutated in *gld* mice and is part of a TNF family gene cluster., Immunity. 1:131 (1994)
15. S.-T. Ju, D. J. Panka, H. Cul, R. Ettinger, M. El-Khatib, D. h. Sherr, B. Z. Stanger and A. Marshak-Rothstein, Fas(CD95)/FasL interactions required for programmed cell death after T-cell activation, Nature. 373:444 (1995)

16. J. Dhein, H. Walczak, C. Bäumler, K.-M. Debatin and P. H. Krammer, Autocrine T-cell suicide mediated by APO-1/(Fas/CD95), Nature. 373:438 (1995)
17. T. Brunner, R. J. Mogil, D. LaFace, N. J. Yoo, A. Mahboubi, F. Echeverri, S. J. Martin, W. R. Force, D. H. Lynch, C. F. Ware and D. R. Green, Cell-autonomous Fas (CD95)/Fas-ligand interaction mediates activation-induced apoptosis in T-cell hybridomas, Nature. 373:441 (1995)
18. T. Takahashi, M. Tanaka, C. I. Brannan, N. A. Jenkins, N. G. Copeland, T. Suda and S. Nagata, Generalized Lymphoproliferative Disease in Mice, caused by a Point Mutation in the Fas Ligand, Cell. 76:969 (1994)
19. I. Gillette-Ferguson and C. L. Sidman, A specific intercellular pathway of apoptotic cell death is defective in the mature peripheral T cells of autoimmune *lpr* and *gld* mice, Eur. J. Immun. 24:1181 (1994)
20. V. Kronin, K. Winkel, G. Suss, A. Kelso, W. Heath, J. Kirberg, H. von Boehmer and K. Shortman, A subclass of dendritic cells regulates the response of naive CD8 T cells by limiting their IL-2 production, J. Immunol. 157:3819 (1996)
21. D. R. Kaplan, J. E. Hambor and M. L. Tykocinski, An immunoregulatory function for the CD8 molecule, Proc. Natl. Acad. Sci. USA. 86:8512 (1989)
22. J. E. Hambor, D. R. Kaplan and M. L. Tykocinski, CD8 functions as an inhibitory ligand in mediating the immunoregulatory activity of $CD8^+$ cells, J. Immunol. 145:1646 (1990)
23. S. R. Sambhara and R. G. Miller, Programmed cell death of T cells signaled by the T cell receptor and the alpha3 domain of class I MHC., Science. 252:1424 (1991)
24. W. Fung-Leung, M. W. Schilham, A. Rahemtulla, T. M. Kundig, M. Vollenweider, J. Potter, W. Van Ewijk and T. W. Mak, CD8 is needed for development of cytotoxic T cells but not helper T cells., Cell. 65 :443 (1991)
25. W. Jiang, W. J. Swiggard, C. Heufler, M. Peng, A. Mirza, R. M. Steinman and M. C. Nussenzweig, The receptor DEC-205 expressed by dendritic cells and thymic epithelial cells is involved in antigen processing., Nature. 375:151 (1995)
26. G. Kraal, M. Breel, M. Janse and G. Bruin, Langerhans' cells, veiled cells, and interdigitating cells in the mouse recognized by a monoclonal antibody, J. Exp. Med. 163:981 (1986)

41

CONSTRUCTING ARTIFICIAL ANTIGEN-PRESENTING CELLS FROM DROSOPHILA CELLS

Jonathan Sprent,[1] Zeling Cai,[2] Anders Brunmark,[2] Michael R. Jackson,[2] and Per A. Peterson[2]

[1]The Scripps Research Institute
Department of Immunology, IMM4
10550 N. Torrey Pines Road
La Jolla, California, 92037
[2]R.W. Johnson Pharmaceutical Research Institute
3535 General Atomics Ct., Suite 100
San Diego, California 92121

1. INTRODUCTION

Stimulation of unprimed T cells is controlled by professional antigen-presenting cells (APC) such as dendritic cells (DC).[1-3] The strong APC function of DC is presumed to reflect that these cells express a high density of major histocompatibility complex (MHC) molecules and a variety of costimulatory molecules. In this respect, activation of naive T cells by APC is thought to require two distinct signals: Signal 1 reflects T cell receptor (TCR) contact with specific peptides bound to MHC molecules, and Signal 2 is the consequence of other molecules on T cells, e.g. CD28, interacting with costimulatory molecules, e.g. B7 (B7–1, B7–2), on APC. Since many different accessory molecules on APC can express costimulatory function for T cells under defined conditions, which particular accessory molecules are essential for stimulating naive T cells is unclear. We have addressed this issue by constructing artificial APC from a Drosophila cell line by gene transfection.

2. EXPERIMENTAL APPROACH

Drosophila cells are an unlikely tool for studying APC function because these cells die rapidly at 37° C. In their favor, however, Drosophila cells express high levels of surface MHC molecules after gene transfection. Moreover, the lack of TAP 1,2 peptide transporters in Drosophila cells means that transfected MHC class I molecules fail to bind endogenous peptides in the cytoplasm and can therefore be loaded externally with high concentrations of exogenous peptides.[4,5] Drosophila cells thus have the potential to deliver a very strong dose of Signal 1 to T cells.

The APC function of Drosophila cells was examined with the aid of the 2C line of TCR transgenic mice[6] and defined peptides as antigen.[7–11] $CD8^+$ cells from the 2C line are MHC class I restricted and undergo positive selection to K^b molecules and display allore-activity to two peptides, p2Ca and QL9, presented by L^d molecules. These two peptides are identical except that QL9 has one extra amino acid than p2Ca. Both peptides are derived from a Krebs cycle enzyme and are thus ubiquitous self peptides. The p2Ca peptide is an 8-mer and is exposed naturally on the surface of $H\text{-}2^d$, e.g. B10.D2, cells bound to L^d; p2Ca has intermediate binding affinity for soluble L^d molecules ($4 \times 10^6\ M^{-1}$) and high affinity for 2C TCR molecules ($2 \times 10^6\ M^{-1}$ to $1 \times 10^7\ M^{-1}$). QL9 peptide has even higher affinity for these molecules ($2 \times 10^8\ M^{-1}$ for L^d and $2 \times 10^7\ M^{-1}$ for 2C TCR); whether QL9 peptide is expressed naturally on the cell surface is unclear.

To examine the requirements for stimulating 2C cells, purified naive-phenotype ($CD44^{lo}$) $CD8^+$ 2C cells were exposed to p2Ca or QL9 peptides presented by a panel of transfected Drosophila cells expressing L^d alone (L^d APC), L^d + B7–1 (L^d.B7), L^d + ICAM-1 (L^d.ICAM) or L^d + B7–1 + ICAM-1 (L^d.B7.ICAM). With these transfectants we assessed the consequences of exposing T cells to Signal 1 alone vs. Signal 1 + 2. In this respect, the prevailing view is that Signal 1 causes partial T cell activation associated with upregulation of IL-2R (CD25), whereas Signal 2 controls T cell proliferation through production of IL-2 and other cytokines.[1–3] In the experiments discussed below,[12] T cell activation was initially quantitated by examining cell-surface expression of CD25. Virtually identical results applied to CD69 expression (data not shown).

2.1. Effects of Signal 1 vs. Signal 1 + 2 on CD25 Expression

As shown in Fig. 1, presentation of p2Ca peptide by Drosophila cells expressing L^d alone appeared to be completely nonimmunogenic for 2C $CD8^+$ cells. Thus even high doses of p2Ca peptide caused no detectable expression of CD25 on 2C cells. Similar results occurred with the high-affinity QL9 peptide, although a slight increase in CD25 expression was observed with high concentrations of this peptide (10^{-5} M). These data

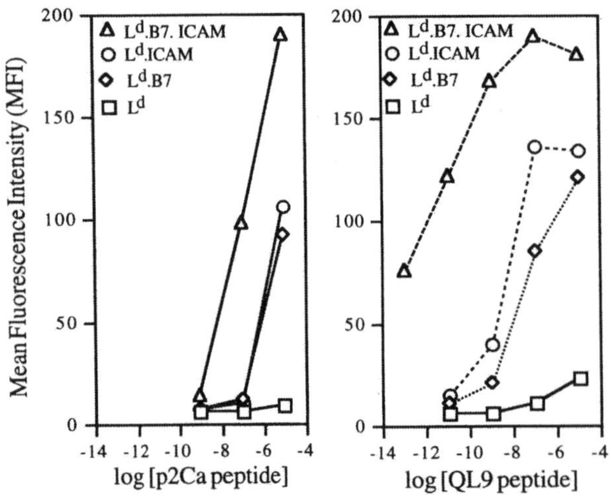

Figure 1. Expression of CD25 on $CD8^+$ 2C cells incubated with p2Ca vs QL9 peptides presented by Drosophila APC. Purified $CD8^+$ 2C cells from lymph nodes were incubated with the panel of transfected Drosophila APC plus varying concentrations of p2Ca or QL9 peptides in bulk (2 ml) cultures for 12 hours and then stained for CD25 expression followed by FACS analysis. The data show the mean fluorescence intensity of CD25 staining on gated $CD8^+$ cells. Data adapted from ref. 12.

indicate therefore that exposure to Signal 1 alone has remarkably little effect on T cells, at least by the parameter of CD25 expression.

Quite different results occurred when the Drosophila APC co-expressed costimulatory molecules (Fig. 1). Thus, strong expression of CD25 occurred when 2C $CD8^+$ cells were exposed to peptides presented by either L^d.B7 or L^d.ICAM APC. This finding applied to both peptides, although QL9 peptide was clearly much more immunogenic than p2Ca. Varying the concentrations of these peptides showed that the APC function of L^d.B7 and L^d.ICAM APC was quite similar. Interestingly, however, B7–1 and ICAM-1 displayed strong synergism when coexpressed. Thus, with limited concentrations of peptides, induction of CD25 expression was far higher with L^d.B7.ICAM APC than with either L^d.B7 or L^d.ICAM APC. With the high-affinity QL9 peptide, upregulation of CD25 expression induced by L^d.B7.ICAM APC was apparent with peptide concentrations as low as 10^{-13} M.

2.2. Proliferative Responses

As for $CD4^+$ cells, proliferative responses of $CD8^+$ cells to antigen depend upon a combination of CD25 (IL-2R) upregulation and contact with IL-2.[13,14] When $CD8^+$ cells are subjected to strong signalling, the cells express CD25 and also synthesize IL-2. Under these conditions, the response of $CD8^+$ cells is thus helper-independent (HI), i.e. the cells proliferate without the need for "help" (exogenous IL-2) provided by other cells, e.g. $CD4^+$ cells. With weaker signalling, however, $CD8^+$ cells express CD25 but fail to synthesize IL-2. Proliferative responses then depend upon exposure to exogenous IL-2. These helper-dependent (HD) responses of $CD8^+$ cells occur when 2C cells are cultured with high concentrations of QL9 peptide (10^{-5} M) presented by Drosophila cells expressing L^d alone (Table 1). In this situation, proliferative responses are undetectable in the absence of IL-2, but are quite high when IL-2 is added, thus indicating that the mild increase in CD25 expression (Fig. 1) is physiologically significant. With lower doses of QL9 peptide ($\leq 10^{-7}$ M), however, proliferative responses in the presence of exogenous IL-2 are virtually undetectable. For the weaker p2Ca peptide, even high doses of this peptide fail to induce proliferative responses with added IL-2.

The above findings apply to APC expressing L^d alone. Very different results occur when the APC express costimulatory molecules. For the stronger QL9 peptide, L^d.B7 and

Table 1. IL-2 dependent proliferative response of $CD8^+$ 2C cells cultured with L^d-transfected Drosophila APC plus p2Ca or QL9 peptides

	$[^3H]$TdR incorporation (cpm × 10^3) with L^d APC plus $CD8^+$ 2C cells	
Peptides added to culture	Without exogenous IL-2	With IL-2 (20 units/ml)
–	97	645
p2Ca (10^{-7} M)	96	1,009
p2Ca (10^{-5} M)	150	1,754
QL9 (10^{-7} M)	112	2,848
QL9 (10^{-5} M)	161	205,026

Doses of 5 × 10^4 purified $CD8^+$ 2C cells were cultured with 2 × 10^5 L^d-transfected Drosophila cells in the presence or absence of the indicated concentrations of peptides for 3 days. $[^3H]$TdR was added during the last 8 hours of culture; rIL-2 was added at a final concentration of 20 units/ml. This data refer to the mean of triplicate cultures and are adapted from ref. 12.

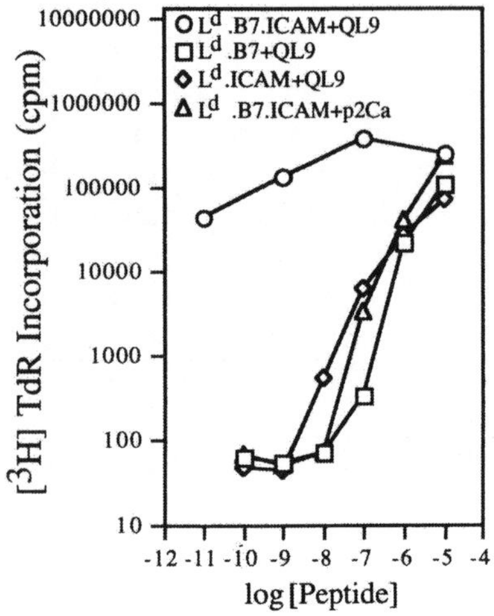

Figure 2. Influence of peptide concentration on the HI proliferative response of CD8$^+$ 2C cells to L^d-transfected Drosophila APC expressing the indicated costimulatory molecules. Cells were cultured as described for Table 1 and responses were measured on day 3; no IL-2 was added to the cultures. The data refer to the mean of triplicate cultures and are adapted from ref. 12.

L^d.ICAM APC both elicit quite strong proliferative responses of 2C CD8$^+$ cells in the absence of added IL-2 (Fig. 2). However, inducing these responses requires relatively high concentrations of peptide. By contrast, co-expression of both B7–1 and ICAM-1 on APC stimulates strong HI responses of 2C cells with very low doses of QL9 peptide, e.g. 10^{-11} M. With the weaker p2Ca peptide, L^d.B7 and L^d.ICAM APC fail to elicit HI responses of 2C cells even with high doses of peptide.[12] By contrast, strong HI responses to p2Ca peptide occur with L^d.B7.ICAM APC, though only with relatively high concentrations of peptide (Fig. 2).

These data on HI proliferative responses correlate closely with the above findings on CD25 expression and indicate that B7–1 and ICAM-1 display striking synergy when coexpressed. This synergy is most conspicuous with limiting concentrations of peptides. Very similar findings apply to IL-2 production.[12] Thus, levels of IL-2 are very low when CD8$^+$ cells are stimulated with either L^d.B7 or L^d.ICAM APC plus peptides but high with L^d.B7.ICAM APC.

3. CONCLUDING COMMENTS

Since CD8$^+$ cells are often viewed as crippled cells that fail to respond to antigen unless supplemented with "help", the capacity of artificial APC to induce strong primary responses of CD8$^+$ cells without the requirement for exogenous cytokines may seem surprising. As discussed earlier, however, pulsing class I-transfected Drosophila cells with peptides produces a very high density of peptide/MHC complexes on these cells and thus delivers a strong dose of Signal 1 to antigen-reactive CD8$^+$ cells. By itself, Signal 1 has little apparent effect on CD8$^+$ cells, at least in terms of stimulating CD25 and CD69 expression. Nevertheless, Signal 1 clearly makes the cells sensitive to Signal 2, i.e. to the effects of costimulation. With a strong dose of Signal 1, the sensitivity of T cells to Signal 2 is greatly increased, with the result that even a limited level of costimulation is sufficient to generate a strong response. This situation applies when 2C CD8$^+$ cells are exposed to high concentrations of the high-affinity QL9 peptide presented by Drosophila APC. Here, Signal 1 is intense, and triggering the cells requires contact with only a single costimulatory molecule on APC, i.e. either B7–1 or ICAM-1. With limiting concentrations of peptide, however, Signal 1 is less intense and the cells show a much greater dependency on costimulation, contact with two different costimulatory molecules, B7–1 + ICAM-1, being required for T cell triggering. This increased dependency on costimulation is especially apparent when the affinity of the peptide is reduced, e.g. when p2Ca rather than QL9 is used.

Although the present data are in close accord with the Signal 1/Signal 2 concept, the strong synergy observed between B7–1 and ICAM-1 suggests that Signal 2 has more than one component. Currently, the prevailing view is that costimulation is largely a reflection of CD28/B7 interaction, and that "costimulation" mediated by other accessory molecules, e.g. between LFA-1 and ICAM-1, is a manifestation of enhanced cell adhesion (conjugate formation) rather than to signal transduction. However, the definition of LFA-1 as an adhesion molecule[15] does not preclude a signalling role for LFA-1. Indeed, there is strong evidence that LFA-1 does contribute to signalling, especially for human T cells.[16] Hence, bearing in mind that the APC function of L^d.B7 and L^d.ICAM APC is virtually identical for 2C cells, it would seem highly likely that ICAM-1 acts at least partly as a signalling molecule, i.e. as a source of Signal 2. Nevertheless, the marked synergy observed when B7–1 and ICAM-1 are co-expressed suggests that the type of Signal 2 mediated by these two molecules is quite different.

The notion that Signal 2 can have at least two components suggests that the signalling events occurring when T cells recognize antigen under physiological conditions on normal APC are highly complex. Thus, professional APC such as DC are known to express a multiplicity of different accessory molecules, and each of these molecules could play an important role in APC function, either by facilitating cell adhesion or through signalling or both. At present, the functional role of the various accessory molecules on APC is still largely obscure. To address this issue, we are currently testing the function of a range of other accessory molecules expressed in Drosophila cells.

In addition to stimulating TCR transgenic T cells, we have recently found that Drosophila APC are capable of generating primary responses by normal (non-transgenic) $CD8^+$ cells.[17] This finding could be relevant to tumor immunotherapy. Thus, studies with a peptide derived from p815 ($H\text{-}2^d$) mastocytoma cells have shown that presentation of this peptide by Ld Drosophila cells to $H\text{-}2^d$ spleen cells in vitro generates strong CTL activity to P815 and rejection of this tumor after adoptive transfer in vivo.[17] Whether this approach is applicable to human tumors is under investigation.

REFERENCES

1. Janeway, C.A., and K. Bottomly. 1994. Signals and signs for lymphocyte responses. *Cell* 76:275–285.
2. Schwartz, R.H. 1992. Costimulation of T lymphocytes: the role of CD28, CTLA-4, and B7/BB1 in interleukin-2 production and immunotherapy. *Cell* 71:1065–1068.
3. Steinman, R.M. 1991. The dendritic cell system and its role in immunogenicity. *Annu. Rev. Immunol.* 9:271–296.
4. Jackson, M.R., E.S. Song, Y. Yang, and P.A. Peterson. 1992. Empty and peptide-containing conformers of class I major histocompatibility complex molecules expressed in Drosophila melanogaster cells. *Proc.Natl.Acad.Sci.USA* 89:12117–12121.
5. Yang, Y., K. Fruh, J. Chambers, J.B. Waters, L. Wu, T. Spies, and P.A. Peterson. 1992. Major histocompatibility complex (MHC)-encoded HAM2 is necessary for antigenic peptide loading onto class I MHC molecules. *J. Biol. Chem.* 267:11669–11672.
6. Sha, W.C., C.A. Nelson, R.D. Newberry, D.M. Kranz, J.H. Russell, and D.Y. Loh. 1988. Selective expression of an antigen receptor on CD8-bearing T lymphocytes in transgenic mice. *Nature* 335:271–274.
7. Kranz, D.M., S. Tonegawa, and H.N. Eisen. 1984. Attachment of an anti-receptor antibody to non-target cells renders them susceptible to lysis by a clone of cytotoxic T lymphocytes. *Proc. Natl. Acad. Sci. USA* 81:7922–7926.

8. Sykulev, Y., A. Brunmark, M. Jackson, R.J. Cohen, P.A. Peterson, and H.N. Eisen. 1994. Kinetics and affinity of reactions between an antigen-specific T cell receptor and peptide-MHC complexes. *Immunity* 1:15–22.
9. Udaka, K., T.J. Tsomides, and H.N. Eisen. 1992. A naturally occurring peptide recognized by alloreactive $CD8^+$ cytotoxic T lymphocytes in association with a class I MHC protein. *Cell* 69:989–998.
10. Udaka, K., T.J. Tsomides, P. Walden, N. Fukusen, and H.N. Eisen. 1993. A ubiquitous protein is the source of naturally occurring peptides that are recognized by a $CD8^+$ T-cell clone. *Proc. Natl. Acad. Sci. USA* 90:11272–11276.
11. Corr, M., A.E. Slanetz, L.F. Boyd, M.T. Jelonek, S. Khilko, B.K. Al-Ramadi, Y. Sang Kim, S.E. Maher, A.L.M. Bothwell, and D.H. Margulies. 1994. T cell receptor-MHC class I peptide interactions: affinity, kinetics, and specificity. *Science* 265:946–949.
12. Cai, Z., A. Brunmark, M.R. Jackson, D. Loh, P.A. Peterson, and J. Sprent. 1996. Transfected Drosophila cells as a probe for defining the minimal requirements for stimulating unprimed $CD8^+$ T cells. *Proc. Natl. Acad. Sci. USA* In Press.
13. Sprent, J., and M. Schaefer. 1990. Antigen-presenting cells for $CD8^+$ T cells. *Immunol. Rev.* 117:213–234.
14. Cai, Z., and J. Sprent. 1994. Resting and activated T cells display different requirements for CD8 molecules. *J. Exp. Med.* 179:2005–2015.
15. Springer, T.A., M.L. Dustin, T.K. Kishimoto, and S.D. Marlin. 1987. The lymphocyte function associated LFA-1, CD2, and LFA-3 molecules: cell adhesion receptors of the immune system. *Annu. Rev. Immunol.* 5:223–252.
16. van Seventer, G.A., Y. Shimizu, K.J. Horgan, and S. Shaw. 1990. The LFA-1 ligand ICAM-1 provides an important costimulatory signal for T cell receptor-mediated activation of resting T cells. *J. Immunol.* 144:4579–4586.
17. Sun, S., Z. Cai, P. Langlade-Demoyen, H. Kosaka, A. Brunmark, M.R. Jackson, P.A. Peterson, and J. Sprent. 1996. Dual function of Drosophila cells as APC for naive $CD8^+$ T cells: implications for tumor immunotherapy. *Immunity* 4:555–564.

42

ROLE OF DENDRITIC CELLS IN INDUCTION OF TOLERANCE AND IMMUNITY *IN VIVO*

Barbara Fazekas de St. Groth, Matthew C. Cook, Adrian L. Smith, Matthew E. Wikstrom, and Antony Basten

Centenary Institute of Cancer Medicine and Cell Biology
Locked Bag #6
Newtown, Sydney, Australia

1. INTRODUCTION

Despite the recent revolution in understanding the molecular mechanisms of T cell activation, the means by which naive $CD4^+$ T cells become tolerant or immune in response to antigen remains controversial. Evidence from a number of experimental systems has suggested that the process of determining the phenotype (but not the specificity) of immune responses is initiated by APCs[1,2]. However not only is the precise signalling pathway between APCs and naive T cells still undefined, but the role of each particular type of APC in the initiation of diverse immune responses *in vivo* is unclear.

The study of interactions between purified APC subpopulations and naive T cells is most easily achieved *in vitro*. However the validity of the conclusions drawn from *in vitro* experiments remains to be established. *In vitro* studies have demonstrated only quantitative differences in stimulation of naive T cells by APCs as disparate as splenic DC and transfected L cells[3-7]. In contrast, a number of investigators have shown clear qualitative effects attributable to the activity of different APCs *in vivo* [8-13]. There are two principal difficulties in interpretation of *in vitro* experiments. Firstly, the true phenotypes of tolerance and immunity *in vitro* may not be intuitively obvious. While it might be expected that naive cells in the process of becoming tolerant would manifest little reactivity in any of the standard *in vitro* assays such as cytokine production and proliferation, *in vivo* experiments have shown that induction of T cell tolerance can be accompanied by an initial phase of activation and proliferation which is difficult to distinguish from the initial phase of an immunogenic response[14]. Thus induction of immunity and tolerance *in vitro* may have been confused with each other. Secondly, naive T cells are segregated from most APCs *in vivo*, with the exception of interdigitating dendritic cells (IDC) which form a matrix in the T cell areas of the spleen and lymph nodes. In particular, the probability of contact between naive T and B cells is often overestimated, since B cells enter the T cell areas only after activation by specific antigen[15].

2. EXPERIMENTAL MODELS TO EXAMINE INTERACTIONS BETWEEN APCS AND NAIVE T CELLS *IN VIVO*

Our recent efforts have been concentrated on determining the outcome of several different APC-naive T cell interactions *in vivo*. A number of experimental protocols have been commonly used to examine the function of APC subsets *in vivo*. Antibody-mediated targeting of antigen to particular APC subsets has been achieved by coupling antigen to antibodies directed towards APC surface molecules, or by using xenogeneic anti-APC antibodies themselves as the antigen. The principal difficulty with this experimental technique is uncertainty as to the distribution of the antigen, which is inferred rather than measured experimentally. Thus although the data of both Parker[10] and Finkelman[9] support the conclusion that antigen-presentation by naive B cells induces T cell tolerance whereas activated B cells induce immunity, more direct experimental systems have suggested that activated B cells are actually tolerogenic when they are the sole APC *(vide infra)*.

Adoptive transfer of purified antigen-pulsed APCs has been effective in inducing immunity in response to mature DCs injected subcutaneously[16]. However its usefulness in addressing the question of whether other APC subpopulations can also generate true immunity is limited by the possibility that antigen may be eluted from the purified injected APCs and presented by potent immunogenic endogenous DCs. In order to avoid this technical difficulty, antigen presentation can be restricted to a subpopulation of injected cells on the basis of expression of an appropriate MHC allele. Ronchese and Hausmann used semi-allogeneic transfer into immunodeficient hosts to demonstrate that neither naive nor activated B cells were capable of priming naive T cells when the participation of endogenous non-B cell APC was rigorously excluded[13], in contrast to prior conclusions reached using antibody-mediated targeting. Fuchs and Matzinger[11] extended this argument by transferring allogeneic B cells into immunocompetent recipients and demonstrated that both naive and activated B cells rendered naive T cells tolerant.

2.1. An Antigen Receptor Transgenic Model of Presentation by B Cells

We have extended these studies of B cells as the sole initiating APC by transferring antigen-specific transgenic B and T cells into syngeneic or semi-allogeneic *scid/scid* hosts (figure 1). This model enables us to visualise the T and B cells as they respond to antigen. Since the immunogenic peptide of moth cytochrome C requires IE^k or IE^b for presentation, IE-negative hosts such as $H\text{-}2^b$ mice are unable to present cytochrome peptide or protein to purified T cells derived from anti-cytochrome TcR transgenic mice[17]. Thus in our protocol only the hen egg lysozyme (HEL)-specific transgenic $H\text{-}2^{bk}$ B cells[18] can act as APC in the $H\text{-}2^b$ hosts. In $H\text{-}2^{bk}$ hosts, macrophages and dendritic cells also express IE and can act as APC for cytochrome. The recipient *scid/scid* mice were immunised intravenously with PBS, moth cytochrome (MCC) peptide 87–103, or a fusion protein bearing both HEL and cytochrome determinants (HELCYT), to activate the B cells specifically while simultaneously delivering cytochrome for presentation to the T cells[19]. A response was detected by division of T cells labelled with the intracellular dye, CFSE, whose fluorescence is halved by each cell division[15]. After intravenous injection of antigen, activated but not naive B cells were capable of stimulating proliferation of naive T cells (figure 2). However naive B cells pulsed with cytochrome peptide *in vitro* prior to adoptive transfer also induced proliferation of naive T cells, presumably because they presented far higher concentrations of antigen than would be achieved *in vivo*. T cells taken on day 14 from $H\text{-}2^b$ recipi-

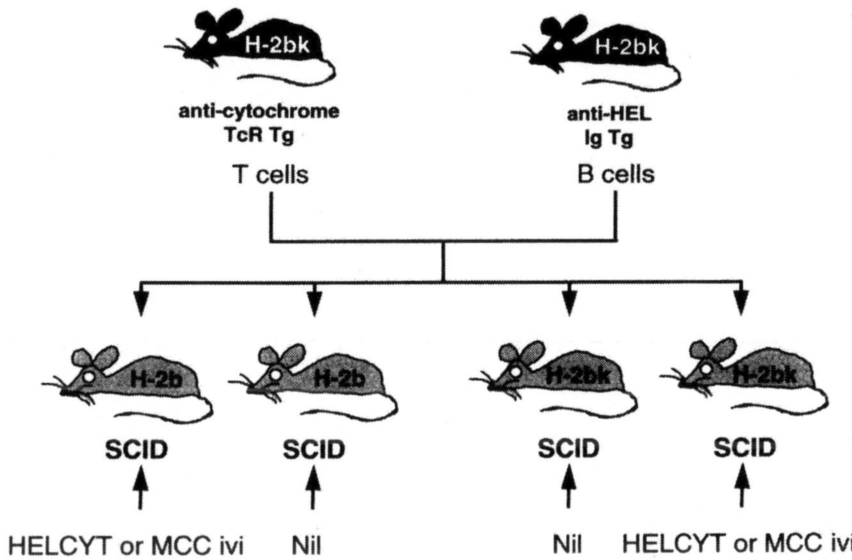

Figure 1. Experimental protocol for restricting antigen-presentation to B cells in immunodeficient hosts.

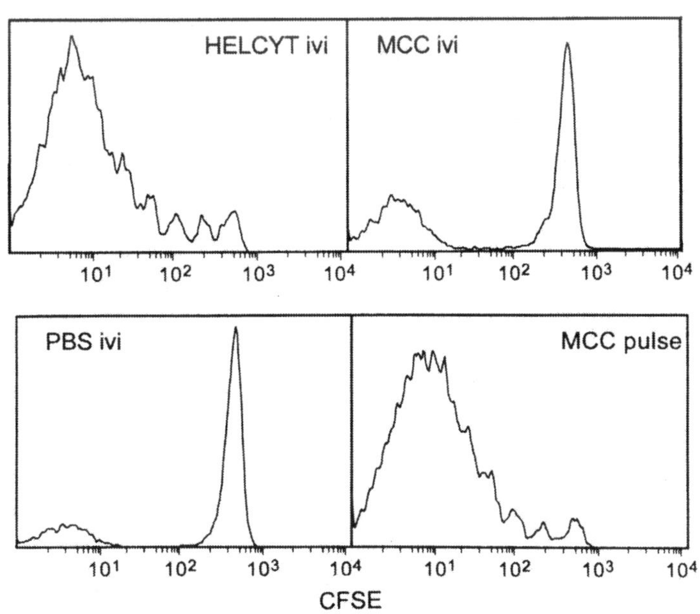

Figure 2. *In vivo* T cell proliferation in response to antigen presented by B cells. T and B cells were isolated from transgenic H-2^{bk} donors, percoll purified and CFSE-labelled before adoptive transfer into H-2^b *scid/scid* hosts. Antigen was administered intravenously 4 hours after cell transfer, except in the last group in which the B cells were pulsed with peptide *in vitro* before transfer. FACS profiles are of transgene-expressing T cells 7 days after antigen challenge. Each cell division is indicated by a halving of the CFSE intensity.

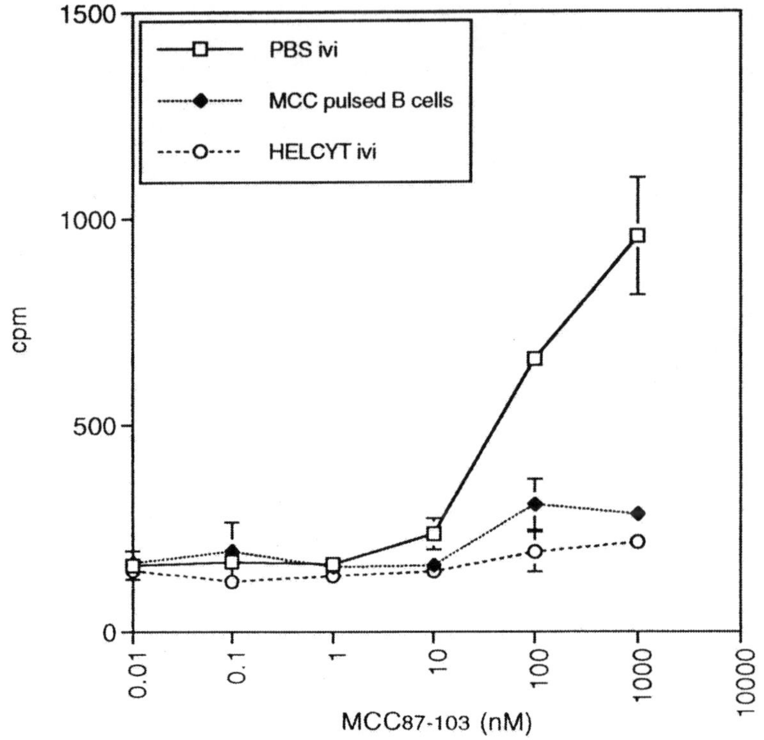

Figure 3. Proliferation of splenic T cells recovered from H-2^b *scid/scid* hosts 14 days after challenge with intravenous PBS, HELCYT or peptide-pulsed B cells. 10^4 transgene-expressing T cells were incubated with 5×10^5 H-2^{bk} irradiated splenocytes in the presence of the indicated concentrations of MCC peptide for 4 days.

ents immunised with HELCYT or peptide-pulsed B cells were profoundly anergic on restimulation *in vitro* (figure 3), whereas those from cytochrome peptide-immunised mice were indistinguishable from naive controls (not shown). In contrast, T cells in H-2^{bk} recipients made a much larger and more prolonged response to immunisation with HEL-cyt and T cell priming was readily detectable *in vitro* at 14 days (not shown). Thus the presence of H-2^{bk} host-derived (non-B cell) APCs converted an intrinsically tolerogenic response stimulated by antigen-activated B cells into a larger immunogenic response.

These experimental data confirm the conclusion of Fuchs and Matzinger[11] that both naive and activated B cells are tolerogenic for naive T cells. However, we also demonstrated that induction of tolerance was accompanied by a short lived immune response manifested by proliferation, cytokine production and antibody secretion. Thus the induction of tolerance did not appear to be the passive process envisaged by Fuchs and Matzinger, who argued strongly that both *in vivo* and *in vitro* demonstrations of naive T cell activation by B cells were essentially artifactual[11]. Rather, activation during the initiation of tolerance was remarkably similar in appearance to activation during the initiation of immunity. However, because the proliferation phase was curtailed when tolerance was being induced, the total clone size reached would be insufficient to be detectable in any but TcR transgenic models.

One further conclusion can be drawn from our experiments. Soluble antigen administered in the presence of a large number of specific B cells was clearly not targeted exclusively to those B cells, since endogenous non-B cells (presumably DC) altered the phenotype of the response when they expressed the MHC required for presentation. Thus it appears unlikely that antigen in general will be presented solely by B cells under physiological conditions, unless it is a B cell-specific self antigen which for some particular reason cannot be re-processed and presented by professional APCs. These data therefore suggest a model in which DC initiate all immune responses, including those that induce tolerance. Such a model is consistent with the histological organisation of peripheral immune tissue, in which only IDCs are constitutively located in the T cell areas, whereas B make contact with T cells only after activation[15].

2.2. Models of Antigen Presentation Using MHC Transgenic Mice

The experiments described above suggested that although B cells and naive T cells were capable of interacting *in vivo*, the process was inefficient in comparison to sequential T cell interaction with DC followed by B cells. Although *scid/scid* hosts reconstitute a fairly normal histological appearance in the spleen and LN within a few days of adoptive transfer of T and B cells, their anatomy is clearly disturbed at the time of adoptive transfer. Thus there may have been more initial contact between T and B cells than would occur in a normal host. Our most recent experimental approach has therefore focussed on reconstituting the interaction between APCs and T cells in the most physiological possible manner, while still taking advantage of the semi-allogeneic model of restricting antigen-presentation to particular subsets of APCs. This approach requires a normal host that is tolerant of semi-allogeneic donor APCs, such as the transgenic 36–2 line[20,21], in which expression of the IEα^d transgene, which pairs with endogenous IEβ^b from the H-2^b host, is limited to the thymus. 36–2 mice are thus effectively tolerant of IE^{db} without expressing it in the periphery. The 107–1 transgenic line[20,21] in which the IEα^d transgene is expressed with a wild-type distribution, can be used both as an APC donor and a breeding partner for the IE-restricted anti-cytochrome TcR transgenic line. Double transgenic mice then serve as donors of positively-selected anti-cytochrome T cells in adoptive transfer (figure 4). 107–1 mice can also be bred with anti-HEL immunoglobulin transgenic mice to serve as donors of IE-positive HEL-specific B cells.

2.2.1. B Cell Presentation. We have used this experimental system to examine whether activated B cells alone can stimulate T cell proliferation in mice with normal immune systems. Some specific T cell proliferation was seen in 36–2 hosts of adoptively transferred purified naive 107–1 × Ig double transgenic B cells and 107–1 × TcR double transgenic T cells immunised with HELCYT, although clearly less than when the host itself was from the 107–1 line (data not shown). This confirms our observations in the *scid/scid* model, namely that B cell presentation is possible but inefficient in comparison with presentation by both DCs and B cells. Our conclusions in these B cell reconstitution experiments must be clearly distinguished from experiments using B cell deficient mice to examine the effect of the addition of B cells to endogenous non-B cell APCs[8,13,22–27]. Such experiments seek to answer a different question, namely whether B cells are required *in addition to* non-B cell APCs for induction of an optimal T cell response. Although the data are not entirely consistent, the work of Sunshine *et al.*,[8] Ronchese and Hausmann[13], and Epstein *et al.*[27] clearly establishes that T cell priming can occur in the absence of B cells. In addition, B cells appear to amplify or modulate the immune response generated by DC, particularly when that response is submaximal[23,26,28]. We have repeated these B cell addition experiments in the 36–2/107–1 model. A comparison of transgenic T cell priming in

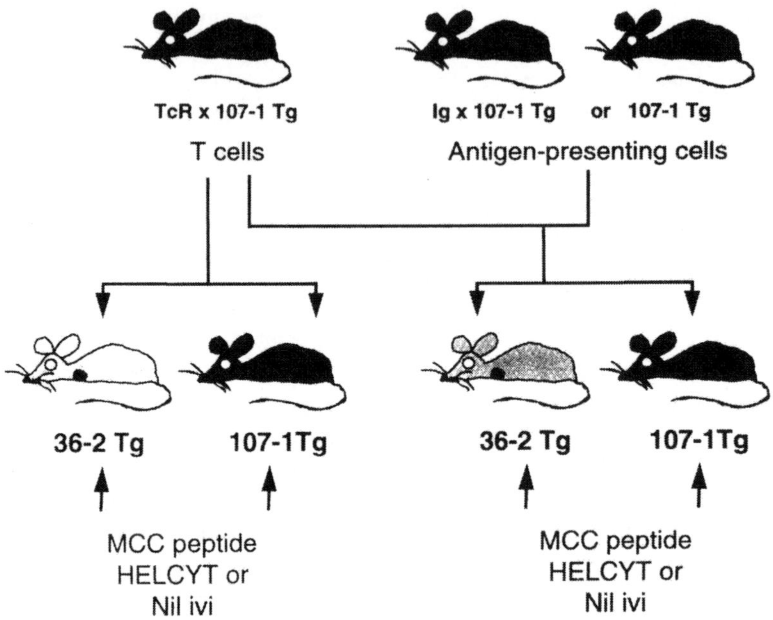

Figure 4. Experimental protocol for restricting antigen-presentation to purified donor cells in immunosufficient hosts.

107–1 hosts adoptively transferred with either 107–1 or wild type H-2^b transgenic B cells revealed that the presence of specific B cells expressing the appropriate MHC amplified the proliferative response of naive transgenic T cells but was not required to generate a detectable response (data not shown).

2.2.2. Dendritic Cell Presentation. In a second set of experiments using the 36–2/107–1 model, we attempted to determine the particular APC responsible for the induction of T cell proliferation and deletion in response to intravenous injection of peptide, since our *scid/scid* experiment (figure 2) showed that B cells were not capable of inducing such a response. In an initial experiment, spleen cells from 107–1 donors were fractionated on a Nycodenz gradient and 80 × 10^6 cells with a density <= 1.077g/cm^2 (comprising 24% CD11c$^+$IE$^+$B220$^-$ and 56% B220$^+$IE$^+$CD11c$^-$ cells) were transferred into 36–2 recipients which had previously been injected intravenously with CFSE-labelled Percoll-purified small T cells from 107–1 × TcR double transgenic donors. One day later, the mice received 15 ug moth cytochrome peptide 87–103 intravenously. A small percentage of the cytochrome-specific transgenic T cells in the spleen responded to the injected peptide by dividing between days 0 and 3 (figure 5A). The cell division profile was quite distinct from that seen in intact 107–1 x TcR double transgenic mice (figure 5B), indicating that antigen-presentation was severely limited in the 36–2 adoptive hosts, since the number of dividing cells declined with each cell division. In contrast, 95% of possible responder cells underwent at least 2 cell divisions in the intact transgenic controls. The crucial qualitative difference between the intact double transgenic and the adoptively transferred mice was in the fate of divided cells. Whereas over 90% of divided cells died by day 7 in double transgenic mice, the number of divided cells actually increased between days 3 and 7 in adoptively transferred 36–2 hosts, a pattern that is associated with generation of immunity in response to subcutaneous peptide rather than tolerance.

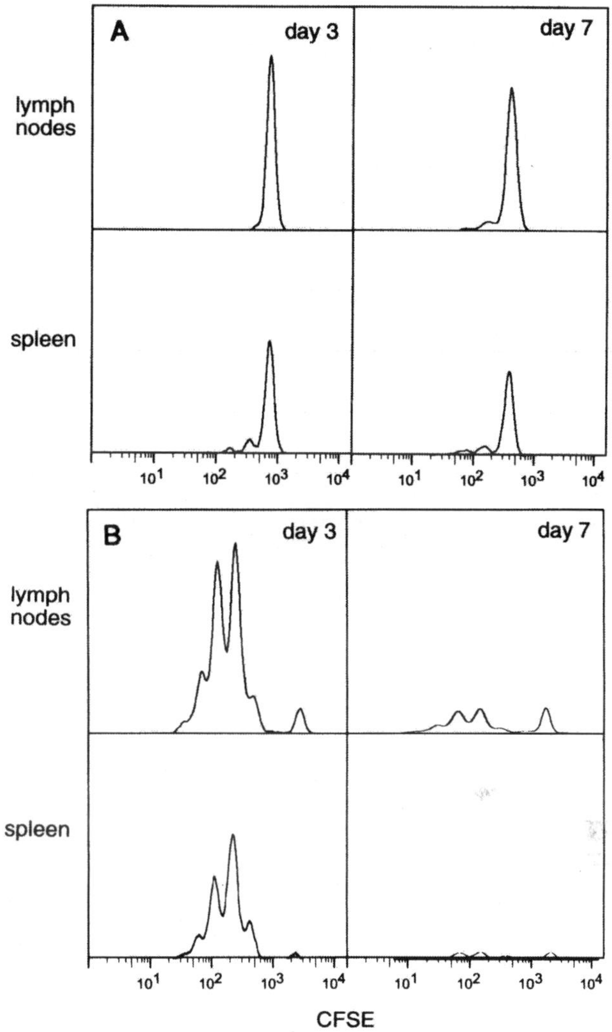

Figure 5. *In vivo* T cell proliferation response to intravenous MCC peptide presented by light density splenocytes. A: Percoll purified T cells were isolated from transgenic 107–1 donors, CFSE-labelled and adoptively transferred into 36–2 hosts together with light density splenocytes from 107–1 donors. B: transgenic T cells isolated from H-2^{bk} mice, CFSE-labelled and adoptively transferred into syngeneic hosts. FACS profiles are of transgene-expressing T cells 3 and 7 days after intravenous peptide challenge. Each cell division is indicated by a halving of the CFSE intensity.

Further experiments using subfractionated spleen cells are under way. In particular, we are collaborating with Prof. K. Shortman (WEHI) in the purification and adoptive transfer of $CD8^+$ and $CD8^-$ splenic DC using the model described above. Our hypothesis is that $CD8^+$ lymphoid-derived DC are tolerogenic, whereas classical myeloid-derived $CD8^-$ DC are immunogenic. When both are present, as in the preliminary adoptive transfer experiment described above, immunity is dominant over tolerance. An alternative possibility is that the phenotype of a single type of IDC changes in response to adjuvant-derived stimuli and that this change regulates the tolerance-immunity decision.

3. CONCLUSIONS

Our data suggest that dendritic cells are the crucial inducer of both tolerance and immunity *in vivo*. Moreover, the roles played by both DC and B cells can change dramatically, depending on the form of the antigen and the other APC present. Whether heterogeneity in DC function will correlate with functional changes within a single population or with subsets of different origin remains to be determined.

ACKNOWLEDGMENTS

This work was supported by the National Health and Medical Research Council of Australia, the Wellcome Trust, and the Medical Foundation of the University of Sydney.

REFERENCES

1. Janeway, C. A. J. The immune system evolved to discriminate infectious nonself from noninfectious self. *Immunol. Today* 13:11–16 (1992).
2. Fazekas de St. Groth, B. Regulation of the immune response - lessons from transgenic models. *Aust. N.Z. J. Med.* 25:761–767 (1995).
3. Seder, R. A., Paul, W. E., Davis, M. M., and Fazekas de St. Groth, B. The presence of interleukin 4 during in vitro priming determines the lymphokine-producing potential of $CD4^+$ T cells from T cell receptor transgenic mice. *J. Exp. Med.* 176:1091–1098 (1992).
4. Hsieh, C.-S., Heimberger, A. B., Gold, J. S., O'Garra, A., and Murphy, K. M. Differential regulation of T helper phenotype development by interleukins 4 and 10 in an αβ T-cell-receptor transgenic system. *P. N. A. S. (U. S. A.)* 89:6065–6069 (1992).
5. Duncan, D. D., and Swain, S. L. Role of antigen-presenting cells in the polarized development of helper T cell subsets: evidence for differential cytokine production by Th0 cells in response to antigen presentation by B cells and macrophages. *Eur. J. Immunol.* 24:2506–2514 (1994).
6. Cassell, D. J., and Schwartz, R. H. A quantitative analysis of antigen-presenting cell function: activated B cells stimulate naive CD4 T cells but are inferior to dendritic cells in providing costimulation. *J. Exp. Med.* 180:1829–1840 (1994).
7. Constant, S., Sant'Angelo, D., Pasqualini, T., Taylor, T., Levin, D., Flavell, R., and Bottomly, K. Peptide and protein antigens require distinct antigen-presenting cell subsets for the priming of $CD4^+$ T cells. *J. Immunol.* 154:4915–4923 (1995).
8. Sunshine, G. H., Jimmo, B. L., Ianelli, C., and Jarvis, L. Strong priming of T cells adoptively transferred into *scid* mice. *J. Exp. Med.* 174:1643–1656 (1991).
9. Goroff, D. K., Holmes, J. M., Basin, H., Nisol, F., and Finkelman, F. D. Polyclonal activation of the murine immune system by an antibody to IgD: XI contribution of membrane IgD cross-linking to the generation of an in vivo polyclonal antibody response. *J. Immunol.* 146:18–25 (1991).
10. Eynon, E. E., and Parker, D. C. Small B cells as antigen-presenting cells in the induction of tolerance to soluble protein antigens. *J. Exp. Med.* 175:131–138 (1992).
11. Fuchs, E. J., and Matzinger, P. B cells turn off virgin but not memory T cells. *Science* 258:1156–1159 (1992).
12. Day, M. J., Tse, A. G. D., Puklavec, M., Simmonds, S. J., and Mason, D. W. Targeting autoantigen to B cells prevents the induction of a cell-mediated autoimmune disease in rats. *J. Exp. Med.* 175:655–659 (1992).
13. Ronchese, F., and Hausmann, B. B lymphocytes in vivo fail to prime naive T cells but can stimulate antigen-experienced T lymphocytes. *J. Exp. Med.* 177:679–690 (1993).
14. Webb, S., Morris, C., and Sprent, J. Extrathymic tolerance of mature T cells: clonal elimination as a consequence of immunity. *Cell* 63:1249–1256 (1990).
15. Fulcher, D. A., Lyons, A. B., Korn, S. L., Cook, M. C., Koleda, C., Parish, C., Fazekas de St. Groth, B., and Basten, A. The fate of self-reactive B-cells depends primarily on the degree of antigen receptor engagement and availability of T-cell help. *J. Exp. Med.* 183:2313–2328 (1996).
16. Inaba, K., Metlay, J. P., Crowley, M. T., and Steinman, R. M. Dendritic cells pulsed with protein antigens in vitro can prime antigen-specific, MHC-restricted T cells in situ. *J. Exp. Med.* 172:631–640 (1990).
17. Fazekas de St. Groth, B., Patten, P. A., Ho, W. Y., Rock, E. P., and Davis, M. M. An analysis of T cell receptor-ligand interaction using a transgenic antigen model for T cell tolerance and T cell receptor mutagenesis. Academic Press. San Diego. (1992).
18. Goodnow, C. C., Crosbie, J., Adelstein, S., Lavoie, T. B., Smith-Gill, S. J., Brink, R. A., Pritchard-Briscoe, H., Wotherspoon, J. S., Loblay, R. H., Raphael, K., Trent, R. J., and Basten, A. Altered immunoglobulin ex-

pression and functional silencing of self-reactive B lymphocytes in transgenic mice. *Nature* 334:676–682 (1988).

19. Fazekas de St. Groth, B., Cook, M. C., and Smith, A. L. The role of T cells in the regulation of B cell tolerance. *Int. Rev. Immunol.* in press (1996).
20. Widera, G., Burkly, L. C., Pinkert, C. A., Böttger, E. C., Cowing, C., Palmiter, R. D., Brinster, R. L., and Flavell, R. A. Transgenic mice selectively lacking MHC Class II (I-E) antigen expression on B cells: an in vivo approach to investigate Ia gene function. *Cell* 51:175–187 (1987).
21. Marrack, P., Lo, D., Brinster, R., Palmiter, R., Burkly, L., Flavell, R. H., and Kappler, J. The effect of thymus environment on T cell development and tolerance. *Cell* 53:627–634 (1988).
22. HayGlass, K. T., Naides, S. J., Scott, C. F. J., Benacerraf, B., and Sy, M.-S. T cell development in B cell-deficient mice. IV. The role of the B cells as antigen-presenting cells in vivo. *J. Immunol.*:823–829 (1986).
23. Ron, Y., and Sprent, J. T cell priming in vivo: a major role for B cells in presenting antigen to T cells in lymph nodes. *J. Immunol.* 138:2848–2856 (1987).
24. Janeway, C. A., Ron, J., and Katz, M. E. The B cell is the initiating antigen-presenting cell in peripheral lymph nodes. *J. Immunol.* 138:1051–1055 (1987).
25. Liu, Y., Wu, Y., Ramarathinam, L., Guo, Y., Huszar, D., Trounstine, M., and Zhao, M. Gene-targeted B-deficient mice reveal a critical role for B cells in the CD4 T cell response. *Int. Immunol.* 7:1353–1362 (1995).
26. Constant, S., Schweitzer, N., West, J., Ranney, P., and Bottomly, K. B lymphocytes can be competent antigen-presenting cells for priming $CD4^+$ T cells to protein antigens in vivo. *J. Immunol.* 155:3734–3741 (1995).
27. Epstein, M. M., Di Rosa, F., Jankovic, D., Sher, A., and Matzinger, P. Successful T cell priming in B cell-deficient mice. *J. Exp. Med.* 182:915–922 (1995).
28. Kurt-Jones, E. A., Liano, D., HayGlass, K. A., Benacerraf, B., Sy, M.-S., and Abbas, A. K. The role of antigen-presenting B cells in T cell priming in vivo. Studies of B cell-deficient mice. *J. Immunol.* 140:3773–3778 (1988).

B7–2^{+} LOW DENSITY APCS ARE AS EFFECTIVE AS B7–2^{-} SMALL RESTING B CELLS IN INDUCING SPECIFIC UNRESPONSIVENESS TO MINOR HISTOCOMPATIBILITY (miH) ANTIGEN(S) *IN VIVO*

Masanori Niimi, Mary-Anne Rawsthorne, Peter J. Morris, and
Kathryn J. Wood

Nuffield Department of Surgery
University of Oxford
John Radcliffe Hospital
Oxford OX3 9DU, United Kingdom

1. INTRODUCTION

Dendritic cells and small resting B cells are thought to be immunogenic and tolerogenic in the periphery, respectively, as dendritic cells but not resting B cells can provide costimulatory signal(s) (1–7). Fuchs and Matzinger demonstrated that resting B cells but not dendritic cells could induce tolerance to H-Y mismatched skin grafts (8). We have extended this study by comparing the ability of dendritic cells and resting B cells to induce tolerance to cardiac grafts mismatched for miH antigen(s).

2. MATERIALS AND METHODS

Low density adherent cells (LODACs) were prepared as described by Steinman, and were 98.2% class II^{+}, 80.1% B7–1^{+} and 93.7% B7–2^{+}. Resting B cells were prepared by complement lysis of T cells, purification using Sephadex G-10 columns and Percoll gradients and were 95.7% class II^{+}, 1.2% B7–1^{+} and 2.5% B7–2^{+}.

Mice were pretreated intravenously with either 1×10^7 resting B cells or 5×10^5 LODACs one week before skin grafting or two weeks before heart grafting.

Dendritic Cells in Fundamental and Clinical Immunology
edited by Ricciardi-Castagnoli, Plenum Press, New York, 1997

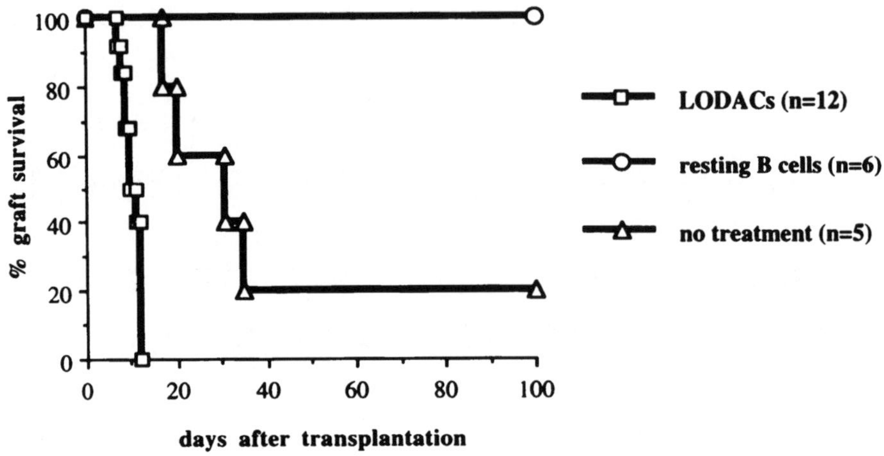

Figure 1. Resting B cells but not LODACs induce indefinite survival of H-Y mismatched skin grafts.

3 RESULTS

3.1. Skin Graft

3.1.1. Primary Skin Grafts. Resting B cells induced indefinite prolongation of H-Y disparate *skin* grafts in C57BL/10 ($H2^b$) female mice (MST > 100 days), as has been reported previously (8). In contrast, LODACs had no effect on the survival of male skin grafts (MST = 10.5 days) (Figure 1).

Next, we explored the ability of resting B cells to prolong graft survival in donor recipient combinations mismatched for either multiple miH antigens or MHC and miH antigens. In both situations, pretreatment with resting B cells and LODACs had no beneficial effect on skin graft survival (MST=14 and 12.5 days, respectively C3H to CBA; 11 and 10 days, respectively C57BL/10 to C3H).

3.1.2. Sencondary Skin Grafts. To investigate whether resting B cells were able to induce operational tolerance to skin grafts mismatched for a single miH antigen, H-Y, recipients were grafted with a 2nd H-Y mismatched skin graft 100 days after transplantation. 2nd grafts were rejected with a MST of 20 days, although the primary skin grafts continued to survive and were unaffected by the rejection of the second challenge graft.

3.2. Cardiac Graft

3.2.1. Primary Cardiac Grafts. In contrast to the effects for H-Y mismatched skin grafts, surprisingly, both resting B cells and LODACs induced indefinite survival of H-Y mismatched cardiac grafts (MSTs > 300 and >200 days, respectively) (Figure 2). [Naive female mice rejected male hearts (MST = 42 days).] Resting B cells and LODACs were also equally effective in their ability to induce unresponsiveness for cardiac grafts mismatched for multiple miH antigens (C3H ($H2^k$) to CBA ($H2^k$)), (MSTs >107 and >99 days, respectively) (Figure 3). In marked contrast, in a fully allogeneic combination (C57BL/10

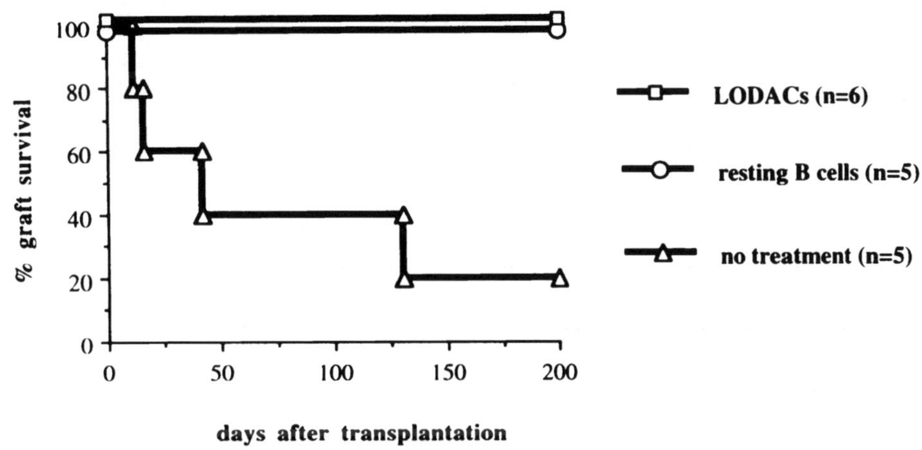

Figure 2. Resting B cells and LODACs induce indefinite graft survival of H-Y mismatched cardiac grafts.

to C3H), LODACs accelerated the rejection of cardiac grafts (MST=8 days) compared with no treatment and resting B cells (MST= 10 days and 22 days, respectively).

3.2.2. Secondary Heart Grafts. When recipients with long term surviving cardiac grafts (>300 days) were rechallenged with second grafts to determine if operational tolerance had been induced, 2nd H-Y mismatched cardiac grafts were accepted indefinetly.

4. DISCUSSION

The data obtained in this study confirms the ability of resting B cells to induce unrepsonsiveness to skin grafts mismatched for a single miH antigen. However, the data presented demonstrates that operational tolerance to the H-Y antigen was not induced by

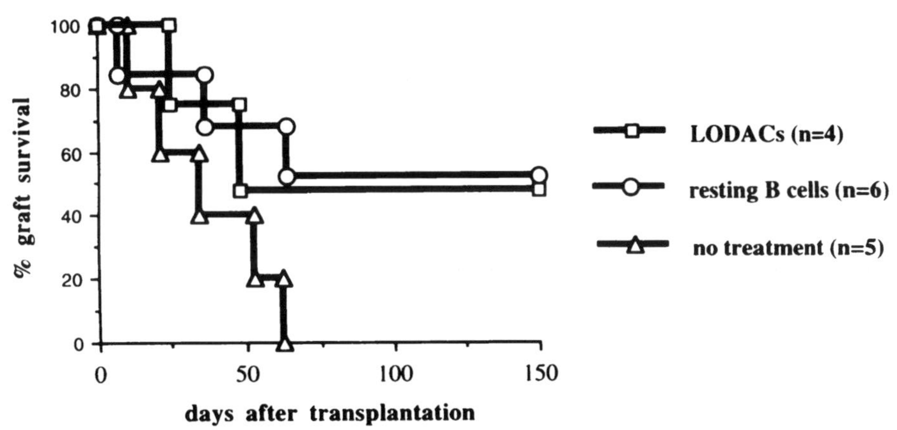

Figure 3. Resting B cells and LODACs induced indefinte survival of cardiac grafts mismatched for multiple miH antigens in 50% of the recipients.

pretreatment with resting B cells, or second skin grafts were rejected by recipients bearing long term surviving primary grafts. In contrast, when cardiac grafts were transplanted both primary and secondary challenge grafts survived indefinitely after pretreatment with resting B cells. These data support the hypothesis that the cardiac graft plays an important role in the maintenance of unresponsiveness in vivo (9).

Interestingly, our study also showed that LODACs were as effective as resting B cells in inducing prolonged survival of cardiac but not skin grafts mismatched for miH antigen(s). LODACs are thought to be immunogenic according to two signal hypothesis as they express both B7–1 and B7–2. However, this may not be the case in every situation, particularly, as shown here for vascularised organ grafts.

5. REFERENCES

1. Koulova L, Clark EA, Shu G, Dupont B. The CD28 ligand B7/BB1 provides costimulatory signal for alloactivation of $CD4^+$ T cells. J Exp Med 1991;173:759–62.
2. Linsley P, Brady W, Urnes M, Grosmaire L, Damle N, Ledbetter J. CTLA-4 is a second receptor for the B cell activation antigen B7. J Exp Med 1991;174:561–569.
3. Freeman GJ, Gribben JG, Boussiotis VA, et al. Cloning of B7–2: A CTLA-4 counter-receptor that costimulates human T cell prloiferation. Science 1993;262:909–11.
4. Azuma M, Ito D, Yagita H, et al. B 70 antigen is a second ligand for CTLA-4 and CD28. Nature 1993;366:76–9.
5. Harding FA, McArthur JG, Gross JA, Raulet DH, Allison JP. CD28-mediated signalling co-stimulates murine T cells and prevents induction of anergy in T-cell clones. Nature 1992;356:607–609.
6. Schwartz RH. A cell culture model for T lymphocyte clonal anergy. Science 1990;248:1349–1355.
7. Bluestone JA. New perspectives of CD28-B7-mediated T cell costimulation. Immunity 1995;2:555–559.
8. Fuchs EJ, Matzinger P. B cells turn off virgin but not memory T cells. Science 1992;258:1156–1159.
9. Hamano K, Rawsthorne M, Bushell A, Morris PJ, Wood KJ. Donor microchimerism is not responsible for the maintenace of tolerance to donor alloantigens in recipients tolerant of cardiac allografts. Transplant Proc 1995;27(1):151–152.

44

AN ATTEMPT TO INDUCE TOLERANCE WITH INFUSION OF DONOR BONE MARROW IN ORGAN ALLOGRAFT RECIPIENTS*

Abdul S. Rao,[1] D. Phil,[2,3] Paulo Fontes,[2] Anand Iyengar,[2]
Adelouahab Aitouche,[2] Ron Shapiro,[2] Adriana Zeevi,[3] Forrest Dodson,[2]
Robert Corry,[2] Cristiana Rastellini,[2] John J. Fung,[2] and Thomas E. Starzl[2]

[1]Thomas E. Starzl Transplantation Institute
[2]Department of Surgery and
[3]Department of Pathology
University of Pittsburgh Medical Center
Pittsburgh, Pennsylvania

1. THE ROLE OF PASSENGER LEUKOCYTES IN ALLOGRAFT REJECTION

Early observations by Snell and Steinmuller provided unequivocal evidence that non-parenchymal cells, resident within the organ play an important role in allograft rejection[1–3].

However, despite these findings and the subsequent employment of the term "passenger leukocytes" to illustrate their migratory capacity[4], identity of the donor cells involved in allograft rejection (at that time) was nevertheless enigmatic. These observations nonetheless had two unique corollaries for transplantation (Tx). Firstly, that the severity of rejection might be mitigated by functionally modulating or depleting donor leukocytes prior to organ tx, and secondly, that the majority of the transplanted tissue is non-immunogenic and may therefore play an infinitesimal role in graft rejection. The realization of the importance of these findings prompted the initiation of numerous studies intended to modulate the immune responses by manipulating resident non-parenchymal cells prior to transplantation[5–7]. In spite of the novelty of these experiments, the success was however, variable.

1.1. Importance of Dendritic Cells (DC) in Primary Immune Responses

Whilst the role of passenger leukocytes in initiating allograft rejection was undeniable, the identity of the cell(s) which instigated this immune response was however,

* Aided by Project Grant No. AI 40329 and DK 29961 from the National Institute of Health, Bethesda, Maryland.

equivocal. Given their unique capacity to present antigens to naive T cells, it was postulated that perhaps cells of dendritic leukocyte lineage may play a vital role in allosensitization[8]. Supporting evidence for the latter argument came from the initial studies by Lechler and Batchelor, who demonstrated that allogeneic kidneys were indefinitely accepted in rodents if they were initially "parked" in an immunosuppressed recipient prior to retransplantation into a secondary naive syngeneic animal[9]. More importantly, the allografts were acutely rejected if DC obtained from donor-strain animals were infused into the secondary recipients indicating that the latter manipulation perhaps, led to the restoration of immunogenicity of the leukocyte-depleted allograft resulting in its prompt rejection. This and numerous subsequent studies[10–12] have unambiguously established the identity of donor DC as the critical passenger leukocytes primarily responsible for allograft rejection, supporting the concept that their elimination prior to Tx may possibly culminate in prolonged graft survival.

1.2. DC in Transplantation Tolerance

A diametrical yet entirely compatible role for DC has been proposed by recent studies in rodents[13, 14] and sub-human primates[15]. Our laboratory has shown previously that livers transplanted orthotopically across most mouse strain combinations are spontaneously accepted[16]. Interestingly, DC isolated from murine livers exhibit phenotype and function distinctive of that of immature cells, suggesting perhaps that upon migration from the graft into the host, these cells may provide "deviant" signals resulting in induction of donor-specific hyporeactivity rather than allosensitization[13]. It is interesting to note that organs other than livers which are promptly rejected following Tx also contain, albeit few, resident immature DC[17]. It is therefore entirely conceivable that the inherent tolerogenicity uniquely ascribed to murine livers may be contingent on the presence within it an apropos quality and quantity of resident leukocytes bearing an immature phenotype.

1.3. Chimerism and Its Role in Organ Allograft Acceptance

The ubiquitous detection of the presence of donor cell chimerism in the tissues of successful long-term human kidney and liver recipients provided irrevocable evidence for the role of microchimerism in Tx tolerance[18,19]. These observations were supported by contemporaneous studies in rodents prompting evolvement of a hypothesis that the establishment of microchimerism by migratory resident leukocytes may play a seminal role in the induction of donor-specific tolerance[13, 15, 20, 21]. Whilst the role of migratory donor cells in the induction of donor-specific hyporeactivity was undeniable, the mechanism responsible for the perpetuation of chimerism and therefore of tolerance, years after organ Tx was nevertheless unclear. However, recent studies in rodents[22,23] and humans[24], have confirmed the presence of pluripotent stem cells in the interstitium of the grafted organ providing explication for the long-term persistence of donor cell chimerism.

2. AUGMENTATION OF CHIMERISM IN HUMAN ALLOGRAFT RECIPIENTS

Recognizing the seminal role played by migratory donor leukocytes in allograft acceptance and the induction of donor-specific tolerance almost four years ago, we embarked on a prospective clinical trial to augment the phenomenon in organ recipients by perioperative donor bone marrow (BM) infusion. In addition to verifying the safety of this procedure, three long-term goals of this study were to: (i) prevent delayed graft rejection, (ii) abrogate or mitigate chronic rejection and (iii) to decrease drug-dependence. Since June 1992, 189 recipients of liver (*n*=55), kidney (*n*=36), kidney + pancreas (*n*=33), kidney + islets (*n*=7), heart (*n*=24), lungs (*n*=18), small bowel (*n*=13) and multi-organ (*n*=3) have received a single peri-operative infusion of 3–6×10^8 BM cells/kg body weight. Additionally, subsequent to the implementation of a modified protocol in April 1996, ten recipients of liver (*n*=5), kidney (*n*=4), kidney + delayed islets (*n*=1) have received multiple infusions of 1×10^8 cells/kg body weight/day for five consecutive days (day 0–4) post-Tx.

Immunosuppression (IS) was with tacrolimus and steroids; CellCept was added to this regimen in 32 study and 15 control patients. BM was isolated from the vertebral bodies (VB) of cadaveric donors by a method described previously [25] and infused intravenously without cryopreservation. The recipients were not conditioned nor was the BM modified prior to infusion. The unavailability of consent to retrieve VB from the cadaveric donors resulted in accrual of 115 recipients of liver (*n*=32), kidney (*n*=19), kidney + pancreas (*n*=20), kidney + delayed islets (*n*=2), heart (*n*=20), lung (*n*=10), small bowel (*n*=7) and multiorgan (*n*=1) who were followed as contemporaneous controls. Episodes of acute rejection were treated with dose adjustments of routine IS, whereas OKT3 was reserved for the treatment of steroid-resistant rejection.

2.1. *In Vitro* Monitoring

Using peripheral blood mononuclear cells (PBMC), the immune status of the recipients was monitored serially by mixed leukocyte reaction (MLR), limiting dilution assay and proliferation against recall antigens (ConA and PHA). Using primers specific for either HLA-allele or the sex-determining region of the Y (SRY) chromosome (in malem-female recipients), the presence of donor DNA was determined periodically in the study and control patients. Serial quantitative evaluations of donor cell chimerism in a selected cohort of study and control patients were performed using a modified limiting dilution PCR (LDA-PCR) assay. The multilineage character of chimerism was also ascertained by PCR detection of donor DNA in lineage$^+$ cells sorted from recipients' PBMC. The evidence for the presence of donor DC progenitors and therefore of engraftment was obtained by propagation of recipients' PBMC in rhGM-CSF and rhIL-4-enriched cultures. Subsequent to enrichment for lineage^{null}/MHC class II$^+$ population, the presence of donor DNA within the sorted cells was confirmed by PCR analysis.

2.2. Clinical Outcome

The ancillary BM infusion was safe and no complications that could be uniquely attributed to this procedure were witnessed in any of the 199 study patients. All but 18/199 (9%) of BM-augmented recipients are alive compared to 13/115 (11%) of controls, who have died during the course of this follow-up (Table 1). Additionally, grafts in nine study and five control patients have been lost during their successive follow-up (Table 1). It is noteworthy that no deaths or graft losses in the augmented group were related to BM infusion. No evidence of any inimical complication was witnessed in any of the ten patients who have received multiple BM infusion during the course of their follow-up (5–154 days). All of the surviving patients have adequate graft function.

Whilst all patients remain IS-dependent, it is nonetheless interesting to note that steroid-free existence has been achieved in 61% study and 40% control patients who are at

Table 1. BM-Augmented (study) and non-augmented (control) transplant recipients: follow-up and patient graft survival

Organs Tx	*n*	Follow-up POD (X±SD)	Patient Survival (%)	Graft Survival (%)
Liver				
Study	60	610±366	53/60 (88%)	51/60 (85%)
Control	32	798±354	28/32 (88%)	28/32 (88%)
Kidney				
Study	81	588±311	80/81 (99%)	75/81 (93%)
Control	41	617±311	38/41 (93%)	37/41 (90%)
Heart				
Study	24	481±330	21/24 (88%)	21/24 (88%)
Control	20	429±226	18/20 (90%)	18/20 (90%)
Lung				
Study	18	380±269	14/18 (78%)	14/18 (78%)
Control	10	478±188	8/10 (80%)	8/10 (80%)
Small Bowel				
Study	13	289±234	10/13 (77%)	9/13 (69%)
Control	07	302±219	5/7 (71%)	5/7 (71%)
Multi-organ				
Study	03	508±136	3/3 (100%)	3/3 (100%)
Control	01	248	1/1 (100%)	1/1 (100%)

least 12 months post-Tx (figure 1). Equally significant however, is the observations that a statistically higher number of kidney recipients in the study group (71%) were weaned off steroids as compared to the controls.

2.2.1. Incidence of Acute and Chronic Rejection. The tempo, severity and cumulative incidence of acute cellular rejection was comparable (~60%) in patients in the study and control groups. Graft versus host disease was witnessed in only two (1%) BM-augmented recipients (both of liver) which for its resolution required minor dose adjustments of routine IS. Whilst histopathological changes pathognomonic of chronic rejection have not manifested in any BM-augmented patient, at a comparable duration of follow-up 2/7 (29%) surviving non-augmented lung recipients have however, exhibited evidence for the development of obliterative bronchiolitis.

2.2.2. Multilineage Chimerism and Evidence for Infused Donor Cell Engraftment. The incidence of chimerism was much higher (94%) in the BM-augmented patients as compared to the controls (56%). Additionally, using LDA-PCR, the levels of chimerism were found to be at least 10–100 fold higher in a selected cohort of study patients as compared to the controls. Furthermore, the presence of donor DNA in sorted lineage$^+$ cells provided unequivocal evidence of the multilineage nature of donor cell chimerism in BM-augmented patients. It must be emphasized that our ability to identify donor DNA in cultured/sorted (lineage$^+$/class II$^+$) cells generated from the PBMC of 4/5 evaluated study patients who were at least one year post-Tx provide sustenance to our claim that infused marrow may have indeed engrafted.

2.2.3. In Vitro Immune Modulation. A previously established criteria was used to access the extent of donor-specific immune modulation in control and study patients.[26] At a two-year follow-up, a higher proportion of BM-augmented liver (50%) and lung (60%) re-

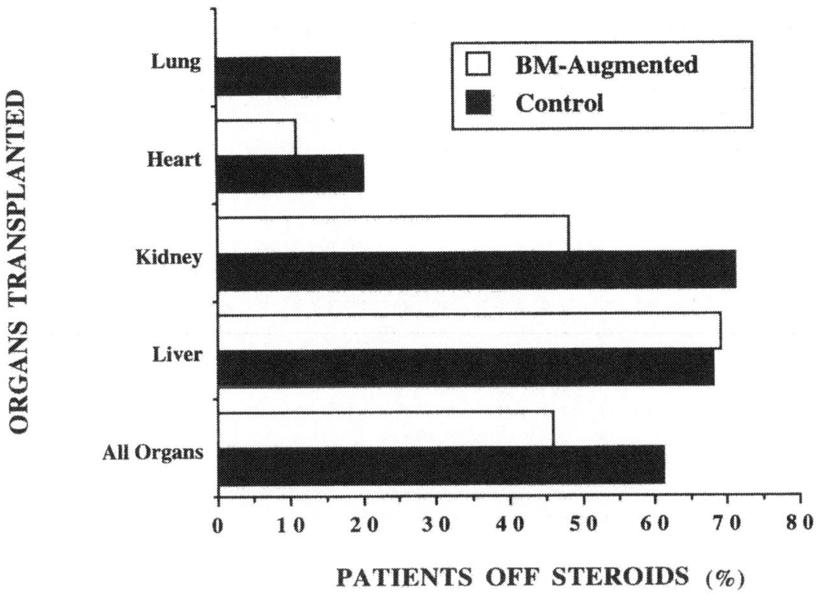

Figure 1. The incidence of steroid-free existence in BM-augmented and non-augmented (control) organ recipients who are at least 12 months post-Tx.

cipients exhibited donor-specific hypo or intermediate (DSHI) responses as compared to the controls (Figure 2). On the contrary, there was a comparable degree of immunomodulation in study and control kidney and heart recipients at a similar duration of follow-up.

3. CONCLUSION

Whilst in rodents and large animals, donor-specific tolerance could be achieved in months or years, the expectation of a similar outcome in humans is perhaps erroneous.

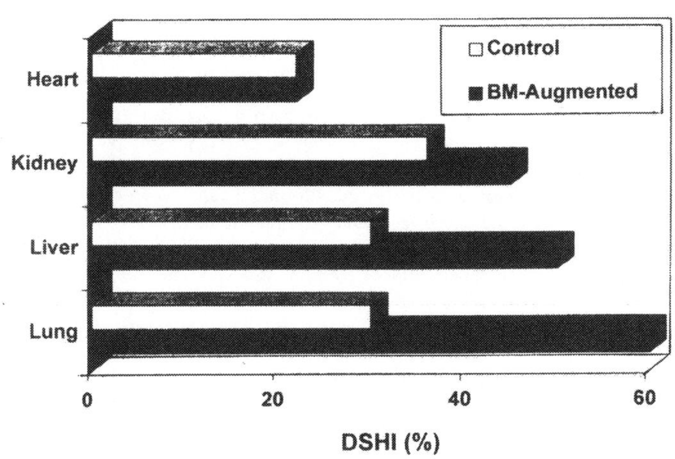

Figure 2. The incidence of evolvement of donor-specific hypo or intermediate (DSHI) reactivity determined by MLR assay in BM-augmented and control organ allograft recipients at a two-year follow-up.

Given our experience with long-term successful organ transplant recipients who are currently being weaned off IS under a meticulously planned protocol, it is anticipated that only a select cohort of patients in this study will ever achieve a drug-free state perhaps years if not a decade after Tx.

REFERENCES

1. Snell GD. The homograft reaction. Annu Rev Microbio 1957; 11:439.
2. Steinmuller D. Immunization with skin isografts taken from tolerant mice. Science 1967; 158:127.
3. Steinmuller D. Passenger leukocytes and the immunogenicity of skin allografts. J Invest Dermatol 1980; 75:107.
4. Elkins WL, Guttmann RD. Pathogenesis of a local graft versus host reaction: Immunogenicity of circulating host leukocytes. Science 1968; 159:1250.
5. Lafferty KJ, et al. Thyroid allograft immunogenicity is reduced after a period in organ culture. Science 1975; 188:259.
6. Silvers WK, et al. The influence of removing passenger cells on the fate of skin and parathyroid allografts. Diabetes 1982; 31:60.
7. Lacy P, Davie J, Finke E. Prolongation of islet allograft survival following in vitro culture (24°C) and a single injection of ALS. Science 1979; 204:312.
8. Steinman RM. The dendritic cell system and its role in immunogenicity. Annu Rev Immunol 1991; 9:271.
9. Lechler RI, Batchelor JR. Restoration of immunogenicity to passenger cell-depleted kidney allografts by the addition of donor strain dendritic cells. J Exp Med 1979; 150:455.
10. McKenzie JL, Beard ME, Hart DJN. The effect of donor pretreatment on interstitial dendritic cell content and rat cardiac allograft survival. Transplantation 1984; 38:371.
11. Faustman DL, et al. Prevention of rejection of murine islet allografts by pretreatment with anti-dendritic cell antibody. Proc Natl Acad Sci USA 1984; 81:3864.
12. Austyn JM, Larsen CP. Migration pattern of dendritic leukocytes. Transplantation 1990; 19:1.
13. Starzl TE, et al. Cell migration and chimerism after whole organ transplantation: The basis of graft acceptance. Hepatology 1993; 17:1127.
14. Streilein JW, et al. Immunoregulatory mechanisms of the eye. J Neuroimmunol 1992; 39:185.
15. Thomas JM, et al. Further studies of veto activity in rhesus monkey bone marrow in relation to allograft tolerance and chimerism. Transplantation 1994; 57:101.
16. Qian S, et al. Murine liver allograft transplantation: Tolerance and donor cell chimerism. Hepatology 1994; 19:916.
17. Austyn JM, et al. Isolation and characterization of dendritic cells from mouse heart and kidney. J Immunol 1994; 152:2401.
18. Starzl TE, et al. Cell migration, chimerism, and graft acceptance. Lancet 1992; 339:1582.
19. Starzl TE, et al. Cell migration and chimerism after whole organ transplantation: The basis of graft acceptance. Hepatology 1993; 17:1152.
20. Murase N, et al. Variable chimerism, graft versus host disease, and tolerance after different kinds of cell and whole organ transplantation from Lewis to Brown-Norway rats. Transplantation 1995;60:158.
21. Demetris AJ, et al. Hematolymphoid cell trafficking, microchimerism, and GVHD reactions after liver, bone marrow, and heart transplantation. Transplantation Proc 1993; 25:3337.
22. Lu L, et al. Growth of donor-derived dendritic cells from the bone marrow of murine liver allograft recipients in response to granulocyte/macrophage colony-stimulating factor. J Exp Med 1995; 182:379.
23. Murase N, et al. Multilineage hematopoietic reconstitution of supralethally irradiated rats by syngeneic whole organ transplantation: With particular reference to the liver. Transplantation 1996; 61:1.
24. Rugeles MT, et al. Evidence for the presence of multilineage chimerism and progenitors of donor dendritic cells in the peripheral blood of bone marrow-augmented organ transplant recipients. J Immunol (submitted), 1996.
25. Fontes P, et al. Bone marrow-augmentation of donor-cell chimerism in kidney, liver, heart and pancreas islet transplantation. Lancet 1994; 344:151.
26. Zeevi A, et al. Immune status of recipients following bone marrow augmented solid organ transplantation. Transplantation 1995; 59:616.

45

BLOCKING OF THE B7-CD28 PATHWAY INCREASES THE CAPACITY OF FasL⁺ (CD95L⁺) DENDRITIC CELLS TO KILL ALLOACTIVATED T CELLS

Lina Lu,[1,2] Shiguang Qian,[1,2] Thomas E. Starzl,[1,2] David H. Lynch,[4] and Angus W. Thomson[1,2,3*]

[1]Thomas E. Starzl Transplantation Institute
[2]Department of Surgery and
[3]Department of Molecular Genetics and Biochemistry
University of Pittsburgh
Pittsburgh, Pennsylvania 15213
[4]Immunex Research and Development Corporation
Seattle, Washington 98101

1. INTRODUCTION

Dendritic cells (DC) are antigen presenting cells of hemopoietic origin, uniquely well-equipped to activate naive T cells.[1] Evidence also exists however, for DC tolerogenicity.[2] During primary activation, mature T cells change from an activation-induced cell death (AICD) -resistant to an AICD-sensitive phenotype.[3] The complete molecular basis for this transition remains to be determined, but CD95 (Fas/Apo-1) and CD95L (Fas ligand; FasL) appear to play an important role in the homeostatic regulation of T cell responses.[4] It seems that CD95L can mediate opposite effects (T cell activation or apoptosis) depending on the state of activation of the responding T cells. Previously, we have shown that mouse bone marrow (BM)-derived DC progenitors deficient in expression of cell surface costimulatory molecules B7–1 (CD80) and B7–2 (CD86) are capable of inducing alloantigen-specific T cell hyporesponsiveness[5] and prolonging allograft survival.[6] Others have shown recently that a major subpopulation of freshly-isolated mouse lymphoid tissue DC is FasL$^+$, and can induce apoptosis in allogeneic T cells.[7] In the course of studies on DC propagated *in vitro* from mouse BM, we observed that these cells expressed Fas L. Our aim was to determine whether these cells could induce T cell apoptosis and whether costimulatory molecule expression could affect this activity.

* Address correspondence to: Angus W. Thomson, D.Sc., Thomas E. Starzl Transplantation Institute, 15th Floor, Biomedical Science Tower, University of Pittsburgh Medical Center, 200 Lothrop Street, Pittsburgh, PA 15213.

2. MATERIALS AND METHODS

2.1. Animals

10–12-week-old male B10.BR/SnJ (BR; H-2^k), C57BL/10J (B10, H-2^b), C57BL/6J (B6; H-2^b) and B6.Smn.C3H-*gld* (B6.*gld*, H-2^b; FasL-deficient) mice were purchased from The Jackson Laboratory (Bar Harbor, ME).

2.2. Propagation and Purification of BM-Derived DC

The DC were propagated from normal B10.BR mouse BM in GM-CSF+IL-4 as described.[5,8] They were purified over metrizamide (to approximately 90% purity; in some experiments, DEC-205$^+$, B7–2$^+$ DC (purity >95%) sorted using a FACStar® plus cell sorter (Becton Dickinson & Co, Mountain View, CA) were used.[9]

2.3. Flow Cytometric Analysis of FasL Expression Using Fas-Fc Fusion Protein

FasL expression on DC was analyzed by FACS. Metrizamide-purified BM DC were first incubated with 0.1 mg/ml mouse Ig (PharMingen, San Diego, CA), 1% v/v normal goat serum (NGS; Vector Laboratories Inc., Burlingame, CA) and 50 µg/ml anti-FcR mAb 2.4 G2 (PharMingen) for 10 min at 4°C to block Fc receptor binding. Mouse Fas-Fc fusion protein (Immunex Corporation, Seattle, WA) or purified human IgG was then added at 50µg/ml and the cells incubated for 30 min. They were then washed and stained with anti-human IgG-FITC (Caltag Laboratories, South San Francisco, CA) at the same concentration.

2.4. Immunoperoxidase Staining for FasL

Fas-Fc fusion protein was also used for staining of cytospins and was added (l0µg/ml) to unfixed preparations of DC after incubation with 0.1 mg/ml mouse Ig, NGS (1%) and 50 mg/ml anti-FcR mAb 2.4G2 (Pharmingen) for 10 min at 4°C. Peroxidase-conjugated mouse anti-human IgG was then added, and the Fas-Fc fusion protein localized by the addition of chromogenic AEC substrate for 5 min.

2.5. Reverse Transcriptase PCR and DNA Sequencing

Total cellular RNA was extracted with TRIzol (Life Technologies, Gaithersburg, MD) and 2 µg was used for cDNA synthesis using SuperScript II RT (Life Technologies) and both Fas L and β-actin primers. The FasL amplified product from the purified DC cDNA was cloned into Bluescript II SK (Stratagene, La Jolla, CA) and sequenced with an automated sequencer (Applied Biosystems Inc., Foster City, CA).

2.6. Mixed Leukocyte Reactions (MLR)

These were performed using nylon wool purified splenic T cells as responders, as described previously in detail.[10]

2.7. Costimulatory Molecule Blockade

In some experiments, mouse CTLA4-Ig fusion protein (a gift from Dr. P.S. Linsley, Bristol-Myers-Squibb, Seattle, WA) or control human Ig was added at the start of the cell proliferation or DNA fragmentation assays.

2.8. Radiometric Assay for DNA Fragmentation

Splenic T cells activated either with Con A (3μg/ml; 48 h) or irradiated allogeneic spleen cells (1:1 ratio; 72 h) were labeled for 18 h with 5μCi/ml [^{3}H] thymidine at 37°C in complete medium. The target cells (10^5) were then combined with various numbers of DC or spleen cells (as effectors) in a total volume of 200μl per well in complete medium in 96 well plates and incubated at 37°C for 4 - 18 h. Unfragmented labeled high molecular weight DNA was collected by filtration through glass fiber filters (Pharmacia) and counted in a liquid scintillation counter. Data are expressed as % DNA fragmentation = 100 x (1-cpm in the presence of effectors/cpm in the absence of effectors).

2.9. Spectrofluorometric Assay for DNA Fragmentation

The protocol was as described previously in detail by McCarthy *et al.* (11).

2.10. In Situ Nick-End Labeling

DNA strand breaks in fixed cytospin preparations were identified by *in situ* terminal deoxynucleotidyl transferase (TdT)-mediated (dUTP) nick end labeling (TUNEL) (12).

3. RESULTS

3.1. Culture, Purification, and Allostimulatory Activity of Bone Marrow (BM)-Derived DC

The DC propagated in GM-CSF+IL-4 and purified as described in the Materials and Methods (DEC 205$^+$; MHC class II^{hi}, B7–2$^+$[CD86$^+$], CD40$^+$, CDllc$^+$) were highly efficient inducers of primary allogeneic T cell responses. Blockade of the B7-CD28 costimulatory pathway however, by addition of CTLA4-Ig at the start of cultures, significantly inhibited the MLR.

3.2. FasL Expression on BM-Derived DC

FasL expression was detected on DC by immunoperoxidase labeling, and by flow cytometric analysis using the mouse Fas-Fc fusion protein (Fig. 1).

3.3. RT-PCR Analysis and cDNA Cloning

The expression of message for Fas L in purified DC was confirmed by RT-PCR analysis. Levels of FasL mRNA in DC were similar to those expressed by Con A-activated T cells. The FasL amplified product from the DC cDNA was cloned into Bluescript II SK, sequenced and found to be identical to the mouse FasL sequence deposited in GenBank, Accession No. U06948.

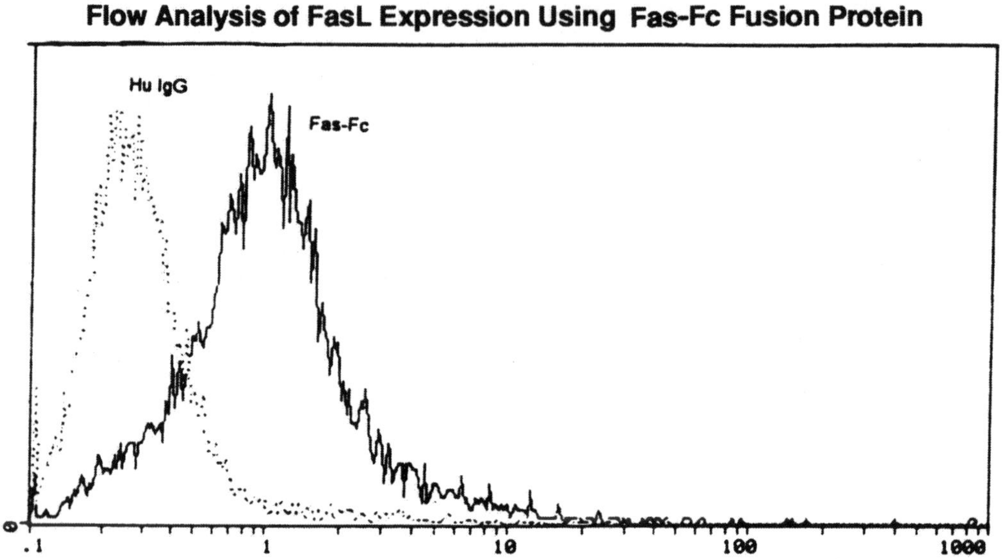

Figure 1. Immunofluorescence staining for FasL on in vitro generated DC using Fas-Fc fusion protein and detected by flow cytometric analysis.

3.4. Induction of DNA Fragmentation in Fas$^+$ Jurkat T Cells and Its Inhibition by Fas-Fc Fusion Protein

DC propagated from the BM of "wild-type" B6 mice but not those from *gld* (FasL-deficient) mice induced dose-related levels of DNA fragmentation. Similar data were obtained using DC propagated from B10.BR BM. Killing was partially blocked however, by the addition of Fas-Fc fusion protein at the start of the assay (data not shown).

3.5. Induction of DNA Fragmentation in Activated T Cells and Its Enhancement by CTLA4-Ig

Con A-stimulated syngeneic (Bl0.BR) or allogeneic (B10) T cell blasts or allo-activated T cells were labeled with [^{3}H] thymidine and used as targets. Cultured DC or fresh bulk spleen cells from B10.BR mice were also used as effectors. The DC induced only a low level of DNA fragmentation. DC from B10.BR.BM induced similar levels of killing in B10 (allogeneic) or B10.BR (syngeneic) Con A blasts, indicating that the induction of killing by DC was not MHC restricted. The extent of DNA fragmentation was equal to or higher than that achieved with fresh bulk splenic effector cells. When the DC were incubated with T cells in the presence of CTLA4 - Ig, the effect was increased up to 4-fold. Similar results were obtained using alloactivated T cells as targets (Fig. 2).

3.6. Death of Activated T Cells Induced by DC Is Mediated by the Fas/FasL Pathway

The allostimulatory activity of DC from FasL-deficient mice was significantly higher and associated with lower DNA fragmentation of activated T cell targets compared

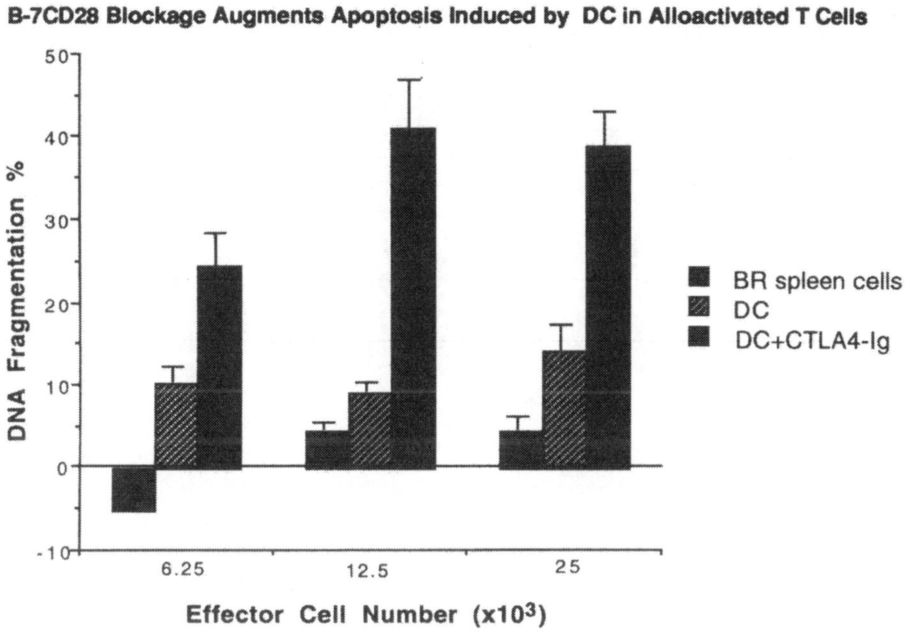

Figure 2. DNA fragmentation induced in alloactivated (B10) T cells by normal (BR) mouse BM-derived DC in the absence or presence of CTLA4-Ig. Addition of CTLA4-Ig (200ng/ml) at the start of the cultures markedly increased the death-inducing activity of the DC. Results are means ± 1SD and are representative of 3 separate experiments.

with DC from wild type B6 mice. However, increased levels of killing were induced by *gld* DC when CTLA4-Ig was added to the cultures (Fig. 3). It appears therefore, that Fas/FasL is not the only pathway by which DC can induce apoptosis of activated T cells.

4. DISCUSSION

This study shows for the first time, that co-stimulatory (B7) and death-signaling (FasL) molecules expressed on *in vitro* propagated DC play counter-regulatory roles in determining activated T cell survival and proliferation. That the death-inducing capacity of the *in vitro* generated DC is linked to cell surface expression of FasL was shown by immunocytochemical techniques and by the capacity of Fas-Fc fusion protein to block T cell killing by the DC. These observations are consistent with a recent report[7] that a major FasL$^+$ subpopulation of freshly-isolated mouse lymphoid tissue DC (CD8$^+$ FasL$^+$) can induce apoptosis in activated T cells. In contrast to the DC studied by Süss and Shortman,[7] however, the FasL$^+$ DC that we generated *in vitro* were CD8$^-$. Thus it appears that the apoptosis-inducing activity of DC and the expression of CD8 (also present on mouse "veto" cells [13]) are not strictly inter-related. Others have shown that monocytes/macrophages, to which DC are related, and with which they may share a common precursor, can promote apoptosis in activated T cells via expression of Fas L.[14,15] Thus, cultured M-CSF-derived macrophages induce selective depletion of allospecific T cells (15). An alternative/additional mechanism by which macrophages have been shown to inhibit T cell

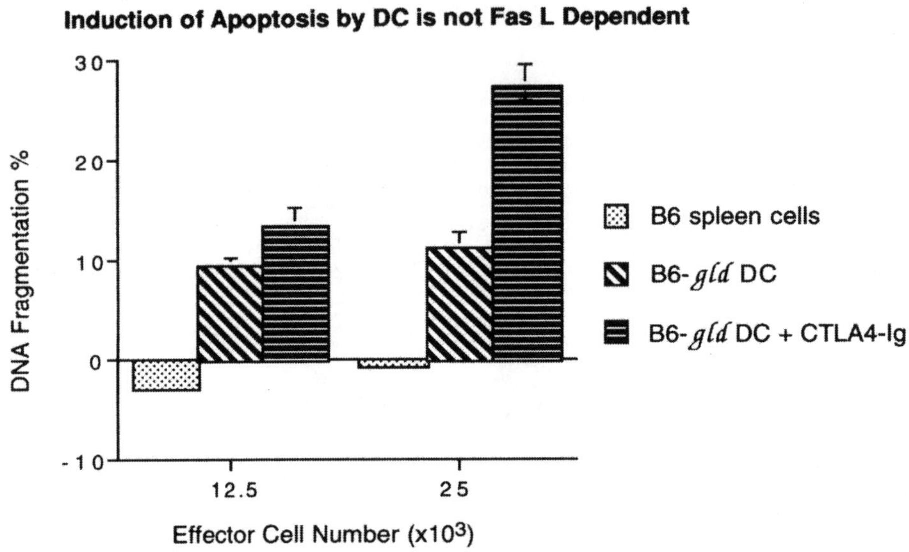

Figure 3. DNA fragmentation in Con A blasts following 4h incubation with FasL-deficient (B6.*gld*) DC in the absence or presence of CTLA4-Ig. Blockade of the B7-CD28 costimulatory pathway by CTLA4-Ig was associated with increased T cell death. Results are means ± 1SD and are representative of 2 separate experiments.

proliferation/induce apoptosis is by the secretion of NO.[9,16,17] NO may also be induced in DC by IFN-γ, endotoxin or interaction with allogeneic T cells.[9]

The Fas (APO-1; CD95) FasL system has emerged as an important pathway regulating the induction of T cell apoptosis[18–20] and has been implicated in the deletion of graft-infiltrating cells in immunologically privileged sites.[21] Both Fas and FasL are induced on T cells upon activation, resulting in cell death by suicide or fratricide. In the present study, the expression of FasL on *in vitro* generated mouse DC was linked to the ability of these cells to induce low levels of killing in activated T cells. No T cell death significantly above background levels could be induced however, by DC propagated from FasL-deficient (B6.*gld*) mice. The capacity of CTLA4-Ig to markedly enhance the ability of potent, antigen-presenting DC to induce DNA fragmentation in T cells implicates a role for CD28 costimulation in preventing T cell death. Others studying both human and mouse systems, have shown recently that CD28 costimulation enhances T cell survival following their activation by TCR cross-linking. The resistance conferred by costimulation has been linked to increased expression of the survival gene bcl-x_L within the T cells.[22]

Our finding that DC generated from BM of FasL-deficient mice can induce apoptosis in activated T cells in the presence of CTLA4-Ig suggests that Fas-independent pathways are also involved in DC-induced apoptosis. Indeed, it has also been shown that Fas is not necessary for activation-induced T cell death (23). We are investigating the expression of TNF family members[23] on DC and in addition, the influence of other costimulatory pathways, such as CD40/CD40L and IL-12/IL-12R on the resistance of activated T cells to apoptosis.

The *in vivo* relevance of FasL expression on DC remains to be elucidated and such studies are clearly warranted. The potential of DC to induce death of activated T cells in the absence of costimulation may have key implications for interpretation of the documented tolerogenic properties of DC. Since cells engineered to express FasL, have immunosuppressive properties,[24] the therapeutic potential of Fas L^+ APC, in particular appropriately "engineered" DC is worthy of evaluation.

ACKNOWLEDGMENTS

We thank Ms. Alison Logar for flow cytometric analysis, Dr. Youping Li for immunocytochemical staining, Dr. William A. Rudert for RT-PCR analysis, Dr. Peter S. Linsley (Bristol-Myers-Squibb, Seattle, WA) for mouse CTLA4-Ig, Dr. Pam Hershberger for valuable discussion and Miss Shelly Lynn Conklin for expert secretarial assistance. The work was supported by National Institutes of Health grants DK 49745-OIA1 and DK 29961-14.

REFERENCES

1. Steinman, R.M. 1991. The dendritic cell system and its role in immunogenicity. *Annu Rev Immunol* 9:271–296.
2. Thomson, A.W., L. Lu, N. Murase, A.J. Demetris, A.S. Rao, and T.E. Starzl. 1995 Microchimerism, dendritic cell progenitors and transplantation tolerance. *Stem Cells* 13:622–639.
3. Green, D.R., and D.W. Scott. 1994. Activation-induced apoptosis in lymphocytes. *Curr. Opin. Immunol.* 6:476–487.
4. Lynch, D.H., F. Ramsdell, and M.R. Alderson. 1995. Fas and FasL in the homeostatic regulation of immune responses. *Immunol. Today* 16:569–574.
5. Lu, L., D. McCaslin, T.E. Starzl, and A.W. Thomson. 1995. Mouse bone marrow-derived dendritic cell progenitors (NLDC 145^+, MHC class II^+, B7-1^{dim}, B7–2^-) induce alloantigen-specific hyporesponsiveness in murine T lymphocytes. *Transplantation* 60:1539–1545.
6. Fu, F., Y. Li, S. Qian, L. Lu, F. Chambers, T.E. Starzl, J.J. Fung, and A.W. Thomson. 1996. Costimulatory molecule-deficient dendritic cell progenitors (MHC class II^+, B7-^{dim}, B7–2^-) prolong cardiac allograft survival in non-immunosuppressed recipients. *Transplantation* 62:659-665.
7. Süss, G., and K. Shortman. 1996. A subclass of dendritic cells kills CD4 T cells via Fas/Fas ligand-induced apoptosis. *J. Exp. Med.* 183:1789–1796.
8. Lu, L., W.A. Rudert, S. Qian, D. McCaslin, F. Fu, A.S. Rao, M. Trucco, J.J. Fung, T.E. Starzl, and A.W. Thomson. 1995. Growth of donor-derived dendritic cells from the bone marrow of liver allograft recipients in response to granulocyte/macrophage colony-stimulating factor. *J.Exp.Med.* 182:379–387.
9. Lu, L., C.A. Bonham, F.G. Chambers, S.C. Watkins, R.A. Hoffman, R.L. Simmons, and A.W. Thomson. 1996. Induction of nitric oxide synthase in mouse dendritic cells by interferon γ, endotoxin and interaction with allogeneic T cells: nitric oxide production is associated with dendritic cell apoptosis. *J. Immunol.* In press.
10. Lu, L., J. Woo, A.S. Rao, Y. Li, S.C. Watkins, S. Qian, T.E. Starzl, A.J. Demetris, and A.W. Thomson. 1994. Propagation of dendritic cell progenitors from normal mouse liver using GM-CSF and their maturational development in the presence of type-1 collagen. *J Exp. Med.* 179:1823–1834.
11. McCarthy, S.A., R.N. Cachione, M.S. Mainwaring, and J.S. Cairns. 1992. The effects of immunosuppressive drugs on the regulation of activation-induced apoptotic cell death in thymocytes. *Transplantation* 54:543–547.
12. Gavrieli, Y., Y. Sherman, and S.A. Ben-Sasson. 1992. Identification of programmed cell death *in situ* via specific labeling of nuclear DNA fragmentation. *J. Cell Biol.* 119:493–501.
13. Vremec, D., M. Zorbas, R. Scollay, D.J. Saunders, C.F. Ardavin, L. Wu, and K. Shortman. 1992. The surface phenotype of dendritic cells purified from mouse thymus and spleen: investigation of the CD8 expression by a subpopulation of dendritic cells. *J. Exp. Med.* 176:47–58.
14. Wu, M.X., Z. Ao, M. Hegen, C. Morimoto, and S.F. Schlossman. 1996. Requirement of Fas (CD95), CD45, and CD11a/CD18 in monocyte-dependent apoptosis of human T cells. *J. Immunol.* 157: 707–713.
15. Munn, D.H., J. Pressey, A.C. Beall, R. Hudes, and M.R. Alderson. 1996. Selective activationinduced apoptosis of peripheral T cells imposed by macrophages. *J. Immunol.* 156:523–532.
16. Hauschildt, S., Bessler, W.G., and P. Scheipers. 1993. Engagement of major histocompatibility complex class II molecules leads to nitrite production in bone marrow-derived macrophages. *Eur. J. Immunol.* 23 :2988–2992.

17. Tian, L., R.J. Noelle, and D.A. Lawrence. 1995. Activated T cells enhance nitric oxide production by murine splenic macrophages through gp39 and LFA-1. *Eur. J. Immunol.* 25:306-309.
18. Ju, S.-T., D.J. Panka, H. Cui, R. Ettinger, M. El-Khatib, D.H. Sherr, B. Z. Stanger, and A. Marshak-Rothstein. 1995. Fas (CD95)/FasL interactions required for programmed cell death after T-cell activation. *Nature* 373:444–448.
19. Nagata, S. 1994. Fas and Fas ligand: a death factor and its receptor. Adv. *Immunol.* 57: 129144.
20. Alderson, M.R., T.W. Tough, T. Davis-Smith, S. Braddy, B. Falk, K.A. Schooley, R.G. Goodwin, C.A. Smith, F. Ramsdell, and D.H. Lynch. 1995. Fas ligand mediates activation-induced cell death in human T lymphocytes. *J. Exp. Med.* 181:71–77.
21. Griffith, T.S., T. Brunner, S.M. Fletcher, D.R. Green, and T.A. Ferguson. 1995. Fas ligand induced apoptosis as a mechanism of immune privilege. *Science* 270:1189–1192.
22. Boise, L.H., A.J. Minn, P.J. Noel, C.H. June, M.A. Accavitti, T. Lindsten, and C.B. Thompson. 1995. CD28 costimulation can promote T cell survival by enhancing the expression of Bcl XL. *Immunity* 3: 87–98.
23. Zheng, L., G. Fisher, R.E. Miller, J. Peschon, D.H. Lynch, and M.J. Lenardo. 1995. Induction of apoptosis in mature T cells by tumour necrosis factor. *Nature* 377:348–351.
24. Lau, H.T., M Yu, A. Fontana, and C.J. Stoeckert, Jr. 1996. Prevention of islet allograft rejection with engineered myoblasts expressing FasL in mice. *Science* 273: 109–112.

46

GENERATION OF NITRIC OXIDE BY MOUSE DENDRITIC CELLS AND ITS IMPLICATIONS FOR IMMUNE RESPONSE REGULATION

C. Andrew Bonham,[1] Lina Lu,[1] Rosemary A. Hoffman,[1] Richard L. Simmons,[1] and Angus W. Thomson[1,2]

[1]Department of Surgery
[2]Department of Molecular Genetics and Biochemistry
University of Pittsburgh
Pittsburgh, Pennsylvania 15213

1. INTRODUCTION

Nitric oxide (NO) is a ubiquitous compound produced by the oxidative metabolism of L-arginine.[1] A variety of functions have been ascribed to it, including neurotransmission, and mediation of smooth muscle relaxation.[2,3] In the immune system, NO is involved in tumoricidal and microbicidal activity, and inhibition of T cell proliferation.[4-7] There are at least three isoforms of nitric oxide synthase (NOS), the cell type-specific enzyme which produces NO.[8] In cells of the immune system, such as macrophages and neutrophils, NOS activity is induced by a number of proinflammatory mediators such as IFN-γ, IL-1, TNF-α, and microbial cell wall products, such as BCG and LPS.[9,10] Dendritic cells (DC) are the most potent known antigen presenting cells capable of activating naive T cells.[11] Mature DC in secondary lymphoid tissues express high levels of major histocompatibility complex (MHC) and costimulatory molecules. DC residing in nonlymphoid tissue, such as the Langerhans cells in the dermis, typically express lower levels of these molecules. However, in response to stress (such as LPS or inflammatory cytokines), these 'immature' cells migrate to T dependent areas in lymphoid tissue, and upregulate expression of MHC and costimulatory molecules.[12,13] Macrophages, which may share a common ontogeny with DC, respond to these cytokines in part by the generation of nitric oxide.[9] However, there are no reports of NO production by DC stimulated by cytokine or LPS. In this study we present evidence that purified mouse bone marrow-derived DC express an inducible NOS which produces NO in response to IFN-γ, LPS, or upon interaction with allogeneic T cells. The synthesis of NO is associated with apoptosis in the DC, and inhibition of proliferation of allogeneic T cells.

Dendritic Cells in Fundamental and Clinical Immunology
edited by Ricciardi-Castagnoli, Plenum Press, New York, 1997

2. MATERIALS AND METHODS

2.1. Animals

Ten to twelve-week old B10.BR (H-2^k, I-E^k) and C57BL/10J (B10; H-2^b, I-A^b) mice were obtained from The Jackson Laboratory, Bar Harbor, Maine, and maintained in the specific pathogen-free facility of the University of Pittsburgh Medical Center.

2.2. Reagents

Recombinant (r) mouse GM-CSF and IL-4 were gifts of Dr. S. K. Narula (Schering-Plough; Kenilworth, NJ). r mouse IFN-γ was from Gibco BRL (Grand Island, NY), LPS (Escherichia coli 011:B4 L3012) from Sigma Chemical Co. (St. Louis, MO), and N^G-momomethyl-L-arginine (NMMA) from Cyclops Biochemical Corp. (Salt Lake City, UT). The NO generating compound *S*-nitoso-*N*-acetyl-penicillamine (SNAP) was synthesized as described.[14]

2.3. Culture and Purification of DC

Mouse bone marrow cells were cultured *in vitro* with the method modified[15] after that first described by Inaba etal.[16] Briefly, 2×10^6 cells were cultured in 24-well plates in RPMI-1640 (Gibco BRL) supplemented with 10% v/v fetal calf serum (Nalgene, Miami, FL), 1000 U/ml r mouse GM-CSF and 1000 U/ml IL-4. Nonadherent cells recovered after 5–6 days of culture were purified on a metrizamide gradient as described elsewhere.[17] Purity of the DC population was verified by flow cytometric analysis with an extensive panel of mAbs as previously described (17). DC antigen presentation was determined in 3 day primary MLR with purified naive splenic T cells as responders.

2.4. Sorting of DC

Cultured bone marrow-derived DC were incubated in 10% v/v normal goat serum (Vector Laboratories, Burlingame, CA) in HBSS (Gibco BRL), at 4° C for 30 min to inhibit nonspecific antibody binding. The cells were then washed and resuspended in buffer at 2×10^7 cells/ml. A saturating concentration of the DC specific antibody NLDC-145[18] (a gift of Dr. R. M. Steinman, The Rockefeller University, New York, NY) or rat IgG2a isotype control (Sigma) was added, and the cells were incubated for 45 min at 4°C. After washing twice with HBSS, the secondary antibody, PE conjugated goat anti-rat IgG (CALTAG, San Francisco, CA) was added as for the primary antibody. After washing, the cells were stained with FITC-conjugated anti-B7–2 mAb or FITC-conjugated rat IgG2a isotype control (both from PharMingen, San Diego, CA). After the final wash, cells were resuspended at 5×10^6/ml in PBS-0.1% BSA. DEC-205^+/B7–2^+ and DEC-205^+/B7–2^- cells were sorted with a FACStar plus® cell sorter (Becton Dickinson & Co.) based on forward and side scatter and FITC/PE fluorescent intensity to obtain a purity >95%.

2.5. Measurement of NO

NO production was measured by assaying the stable end product NO_2^- using the diazotization reaction of Griess.[19] 100μl aliquots of culture supernatant were incubated with an equal volume of Griess reagent (1% sulfanilamide; 0.1% naphthylethylene diamine dihydrochloride; 2.5% H_3PO_4) for 10 min at room temperature. The absorbence was measured at 570 nm in an automated plate reader, with concentration determined with reference to a standard curve.

2.6. Intracellular Staining of NOS

2×10^6 cells/ml were fixed and stained by a modification of the paraformaldehyde-saponin procedure as described elsewhere.[20] Polyclonal rabbit anti-mouse inducible (i) NOS (Transduction Laboratories, Lexington, KY) was added at 2μg/ml at room temperature for 30 min. The cells were washed and incubated with Cy3.18-conjugated goat anti-rabbit Ab (Jackson Immunoresearch Labs,. Inc.) for 1 h. Slides were washed and mounted with buffered glycerol. Cells were imaged with a Nikon FXL microscope equipped with differential interference contrast (DIC) and epifluorescent optics, and the images were digitally collected and superimposed to obtain a combined fluorescent and DIC morphologic image.

2.7. Mixed Leukocyte Reactions (MLR)

Sorted DC or freshly-isolated splenocytes were γ-irradiated (20 Gy) and cultured with 2×10^5 nylon wool column-purified T cells in 0.2 ml RPMI-1640 complete medium 10% FCS in 96 well, round bottom plates in 5% CO_2 in air at 37° for 3–5 d. NMMA (0.5mM) was added to selected wells to inhibit NO release. 10μl [^{3}H]TdR (1μCi) was added to each well for the final 18 h of culture. Cells were harvested with a multiple cell harvester onto glass fiber disks, and the degree of thymidine incorporation was determined in a liquid scintillation counter. Culture supernatants were assayed for nitrite production.

2.8. Identification of Apoptosis

DNA strand breaks were detected using DNA polymerase and biotin-labeled dUTP in the in situ nick translation as described elsewhere.[21] Incorporated biotin-dUTP was detected by peroxidase-labeled avidin followed by an enzyme reaction using DAB as substrate.

3. RESULTS

3.1. Induction of NO Production by DC Stimulated with LPS and IFN-γ

Sorted GM-CSF + IL-4 stimulated DEC-205$^+$/B7–2$^+$ bone marrow-derived DC free of contaminating macrophages, B cells, and granulocytes, were exposed to LPS (10μg/ml), IFN-γ (500U/ml), or the two agents combined for 48 h. These mediators are known to induce NO synthesis in macrophages and neutrophils at similar concentrations.[9] NO production was determined by measuring the accumulation of the stable end product NO_2^- in culture supernatants. As shown in Figure 1A, NO production by DC was stimulated by either agent alone or in combination. No significant NO release was measured in cultures of similar numbers of IFN-γ or LPS-stimulated mouse splenocytes. Furthermore, addition of the NOS inhibitor NMMA (0.5μM) at the start of culture blocked NO release.

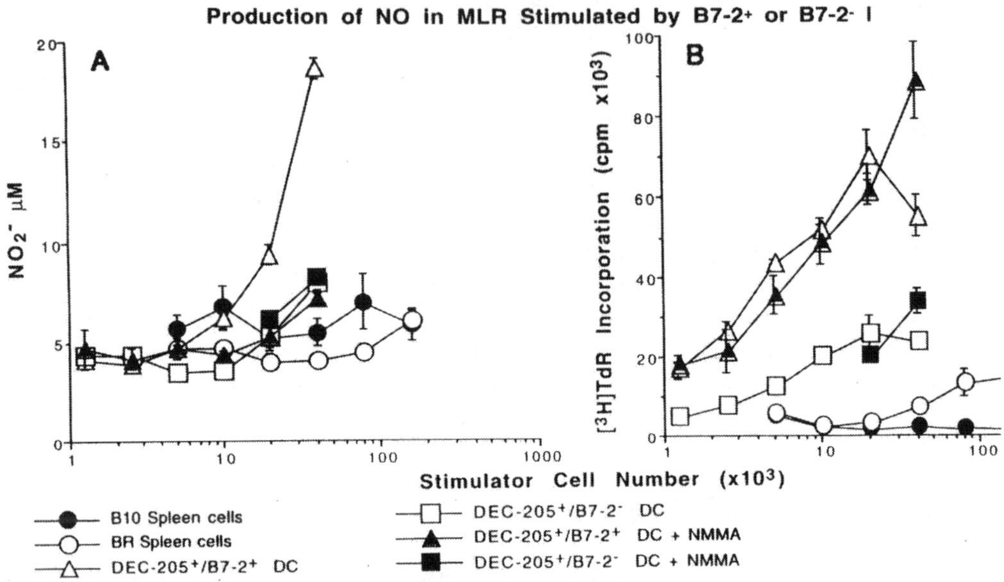

Figure 1. (A), NO_2^- concentrations in supernatants of mouse bone marrow-derived DC or splenocytes cultured in medium alone or with IFN-γ (500U/ml), LPS (10µg/ml) or both combined for 72 h. Selected cultures also contained NMMA (0.5µM) to inhibit NO production by iNOS. All stimulated DC cultures contained high levels of NO_2. NMMA inhibited accumulation of NO2- in these cultures. (B), NO_2^- accumulation in cultures of B7–2$^+$ sorted DC (Δ), B7–2$^-$ sorted DC (□), or spleen cells (○) was measured after stimulation of the cells for 48 h with IFN-γ and LPS. NO2- accumulated in all DC cultures regardless of the presence or absence of costimulatory molecules.

3.2. Induction of NO Production in Costimulatory Molecule Deficient DC

B7–2$^+$ DC are potent antigen presenting cells. In contrast, B7–2$^-$ DC are at best weak stimulators of naive T cells, and have been demonstrated to induce allospecific hyporesponsiveness.[22] However, when cultured for 48 h in IFN-γ and LPS, both DEC-205$^+$/B7–2$^+$ and DEC-205$^+$/B7–2$^-$ cells released comparable amounts of NO (Figure 1B).

3.3. Intracellular Staining of NOS in Stimulated DC

Purified GM-CSF + IL-4 stimulated DC were exposed to LPS and IFN-γ for 18 h, then stained with rabbit anti-mouse iNOS followed by Cy3.18-labeled goat anti-rabbit IgG. A subpopulation of approximately 15–20% of the DC were iNOS positive. Although convincing DC morphology was demonstrated for iNOS positive cells, many cells exhibited characteristics of apoptosis (nuclear and cytoplasmic condensation, and membrane blebbing) (data not shown).

3.4. NO Production by DC Interacting with Allogeneic T Cells and Its Effects in MLR

Purified, γ-irradiated B10.BR DC were cultured with naive B10 T cells in a 3 d primary MLR. NO accumulation in the culture supernatants was proportional to the number of DC stimulators present in the culture, and similar to that found in cultures of DC stimulated by IFN-γ. At high DC:T cell ratios, suboptimal T cell activation was associated with

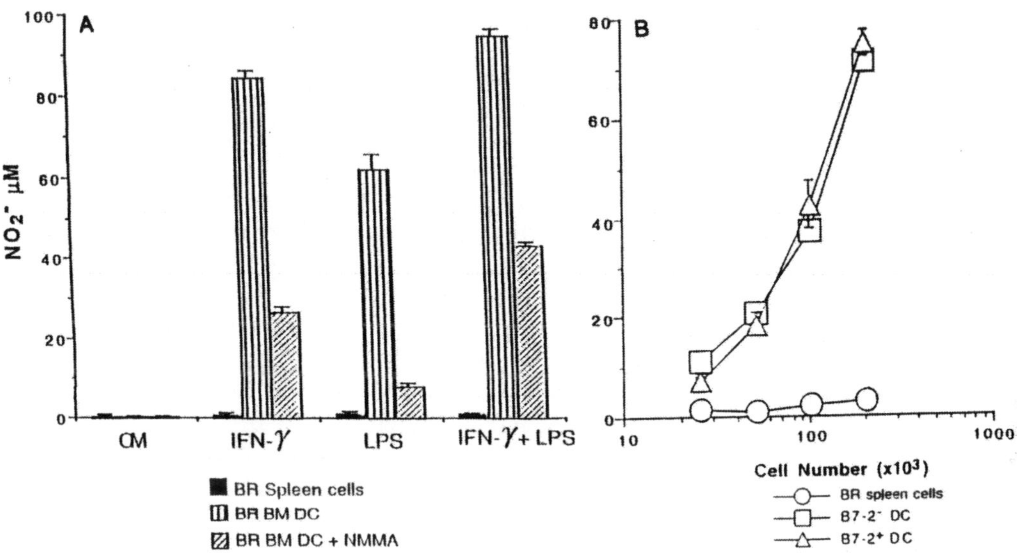

Figure 2. (A), NO production in primary MLR. B10.BR γ-irradiated stimulators were cultured with B10 T cells for 72 h. NO_2^- accumulation in cultures stimulated by DEC-205⁺/B7–2⁺ DC (Δ) was associated with high numbers of stimulators (DC). Addition of NMMA (solid icons) inhibited NO release by DC. Cultures stimulated by DEC-205⁺/B7–2⁻ sorted DC (□) (with or without NMMA) failed to accumulate appreciable levels of NO_2^-, as did cultures stimulated by allogeneic (O) or syngeneic (●) splenocytes. (B), Allostimulatory activity of sorted DC was measured by thymidine incorporation by purified naive B10 splenic T cells in 2 or 3 d MLR. NO_2^- accumulation, as noted in Fig. 2A, correlated with reduced T cell proliferation. NMMA restored the allostimulatory activity at high DC concentrations by inhibiting NO release.

an increasing NO concentration in the culture (Figure 2). Addition of NMMA at initiation of the MLR inhibited the release of NO, abrogating its effect on T cell proliferation as seen by the increased DNA synthesis exhibited by these cells. Syngeneic T cell responders did not stimulate DC NO synthesis, nor did they respond in the MLR. These findings indicate that NO production following allogeneic DC:T cell interaction inhibits T cell proliferation.

3.5. NO Induces Apoptosis in DC

Macrophages may undergo apoptosis when induced to release NO by IFN-γ or LPS.[23] Apoptosis in DC stimulated by IFN-γ and LPS for 48 h was confirmed by detection of DNA strand breaks by in situ nick translation. Addition of NMMA to these cultures inhibited apoptosis, supporting the role for endogenous NO production in DC apoptosis. Confirmatory evidence for NO-mediated DC apoptosis was provided by culturing cells in the presence of the NO donor SNAP in the continued presence of GM-CSF. Apoptosis in DC was induced by NO in a dose dependent manner (data not shown).

4. DISCUSSION

In this study we present evidence for the presence of an inducible NOS in highly purified populations of DEC-205$^+$ DC propagated from mouse bone marrow in response to GM-CSF + IL-4. Culture of DC in the presence of LPS, IFN-γ, or both induced the release of NO. Both B7–2$^+$ and B7–2$^-$ DC were capable of generating NO in response to these molecules. NMMA, known for its ability to inhibit iNOS, prevented the release of NO by DC. The iNOS protein was visualized intracellularly with a rabbit polyclonal antibody. Interestingly, only a subpopulation of the IFN-γ and LPS-stimulated DC were shown to express iNOS in their cytoplasm. As in macrophages, DC iNOS activity may correlate inversely with their life span in culture. The inducibility of iNOS in DC may be affected by the phase of the cell cycle.

When cultured in the presence of allogeneic T cells, DC release NO. Addition of NMMA to the MLR blocks NO production. At high DC:T cell ratios, T cell proliferation declines in correlation with rising levels of NO. NMMA abrogates this effect. In contrast, B7–2$^-$ DC do not release NO when cultured with allogeneic T cells, suggesting that a prerequisite for induction of iNOS in DC may be successful activation of T lymphocytes. Activated T helper 1 (Th1) cells produce IFN-γ, and have been shown to induce NO synthesis in Kupffer cells.[24] Similarly, DC may be induced to release NO by IFN-γ produced by T cells interacting with the DC. Furthermore, signaling through cell membrane components such as gp 39 or LFA-1 upon direct contact of DC with activated T cells may stimulate NO production.[25] An analogous mechanism has been described for macrophages induced to release NO by isolated T cell membranes.[26]

The NO donor SNAP induced apoptosis in DC in a dose-dependent fashion. Additionally, endogenous production of NO by DC was noted to correlate with DC apoptosis. Morphologically, many of the cells staining positive for iNOS exhibited signs of programmed cell death. The implication of this observation is that DC producing NO may themselves be targets of its effects.

The significance of NO production by DC *in vivo* is not clear. The observations presented here indicate that unstimulated and stimulated (by LPS or IFN-γ) DC may have opposite effects on T cell responses. Whereas the initial encounter between DC and T cells may result in vigorous immune activation, as the event unfolds, IFN-γ released by stimulated T cells may induce DC NO production. Rising levels of NO released by DC may serve to regulate the immune reaction by inhibiting further T cell proliferation. Furthermore, T cells and thymocytes may actually undergo apoptosis upon exposure to NO.[27] Apoptosis in a subpopulation of DC may serve to eliminate a powerful source of antigen presentation and T cell activation. In addition, NO may prevent maturation of costimulatory molecule deficient DC, as has been shown by the ability of alveolar macrophages to inhibit maturation of GM-CSF cultured DC in a NO-dependent manner.[28] The existence of a complex feedback loop mediated by NO may have implications for regulation of the immune response.

5. ACKNOWLEDGMENTS

We thank Dr. Simon C. Watkins for invaluable assitance with immunohistochemistry and image analysis, and Dr. Youping Li for assistance with staining for apoptosis.

This work was supported in part by the National Institutes of Health Grants R01-DK49745–01A1 and R37-AI-16869–16.

6. REFERENCES

1. Albina JE, Mills CD, Henry WL Jr, Caldwell MD. Regulation of macrophage physiology by L-arginine: role of the oxidative L-arginine deaminase pathway. J Immunol 1989; 143: 3641.
2. Ignarro LJ, Buga GM, Wood KS, Byrns RE, Chaudhuri G. Endothelium-derived relaxing factor produced and released from artery and vein is nitric oxide. Proc Natl Acad Sci USA 1987; 84: 9265.
3. Bredt DS, Snyder SH. Nitric oxide mediates glutamate-linked enhancement of cGMP levels in the cerebellum. Proc Natl Acad Sci USA 1989; 86: 9030.
4. Stuehr DJ, Nathan CF. Nitric oxide: a macrophage product responsible for cytostasis and respiratory inhibition in tumor target cells. J Exp Med 1990; 169: 1543.
5. Granger DL, Hibbs JB Jr, Perfect JR, Durack DT. Specific amino acid (L-arginine) requirement for the microbiostatic activity of murine macrophages. J Clin Invest 1988; 81: 1129.
6. Albina JD, Abate JA, Henry WL Jr. Nitric oxide production is required for murine resident peritoneal macrophages to suppress mitogen-stimulated T cell proliferation. Role of IFN-gamma in the induction of the nitric oxide synthesizing pathway. J Immunol 1991; 147: 144.
7. Denham S, Rowland IJ. Inhibition of the reactive proliferation of lymphocytes by activated macrophages: the role of nitric oxide. Clin Exp Immunol 1992; 87: 157.
8. Stuehr DJ, Griffith OW. Mammalian nitric oxide synthases. Adv Enzymol Relat Areas Mol Biol 1992; 65: 287.
9. Stuehr DJ, Marletta MA. Induction of nitrite/nitrate synthesis in murine macrophages by BCG infection, lymphokines, or IFN-γ. J Immunol 1987; 139: 518.
10. Deng W, Thiel B, Tannenbaum CS, Hamilton TA, Stuehr DJ. Synergistic cooperation between T cell lymphokines for induction of the nitric oxide synthase gene in murine peritoneal macrophages. J Immunol 1993; 151: 322.
11. Steinman RM. The dendritic cell system and its role in immunogenicity. Annu Rev Immunol 1991; 9: 271.
12. Larsen CP, Ritchie SC, Hendrix R, et al. Regulatory of immunostimulatory function and costimulatory molecule (B7–1 and B7–2) expression on murine dendritic cells. J Immunol 1994; 152: 5208.
13. Roake JA, Rao AS, Morris PJ, Larsen CP, Hankins DF, Austyn JM. Dendritic cell loss from nonlymphoid tissues after systemic administration of lipopolysaccharide, tumor necrosis factor, and interleukin 1. J Exp Med 1995; 181: 2237.
14. Field L, Dilts RV, Ravichandran R, Lenhert PG, Carnahan GE. An usually stable thionitrite from *N*-acetyl-D,L-penicillamine; X-ray crystal and molecular structure of 2-(acetylamino)-2-carboxy-1, 1-dimethylethyl thionitrite. J C S Chem Comm 1978; 1978: 249.
15. Lu L, Rudert WA, Qian S, et al. Growth of donor-derived dendritic cells from the bone marrow of murine liver allograft recipients in response to granulocyte/macrophage colony-stimulating factor. J Exp Med 1995; 182: 379.
16. Inaba K, Inaba N, Romani N, et al. Generation of large numbers of dendritic cells from mouse bone marrow cultures supplemented with granulocyte/macrophage colony-stimulating factor. J Exp Med 1992; 176: 1693.
17. Woo J, Lu L, Rao AS, et al. Isolation, immunophenotype and allostimulatory activity of murine liver dendritic cells. Transplantation 1994; 58: 484.
18. Swiggard WJ, Mirza A, Nussenzweig MC, Steinman RM. DEC-205, a 205-kDa protein abundant on mouse dendritic cells and thymic epithelium that is detected by the monoclonal antibody NLDC-145: purification, characterization, and N-terminal amino acid sequence. Cell Immunol 1995; 165: 302.
19. Billiar TR, Curran RD, Stuehr DJ, West MA, Bentz BG, Simmons RL. An L-arginine-dependent mechanism mediates Kupffer cell inhibition of hepatocyte protein synthesis in vitro. J Exp Med 1989; 169: 1467.
20. Sander B, Andersson J, Andersson U. Assessment of cytokines by immunofluorescence and the paraformaldehyde-saponin procedure. Immunol Rev 1991; 119: 65.
21. Gavrieli Y, Sherman Y, Ben-Sasson SA. Identification of programmed cell death in situ via specific labeling of nuclear DNA fragmentation. J Cell Biol 1992; 119: 493.
22. Lu L, McCaslin D, Starzl TE, Thomson AW. Bone marrow-derived dendritic cell progenitors (NLDC-145$^+$, MHC class II$^+$, B7–1^{dim}, B7–2$^-$) induce alloantigen-specific hyporesponsiveness in murine T lymphocytes. Transplantation 1995; 60: 1539.
23. Albina JE, Cui S, Mateo RB, Reichner JS. Nitric oxide-mediated apoptosis in murine peritoneal macrophages. J Immunol 1993; 150: 5080.
24. Roland CR, Walp L, Stack RM, Flye MW. Outcome of Kupffer cell antigen presentation to a cloned murine Th1 lymphocyte depends on the inducibility of nitric oxide synthase by IFN-γ. J Immunol 1994; 153: 5453.

25. Tian L, Noelle RJ, Lawrence DA. Activated T cells enhance nitric oxide production by murine splenic macrophages through gp 39 and LFA-1. Eur J Immunol 1995; 25: 306.
26. Fao X, Stout RD. T cell-mediated cognate signaling of nitric oxide production by macrophages. Requirements for macrophage activation by plasma membranes isolated from T cells. Eur J Immunol 1993; 23: 2916.
27. Fehsel K, Kroncke K-D, Meyer KL, Humber H, Wahn V, Kolb-Bachofen V. Nitric oxide induces apoptosis in mouse thymocytes. J Immunol 1995; 155: 2858.
28. Holt PG, Oliver J, Bilyk N, et al. Down-regulation of the Ag presenting cell function(s) of pulmonary dendritic cells in vivo by resident alveolar macrophages. J Exp Med 1993; 177: 397.

47

DENDRITIC CELLS IN THE AUTOIMMUNE INSULITIS IN NOD MOUSE MODELS OF DIABETES

Judith G. M. Rosmalen,[1] Pieter J. M. Leenen,[1] Jonathan D. Katz,[2]
Jane S. A. Voerman,[1] and Hemmo A. Drexhage[1]

[1]Department of Immunology
Erasmus University Rotterdam
The Netherlands
[2]IGBMC, BP 163
Illkirch, C.U. de Strasbourg, France

1. INTRODUCTION

Dendritic cells (DC) are thought to play important roles as antigen presenting cells in the autoimmune insulitis preceding insulin dependent diabetes mellitus (IDDM). Not only are DC the first cells appearing in the pancreatic inflammatory process in animal models of diabetes (NOD mouse[1,2] and BB rat[3,4]), DC are also abundantly present in progressed human insulitis[5]. Although insulitis in the NOD mouse starts around 3 weeks, overt diabetes will not develop until an age of 20 weeks. The lag between occult insulitis and diabetes, which can also be observed in human IDDM, was always considered to reflect the time needed for accumulation of a sufficient number of destructive T cells in the islets.

A new NOD mouse model of diabetes has been developed based on the transgenic expression of the rearranged T cell receptor (TCR) genes from the diabetogenic T cell clone BDC2.5[6]. These TCR transgenic NOD mice have been shown to develop extensive insulitis followed by diabetes. Remarkably, in this NOD mouse model the inflammatory insulitis but not the onset of diabetes is dramatically accelerated. Since virtually all T cells in the transgenic NOD mouse express the potentially diabetogenic BDC2.5 T cell receptor, we reason that clonal expansion of specific T cells is unnecessary in this model. Therefore, we hypothesize that there might be less need for 'professional' presentation of pancreatic islet antigens by DC. To investigate the involvement of DC and other lymphoid cells in the inflammatory process, we performed a detailed immunohistochemical analysis of the islet infiltrates in both TCR transgenic and regular NOD mice.

Table 1. Antibodies used for immunohistochemical study

MoAb	Specificity	Source
RA3.6B2	B220, CD45R	R. Coffman
BM8	Mϕ with phagocytic capacity[1]	BMA
ER-MP23	Mouse Mϕ galactose-/N-acetyl galactosamine-specific C-type lectin[1]	P. Leenen
KT3	CD3	ATCC
N418	CD11c[1]	ATCC
NLDC-145	Dendritic cells; DEC205[1]	ATCC

[1]Overview in: Leenen PJM, de Bruijn MFTR, Voerman JSA, Campbell PA, van Ewijk W (1994). Markers of mouse macrophage development detected by monoclonal antibodies. J Immunol Meth 174:5-19.

2. RESULTS AND DISCUSSION

2.1. Insulitis in Regular NOD Mice

In regular NOD mice, six stages of insulitis can be recognized (Table 1)[1]. The insulitis process starts with a swelling of the vessels near the islets and a perivascular increase of different DC and Mϕ populations, characterized by CD11c, ER-MP23 or BM8 expression. At 7 weeks of age, ER-MP23$^+$ Mϕ and CD11c$^+$ DC accumulate around the islets and are the first hematopoietic cells detectable at these locations. Lymphocytes, entering after this accumulation of Mϕ and DC, were initially situated around the islets. In later stages of the insulitis process, ER-MP23$^+$ Mϕ as well as lymphocytes infiltrate the islets. However, only in the diabetes-prone female NOD mice, BM8$^+$ Mϕ are found to infiltrate the islets. Therefore, especially the BM8$^+$ Mϕ are thought to play a role in the destruction of the β cells. True insulitis involving β cell destruction occurs from 15 weeks.

2.2. Insulitis in BDC2.5 TCR transgenic NOD mice

In the TCR-transgenic NOD mice, we find a fast developing inflammatory process starting already at 3 weeks of age. In contrast to the distinct stages of insulitis in the regular NOD mouse, a simultaneous instead of sequential recruitment of DC/Mϕ, T and B cells can be seen. Islets are surrounded by large lymphocytic infiltrates primarily consisting of

Table 2. Insulitis in regular NOD mice

Stage (age)	Characteristic features
0 (<wk 3)	Intact islet as observed in non-diabetes prone mice
I (wk 3)	Intact islet, but swollen vessels and perivascular increase of CD11c$^+$, ERMP23$^+$, MOMA-1$^+$ and BM8$^+$ Mϕ and DC
II (wk 4-7)	Islet surrounded by DC and Mϕ with above mentioned phenotypes; infiltration of islets by CD11c$^+$ DC; lymphocytes are still absent
III (wk 7-10)	Parainsular recruitment of CD4$^+$ and CD8$^+$ T lymphocytes and some B lymphocytes to the accumulation of DC and Mϕ
IV (wk 10-17)	Lymphocytes surround the islet (peri-insulitis)
V (>wk 17)	Lymphocytes and ER-MP23$^+$ and BM8$^+$ Mϕ infiltrate the islet
VI (>>wk 17)	Endstage; no insulin-producing cells detectable

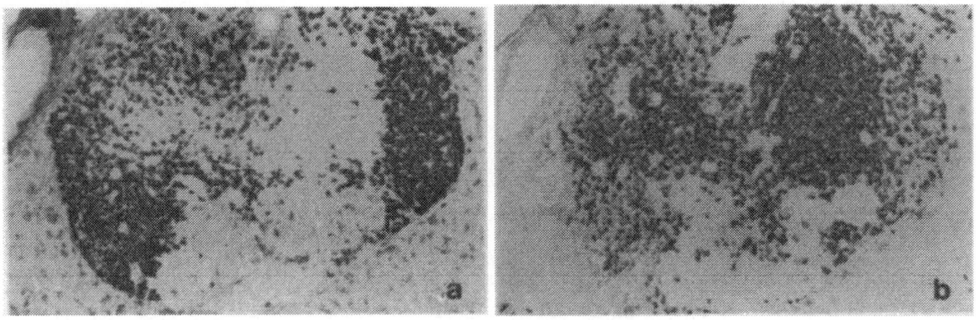

Figure 1. B and T cell staining of an infiltrated islet in a female BDC2.5 TCR transgenic NOD mouse at 3 weeks of age. Note that B and T cells are localized in distinct areas within the infiltrate. Original magnification 90X. (a) B cell staining using anti-B220 (Table 2). (b) T cell staining using anti-CD3 (Table 2).

(transgenic) T cells and B cells in partially segregated areas (Figure 1). The T cell areas are characterized by the presence of high endothelial venules (HEV). Two distinct DC subsets are found to be present in separate locations. CD11c$^+$ DC are found especially at the interface of intact β cells and T cells as well as in the inflamed islet mass; NLDC-145$^+$ DC are found within the T cell areas (Figure 2). In contrast to the regular NOD insulitis, BM8$^+$ scavenger Mϕ are scarce in inflamed islets in the TCR transgenic mice. The peri-insulitis slowly protrudes into the islets, and almost no insulin-producing β cells can be detected in the TCR transgenic NOD mice from 20 weeks of age. In conclusion, insulitis and subsequent β cell destruction in the regular NOD mouse and in the BDC2.5 TCR transgenic NOD mouse follow different kinetics. In both models of type 1 diabetes, however, DC accumulate in the earliest stage of inflammation. Furthermore, DC remain present throughout the process of β cell destruction, suggesting an important role for DC in the initiation and progression of the autoimmune process.

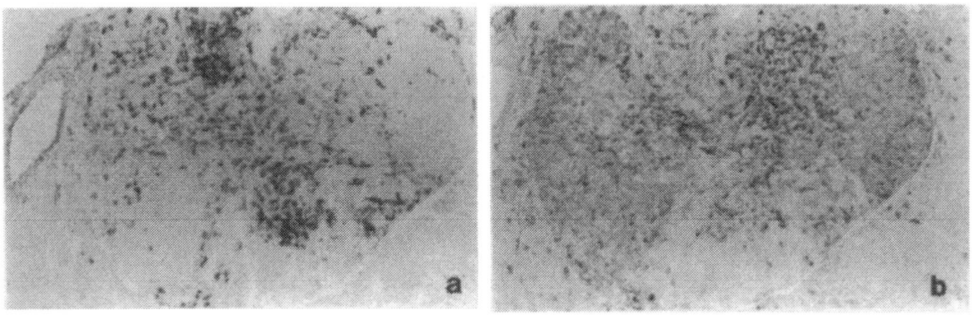

Figure 2. DC staining of the same infiltrated islet as in Figure 1. CD11c$^+$ and NLDC-145$^+$ DC are localized in partially distinct areas within the infiltrate. Original magnification 90X. (a) CD11c staining (Table 2). CD11c$^+$ DC are especially found at the interface of intact β cells and T cells as well as in the T cells areas. (b) NLDC-145 staining (Table 2). NLDC-145$^+$ DC are especially found within the T cell areas.

ACKNOWLEDGMENTS

Work in our laboratory is supported by the Dutch Diabetic Fund.

REFERENCES

1. Jansen A, Homo-Delarche F, Hooijkaas H, Leenen PJ, Dardenne M, Drexhage HA (1994). Immunohistochemical characterization of monocytes-macrophages and dendritic cells involved in the initiation of the insulitis and β cell destruction in NOD mice. Diabetes 43:667–675
2. Lee KU, Amano K, Yoon JW (1988). Evidence for initial involvement of macrophage in development of insulitis in NOD mice. Diabetes 37:989–991
3. Voorbij HAM, Jeucken PHM, Kabel PJ, De Haan M, Drexhage HA (1989). Dendritic cells and scavenger macrophages in pancreatic islets of prediabetic BB rats. Diabetes 38: 1623–1629
4. Hanenberg H, Kolb-Bachofen V, Kantwerk-Funke G, Kolb H (1989). Macrophage infiltration precedes and is prerequisite for lymphocytic insulitis in pancreas of prediabetic BB rats. Diabetologia 32:126
5. Jansen A, Voorbij HAM, Jeucken PHM, Bruining GJ, Hooijkaas H, Drexhage HA (1993). An immunohistochemical study on organized lymphoid cell infiltrates in fetal and neonatal pancreases. A comparison with similar infiltrates found in the pancreas of a diabetic infant. Autoimmunity 15:31–38
6. Katz, JD, Wang B, Haskins K, Benoist C, Mathis D (1993). Following a diabetogenic T cell from genesis to pathogenesis. Cell 74:1089–1100

48

DENDRITIC CELLS IN INFLAMMATORY RESPONSES IN THE CNS

M. K. Matyszak* and V. H. Perry

Department of Pharmacology
Oxford University
Mansfield Rd., Oxford OX1 3QT

1. INTRODUCTION

There are a number of bone marrow derived mononuclear phagocytes associated with the central nervous system. These include: microglia in the brain parenchyma, perivascular macrophages and macrophages in the meninges and in the choroid plexus[1]. MHC class II is not expressed on microglia in the normal rodent brain, but it can be rapidly upregulated during an inflammatory response. Other macrophages associated with the CNS express MHC class II constitutively. Although the expression of MHC antigens is a prerequisite for antigen presentation, it is not by itself sufficient to stimulate T cell responses. *In vitro* studies by Fontana and his colleagues showed that both astrocytes and microglia purified from the brains of new born animals, cultured and stimulated with INF-γ can present an antigen to already primed T cells[2]. In contrast, recent *in vitro* studies have shown that MHC class II+ microglia rapidly isolated from the adult rat brain are poor accessory cells, even when presenting an antigen to primed T cells, but other macrophages associated with the CNS can efficiently stimulate already primed T cells[3]. Using bone marrow chimeras Hickey and Kimura showed that perivascular macrophages are important accessory cells for the induction of immune responses in the CNS following peripheral activation of T cells[4]. However, it is still debatable whether any of the MHC class II+ cells in the CNS can induce a primary immune response.

2. IS THERE A POPULATION OF DENDRITIC CELLS ASSOCIATED WITH THE CNS?

Progenitor dendritic cells have been found in most non-neuronal tissues[5]. Early studies indicated that these cells are absent from the CNS. However, since dendritic cell spe-

* Correspondence to: Dr. M.K. Matyszak, Department of Pharmacology, Oxford University, Mansfield Rd., Oxford OX1 3QT. Tel: (01865) 271587; Fax: (01865) 271853, e-mail: Matyszak@vax.ox.ac.uk.

cific antibodies were not used in those studies, we have recently carried out a detailed immunohistochemical analysis with OX62 mAb on rat neuronal tissues. OX62 mAb recognises the α-like integrin subunit present on dendritic cells. OX62+ cells purified from either the lymphoid tissues or peripheral blood are able to stimulate primary responses[6,7]. Our results showed dendritic cells are absent from the CNS parenchyma, but we detected a population of OX62+ cells in the meninges and the choroid plexus[8]. In the choroid plexus these cells constituted only 2% of the total number of MHC class II positive cells. However, immunohistochemistry alone is insufficient to demonstrate the accessory function of these cells and further functional assessment is needed.

3. INFLAMMATORY RESPONSES TO BACILLUS CALMETTE-GUÉRIN IN THE BRAIN PARENCHYMA

Bacillus Calmette-Guérin (BCG) sequestered in the non-neuronal tissues, for example the peritoneal cavity, induces early recruitment of immature dendritic cells into the site of BCG deposits[9]. Dendritic cell recruitment was shown to precede the entry of neutrophils and monocytes. Once within the inflammatory site, they phagocytose the foreign antigen and migrate into local lymphoid tissues where they can present the antigen to naive T cells, stimulating an antigen specific immune response. We studied the recruitment of dendritic cells into the CNS parenchyma after a single intracerebral (i.c.) injection of heat-killed BCG[10]. Animals were injected stereotaxically with 10^5 organisms of heat-killed BCG. BCG in the brain parenchyma induced a rapid acute inflammatory response which was comparable with that in the skin. The neutrophil response resolved within the first week. The monocyte response was delayed for approximately 24–48 hours and was less pronounced when compared with skin lesions induced by intradermal injection of the same dose of BCG. Staining with OX62 mAb revealed no OX62+ cells at the site of BCG deposits in the brain at early time points. There was also only an occasional OX62+ cell at the height of acute inflammatory response.

By four weeks post injection there was little sign of an inflammatory response in the CNS and the blood-brain barrier re-sealed. However, the myelomonocytic response in the CNS parenchyma failed to clear the BCG. Immunogold labelling of ultrathin CNS sections with anti-BCG antibody revealed the presence of BCG in the lysosomes and lipofuscin granules in mononuclear phagocytes within the parenchyma and in perivascular macrophages[11]. Long-term experiments showed that BCG can persist in the brain for months undetected by the immune system.

It has been previously shown that proteins such as ovalbumin, injected into the brain parenchyma, drain into the deep cervical lymph nodes. Anti-ovalbumin antibodies were detected in the serum of these animals[12]. We have tested sera from animals which received an i.c. injection of 10^5 organisms BCG for the presence of antibodies to purified protein derivatives (PPD), the major protein component of the mycobacteria cell wall. Blood samples were collected from experimental animals at different times after i.c. injection starting from day seven to six weeks and the sera were tested for anti-PPD antibodies. Control groups of animals which received intradermal injection of 10^5 organisms of BCG were tested in parallel. Antibody responses were measured using immunoblots. There was a detectable level of antibodies to PPD in 50% of animals, seven days after intradermal injection of BCG. By 14 days onwards all animals which received intradermal injection tested positive for antibody to PPD. However, none of the animals which received an i.c. injection of BCG was tested positive at any time after the injection (Matyszak and Perry, unpublished observation).

We have also studied T cell responses to PPD using the tuberculin test. Animals were injected i.c. with heat-killed BCG and two weeks later they received a subcutaneous challenge with 2 μg of PPD. Animals were killed 72 hours later, and the PPD injected skin was assessed immunohistochemically. No inflammatory lesions were detected at the site of the PPD inoculation in any of the animals studied. In contrast, control animals which received an intradermal injection followed by PPD challenge all tested positive with the tuberculin test (Matyszak and Perry, unpublished).

It is apparent that under our experimental conditions, the BCG sequestered in the CNS parenchyma is not recognised by the immune system. The resident microglia and perivascular macrophages do not have the capacity to initiate a primary T cell response which would subsequently lead to the induction of a DTH response in the brain. At present it is not clear whether the failure to sensitise the immune system to pathogens sequestered in the CNS parenchyma is at the level of antigen presentation or is due to the inability of resident cells in the CNS to transport the BCG into the local lymphoid organs.

4. DENDRITIC CELLS IN IMMUNE MEDIATED RESPONSES IN THE CNS

4.1. Immune Responses to BCG Sequestered in the CNS Parenchyma

As discussed above, pathogens such as BCG can persist sequestered in the CNS parenchyma for months, unrecognised by the immune system. However, we have shown that if an animal with intraparenchymal BCG deposits is at some later time challenged subcutaneously with BCG, an immune response is evoked in the CNS at the site of original injection[13]. The CNS lesion is composed predominantly of T cells and macrophages. Dendritic cells (OX62+ cells) were also detected in these lesions[8]. OX62+ cells were detected in the brain as early as day 8 after subcutaneous challenge with BCG and their entry correlated with the infiltration of the first T cells. At day 8 the number of OX62+ cells in CNS lesions was relatively low. However, by day 12 their number increased approximately ten-fold (Fig.1a). The presence of dendritic cells was further confirmed with electron microscopy. Dendritic cells were characterised by a pale cytoplasm with perinuclear accumulation of the organelles and organelle-free rounded processes. Perinuclear organelles were rich in large mitochondria and rough endoplasmic reticulum. Many of these cells were seen in close contact with lymphocytes (Fig.1b). After the first two weeks the number of DCs decreased rapidly. At three weeks there were approximately 50% less OX62+ cells in these lesions than at day 12, although there was no significant change in the number of T cells; and in many lesions there was a noticeable increase in the total number of MHC class II+ inflammatory cells. OX62+ cells which remained in the CNS lesions persisted for months.

4.2. Dendritic Cells in Experimental Allergic Encephalomyelitis (EAE) Lesions

We have also studied the presence of OX62+ cells in acute EAE lesions[8]. EAE was induced by subcutaneous injection of myelin basic protein in complete Freund's adjuvant in Lewis rats. This protocol gives rise to a monophasic disease characterised by multiple perivascular lesions, most commonly in the brain stem and the spinal cord. Animals often suffer from clinical signs varying from limb weakness to paralysis. Most animals recover over the period of four to five days and they become refractory to subsequent episodes.

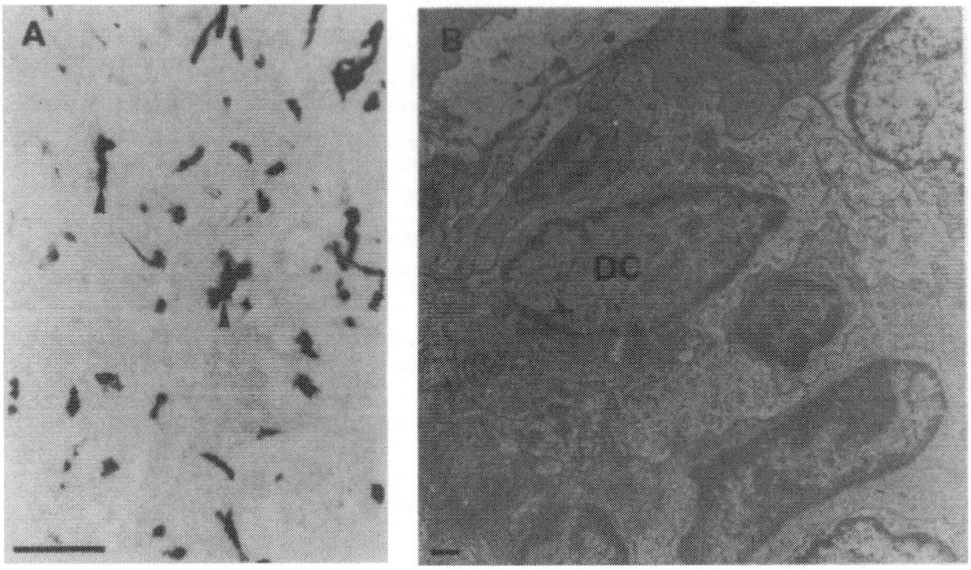

Figure 1. Presence of dendritic cells in delayed-type hypersensitivity lesions directed against BCG in the brain. **A.** OX62+ cells in a typical DTH lesion, 12 days after subcutaneous injection of BCG. **B.** An electron micrograph of a DTH lesion with a dendritic cell in the middle of the field. The DC shows extensive round processes which are organelle free and an organelle-rich perinuclear area. Two lymphocytes are in close contact with the DC. Scale bars, 50 μm (A), 1 μm (B).

EAE lesions were composed predominantly of T cells and inflammatory macrophages. There was also a wide spread activation of microglia. There was a small number of OX62+ cells in many but not all EAE lesions. OX62+ cells were most numerous in perivascular cuffs (approximately 2% of all leucocytes) (Fig. 2). In the brain parenchyma there were only few of these cells and their number declined rapidly with distance from the perivascular cuff. The number of OX62+ cells in an EAE lesions was significantly smaller than in DTH lesions described above.

5. WHAT IS THE POSSIBLE FUNCTION OF DENDRITIC CELLS IN IMMUNE MEDIATED INFLAMMATORY RESPONSES IN THE CNS?

Dendritic cells can contribute to the pathogenesis of inflammatory responses in the CNS in a number of ways. They may contribute to the chronicity of the disease by *in situ* activation of T lymphocytes. This may be of special importance in the light of recent studies which suggest that microglia suppress T cell activation[3]. Furthermore, studies by Ibrahim *et al.*[14] suggest that dendritic cells are activated by tissue injury. In chronic responses in the CNS directed against BCG there is extensive damage to the blood-brain barrier and neuronal tissue, including myelin loss. Studies are now underway in our laboratory as to whether damaged CNS tissue may be taken up by dendritic cells in the CNS lesion and subsequently presented to immune cells, leading to the induction of an autoimmune response. The results will help in better understanding of the pathogenesis of some CNS diseases, particularly multiple sclerosis.

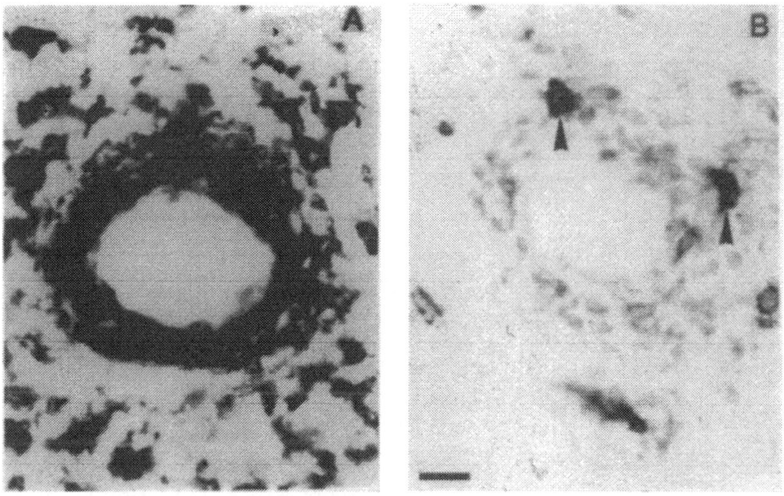

Figure 2. A typical EAE lesion, 12 days after subcutaneous injection of MBP in CFA. **A**. OX6 staining showing the thick perivascular cuff. **B**. the same perivascular cuff as in (A) stained with OX62 mAb. Note that OX62+ cells constitute only a small subpopulation of MHC class II+ cells in an EAE lesion. Scale bar, 10 μm.

REFERENCES

1. Perry, V.H., and S. Gordon. 1991. Macrophages and the nervous system. *Int-Rev-Cytol.* 125:203.
2. Fontana, A., K. Frei, S. Bodmer, and E. Hofer. 1987. Immune mediated encephalitis: On the role of antigen-presenting cells in brain tissue. *Immunol. Rev.* 100:185.
3. Ford, A.L., A.L. Goodsall, W.F. Hickey, and J.D. Sedgwick. 1995. Normal adult ramified microglia separated from other central nervous system macrophages by flow cytometric sorting. Phenotypic differences defined and direct ex vivo antigen presentation to myelin basic protein-reactive CD4+ T cells compared. *J. Immunol* 154, 4309.
4. Hickey, W.F., and H. Kimura. 1988. Perivascular microglial cells of the CNS are bone marrow-derived and present antigen in vivo. *Science* 239, 290.
5. Steinman, R.M. 1991. The dendritic cell system and its role in immunogenicity. *Annu. Rev. Immunol.* 9:271.
6. Brenan, M., and M. Puklavec. 1992. The MRC OX-62 antigen: A useful marker in the purification of rat veiled cells with the biochemical properties of an integrin. *J. Exp. Med.* 175:1457.
7. Chen-Woan, M., C.P. Delaney, V. Fournier, Y. Wakizaka, N. Murase, J. Fung, T.E. Starzl, and A.J. Demetris. 1996. In vitro characterization of rat bone marrow-derived dendritic cells and their precursors. *J. Leucocyte Biol.* 59:196.
8. Matyszak, M.K., and V.H. Perry. 1996. The potential role of dendritic cells in immune mediated inflammatory responses in the central nervous system. *Neuroscience*. 74, 599.
9. van Vugt, E., M.A. Verdaasdonk, R.H. Beelen, and E.W. Kamperdijk. 1992. Induction of an increased number of dendritic cells in the peritoneal cavity of rats by intraperitoneal administration of Bacillus Calmette-Guerin. *Immunobiology* 186, 230.
10. Matyszak, M.K., and V.H. Perry. 1996. A comparison of leucocyte responses to heat-killed bacillus Calmette-Guérin in different CNS compartments. *Neuropathol. Appl. Neurobiol.* 22:44.
11. Matyszak, M.K., M.J. Townsend, and V.H. Perry. 1996. Ultrastructural studies of an immune mediated inflammatory response in the CNS parenchyma directed against a non-CNS antigen. *Neuroscience*, in press.
12. Cserr, H.F., and P.M. Knopf. 1992. Cervical lymphatics, the blood-brain barrier and the immunoreactivity of the brain: a new view. *Immunol. Today* 13, no. 12:507.
13. Matyszak, M.K., and V.H. Perry. 1995. Demyelination in the central nervous system following a delayed-type hypersensitivity response to bacillus Calmette-Guérin. *Neuroscience* 64, no. 4:967.
14. Ibrahim, M.A.A., B.M. Chain, and D.R. Katz. 1995. The injured cell: the role of the dendritic cell system as a sentinel receptor pathway. *Immunol. Today* 16, no. 4:181.

TH-1/TH-2 SWITCH REGULATION IN IMMUNE RESPONSES TO INHALED ANTIGENS

Role of Dendritic Cells in the Aetiology of Allergic Respiratory Disease

P. G. Holt, C. Macaubas, D. Cooper, D. J. Nelson, and A. S. McWilliam

Division of Cell Biology
TVW Telethon Institute for Child Health Research
PO Box 855, West Perth
Western Australia 6872

1. INTRODUCTION

Until comparatively recently, allergic (atopic) disease was viewed as a manifestation of hyperreactivity to essentially non-pathogenic soluble protein antigens which are ubiquitous in the natural environment. In this context, normality would equate to non-responsiveness, resulting from either tolerance or ignorance of the antigens. However, it is now clear (reviewed in[1]) that active T cell immunity to at least one class of these antigens (airborne "inhalant" allergens) is essentially universal amongst adults, clinical reactivity being a function of the cytokine profiles of $CD4^+$ Th-cells which dominate relevant Th-memory populations. Thus, atopics who respond to allergen exposure via IgE production, eosinophilia etc., manifest Th-2-skewed memory, whereas T cell memory in non-responsive normal adults is dominated by Th-1-cytokines such as IFNγ. The situation with respect to ingested environmental allergens (ubiquitous in the diet) appears both qualitatively and quantitatively different, as T cell reactivity to this class of antigens is considerably less frequent in the adult population[1,2].

2. ANIMAL MODEL STUDIES

2.1. Antigen Inhalation

The mucosal immunology literature has provided significant insights into the mechanism(s) involved in regulation of T cell responses to these two classes of antigens. Our laboratory has had a long-standing interest in the role of immune deviation mechanisms in regulation of primary immunity to antigens presented via respiratory mucosal surfaces. Our studies have defined a "default" response in normal immunocompetent animals undergoing aerosol antigen exposure that involves an initial burst of Th-2-dependent IgE production in regional lymph nodes draining the airway mucosa, which (if exposure continues) is terminated by the development of Th-1-skewed memory dominated by MHC class I restricted CD8$^+$ T cells[1,3,4]. The latter cells respond *in vitro* in an antigen-specific fashion to the eliciting (exogenous) antigen, provided IL-2 help from MHC class II restricted CD4$^+$ T-cells is present[4]. In the model we have described, the latter cells are responsive to the same antigen[4]; moreover, during this period, IL-4 dependent IgE production occurs within the same lymphoid microenvironment, and it is thus likely that this CD4$^+$ help emanates from IL-2/IL-4 producing Th-0 cells[1]. With continuing airborne antigen exposure, the CD8$^+$ population becomes progressively less dependent upon exogenous IL-2 help, and develops the capacity to selectively suppress Th-2 responses in adoptive recipients[2,4]; flow cytometric analytic studies on this CD8$^+$ population indicate an important contribution from CD8$^+$ TcRγ/δ cells in this process[5,6]. We have hypothesized[4] that with long-term antigen exposure, mimicking the situation pertaining to ubiquitous environmental antigens, these Th-1-skewed CD8$^+$ regulator T cells may "select" for CD4$^+$ Th-1 cells of the same specificity, which would eventually dominate the memory compartment.

Two key observations from these models[1] which are relevant to more recent studies in humans; (a) once Th-2 memory is consolidated, antigen inhalation *per se* appears incapable of deviation of the ongoing response towards Th-1, and (b) this immune deviation mechanism only functions in immunocompetent adult animals being essentially inoperative during infancy.

2.2. Antigen Feeding

The parallel process of Oral Tolerance which underlies regulation of immune responses to ingested antigens, initially described over 80 years ago, has enjoyed a rebirth through the recent autoimmunity literature, which is driven by the demonstration that controlled feeding of autoantigens has significant therapeutic potential in several disease models. The nature of the underlying mechanisms are controversial, particularly regarding the induction of tolerance to exogenous antigens, with reports of both Th-2-skewed and Th-1-skewed immunity being dominant in the process[7]. Part of this controversy may eventually be resolved by further studies on the role of antigen dosage in selection of different classes of regulatory mechanisms, in view of recent demonstrations that low-level exposure triggers exclusively immune deviation mechanisms whereas high dose feeding leads to development of T cell anergy and eventually T cell deletion[8].

It is also important to note that, analogous to the situation reported above for inhaled antigens, Oral Tolerance mechanisms function extremely poorly in the early postnatal period[9].

3. DEVELOPMENT OF T-CELL IMMUNITY TO ENVIRONMENTAL ALLERGENS IN HUMANS

3.1. Seroepidemiology

There is now a considerable body of evidence (reviewed in[1]) that antibody production against environmental allergens is initiated in early postnatal life. Responses to dietary antigens proceed most rapidly, reflecting the relatively high levels of stimulation from gastrointestinal exposure; IgG1 responses typically peak in both affinity and titre during infancy, whereas IgG4 responses often demonstrate slow affinity maturation with age and are maintained at low titres throughout adulthood. IgG subclass antibody production against common inhalant allergens is also usually initiated within 3 months of birth[10], even though inhalation exposure is only in the nanogram-microgram range, and low levels of antibody production are common throughout life in both atopics and non-atopics.

Prospective studies on age-related changes in allergen-specific IgE titres in panels of infants have been particularly instructive. These studies[1,11] have demonstrated that IgE antibody responses to food antigens are initiated in most children within 3 months of birth, but are almost invariably biphasic and disappear over the ensuing 2–3 years, consistent with the onset of some form of selective tolerance. IgE responses to inhalant allergens are usually not measurable until 1–2 years of age, but again occur at low titre in most young children, and are also typically biphasic and disappear over the next 3–4 years. However, a subset of these allergen-specific responses in a subset of children do not switch off, but instead are progressively amplified during childhood and persist into adulthood, reflecting an apparent failure of the regulatory mechanisms which terminate the same responses in normal children. These persisting IgE responses are associated with the manifestation of allergic diseases such as rhinitis and asthma, in late childhood and adulthood.

3.2. Allergen-Specific T-Cell Function in Early Life

Recent studies (reviewed in[12]) from a number of laboratories suggest that initial weak priming of the T cell system to these antigens occurs transplacentally, as evidenced by the presence of T cells in cord blood which proliferate in response to both dietary and inhalant allergens; in contrast, responses to vaccine antigens to which pregnant women are rarely exposed, are not seen[2]. Further studies in older age groups indicate that as per antibody production, lymphoproliferative responses to food antigens typically peak during infancy, and by adulthood <20% of individuals display (weak) responses; the spread of T cell epitopes recognised *in vitro* is also inversely related to age[13].

These findings are consistent with a dominant role for T cell deletion mechanisms in regulation of T cell responses to dietary antigens. In contrast, both the magnitude and frequency of T cell responses to inhalant allergens increase with age, to the extent that virtually 100% of adults respond vigorously[2]; taken in conjunction with the T cell cloning literature indicating that these latter responses segregate into Th-1 versus Th-2 skewed cytokine patterns, the principal regulatory mechanism would appear to be classical immune deviation[1,2]. This disparity is consistent with findings from the animals models cited above, viz. that low-level inhalant allergen exposure (ng-μg range) selectively triggers immune deviation mechanisms, whereas food antigen exposure would cover a wider exposure range up to that required to trigger T cell deletion—the much lower overall prevalence of food allergy may be a reflection of the availability of a series of overlapping T cell control mechanisms, as opposed to reliance solely on immune deviations for regulation of responses to inhalants.

4. TH-1/TH-2 SWITCH REGULATION IN EARLY LIFE: ARE THE DICE LOADED?

Epidemiologists since the early 1970s have recognised that early postnatal exposure to relatively high levels of allergen, birth during the pollen season being the best studied example, is associated with markedly increased risk of pollen/allergy in adulthood. (These findings echo the failure of tolerance/immune deviation mechanisms in the respiratory and gastrointestinal tracts of rodents cited above.) This can now be restated in more precise immunological terms: early postnatal stimulation of the weakly primed immune system, predisposes to positive selection for Th-2-skewed memory.

Why is this so? A likely explanation can be found in the rapidly evolving literature on the immunology of pregnancy. It is now evident[14] that the cytokine milieu at the feto-maternal interface is constitutively skewed towards Th-2, an adaptation which is theorised to provide protection of the fetoplacental unit against the toxic effects of Th-1 cytokines[15]. The Th-1/Th-2 "balance" typical of immunocompetent adults is not established until after infancy in mice, as shown by the preferential expansion of Th-2-memory cells in immunised animals[16]. It is also clear that this phenomenon is not restricted to mice, as human infants are also deficient (relative to adults) in production of Th-1 cytokines such as IFNγ (reviewed in[17]) Interestingly, this skew away from Th-1 appears most marked in infants genetically at high risk for atopy[18–21].

5. TH-1/TH-2 SWITCH REGULATION IN INFANCY: ANTIGEN PRESENTING CELLS (APC) AS THE LIMITING FACTOR?

It has been known since the mid 1980s that the reduced capacity of human infant T cells to produce IFNγ in response to polyclonal stimuli is due in part to deficient accessory cell function, as supplementation of neonatal T cell cultures with adult APCs markedly increases cytokine production[17]. More importantly, it has recently been demonstrated that the phenomenon of neonatal tolerance is explicable on the basis of selective priming of Th-2-memory cells during neonatal antigen stimulation, leading to subsequent down-regulation of Th-1-dependent responses following rechallenge in adulthood via immune deviation.[22,23] Additionally, the tolerance phenomenon is bypassed if neonates are primed with antigen-pulsed syngenic Dendritic Cells (DC), which promotes adult-equivalent levels of Th-1-memory development[23].

These results have potentially important ramifications with respect to Th-memory development against environmental allergens in humans. The findings discussed above in the human system argue strongly that the principal mechanism in humans responsible for "protection" against the postnatal development of pathogenic Th-2-skewed inhalant allergen-specific memory is immune deviation, a process involving antigen-driven (and hence APC-dependent) T cell selection during early childhood, when the weakly primed T cell system confronts inhaled environmental allergens via the respiratory mucosa. We have presented evidence that the principal APCs regulating this process are airway intraepithelial DC[24] which traffic in large numbers per day between the mucosa and regional lymph nodes[25]. However, this DC network is virtually non-existent at birth, and is established slowly between birth and weaning, a period during which MHC class II expression, GM-CSF responsiveness, chemokine responsiveness, and APC function are all markedly reduced relative to immunocompetent adults[26,27].

We speculate that the kinetics of postnatal maturation of the function(s) of this airway intraepithelial DC network may thus be an important factor in the outcome of Th1/Th-2 memory cell selection in early immune responses to airborne environmental allergens, and further, that variations in the efficiency of this maturation process may be a key determinant of genetic risk of atopy.

6. ACKNOWLEDGMENTS

The authors' laboratory is supported by Glaxo Wellcome, and the National Health and Medical Research Council of Australia.

7. REFERENCES

1. Holt PG. Immunoprophylaxis of atopy: light at the end of the tunnel? Immunol Today 1994; 15:484–9.
2. Holt PG, O'Keeffe PO, Holt BJ, Upham JW, Baron-Hay MJ, Suphioglu C, Knox B, Stewart GA, Thomas WR, Sly PD. T-cell "priming" against environmental allergens in human neonates: sequential deletion of food antigen specificities during infancy with concomitant expansion of responses to ubiquitous inhalant allergens. *Ped Allergy Immunol* 1995; 6:85–90.
3. Sedgwick JD, Holt PG. Induction of IgE-secreting cells and IgE isotype-specific suppressor T cells in the respiratory lymph nodes of rats in response to antigen inhalation. Cell Immunol 1985; 94:182–94.
4. McMenamin C, Holt PG. The natural immune response to inhaled soluble protein antigens involves major histocompatibility complex (MHC) class I-restricted CD8+ T cell-mediated but MHC class II-restricted CD4+ T cell-dependent immune deviation resulting in selective suppression of IgE production. J Exp Med 1993; 178:889–99.
5. McMenamin C, Pimm C, McKersey M, Holt PG. Regulation of $CD4^+$-TH-2-dependent IgE responses to inhaled antigen in mice by antigen-specific γ/δ T-cells. Science 1994; 265:1859–1861.
6. McMenamin C, McKersey M, Kuhnlein P, Hunig T, Holt PG. γ/δ T-cells downregulate primary IgE responses in rats to inhaled soluble protein antigens. *J Immunol* 1995; 154:4390–94.
7. Thomas WR, Cooper D, Holt PG. Immunity at body surfaces. *The Immunologist* 1995; 3:130–132.
8. Friedman A, Weiner HL. Induction of anergy or active suppression following oral tolerance is determined by antigen dosage. Proc. Natl. Acad. Science. USA1994; 91:6688–92.
9. Strobel S, Ferguson A. Immune responses to fed protein antigens in mice. 3. Systemic tolerance or priming is related to age at which antigen is first encountered. Pediatr Res 1984; 18:588–94.
10. Mariani F, Price JF, Kemeny DM. The IgG subclass antibody response to an inhalant antigen (*Dermatophagoides pteronyssinus*) during the first year of life: evidence for early stimulation of the immune system following natural exposure. Clin Exp Allergy 1992; 22:29–33.
11. Hattevig G, Kjellman B, Bjërkstén B. Appearance IgE antibodies to ingested and inhaled allergens during the first 12 years of life in atopic and non-atopic children. Ped Allergy Immunol 1993; 4:182–6.
12. Holt PG. Primary allergic sensitisation to environmental antigens: perinatal T-cell priming as a determinant of responder phentotype in adulthood. *J Exp Med* 1996; 183:1297–1301.
13. Yabuhara A, Macaubas C, Prescott SL, Venaille T, Holt BJ, Habre W, Sly PD, Holt PG. Th-2-polarised immunological memory to inhalant allergens in atopics is established during infancy and early childhood. *Clin Exp Allergy* Submitted:
14. Wegmann TG, Lin H, Guilbert L, Mosmann TR. Bidirectional cytokine interactions in the maternal-fetal relationshiip: is successful pregnancy a Th2 phenomenon? Immunol Today 1993; 14:353–56.
15. Krishnan L, Guilbert LJ, Wegmann TG, Belosevic M, Mosmann TR. T helper 1 response against *Leishmania major* in pregnant C57BL/6 mice increases implantation failure and fetal resportions. J Immunol 1996; 156:653–62.
16. Chen N, Gao Q, Field EH. Expansion of memory Th2 cells over Th1 cells in neonatal primed mice. Transplantation 1995; 60:1187–93.
17. Holt PG. Postnatal maturation of immune competence during infancy and early childhood. Pediatr Allergy Immunol 1995; 6:59–70.
18. Holt PG, Clough JB, Holt BJ, Baron-Hay MJ, Rose AH, Robinson BWS, Thomas WR. Genetic 'risk' for atopy is associated with delayed postnatal maturation of T-cell competence. Clin Exp Allergy 1992; 22:1093–9.
19. Warner JA, Miles EA, Jones AC, Quint DJ, Colwell BM, Warner JO. Is deficiency of interferon gamma production by allergen triggered cord blood cells a predictor of atopic eczema? Clin Exp Allergy 1994; 24:423–30.
20. Tang MLK, Kemp AS, Thorburn J, Hill DJ. Reduced interferon-γ secretion in neonates and subsequent atopy. *Lancet* 1994; 344:983–86.
21. Rinas U, Horneff G, Wahn V. Interferon-γ production by cord-blood mononuclear cells is reduced in newborns with a family history of atopic disease and is independent from cord blood IgE-levels. Pediatr Allergy Immunol 1993; 4:60–4.

22. Singh RR, Hahn BH, Sercarz EE. Neonatal peptide exposure can prime T cells, and upon subsequent immunization induce their immune deviation: implications for antibody vs. T cell-mediated autoimmunity. J Exp Med 1996; 183:1613–1622.
23. Ridge JP, Fuchs EJ, Matzinger P. Neonatal tolerance revisited: turning on newborn T cells with dendritic cells. *Science* 1996; 271:1723–1726.
24. Schon-Hegrad MA, Oliver J, McMenamin PG, Holt PG. Studies on the density, distribution, and surface phenotype of intraepithelial class II major histocompatibility complex antigen (Ia)-bearing dendritic cells (DC) in the conducting airways. J Exp Med 1991; 173:1345–56.
25. Holt PG, Haining S, Nelson DJ, Sedgwick JD. Origin and steady-state turnover of class II MHC-bearing Dendritic Cells in the epithelium of the conducting airways. J Immunol 1994; 153:256–61.
26. Nelson DJ, McMenamin C, McWilliam AS, Brenan M, Holt PG. Development of the airway intraepithelial Dendritic Cell network in the rat from class II MHC (Ia) negative precursors: differential regulation of Ia expression at different levels of the respiratory tract. J Exp Med 1994; 179:203–12.
27. Nelson DJ, Holt PG. Defective regional immunity in the respiratory tract of neonates is attributable to hyporesponsiveness of local Dendritic Cells to activation signals. *J Immunol* 1995; 155:3517–3524.

LANGERHANS CELLS IN THE TGFβ1 NULL MOUSE

Teresa A. Borkowski,[1] John J. Letterio,[2] Crystal L. Mackall,[3] Atsushi Saitoh,[1] Andrew G. Farr,[4] Xiao-Jing Wang,[5] Dennis R. Roop,[5] Ronald E. Gress,[3] and Mark C. Udey[1]

[1]Dermatology Branch
[2]Laboratory of Chemoprevention
[3]Experimental Immunology Branch
National Cancer Institute
National Institutes of Health
Bethesda, Maryland
[4]Department of Biological Structure
University of Washington
Seattle, Washington
[5]Department of Cell Biology
Baylor College of Medicine
Houston, Texas

1. INTRODUCTION

1.1. TGFβ1 and Dendritic Cell gp40 Expression

We recently identified a cell surface protein (gp40) that is homologous to human Ep-CAM, a putative homophilic adhesion molecule, and that is abundantly expressed by some murine dendritic cells (including Langerhans cells) and in certain epithelia (1). While characterizing various dendritic cells with regard to gp40 expression, we determined that the pleiotropic cytokine TGFβ1 was uniquely able to induce cell surface expression of gp40 on dendritic cells propagated from murine bone marrow in fetal calf serum- and GM-CSF-supplemented media. To determine whether or not this finding was potentially physiologically significant, we studied Langerhans cells in TGFβ1 null mice that had previously been generated and described by Kulkarni, Karlsson and coworkers (2).

1.2. TGFβ1 Null Mice

Although the TGFβ family is comprised of three isoforms that are frequently coexpressed *in vivo*, that bind to the same receptors and that trigger identical biocemical changes in responsive cells *in vitro* (3, 4), mice deficient in one of the isoforms (TGFβ1) exhibit a catastrophic phenotype (2, 5). TGFβ1 null mice are normally formed at birth, but experience ~50% intrauterine lethality (6). TGFβ1 null mice are indistinguishable from their wildtype littermates for the first 7–10 days of life but, thereafter, they exhibit progressive growth retardation, ultimately waste and invariably die by 4 weeks of age. The demise of TGFβ1 null animals is associated with, and preceded by, development of a multiorgan inflammatory syndrome characterized by lymphocyte and macrophage predominant inflammation involving lungs, heart, liver and other organs (2, 5). The skin of TGFβ1 null mice is not prominently involved, however. Abnormalities in cytokine synthesis (5), MHC antigen expression (7), autoantibody production (8, 9), and nitric oxide metabolism (10) have been described in TGFβ1 null mice, but the causal relationship between these findings and the multiorgan inflammatory syndrome that is characteristic of the genotype has not been established.

2. METHODOLOGY AND RESULTS

2.1. Epidermal Leukocytes in TGFβ1 Null Mice

To determine if TGFβ1 regulates gp40 expression by Langerhans cells in vivo, we prepared epidermal cell suspensions by limited trypsinization and examined viable cells for simultaneous expression of I-A antigens and gp40. Surprisingly, we did not detect I-A-bearing epidermal cells (Langerhans cells) in suspensions from TGFβ1 null mice. These results were confirmed in >10 individual animals ranging in age from 8–18 days. Younger animals were not studied because Langerhans cell frequencies do not approximate adult levels until mice are ~ 7 days old. Older TGFβ1 null mice were not studied because they rapidly deteriorate after 18–21 days of age. Although Langerhans cells were absent from TGFβ1 null skin (as assessed via multicolor flow cytometry, by single color immunofluorescence in ammonium isothiocyanate-dissociated epidermal sheets and using assays of accessory cell activity in epidermal cell suspensions), dendritic epidemal T cells were present in normal numbers. In addition, essentially all the leukocytes in TGFβ1 null epidermis were dendritic epidermal T cells. This result excludes the existence of a significant population of incompletely differentiated Langerhans cells in TGFβ1 epidermis.

2.2. Lymph Node Dendritic Cells in TGFβ1 Null Mice

To determine if the absolute deficiency of Langerhans cells in TGFβ1 null mice was reflective of a more generalized abnormality in dendritic cells, we analyzed skin draining lymph nodes from TGFβ1 null and littermate controls for dendritic cells using several dendritic cell-selective markers. Using a sensitive 3 step immunohistochemical technique, we readily identified $CD11c^+$ dendritic cells in T cell-dependent areas of lymph nodes from both TGFβ1 null and control animals. Of interest, gp40-bearing dendritic cells were present in normal lymph nodes and absent from lymph nodes from TGFβ1 null mice.

2.3. Dissociation of Langerhans Cell Deficiency from Inflammation and Wasting in TGFβ1 Null Mice

The skin of TGFβ1 null mice is not obviously inflamed. To rigorously exclude inflammation in skin or other organs as a cause of the Langerhans cell deficiency that we observed, several litters including TGFβ1 null and control mice were treated with the immunosuppressive rapamycin (4 mg/kg i.p.) beginning on postnatal day 10 and 3x per week thereafter. Rapamycin suppressed the inflammatory syndrome in TGFβ1 null mice and wasting was abrogated. However, treatment of TGFβ1 null mice with rapamycin for up to 42 days after birth did not result in repopulation of epidermis with Langerhans cells. Importantly, rapamycin had no effect on Langerhans cells in wildtype or TGFβ1 null heterozygotes. Thus, the Langerhans cell deficiency in TGFβ1 null mice can be dissociated from the inflammatory and wasting component of the phenotype.

2.4. Langerhans Cell Progenitors in TGFβ1 Null Mice

To examine TGFβ1 null bone marrow for Langerhans cell precursors, Ly-5.1$^+$ T cell-depleted bone marrow was administered to lethally-irradiated Ly-5.2$^+$ C57BL/6 mice and recipients were examined for Langerhans cells at various times thereafter. In 1 of 2 experiments harvested 8 weeks after transplantation, the frequency of Langerhans cells in recipients reconstituted with TGFβ1 null marrow was lower than in recipients given wildtype cells. In 1 of 2 experiments terminated at 8 weeks and 3 of 3 experiments terminated at 16 weeks, Langerhans cell numbers in recipients engrafted with TGFβ1 null and control precursors were comparable.

2.5. Localization of Langerhans Cells in TGFβ1 Null or TGFβ1-Unresponsive Epidermis

To determine if normal Langerhans cells could infiltrate TGFβ1 null epidermis, we placed full-thickness skin grafts from TGFβ1 null and control mice onto BALB/c nude mice and prepared epidermal sheets 12 weeks later. Similar numbers of I-A^d-bearing Langerhans cells were detected in grafts from TGFβ1 null and control mice. It is possible that TGFβ1 null skin does not remain TGFβ1 deficient after placement on nude recipients (11). Therefore, we enumerated Langerhans cells in transgenic mice expressing a dominant negative type II TGFβ receptor driven by a truncated loricrin promoter (12). The construct is expressed in suprabasal as well as basal keratinocytes beginning on gestation day 16. These mice exhibit a hyperproliferative epidermis transiently, and primary keratinocytes derived from transgenic animals are resistant to the growth inhibitory effects of TGFβ1 *in vitro*. Whether or not keratinocytes from these animals are resistant to other effects of TGFβ1 is unknown. Nonetheless, normal numbers of Langerhans cells were present in dominant negative TGFβ type II receptor animals ranging in age from 5–14 days.

2.6. Cytokine Production in TGFβ1 Null Skin

TGFβ1 may influence the proliferation, differentiation or survival of Langerhans cells or their precursors (13), or regulate trafficking by modulating production of proinflammatory cytokines. We prepared total RNA from the skin of TGFβ1 null and control mice and assessed cytokine mRNA levels. Although increased levels of many cytokine mRNAs were detected in the skin of TGFβ1 null mice, levels of IL-1α, IL-1β and TNFα mRNAs were similar in TGFβ1 null and control skin. Overproduction of proinflammatory cytokines that activate and mobilize Langerhans cells (14) does not obviously contribute to the Langerhans cell deficiency that is observed in TGFβ1 null mice.

3. CONCLUSION

Although TGFβ1 is clearly essential for normal Langerhans cell development and/or localization, the precise explanation for the Langerhans cell abnormality that is evident in TGFβ1 null mice remains to be elucidated. Within epidermis, TGFβ1 is produced by keratinocytes as well as Langerhans cells (15, 16, 17). Thus, the requirement of Langerhans cells for TGFβ1 could be satisfied by endogenous or exogenous cytokine. This likely explains the results of the bone marrow transplant studies as well as the skin grafting/transgenic experiments. While our results do not absolutely exclude local or extracutaneous overproduction of proinflammatory cytokines as a cause of the Langerhans cell deficiency, neither are they supportive of this concept.

4. REFERENCES

1. T. A. Borkowski, A. J. Nelson, A. G. Farr, M. C. Udey, *Eur. J. Immunol.* **26**, 110–114 (1996).
2. A. B. Kulkarni, et al., *Proc. Natl. Acad. Sci.* **90**, 770–774 (1993).
3. M. B. Sporn, A. B. Roberts, *J. Cell Biol.* **119**, 1017–1021 (1992).
4. D. M. Kingsley, *Genes Devel.* **8**, 133–146 (1994).
5. M. M. Shull, et al., *Nature* **359**, 693–699 (1992).
6. M. C. Dickson, et al., *Development* **121**, 1845–1854 (1995).
7. A. G. Geiser, et al., *Proc. Natl. Acad. Sci.* **90**, 9944–9948 (1993).
8. H. Dang, et al., *J. Immunol.* **155**, 3205–3212 (1995).
9. L. Yaswen, et al., *Blood* **87**, 1439–1445 (1996).
10. Y. Vodovotz, et al., *J. Exp. Med.* **183**, 2337–2342 (1996).
11. A. B. Glick, et al., *Genes Devel.* **8**, 2429–2440 (1994).
12. X.-J. Wang, et al., *Proc. Natl. Acad. Sci.* , (In press).
13. H. Strobl, et al., *J. Immunol.* **157**, 1499–1507 (1996).
14. K. Schwarzenberger, M. C. Udey, *J. Invest. Dermatol.* **106**, 553–558 (1996).
15. A. B. Glick, et al., *Proc. Natl. Acad. Sci.* **90**, 6076–6080 (1993).
16. M. S. Gruschwitz, O. P. Hornstein, *J. Invest. Dermatol.* **99**, 114–116 (1992).
17. F. Granucci, et al., *Eur. J. Immunol.* **24**, 2522–2526 (1994).

MATURATION AND MIGRATION OF MURINE DENDRITIC CELLS IN SITU

Observations in a Skin Organ Culture Model

Hella Stössel, Franz Koch, and Nikolaus Romani

Department of Dermatology
University of Innsbruck
Innsbruck, Austria

1. ABSTRACT

Dendritic cells reside in tissues such as skin in an immature state. Upon antigenic challenge they begin to mature and migrate to the draining lymph nodes. These processes are still poorly understood. One way to study in situ aspects of maturation and migration are skin organ culture models. In an attempt to learn more about the relationship between maturation and migration we investigated the expression of several marker molecules by immunohistochemistry. Sheets from normal murine ear skin and from skin that had been cultured for three days were compared.

During culture the numbers of epidermal Langerhans cells decreased and accumulations of strongly MHC class II-positive cells ("cords") were found in the dermis. As compared to untreated skin, the few Langerhans cells remaining in the epidermal sheets after 2–3 days expressed increased levels of MHC class II and had also upregulated B7–2 (CD86) as described. They did not express the antigen recognized by mAb 2A1, a marker for mature dendritic cells. Double-staining of dermal sheets after 3 days of culture showed that dendritic cells in the "cords" expressed high levels of MHC class II and CD86 but were also reactive with mAb 2A1. This pattern is identical to those dendritic cells that had emigrated into the culture medium over the period of 3 days. Invariant chain (mAb In1) was detected at all stages of culture as opposed to isolated epidermal Langerhans cells in suspension where invariant chain expression disappears after 3 days of culture.

We conclude that the up- (class II, B7–2, 2A1) and down-regulation (invariant chain) of dendritic cell molecules during migration does not happen in a synchronized manner. The molecule recognized by mAb 2A1 seems to appear late in maturation.

Dendritic Cells in Fundamental and Clinical Immunology
edited by Ricciardi-Castagnoli, Plenum Press, New York, 1997

2. INTRODUCTION

Dendritic cells exert their immunogenic functions at different stages of their life cycle. This pathway of differentiation in dendritic cells was termed "maturation". It was discovered in epidermal Langerhans cells[1] and its main features were investigated and described to the greatest part in this very model[2–5]. After their generation in the bone marrow dendritic cells populate many lymphoid and non-lymphoid organs. Upon experimental isolation from the tissues they are considered *immature* dendritic cells. At this stage they are highly capable of processing protein antigens for the MHC class II pathway. The basis for this ability is their strong macropinocytic activity[6], their expression of pattern-recognition receptors such as DEC-205[7], the presence of many organelles where processing can take place[4,8], and the intense biosynthesis of MHC products[5]. After short term culture dendritic cells turn into *mature* dendritic cells that are well equipped to sensitize resting, naive T cells, the reason for this being their expression of high levels of known—and perhaps still unknown adhesion and costimulator molecules[9] as well as of soluble immungenic factors such as IL-12[10].

Most of the data on which the "concept of dendritic cell maturation" is based were corroborated in *in vitro* experiments. We were interested in *in vivo* aspects of dendritic cell maturation and therefore chose to study the murine skin organ culture model established by Larsen et al.[11] Specifically, we wanted to monitor the maturation process along part of the migratory pathway of dendritic cells.

3. MATERIAL AND METHODS

Murine ears were split in halves and were cultured in RPMI-1640 supplemented with 10% FCS for 2–3 days as described[11,12]. At these time points the whole skin explants were processed into epidermal and dermal sheets by means of ammoniumthiocyanate. Sheets were stained immunohistochemically applying a five-step technique. Briefly, sheets were incubated sequentially with primary rat mAb's (see below), biotinylated anti-rat Ig, streptavidin-Texas Red, rat Ig (blocking step), and FITC-conjugated anti-MHC class II (mAb 2G9, Pharmingen). Stained sheets were evaluated under a standard fluorescence microscope. The following primary mAb's were used: In-1, anti-invariant chain, gift of Dr. N. Koch, Bonn; GL1, anti-B7–2 / CD86, Pharmingen; and 2A1[13], marker for mature dendritic cells; gift of Dr. R.M. Steinman, New York).

4. RESULTS

Langerhans cells in untreated and uncultured skin express the invariant chain, but do not express B7–2 nor 2A1 molecules. Epidermis from skin cultured for two to three days showed a markedly reduced number of Langerhans cells; accumulations of strongly MHC class II-positive cells appeared in the dermis ("cords"). Both features had been originally described by Larsen et al.[11] When we double-labeled sheets from cultured skin we observed a non-synchronous development of phenotypical markers (Table 1). MHC class II expression was clearly enhanced on dendritic cells in both epidermis and dermis from cultured skin. B7–2 / CD86 molecules appeared on epidermal and dermal dendritic cells (in "cords"); 2A1 molecules showed a slower time course. They were only detected on dendritic cells in dermal "cords" but not (yet) on Langerhans cells in the epidermis from 3-

Table 1. Staining patterns of migrating epidermal Langerhans cells in situ at different stages along the transcutaneous migration pathway in comparison with isolated Langerhans cells in suspension

	MHC class II mAb 2G9	Invariant chain mAb In1	Costimulator CD86 mAb GL1	Maturation marker mAb 2A1
Freshly isolated Langerhans cells	++	+	–	–
Langerhans cells in uncultured epidermis	+	+	–	–
Langerhans cells in epidermis from 3-day cultured skin	+++	+	+	–
Langerhans cells in dermis from 3-day cultured skin ("cords")	+++	+	++	+++
Cultured Langerhans cells	+++	–	++	+++

Fluorescence staining intensities are presented in an arbitrary scale (– to +++).

day cultured skin. Expression of the invariant chain remained throughout the cultures on dendritic cells of both epidermis and dermis.

5. DISCUSSION

We were looking at three stages of dendritic cell development in the skin explant model: (1) untreated and uncultured skin, representing resident, stationary dendritic cells, (2) epidermis from skin cultured for two to three days, and (3) dermis from skin cultured for two to three days. We imply a time sequence between steps (2) and (3) in that we assume that the Langerhans cells found in the epidermis will eventually end up in the dermal "cords"—and finally in the culture medium.

In the *in vitro* model where dendritic cell maturation was originally described, namely the short term culture of isolated epidermal Langerhans cells, immature Langerhans cells were invariably MHC class II + / invariant chain + / B7–2 - / 2A1 -. After maturation in culture this pattern changed to MHC class II +++ / invariant chain - / B7–2 ++ / 2A1 +++[14]. Our data show that these markers do not appear or disappear in a synchronous fashion in vivo and that there are intermediate phenotypes. Still within the epidermis, the initial phenotype of MHC class II + / invariant chain + / B7–2 - / 2A1 - changes to MHC class II +++ / invariant chain + / B7–2 + / 2A1 -. This means that dendritic cells have upregulated class II and B7–2, indicating maturation. The lack of 2A1 and the persistence of the invariant chain, however, suggest that maturation is not complete. One step further down the migratory pathway, that is within the dermal "cords", dendritic cells acquire the 2A1 maturation marker. However, they still express the invariant chain. From these observations we conclude that maturation starts with the onset of migration. It is not fully completed, however, within the cutaneous microenvironment.

The persistence of invariant chain expression on maturing dendritic cells may help explain seemingly contradictory data about the relative processing capacities of immature versus mature dendritic cells[15,16]. There may be conditions where dendritic cells that are mature, as measured by the expression of CD86 and 2A1, can still process. We have not tested the processing capacity of migrating dendritic cells nor have we attempted to experimentally induce a higher degree of maturity (i.e. disappearance of invariant chain) in our model.

Turley, Mellman and colleagues (unpublished data; Symposium on Dendritic Cells; Venezia, October 1996) have dissected the maturational pathway of murine dendritic cells

Table 2. Classification of developmental stages of murine cutaneous dendritic cells

Early Immature	Intermediate Immature	Late Mature
resident, in situ	after isolation *in vitro*; after "activation" *in situ* (e.g. in skin organ culture or after application of contact sensitizers)	after 3-day culture of isolated Langerhans cells; after 2-3 days of skin organ culture in situ
	excellent antigen processing properties; poor T cell sensitizing capacity	processing capacity absent or low; strong T cell sensitizing capacity

using mainly the pattern of intracellular organelles as criteria. They distinguish the "early", "intermediate", and "late" stage of development. Using confocal microscopy in the same skin organ culture model they describe "early" as being 2A1-negative, "intermediate" as being 2A1-positive in a dispersed pattern, and "late" as being 2A1-positive in the typical clustered, perinuclear pattern. We were unable to find a dispersed pattern of 2A1 staining, perhaps due to the lack of confocal microscopy in our experiments. However, like these authors we identified the 2A1-negative "early" stage and the 2A1-positive (perinuclear pattern) "late" stage. At this point it is important to remember that the concept of mature and immature dendritic cells is largely based on *in vitro* experiments with dendritic cells isolated from different tissues. Mellman's "early – intermediate – late" classification scheme is to a large part based on observations *in situ* and *in vivo*. It therefore adds a stage of development that could not be considered in previous *in vitro* experiments, namely the stage of resting dendritic cells in tissues. Table 2 shows an attempt to relate the "mature - immature" classification with the "early – intermediate – late" classification. It is likely that isolation techniques such as the trypsinization of epidermis, that is necessary in order to obtain Langerhans cells, induce the transition from the "early" to the "intermediate" stage.

The skin organ culture model will be helpful to further analyze dendritic cell maturation in vivo.

6. REFERENCES

1. Schuler, G. and R. M. Steinman. 1985. Murine epidermal Langerhans cells mature into potent immunostimulatory dendritic cells *in vitro*. *J. Exp. Med.* 161:526–546.
2. Heufler, C., F. Koch, and G. Schuler. 1988. Granulocyte-macrophage colony-stimulating factor and interleukin-1 mediate the maturation of murine epidermal Langerhans cells into potent immunostimulatory dendritic cells. *J. Exp. Med.* 167:700–705.
3. Romani, N., S. Koide, M. Crowley, M. Witmer-Pack, A. M. Livingstone, C. G. Fathman, K. Inaba, and R. M. Steinman. 1989. Presentation of exogenous protein antigens by dendritic cells to T cell clones: intact protein is presented best by immature epidermal Langerhans cells. *J. Exp. Med.* 169:1169–1178.
4. Stössel, H., F. Koch, E. Kämpgen, P. Stöger, A. Lenz, C. Heufler, N. Romani, and G. Schuler. 1990. Disappearance of certain acidic organelles (endosomes and Langerhans cell granules) accompanies loss of antigen processing capacity upon culture of epidermal Langerhans cells. *J. Exp. Med.* 172:1471–1482.
5. Kämpgen, E., N. Koch, F. Koch, P. Stöger, C. Heufler, G. Schuler, and N. Romani. 1991. Class II major histocompatibility complex molecules of murine dendritic cells: Synthesis, sialylation of invariant chain, and antigen processing capacity are down-regulated upon culture. *Proc. Natl. Acad. Sci. USA* 88:3014–3018.
6. Sallusto, F., M. Cella, C. Danieli, and A. Lanzavecchia. 1995. Dendritic cells use macropinocytosis and the mannose receptor to concentrate macromolecules in the major histocompatibility complex class II compartment: Downregulation by cytokines and bacterial products. *J. Exp. Med.* 182:389–400.

7. Jiang, W., W. J. Swiggard, C. Heufler, M. Peng, A. Mirza, R. M. Steinman, and M. C. Nussenzweig. 1995. The receptor DEC-205 expressed by dendritic cells and thymic epithelial cells is involved in antigen processing. *Nature* 375:151–155.
8. Kleijmeer, M. J., V. M. J. Oorschot, and H. J. Geuze. 1994. Human resident Langerhans cells display a lysosomal compartment enriched in MHC class II. *J. Invest. Dermatol.* 103:516–523.
9. Inaba, K., M. Witmer-Pack, M. Inaba, K. S. Hathcock, H. Sakuta, M. Azuma, H. Yagita, K. Okumura, P. S. Linsley, S. Ikehara, S. Muramatsu, R. J. Hodes, and R. M. Steinman. 1994. The tissue distribution of the B7–2 costimulator in mice: Abundant expression on dendritic cells in situ and during maturation in vitro. *J. Exp. Med.* 180:1849–1860.
10. Heufler, C., F. Koch, U. Stanzl, G. Topar, M. Wysocka, G. Trinchieri, A. Enk, R. M. Steinman, N. Romani, and G. Schuler. 1996. Interleukin-12 is produced by dendritic cells and mediates T helper 1 development as well as interferon-gamma production by T helper 1 cells. *Eur. J. Immunol.* 26:659–668.
11. Larsen, C. P., R. M. Steinman, M. Witmer-Pack, D. F. Hankins, P. J. Morris, and J. M. Austyn. 1990. Migration and maturation of Langerhans cells in skin transplants and explants. *J. Exp. Med.* 172:1483–1493.
12. Ortner, U., K. Inaba, F. Koch, M. Heine, M. Miwa, G. Schuler, and N. Romani. 1996. An improved isolation method for murine migratory cutaneous dendritic cells. *J. Immunol. Methods* 193:71–79.
13. Inaba, K., M. Inaba, N. Romani, H. Aya, M. Deguchi, S. Ikehara, S. Muramatsu, and R. M. Steinman. 1992. Generation of large numbers of dendritic cells from mouse bone marrow cultures supplemented with granulocyte/macrophage colony-stimulating factor. *J. Exp. Med.* 176:1693–1702.
14. Larsen, C. P., S. C. Ritchie, T. C. Pearson, P. S. Linsley, and R. P. Lowry. 1992. Functional expression of the costimulatory molecule, B7/BB1, on murine dendritic cell populations. *J. Exp. Med.* 176:1215–1220.
15. Liu, L. M. and G. G. MacPherson. 1995. Antigen processing: cultured lymph-borne dendritic cells can process and present native protein antigens. *Immunology* 84:241–246.
16. Koch, F., B. Trockenbacher, E. Kämpgen, O. Grauer, H. Stössel, A. M. Livingstone, G. Schuler, and N. Romani. 1995. Antigen processing in populations of mature murine dendritic cells is caused by subsets of incompletely matured cells. *J. Immunol.* 155:93–100.

52

ANTIGEN PRESENTATION AND IL-12 PRODUCTION BY DENDRITIC CELLS IN VIVO

Jean-Charles Guéry,[2] Francesco Ria,[1*] Francesca Galbiati,[1] and Luciano Adorini[1]

[1]Roche Milano Ricerche Via Olgettina, 58
I-20132 Milano, Italy
[2]INSERM U28
Hôpital Purpan
31059 Toulouse, France

1. INTRODUCTION

IL-12 is a key regulatory cytokine produced by APC which drives the development of interferon-γ (IFN-γ)-producing Th1 cells and promotes cell-mediated immunity[1]. The biologically active IL-12 molecule is a covalently linked heterodimer of approximately 75 kDa composed of two chains, p40 and p35 encoded by separate genes[2,3].

We have studied antigen presentation and IL-12 production by APC obtained from an inflammatory site, since IL-12, a pro-inflammatory cytokine, is likely to be produced in higher amounts at the site of inflammation by professional APC. Draining lymph node cells from mice immunized subcutaneously with hen egg-white lysozyme (HEL) in adjuvant display HEL peptide-MHC class II complexes able to stimulate, in the absence of any further antigen addition, specific T hybridoma cells[4]. We have analyzed the phenotype of the APC bearing antigenic complexes and shown that following immunization with HEL in adjuvant, dendritic cells (DC) are the only lymph node APC population expressing detectable HEL peptide-class II complexes[5].

Here we show that DC not only present efficiently antigenic complexes formed *in vivo*, but also secrete substantial amounts of bioactive IL-12p75. DC do not produce IL-12 constitutively, but only after induction, in our case, by $CD4^+$ T hybridoma cells. IL-12 production is increased when DC present antigenic complexes, indicating that ligand-TCR interaction can upregulate IL-12 production. In addition, IL-12 secretion by DC requires CD40-CD40L interaction with $CD4^+$ T cells. Therefore, our data indicate that during antigen presentation positive feedback is established through cognate interaction between T cells and DC, resulting in increased IL-12 production.

---

* Present address: Istituto di Patologia Generale, Università Cattolica del Sacro Cuore, 00168 Roma, Italy.

Dendritic Cells in Fundamental and Clinical Immunology
edited by Ricciardi-Castagnoli, Plenum Press, New York, 1997

2. RESULTS

2.1. Antigen-Presentation and IL-12 Production by DC-Enriched Lymph Node Cells

Lymph node APC from HEL-IFA immunized BALB/c mice were separated into three populations (small B cells, large B cells and DC-enriched) by a combination of Percoll centrifugation and magnetic cell sorting as described[5]. We compared the relative capacity of these different APC population to present antigenic complexes formed in vivo and to produce IL-12 upon coculture with the HEL107–116-specific, I-E^d-restricted T cell hybridoma 1H11.3. In agreement with our previous observation[5], DC but not B cells, have the capacity to induce IL-2 and IFN-γ secretion by T hybridoma cells without antigen addition in vitro (Fig. 1). This is not due to a defect in the antigen presenting capacity of B cells since they present efficiently antigenic peptide added in vitro, however less efficiently than DC. Interestingly, coculture of APC with the T hybridoma cells reveals a similar pattern in IL-12 production (Fig. 1). The capacity of the DC-enriched population to express antigenic complexes formed in vivo following administration of HEL in adjuvant is thus paralleled by its ability to secrete IL-12.

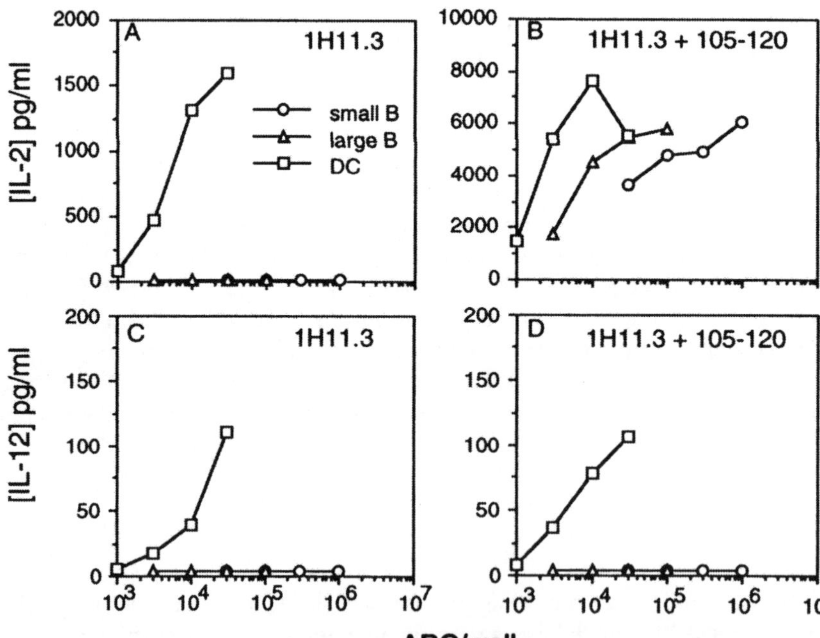

Figure 1. Antigen presentation and IL-12 production by lymph node DC following administration of protein antigen in adjuvant. BALB/c mice were immunized into the hind footpads with 10 nmoles/mouse HEL in IFA. Lymph node APC were separated in DC-enriched, large and small B cells by a combination of Percoll centrifugation and magnetic cell sorting on MiniMACS column using B220-conjugated microbeads. Graded numbers of APC were cultured with the HEL-specific I-E^d-restricted 1H11.3 T cell hybridomas (5×10^4 cells/well) alone (A, C) or in the presence of 0.3 μM HEL105–120 (B, D). After 24 h, IL-2 (A, B) and IL-12p75 (C, D) were measured in culture supernatants by two sites sandwich ELISA.

2.2. Ligand-TCR Interaction Upregulates IL-12p75 Production by Lymph Node DC

We compared the relative capacity of different APC populations (large B cells, macrophages, DC-enriched population from spleen or immune lymph node) to produce IL-12 upon activation with T cell-independent or T cell-dependent stimuli. The DC-enriched populations were obtained from immune LNC or collagenase-treated SC by a combination of gradient centrifugation and negative selection using both B220- and Thy1.2-coated magnetic particles. Large B cells were obtained from low buoyant density cell population from either spleen or immune lymph nodes positively selected using B220-conjugated microbeads. As a source of macrophages (Mϕ), peritoneal exudate cells were used. IL-12 production in 24h culture supernatants from these different APC populations stimulated by SAC and IFN-γ was measured using two sites sandwich ELISA. As shown in Fig. 2A, peritoneal Mϕ and the DC-enriched population from spleen and lymph node have a similar capacity to secrete IL-12p75 heterodimers. Conversely, large B cells isolated from immune lymph node (Fig. 2A) or spleen (not shown) fail to secrete IL-12p75. We next examined the capacity of these APC to produce IL-12 upon antigen-specific interaction with class II-restricted T cells. APC were cultured with T hybridoma cells in the presence or absence of antigenic peptide, and the concentration of IL-12p75 in 24h culture supernatants determined as above. Results in Fig. 2B and C show that Mϕ and the DC-enriched populations from spleen or lymph nodes, but not B cells, are able to produce comparable amounts of IL-12p75 upon activation with SAC + IFN-γ. Conversely, IL-12p75 is measurable in supernatant of lymph node DC cultured with 1H11.3 cells, while it is undectable or just above the detection limit using the other APC populations. Interestingly, addition of HEL peptide 105–120 to cultures upregulates IL-12p75 production by lymph node DC. A modest upregulation of IL-12p75 production by the splenic DC-enriched population is also observed. Therefore, upregulation of IL-12 production by antigenic peptide requires ligand-TCR interaction between DC and T cells.

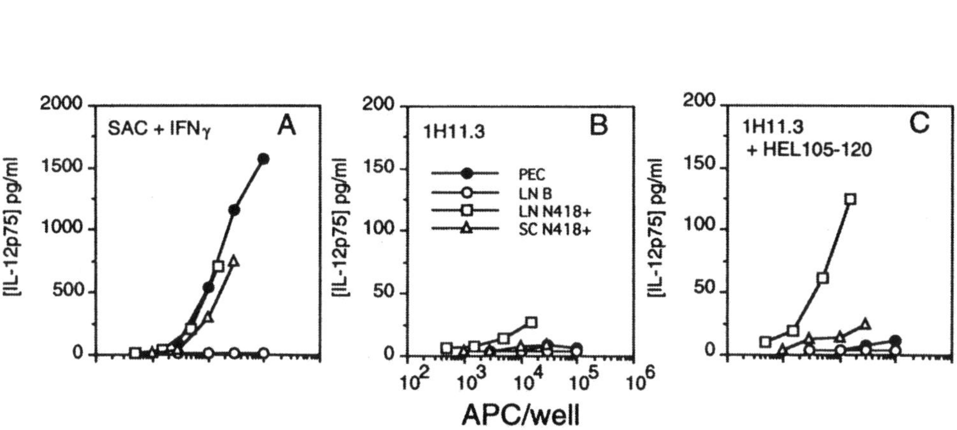

Figure 2. Ligand-TCR interaction increases IL-12p75 production by lymph node DC. In panel A, the indicated numbers of Mϕ, B cells and DC-enriched N418[+] APC from BALB/c mice were cultured in the presence of SAC (1:5000) and IFN-γ (50 U/ml) in complete medium at 37°C. In panels B and C, APC were cultured as indicated with 1H11.3 cells alone (B) or in the presence of 0.3 μM HEL105–120 (C). The concentration of IL-12p75 in 24 h culture supernatants was determined by two-sites sandwich ELISA.

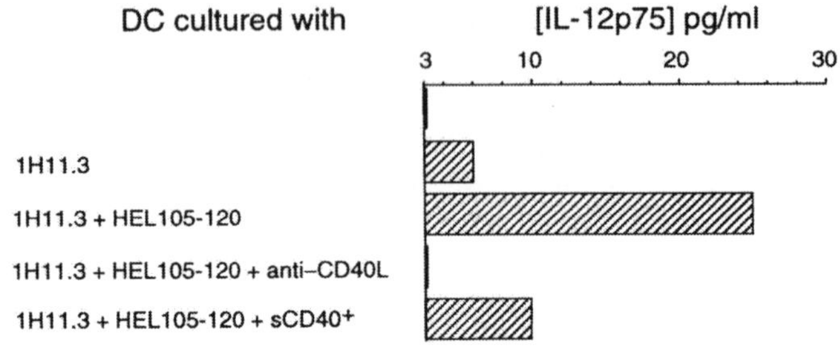

Figure 3. CD40/CD40 ligand interaction are required for T cell-dependent production of IL-12 by lymph node DC. Lymph node DC were enriched from CFA-primed BALB/c mice as in Fig.1. DC (75% N418$^+$) were cultured at 30 x 10^4 N418$^+$ cells/well with 5 x 10^4 1H11.3 T cells/well with or without 0.3 μM HEL105–120. Anti-CD40L mAb (10 μg/ml) or soluble hCD40-Hμ (10 μg/ml) purified from culture supernatant by affinity chromatography on anti-human IgM column (sCD40$^+$), were added to the indicated cultures. IL-12 p75 was measured in 20h supernatants from duplicate cultures by two-sites sandwich ELISA.

2.3. Blocking CD40-CD40L interaction prevents IL-12 production

Since it has been recently shown that activated T cells induce IL-12 production by monocytes via CD40-CD40L interaction[6,7], we examined whether this interaction plays a role in our model. CD40 and CD40L are expressed on DC and T hybridoma cells, respectively (not shown). Expression of CD40L is rapidly upregulated on T hybridoma cells following activation by ionomycin and phorbol esters. As shown in Fig. 3, addition of HEL105–120 to cultures of DC and 1H11.3 T cells, increases IL-12 heterodimer production which is prevented by addition of anti-class II mAb (not shown). Addition of soluble hCD40-Hμ chimeric molecules or anti-CD40L mAb prevents almost completely upregulation of IL-12p75 production. In addition, ligation of CD40 on DC by plate-bound anti-CD40 mAb results in a two-fold increased IL-12 secretion, which was correlated with an upregulation of IFN-γ and IL-2 secretion by T hybridoma cells (not shown). Taken together, these data demonstrate that upregulation of IL-12p75 by lymph node DC mediated by ligand-TCR interactions with antigen-specific T cells does not involve soluble factors such as IFN-γ but rather requires cognate interaction involving the T cell-associated CD40L and CD40 on DC.

3. DISCUSSION

We have previously shown that lymph node DC isolated from mice immunized with protein antigen in adjuvant are the main APC population able to express detectable antigenic complexes[5]. Since this protocol of immunization preferentially induces a Th1-type immune response[8], it was of interest to analyze whether in addition to their antigen presenting capacity DC from immune lymph nodes were able to secrete IL-12 p75, a proinflammatory cytokine involved in Th1 differentiation[1]. We now show that following administration of protein antigen in adjuvant, lymph node DC analyzed ex vivo not only present peptide-class II complexes formed in vivo, but also produce substantial amounts of IL-12 p75. DC do not produce IL-12 constitutively, but only after induction, in our case, by T hybridoma cells, and this is strongly upregulated by antigen-specific interaction. These data emphasize the critical role of lymph node DC in both activation and differentiation in vivo of $CD4^+$ T cells following antigen priming. Conversely, B cells fail to activate T cells and to secrete IL-12 p75, even when stimulated with SAC and IFN-γ. In vivo loaded DC, but not B cells, induce IFN-γ production by antigen-specific T cells. Interestingly, both DC and B cells induce IL-2 secretion by T cells when antigenic peptide is added in vitro, but IFN-γ production is still only induced by DC. This antigen-specific IFN-γ production is likely due to IL-12 p75 secretion by DC, since addition of IL-12 p75 in vitro induces IFN-γ secretion by T hybridoma cells stimulated by antigen and B cells (J-C. Guéry, unpublished).

Lymph node DC and peritoneal exudate Mϕ produce comparable amount of IL-12 upon activation with bacterial products and IFN-γ, but upregulation of IL-12 secretion upon antigen-specific interaction with T cells is strongly induced only in DC and is negligible in Mϕ. This indicates an heterogeneity between professional APC populations in their capacity to upregulate IL-12 production in response to signals from antigen-specific cells. Heterogeneity is also present between different DC populations, as demonstrated by the observation that DC from immune lymph node, as compared to DC enriched from steady state spleen, are far more efficient in secreting IL-12 p75 upon interaction with antigen-specific T cells.

Several lines of evidence indicate that cognate interactions with antigen-specific T cells play an important role in IL-12 secretion by DC from immune lymph nodes, in agreement with results obtained with mouse splenic[9] and human[10] DC. First, IL-12 production is increased when DC present antigenic complexes to specific T cells, indicating that ligand-TCR interaction can upregulate IL-12 production. Second, CD40 and CD40L are expressed by DC and T hybridoma cells, respectively. Ligation of CD40 on DC by plate-bound anti-CD40 mAb increased two-fold IL-12 production, which was correlated to increased IL-2 and IFN-γ production by T cells (J-C. Guéry, unpublished results). Third, soluble CD40 molecules and anti-CD40L mAb inhibit upregulation of IL-12 secretion by lymph node DC.

In conclusion, our results demonstrate that lymph node DC present very efficiently antigenic peptides derived from processing of protein antigen endocytosed at the site of inflammation and simultaneously produce IL-12, a key cytokine in the development of cell-mediated immunity. The efficient presentation of antigenic complexes derived from proteins present in inflammatory sites and the production of IL-12 may account for the unique capacity of lymph node DC to prime naive $CD4^+$ T cells in vivo.

4. REFERENCES

1. Trinchieri, G. (1995) *Annu. Rev. Immunol.* 13, 251–276
2. Wolf, S.F., Temple, P.A., Kobayashi, M. *et al.* (1991) *J. Immunol.* 146, 3074–3081
3. Stern, A.S., Podlaski, F.J., Hulmes, J.D. *et al.* (1990) *Proc. Natl. Acad. Sci. USA* 87, 6808–6812
4. Guéry, J.-C., Sette, A., Leighton, J., Dragomir, A. and Adorini, L. (1992) *J. Exp. Med.* 175, 1345–1352
5. Guéry, J.C., Ria, F. and Adorini, L. (1996) *J. Exp. Med.* 183, 751–757
6. Shu, U., Kiniwa, M., Wu, C.Y. *et al.* (1995) *Eur. J. Immunol.* 25, 1125–1128
7. Stuber, E., Strober, W. and Neurath, M. (1996) *J. Exp. Med.* 183, 693–698
8. Guéry, J-C., Galbiati, F., Smiroldo, S. and Adorini, L. (1996) *J. Exp. Med.* 183, 485–497
9. Koch, F., Stanzl, U., Jennewein, P. *et al.* (1996) *J. Exp. Med.* 184, 741–746
10. Cella, M., Scheidegger, D., Palmer-Lehmann, K. *et al.* (1996) *J. Exp. Med.* 184, 747–752

IL-10 PREVENTS THE GENERATION OF DENDRITIC CELLS FROM CD14+ BLOOD MONOCYTES, PROMOTES THE DIFFERENTIATION TO MATURE MACROPHAGES AND STIMULATES ENDOCYTOSIS OF FITC-DEXTRAN

P. Allavena,[1] L. Piemonti,[1] D. Longoni,[1] S. Bernasconi,[1] A. Stoppacciaro,[2] L. Ruco,[2] and A. Mantovani[1,3]

[1]Department of Immunology and Cell Biology
"Mario Negri" Institute
20157 Milano, Italy
[2]Department of Expimental Medince and Pathology
University of Rome
00161Rome, Italy
[3]Division of General Pathology
University of Brescia
25123 Brescia, Italy

INTRODUCTION

Dendritic cells (DC) constitute an heterogeneous system of leukocytes specialized in antigen (Ag) capture, processing and presentation to immunocompetent cells[1,2]. Their hematopoietic differentiation lineage of DC has not clearly been defined. At least some DC are closely related to the mononuclear phagocyte system, but there is also evidence of DC belonging to a separate differentiation pathway[1,2]. Recently large quantities of DC have been differentiated from monocyte-enriched PBMC using GM-CSF and IL-4[3,4]. We have demonstrated that IL-4 can be substituted with IL-13, with identical results[5].

IL-10 is a potent immunosuppressive cytokine produced by mononuclear phagocytes, T cells and neoplastic lines. IL-10 has profound effects on the function of mononuclear phagocytes: inhibits the production of proinflammatory cytokines reduces the expression of MHC Class II and APC function[6] . On differentiated DC IL-10 reduces the expression of B7.1[7] and of B7.2[8] and inhibits tumor antigen presentation by epidermal APC[9]. In the present study we investigated how IL-10 affects differentiation, maturation and function of DC obtained by in vitro culture of CD14+ monocytes with GM-CSF and IL-13.

Table 1. Inhibition of DC differentiation from monocytes by IL-10

	Phenotype % positive cells (RFI)°				MLR % APC		
Cells cultured with	CD1a	MHCII	CD16	CD14	10	3	1
GM-CSF+IL-13	79 (29)	87 (44)	9 (<1)	4 (<1)	11769*	9714	5930
GM-CSF+IL-13 +IL-10							
(20 ng/ml)	20 (<1)	91 (25)	83 (5)	81 (8)	7745	2286	1227
(2 ng/ml)	38 (1)	88 (39)	82 (4)	82 (10)	10966	8113	2774
(0.2 ng/ml)	44 (6)	81 (44)	36 (1)	27 (2)	13475	7767	3629

Monocytes were cultured for 7 days with GM-CSF (50 ng/ml) + IL-13 (10 ng/ml) with or without IL-10.
Phenotype analysis was performed with FACSTAR.
°RFI: Relative fluorescence intensity.
MLR: Different number of APC were cultured with allogenic cord blood lymphocytes (10^5/ml) for 5 days.
*Results are expressed as cpm of ^{3}HTdr.

RESULTS

1. IL-10 Inhibits the Differentiation of Monocyte-Derived DC

1.1. Morphological Analysis. Highly purified blood monocytes were cultured for 7 days with GM-CSF + IL-13 with or without IL-10. Cultures in GM-CSF+IL-13 developed into cells with abundant cytoplasm, irregular membrane ruffling and cytoplasm protrusions, characteristic of DC. Cultures in GM-CSF+IL-13 in the presence of IL-10 did not develop into cells with DC morphology, about 30–40% of the cells were very large, with a macrophage-like morphology.

1.2. Phenotype Analysis and APC Function. Table 1 shows the phenotype analysis and the ability of DC to stimulate cord blood T cells in MLR. Cells cultured with GM-CSF+IL-13 after 7 days expressed high levels of CD1a and were negative for CD14 and CD16. Cells cultured with GM-CSF+IL-13 in the presence of IL-10 were negative for CD1a but expressed CD14, and high levels of the macrophage markers CD16 and CD68. Optimal inhibition of DC differentiation was observed with 20 ng/ml of IL-10 and the effect of IL-10 was dose-dependent. IL-10 did not modify the number of large cells positive for MHC class II but fluorescence intensity was greatly decreased.

Cultures differentiated in the presence of 20 ng/ml of IL-10 were less potent APC than control cultures, showing 10 to 100 fold less capacity of T cell stimulation. At lower doses, the effect of IL-10 was not statistically different.

The inhibition of DC differentiation was best observed when IL-10 was added to the cocktail of GM-CSF and IL-13 at the beginning of the 7 day culture. When IL-10 was added at day +3 or +6, a smaller inhibition was observed (not shown).

1.3. Endocytic Acivity. Monocyte-derived DC cultured with GM-CSF and IL-13 showed high endocytic acivity, measured as uptake of FITC-dextran. Interestingly, cells cultured with GM-CSF+IL-13 in the presence of IL-10 had much higher endocytosis of FITC-dextran (5.7±1.4 fold increase, mean ±SD of 4 experiments), a representative experiment is shown in Figure 1 A. It is worth noting that IL-10 increased the fluorescence intensity of the mannose receptor (MR) expression, recognized by PAM-1 mAb on these cells (cultures in GM-CSF+IL-13, RFI = 12; GM-CSF+IL-13 + IL-10; RFI = 24). Endocytosis of FITC-dextran reflects uptake via the MR and macropinocytosis.Experiments of

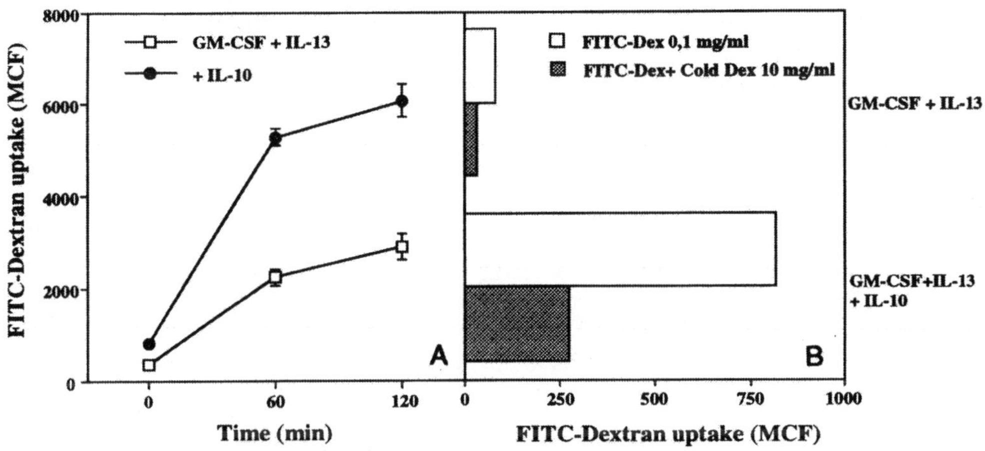

Figure 1. Effect of IL-10 on the endocytosis of FITC-dextran. Monocytes were cultured for 7 days in GM-CSF (50 ng/ml) + IL-13 (10 ng/ml) with or without IL-10 (20ng/ml). A) Endocytosis was evaluated as uptake of FITC-dextran (1 mg/ml) at 1 and 2 h and measured using a FACSTAR. B) Cold competition with unlabeled dextran (10 mg/ml), at 2 h. Results are expressed as mean channel fluorescence (MCF).

competition with unlabeled dextran indicated that IL-10 increases both MR-dependent and independent endocytosis (Fig.1 B).

2. Promotion of Differentiation to Mature Macrophages by IL-10

2.1. Cytochemical Analysis. To confirm that the abrogation of DC differentiation from monocytes by IL-10 was accompanied by the appearance in culture of large macrophage-like cells, we performed a cytochemical analysis. While control DC did not express non-specific esterase (ANAE, 5% of the cells), cultures treated with IL-10 were 72% positive, similarly to macrophages (69%) differentiated in vitro from monocytes cultured with 40% human serum.

Collectively these results indicate that in the presence of IL-10 a substantial proportion of monocytes are driven to become mature mononuclear phagocytes.

3. Effect of IL-10 on Differentiated DC

It was important to evaluate whether IL-10, in addition to blocking differentiation of monocyte-derived DC, also affected functions of differentiated DC and induced differentiated DC to acquire macrophage like characteristics. As mentioned above, exposure of differentiated DC (day 6 of culture) to IL-10 for 24 h caused only a modest reduction of MHC Class II and CD1a expression, and little inhibition of their ability to stimulate cord blood T cells in MLR. Culture of differentiated DC for as long as 5 days with 20 ng/ml of IL-10 did not result in the expression of the macrophage markers CD14, CD16 and CD68 although there was some reduction of CD1a expression and of stimulation of cord blood T cells (not shown). Thus these results indicate that differentiated DC do not acquire a phenotype of macrophages when exposed for relatively long time to IL-10.

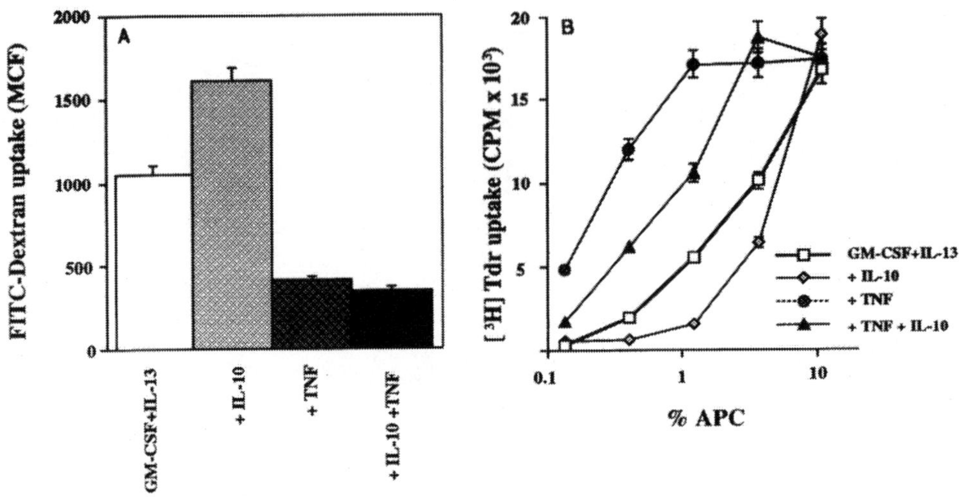

Figure 2. Effect of IL-10 and TNF on the endocytic activity of differentiated DC. DC were obtained by culturing monocytes for 7 days in GM-CSF (50 ng/ml) + IL-13 (10 ng/ml). A) Endocytosis of FITC-dextran. IL-10 (20 ng/ml) and/or TNF (20ng/ml) were concomitantly added in the last 24 h of culture. Endocytosis was evaluated as uptake of FITC-dextran (1 mg/ml) at 2 h, and measured using a FACS. The results are expressed as mean channel fluorescence. B) MLR. Responder cells were cord blood T lymphocytes depleted of monocytes and B lymphocytes; they were plated at 1×10^5 cells/well. APC were mixed at the indicated concentrations. 3H-Tdr was added in the last 18 h of a 5 days experiment.

3.1. IL-10 Augments Endocytosis of Differentiated DC. Twenty-four hour exposure of differentiated DC to IL-10 augmented by 50% the endocytosis of FITC-dextran, while as expected, TNF reduced it by 70% (Fig.2, A). IL-10 added concomitantly with TNF did not counteract its inhibitory function. Panel B shows that TNF dramatically augmented the accessory function of DC; in contrast to what observed with endocytosis, IL-10 partially inhibited the TNF-induced stimulation of MLR.

4. Concluding Remarks

The present study demonstrates that IL-10 completely blocks the GM-CSF+IL-13-driven differentiation of CD14+ monocytes to DC as assessed by morphology, expression of CD1a and stimulatory capacity in MLR for naive cord blood T cells. Moreover, in the presence of IL-10, a substantial proportion of the cells had the morphology of mature macrophages and showed the cytochemical (non-specific esterase) and phenotypical (CD14+, CD16+, CD68+) features of mature macrophages. Hence IL-10, in concert with GM-CSF and IL-13, does not freeze CD14+ precursors at an immature stage, but allows or promotes their differentiation to mature macrophages.

When differentiated DC, obtained by 6-day culture with GM-CSF and IL-13, were exposed to IL-10 for 24 h, or as long as 5 days, cells did not acquire the morphological and phenotypic features of mature macrophages. On the other hand, monocytes-derived macrophages obtained with GM-CSF + IL-13 + IL-10 did not acquire a DC phenotype when cultures were prolonged in the absence of IL-10 (not shown). Thus, IL-10 acts only on monocytes prior to their committment and does not shunt already differentiated DC to mature macrophages, or vice versa.

In the present study, we have also found that IL-10 stimulates expression of the MR and the endocytic activity of DC. Hence IL-10 is not merely an inhibitor of accessory cell function but rather a selective modulator. We speculate that augmented scavenging of antigen and decreased antigen presenting capacity of differentiated DC, associated with blocking of differentiation from monocytes, contributes to the immunosuppressive and, possibly, tolerogenic activity of IL-10.

ACKNOWLEDGMENTS

Supported in part by the Italian Association for Cancer Research (AIRC) and Istituto Superiore di Sanit' (project AIDS).

REFERENCES

1. Steinman RM: The dendritic cell system and its role in immunogenicity. Annu Rev Immunol 9:271, 1991
2. Caux C, Liu YJ, Banchereau J: Recent advances in the study of dendritic cells and follicular dendritic cells. Immunol Today 16:2, 1995
3. Sallusto F, Lanzavecchia A: Efficient Presentation of Soluble Antigen by Cultured Human Dendritic Cells Is Maintained by Granulocyte/ Macrophage Colony-Stimulating Factor Plus Interleukin-4 and Downregulated by Tumor Necrosis Factor-alpha. J Exp Med 179:1109, 1994
4. Romani N, Gruner S, Brang D, Kampgen E, Lenz A, Trockenbacher B, Konwalinka G, Fritsch PO, Steinman RM, Schuler G: Proliferating dendritic cell progenitors in human blood. J Exp Med 180:83, 1994
5. Piemonti L, Bernasconi S, Luini W, Trobonjaca Z, Minty A, Allavena P, Mantovani A: IL-13 supports differentiation of dendritic cells from circulating precursors in concert with GM-CSF. Eur Cytokine Netw 6:245, 1995
6. Mosmann TR: Properties and functions of interleukin-10. Adv Immunol 56:1, 1994
7. Chang CH, Furue M, Tamaki K: B7–1 expression of Langerhans cells is up-regulated by proinflammatory cytokines, and is down-regulated by interferon-gamma or by interleukin-10. Eur J Immunol 25:394, 1995
8. Buelens C, Willems F, Delvaux A, Picrard G, Delville JP, Velu T, Goldman M: Interleukin-10 differentially regulates B7–1 (CD80) and B7- 2 (CD86) expression on human peripheral blood dendritic cells. Eur J Immunol 25:2668, 1995
9. Beissert S, Hosoi J, Grabbe S, Asahina A, Granstein RD: IL-10 inhibits tumor antigen presentation by epidermal antigen-presenting cells. J Immunol 154:1280, 1995

54

HUMAN DENDRITIC/LANGERHANS CELLS CONTROL GROWTH AND DIFFERENTIATION OF CD40 ACTIVATED B CELLS

Bertrand Dubois, Béatrice Vanbervliet, Jérôme Fayette, Catherine Massacrier, Francine Brière, Jacques Banchereau, and Christophe Caux

Schering-Plough
Laboratory for Immunological Research
Dardilly, France

INTRODUCTION

During an immune response, dendritic cells (DC) capture the antigen at site of injury, and migrate through the afferent lymph stream to the lymph-nodes where they efficiently activate naive T cells. This T cell activation is followed by B cell recruitment, which occurs in the extrafollicular area, where DC home[1]. Accordingly, we wondered herein whether DC might directly interact with B cells, using DC generated *in vitro* from $CD34^+$ progenitors, called Dendritic-Langerhans cells[2,3] (D-Lc). As both DCs[4] and B cells[5] express functional CD40, we used CD40-ligand transfected L cells[6] as surrogate activated T cells, to study the effect of DCs on B cell activation. We show that D-Lc enhanced the proliferation of CD40-activated naive B cells and Ig commited B lymphocytes (3–8 fold). In addition, D-Lc induced a 10 to 100 fold enhancement of IgG and IgA secretions by Ig committed B cells (essentially memory B cells). Most notably, in the presence of IL-2 , D-Lc stimulated CD40-activated naive B cells to produce high amounts of IgM. Using transwells, we showed that the effect on B cell proliferation is soluble and independent of CD40 triggering on D-Lc. In contrast, the IgM production is partly dependent of soluble molecule(s) secreted after CD40 engagement on D-Lc. The use of anti-cytokine blocking antibodies indicate that D-Lc derived IL-12 is absolutely required (although not sufficient) for the differentiation of naive B cells into IgM secreting cells in presence of IL-2.

Thus, in addition to priming naive T cells in the extrafollicular areas of secondary lymphoid organs, DC may also directly regulate B cell differentiation and participate in the IL-2 dependent primary B cell activation through the release of IL-12.

Dendritic Cells in Fundamental and Clinical Immunology
edited by Ricciardi-Castagnoli, Plenum Press, New York, 1997

MATERIALS AND METHODS

1. Generation of D-Lc

Umbilical cord blood samples were obtained according to institutional guidelines. Cells bearing CD34 antigen were isolated from non adherent mononuclear fractions through positive selection, using anti CD34 Monoclonal Antibody (MAb) Immu-133.3 (Immunotech, Marseille, France) and a Magnetic Cell Sorter (Mini Macs, Miltenyi Biotech). Cultures were established in the presence of GM-CSF (100 ng/ml) and TNF α (2,5 ng/ml) in complete medium (10% FCS). After 12 days, cultures contained 70 to 90% CD1a$^+$ D-Lc[2]. D-Lc were further separated according to CD1a expression by FACS sorting (>98% pure).

2. Purification of B Cells

Total B cells were purified from tonsil through Ficoll-Rosetting and depletion with MAbs directed against non B cells (T cells, monocytes and NK cells). In the isolated

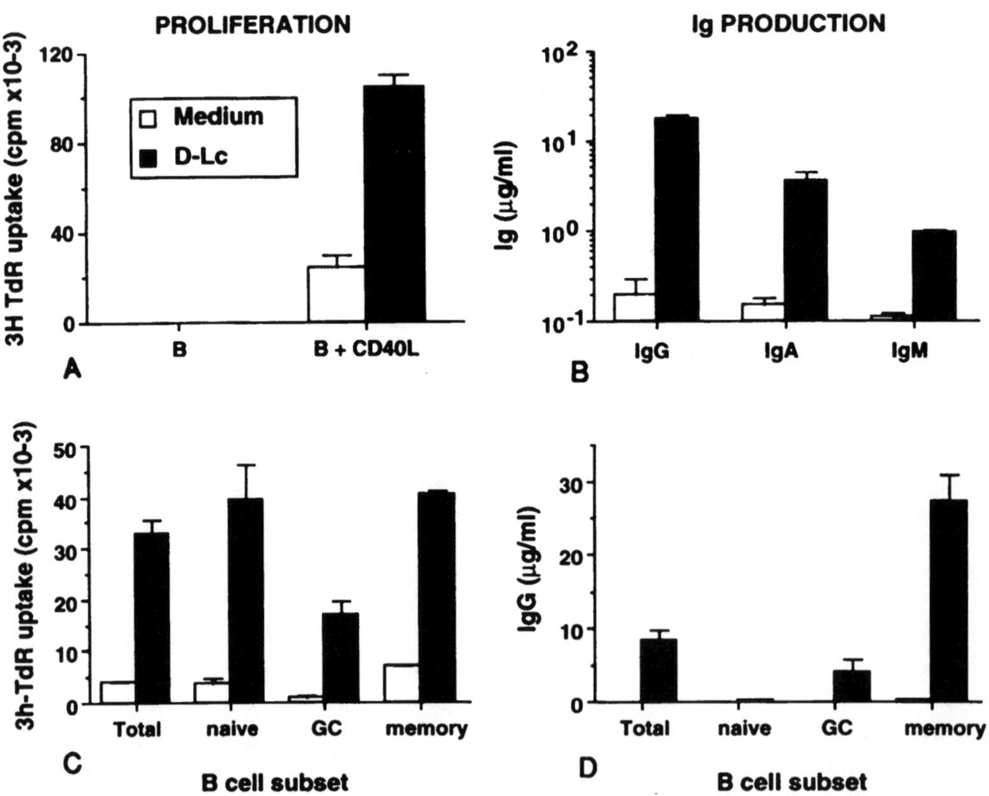

Figure 1. CD34$^+$ progenitors derived D-Lc were used, after irradiation (3000 rad) as stimulators of CD40 dependent B cell proliferation and differentiation. (A,B) Total tonsillar B cells (10^4 per well) were cultured over irradiated CD40-L transfected L cells (7500 rad, $2{,}5.10^3$ per well) in the presence or absence of D-Lc (10^4 per well). (C,D) Total B cells, B cells from Germinal Center or memory B cells (purification detailed in materials & methods) were cultured under CD40 triggering in presence or absence of 10^4 D-Lc. (A,C) B cell proliferation at day 6. (B,D) production of Ig measured at day 13.

population, >99% expressed CD19 and CD20 and <1% expressed CD2 or CD14 Ags. Isolation of $sIgD^+$ and $sIgD^-$ B cells subpopulations was performed using a preparative magnetic cell sorter (MACS, Becton Dickinson & Co). The separation based on sIgD expression has been described in details elsewhere[7]. IgD was expressed on >99% of the $sIgD^+$ B cell subpopulation and <1% of $sIgD^-$ B cell subpopulation, as assessed by fluorescence analysis using a FACScan (Becton Dickinson). For certain experiments, $sIgD^-$ B cells were further separated according to CD38 and CD39 expression into $CD38^+CD39^-$ germinal center (GC) B cells and $CD38^-CD39^+$ memory B cells using anti-CD38 and anti-CD39 mAbs and bead depletion as described earlier[8].

3. Coculture of B Cells and D-Lc

2.5×10^3 irradiated CD40L L cells (7500 rad) were seeded together with 10^4 B lymphocytes (either total B cells or B cells subsets) in the presence or absence of *in vitro* generated D-Lc (10^4) in 96 wells culture plate (Nunc). B cell proliferation was monitored by tritiated Thymidine ([^{3}H]TdR) incorporation after 6 days of coculture. Cells were incubated for the last 16 hours with 1 µCi of [^{3}H]TdR. For determination of Ig production, supernatants were recovered after 13 days and used for indirect ELISA[9].

In other experiments, B cells and D-Lc were cultured in separate compartments using transwells (Costar, Wilmington, MA). 10^5 D-Lc cultured in the presence or absence of CD40 triggering (2.5×10^4 CD40L L cells or CD32 L cells used as control) in the lower compartment (in a total volume of 0.8 ml) were assayed for their ability to stimulate growth and differentiation of 1.5×10^4 B cells activated by 3.75×10^3 CD40L L cells in the top of the transwells (in a total volume of 0.2 ml). DNA synthesis of B cells was performed by transfering, at day 6, the cells present in the top of the transwells into flat bottom 96 wells plates and pulsing them with [^{3}H]TdR for the last 16 hours of the culture period.

RESULTS

1. D-Lc Induced Memory B Cells to Produce High Amounts of Igs

Tonsillar B cells (10^4 per well) were cultured over irradiated CD40-L L cells ($2{,}5.10^3$ per well, 7500 rad) in presence or absence of $CD34^+$ progenitors derived D-Lc (3000 rad). Although, D-Lc had no effect on B cell proliferation in the absence of any activator (right panel of figure 1A), D-Lc enhanced the proliferation of CD40-activated total B cells (3–8 fold enhancement of [^{3}H]TdR uptake) (left panel of figure 1A). Most notably, in the absence of exogenous cytokines, the addition of D-Lc dramatically increased the production of IgG, IgA and, to a lesser extent IgM, by CD40-activated B cells. In the experiment shown in figure 1B (note the log scale), addition of 10^4 D-Lc increased the production of IgG by 100 fold, IgA (by 20 fold) and IgM (by 10 fold) over the background (≤ 300 ng/ml).

Tonsillar B cells consist of naive B cells, expressing surface IgM and IgD as well as germinal center (GC) B cells and memory B cells[8] that have mostly undergone isotype switch, thus resulting in a loss of IgD expression. Therefore we analysed whether either purified $sIgD^+$ B cells (naive B cells) or $sIgD^-$ B cells (B cells from germinal centers and memory B cells) were preferential targets for D-Lc effects. As shown in figure 1C, D-Lc stimulates the proliferation of both CD40-activated naive (IgD^+) and Ig committed B cells.

As shown in figure 1D, CD40-activated naive $sIgD^+$ B cells did not produce IgG whether or not D-Lc are added. Separation of $sIgD^-$ B cells into $CD38^+CD39^-$ GC cells and $CD38^-$ $CD39^+$ memory B cells (using Mabs and bead depletion)[8], allowed identification of the responding cell population. As shown in figure 1D, memory B cells represent the main source of IgG as, when cultured in the presence of D-Lc, 10^4 memory B cells produced 27 μg/ml while GC cells produced less than 0.5 μg/ml and 4 μg/ml respectively, under such culture conditions.

2. In Presence of IL-2, D-Lc Induced Naive B Cells to Secrete High Levels of IgM

In view of the weak stimulation of IgM production (figure 1B), although D-Lc strongly stimulated the growth of naive B cells (figure 1C), we wondered whether addition of exogenous cytokines could enhance this production. In IL-2 supplemented cultures, addition of D-Lc induced CD40-activated B cells to secrete high levels of IgM (figure 2A), which reached 50 μg/ml (mean increase: 161 fold, n=16). Separation of B cells into $sIgD^+$ and $sIgD^-$ B cells showed that production of IgM was mainly a property of the naive $sIgD^+$ B cell population (figure 2B). Note, however that $sIgD^-$ B cells can be induced to produce significant levels of IgM.

3. Involvment of IL-12 in the IL-2 Dependent IgM Production

Experiments were designed to determine whether the effects of D-Lc on B cells were mediated through soluble factors and/or cell/cell interactions and whether D-Lc had to be activated through their CD40 antigen. D-Lc were separated from B cells by a permeable membrane using Transwells. As shown in figure 3A, D-Lc cultured in the lower compartment, with or without CD40 triggering, can enhance CD40-dependent B cell proliferation as much as if they were cocultured in direct contact. In the presence of IL-2, CD40-activated D-Lc, but not unactivated D-Lc, can support IgM production by B cells

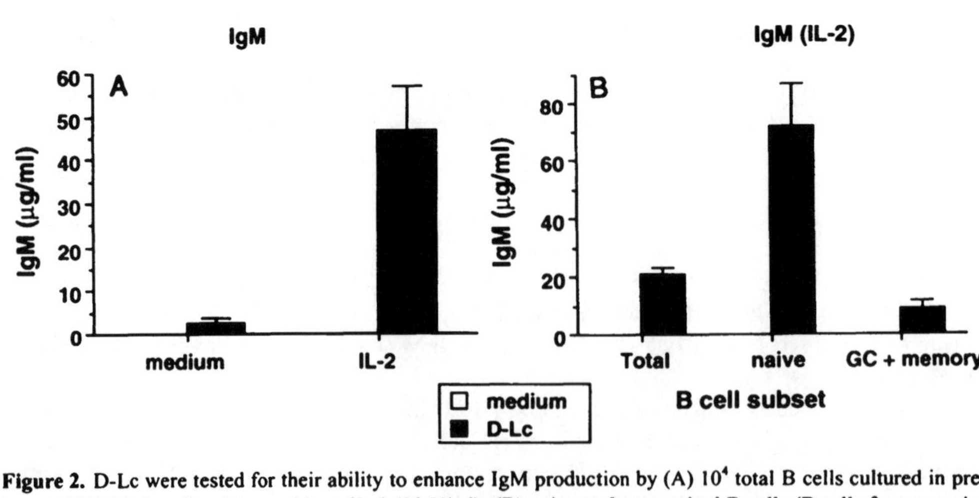

Figure 2. D-Lc were tested for their ability to enhance IgM production by (A) 10^4 total B cells cultured in presence of CD40L L cells with or without IL-2 (20 U/ml); (B) naive or Ig commited B cells (B cells from germinal centers and memory B cells as detailed in materials & methods) in presence of IL-2. IgM production was measured after 13 days of culture.

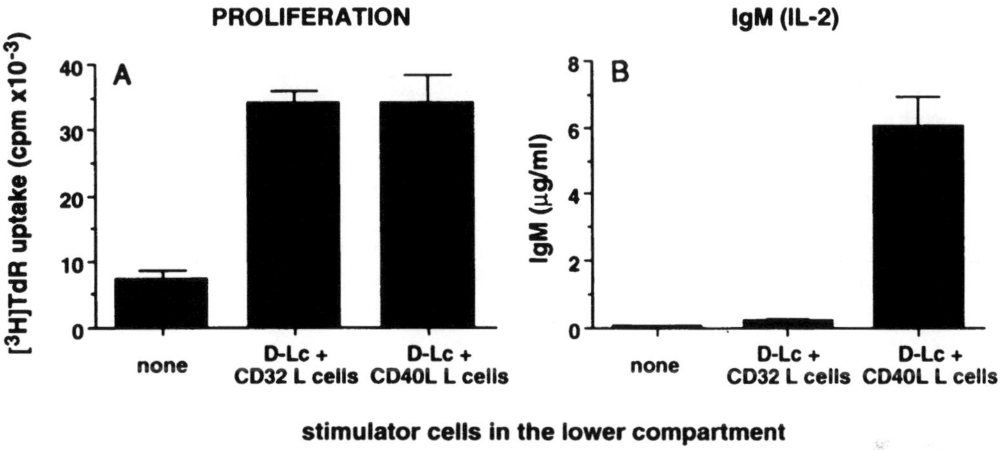

Figure 3. 1.5×10^4 sIgD$^+$ B cells were cultured in the presence of 3.75×10^3 CD40L L cells in the top of the transwell. 10^5 D-Lc were seeded in the lower compartment with 2.5×10^4 CD40L L cells or CD32 L cells used as controls. DNA synthesis of B cells (A) of the upper compartment, in absence of cytokine, was determined after 6 days of coculture. IgM production in the presence of IL-2 (B) was determined by ELISA after 13 days.

activated through CD40 (Figure 3B). However, IgM levels obtained were lower than those observed in direct contact coculture (6.1±0.8 µg/ml versus 15.5±2.5 µg/ml), suggesting that cell/cell contacts may also contribute. Note that D-Lc do not deliver any signal when B cells are not activated. Thus, the effect of D-Lc on B cell proliferation appears to be mediated by soluble factor(s), which production is independent of CD40 engagement on the D-Lc. In contrast, the effect on B cell differentiation, also mediated through soluble mediator(s), requires CD40 ligation on D-Lc.

We then screened a number of anti-cytokine antibodies for their ability to block either the D-Lc dependent B cell proliferation or the IgM production in presence of IL-2. Among those tested, an antibody directed against IL-12 appeared to block 95% of the IgM production in presence of D-Lc and IL-2 (data not shown). Note that this antibody has no effect on B cell proliferation induced by D-Lc. However, the production of IL-12 by D-Lc under CD40 triggering is not sufficient to account for the IL-2 dependent IgM production. Accordingly, IL-12 has little if any effect on IgM production by naive B cells cultured in presence of IL-2 and CD40L L cells without D-Lc (data not shown).

DISCUSSION

We have shown that D-Lc generated *in vitro* enhanced the proliferation and differentiation of both naive and memory B cells in presence of CD40 triggering but in absence of T cells. D-Lc generated *in vitro* are composed of LC and other DCs that may contain the Interdigitating Dendritic Cells population of the extrafollicular area of secondary lymphoid organs[3]. The latter cells are particulary important in the initiation of primary and reactivation of secondary humoral response[1]. Thus in addition to activating T cells in this area, DCs could directly modulate B cell activation, during primary and secondary humoral response.

The mechanism by which D-Lc regulates B cell responses is not yet totally elucidated. However, Transwells experiments show that D-Lc enhance growth of CD40-activated B cells through the production of soluble factor(s) independently of CD40 triggering. In contrast, the effect on IgM secretion in the presence of IL-2 is partly dependent on the release by D-Lc of soluble mediator(s) after CD40 engagement. IL-12 play an important role in this IgM production, although not sufficient, thus in addition to its role in T cell priming[10], IL-12 may also participate in the activation of naive B cells. The mechanism of induction of IL-2 responsiveness is not yet clear. As D-Lc expressed CD25 after CD40 engagement[4], IL-2 could signal D-Lc which in return induce B cells to produce IgM. Alternatively, such an effect might be due to the ability of D-Lc to upregulate CD25 expression on CD40-activated naive B cells (data not shown).

Many authors have underlined the critical role of DCs in humoral response[11–13]. The direct modulation of B cell responses together with T cell priming may explain the primordial role of DCs in this process.

REFERENCES

1. Steinman, R. M.. The dendritic cell system and its role in immunogenicity. Ann Rev Immunol. 9:271–296 (1991).
2. Caux, C., C. Dezutter-Dambuyant, D. Schmitt, and J. Banchereau. GM-CSF and TNF-alpha cooperate in the generation of dendritic Langerhans cells. Nature. 360:258–261 (1992).
3. Caux, C., B. Vanbervliet, C. Massacrier, C. Dezutter-Dambuyant, B. de Saint-Vis, C. Jacquet, K. Yoneda, S. Imamura, D. Schmitt and J. Banchereau. $CD34^+$ Hematopoietic Progenitors from Human Cord Blood Differentiate Along Two Independent Dendritic Cell Pathways in Response to GM-CSF+TNFα. J Exp Med. 184: 695–706 (1996).
4. Caux, C., C. Massacrier, B. Vanbervliet, B. Dubois, C. Van Kooten, I. Durand, and J. Banchereau. Activation of human dendritic cells through CD40 cross-linking. J. Exp. Med. 180:1263–1272 (1994).
5. Banchereau, J., F. Bazan, D. Blanchard, F. Briere, J. P. Galizzi, C. van Kooten, Y. J. Liu, F. Rousset, and S. Saeland. The CD40 Antigen ans its Ligand. Annu. Rev. Immunol. 12:881–922 (1994).
6. Garrone, P., E.M. Neidhardt, E. Garcia, L. Galibert, C. van Kooten, and J. Banchereau. Fas ligation induces apoptosis of CD40-activated human B lymphocytes. J. Exp. Med. 182:1265–1273 (1995).
7. Defrance, T., B. Vanbervliet, F. Brière, I. Durand, F. Rousset, and J. Banchereau. Interleukin 10 and Transforming Growth Factor ß cooperate to induce anti-CD40-activated naive human B cells to secrete Immunoglobulin. J. Exp. Med. 175:671–682 (1992).
8. Liu, Y.J., C. Barthélémy, O. de Bouteiller, C. Arpin, I. Durand, and J. Banchereau. Memory B cells from human tonsils colonize mucosal epithelium and directly present antigen to T cells by rapid upregulation of B7.1 and B7.2. Immunity 2:238–248 (1995).
9. Defrance, T., B. Vanbervliet, J. Pène, and J. Banchereau. Human recombinant IL-4 induces activated B lymphocytes to produce IgG and IgM. J. Immunol. 141:2000–2005 (1988).
10. Seder, R.A., R. Gazzinelli, A. Sher and W.E. Paul. IL-12 acts directly on $CD4^+$ T cells to enhance priming for IFNγ production and diminishes IL-4 inhibition in such priming. Proc. Natl. Acad. Sci. USA. 90:10188–10192 (1993).
11. Inaba, K., J. P. Metlay, M. T. Crowley, and R. M. Steinman. Dendritic cells pulsed with protein antigens in vitro can prime antigen-specific, MHC-restricted T cells in situ. J Exp Med. 172:631–640 (1990).
12. Sornasse, T., V. Flamand, B. G. De, H. Bazin, F. Tielemans, K. Thielemans, J. Urbain, O. Leo, and M. Moser. Antigen-pulsed dendritic cells can efficiently induce an antibody response in vivo. J Exp Med. 175:15–21 (1992).
13. Flamand, V., T. Sornasse, K. Thielemans, C. Demanet, M. Bakkus, H. Bazin, F. Tielemans, O. Leo, J. Urbain, and M. Moser. Murine dendritic cells pulsed in vitro with tumor antigen induce tumor resistance in vivo. Eur J Immunol. 24:605–610 (1994).

55

ACTIVATION OF DENDRITIC CELLS BY SURROGATE T CELL INTERACTIONS LEADS TO ENHANCED COSTIMULATION, SECRETION OF TH1–ASSOCIATED CYTOKINES, AND CTL INDUCTIVE CAPACITY

Cara C. Wilson,[1] Thomas Tueting,[2] Debora Ma,[2] Cathy Haluszczak,[2] Michael Lotze,[2] and Walter Storkus[2]

[1]Department of Medicine and
[2]Department of Surgery
University of Pittsburgh Medical Center and the University of Pittsburgh Cancer Institute
Pittsburgh, Pennsylvania 15261

1. INTRODUCTION

Dendritic cells (DC), as professional antigen presenting cells (APC), have the capacity to capture antigen and migrate to lymphoid organs, where antigen presentation to and priming of naive T cells takes place (reviewed in 1). It is currently believed that effective induction of antigen-specific T cell responses requires a dynamic dialogue of multiple, sequential interactions between the DC as APC and a given T cell clone in order to prime the cell for expansion and effector function. The first interaction (signal 1) is felt to involve the MHC I/II-peptide on APC with the T cell receptor (TCR) on the effector lymphocyte, with a second signal being the costimulatory interaction between CD80 (B7.1)/CD86(B7.2) on DC with CD28 on the T cell. Signals 1 and 2 combine to activate the T cell, upregulating the expression of CD40L which can interact with DC-expressed CD40 (reviewed in 2). This DC-T cell interaction results in the generation of an antigen-specific T cell responses. The ability of DC to prime naive T cells has been attributed to several factors, most notably the high levels of expression of MHC and costimulatory molecules such as B7.1 and B7.2. It has also been shown that DC are capable of producing several cytokines, among them IL-12, a cytokine critical for the development of Th1 responses (3,4). Recently, ligation of CD40 on both human (5) and murine (6) DC has been reported to result in the upregulation of costimulatory molecules and the induction of IL-12. These findings suggest that the interaction of T cells with DC result in signals to DC which are critical for induction of the cellular immune response. Whereas many of the signals required by T cells for activation and proliferation have been described, the signaling requirements for optimal DC activation remain poorly defined.

We have evaluated the impact of certain T-DC interactions on human DC cultured from peripheral blood in the presence of rIL-4 and rGM-CSF. By mimicking T cell ligands using monoclonal antibodies specific for DC-expressed counter-receptors (i.e. anti-CD40, -CD80, -CD86, -MHCI/II) and crosslinking them with anti-mouse Ig or by adherence to plastic (providing a solid-state cross-linking matrix), we can dissect out the signals, singly or in combination, which have important effects on DC activation. By investigating the impact of T cell-mediated signals on DC phenotype and function (cytokine secretion and T cell stimulatory capacity) we hope to better understand how these signals influence the nature and magnitude of antigen-specific T cell immune responses. Further, it is hoped that such knowledge will provide the means by which to optimally activate DC for implementation in clinical vaccine trials.

2. MATERIAL AND METHODS

2.1. Media and Reagents for Cell Culture

The medium used for all DC cultures was serum-free AIM-V (GIBCO-BRL, Gaithersburg, MD) supplemented with 1000 U/ml rhIL-4 and 1000 U/ml rhGM-CSF (Schering-Plough Research Institute, Kenilworth, NJ). T cells for functional assays were maintained in RPMI 1640 media (GIBCO-BRL) supplemented with 100 U/ml penicillin, 100 μg/ml streptomycin, and 10% heat-inactivated human AB serum with or without 10–50 U/ml rIL-2.

2.2. Dendritic Cell Generation

Heparinized peripheral blood was obtained by venipuncture from normal donors with their consent. PBMC were obtained after density centrifugation on Ficoll-Hypaque gradients per the manufacturer s instructions (LSM, Organon-Teknika, Durham, NC). After 4–5 washes in Hank s Buffered Saline (HBSS from GIBCO-BRL, Gaithersburg, MD), 10^8 cells per 10 ml serum-free AIM-V medium were plated in T75 flasks (COSTAR, Cambridge, MA) and incubated at 37° C for 1–1.5h to allow for cell attachment. After removal of nonadherent cells with 4 gentle washes with HBSS, the plastic-adherent cells were cultured for 5–7 days in AIM-V medium supplemented with 1000 U/ml rIL-4 and 1000 U/ml rGM-CSF (7,8). Following a 5–7 day culture period, nonadherent cells were harvested and DC were further purified by discontinuous density centrifugation on a layer of Nycoprep 1.064(Nycomed):LSM(Organon-Teknika) 9:1, 1000g × 10 minutes. Gradient interface cells were harvested, washed with HBSS, and used in phenotypic and functional assays. Cells generated in this fashion were determined to be > 90% DC based on morphology and the expression of a CD3/CD14/CD16/CD20 negative, MHC class II +, CD40+, CD80+, CD86+ phenotype assessed by indirect immunofluorescence assays monitored by flow cytometry. Day 7 yields were approximately 8–15% of starting normal donor PBMC numbers.

2.3. Crosslinking of Surface Markers on Dendritic Cells

Cultured dendritic cells (day 5–7) were incubated with primary antibodies specific for CD40 (628.5, kindly provided by Dr. J. Ledbetter, Bristol-Meyer Squibb, Seattle,WA),

MHC I (W632, American Type Culture Collection, Rockville, MD), MHC II (L243, ATCC, Rockville, MD) at concentrations of 1–10μg/10^6 DC at 4 C for 30 minutes. After 2 washes with HBSS, 1–5μg/ml anti-mouse Ig G (Fc specific) F(ab $)_2$ antibody (Sigma) was added for 30 min at 4 C. DC were washed and placed in serum-free AIM-V medium to incubate at 37 C for 24–72h prior to harvesting for flow cytometric analysis and functional assays.

2.4. Analysis of DC Cell Surface Markers by Flow Cytometry

DC were phenotyped using a panel of commercially-available, directly fluorochrome-labelled antibodies with the following specificities: CD3, CD14, CD16, CD20, HLA-DR (Becton-Dickinson, Mountainview, CA); CD40, CD80, CD86 (Ancell Corporation, Bayport, MN). Cells were stained using standard procedures (9), fixed in 4% formalin and analyzed by flow cytometry (FACScan, Becton-Dickinson, Mountainview, CA).

2.5. Detection of Intracellular Cytokines by Flow Cytometry

Cultured DC activated for 24–48h with cross-linked mAb were harvested and incubated in AIM-V with Brefelden A (Sigma) at 1 μg/ml at 37°C for 3–4h. Cells were stained for surface DC markers, permeabilized with saponin buffer (0.1% saponon, 0.1% sodium azide, 0.1% BSA, 10mM Hepes), then stained with directly-conjugated anti-cytokine mAbs as descriibed in (10). The specific antibodies used for intracellular staining, anti-IL-12 p75 (20C2, Hoffmann-La Roche) and anti-IFNalpha (MMHA-2, PBL Biomedical Labs, West Caldwell, NJ) were directly FITC-conjugated per the fluorochrome manufacturer's guidelines (Sigma). Labelled cells were then assessed by FACScan analysis.

2.6. Measurement of hIL-12 and hIFN-α in DC Culture Supernatant by ELISA

Cell-free supernatant was collected from DC cultures 24–72h following cross-linking and stored at -20° C. hIL-12 p75 heterodimer and hIFN-α subtypes 1 and 2 were detected using sandwich ELISAs. The capture mAb for IL-12p75 ELISA was the p75-specific mAb derived from clone 20C2, and the detection mAb was p40-specific (clone 4D6, POD-conjugated). Both antibodies were kindly provided by Dr. Maurey Gately, Hoffmann-La Roche, Inc., Nutley, N.J. The capture mAb for the IFN-α ELISA was a polyclonal sheep anti-h IFN-α Ab (kindly provided by P. Zavodny, Schering-Plough Research Institute, Kenilworth, NJ) and the detection antibody was the murine anti-hIFN-α mAb, MMHA-2 (PBL Biomedical Labs, West Caldwell, NJ). A POD-conjugated goat anti-mouse antibody was used as a detector. TMB substrate was added and color developed for 15–30 min, then plates read for absorbance at 450nm.

2.7. Primary Induction of Allogeneic Cytotoxic T Lymphocytes

Donor HLA-A2+/A3- DC (day 7) were generated as described above (section 2.2), cross-linked as outlined in section 2.3 for 48h, irradiated x 2500 rad, and co-cultured with allogeneic HLA-A2- responder peripheral blood T cells in a R:S ratio of 20:1 in 24 well plates in complete RPMI with 10% human AB serum. 10 IU/ml rhIL-2 was added to the cultures on day 5. On day 7, cultured T cells were harvested and plated as effector cells in a standard 4h chromium release assay at an effector:target ratio of 10:1. Targets consisted of chromium-labelled EBV-B-LCL generated from the DC donor, EBV BCL expressing only the relevant alloantigen HLA-A2 (C1R.A2) or an EBV BCL expressing only the irrelevant allospecificity HLA-A3 (C1R.A3) (9).

3. RESULTS

3.1. CD40 and MHCII Ligation on DC Results in Upregulation of Costimulatory and MHC Molecules

DC cultured in GM-CSF and IL-4 for 5–7 days are characterized by a large percentage of cells expressing HLA-DR, CD86, and CD40, and with more moderate numbers of cells expressing CD80 and CD1a. As APC expression of MHC and costimulatory molecules has been shown to be critical for T cell priming, we investigated the expression levels of these surface markers 48–72 hours following cross-linking of MHC I, MHC II, and CD40 on DC. Ligation of CD40 on DC gave up to 3 fold increases in expression over baseline of costimulatory molecules CD80, CD86, and CD40 itself, as well as a similar increase in expression of MHC II. MHC II ligation also led to upregulation of certain APC-expressed molecules, primarily CD86, CD40, and MHCII, with less of an effect on CD80 expression. Co-ligation of CD40 with MHC II did not enhance expression of CD80 or CD86 over the maximum level of expression observed with ligation of either molecule alone, and tended to lead to a slight decrease in the relative expression of MHCII and CD40. These points are illustrated in a representative experiment in Figure 1.

3.2. Co-Ligation of MHC I with either CD40 or MHC II May Suppress Upregulatory Effects on Certain Surface Markers

Crosslinking of anti-MHC I mAbs alone on DC had no effect on surface expression of MHC and costimulatory molecules in multiple experiments. However, when MHC I

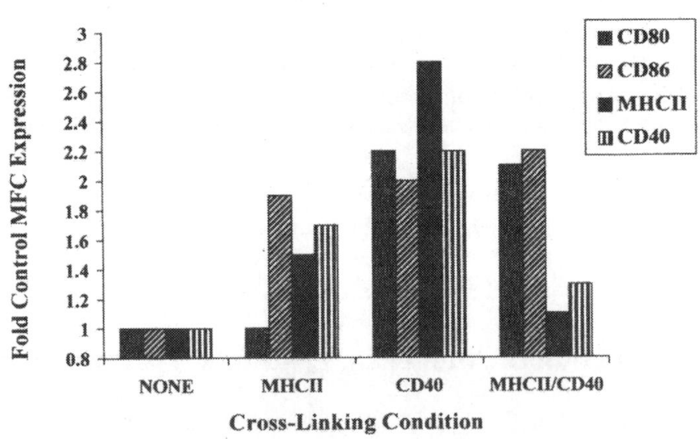

Figure 1. CD40 and MHC II ligation on DC results in upregulation of costimulatory and MHC molecules. Cultured DC were incubated with mAb specific for MHC II and CD40, alone and in combination, then cross-linked using anti-mouse Ig as described in Materials and Methods. Following 48–72h incubation, DC were stained with fluorochrome-labelled mAb against CD80, CD86, MHC II (DR), and CD40. Cells were then analyzed for expression of these surface markers by flow cytometry.

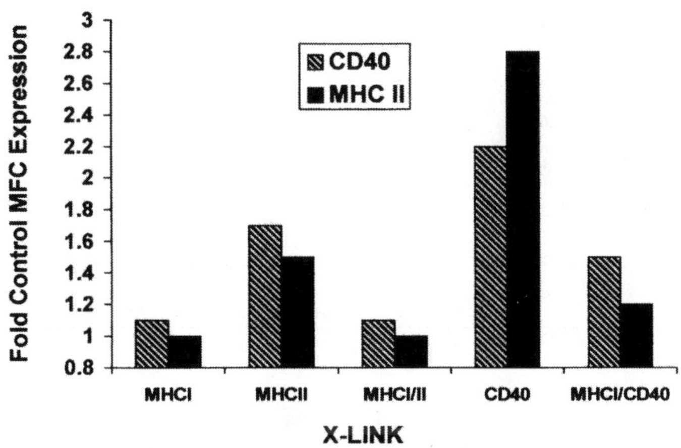

Figure 2. Co-ligation of MHC I with CD40 or MHC II suppresses upregulatory effects on DC surface marker expression. Cultured DC were incubated with mAb specific for MHC II and CD40, alone and in combination with anti-MHC I mAb, then cross-linked using anti-mouse Ig as described in Materials and Methods. Following 48–72h incubation, DC were stained with fluorochrome-labelled mAb against MHC II (DR), and CD40. Cells were then analyzed for expression of these surface markers by flow cytometry.

and either CD40 or MHC II were co-ligated on the same cells, marked decreases in the expression of both CD40 and MHCII, relative to that seen after ligation of either CD40 or MHC II alone, were noted. This point is illustrated in Figure 2. Interestingly, such suppressive effects of co-ligation were not observed in analysis of CD80 and CD86 expression which was enhanced by CD40 or MHC II ligation in the presence or absence of anti-class I (data not shown).

3.3. IL-12 Production and Secretion Is Induced by the Crosslinking of mAb Against CD40 and MHC II on DC

CD86+ DC were evaluated for intracellular expression of IL-12 p75 heterodimer at 24h post-cross-linking by flow cytometric analysis, as illustrated in Fig. 3a. When no cross-linking was employed, < 5% of cultured DC constitutively expressed IL-12. However, following ligation of CD40, MHC II, or a combination, the percentage of CD86+ DC staining for IL-12 increased several-fold, with up to 40% of DC expressing cytoplasmic IL-12 following co-ligation of CD40 and MHC II. The secretion of hIL-12 p75 was also evaluated after cross-linking DC for 24–72h (Figure 3b). The secretion patterns of IL-12 following ligation of CD40 and MHC II were in keeping with the patterns of intracellular staining for each given condition, with CD40 and MHC II ligation leading to 4–8 fold increases in secretion over background levels. The co-ligation of CD40 and MHC II induced the largest amount of secreted IL-12.

3.4. Activation of DC by Cross-Linking Leads to the Secretion of Pre-Formed IFN-α

Although freshly-isolated peripheral blood-derived DC have been shown to produce IFN-α following stimulation with viruses such as HIV (Ferbas et al. JI, 1994), to our

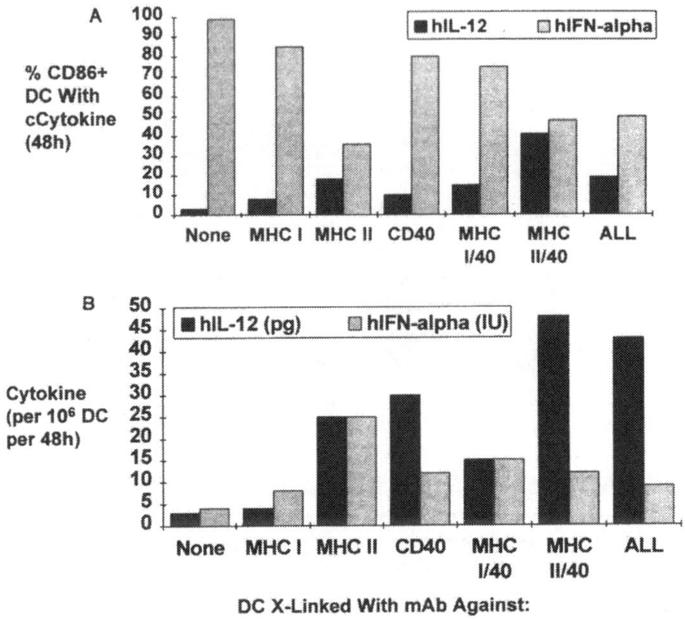

Figure 3. A and B. Induction of IL-12 and secretion of IFN-alpha by DC following ligation of CD40 and MHC II. DC were incubated with mAb specific for MHC I, MHC II, and CD40, then cross-linked using anti-mouse Ig as described in Materials and Methods. In figure 3A, 24h following cross-linking, DC were harvested and first stained for expression of surface CD86, followed by permeabilization and staining for intracellular expression of IL-12 and IFN-alpha as described. Results are depicted as a percentage of CD86+ DC which also express the given cytokine. In figure 3B, cell-free supernatants from DC cultured for 48h following cross-linking were collected and analyzed for presence of IL-12 or IFN-alpha by ELISA, as described.

knowledge this is the first time that IFN-α production has been evaluated in cultured human blood-derived DC following activation. Based on intracellular staining at 24h post-treatment, we find that a high percentage of untreated DC (i.e. no cross-linking) express IFN-α, and the percentage of cells expressing this cytokine decrease to varying degrees following crosslinking, with the most marked decrease following MHC II ligation (Figure 3a). In the case of secreted IFN-α, the amount of cytokine secreted varies inversely with the levels expressed intracellularly under a given condition, with the highest amounts of secreted IFN-α detected in the supernatants of DC activated by MHC II ligation. These data suggest that ligation of MHC II on DC may give a signal to secrete pre-formed IFN-α, thereby depleting intracellular stores of this cytokine. We are currently investigating which culture conditions or other factors might contribute to the production of IFN-α in cultured DC.

3.5. DC Activation via Cross-Linking Enhances Primary Allospecific CTL Induction *in Vitro*

DC are potent stimulators of allospecific T cell responses, yet primary induction of allospecific CTL may require multiple rounds of stimulation in vitro before killing is observed (unpublished observations). We investigated whether this process of CTL induction could be enhanced by activating DC prior to T cell priming (Figure 4). We observed

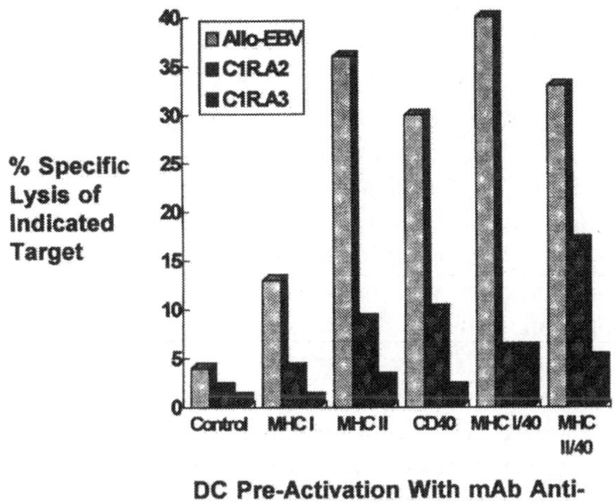

Figure 4. DC activation via cross-linking enhances primary allospecific CTL induction in vitro. Donor HLA-A2+/A3- DC (day 7) were generated as outlined in Materials and Methods. These cells were then cross-linked as indicated for 48h, irradiated, and used to stimulate allogeneic HLA-A2- responder peripheral blood T cells for 7 days. T cells were harvested and used in 4h cytotoxicity assays against EBV-BCL generated from the DC donor, EBV-BCL expressing only the HLA-A2 alloantigen (C1R.A2), or EBV-BCL expressing an irrelevant allospecificity (C1R.A3).

that although little response was noted 1 week following induction by unactivated DC, DC activated by ligation of MHC II and CD40 (alone and in combination) were potent inducers of allospecific CTL in this same time period. As might be expected, responses were strongest against the fully mismatched target (B-LCL from the DC donor), but positive responses could still be observed against targets expressing the single relevant alloantigen HLA-A2 in those T cell cultures primed by cross-linked DC. It is of interest to note that strong allo-CTL were also induced by DC following co-ligation of MHC I and CD40 (but not MHC I alone), and we are further investigating these responses in terms of MHC class I or II restriction and CD4+ versus CD8+ T cell phenotype.

4. DISCUSSION

Significant progress has been made in characterizing DC and defining their role as the most potent APC for inducing antigen-specific T cell responses (reviewed in 1,11). Progress has also been made in the recent past which supports the use of DC as biologic adjuvants in cancer vaccines (12,13). In order to increase the clinical utility of DC, it will be important to better understand the immunobiology of DC and their interactions with T cells.

DC are capable of priming both MHC class I- and II-restricted, antigen-specific T cell responses, however, in most antigen-limiting conditions the generation of CD8+ effector T cells is probably preceded by the induction of Th1 CD4+ antigen-specific T cells. DC have been shown to secrete IL-12, a potent Th1-promoting cytokine (14,15), following interaction with CD4+ T cells (16), and more recently have been shown to secrete large amounts of IL-12 following CD40 ligation (5,6). As activated CD4+ T cells have been shown to be the major expressors of CD40L, it follows that both MHC II/peptide-

TCR as well as CD40-CD40L interactions would be critical events leading to the generation of cellular immunity. We have confirmed the findings, with respect to IL-12 production by DC following CD40 ligation, reported by Cella et al and Koch et al., and furthermore, we propose a more prominent role in T cell induction for signaling events occurring through MHC II molecules on DC. We also report that MHC II ligation augments the production of IL-12 by DC, leads to upregulation of costimulatory molecules, and may serve as a secretory signal for the release of IFN-α. The increased expression of costimulatory molecules following CD4+ T cell-mediated signals should serve to enhance DC APC function as well as to further activate the T cell. The production and secretion of IL-12 by DC following interaction with CD4+ T cells could serve to enhance both T helper development and function, ultimately leading to enhancement of effector CD8+ CTL generation most likely through IFN-γ. As viruses and other foreign antigens have been shown to induce the production of IFN-α by DC (17) and other APC, the subsequent ligation of MHC II on such an activated DC by CD4+ T cells during antigen presentation could lead to increased secretion of this cytokine. IFN-α has been shown to have a number of immunomodulatory effects, including increasing the frequency of IFN-γ-producing CD4+ T cells (18), stimulating IL-2 production in activated T cells (19), and suppressing Th2 cytokines and the development of Th2 clones (20,21). In a recent report by Tough et al.(22), IFN-α was also shown to induce the proliferation of bystander CTL in vivo. The ability of DC-CD4+ T cell interactions to augment these effects could significantly impact on the resulting immune response.

In many instances we have noted a suppressive effect on surface marker expession and even cytokine production following co-ligation of CD40 and MHC I on DC. Although these data are preliminary, it is interesting to postulate a possible down-regulatory role for TCR-MHC I ligation following DC-CD8+ T cell interaction. Once effector cells have been primed, this might be a means of turning down, or regulating, stimulatory T cell signals from DC.

We are currently investigating the impact of activating DC by cross-linking on their ability to promote enhanced induction of antigen-specific T effector cells *in vitro*. Preliminary studies show that DC activated by ligating CD40 or MHC II are superior at priming allogeneic-specific cytotoxic T cells. We also have evidence that DC activated by MHC II ligation are better stimulators of MHC class I-restricted, peptide-specific CTL (unpublished results). We hope to apply these concepts toward the development of more effective anti-viral and anti-tumor DC-based vaccines and therapies.

REFERENCES

1. Caux C, Liu YJ, Banchereau J. Recent Advances in the Study of Dendritic Cells and Follicular Dendritic Cells. Immunol Today. 1995;16:2–4.
2. Grewal IS, Flavell RA. A Central Role of CD40 Ligand in the Regulation of CD4+ T-cell Responses. Immunol Today. 1996;17:410–414.
3. Macatonia SE, Hosken NA, Litton M, Vieira P, Hsieh CS, Culpepper JA, Wysocka M, Trinchieri G, Murphy KM, O'Garra A. Dendritic Cells Produce IL-12 and Direct the Development of Th1 Cells from Naive CD4+ T cells. J Immunol 1995;154:5071–5079.
4. Scheicher C, Mehling M, Dienes HP, Reske K. Uptake of Microparticle-absorbed Protein Antigen by Bone Marrow Dendritic Cells Results in Upregulation of Interleukin-1α and Interleukin-12 p40/p35 and Triggers Prolonged, Efficient Antigen Presentation. Eur J Immunol 1995;25:1566–1572.
5. Cella M, Scheidegger D, Palmer-Lehmann K, Lane P, Lanzavecchia A, Alber G. Ligation of CD40 on Dendritic Cells Triggers Production of High Levels of Interleukin-12 and Enhances T Cell Stimulatory Capacity: T-T Help via APC Activation. J Exp Med 1996;184:747–752.

6. Koch F, Stanzl U, Jennewein P, Janke K, Heufler C, Kämpgen, Romani N, Schuler G. High Level IL-12 Production by Murine Dendritic Cells: Upregulation via MHC Class II and CD40 Molecules and Down-regulation by IL-4 and IL-10.J Exp Med 1996, 184:741–746.
7. Sallsto F, Lanzavecchia A. Efficient Presentation of Soluble Antigen by Cultured Human Dendritic Cells is Maintained by Granulocyte/Macrophage Colony-stimulating Factor Plus Interleukin 4 and Downregulated by Tumor Necrosis Factor Alpha. J Exp Med. 1994;179:1109–1118.
8. Kiertscher SM, Roth MD. Human CD14+ leukocytes Acquire the Phenotype and Function of Antigen-Presenting Dendritic Cells when Cultured in GM-CSF and Il-4. J Leuk Biol. 1996;59:208–218.
9. Storkus WJ, Alexander J, Payne JA, Dawson JR, Cresswell P. Reversal of NK susceptibility in target cells expressing transfected class I HLA genes. Proc Natl Acad Sci USA 1989, 86:2361–2367.
10. Jung T, Schauer U, Heusser C, Neumann C, Reiger C. Detection of intracellular cytokines by flow cytometry. J Immunol Methods 1993, 159:197–207.
11. Steinman RM. The Dendritic Cell System and Its Role in Immunogenicity. Annu Rev Immunol 1991;9:271–296.
12. Young JW, Inaba K. Dendritic Cells as Adjuvants for Class I Major Histocompatibility Complex-Restricted Antitumor Immunity. J Exp Med 1996;183:7–11.
13. Mayordomo JI, Zorina T, Storkus WJ, Zitvogel L, Celluzzi C, Falo LD, Melief CJ, Ildstad ST, Kast WM, DeLeo A, Lotze MT. Bone Marrow Derived Dendritic Cells Pulsed with Synthetic Tumor Peptides Elicit Protective and Therapeutic Anti-tumor Immunity. Nature Med 1995;1:1297–1302.
14. Manetti R, Parronchi P, Guidizi MG, Piccinini MP, Maggi E, Trinchieri G, Romagnani S. Natural Kille Cell Stimulatory Factor (interleukin 12 [IL-12]) Induces T Helper Type 1 (th1)-specific Immune Responses and Inhibits the Development of IL4-producing Th Cells. J Exp Med 1993;177:1199–1204.
15. Trinchieri G. Interleukin-12 and its Role in the Generation of Th1 Cells. Immunol Today 1993;14:335–338.
16. Heufler C, Koch F, Stanzl U, Topar G, Wysocka M, Trinchieri G, Enk A, Steinman RM, Romani N, Schuler G. Interleukin-12 is produced by dendritic cells and mediates TH1 development as well as Interferon-gamma production by TH1 cells. Eur J Immunol 1996, 26:659–668.
17. Ferbas JJ, Toso JF, Logar AJ, Navratil JS, Rinaldo CR. CD4+ Blood Dendritic Cells are Potent Producers of IFN-α in Response to In Vitro HIV-1 Infection. J Immunol 1994;152:4649–4662.
18. Brinkmann V, Geiger T, Alkan S, Heusser CH. Interferon Alpha Increases the Frequency of Interferon Gamma-producing Human CD4+ T Cells. J Exp Med. 1993;178:1655–1663.
19. Holáň VK, Kohno K, Minowada J. Natural Human Interferon-α Augments Interleukin-2 Production by a Direct Action on the Activated IL-2-producing T Cells. J Interferon Res 1991;11:319.
20. Schandene L, Del Prete GF, Cogan E, Stordeur P, Crusiaux A, Kennes B, Romagnani S, Goldman M. Recombinant Interferon-alpha Selectively Inhibits the Production of Interleukin-5 by Human CD4+ T Cells. J Clin Invest 1996;97:309–315.
21. Manetti R, Annunziato F, Tomasevic L, Gianno V, Parronchi P, Romagnani S, Maggi E. Polyinosinic Acid: Polycytidylic Acid Promotes T helper Type 1-Specific Immune Responses by Stimulating Macrophage Production of Interferon-alpha and Interleukin-12. Eur J Immunol 1995;25:2656–2660.
22. Tough DF, Borrow P, Sprent J. Induction of Bystander T cell Proliferation by Viruses and Type I Interferon in vivo. Science 1996;272:1947.

56

SOLUBLE CD16/FcγRIII INDUCES MATURATION OF DENDRITIC CELLS AND PRODUCTION OF SEVERAL CYTOKINES INCLUDING IL-12

H. de la Salle,[1] J. Galon,[2] H. Bausinger,[1] D. Spehner,[1] A. Bohbot,[3] J. Cohen,[4] J. -P. Cazenave,[5] W. -H. Fridman,[2] C. Sautès,[2] and D. Hanau[1]

[1]CJF 94–03 INSERM, AND
[5]U. 311, E.T.S. INSERM, Strasbourg
[2]INSERM U. 255
Institut Curie, Paris
[3]Service d'Onco-Hématologie
Hôpital de Hautepierre, Strasbourg
[4]Laboratoire d'Immunologie
Hôpital Robert Debré
Reims, France

ABSTRACT

FcγRIII (CD16), a low affinity FcR which binds IgG-containing immune-complexes, exists under membrane-associated forms and under a soluble form (sFcγRIII). The latter, present in biological fluids (serum, saliva), is generated by proteolytic cleavage of the two membrane-associated FcγRIII isoforms, FcγRIII-A (expressed by macrophages and NK cells) and FcγRIII-B (expressed exclusively by neutrophils). Herein we demonstrate that dendritic cells (DCs), generated by culturing monocytes with GM-CSF and IL-4, bind biotinylated recombinant sFcγRIII. This binding is specific and involves the complement receptor CR3 (CD11b/CD18) and CR4 (CD11c/CD18). Indeed, preincubation of DCs with anti-CD11b and anti-CD11c mAbs decreased by 52 % and 62 % respectively the binding with sFcγRIII. Moreover, electron microscopy showed that binding of gold-labeled sFcγRIII to DCs maintained at 4°C occured within clathrin-coated pits. Once internalized, at 37°C, sFcγRIII entered the endocytic pathway and reached the MHC class II compartments. Furthemore, DCs incubated for 48 h with multivalent sFcγRIII expressed increased levels of CD40, CD80, CD86, CD54, CD58, HLA class I and class II molecules and decreased levels of CD23 and CD32. These effects result in an increased capacity of DCs to trigger proliferative responses by CD4$^+$ CD45RA$^+$ allogeneic T cells. RT-PCR amplification demonstrated that incubation of DCs for 20 h in the presence of multivalent sFcγRIII induced the appearance of GM-CSF and IL-12 p40 mRNA. Among the cytokines constitutively expressed, IL-1β and IL-8 were strongly up-regulated whereas IL-6 and IL-12 p35 mRNA were increased to a lesser extent and the expression of MIP-1α mRNA remained constant. Finally, ELISA tests demonstrated that DCs incubated with multivalent sFcγRIII secreted the cytokines IL-1β, IL-6, IL-8, GM-CSF and IL-12 p75. Thus, while becoming internalized sFcγRIII could affect the capacity of DCs to present antigens and, via the induction of accessory molecules and the release of the IL-12 p75 protein, could initiate Th1 type immune response.

1. INTRODUCTION

Over the last few years it has been shown that maturation of DCs can be induced *in vitro* by proinflammatory stimuli or interactions with lymphocytes involving particular receptor-ligand pairs. Thus DCs[1] can be triggered by proinflammatory stimuli such as TNF-α, IL-1, and LPS to mature and to upregulate (i) the expression of adhesion factors and costimulatory molecules and (ii) their capacity to stimulate allogeneic T cells. Likewise binding of CD40 present at the DC surface with the CD40 ligand (CD40L) induces (i) an increase in the level of adhesion factors and costimulatory molecules, (ii) an enhancement of their stimulatory capacity in autologous and allogeneic MLR and (iii) the production of massive amounts of IL-12[2].

Soluble FcγRIII (sCD16) is the major species of soluble FcγR found in blood. It is produced by proteolytic cleavage of membrane-associated FcγRIII found on the surface of NK cells, macrophages, activated monocytes and neutrophils. Recent experiments have shown that sCD16 binds to the complement receptors CR3 (CD11b/CD18) and CR4 (CD11c/CD18) and that this interaction induces the secretion of proinflammatory cytokines such as IL-6 and IL-8 by monocytes[3]. The engagement of CR3 by specific mAbs also triggers activation signals and TNF-α production by monocytes[4]. Since differentiated DCs obtained from monocytes cultured in the presence of GM-CSF and IL-4 express high levels of CR3 and CR4, we investigated in the present work whether such DCs bind sCD16. We show that DCs not only bind sCD16 but also that this binding induces (i) an increase in the level of adhesion factors and costimulatory molecules as well as of the MHC class I and class II molecules, (ii) an enhancement of the stimulatory capacity of DCs in allogeneic MLR and (iii) the production of various cytokines by DCs, among which IL-12.

2. MATERIALS AND METHODS

2.1. *In Vitro* Differentiation of Peripheral Blood Monocytes into DCs

Human blood monocytes were isolated by continuous flow centrifugation leukapheresis and counterflow centrifugation elutriation as previously described[5]. They were cultured in RPMI 1640–10 % FCS supplemented with 50 ng/ml GM-CSF (Shering-Plough, Levallois-Perret, France) and 200 U/ml IL-4 (PeproTech, Rocky Hill, NJ) for 7 days. Because at day 7 a small percentage of the cells occasionally were $CD14^+$, cells were routinely treated with magnetic beads coated with an anti-CD14 mAb and the disappearance of the CD14 reactivity checked by FACS analysis.

2.2. Activation of DCs by sCD16, TNF-α, and CD40L

On day 7, DCs were cultured, for 24 to 48 h, in the presence of either gold-conjugated recombinant sCD16 (sCD16-Au) (< 0.1 IU/ml endotoxins), polymeric recombinant sCD16 (< 0.6 IU/ml endotoxins), gold-conjugated BSA (BSA-Au) (< 0.1 IU/ml endotoxins) (British BioCell International, Cardiff, UK), gold-conjugated F(ab')2 fragments of the anti-CD11b or anti-CD11c mAb, 20 ng/ml recombinant human TNF-α (Genzyme, Boston, MA) or irradiated J558L cells transfected with CD40L (a generous gift of Dr. Peter Lane, Basel Institute for Immunology)[6] at a 1 : 1 ratio. Polymeric recombinant sCD16 was purified from supernatants of CHO cells cotransfected with pKC3 and C4bp. Recombinant sCD16 was obtained as described[3]. Gold-conjugation was performed as previously described[7].

2.3. Immunofluorescence Analysis

Cell staining was performed using mouse mAbs followed by FITC-conjugated, affinity isolated, F(ab')2 fraction of a sheep anti-mouse IgG antibody (Silenius, Hawthorn Victoria, Australia). The following mAbs were used: BL6 (IgG1, anti-CD1a, Immunotech, Marseille, France), MAB89 (IgG1, anti-CD40, Immunotech), MAB104 (IgG1, anti-CD80, Immunotech), B70/B7–2 (IgG2b, anti-CD86, Pharmingen, San Diego, CA), IV.3 (IgG2b, anti-CD32, Medarex, Annandale, NJ), Bear1 (IgG1, anti-CD11b, Immunotech), BU15 (IgG1, anti-CD11c, Immunotech), MHM6 (IgG1, anti-CD23, Dako, Copenhague, Danemark), 84H10 (IgG1, anti-CD54, Immunotech), AICD 58 (IgG2a, anti-CD58, Immunotech), W6/32 (IgG2a, anti-HLA class I, Dako), L243 (IgG2a, anti-HLA-DR, Becton Dickinson, San José, CA), and appropriate isotype controls. In some experiments DCs were incubated successively (i) immediately with biotinylated sCD16 (prepared as previously described[3]) or after a preincubation with either anti-CD11b or anti-CD11c mAb and (ii) with streptavidin-phycoerythrin. Samples were analysed on a FACScan (Becton Dickinson) using propidium iodide to exclude dead cells.

2.4. Immunogold Labeling Procedure

DCs were cooled to 4°C for 15 min, incubated at 4°C for 60 min in the presence of sCD16-Au and fixed at 4°C for electron microscopy. Specificity of the binding was controlled by adding a 40-fold excess of unlabeled sCD16 during the incubation period. In another experiment, DCs were warmed to 37°C for 5–10 min and fixed, at 37°C, for electron microscopy. Finally, DCs were directly incubated at 37°C for 15 min and sCD16-Au was added for 15, 30 or 45 min before fixation and pellets were embedded either in Epon or in Lowicryl[8]. Lowicryl ultrathin sections were successively incubated with (i) a rabbit antiserum directed against the DR α chain (a generous gift of Dr. Jacques Neefjes, The Netherlands Cancer Institute) and (ii) 20-nm gold-conjugated goat anti-rabbit antibodies. Finally, ultrathin sections were examined under a Philips CM 120 BioTWIN electron microscope (120 kV).

2.5. Cellular Cytokine mRNA Expression

mRNA was extracted from DCs according to manufacturer's instructions (Bioprobe Systems, Montreuil-Sous-Bois, France). Total RNA was reverse transcribed using AMV reverse transcriptase, random primers and dNTPs in presence of RNAase inhibitor (all from Boehringer Mannheim, Meylan, France). The cDNA were amplified using cytokine-specific primers, dNTPs and DNA Taq polymerase (Promega, Madison, WI). Thirty cycles were performed each consisting of 30 sec at 94°C and 55°C and 1 min at 72°C. The specificity of the products was tested by hybridization using internal oligonucleotide probes.

2.6. Detection of Cytokines in Culture Supernatants by ELISA

After 24 h of culture, DC supernatants were collected and assayed for IL-1β, IL-6, IL-8, GM-CSF, IL-15 and IL-12p75 by sandwich ELISA. Tests samples were added and incubated according to the manufacturer's instructions: R&D Systems (Minneapolis, MN) for IL-1β, IL-6, IL-8, and IL-12p75, Genzyme (Cambridge, MA) for IL-15, and Medgenix (Fleurus, Belgium) for GM-CSF. Sensitivity of the ELISA were 0.3 pg/ml for IL-1β, 0.7 pg/ml for IL-6, 3 pg/ml for IL-8, 3 pg/ml for GM-CSF, 10 pg/ml for IL-15, and 5 pg/ml for IL-12p75.

2.7. Mixed Lymphocyte Reaction

DCs were cultured for 48 h in medium or with sCD16-Au, 20 ng/ml recombinant human TNF-α (Genzyme), or with irradiated J558L cells transfected with CD40L and used as stimulators in allogeneic MLR. Graded numbers of stimulator cells were cocultured with 1.5×10^5 $CD4^+$ $CD45RA^+$ responder T cells in round-bottom, 96-well microplates. Thymidine incorporation was measured on day 5 by an 18-h pulse with (^3H)thymidine.

3. RESULTS

3.1. Soluble CD16 Binds Specifically to CR3 and CR4 Present on the DC Surface

Binding of sCD16 to DCs was demonstrated by FACS analysis using biotinylated sCD16 followed by streptavidin. Preincubation of DCs with 10 μg/ml of the anti-CD11b mAb Bear-1 (IgG1) or with the anti-CD11c mAb BU15 (IgG1) reduced by 52 and 62 % respectively the intensity of the sCD16 labeling. No modification of the intensity of the sCD16 labelling was observed when DCs were preincubated with an identical concentration of a control mouse IgG1. Finally, in order to investigate the specificity of the binding, competition experiments were performed and examined under the electron microscope. Adding a 40-fold excess of unlabeled sCD16 — during the incubation period, at 4°C, with sCD16-Au — dramatically reduced the intensity of the DC surface labeling by gold particles.

3.2. Receptor-Mediated Endocytosis Allows Soluble CD16-Au to Reach the MHC Class II Compartments Through the Endocytic Pathway

In a first approach, meant to define the fate of sCD16 that binds to CR3 and CR4 at the DC surface, DCs were incubated for 60 mins at 4°C in the presence of sCD16-Au. Under the electron microscope numerous gold particles were already located in coated pits. These data rule out the possibility that internalization of sCD16-Au was caused by CD16-

Au-induced CR3/CR4 crosslinking and imply that a spontaneous and continuous internalization of CR3 and CR4 in DCs occurs. This conclusion was further confirmed by incubating DCs in the presence of gold-conjugated F(ab')2 fragments of the anti-CD11b or anti-CD11c mAb which showed, likewise, gold particles located in coated pits. Moreover, when DCs were incubated for 60 mins at 4°C and then for 5–10 mins at 37°C, gold particles were also found in coated vesicles and in tubulo-vesicular endosomal structures. After 15 to 45 min of internalization at 37°C, the intracellular compartments—where MHC class II molecules accumulate—were also labelled with internalized sCD16-Au, as observed under the electron microscope in Epon as well as Lowicryl thin sections. Indeed, in Epon thin sections sCD16-Au were located in vesicles characterized by concentric arrays of internal membranes or showing membrane infoldings whereas, on Lowicryl thin sections, sCD16-Au clearly colocalized in compartments which could be heavily labelled by a polyclonal anti-DRα chain antibody.

3.3. Binding of Soluble CD16 Onto DCs Results in Upregulation of Adhesion Factors, Costimulatory and HLA-A, B, C and -DR Molecules and in Downregulation of the Low Affinity FcR CD23 and CD32

As internalization of sCD16 in DCs is likely to facilitate the uptake of immune complexes by DCs we wondered whether binding/internalization of multivalent sCD16, i.e., sCD16-Au, would induce modulation of accessory molecules on the DC surface, i.e., "phenotypic" maturation of the DCs. Therefore, "immature" DCs were incubated for 48 h - in the presence of GM-CSF and IL-4 - with either TNF-α (known to induce the maturation of monocyte-derived DCs (1)) or sCD16-Au or remained "untreated". DCs were stained with a panel of mAbs revealed by the FITC-conjugated F(ab')2 fraction of a sheep anti-mouse IgG antibody. As shown in Fig. 1, DCs incubated in the presence of sCD16-

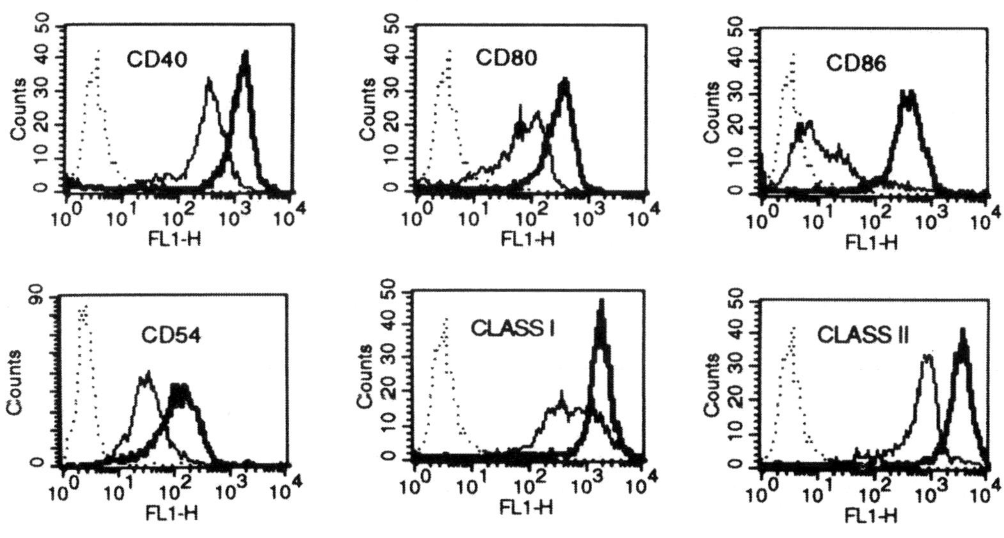

Figure 1. Binding of sCD16-Au to DCs induces upregulation of accessory and MHC molecules. Dotted lines: isotype controls; fine lines: unstimulated DCs; bold lines: DCs stimulated for 48 h with sCD16-Au.

Table I. Relative level of DC cytokine mRNAs in the presence of sCD16-Au with respect to BSA-Au. RT-PCR amplification products were scanned and the band intensities quantified. Results were normalized by comparison with the intensity of a β-actin band amplified from the same cDNAs

Cytokine mRNA	Fold increase sCD16/BSA
βactine	0,92
IL-1β	9,9
IL-6	4,7
IL-8	10,1
MIP-1α	1,1
IL-12p35	3,6
IL-12p40	9,8
GM-CSF	8,9

Au upregulated the expression of CD54, CD58, CD40, CD80 and particularly of CD86. MHC class I and MHC class II were also upregulated, while the expression of the two low affinity FcR for Ig present on the DC surface, i.e., CD23 and CD32, was downregulated. TNF-α had similar effects, although less pronounced, on the expression of CD80, CD86, MHC class I and MHC class II.

3.4. Binding of sCD16 to DCs Induces Cytokine Production

Since sCD16-Au was able to induce, at least according to "phenotypic" criteria, maturation of DCs, we then asked whether such "maturation" of DCs was concomitant with the production of cytokines. Therefore, DCs were incubated for 24 h in medium or in the presence of either sCD16-Au, polymeric recombinant sCD16 or, for control purposes, with BSA-Au. In a first step DC mRNAs were subjected to RT-PCR analysis. As shown in Table I, incubating DCs for 24 h in the presence of sCD16-Au induced the appearance of GM-CSF and IL-12 p40 mRNA. Among the cytokines constitutively expressed, IL-1β and IL-8 were strongly up-regulated whereas IL-6 and IL-12 p35 mRNA were increased to a lesser extent and the expression of MIP-1α mRNA remained constant. Cytokine production was then measured in 24 h DC supernatants using specific ELISA assays. As shown in Table II, DCs secreted the cytokines IL-1β, IL-6, IL-8, GM-CSF and IL-12 p75 whereas IL-15 was not detected.

3.5. Stimulatory Capacity of Soluble CD16-Treated DCs in Allogeneic MLR

Since DCs treated with sCD16 undergo phenotypic alterations characteristic of mature DCs we investigated whether DCs also acquire the ability to stimulate $CD4^+$ $CD45RA^+$ allogeneic T cells. We found that, whatever the cell ratio, DCs incubated with sCD16-Au were better stimulators of $CD4^+$ $CD45RA^+$ T lymphocytes than TNF-α treated DCs which in turn were superior to untreated cells. Preliminary experiments show that DCs incubated with sCD16-Au were as good stimulators of $CD4^+$ $CD45RA^+$ allogeneic T cells as CD40L-treated DCs (data not shown).

Table II. Induction of cytokine secretion in DCs stimulated either by sCD16-Au or BSA-Au

Stimulation	IL-1β (pg/ml)	IL-6 (pg/ml)	IL-8 (ng/ml)	GM-CSF (pg/ml)	IL-12p75 (pg/ml)
0	< 0.3	24.3±7.0	4.7±2.6	3.0±1.2	< 5
BSA-Au	< 0.3	235±63	18.0±10.4	3.0±1.2	< 5
sCD16-Au	510±10	1551±28	80.3±11.4	394.4±59.9	1025±25

4. DISCUSSION

In the present study we demonstrate that DCs not only bind sCD16 but also that this binding induces maturation of DCs and production of several cytokines including IL-12. Maturation of DCs can be induced *in vitro* by proinflammatory stimuli such as TNF-α, IL-1, and LPS[1] or by ligation of CD40[1, 9, 10]. This ligation also induces IL-12p75 production by monocytes treated for 48 hours in GM-CSF and IFN-γ[11] and by human and murine DCs[2,10]. While monocytes produce only low levels of IL-12p75 (50–100 pg/ml)[11] CD40L-stimulated DCs release high levels of bioactive IL-12 (up to 15 ng/ml for human monocyte-derived DCs[2] and between 263–4726 pg/ml for murine spleen DCs[10]). Surprisingly, the stimuli which share with CD40L the capacity to trigger DC maturation, i.e., LPS and TNF-α, are less efficient or ineffective in driving IL-12p75 production by DCs. On the other hand, pathogens that have been reported to induce IL-12 production by murine macrophages—such as *Staphylococcus aureus, Pseudomonas aeruginosa, Candida albicans, and Leishmania major*—appear less effective or ineffective at all on DCs[2]. This selectivity of DCs in their capacity to produce IL-12, under the influence of different stimuli, appears to be dependent on the ligation of some of the DC surface molecules: CD40[2,10], MHC class II molecules[10] and, as demonstrated in the present work, CR3 and CR4. Indeed, incubating DCs with gold-conjugated F(ab')2 fragments of an anti-CD11b or anti-CD11c mAb instead of CD16-Au induces the same level of secretion of IL-12p75 (data not shown).

Presentation of exogenous soluble antigen to Th cells is a complex process that requires uptake of proteins by APC, their digestion into immunogenic peptides, their intracellular association with MHC class II molecules, and the transport of the resultant immunogenic complexes to the plasma membrane for recognition by Th cells. Targeting of antigen to specific cell surface receptors enhances the efficiency of presentation by 100 to more than 1000 fold compared with that obtained with soluble antigen. Since sCD16/FcγRIII binds IgG-containing immune complexes and targets the DC surface receptors CR3 and CR4, it is likely that its internalization into DCs enhances the efficiency of presentation of an antigen complexed to IgG. This is what is expected given that sCD16-Au reaches, via receptor-mediated endocytosis and the endocytic pathway, the MHC class II compartments. These events could take place in nonlymphoid organs where memory T cells recirculate. Activation of these cells would be further stimulated since binding of sCD16 to DCs induces DC maturation - leading in particular to an increase in B7 molecules (CD80 et CD86) - and secretion of cytokines, among which IL-12. The latter cytokine is expected to favor the generation and maturation of Th type 1 cells all the more so since the interaction of B7 from DCs with CD28 molecules on T cells synergizes with IL-12, in inducing proliferation and cytokine production of T cells[12]. Moreover, the interaction of CD40, highly expressed on the surface of mature DCs, with the CD40L of activated T cells and the interaction of the TCR with the MHC class II molecules from DCs should also amplify IL-12 secretion[2, 10] until negative feedback mechanisms mediated by IL-10 and possibly by IL-4 interrupt the IL-12 production. However, other cytokines secreted by DCs under the influence of sCD16 can also affect T cells. Thus IL-1β, in association with IL-6, can costimulate T cell activation[13], while IL-8 will chemoattract neutrophils, basophils and a subpopulation of lymphocytes, and GM-CSF will promote survival and influence the distribution and functional activity of DCs[14]. As we have also recently found that DCs are themselves able to secrete IL-10 under the influence of certain stimuli, it is apparent that according to the type of ligation occurring on the surface of DCs, these cells can contribute to the development of either a strong cell mediated immunity or a hapten-specific tolerance[15].

5. ACKNOWLEDGMENT

This work was supported by INSERM, the Agence Française du Sang, Roussel-Uclaf, Institut Curie and the Etablissement de Transfusion Sanguine de Strasbourg.

6. REFERENCES

1. Sallusto F., Lanzavecchia A. J. Exp. Med., 1994, 179: 1109–1118.
2. Cella M., Scheidegger D., Palmer-Lehman K., Lane P., Lanzavecchia A., Albert G. J. Exp. Med., 1996, 184: 747–752.
3. Galon J., Gauchat J.-F., Mazières N., Spagnoli R., Storkus W., Lötze M., Bonnefoy J.-Y., Fridman W.-H., Sautès C. J. Immunol., 1996, 157: 1184–1192.
4. Fan S., Edgington. T. J. Immunol., 1993, 150: 2972–2980.
5. Faradji A., Bohbot A., Schmitt-Goguel M., Siffert J.C., Dumont S., Wiesel M.L., Piemont Y., Eischen A., Bergerat J.P., Bartholeyns J., Poindron P., Witz J.P., Oberling F. J. Immunol. Methods, 1994, 174: 297–309.
6. Lane P., Burdet C., Mc Connell F., Lanzavecchia A., Padovan E. Eur. J. Immunol., 1995, 25: 1788–1793.
7. Hanau D., Fabre M., Schmitt D.A., Garaud J.-C., Pauly G., Tongio M.-M., Mayer S., Cazenave J.-P. Proc. Natl. Acad. Sci. , 1987, 84, 2901–2905.
8. Gachet C., Hanau D., Spehner D., Brisson C., Garaud J.-C., Schmitt D.A., Ohlmann P., Cazenave J.-P. J. Cell Biol., 1992, 120: 1021–1030.
9. Caux C., Massacrier C., Vanbervliet B., Dubois B., Van Kooten C., Durand I., Banchereau J. J. Exp. Med., 1994, 180: 1263–1272.
10. Koch F., Stanzl U., Jennewein P., Janke K., Heufler C., Kämpgen E., Romani N., Schuler G. J. Exp. Med., 1996, 184: 741–746.
11. Shu U., Kiniwa M., You Wu C., Maliszewski C., Vezzio N., Hakimi J., Gately M., Delespesse G. Eur. J. Immunol., 1995, 25: 1125–1128.
12. Kubin M., Kamoun M., Trinchieri G. J. Exp. Med., 1994, 180: 211–222.
13. Cumberbatch M., Dearman R.J., Kimber I. Immunology, 1996, 87: 513–518.
14. Markowicz S., Engleman E.G. J. Clin. Invest., 1990, 85: 955–961.
15. Enk A.H., Saloga J., Becker D., Mohamadzadeh M., Knop J. J. Exp. Med., 1994, 179: 1397–1402.

INDUCTION OF FcεRIα mRNA AND PROTEIN SYNTHESIS BY INTERLEUKIN 4 IN CD34$^+$ CELLS-DERIVED CD1a$^+$ DENDRITIC CELLS

Roland Magerstaedt,[1,2] Stefan Kraft,[1] Isolde Strobel,[1] Maren Jürgens,[1] Daniel Hanau,[3] Joerg Wessendorf,[1] and Thomas Bieber[1]

[1]Department of Dermatology
[2]Genzentrum of Ludwig-Maximilians-University of Munich
Munich, Germany
[3]Laboratoire d'Histocompatibilite
Etablissement Regional deTransfusion
Sanguine, Strasbourg, France

INTRODUCTION

IgE-bearing human epidermal Langerhans cells (LC) are suspected of playing a key role in the pathogenesis of atopic dermatitis (AD). We and others have shown that these cells bind IgE via the high affinity receptor Fc epsilon RI (FcεRI) (1, 2). Recent findings have shown that the expression of this receptor on LC and related dendritic cells in the skin is highly upregulated in lesions skin of AD when compared with nonatopic individuals and this is correlated to a high IgE serum levels (3). Functionally, LC from normal skin differ from those of atopic skin in terms of calcium mobilization upon receptor ligation (4). Beside qualitative alterations in the activation cascade, unresponsiveness also may be related to inefficient triggering in normal LC displaying low amounts of receptors. Thus, the variations in receptor expression may be crucial for the outcome of the signal transduction initiated by FcεRI-cross-linking on LC, implying the need of a better understanding of the mechanisms regulating the receptor display on LC. However, detailed functional, biochemical and molecular biological analysis of FcεRI + LC is hampered by the limited number of LC routinely obtained from skin biopsies or surgical samples. Recently several cytokines, especially granulocyte/macrophage colony-stimulating factor (GM-CSF) and tumor necrosis factor alpha (TNF-α), have been identified that speed the development of dendritic cells from blood and bone marrow precursors in suspension cultures (5, 6, 7, 8, 9). We took advantage of this progress to establish a model for the study of the regulation of FcεRI on LC and other dendritic cells.

MATERIAL AND METHODS

After gradient density centrifugation, $CD34^+$ cells were isolated from cord blood as described by Caux et al. by the means of anti-CD34 microbeads and the MACS system (Miltenyi Biotec, Bergisch Gladbach, Germany). Cells were then seeded at a density of $8 \times 10^4 - 1 \times 10^5$ cells/ml in Iscove's modified Dulbecco's medium (IMDM) containing 10% heat-inactivated FCS, 100 U/ml penicillin and 100 ng/ml streptomycin in 24 well tissue culture plates. Cultures were maintained by replacing one half of the culture medium with fresh cytokine-supplemented medium every 4 days for a maximum 8 weeks. Human recombinant cytokines were used at the following final concentrations: GM-CSF (300U/ml), TNF-α (50U/ml), IL-4, (250–300U/ml) to support the proliferation of DCs cytokines in suspension culture.

At different time points of the culture (24h, 48h, 72h, 96h, 124h, 144h, 250h), the cells were harvested and the surface and intracellular expression of several molecules was analysed by flow cytometry as described in detail (10).

For analysis of mRNA transcripts of FcεRI subunits, highly purified in vitro generated LC/DC were obtained by positive selection using anti-CD1a antibody (IOT6a) bound to sheep-anti-mouse coated 280nm magnetic beads according to the manufacturer's protocols (Dynal, Oslo, Norway). Highly purified in vitro generated LC/DC were not contaminated by basophils as shown by the failure to detect PCR-products using specific primers for tryptase in the same samples. Total RNA was extracted and cDNA was obtained by RT as described (1). Specific primer sequences for each gene were as described (1).

Analysis of calcium mobilization was performed as described in detail (4).

RESULTS AND DISCUSSION

Characterisation by extensive phenotyping of the cells at day 9–12, revealed that dendritic cells were CD1a, CD1c, HLA-DR, HLA-DQ, CD4, CD11a, CD11b, CD11c, CD13, CD33, CD83 (11), CD32, CD36, CD23, CD45RA, CD45R0, B7.1, B-7.2, CD40 and CD54. They also expressed the intracellular marker p55 (12) and Birbeck granules (25%).

Spontaneous surface expression of FcεRIα was found in a limited number of cases (9/60) of cultures performed with GM-CSF and TNF-α.

Since IL-4 is known as a key cytokine in the biology of IgE, we asked whether its addition to the culture may induce/enhance the expression of FcεRIα. However, under these conditions, no effect was observed when analysing the cell surface of $CD1a^+$ cells. In contrast, we could show that IL-4 was able to induce the presence of intracellular FcεRIα moieties in $CD1a^+$ cells. This induction was confirmed by RT-PCR experiments where transcripts for FcεRIα could be detected in those cultures exposed to IL-4. Low but significant amounts of FcεRIγ were reproducibly found in all conditions.

Finally, we asked whether ligation of spontaneous expressed FcεRI may lead to activation of $CD1a^+$ as determined by calcium mobilization. As already shown for normal epidermal LC, cross-linking of FcεRI on in vitro generated $CD1a^+$ DC did not provoke measurable calcium influx.

Taken together, our results suggest that IL-4 plays a crucial role in the regulation of the FcεRI expression on dendritic cells by inducing at least the transcription, protein synthesis and intracellular localization of FcεRIα. However, in contrast to bone marrow derived mast cells (13), other yet-to-be-defined factors are required for the surface expression and the function of the receptor on these cells.

ACKNOWLEDGMENTS

We are grateful to Dr. E. Langhoff for providing us with the anti-p55 Dr. J. Kochan for anti- FcεRIα and FcεRIγ and Dr. S. Tedder anti-CD83 antibodies.

This work was supported by grants from the Deutsche Forschungsgemeinschaft (SFB 217/D4), from INSERM (CRE 93 06 06 and CJF 9403) and from the DAAD (312/pro-gg).

REFERENCES

1. Bieber, T., de la Salle, H., Wollenberg, A., Hakimi, J., Chizzonite, R., Ring, J., Hanau, D., and de la Salle, C., Human Epidermal Langerhans cells Express the High affinity Receptor for Immunoglobulin E (FcεRI), J. Exp. Med. 175: 1285(1992)
2. Wang, B., Rieger, A. Kilgus, O. Ochiai, K. , Maurer, D., Fodinge, D. Kinet, J.P. and Stingl, G., Epidermal Langerhans cells from normal human skin bind monomeric IgE via FcεRI. J. Exp. Med. 175: 1285(1992)
3. Wollenberg, A., Wen, S.P. and Bieber, T., Langerhans cell phenotyping: a new tool for differential diagnosis of inflammatory skin diseases. Lancet 346: 1626 (1995)
4. Jurgens, M., Wollenberg, A., Hanau, D., de la Salle, H. and Bieber, T., Activation of human epidermal Langerhans cells by engagement of the high affinity receptor for IgE, Fc epsilon RI. J. Immunol. 155:5184 (1995)
5. Caux, C., Dezutter-Dambuyant, C., Schmitt, D., and Banchereau, J., GM-CSF and TNF-alpha cooperate in the generation of dendritic Langerhans cells, Nature 360: 258 (1992)
6. Szabolcs, P., Feller, E.D., Moore, Mas and Youg, J.W., Progenitor recruitment and in vitro expansion of immunostimmulatory dendrtic cells from human CD34⁺ bone marrow cells by c-kit-ligand, GM-CSF and TNF-alpha. Adv Exp Med Biol 378.17:17
7. Strunk, D., Rappersberger, K., Egger, C., Strobel, H., Kromer, E., Elbe, A., Maurer, D. and Stingl, G., Generation of human dendritic cells/Langerhans cells from circulating CD34⁺ haematopoetic progenitor cells. Blood 87: 1292 (1996)
8. Mackensen, A., Herbst, B., Kohler, G., Wolffvorbeck, G., Rosenthal, F.M., Veelken, H., Kulmburg, P., Schaefer, H.E., Mertelsmann, R. and Lindemann, A., Delineation of the dendritic cell lineage by generating large numbers of Birbeck granule-positive Langerhans cells from human peripheral blood progenitor cells in vitro. Blood. 86: 2699 (1995)
9. Reid, C. D., Stackpoole, A., Meager, A. and Tikerpae, J., Progenitor recruitment and in vitro expansion of immunostimulatory dendritic cells from human CD34⁺ bone marrow cells by c-kit-ligand, GM-CSF, and TNF alpha Interactions of tumor necrosis factor with granulocyte-macrophage colony-stimulating factor and other cytokines in the regulation of dendritic cell growth in vitro from early bipotent CD34⁺ progenitors in human bone marrow. J. Immunol. 142, 2681 (1992)
10. Wollenberg, A. , Kraft, S., Hanau, D., Bieber,T., Immunomorphological and ultrastructural characterization of Langerhans cells and a novel, inflammatory dendritic epidermal cel (IDEC) population in lesional skin of atopic eczema. J. Invest Dermatol 106:446 (1996)
11. Zhou, L.-J.and Tedder, T.F. Human blood dendritic cells selectively express CD83, a member of the immunoglobulin superfamily. J.Immunol.154, 3821(1995)
12. Moisialos, G., Birkenbach, M., Ayehunie, S., Matasumura, F., Pinkus, G.S., Kieff, E. and Langhoff, E. Circulating human dendritic cells differentially express high levels of a 55-KD actin bundling protein. Am. J. Paththol. 148, 593 (1996)
13. Toru, H., C. Ra, S. Nonoyama, K., Suzuki, J. Yata and T. Nakahata Induction of the high-affinity IgE receptor (Fc epsilon RI) on human mast cells by IL-4. Int.Immunol 8:1367–1373 (1996)

58

FACS-SORTED SPLEEN AND PEYER'S PATCH DENDRITIC CELLS INDUCE DIFFERENT RESPONSES IN TH0 CLONES

Michael P. Everson,[1,3] David G. Lemak,[1] Jerry R. McGhee,[2] and Kenneth W. Beagley[1,2,4]

[1]Department of Medicine
[2]Department of Microbiology
The University of Alabama at Birmingham
Birmingham, Alabama 35294
[3]Birmingham Department of Veterans Affairs Medical Center
Birmingham, Alabama 35233
[4]Discipline of Pathology
The University of Newcastle
Newcastle, Australia

1. INTRODUCTION

Helper T cells in mice have been divided into two functionally distinct subsets based upon the pattern of cytokines secreted,[1,2] with Th type 1 (Th1) cells producing IL-2 and IFN-γ upon activation and predominantly mediating cell-mediated immunity (CMI), and Th type 2 (Th2) cells producing IL-4, IL-5, IL-6, and IL-10 upon activation and primarily mediating humoral immunity by providing T cell help via cytokines for Ig isotype and subclass responses.[1–4] For T cell activation, dendritic cells (DC) are the most potent inducers of primary in vivo T cell responses and are the principle *in vitro* stimulators of naive T cell activation in both mice and humans.[5,6] Previous studies showed that DC from different lymphoid tissues could possess similar[7] or different[8] functions. These studies supported the notion that PP DC induce preferential IgA production (predominantly found at mucosal sites) and SP DC support IgM production by controlling the cytokines produced by T cells in mucosal and systemic tissues, respectively. Recent studies suggest that cytokines present during primary activation of naive T cells play a role in determining the pattern of cytokines produced during subsequent antigenic challenge.[9] Perhaps more important to these stimulatory events are the APC or accessory cells (AC) resident in the local tissues that can stimulate initial cytokine secretion in this inductive milieu. In this regard, the present studies have addressed tissue specificity of the AC component of T cell activation using a single type of AC, namely, DC isolated from two different anatomical sites. We asked whether SP DC and PP DC, isolated under identical conditions, induce production of similar levels of T cell-derived cytokines that are involved in regulation of immune responses.

2. MATERIALS AND METHODS

2.1. Mice

Male C3H/HeN mice (8–20 wks of age) were obtained from the Frederick Cancer Research Center (Frederick, MD) and used for preparation of SP and PP DC.

2.2. Antibodies and FACS (Fluorescence-Activated Cell Sorter) Sorting

The following were used as primary antibodies for flow cytometric analysis and sorting: N418, biotinylated or unlabelled hamster anti-mouse CD11c, DC restricted in mice,[10] HB 224, (American Type Culture Collection [ATCC], Rockville, MD); 33D1, anti-mouse dendritic cell,[11] TIB 227, (ATCC); F4/80, biotinylated anti-mouse macrophage,[12] HB 198, (ATCC); and 11–5.2.1.9, fluoresceinated anti-mouse Ia^k, TIB 94, (ATCC). Secondary antibodies were: biotinylated or fluoresceinated rabbit anti-hamster IgG (Accurate Chemical, Westbury, NY) to detect N418 binding and RG7/7.6 (fluoresceinated mouse anti-rat kappa light chain, TIB 172, ATCC) or fluoresceinated mouse anti-rat IgG (Boehringer-Mannheim, Indianapolis, IN) to detect 33D1 binding. A tertiary reagent used with biotinylated antibodies was PE-avidin (UAB Immunochemical Facility, Birmingham, AL). Other antibodies included 2.4G2, anti-mouse Fc receptor,[13] HB 197, (ATCC) used to block non-specific binding; and those antibodies used in ELISA protocols as described below. For FACS analysis and sorting, cells were washed and stained in heat-inactivated FCS (5% v/v; Intergen, Purchase, NY) in PBS. Cells were pre-treated with 2.4G2 (anti-mouse Fc receptor) for 30 min on ice to block non-specific binding through Fc receptors. Cells were stained with primary and secondary antibodies (and PE-avidin as required) for 30 min on ice with one wash between stains. Normal mouse serum (1% v/v) was mixed with biotinylated rabbit anti-hamster IgG (0.5% v/v) for 30 min at room temperature prior to its use to aid in blocking non-specific binding. Cells were sorted by fluorescence on a FACStar Plus cell sorter (Becton Dickinson, San Jose, CA).

2.3. Cells

For DC purification, SP and PP tissues were digested with Dispase® and were applied to dense BSA gradients as previously described. Buoyant cells were maintained overnight at 1×10^7 cells/ml complete medium at 4°C as previously described[14] prior to FACS sorting as indicated above using N418 mAb. Macrophage (F4/80) or B cell (B220) contamination was <10%. Confirmatory staining with 33D1 correlated well with N418 staining (not shown). Greater than 96% of SP and PP DC stained positively for cell surface Ia molecules. Pigeon cytochrome *c*-specific Th0 T cell clones (L9A.1 and 16B.2) previously shown to be capable of producing Th1 and Th2 cytokines were kindly provided by Dr. Dan Mueller (Univ. of Minnesota, Minneapolis, MN) and maintained in culture as previously described.[15]

2.4. Supernatant Generation and Cytokine Determinations

Complete medium (RPMI 1640 supplemented with 50 μg/ml gentamicin, 2 mM L-glutamine, 5% heat-inactivated FCS, 50 μM 2-ME, 100 U/ml penicillin, and 100 μg/ml streptomycin) was used to generate culture supernatants from T cell clones. Cells were cultured in 24-well, flat-bottomed tissue culture wells at 2.0–32.0 × 10^4 Ag-specific T cells/ml with graded numbers (8.0–50.0 × 10^3/ml) of SP or PP DC. Culture supernatants were harvested at various timepoints and filter-sterilized (Millex-GV, Millipore, Bedford, MA) prior to determination of cytokine levels. IFN-γ levels were determined in triplicate using WEHI-279 cells[16] (provided by K. Bottomly, Yale University, New Haven, CT) by comparison of supernatant dilutions with recombinant cytokine dose-response curves as previously described.[17] IL-4 determinations by enzyme-linked immunosorbent assays used procedures similar to those previously described[18] and purified rat anti-mouse IL-4 mAbs (11B11 and BVD6–24G2; Pharmingen, San Diego, CA). Statistical comparisons were performed using two-tailed Student's *t* tests and reported as mean ± SD.

3. RESULTS

3.1. Antigen-Specific Stimulation of T Cell Cytokine Production

Production of T cell cytokine patterns was determined using the pigeon cytochrome *c*-specific Th0 T cell clones L9A.1 and 16B.2. Graded numbers of purified, N418⁺ SP and PP DC were cocultured with either 16B.2 (Fig. 1) or L9A.1 (Fig. 2) Th0 clones in the presence of antigen (50 μg/ml pigeon cytochrome *c*). As with polyclonal activation of T cells shown in a previous study,[19] purified SP DC induced higher levels of IFN-γ than purified PP DC in Ag-stimulated T cell activation cultures (Fig. 1). Patterns of IFN-γ similar to these were induced by SP and PP DC using the A.E7 Th1 clone,[20] thus confirming these findings (data not shown). In contrast, PP DC stimulated pigeon cytochrome *c*-specific Th0 clone L9A.1 cells to produce elevated levels of IL-4 relative to the levels stimulated by SP DC (Fig. 2). Although these data do not address the possibility of variable antigen processing and presentation by DC from different tissues, the fact that they parallel the data generated previously using mitogens suggests that the different patterns of cytokines produced by DC-activated T cells is a function of the DC tissue of origin and not due to differential antigen handling.

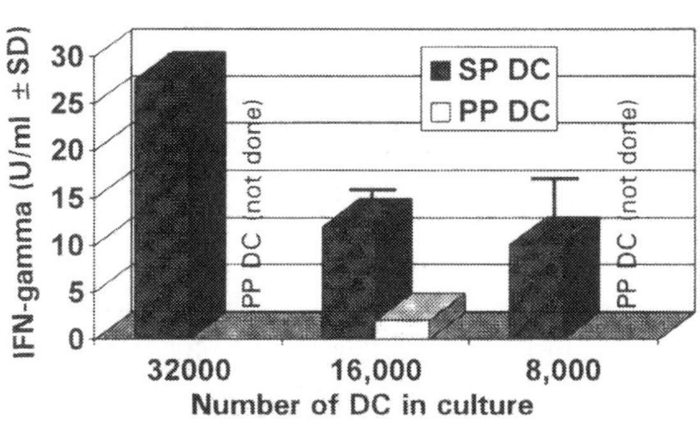

Figure 1. For IFN-γ production (Fig. 1), graded numbers of N418⁺ FACS-sorted SP DC or PP DC were cultured for 16 hours with 3.2 × 10^6 16B.2 T cells. Supernatants tested for IFN-γ that fell below the limits of detection (4 U/ml) were graphed as 2 U/ml; supernatants not depicted here that had <4 U/ml IFN-γ included 3.2 × 10^6 16B.2 T cells alone as well as 3.2 × 10^4, 1.6 × 10^4, and 8.0 × 10^3 SP DC alone and 1.6 × 10^4 PP DC alone (other PP DC numbers not included due to limited cell yield).

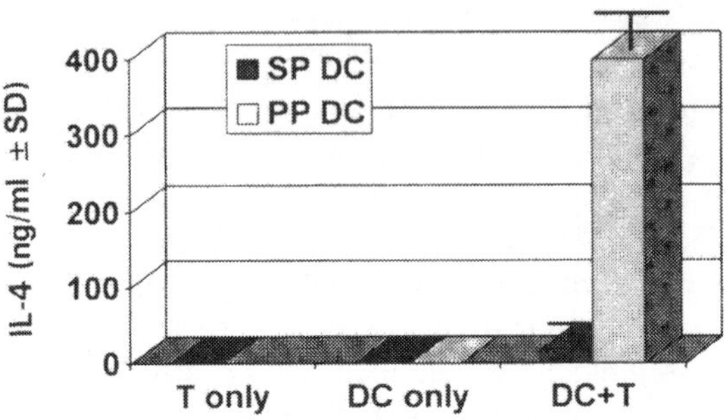

Figure 2. For IL-4 production (Fig. 2), N418⁺ SP or PP DC (5×10^4/well) were cultured for 60 hours with 2×10^4 L9A.1 T cells/well and 50 μg/ml pigeon cytochrome *c*.

4. DISCUSSION

Previous studies provide evidence for diversity of functional capabilities of a given type of APC. An example of macrophage diversity as a function of buoyant density is reflected by the production of different quantities of tumor necrosis factor-α in pulmonary fibrosis by different density-defined alveolar macrophage subpopulations.[21] Examples of DC diversity as a function of tissue origin are presented here and in the work of Spalding and colleagues.[8] For example, Spalding demonstrated that SP and PP DC induced different profiles of Ig regardless of the T cell and B cell tissue origin. Experiments described previously (manuscript submitted) indicate that periodate-modified whole SP cells proliferate and produce IL-2 from 3- to 20-fold more readily than identically treated whole PP cells, respectively. Although provocative, these experiments did not use DC enriched by identical techniques, thus possibly introducing bias into the measures of their functions. To further investigate this tissue-specific diversity, it was therefore necessary to use a unified procedure, such as FACS sorting, for enriching DC from SP and PP.

The possibility that different APC may induce divergent cytokine production profiles has been suggested by others and has only recently been demonstrated for different cell populations, e.g., B cells versus macrophages.[22] The studies presented here complement these past observations and suggest that the same cell type, i.e., DC, from distinct tissues (SP versus PP) can induce divergent T cell-derived cytokine production profiles.

Other workers determined that SP DC could be isolated by flow cytometry using the hamster anti-mouse DC mAb raised against the CD11c molecule (DC restricted in mice) (N418, American Type Culture Collection, Rockville, MD).[10] We then tested PP DC for N418 staining and determined that both SP and PP DC could be isolated using this Ab and a FACS. DC staining with N418 correlated well with the DC-specific 33D1 mAb for both SP and PP DC. Ia^+, $N418^-$ cells were also found, however, these cells lacked stimulatory capacity for proliferation and IL-2 induction in oxidative mitogenesis reactions and were not tested further. To confirm that the AC that were to be used in these experiments were indeed DC and not some uncharacterized, contaminant population, we performed multiple control experiments (these experiments used 2.4G2-pre-treated spleen cells—see below for explanation for this pre-treatment). It was demonstrated that: 1) >95% of the $N418^+$ cells were Ia^+, and 2) >97% of the stimulatory activity in oxidative mitogenesis reactions was found in the fraction of buoyant SP cells that stained strongly for N418 and Ia, thus strongly supporting the FACS data that DC (the cell population responsible for stimulation in oxidative mitogenesis reactions[23]) stain with N418, 33D1, and anti-Ia. Moreover, the non-DC population did not augment DC-induced IL-2 production when mixed with half-maximal numbers of DC. This experiment controlled for any putative unknown contaminant cell type being responsible for augmenting DC function. These results firmly support and confirm the view that the induction of cytokines in our cultures is mediated by DC and not some contaminant population.

The exact mechanism by which DC from different tissues induce different cytokine secretion profiles is unclear. Nonetheless, several possibilities can be considered. Firstly, the possibility that DC from different tissues may themselves produce different cytokines must be considered. It has been shown that SP DC stained positively for the p40 subunit of IL-12 intracellularly[24] and would thus favor induction of Th1 cells. The cytokines produced by PP DC have not been defined. Secondly, T cell activation not only requires ligation of the TCR:CD3 complex by MHC plus peptide, but also binding of accessory molecules on the T cell surface (e.g., LFA-1, ICAM-1, CD2, CD28) with the appropriate counter receptors on the surface of the APC. Expression of different combinations of these counter receptor molecules on DC from different tissues could result in different signals being delivered to T cells which could affect the profile of elaborated cytokines. Finally, the local microenvironment from which the DC are isolated must be considered. Lymphoid cells in gut-associated lymphoid tissues are continually in contact with bacterial products such as endotoxin, food-derived antigens, and breakdown products of epithelial cell turnover. As a result of exposure to the constituents of this microenvironment, gut DC may have an activated phenotype; such conditions are unlikely to be found in the spleen. Also, it has been suggested that exposure to these environmental products may influence induction of cytokine secretion or modification of counter receptor expression. Taken together, these findings support the current concept that different tissues, each with their distinct microenvironment of cytokines, hormones, and cellular elements, are involved in the selection, promotion, and/or maintenance of different immune responses. With regard to DC induction of T cell-derived cytokines, it is apparent that the tissue of DC origin determines the cytokine profiles produced and that DC from different tissues favor either cellular versus humoral immune responses by influencing T-cell cytokine production.

5. ACKNOWLEDGMENTS

The authors are greatly appreciative of the expert technical assistance provided by Mrs. Debbie McDuffie during the course of these studies. This work was supported in part by grants from NIH (AI 18958) (J.R.M., K.W.B., M.P.E.) and the Office of Research and Development, Medical Research Service, Department of Veterans Affairs (M.P.E.).

6. REFERENCES

1. T.R. Mosmann and R.L. Coffman, TH1 and TH2 cells: different patterns of lymphokine secretion lead to different functional properties, *Annu. Rev. Immunol.* 7:145 (1989).
2. T.R. Mosmann, J.H Schumacher, et al., Diversity of cytokine synthesis and function of mouse CD4⁺ T cells, *Immuno. Rev.* 123:209 (1991).
3. C. Pfeiffer, J. Murray, J. Madri, and K. Bottomly, Selective activation of Th1- and Th2-like cells *in vivo* - response to human collagen IV, *Immunol. Rev.* 123:65 (1991).
4. S.L. Swain, L.M. Bradley, et al., Helper T-cell subsets: phenotype, function and the role of lymphokines in regulating their development, *Immunol. Rev.* 123:115 (1991).

5. K. Inaba, J. Metlay, M.T. Crowley, and R.M. Steinman, Dendritic cells pulsed with protein antigens in vitro can prime antigen-specific, MHC-restricted T cells in situ, *J. Exp. Med.* 172:631 (1990).
6. W.C. Van Voohris, J. Valinsky, et al., Relative efficacy of human monocytes and dendritic cells as accessory cells for T cell replication, *J. Exp. Med.* 158:174 (1983).
7. D.M. Spalding, W.J. Koopman, J.H. Eldridge, J.R. McGhee, and R.M. Steinman, Accessory cells in murine Peyer's patch.Identification and enrichment of a functional dendritic cell, *J.Exp.Med.* 157:1646(1983).
8. D.M. Spalding and J.A. Griffin, Different pathways of differentiation of pre-B cell lines are induced by dendritic cells and T cells from different lymphoid tissue, *Cell* 44:507 (1986).
9. C.S. Hsieh, S.E. Macatonia, C.S. Tripp, S.F. Wolf, A. O'Garra, and K.M. Murphy, Development of TH1 CD4+ T cells through IL-12 produced by Listeria-induced macrophages, *Science* 260:547 (1993).
10. J.P. Metlay, M.D. Witmer-Pack, et al., The distinct leukocyte integrins of mouse spleen dendritic cells as identified with new hamster monoclonal antibodies, *J. Exp. Med.* 171:1753 (1990).
11. M.C. Nussenzweig, M.C., R.M. Steinman, M.D. Witmer, and B. Gutchinov, A monoclonal antibody specific for mouse dendritic cells, *Proc. Natl. Acad. Sci. USA* 79:161 (1982).
12. J.M. Austyn and S. Gordon, F4/80, a monoclonal antibody directed specifically against the mouse macrophage, *Eur. J. Immunol.* 11:805 (1981).
13. J.C. Unkeless, Characterization of a monoclonal antibody directed against mouse macrophage and lymphocyte Fc receptors, *J. Exp. Med.* 150:580 (1979).
14. M.P. Everson, D.M. Spalding, and W.J. Koopman, Exquisite sensitivity of dendritic cells to ultraviolet radiation and temperature changes, *Transplantation* 48:666 (1989).
15. D.L. Mueller, L. Chiodetti, P.A. Brown, and R.H. Schwartz, Clonal anergy blocks the response to IL-4, as well as the production of IL-2, in dual-producing T helper cell clones, *J. Immunol.*147:4118(1991).
16. D.S. Reynolds, W.H. Boom, and A.K. Abbas, Inhibition of B lymphocyte activation by interferon-γ, *J. Immunol.* 139:767 (1987).
17. M.P. Everson, D.M. Spalding, and W.J. Koopman, Enhancement of IL-2-induced T cell proliferation by a novel factor(s) present in murine spleen dendritic cell-T cell culture supernatants, *J. Immunol.* 142:1183 (1989).
18. J.H. Schumacher, A. O'Garra, B. Shrader, A. van Kimmenade, M.W. Bond, T.R. Mosmann, and R.L. Coffman, The characterization of four monoclonal antibodies specific for mouse IL-5 and development of mouse and human IL-5 enzyme-linked immunosorbent, *J. Immunol.* 141:1576 (1988).
19. M.P. Everson, D.S. McDuffie, D.G. Lemak, W.J. Koopman, J.R. McGhee, and K.W. Beagley, Dendritic cells from different tissues induce production of different T-cell cytokine profiles, *J.Leuk.Biol.* 59:494–498(1996).
20. T.T. Hecht, D.L. Longo, and L.A. Matis, The relationship between immune interferon production and proliferation in antigen-specific, MHC-restricted T cell lines and clones, *J. Immunol.* 131:1049 (1983).
21. M.P. Everson and D.B. Chandler, Changes in distribution, morphology, and tumor necrosis factor-α secretion of alveolar macrophage subpopulations during the development of bleomycin-induced pulmonary fibrosis, *Am. J. Pathol.* 140:503 (1992).
22. C.A. Janeway, Jr., S. Carding, B. Jones, J. Murray, P. Portoles, R. Rasmussen, J. Rojo, K. Saizawa, J. West, and K. Bottomly, $CD4^+$ T cells: specificity and function, *Immunol. Rev.* 101:39 (1988).
23. J.M. Austyn, R.M. Steinman, D.E. Weinstein, A. Granelli-Piperno, and M.A. Pallidino, Dendritic cells initiate a two-stage mechanism of T lymphocyte proliferation, *J. Exp. Med.* 157:1101 (1983).
24. S.E. Macatonia, N.A. Hosken, et al., Dendritic cells produce IL-12 and direct the development of Th1 cells from naive $CD4^+$ T cells, *J. Immunol.* 154:5071 (1995).

59

DENDRITIC CELLS, OBTAINED FROM PERIPHERAL BLOOD PRECURSORS IN THE PRESENCE OF PGE_2, PROMOTE Th2 RESPONSES

Paweł Kaliński, Catharien M. U. Hilkens, Alies Snijders, Frank. G. M. Snijdewint, and Martien L. Kapsenberg

Department of Cell Biology and Histology
Academic Medical Centre
University of Amsterdam
Meibergdreef 15
1105 AZ Amsterdam, The Netherlands

ABSTRACT

In order to investigate the impact of an inflamatory mediator PGE_2 on the functions of maturing DC we used an *in vitro* model of DC generation from peripheral blood monocytes. Addition of PGE_2 (10^{-9}M-10^{-6}M) to the cultures performed in the presence of GM-CSF and IL-4 did not alter the morphology nor high levels of expression of class II MHC and co-stimulatory molecules on arising DC, although at concentrations above 10^{-8} M, the acquisition of CD1a was selectively prevented. Control DC and the DC maturing in the presence of PGE_2 (PGE_2-DC) induced a similar proliferation of naive Th cells. Control DC produced high amounts of IL-12, and only trace amounts of IL-10, whereas PGE_2-DC produced no IL-12 and high levels of IL-10, when stimulated after the removal of PGE_2. The deficient IL-12 production by PGE_2-DC was observed after stimulation both in the absence and in the presence of IFNγ, and was not compensated during further 48 h culture in the absence of PGE_2. Compared to control DC, PGE_2-DC induced development of Th cells secreting elevated amounts of IL-4 and IL-5, from naive precursors. These data indicate that elevated tissue levels of PGE_2 may promote type 2 Th responses by impairing the ability of locally maturing DC to produce IL-12. Since Th2 responses mediate protection in Th1-related autoimmune disorders, the use of PGE_2-DC in immunotherapy of such disorders may be considered.

INTRODUCTION

The release of inflammatory cytokines at the site of pathogen entry increases the turnover of local DC and, by increasing their immunostimulatory potential, contributes to

Table I. Comparison of the surface phenotype and the functions of DC obtained from peripheral blood monocytes in the absence (control DC) and the presence (PGE_2-DC) of PGE_2

	Control DC	PGE_2-DC
HLA-A,B,C,	+++	+++
HLA-DR	+++	+++
HLA-DP	+++	+++
HLA-DQ	+++	+++
CD80	++	+/++
CD86	++	++
CD40	+++	++/+++
CD1a	+++	+++/– *
CD14	–	–/+++ *
Stimulation of naive Th cells	+++	+++
IL-12 production	+++	–
IL-10 production	–/+	+++
Cytokine profile induced in naive Th cells	Th1/Th0	Th2/Th0

–: negative, +: low, ++: high, +++: very high.
*Concentrations of PGE_2 above 10^{-8} M selectively prevented the disapearance of CD14 and the acquisition of CD1a.

the initiation of antigen-specific immune responses. GM-CSF, a factor involved in the recruitment of DC precursors to the tissues (1), promotes the generation of $CD1^+$ DC from human bone marrow or peripheral blood precursors *in vitro*, when combined with TNFα or IL-4 (rev: 2). This *in vitro* model facilitates studies on the impact of additional microenvironmental factors on the functional maturation of human DC.

Elevated production of an inflamatory mediator prostaglandin E_2 (PGE_2) is observed in several Th2-related diseases (3–5). At concentrations above 10^{-7} M, PGE_2 favors the *in vitro* production of type 2 cytokines in Th cells directly, via the selective downregulation of type 1 cytokines (6). The presence of much lower concentrations of PGE_2 during stimulation downregulates the production of the major Th1-promoting factor IL-12 in whole blood cultures (7), which raises the possibility that PGE_2 may also promote Th2-type responses indirectly.

This study addresses the question whether the pre-exposure of DC to PGE_2 in the tissues may bear functional consequences for the subsequent priming of naive Th cells in lymph nodes. To this aim, we tested if the exposure of DC to PGE_2 during their *in vitro* development from peripheral blood precursors, modifies their subsequent ability to induce a particular cytokine profile in developing naive Th cells. The data obtained are summarized in Table I.

RESULTS AND DISCUSSION

APC maturing in the absence and in the presence of PGE_2 develop DC phenotype and potent stimulatory activity for naive Th cells.

Adherent monocytes cultured for 6 days in GM-CSF and IL-4 (8), in the absence or in the presence of PGE_2 (10^{-9}–10^{-6} M) developed into nonadherent cells of dendritic morphology. DC obtained in the absence (control DC) and in the presence of PGE_2 (PGE_2-DC) both showed high expression of MHC class I, HLA-DR, -DP, -DQ, and the costimulatory molecules CD80, CD86 and CD40, although the expression of CD80 and CD40 on PGE_2-DC was often slightly lower. At concentrations above 10^{-8} M, PGE_2 prevented the disappearance of CD14 and the acquisition of CD1a, without any major influence on the expression of other markers. After removal of all the factors which had been present during the initial 6 days of DC cultures, both PGE_2-DC and control DC obtained from the same donors showed comparable capacity to induce the proliferation of allogeneic $CD4^{+}CD45RA^{high}$ naive Th cells as well the responses of autologous naive Th cells to superantigen, while monocytes from the same donors were clearly less effective.

PGE_2-DC Have a Stably Impaired Ability to Produce IL-12

Control DC produced substantial levels of IL-12 p70 (mean 42 pg/ml and 35 pg/ml; n = 11 donors) upon stimulation with SAC or LPS, respectively. Addition of IFNγ upregulated IL-12 production by DC to nanogram per ml concentrations and allowed for high production of IL-12 in response to a soluble trimeric CD40 ligand (9). DC maturing in the presence of even the lowest PGE_2 concentrations tested (10^{-9} M) showed 90% inhibition of the subsequent IL-12 production, while the DC maturing in the presence of 10^{-7} M PGE_2 did not produce any detectable IL-12. This concentration of PGE_2 was therefore used in subsequent experiments. The deficient IL-12 production by PGE_2-DC was observed in all experimental conditions; in all donors and after all modes of stimulation. Importantly, this deficit was not compensated in the presence of high amounts of IFNγ, which strongly upregulates IL-12 production in APC in most of the experimental conditions described thus far.

PGE_2-DC produced high amounts of IL-10. However, the addition of IL-10 neutralizing antibodies did not restore IL-12 production in SAC-stimulated PGE_2-DC, indicating that their inability to make IL-12 does not result from elevated IL-10 production. Additional support for this comes from the observation that PGE_2-DC did not secrete IL-12 after stimulation with CD40L plus IFNγ, which induced only marginal IL-10 production. Both control DC and PGE_2-DC produced similar amounts of IL-1β, which were 10 - 100 times lower compared to the monocyte-derived macrophages from the same donors, which were obtained in the presence of GM-CSF alone.

The difference in the cytokine profile of the two DC populations persisted for 48 h after removal of PGE_2, in the cultures supplemented with GM-CSF, the combination of GM-CSF and IL-4, or none of these factors.

PGE_2-DC Promote the Production of Type 2 Cytokines in Maturing Th Cells

To study whether the maturation of DC precursors in a PGE_2-rich environment has consequences for the type of immune responses initiated by these cells, we compared the ability of PGE_2-DC and control DC to induce the production of type 1 (IFNγ) and type 2 (IL-4 and IL-5) cytokines in maturing naive Th cells. Autologous naive Th cells were primed with the superantigen (SEA) presented by the two types of DC. Within the period of 14 days, Th cells acquired the CD45R0 memory marker and the number of Th cells primed by either type of DC increased similarly about 200- to 500-fold. After restimulation with CD3 mAb, in the absence of DC, Th cells primed with control DC displayed Th0-like cytokine profiles. In 7 independent experiments, despite a donor to donor variation with respect to the absolute amounts of the cytokines produced, naive Th cells primed with the PGE_2-DC produced as an average 3-fold higher amounts of IL-4 and IL-5, whereas the average IFNγ production was only 70% of the levels induced by control DC. This bias was observed already after 1 round of stimulation of Th cells, and most likely represents the acquisition of a different cytokine profile by naive Th cells, rather than a selective outgrowth of contaminating Th2-like memory cells, since the purity of the starting $CD4^{+}CD45RA^{high}$ population was above 98% and both DC types induced similarly strong proliferative responses, yielding similar numbers of Th cells after 14 days of culture. Additionally, a similar steering effect was observed in an accessory cell-free model, which was established previously in order to eliminate any possible selection by providing the conditions in which naive and memory Th cells have similar proliferation rates (10). In this case the supernatants from SAC-activated DC and PGE_2-DC were added to the naive Th cells stimulated with immobilized CD3 mAb in the presence of CD28 mAb. After restimulation, the Th cells primed in the presence of PGE_2-DC supernatants produced less IFNγ but more IL-4 and IL-5, compared to the cells primed in the presence of supernatants from control DC, indicating the involvement of soluble factors in the differential steering effects of the two DC types. These differences were abolished when IL-12 neutralizing Ab were added to the cultures, indicating that the absence of IL-12 production by PGE_2-DC is responsible for their ability to induce high levels of type 2 cytokines. Although these data underlines the dominant role of IL-12 in the differential steering of Th development by PGE_2-DC versus control DC, they do not exclude the possible participation of other soluble of cell surface-related APC factors.

The impact of exogenous PGE_2 on the cytokine profile of maturing DC was observed at a wide concentration range, with a strong effect even at 10^{-9} M. This implicates its physiological relevance, since these and higher PGE_2 levels are found in inflamed tissues (3,11) or are secreted by tumor cells, or tumor stroma (12,13). Although the distinct effect on cytokine profiles was seen already at a concentration as low as 10^{-9} M, even the highest concentrations of PGE_2 tested (10^{-6} M) did not prevent the acquisition of the dendritic morphology, high expression of class I and II MHC and co-stimulatory molecules by DC, nor their high activity in stimulating naive Th cells. Concentrations above 10^{-8} M selectively prevented the appearance of CD1a marker and the disappearance of CD14. Such $CD1^{-}$ cells obtained in high concentrations of PGE_2 resembled $CD1^{+}$ dendritic cells in respect to their morphology, potent stimulatory capacity for naive Th cells and low production of the inflammatory cytokine IL-1ß. Their relation to $CD14^{+}$ DC observed in the epithelia of upper airways (14), in the lung (15), or the dermis (16) has yet to be established.

The stability of PGE_2-induced IL-12 deficiency implies that the modulation of the cytokine pattern of DC in tissue compartments will be preserved after emigration of DC to the draining lymph nodes, where they thus contribute to the development of Th2-biased responses. Such a stability is important for the possible therapeutic application of PGE_2-DC. Since the induction of antigen-specific Th2 responses may prevent, or even ameliorate already ongoing, autoimmune processes in EAE and IDDM, murine models of human multiple sclerosis and diabetes (rev: 17), PGE_2-DC may be considered for use in antigen-specific therapies of Th1-related autoimmune disorders.

The present data suggest that elevated levels of PGE_2 in the tissue promote type 2 Th responses via a stable impairment of IL-12 production in locally maturing DC.

Such stable modulation of IL-12 production in maturing DC, by a factor present in a local inflammatory environment, indicates that the tissue factors induced at the site of pathogen entry may influence the character of the primary immune response, by the modulation of the function of local APC.

ACKNOWLEDGMENTS

This work was partially supported by grants from The Netherlands Organization for Scientific Research (No. 900–505–251 to CMUH), and Yamanouchi Europe B.V., The Netherlands (to FGMS).

REFERENCES

1. Kaplan, G., G. Walsh, L.S. Guido, P. Meyn, R.A. Burkhardt, R.M. Abalos, J. Barker, P.A. Frindt, T.T. Fajardo, R. Celona, and Z.A. Cohn. 1992. Novel responses of human skin to intradermal recombinant granulocyte/macrophage colony-stimulating factor: Langerhans cell recruitment, keratinocyte growth and enhanced wound healing. *J. Exp. Med.* 175:1717–1728.
2. Peters, J.H., R.K.H. Gieseler, B. Thiele, and F. Steinbach. 1996. Dendritic cells: from ontogenetic orphans to myelomonocytic descendants. *Immunol.Today* 17:273- 278.
3. Fogh, K., T. Herlin, and K. Kragballe. 1989. Eicosanoids in skin of patients with atopic dermatitis: prostaglandin E2 and leukotriene B4 are present in biologically active concentrations. *J. Allergy Clin. Immunol.* 83:450–457.
4. Misra, N., M. Selvakumar, S. Singh, M. Bharadwaj, V. Ramesh, R.S. Misra, and I. Nath. 1995. Monocyte derived IL-10 and PGE2 are associated with the absence of Th1 cells and in vitro T cell suppression in lepromatous leprosy. *Immunol. Lett.* 48:123–128.
5. Foley, P., F. Kazazi, R. Biti, C. Sorrel, and A.L. Cunningham. 1992. HIV infection of monocytes inhibits the T-lymphocyte proliferative response to recall antigens, via production of eicosanoids. *Immunology* 75:391–397.
6. Snijdewint, F.G., P. Kalinski, E.A. Wierenga, J.D. Bos, and M.L. Kapsenberg. 1993. Prostaglandin E2 differentially modulates cytokine secretion profiles of human T helper lymphocytes. *J. Immunol.* 150:5321–5329.
7. van der Pouw Kraan, T.C., L.C. Boeije, R.J. Smeenk, J. Wijdenes, and L.A. Aarden. 1995. Prostaglandin-E2 is a potent inhibitor of human interleukin 12 production. *J. Exp. Med.* 181:775–779.
8. Sallusto, F. and A. Lanzavecchia. 1994. Efficient presentation of soluble antigen by cultured human dendritic cells is maintained by granulocyte/macrophage colony-stimulating factor plus interleukin 4 and downregulated by tumor necrosis factor α. *J. Exp. Med.* 179:1109–1118.
9. Fanslow, W.C., S. Srinivasan, R. Paxton, M.G. Gibson, M.K. Spriggs, and M.J. Armitage. 1994. Structural characteristics of CD40 ligand that determine biological function. *Seminars Immunol.* 6:267
10. Kalinski, P., C.M. Hilkens, E.A. Wierenga, T.C. van der Pouw-Kraan, R.A. van Lier, J.D. Bos, M.L. Kapsenberg, and F.G. Snijdewint. 1995. Functional maturation of human naive T helper cells in the absence of accessory cells. Generation of IL-4-producing T helper cells does not require exogenous IL-4. *J. Immunol.* 154:3753–3760.
11. Nygard, G., A. Larsson, B. Gerdin, S. Ejerblad, and T. Berglindh. 1992. Viability, prostaglandin E2 production, and protein handling in normal and inflamed human colonic mucosa cultured for up to 48 h in vitro. *Scand. J Gastroenterol.* 27:303–310.
12. Pugh, S. and G.A. Thomas. 1994. Patients with adenomatous polyps and carcinomas have increased colonic mucosal prostaglandin E2. *Gut* 35:675–678.
13. Schrey, M.P. and K.V. Patel. 1995. Prostaglandin E2 production and metabolism in human breast cancer cells and breast fibroblasts. Regulation by inflammatory mediators. *Br. J. Cancer* 72:1412–1419.
14. Graeme-Cook, F., A.K. Bhan, and N.L. Harris. 1993. Immunohistochemical characterization of intraepithelial and subepithelial mononuclear cells of the upper airways. *Am. J. Pathol.* 143:1416–1422.
15. van Haarst, J.M.W., H.C. Hoogsteden, H.J. de Wit, G.T. Verhoeven, C.E.G. Havenith, and H.A. Drexhage. 1994. Dendritic cells and their precursors isolated from human bronchoalveolar lavage: immunocytologic and functional properties. *Am. J. Respir. Cell Mol. Biol.* 11:344–350.
16. Cerio, R., C.E.M. Griffiths, K.D. Cooper, B.J. Nickoloff, and J.T. Headington. 1989. Characterization of factor XIIIa positive dermal dendritic cells in normal and inflamed skin. *Br. J. Dermatol.* 121:421–431.
17. Liblau, S.R., S.M. Singer, and H.O.McDevitt. 1995. Th1 and Th2 CD4+ T cells in the pathogenesis of organ-specific autoimmune diseases. *Immunol. Today* 16:34–42.

DENDRITIC CELLS AND MACROPHAGES INDUCE THE DEVELOPMENT OF DISTINCT T HELPER CELL POPULATIONS *IN VIVO*

Geneviève De Becker, Véronique Moulin, Marcelle Van Mechelen, Françoise Tielemans, Jacques Urbain, Oberdan Leo, and Muriel Moser

Laboratoire de Physiologie Animale
Université Libre de Bruxelles
Rue des Chevaux, 67
B-1640 Rhode-Saint-Genèse, Belgium

INTRODUCTION

We have previously shown that the injection of antigen-pulsed DC induces the synthesis of specific IgG1 and IgG2a antibodies, whereas macrophages favor the production of IgG1 and IgE antibodies specific for the antigen. These data indicate that the isotype and the amplitude of the B cell response can be regulated by the nature of the APC, suggesting that Th cell differentiation is controlled at the level of antigen presentation[1,2]. In this report, we directly evaluate Th1 and Th2 functions in lymph nodes of mice that were immunized by injection of dendritic cells or peritoneal macrophages pulsed with the antigen. We show that macrophages induce Th2 differentiation *in vivo*, whereas DC drive the development of cells able to secrete Th1 and Th2-derived cytokines.

RESULTS

Dendritic Cells and Macrophages Pulsed *in Vitro* with Antigen Sensitize T Cells *in Vivo*

DC and macrophages were purified and pulsed with HGG (100μg/ml) as previously described[2,3], and injected into the fore and hind footpads, according to a protocol described by Inaba *et al.*[4]. The data in Figure 1 show that lymph node cells from DC- but not macrophages-injected mice proliferated *in vitro* in the presence of HGG. Some mice were then challenged by a second injection of antigen-pulsed APC and tested 2 d later. Lymph node cells from mice injected twice with DC or macrophages developed specific antigen responsiveness (Figure 1). The background proliferation of lymph node cells from mice injected twice with HGG-pulsed DC was elevated in all experiments performed and may

Dendritic Cells in Fundamental and Clinical Immunology
edited by Ricciardi-Castagnoli, Plenum Press, New York, 1997

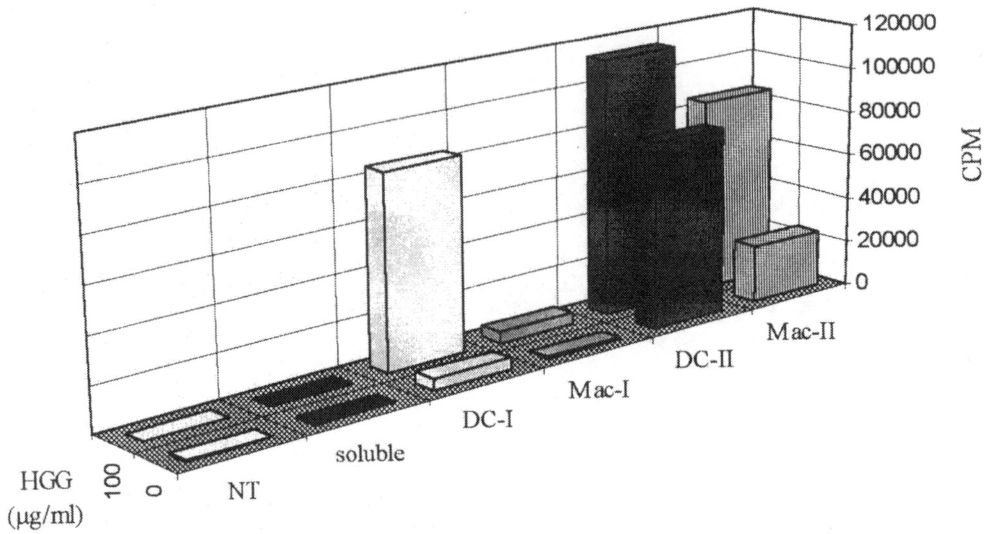

Figure 1. HGG-pulsed DC and macrophages sensitize T cells *in vivo*. BALB/c mice (3-to-4 mice per group) received one (I) or two (II) injection(s) of 3 × 10^5 HGG-pulsed DC (DC) or macrophages (Mac), 50 μg HGG (soluble) into the fore and hind footpads or were left untreated (control). Brachial and popliteal lymph nodes were harvested 5 days (I) or 2 days (II) later and 6 × 10^5 and 3 × 10^5 cells were cultured in 96-well microtiter plates with or without HGG (100 μg/ml). Proliferation was measured by [3H] thymidine incorporation during the last 16h of 72h culture. The data are expressed as the mean CPM of duplicate wells for each culture condition. Five experiments have been performed with similar results.

represent the continuous proliferation of T cells activated *in vivo* and/or may be due to remaining antigen associated with APC from primed mice.

DC But Not Macrophages drive Development of Th1-Type Cells

Antigen-specific T cells from mice injected with APC were tested for the ability to secrete the lymphokines characteristics of Th1 responses. As shown in Figure 2, lymph node T cells from mice immunized by 2 injections of DC pulsed with HGG produced IL-2 and IFN-γ when rechallenged with the same antigen *in vitro*. By contrast, T cells from macrophages-injected animals did not secrete detectable amounts of IL-2 nor IFN-γ in the same conditions.

DC and Macrophages Pulsed with Antigen Prime IL-4 and IL-5-Producing T Cells

We next compared the ability of lymph node T cells from mice immunized with either APC to secrete lymphokines characteristics of Th2 responses. Figure 3 shows that both groups of mice produced IL-4 and IL-5, although macrophages secrete slightly higher levels of cytokines in several independent experiments.

DC but Not Macrophages Express IL-12 mRNA

RNA was prepared from purified DC and macrophages, and PCR amplification was used to detect mRNA specific for IL-12, which has been shown to stimulate the IFN-γ

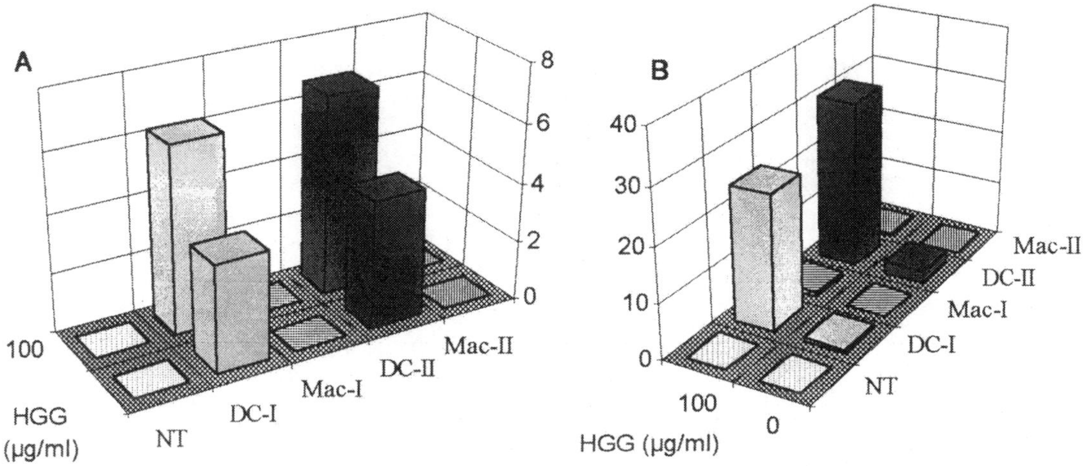

Figure 2. DC, but not macrophages, prime IL-2- and IFN-γ-producing T cells. Balb/c mice (pool of 2 or 3 animals per group) were immunized by two injections of 3 × 10^5 HGG-pulsed DC or macrophages (Mac) into the fore and hind footpads, or were left untreated (NT). Brachial and popliteal lymph nodes were harvested 2 days after second injection and purified T cells were cultured with irradiated syngeneic spleen cells, with or without addition of 100µg/ml HGG. Supernatants were tested for (A) IL-2 content by a bioassay and (B) IFN-γ content by ELISA. Results are expressed in U/ml. This experiment is representative of five independent experiments.

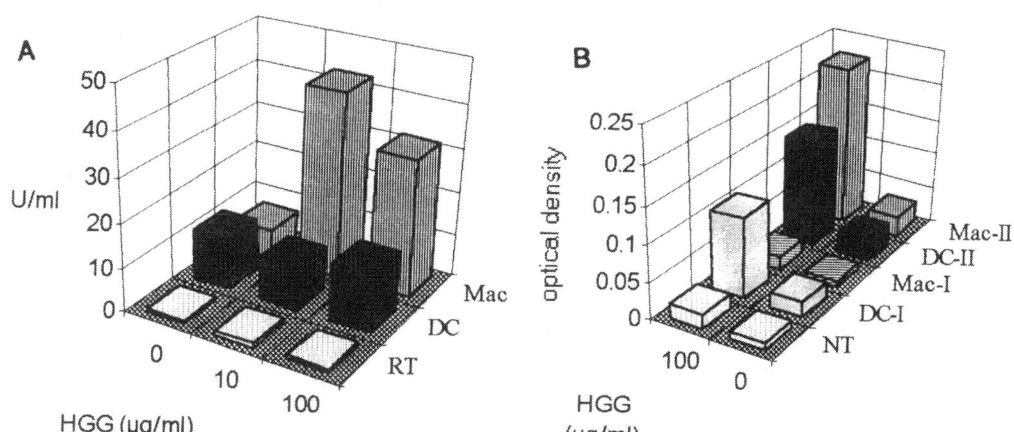

Figure 3. DC and macrophages induce development of IL-4- and IL-5-secreting T cells. Balb/c mice (pool of 3 animals per group) were immunized by 2 injections of 3 × 10^5 HGG-pulsed DC or macrophages (Mac), or were left untreated (NT). T cells were purified from brachial and popliteal lymph nodes harvested 2 days after second injection and cultured (3 × 10^5) with irradiated syngeneic spleen cells (2 × 10^5) in the presence of 10 µg/ml or 100 µg/ml HGG, or in the absence of antigen. (A) IL-4 content was evaluated by increase of MHC-II on splenic cells after 40-h of culture. Anti-IL-4 mAb (11B11) was added to some cultures, as indicated. Results are expressed as U/ml. (B) Supernatants were assayed for IL-5 content by ELISA and the results are expressed as mean absorbance. Four additional experiments were performed with similar results.

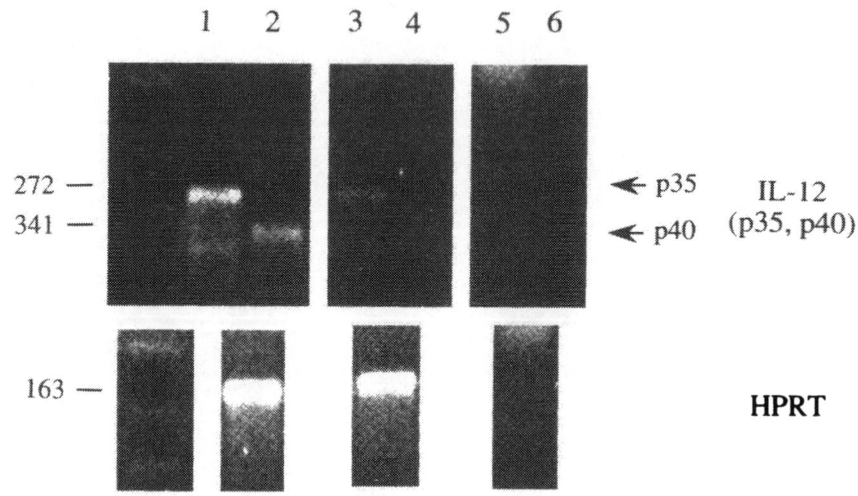

Figure 4. PCR analysis of IL-12 gene expression in splenic DC and peritoneal macrophages. Primers specific for the cytokine sequence were used to amplify RNA isolated from spleen DC and peritoneal macrophages. The housekeeping HPRT gene was used to assess the amount and the integrity of the RNA. After preparation of cDNA, PCR was performed essentially as previously described[6,7]. The PCR products were analyzed by 3% agarose gel electrophoresis and visualized by ethidium bromide staining. The size of the bands (base pair) is given for the amplified products and for the molecular weigth markers. Sample are 1) P35 mRNA in DC; 2) P40 mRNA in DC; 3) P35 mRNA in macrophages; 4) P40 mRNA in macrophages; 5) and 6) negative controls (no RNA) for P35 and P40, respectively.

production by Th1 cells[5]. The data in Figure 4 show that DC, but not macrophages, express mRNA for IL-12 p40.

DISCUSSION

Our data show that the nature of the cell presenting the antigen to naive T lymphocytes could influence the preferential development of Th1 and/or Th2 cells. DC drive the development of T cells secreting Th1-and Th2-derived cytokines, whereas macrophages induce the development of a polarized Th2 phenotype *in vivo*. T cells primed by DC *in vivo* could be Th1 and Th2 cells, each producing a polarized set of lymphokines, or alternatively a Th0 population, able to produce low levels of IL-4 and IFN-γ simultaneously. The analysis of the cytokines produced at the single cell level will help to clarify this issue. The finding that DC can prime Th1- and Th2-like cells is consistent with the data of Ronchese *et al.* showing that IFN-γ and IL-4 producing T cells can be primed on DC *in vivo*[8]. *In vitro*, Macatonia *et al.* showed that dendritic cells induced low levels of IFN-γ and IL-4 during secondary reactivation with antigen of T cells derived from unimmunized αβ TCR transgenic mice[5].

The differentiation of Th cells may be influenced by the strength of the activation signal[9,10]. There is evidence that high antigen doses favor the differentiation of Th1, whereas low dose of peptide induces the development of IL-4 producers. This observation may be relevant to our studies, since the relative affinity of the binding of Ag/MHC complexes as well as adhesion molecules and costimulatory signals on corresponding receptors on T lymphocytes is expected to be of much higher affinity on DC than on macrophages[11]. Furthermore, soluble (IL-12, IL-4)[12,13] and membrane-(B7–1, B7–2, CD40)[13,14,15] bound signals delivered between T cells and APC have been reported to influence Th1 or Th2 commitment. In particular, purified DC but not macrophages express mRNA specific for inducible chain of IL-12 (p40) (Figure 4). The injection of specific blocking antibodies will help to evaluate the role of B7–1, B7–2, CD40 and IL-12 in the *in vivo* priming by dendritic cells or macrophages.

REFERENCES

1. Sornasse, T. *et al.* 1992. *J. Exp. Med.* 172: 15–21.
2. De Becker, G. *et al.* 1994. *Eur. J. Immunol. 24:* 1523–1528.
3. Crowley, M., Inaba K., Witmer-Pack M. & R.M. Steinman. 1989. *Cell. Immunol. 118: 108.*
4. Inaba, K. Metlay J.P., Crowley M.T. & R. M. Steinman.1990. *J. Exp. Med.* 172: 63.
5. Macatonia, S.E., Hsieh C-S, Murphy K.M. & A. O'Garra. 1993. *Intern. Immunol.* 5(9): 1119.
6. De Wit, D. *et al.* 1993. *J. Immunol.* 150: 361.
7. Saiti, R.K. et al. 1985. *Science* 230: 1350.
8. Ronchese, F., Hausmann B., Hubele S. & P. Lane. 1994. *J. Exp. Med.* 179: 809.
9. Constant, S., Pfeiffer C., Woodard A., Pasqualini T. and K. Bottomly. 1995. *J. Exp. Med.* 182: 1591.
10. Kumar, V., Bhardwaj V., Soares L., Alexander J., Sette A. and E. Sercarz. 1995. *Proc. Natl. Acad. Sci.* 92: 9510.
11. Steinman, R.M.. 1991. *Ann. Rev. Immunol.* 9: 271.
12. Macatonia, S.E., Hosken N.A., Litton M., Vieira P., Hsieh C-S, Culpepper J.A., WWysocka M., Trinchieri G., Murphy K.M. and A. O'Garra. 1995. *J Immunol.* 154: 5071.
13. Koch et al. 1996. *J. Exp. Med.,* 184: 741.
14. Freeman, G.J., Boussiotis V.A., Anumanthan A., Bernstein G.M., Ke X-Y, Rennert P.D., Gray G.S., Gribben J.G. and L.M.. Nadler. 1995. *Immunity* 2: 523.
15. Kuchroo, V.K., Prabhu das M., Brown J.A., ranger A.M., zamvil S.S., Sobel R.A., Weiner H.L., Nabavi N. and L.H. Glimcher. 1995. *Cell* 80: 707.

61

A POTENTIAL PATHWAY OF TH2 DEVELOPMENT DURING PRIMARY IMMUNE RESPONSE

IL-10 Pretreated Dendritic Cells Can Prime Naive CD4⁺ T Cells to Secrete IL-4

Liming Liu,[1] Benjamin E. Rich,[2] Junichi Inobe,[1] Wunjun Chen,[1] and Howard L. Weiner[1]

[1]Center for Neurologic Diseases
[2]Harvard Skin Disease Research Center
Brigham and Women's Hospital and Harvard Medical School
221 Longwood Avenue
Boston, Massachusetts 02115

1. INTRODUCTION

Differential cytokine production (Th1 versus Th2) by $CD4^+$ T cells during an immune response plays an important role in determining the biological implications of the response. Th1 cells are characterized by the dominance of IFN-γ production while Th2 cells produce predominantly IL-4. It is now well established that IL-12 is essential for the priming of Th1 cytokine secreting T cells[1] while IL-4 is critical in the priming of Th2 cytokine secreting T cells[2]. While it has been suggested that CD1 specific NK1.1 $CD4^+$ T cells in the spleen[3], or CD4- CD8-, $TCR\alpha\beta^+$ T cells restricted by MHC class I[4] might be the potential sources of IL-4 during primary immune responses, these have not been clearly demonstrated.

Dendritic cells (DC) are the most potent antigen presenting cells (APC) in priming naive T cells and they have been used as APC in Th1 and Th2 differentiation assays[2,5]. Although it has been suggested that IL-10 may have a role in Th2 cell development, there is no direct evidence that IL-10 induces the differentiation of Th2 cells[2,6]. In previous experimental systems used to examine helper T cell differentiation, DC (or other APC) and naive T cells were exposed to antigen in the presence of influential cytokines (IL-10, IL-4 or IL-12)[1,2,6,7]. While the overall effects of added cytokines were evaluated in such systems, the precise cellular targets of these factors could not be identified. Physiologically, DC spend a certain amount time in the peripheral tissues such as skin or gut wall before they migrate to draining lymph nodes and present antigens to T cells there[8,9,10]. The original microenvironment of DC may influence their antigen presentation properties. We hypothesize that DC may prime naive T cells for either Th1 or Th2 development depending on their cytokine environment. To test this hypothesis we developed an *in vitro* priming system with separate preparations of highly enriched DC and naive OVA-specific TCR transgenic T cells, allowing the specific evaluation of the biologic effects of cytokines on DC. Our results demonstrate that DC pretreated with IL-10 (IL-10 DC) can prime naive T cells to secrete IL-4 while DC treated with rIL-4, rIL-12 or medium prime for almost exclusively Th1 cells. Since IL-10 and IL-12 are often present in the early immune response depending on the type of antigen, DC exposed to these cytokines may determine the differential T helper cell development (Th1 versus Th2).

2. MATERIALS AND METHODS

2.1. Animals

BALB/c mice carrying the I-A^d-restricted, DO11.10 TCR-αβ transgene specific for amino acids 323 to 339 of ovalbumin (OVA), were a generous gift of Dr. Dennis Loh[11] and bred in a facility managed by the Division of Veterinary Medicine of Harvard Medical School.

2.2. Preparation of Cytokine Treated DC

Spleen cells from BALB/c mice were cultured in 100x15mm culture dishes for two hours. Non-adherent cells were rinsed away with warm medium and the adherent cells were cultured overnight in fresh complete DMEM in the presence or absence of 2 ng/ml of mouse rIL-4, rIL-10 or rIL-12. 1 mg/ml OVA was added into the cultures 3–4 hours before harvesting non-adherent cells from the overnight cultures. DC were enriched by density separation with 15% Percoll. The enriched DC (40–60%) were further purified to 80–92% as judged by staining with DC specific mAb N418 by removing $B220^+$ cells with BioMag cell sorting. All DC preparations were irradiated prior to subsequent culture.

2.3. Preparation of Naive T Cells from OVA-Specific TCR Transgenic Mice

Spleen cells from OVA-TCR transgenic mice were passed over a nylon wool column and the eluted cells were further purified by removing cells other than $CD4^+$ T cells with BioMag cell sorting using mAb specific for MHC class II, B220, CD8 and F4/80 (a macrophage marker). The remaining highly enriched T cells (90–95% $CD4^+$) were layered over Percoll step-gradients of 50%, 60%, 66% and 70% and centrifuged at 400g. High density cells were recovered from the interface between 66% and 70%. These cells were CD25 negative, CD44 ^{low}, CD45 RB ^{high}, Mel 14 (CD62L)^{high}, a naive T cell phenotype (not shown).

2.4. Priming Naive OVA-TCR Transgenic $CD4^+$ T Cells with Cytokine Treated DC

5×10^6 naive T cells per well were cultured in 24 well plates with 2.5×10^5 cytokine-treated, OVA-pulsed DC for 4–5 days. Alternatively, 2×10^6 naive T cells were cultured with 1 × 10^5 cytokine-treated and OVA-pulsed DC in 48-well plates for 4–5 days. The recovered cells from 4–5 day cultures were washed three times and rested for another 4–5 days with irradiated APC. Viable cells of resting cultures were obtained by density gradient centrifugation using Isopaque-Ficoll. The recovered viable cells were restimulated in X-VIVO 20 with OVA (50 μg/ml) and irradiated splenic cells. Supernatants were harvested 22 h after restimulation and cytokines were measured with ELISA kits (Pharmingen, see Fig. 1A).

3. RESULTS AND DISCUSSION

3.1. IL-10 DC Prime Naive $CD4^+$ T Cells to Secrete IL-4 upon Restimulation

To examine whether exposure to cytokines can influence the ability of DC to prime T cells for differential cytokine secretion, a culture system was established involving separate preparations of purified DC and T cells (See Figure 1A and Materials and Methods). In this system, DC were purified from adherent spleen cells that had been previously exposed to cytokines overnight (rIL-4, rIL-10 or rIL-12) and then antigen (OVA). In the priming cultures, naive $CD4^+$ T cells (high density, Mel-14^{high}, $CD44^{low}$) purified from OVA-TCR transgenic mice were exposed to the treated DC for 4–5 d. These T cells were then allowed to "rest" for 4–5 days in the presence of fresh irradiated splenic APC without further added antigen and then restimulated with fresh splenic APC and OVA for 22 hours. Supernatants were then assayed for IL-2, IL-4, IL-10 and IFN-γ. IL-10 pretreated DC primed naive $CD4^+$ T cells to secrete IL-4, IL-2, IFN-γ and IL-10 (Th0) upon restimulation while IL-4, IL-12 or medium treated DC primed naive T cells secreted exclusively Th1 cytokines, e.g., IL-2, IFN-γ (Fig.1B). When these primed T cells were restimulated one more time *in vitro* , Th2 dominant cytokines were produced in IL-10 DC primed T cells while IL-12 or medium treated DC primed T cells still produced exclusively Th1 cytokines (Fig.1B). The effect of various doses of IL-10 in the preincubation culture was also tested. T cells primed by DC exposed to the higher dose (6 ng/ml) secreted predominantly the Th2 like cytokines such as IL-4 and IL-10 with little IFN-γ and IL-2 upon secondary restimulation while T cells primed by DC exposed to lower doses (0.5 ng or 2 ng/ml) produced Th0 type cytokines (Fig. 1C).

3.2. Mechanisms Involved in Priming of IL-4 Secreting T Cells by IL-10 DC

IL-10 can inhibit macrophage-dependent stimulation of Th1 cells through blocking IL-12 production of macrophages[12]. IL-10 suppresses DC-stimulated IFN-γ production by CD4 T cells[13]. We also noticed in the primary culture that IL-10 DC stimulated much less IFN-γ when compared to IL-4, IL-12 or control pretreated DC (Fig.1B), suggesting that expression of IL-12 in preincubation cultures may be inhibited by IL-10. To confirm this, we measured levels of the p40 subunit of IL-12 in preincubation cultures into which different cytokines were added. Supernatants of IL-10 pretreated cultures contained significantly less IL-12 p40 than that of control culture (not shown). To further determine the role of IL-12 in regulating IL-4 priming capacity by IL-10 DC, rIL-12 was added either directly into the priming cultures or the preincubation cultures and IL-4 production upon restimulation was measured. Addition of rIL-12 into the priming cultures completely

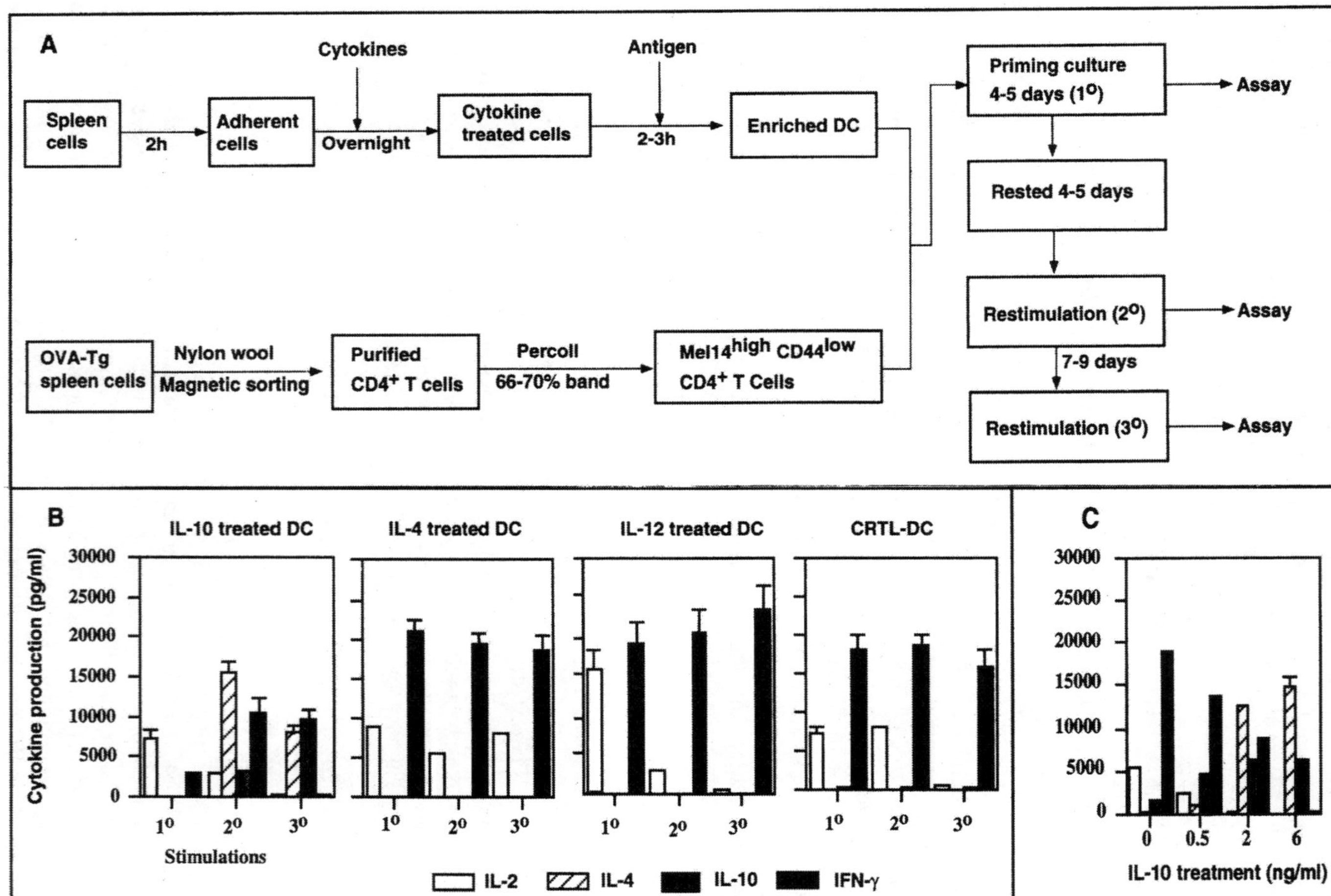

Figure 1. A, Experimental scheme (also see Materials and Methods). B, Cytokine secretion of T cells primed by different cytokine treated DC upon restimulation. C, Cytokine secretion of T cells primed by DC exposed to varying levels of IL-10.

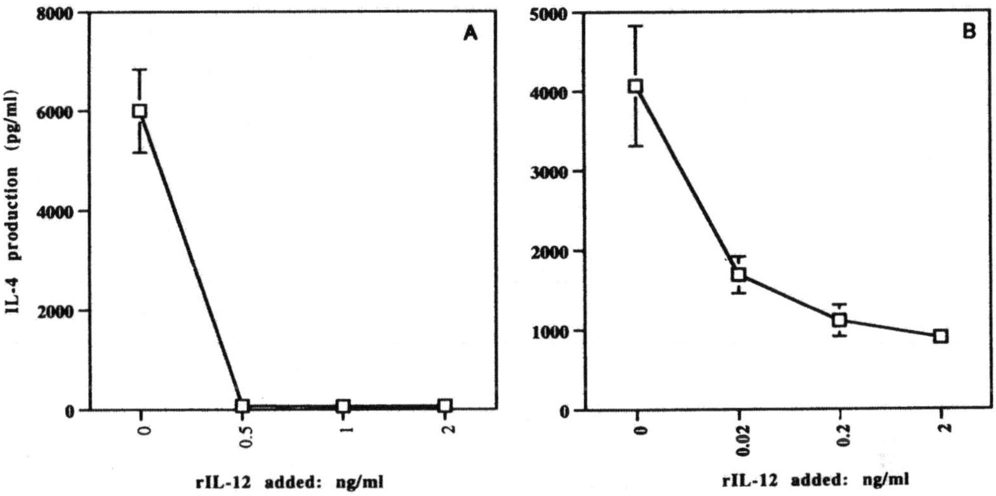

Figure 2. IL-4 secretion of T cells primed by IL-10 DC: A, rIL-12 present in priming cultures. B, rIL-12 present in preincubation cultures.

blocked IL-4 priming capacity of IL-10 treated-DC (Fig. 2A). When we added rIL-12 into the preincubation cultures which contained rIL-10, we observed reduced capacity of IL-10 DC to prime for IL-4 secretion dependent on the dose of IL-12 (Fig. 2B). Thus, the regulation of IL-12 production in preincubation cultures is essential for determining whether Th1 or Th2 type responses predominate.

IL-10 down-regulates expression of B7.2 but not B7.1 on blood DC[14] and also inhibits B7.1 and B7.2 expression on Langerhans' cells[15]. However, no significant difference in the number of DC expressing B7.1 or B7.2 was found in preparations exposed to doses of IL-10 up to 8 ng/ml (0, 0.5, 2, 8 ng/ml) and the staining intensity was also similar (not shown). B7.2 is a critical costimulatory molecule in the initial production of IL-4[16] and in the Th2 pathway[17]. To determine the contribution of B7 molecules to the priming capacity of IL-10-DC, anti-B7.1 or anti-B7.2 mAb were added to the priming culture. Only anti-B7.2 inhibited IL-4 priming capacity of IL-10 DC (Fig. 3a) while anti-B7.1 slightly increased IL-4 priming capacity of IL-10 DC. The effect of anti-B7.2 was not due to toxicity since reduced IL-4 production was accompanied by increased IFN-γ production (Fig. 3B). Neither anti-B7.1 or anti-B7.2 induced IL-4 priming by medium-treated DC (not shown). No significant differences were found in the levels of expression of ICAM-1 or MHC class II on IL-10 treated or control DC (not shown). Significantly, the ability of IL-10 DC to prime T cells to secrete IL-4 is dependent on direct cell to cell contact as well as IL-4 production in the priming culture (not shown).

4. CONCLUSION

Splenic DC exposed to IL-10 prime naive $CD4^+$ T cells to secrete IL-4 leading to Th2 cell development. IL-4, IL-12 medium-pretreated DC induce exclusively Th1 effector T cells. The balance between IL-12 and IL-10 in the preincubation cultures determines whether IL-4 secreting T cells are developed. Direct contact between IL-10 DC and naive T cells is required for the priming of IL-4 secreting cells. Although the expression of B7.1

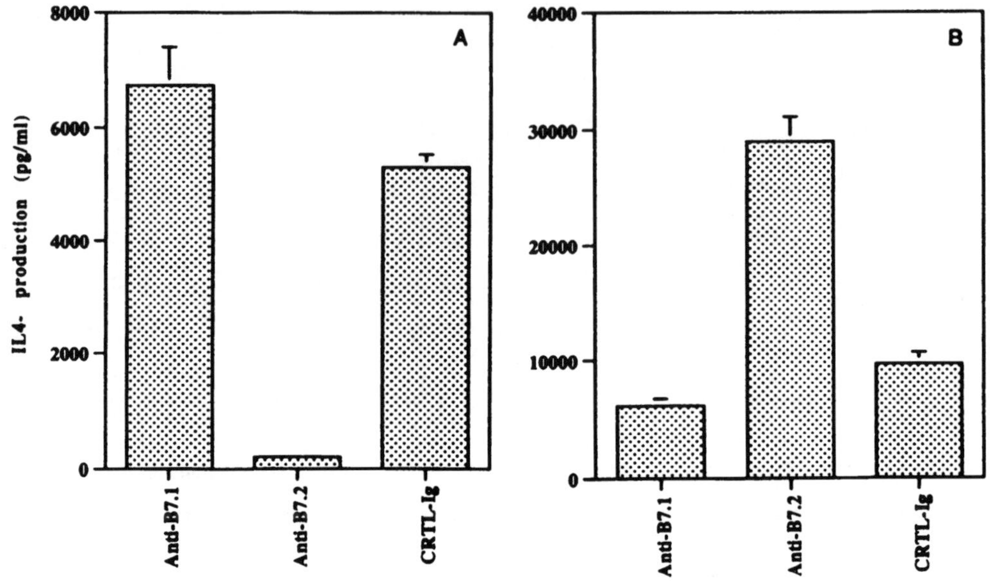

Figure 3. Cytokine secretion of T cells primed by IL-10 DC in the presence of anti-B7 antibodies. A, IL-4. B, IFN-γ.

and B7.2 on both control and IL-10 DC are similar, anti-B7.2 antibody inhibits the priming for IL-4 secreting T cells.

5. REFERENCES

1. Hsieh, C.-S., S.E. Macatonia, C.S. Tripp, S.F. Wolf, A. O'Garra, and K.M. Murphy. 1993. Development of Th1 $CD4^+$ T cells through IL-12 produced by *Listeria* -induced macrophages. *Science (Wash.DC)* 260:547.
2. Seder, R.A., W.E.Paul, M.M.Davis, and B.F.D.T. Groth. 1992. The presence of Interleukin 4 during *in vitro* priming determines the lymphokine-producing potential of $CD4^+$ T cells from T cell receptor transgenic mice. *J. Exp. Med*. 176:1091–1098.
3. Yoshimoto, T., and E.P. Paul. 1994. $CD4^{pos}$. $NK1.1^{pos}$ T cells promptly produce interleukin 4 in response to *in vivo* challenge with anti-CD3. *J. Exp. Med.* 179:1285–1295.
4. Leite-de-Moraes, M.C., A. Herbelin, F. Machavoine, A. Vicari, J.M. Gombert, and M.D.M. Papiernik. 1995. MHC class I- selected CD4-CD8-TCR-alpha beta $^+$ T cells are potential source of IL-4 during primary immune response. *J. Immunol.* 155:4544–50.
5. Hosken, N.A., K. Shibuya, A.W. Heath, K.M. Murphy, and A. O'Garra. 1995. The effect of antigen dose on $CD4^+$ T helper cells phenotype development in a T cell receptor-αβ transgenic model. *J. Exp. Med.* 182:1579.
6. Hsieh, C.-S., A.B. Heimberger, J.S. Gold, A. O'Garra, and K.M. Murphy. 1992. Differential regulation of T helper phenotype development by interleukin 4 and 10 in αβ T cell-receptor transgenic system. *Proc. Natl. Acad. Sci. USA* 89:6065.
7. Swain, S.L., A.D. Weinberg, M. English, and G. Huston. 1990. IL-4 direct the development of Th2-like helper effectors. *J. Immunol.* 145:3796–3806.
8. Bujdoso, R., J. Hopkins, B.M. Dutia, P. Young, and I. McConnell. 1989. Characterization of sheep afferent lymph dendritic cells and their role in antigen carriage. *J. Exp. Med.* 170:1285.
9. Macatonia, S.E., S.C. Knight, A.J. Edwards, S. Griffiths, and P. Fryer. 1987. Localisation of antigen on lymph node dendritic cells after exposure to the contact sensitizer fluorescein isothiocyanate. Functional and Morphological studies. *J. Exp. Med.* 166:1654.

10. Liu, L.M., and G.G. MacPherson. 1993. Antigen acquisition by dendritic cells: intestinal dendritic cells acquire antigen administered orally and can prime naive T cells *in vivo*. *J. Exp. Med.* 177:1299–1307.
11. Murphy, K., A. Heimberger and D. Loh. 1990. Induction by antigen of intrathymic apoptosis of CD4+CD8+TCR^{lo} Thymocytes in vivo. *Science*. 250:1720.
12. O'Garra, A. 1995. Interleukin-10: an inhibitor of macrophage-dependent stimulation of Th1 cells. Interleukin-10., ed. d.W.M.R.a.J.E.d. Vries, Chapter 6. R.G. Landes Company, Texas. 53–55 pp.
13. Macatonia, S.E., T.M. Doherty, S.C. Knight, and A. O'Garra. 1993. Differential effect of Interleukin-10 on dendritic cell-induced T cell proliferation and interferon-g production. *J. Immunol.* 150:3755–65.
14. Buelens, C., F. Willems, A. Delvaux, G. Pierard, J.P. Delville, T. Velu, and M. Goldman. 1995. Interleukin-10 differentially regulates B7–1 (CD80) and B7–2 (CD86) expression on human peripheral blood dendritic cells. *European Journal of Immunology* 25, no. 9:2668–72.
15. Kawamura, T., and M. Furue. 1995. Comparative analysis of B7–1 and B7–2 expression in Langerhans cells: differential regulation by T helper type 1 and T helper type 2 cytokines. *European Journal of Immunology* 25, no. 7:1913–7.
16. Freeman, G.J., V.A. Boussiotis, A. Anumanthan, G.M. Bernstein, X.-Y. Ke, P.D. Rennert, G.S. Gray, J.G. Gribben, and L.M. Nadler. 1995. B7–1 and B7–2 do not deliver identical costimulatory signals, since B7–2 but not B7–1 preferentially costimulates the initial production of IL-4. *Immunity* 2:523–532.
17. Kuchroo, V.K., M.P. Das, J.A. Brown, A.M. Ranger, S.S. Zamvil, R.A. Sobel, H.L. Weiner, N. Nabavi, and L.H. Glimcher. 1995. B7–1 and B7–2 costimulatory molecules activate differentially the Th1/Th2 developmental pathways: application to autoimmune disease therapy. *Cell* . 80:707–718.

62

DENDRITIC CELLS AS IMMUNOGENS FOR HUMAN CTL RESPONSES

Armin Bender,[2] Mark Sapp,[1] Mary Feldman,[1] Anita Reddy,[1] Robert Seder,[3] Gerold Schuler,[2] Ralph M. Steinman,[1] and Nina Bhardwaj[1]

[1]The Rockefeller University
1230 York Avenue, New York 10021
[2]University of Erlangen
Erlangen, Germany
[3]The National Institutes of Health
Bethesda, Maryland

SUMMARY

The cellular requirements for generating potent human CD8+ CTLs to influenza A virus in vitro have been defined. Furthermore, we have developed improved methods for generating large numbers of DCs from non-proliferating progenitors. These developments have enabled the design of new strategies to elicit CTLs in vivo. For example, together with IL-12, antigen-pulsed DCs may be a useful approach for boosting CTL responses against infectious agents and malignancies. Our results also reopen the potential use of inactivated virus preparations as immunogens for CTL responses.

INTRODUCTION

CD8+ CTLs are important mediators for resistance to infections and malignant diseases. Vaccines and immunotherapies that preferentially prime this arm of the immune response are critical for establishing effective host immunity. We have previously identified the cellular requirements for generating potent human CD8+ CTLs to viral antigens using influenza A virus as a model agent. Small numbers of dendritic cells [DCs] induce substantial responses from fresh blood T cells [T:APC ratio of 50:1 – 100:1] within 7 days, while B cells and monocytes are inactive (1). Whereas both DCs and monocytes are infected with influenza virus, the former serve as effective APCs for the induction of CD8+ CTLs while the latter act as targets for the CTLs that are induced. The strong CD8+ T cell response to influenza-infected DCs is manifested by extensive proliferation. Furthermore, the CTLs are generated in the absence of CD4+ helper cells or exogenous cytokines and

Dendritic Cells in Fundamental and Clinical Immunology
edited by Ricciardi-Castagnoli, Plenum Press, New York, 1997

recognize virus infected or peptide pulsed targets in a class I restricted fashion. These findings indicate that DCs are potent adjuvants for the generation of anti-viral CTLs and will be useful agents in the design of vaccines where CTL generation is the desired endpoint. Below we describe additional studies demonstrating the critical role of DCs as immunogens for human CTL responses.

INACTIVATED NON-REPLICATING INFLUENZA VIRUS ELICITS HUMAN CD8+ CYTOLYTIC T CELL RESPONSES

The generation of CTL responses is thought to require replicating virus, primarily to provide enough antigen for processing and presentation on MHC class I molecules of APCs. We have found that equally potent human CD8+ CTL responses are generated when non-replicating, inactivated influenza virus is pulsed onto dendritic cells (2). Furthermore, the CTLs are generated in the apparent absence of CD4+ helper cells or exogenous cytokines. We have confirmed that there is little to no detectable viral protein in DCs pulsed with non-replicating virus. In particular, NS-1 protein, which is only synthesized in the infectious cycle, is evident in <1% of cells by FACS staining. These results indicate that active viral protein synthesis is not a requirement for charging class I molecules on DCs. To be optimally effective, the inactivated virus must retain its fusogenic activity and presumably gain access to the cytoplasm. Fusogenic activity can be blocked by altering the envelope hemagglutinin molecule [HA], which is required for receptor mediated endocytosis, entry into endosomes and subsequent fusion with the endosomal membrane. Fusion can be blocked by pre-treating the influenza virus at pH 5 [which alters the conformation of HA] or using virus with a precursor non-cleaved HA. Neither form of virus elicits CTL responses when pulsed onto DCs. Thus DCs probably do not rely on an exogenous pathway to present non-replicating virus but are capable of efficiently processing small amounts of protein brought in by the non-replicating virus.

IL-12 IN CONJUNCTION WITH DCS ENHANCES ANTI-VIRAL, CD8+ CTL RESPONSES IN VITRO

IL-12 is a heterodimeric cytokine with multiple immunoregulatory activities (3,4). It is produced early during the inflammatory response by macrophages, enhances NK cell cytoxicity several fold and induces the production of IFNγ which is critical for increasing the anti-microbial activity of phagocytic cells (5–8). IL-12 also participates in the development of acquired immune responses. It induces the differentiation of Th1 cells through its ability to prime naive Th0 cells for high IFNγ production (9,10). Furthermore, it is a potent growth factor for CD8+ T cells (9) and enhances cytolytic responses to alloantigens (11–13) and in anti-CD3 redirected assays (14).

Dendritic cells effect the development of Th1 cells from Th0 cells by the production of IL-12 (15), suggesting that IL-12 is a key component by which these APCs induce T cell mediated immune responses. We therefore evaluated the role of IL-12 in the generation of CTLs to influenza infected dendritic cells. Significant levels of IL-12 were not detectable in supernatants of infected-dendritic cells, or during CTL generation (16). Furthermore, anti-IL-12 antibody did not block CTL generation. However, exogenous IL-12 [30–300 pg/ml] enhanced CD8+ T cell proliferative and cytolytic responses. The greatest effect was seen in individuals with weak reactivity to influenza virus or at APC:T cell ratios of 1:100 or less. IL-12 augmented interferon γ production during CTL generation. Neutralizing anti-interferon γ antibody did not block the CTL inducing function of IL-12, however. Thus antigen-pulsed dendritic cells together with IL-12 may be a useful approach for boosting CTL responses against infectious agents and malignancies.

IMPROVED METHODS FOR THE GENERATION OF DENDRITIC CELLS FROM NONPROLIFERATING PROGENITORS IN HUMAN BLOOD

We have developed an improved method for generating mature dendritic cells from nonproliferating progenitors in human blood (17,18). 1% human plasma is used in the place of 10% fetal calf serum and two steps are involved. The first step or "priming" phase is a 6–7 day culture of T cell depleted mononuclear cells in medium supplemented with GM-CSF and IL-4. The second step or "differentiation" phase requires exposure to macrophage conditioned medium. Substantial yields are obtained using this two-step approach. About 1–3 x 10^6 mature dendritic cells are generated from 40 ml of blood vs. <0.1 x 10^6 from non-cytokine treated blood. The dendritic cells derive from progenitors found primarily in a population of CD14+ and adherent blood mononuclear cells. They express several features of mature cells, including a stellate cell shape, nonadherence to plastic, and very strong T cell stimulatory activity. Strong APC function was evident for both the proliferation of allogeneic T cells in the MLR [Figure 1], and the generation by syngeneic T cells of class I restricted, CTL responses to influenza virus. Cells pulsed with nM levels of matrix protein peptide are effective at inducing potent CD8+ CTLs. A panel of den-

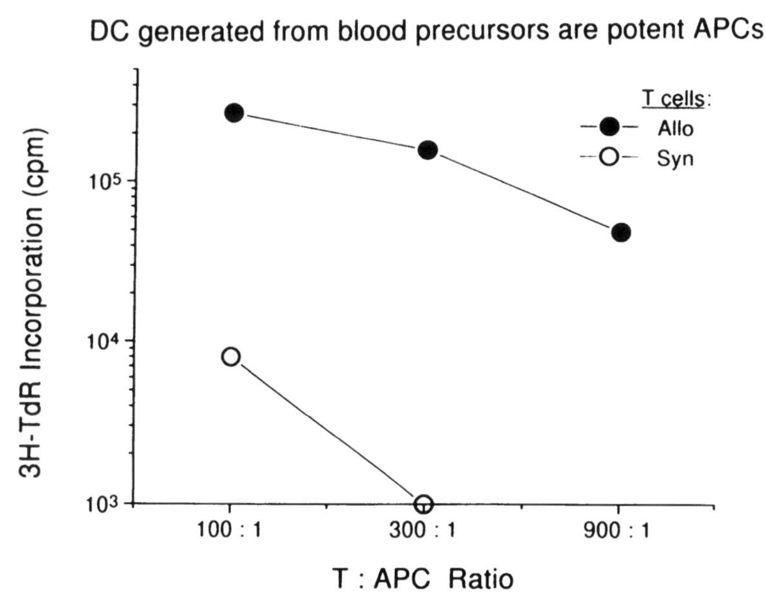

Figure 1. Dendritic cells derived from blood precursors are potent APCs. DCs were derived from PBMCs as described in the text with cytokines and conditioned medium. T cell stimulatory function [primary allogeneic MLR] is shown. Syngeneic MLR responses are included for comparison.

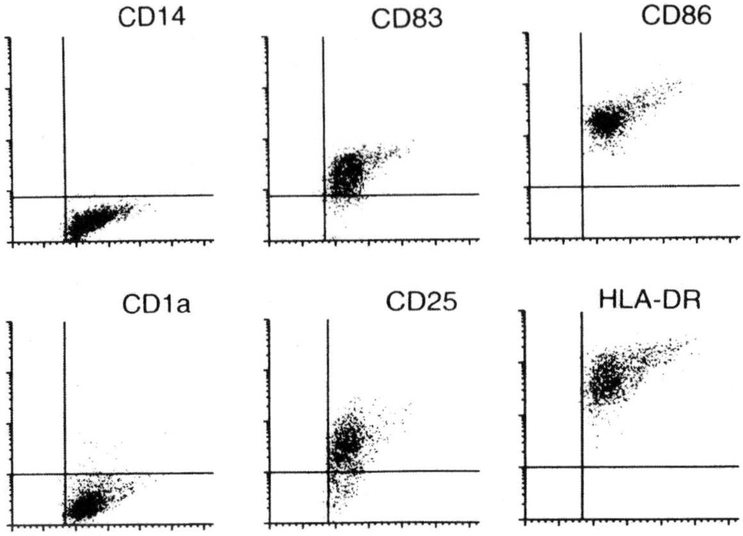

Figure 2. Cytofluorographic analysis of dendritic cells grown in GM-CSF/IL-4 and conditioned medium. Mature DCs were stained with a panel of abs. Dead cells and contaminating lymphocytes were excluded by light scatter properties. Dot plots of the remaining cells are shown.

dritic cell restricted markers was also expressed, including CD83 [Figure 2], p55, and perinuclear CD68. The cells lacked CD14 and CD1a but expressed high levels of DR, CD25 and CD86. These dendritic cell properties were retained for at least 3 days when the cytokines were removed, suggesting that these populations are stable and terminally differentiated. We predict that these cells will be effective adjuvants in vivo for active immunotherapy.

BIBLIOGRAPHY

1. **Bhardwaj, N., A. Bender, N. Gonzalez, L. K. Bui, M. C. Garrett, and R. M. Steinman**. 1994. Influenza virus-infected dendritic cells stimulate strong proliferative and cytolytic responses from human CD8+ T cells. *J. Clin. Invest. 94*:797.
2. **Bender, A., L. K. Bui, M. A. V. Feldman, M. Larsson, and N. Bhardwaj**. 1995. Inactivated influenza virus, when presented on dendritic cells, elicits human $CD8^+$ cytolytic T cell responses. *J. Exp. Med. 182*:1663.
3. **Trinchieri, G.** 1995. Interleukin-12: a proinflammatory cytokine with immunoregulatory functions that bridge innate resistance and antigen-specific adaptive immunity. *Annu. Rev. Immunol. 13*:251.
4. **Trinchieri, G.** 1994. Interleukin-12: A cytokine produced by antigen-presenting cells with immunoregulatory functions in the generation of T-helper cells type 1 and cytotoxic lymphocytes. *Blood 84*:4008.
5. **Scharton, T. M. and P. Scott**. 1993. Natural killer cells are a source of interferon gamma that drives differentiation of CD4+ T cell subsets and induces early resistance to Leishmania major in mice. *J. Exp. Med. 178*:567.
6. **Chehimi, J., N. M. Valiante, A. D'Andrea, M. Rengaraju, Z. Rosado, M. Koyayashi, B. Perussia, S. F. Wolf, S. E. Starr, and G. Trinchieri**. 1993. Enhancing effect of natural killer cell stimulatory factor [NKSF/interleukin-12] on cell-mediated cytotoxicity against tumor-derived and virus-infected cells. *Eur. J. Immunol. 23*:1826.
7. **Chan, S. H., B. Perussia, J. W. Gupta, M. Kobayashi, M. Pospisil, H. A. Young, S. F. Wolf, D. Young, S. C. Clark, and G. Trinchieri**. 1991. Induction of interferon γ production by natural killer cell stimulatory factor: Characterization of the responder cells and synergy with other inducers. *J. Exp. Med. 173*:869.

8. **Kobayashi, M., L. Fitz, M. Ryan, R. M. Hewick, S. C. Clark, S. Chan, R. Loudon, F. Sherman, B. Perussia, and G. Trinchieri**. 1989. Identification and purification of natural killer cell stimulatory factor [NKSF], a cytokine with multiple biologic effects on human lymphocytes. *J. Exp. Med. 170*:827.
9. **Gately, M. K., B. B. Desai, A. G. Wolitzky, P. M. Quinn, C. M. Dwyer, F. J. Podlaski, P. C. Familletti, F. Sinigaglia, R. Chizonnite, U. Gubler, and A. S. Stern**. 1991. Regulation of human lymphocyte proliferation by a heterodimeric cytokine, IL-12 [cytotoxic lymphocyte maturation factor]. *J. Immunol. 147*:874.
10. **Seder, R. A. and W. E. Paul**. 1994. Acquisition of lymphokine-producing phenotype by CD4+ T cells. *Annu. Rev. Immunol. 12*:635.
11. **Gately, M. K., A. G. Wolitzky, P. M. Quinn, and R. Chizzonite**. 1992. Regulation of human cytolytic lymphocyte responses by interleukin-12. *Cell. Immunol. 143*:127.
12. **Chouaib, S., J. Chehimi, L. Bani, N. Genetet, T. Tursz, F. Gay, G. Trinchieri, and F. Mami-Chouaib**. 1994. Interleukin 12 induces the differentiation of major histocompatibility complex class I-primed cytotoxic T-lymphocyte precurosrs into allospecific cytotoxic effectors. *Proc. Natl. Acad. Sci. USA 91*:12659.
13. **Bloom, E. T. and J. A. Horvath**. 1994. Cellular and molecular mechanisms of the IL-12-induced increase in allospecific murine cytolytic T cell activity. Implications for the age-related decline in CTL. *J. Immunol. 152*:4242.
14. **Mehrotra, P. T., D. Wu, J. A. Crim, H. S. Mostowski, and J. P. Siegel**. 1993. Effects of IL-12 on the generation of cytotoxic activity in human CD8+ T lymphocytes. *J. Immunol. 151*:2444.
15. **Macatonia, S. E., N. A. Hosken, M. Litton, P. Vieira, C. Hsieh, J. A. Culpepper, M. Wysocka, G. Trinchieri, K. M. Murphy, and A. O'Garra**. 1995. Dendritic cells produce IL-12 and direct the development of Th1 cells from naive CD4+ T cells. *J. Immunol. 154*:5071.
16. **Bhardwaj, N., R. A. Seder, A. Reddy, and M. V. Feldman**. 1996. IL-12 in conjunction with dendritic cells enhances anti-viral, CD8+ CTL responses in vitro. *J. Clin. Invest. 98*:715.
17. **Bender, A., M. Sapp, G. Schuler, R. M. Steinman, and N. Bhardwaj**. 1996. Improved methods for the generation of dendritic cells from nonproliferating progenitors in human blood. *J. Immunol. Methods 196*:121.
18. **Romani, N., D. Reider, M. Heuer, S. Ebner, B. Eibl, D. Niederwieser, and G. Schuler**. 1996. Generation of mature dendritic cells from human blood: An improved method with special regard to clinical applicability. *J. Immunol. Methods 196*:137.

63

DENDRITIC CELLS AS TARGETS FOR CYTOTOXIC T LYMPHOCYTES

Stella C. Knight,[1] B. A. Askonas,[2] and Steven E. Macatonia[1]

[1]Antigen Presentation Research Group
Imperial College School of Medicine at Northwick Park Institute for Medical Research
Watford Road, Harrow, Middlesex
HA1 3UJ, United Kingdom
[2]Department of Biology
Immunity Infection Section
Imperial College of Science, Technology and Medicine
London SW7 2BB, United Kingdom

1. ABSTRACT

Dendritic cells (DC) carry antigen into lymph nodes where they may cluster with CD4 and CD8+ lymphocytes and activate both subsets in the initiation of immune responses. Since DC do not leave the lymph nodes in the efferent lymph they may die within the lymph nodes. Another possibility is that they are targets for cytotoxic T cells (CTL) when expressing appropriate epitopes. This possibility was tested *in vitro* using human peripheral blood DC to stimulate the development of primary CTL in response to HIV-1 or one of its T-cell epitopdes (e.g. env 111–126) and secondary CTL in response to type A influenza virus. Pooled CTL generated during six day cultures in 60 replicate 20µl hanging drops were tested in a conventional CTL assay. The HIV or HIV peptide stimulated CTL lysed HIV infected DC while the influenza-virus induced CTL killed DC targets infected with this virus. DC were not lysed significantly until they had been exposed to virus for 2–3 days and thus are not highly susceptible to lysis. However, killing of DC after 2–3 days infection with virus may be a feedback mechanism for removing antigen presenting cells after they have stimulated T cell responses. Removal of persistently infected DC by CD8+ CTL may also contribute to the reduction in DC numbers observed in blood and skin in HIV infection.

2. INTRODUCTION

DC as exemplified by Langerhans' cells of the skin are continuously entering the afferent lymphatics and migrating to lymph nodes where it is believed that they may become

DC populations within or surrounding T cell-dependent areas[1,2]. The influx of DC into lymph nodes is significantly increased following exposure to antigens in the periphery and DC within the lymph nodes are then able to stimulate T cell responses to the antigens[3,4]. Despite the continuous influx of DC into the lymph nodes, there is no evidence that significant numbers of DC leave the lymph nodes within the efferent lymphatics except in exceptional circumstances where lymph nodes may be damaged. There must, therefore, be mechanisms for removal of DC within the lymph nodes following the stimulation of immune responses. It has been reported that DC exposed to antigens may become targets for natural killer cells in *in vitro* studies[5]. In this study we examined whether DC could become targets for virus specific CTL.

In some viral infections in mice such as lymphocytic choriomeningitis virus (LCMV) there is evidence that CD8 CTL present in lymph nodes during the virus infection may deplete antigen presenting DC and it has been hypothesised that this may make a major contribution to the development of pathology[6,7]. There is some debate about the role that CTL play in HIV-1 infection. CD8+ CTL are present at a high level in asymptomatic individuals while they are lost as patients develop AIDS. In addition, some individuals who are known to have been exposed to HIV-1 but do not develop the disease (e.g. some Gambian prostitutes) may have cell mediated immunity including CD8+ CTL responses to the virus in the absence of antibody[8]. CTL are thus believed, on this evidence, to be associated with protective effects in the disease, although defects in HIV-1 co-receptor may contribute to resistance in some multiply-exposed individuals[9]. There are also high levels of CTL within lymph nodes, and there are losses in the numbers of DC present in lymph nodes, in the circulation and within the skin[10,11], suggesting that a severe loss of DC numbers appears to occur during HIV infection. In addition to acting as a useful feedback mechanism to remove antigen presenting cells it is possible, therefore, that CTL might cause pathological effects by continuous removal of antigen presenting cells during persistent infection with HIV-1. The possibility that DC exposed to HIV-1 can act as target cells for CD8+ CTL was therefore studied.

3. MATERIALS AND METHODS

3.1. Cell Separation

Ficoll-separated peripheral blood mononuclear cells from normal defibrinated human peripheral blood were incubated overnight in medium (RPMI 1640, Dutch modification, with 100 IU/ml penicillin and 100μg/ml streptomycin and 10% foetal calf serum on petri dishes (NUNC, Glaxo, Uxbridge, Middlesex, UK) at 37°C in a 5% CO_2 incubator. Non-adherent cells were removed and layered onto hypertonic metrizamide gradients (13.7w/v, Nygaard, Oslo, Norway) and centrifuged for 10 min at 600g. Low density cells at the interface were collected, washed in medium and contained 30–35% DC. These cells were treated with a cocktail of antibodies to CD14, CD56, CD19 and CD3 and then incubated on petri dishes coated with rabbit anti-mouse, IgG as previously described. The purified cells were >90% DC[12].

3.2. Antigens

Infectious human immunodeficiency virus-type 1 (HIV-1) was stored at -70°C as culture supernatant from H9 cell lines persistently infected with HIV strain IIIB a T cell tropic line. DC (0.5–1 × 10^6 were incubated with 10^5 tissue culture infectious doses (TCID50) of HIV in 200µl of culture supernatant for two hours at 37°C and washed three times before the cells were used to stimulate primary CD8+ T cell responses. Control DC were incubated in supernatants from uninfected cultures. Alternatively, DC were pulsed for two hours with 20µg/ml of HIV-1 envelope 15 mer peptide (env 111–126) which was previously shown to be T cell epitope of HIV stimulating the development of primary CTL which kill virus-infected target cells[13]. Recombinant influenza A strain virus X31 was a gift from Dr. A. Hay (National Institute for Medical Research, Mill Hill, UK).

3.3. Induction of CD8+ Cytotoxic T Cell Responses

Enriched human T cells were prepared from the cells of the metrizamide gradient by rosetting them overnight with sheep red blood cells treated with neuraminidase (Sigma, Poole, Dorset, UK). T cell rosettes were separated on Ficoll and sheep red cells were lysed using Gey's solution. Enriched human T cells were cultured in 20µl hanging drops at 10^5 T cells per drop and stimulated with 2 × 10^4 autologous DC exposed *in vitro* for 2 hr to HIV, HIV peptide or influenza virus. After 6 days, 60 drops were pooled for the cytotoxic assay. Target cells were prepared by stimulating autologous or homologous lymphocytes for 3 days with phytohaemaglutinin (PHA) (1µg/ml) and recombinant interleukin-2 (IL-2) (30 units/ml) and then infecting with 10^4 TCID 50 of IIIB strain of HIV-1 or 500 HAU influenza virus in 200µl medium and culturing for a further 3 days as previously described[13]. Additionally, DC were cultured per se for 3 days and exposed to the viruses for the last 1 or 2 days. The cells were labelled with ^{51}Cr and used as targets for the CTL[13]. After 6hr in the presence of the CTL population, percentage specific ^{51}Cr release from lysed cells was calculated as 100 × [c.p.m. (sample release) - c.p.m. (spontaneous release)]/c.p.m. total release. Total release in cells lysed by triton was between 1900 and 3500 c.p.m. and the spontaneous release between 10 and 22% of this total.

4. RESULTS

Cultures of lymphocytes stimulated by DC pulsed with X31 influenza virus were harvested after six days to measure cytotoxic activity and this is shown in Fig.1. Specific chromium release from target cells by CTL reached almost 40% using autologous, 3 day influenza virus-infected PHA blast cells as targets with a killer to target cell ratio of 40/1. Using the same conditions, the specific lysis of the infected DC population was almost 50% when the target cells were exposed for 3 days to virus. Low levels of killing were seen using uninfected DC target cells. Lysis by CTL of DC targets pulsed for only 1 or 2 days with X31 virus was considerably lower.

Figure 2 shows that similar results were obtained when a primary CTL response was induced by DC pulsed for two hours with HIV. The DC targets were exposed to HIV virus for 3 days. Thus both X31 and HIV exposed DC could serve as targets for CTL. Primary stimulation of T cells with the ENV peptide 111–126 after pulsing the DC for two hours also generated CTL which could kill both virus-infected DC and PHA-stimulated blast cells with specific ^{51}Cr release of 28% and 34% respectively at a killer to target cell ratio of 40 to 1.

5. DISCUSSION

DC separated from peripheral blood and exposed for 3 days either to HIV or to influenza virus could act as targets for virus-specific CD8+ CTL. Low levels of lysis were

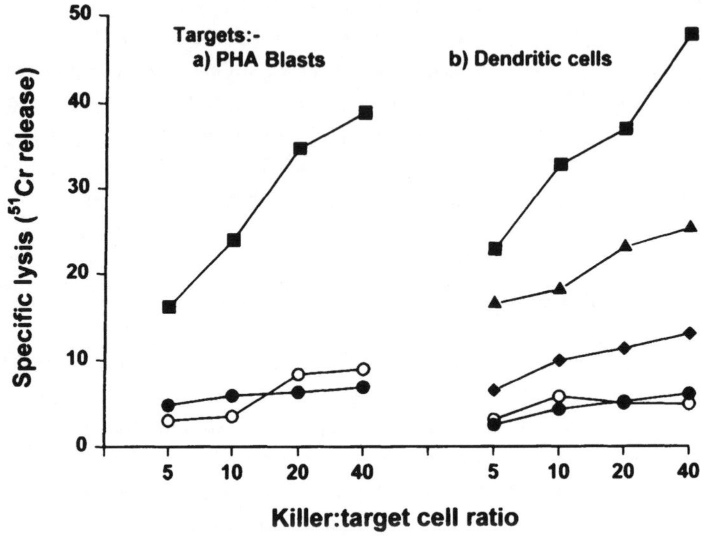

Figure 1. Infected dendritic cells as targets for anti-influenza specific cytotoxic T lymphocytes. Anti-influenza CTL were generated in 20 μl hanging drops by stimulating T cells (10^5) with DC (10^3) pulsed with A/X31 virus in 20 μl hanging drop cultures for 6 days. Pooled CTL from normal donors were tested for their capacity to lyse a) T cell blasts, b) dendritic cells. Untreated targets (O); Targets pulsed with tetanus toxoid for 3 days as a control antigen(●), or targets pulsed for 3 days (■), 2 days (▲), or 1 day (◆) with influenza virus.

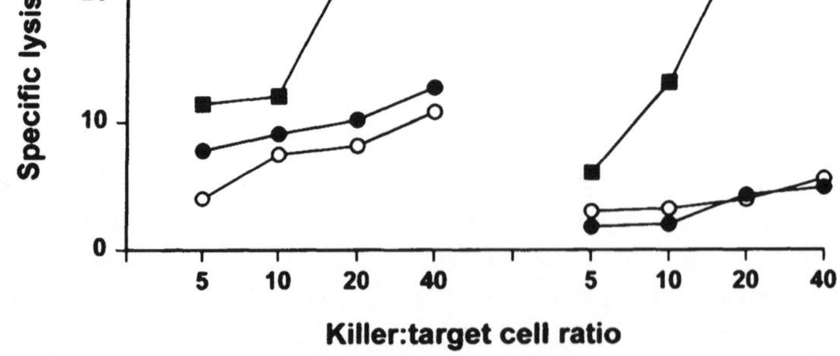

Figure 2. Infected dendritic cells as targets for anti-HIV specific cytotoxic lymphocytes - Primary CTL were generated by stimulation with HIV-1 *in vitro* as for influenza CTL in Figure 1. CTL were assessed for lysis of untreated targets (O), targets exposed to A/X31 influenza virus for 3 days (●) or targets infected for 3 days with HIV-1 (■).

found when target cells were exposed to virus for only 1 or 2 days but after 3 days the viral antigens were expressed sufficiently to allow moderate levels of lysis by specific CTL. Lysis of DC targets was at a similar level to that seen using autologous PHA stimulated blast cells as targets for the CTL; neither population is easily killed by CTL without a long incubation with virus. The present study used DC, pulsed for just two hours with the viruses, to stimulate the primary or secondary CTL responses, harvested after six days of culture. Thus, DC after just a short exposure to virus *in vitro* are able to process and present the virus and stimulate both primary and secondary proliferative and CTL responses. Longer times of exposure of DC to virus results in a loss in the capacity to stimulate T cell proliferation[10] *In vivo,* it is therefore likely that DC carrying antigen into lymph nodes will rapidly stimulate primary T cell responses. Following the activation of CTL they could themselves then be removed by the action of CTL. The capacity of DC to become targets for CTL may require a final maturation of the DC within the lymph nodes. Disappearance of specific antigen-presenting 'DC from lymph nodes draining the site of exposure to a contact sensitizer occurs between two and six days following exposure to antigen[14]. Previous studies have provided evidence that DC that have presented antigen *in vitro* can also become targets for natural killer cells[5].

It seems likely that during a resolving virus infection, removal of antigen presenting cells following the first rounds of exposure to virus would act as a normal feedback inhibition mechanism for inhibition of further immune responses within lymph nodes. However, in cases of persistent virus infection, the CTL could act as a double-edged sword by clearing virus and infected cells but also causing a detrimental deletion of DC. In transgenic mouse models of persistent infection with LCMV, immune suppression is associated with a $CD8^+$ T cell-dependent loss of interdigitating cells from the periarterioler lymphoid sheaths of the spleen; this effect is not seen when the virus strain is given to immunocompetent mice that clear the virus. This immunosuppressive virus also causes high levels of infection in DC in the white pulp[6]. Removal of DC by CTL may occur and contribute to immunosuppression[6,7]. In HIV infection, a reduction in the numbers of DC both in the peripheral blood and in the skin in HIV-infected individuals has been reported[10,11]. Since there is also evidence of a low level but persistent infection of DC with HIV-1[15], removal of DC by CTL could contribute to this finding. Recent studies show that the reduction in DC numbers in blood is inversely proportional to the level of infection in the DC (Patterson, English, Pinching, Helbert and Knight, in preparation). This raises the possibility that the persistent lytic activity of CTL on antigen presenting cells may contribute towards the loss of DC and to pathology in this disease.

6. REFERENCES

1. Lens, J.W., Drexhage, H.A., Benson, W. and Balfour, B.M. *Immunology.* **49**:415–22, 1983.
2. Silberberg-Sinakin,I., Fedorko,M.E., Baer,R.L., Rosenthal, S.A., Berezowsky, V. and Thorbecke, G.J. *Cell.Immunol.* **32(2)**:400–16, 1977.
3. Knight, S.C., Krejci, J., Malkovsky, M., Colizzi, V., Gautam, A. And Asherson, G.L. *Cell.Immunol.* **94**:427–434, 1985.
4. Macatonia, S.E., Taylor, P.M., Knight, S.C. and Askonas, B.A. *J.Exp.Med.* **169**:1255–64, 1989.
5. Shah, P.D., Gilbertson, S.M. and Rowley, D.A. *J.Exp.Med.* **162**:625, 1985.
6. Borrow, P., Evans, C.F. and Oldstone, M.B.A. *J.Virol.* **69**:1059–70, 1995.
7. Zinkernagel, R.M. *Curr.Opin.Immunol.* **7**:462–70, 1995.
8. Rowland-Jones, S.L., McMichael, A *Curr.Opin.Immunol.* **7**:448–55, 1995.
9. Liu, R., Paxton, W.A., Choe, S., Ceradini, D., Martin, S.R., Horuk, R., MacDonald, M.E., Stuhlmann, H., Koup, R.A. and Landau, N.R. *Cell* **86**:367–77, 1996.

10. Macatonia, S.E., Lau, R., Patterson, S., Pinching, A.J. and Knight, S.C. *Immunology*, **71**:38–45, 1990.
11. Belsito, D.V., Sanchez, R., Baer, M.D., Valentine, F. and Thorbecke, G. *N.Engl.J.Med.* **310**:1279–82, 1984
12. Macatonia, S.E., Gompels, M., Pinching, A.J., Patterson, S. And Knight, S.C. *Immunology*, **75**:576–581, 1992
13. Macatonia, S.E., Patterson, S. and Knight, S.C. *Immunology*, **74**:399–406, 1991.
14. Macatonia, S.E., Edwards, A.J. and Knight, S.C. *Immunology*, **59**: 509–14, 1986.
15. Knight, S.C. *AIDS* **10(8)**:807–17, 1996.

64

CUTANEOUS DENDRITIC CELLS PROMOTE REPLICATION OF IMMUNODEFICIENCY VIRUSES

Melissa Pope,[1] Sarah Frankel,[2] Ralph Steinman,[1] David Elmore,[3] David Ho,[4] and Preston Marx[4]

[1]Laboratory of Cellular Physiology and Immunology
The Rockefeller University
1230 York Avenue, New York 10021
[2]The Armed Forces Institute of Pathology
AIDS Division, Washington DC 20306
[3]Charles River Biomedical Resources Foundation Inc.
305 Almeda-Genoa Road
Houston, Texas 77047
[4]The Aaron Diamond AIDS Research Center
The Rockefeller University
455 First Avenue, New York 10016

1. CUTANEOUS DENDRITIC CELL-T CELL MIXTURES PROVIDE A "NATURAL" MILIEU FOR HIV REPLICATION

The body surfaces [such as the linings of the vagina, cervix, anus, and oropharynx] that are potentially exposed to HIV-1 during sexual and perinatal transmission, are similar to the skin in histology and dendritic cell [DC] content. Since large amounts of fresh skin are more readily available, the skin-derived leukocytes have been used as a model for the mucosal leukocytes to study their interaction with virus and potential roles in viral pathogenesis.

DCs and memory T cells [both $CD4^+$ and $CD8^+$ subsets] emigrated from organ cultures of normal human skin[1,2]. The DCs exhibited a characteristic "mature" morphology and phenotype, expressing high levels of MHC proteins and several adhesion and costimulatory molecules. Although some of the T cells were tightly bound to the DCs, the majority of the T cells were not cycling, as determined by the lack of expression of the Ki-67 nuclear antigen[3,4].

When HIV-1 was added to this DC-T cell mix, considerable virus replication took place[2,3]. Much of the new virus was produced by multinucleated syncytia that formed following the fusion of DCs with $CD4^+$ T cells[3]. Virus replication did not require the addition

Dendritic Cells in Fundamental and Clinical Immunology
edited by Ricciardi-Castagnoli, Plenum Press, New York, 1997

of exogenous stimuli or that the T cells be actively proliferating[3,4]. Similar virus growth in multinucleated syncytia has been observed with numerous isolates of HIV-1. These included both macrophage and T cell tropic lab-adapted and primary isolates[2,3], as well as several primary isolates designated as Clade E or B strains [Pope et al, in preparation].

If either the DCs or the T cells were separately exposed to the virus, productive infection only ensued once the two cell types were mixed back together. Usually fewer than 100 copies of proviral DNA were detected in 50,000 DCs or T cells alone. Whereas, thousands of HIV DNA copies were present in the DC-T cell mixtures within days of exposure to virus. Interestingly, this low level infection in the DC population was blocked by AZT, as was the ability of virus-pulsed DCs to initiate infection upon binding to $CD4^+$ T cells[2].

Therefore, we have identified a natural environment created when skin-derived DCs and T cells interact, that supports extensive replication of HIV-1 in the absence of exogenous stimulation of the T cells to proliferate. Similar environments in vivo may provide a niche for amplification of virus and spread from cell to cell.

2. VIRUS-PRODUCING SYNCYTIA IN THE TONSILLAR MUCOSA OF HIV-1-INFECTED INDIVIDUALS

Recent studies have identified a similar environment in vivo, where mature DCs and T cells can continuously interact and where significant virus replication can occur. DCs are located in the surfaces covering the palantine and pharyngeal tonsils. In particular, numerous DCs are positioned within the epithelium lining the deep crypts or surface invaginations typical of these tissues. The mucosa of the tonsil changes from a stratified squamous epithelium on the surface to a much more loosely keratinized lymphoepithelium, which contains many B and T lymphocytes, lining the crypts. It is within this lymphoepithelium that mature DCs can constantly interact with T cells.

During a study of the adenoids of 13 HIV-1-infected individuals, Frankel et al found that virus-producing syncytia could be detected within and just beneath the lymphoepithelium[5]. These adenoids had been removed as they were obstructive, and the patients were basically asymptomatic of HIV disease. The patients fell into various risk groups and were likely originally infected via different routes. Large numbers of S100-positive DCs were identified within the lymphoepithelium of the tonsillar crypts. When stained for viral p24 proteins, many profiles carrying intracellular p24 protein were identified in this area. These virus-positive profiles were multinucleated syncytia that stained for two markers, S100 and p55, that are expressed by mature DCs.

More recent investigations of another 13 tonsils obtained from patients with known medical histories have revealed very similar results [Frankel et al, in preparation]. Again, this was a healthy group of HIV-1-infected individuals with CD4 counts in the range of 200–900/μl. Numerous virus antigen-positive syncytia were located within the lymphoepithelia lining the tonsillar crypts, but rarely within the surface stratified squamous epithelia. Interestingly, many of the syncytia were smaller than those previously described in the adenoid mucosa.

We believe that this DC-T cell environment within the crypt lymphoepithelia is the in vivo counterpart of what we had previously described in vitro[1]. Within this environment, the mature DCs and T cells can interact with each other on a constant basis and create the milieu known to promote HIV-1 replication in vitro[2,3]. This and similar mucosal associated lymphoid tissues [such as in the gut] may provide important "pockets" where new virus can be produced and efficiently spread between cells.

3. DENDRITIC CELLS, T CELLS, AND THE REPLICATION OF SIV

3.1. The SIV-Macaque Model

To better dissect the role of mucosal DCs and T cells in the pathogenesis of primate immunodeficiency viruses, we have chosen to study the SIV-macaque system. SIV is morphologically and genetically similar to HIV[6–12] and exhibits a tropism for cells expressing the CD4 molecule[6,13]. The SIV-macaque model has proven useful to study immune deficiencies[8], since macaques infected with SIV via the intravenous[6,7,14,15] and mucosal routes [oral[16,17] and genital[18]], succumb to a similar disease to that caused by HIV-1 in humans. Monkeys infected with SIV develop anti-SIV CTL and Ab responses, a characteristic skin rash, usually preceeding the onset of an AIDS-like disease, and eventually become immunosuppressed and succumb to opportunistic infections[6–8].

DCs have been identified in the cervical and vaginal mucosae of humans[19–23] and macaques[24,25], with the dendritic processes extending to the epithelial surface[25]. Early after vaginal infection with SIV, SIV-DNA-positive cells that stained for the DC markers S100 and HLA-DR in adjacent tissue sections, were detected[25,26]. More recently, Marx et al demonstrated that animals with progesterone implants exhibited dramatic thinning of the vaginal epithelia and an increased sensitivity to vaginal infection with SIV[27]. In addition, infected cells have been observed in the T cells areas of lymph nodes within the first two weeks of infection[28]. Formal identification of the infected cells in these tissues is still required.

Therefore, it is possible that virus-carrying DCs and/or T cells move from the initial infection site to the draining lymph nodes, targeting the T cell areas. Thereafter, free virus and/or virus-carrying cells could seed the peripheral tissues, including the mucosae, where DCs and T cells continuously interact, initiating sites for chronic virus replication.

3.2. Characterization of DCs and T Cells Isolated from Macaque Skin and Mucosae

Our initial studies focused on the isolation and subsequent characterization of leukocytes from uninfected macaque tissues. Using the organ culture approach developed for human skin[1], we were able to isolate leukocytes from skin and oral and genital mucosal tissues[29]. In the case of the skin and genital mucoase, the leukocytes were almost exclusively T cells and mature DCs, the DCs representing 30–55% of the total leukocyte population. However, the genital cell suspensions were contaminated by up to 50% epithelial cells and/or keratinocytes. The percentage of DCs in the oral mucosal cell suspensions was much lower [<2%] mainly due to the presence of numerous B cells and T cells.

Phenotypic analysis of these cells revealed that the macaque skin and mucosal DCs were very similar to mature human DCs. For example, macaque DCs expressed high levels of MHC class II proteins, the costimulator molecules CD86 and CD40, and the two recently described molecules that are known to be expressed on mature human DCs, CD83[30] and p55[31]. Both $CD4^+$ and $CD8^+$ T cells were present in all suspensions. Some of these T cells were also tightly bound to the DCs forming DC-T cell conjugates. Similar observations have been made with cells isolated from both cynomolgus and rhesus macaques. Therefore, cutaneous and mucosal DCs isolated from uninfected monkeys appeared to be similar to mature human DCs in morphology, phenotype, and T cell binding capacity.

3.3. SIV Replication in DC-T Cell Mixtures

Having characterized the macaque skin and mucosal leukocytes, we were interested in investigating the ability of these cells to support SIV replication in vitro. Cutaneous and mucosal DC-T cell mixtures were exposed to SIVmac251 or SIVmac239 and infections monitored by measuring the production of reverse transcriptase activity and expression of SIV p27 antigens at the cellular level.

The DC-T cell mixtures from skin, oral and vaginal mucosae, but not cervical mucosal tissue, were productively infected by both SIV isolates[29]. Reverse transcriptase activity was detected in the culture supernatants of infected cultures within 7–9 days of exposure to SIV. Examination of cell cytospins immunoperoxidase labeled for SIV p27 protein revealed virus-positive profiles, including many syncytia, 5 days after infection. These suspensions were also infectible with cell line-grown SIV, suggesting that trace contaminants of superantigen or cytokines that may be present in virus stocks grown in activated peripheral blood cells, were not required to stimulate the T cells in these cultures for virus replication to occur.

Therefore, much like the cutaneous DCs and T cells isolated from human skin, the DC-T cell mixtures obtained from the skin and several mucosae of macaques supported the replication of a primate immunodeficiency virus. We are now able to use this system to facilitate extensive studies and elucidate the roles of mucosal DCs and T cells in the initial transmission of infection and in chronic virus replication.

4. SUMMARY

The cutaneous or mucosal DC-T cell environments seem extremely supportive of immunodeficiency virus replication. Apart from very early after SIV infection[28], similar virus producing cells have been difficult to detect in the lymphoid tissues where DCs and T cells are also known to interact. Large amounts of virus can be visualized in the germinal centers of the lymph nodes, much of which represents immune complexed virus that is trapped on the follicular dendritic cell surface[32–38]. However, whether these virus-carrying cells actually make virus or even virus proteins requires further investigation.

We believe that once an individual is systemically infected, free virus and/or virus-infected cells will seed peripheral tissues and when encountering similar DC-T cell environments as described in the oral mucosae, can set up sites of chronic virus replication. For instance, a virus-carrying T cell that migrates to the periphery would, on entering this milieu, interact with the mature DCs and activate virus production. This likely occurs at similar sites around the body, such as the mucosal associated lymphoid tissue of the gut, and is probably independent of the route of infection.

REFERENCES

1. Pope, M., Betjes, M. G. H., Hirmand, H., Hoffman, L. & Steinman, R. M. *J. Invest. Dermatol.* **104**, 11–18 (1995).
2. Pope, M., Gezelter, S., Gallo, N., Hoffman, L. & Steinman, R. M. *J. Exp. Med.* **182**, 2045–2056 (1995).
3. Pope, M., Betjes, M. G. H., Romani, N., et al. *Cell* **78**, 389–398 (1994).
4. Murphy, D. B., Lo, D., Rath, S., et al. *Nature* **338**, 765–768 (1989).
5. Frankel, S. S., Wenig, B. M., Burke, A. P., et al. *Science* **272**, 115–117 (1996).
6. Daniel, M. D., Letvin, N. L., King, N. W., et al. *Science* **228**, 1201–1204 (1985).

7. Letvin, N. L., Daniel, M. D., Sehgal, P. K., et al. *Science* **230**, 71–73 (1985).
8. Desrosiers, R. C. *Ann. Rev. Immunol.* **8**, 557–578 (1990).
9. Chakrabarti, L., Guyader, M., Alizon, M., et al. *Nature* **328**, 543–547 (1987).
10. Franchini, G., Gurgo, C., Guo, H. G., et al. *Nature* **328**, 539–543 (1987).
11. Fukasawa, M., Miura, T., Hasegawa, A., et al. *Nature* **333**, 457–461 (1988).
12. Hirsch, V., Riedel, N. & Mullins, J. I. *Cell* **49**, 307–319 (1987).
13. Kannagi, M., Yetz, J. M. & Letvin, N. L. *Proc. Natl. Acad. Sci. USA* **82**, 7053–7057 (1985).
14. Chalifoux, L. V., Ringler, D. J., King, N. W., et al. *Am. J. Path.* **128**, 104–110 (1987).
15. Kannagi, M., Kiyotaki, M., Desrosiers, R. C., et al. *J. Clin. Invest.* **78**, 1229–1236 (1986).
16. Baba, T. W., Koch, J., Mittler, E. S., et al. *AIDS Res. Hum. Retero.* **10**, 351–357 (1994).
17. Baba, T. W., Jeong, Y. S., Penninck, D., Bronson, R., Greene, M. F. & Ruprecht, R. M. *Science* **267**, 1820–1825 (1995).
18. Miller, C. J., Alexander, N. J., Sutjipto, S., et al. *Journal of Virology* **63**, 4277–4284 (1989).
19. Edwards, J. N. T. & Morris, H. B. *Br. J. Obstet. Gynecol.* **92**, 974–982 (1985).
20. Bjercke, S., Scott, H., Braathen, L. R. & Thorsby, E. *Acta. Obstet. Gynecol. Scand.* **62**, 585–589 (1983).
21. Hackemann, M., Grubb, C. & Hill, K. R. *J. Ultrastruct. Res.* **22**, 443–457 (1968).
22. Morris, H. H. B., Gatter, K. C., Stein, H. & Mason, D. Y. *Br. J. Obstet. Gynecol.* **90**, 400–411 (1983).
23. Younes, M. S., Robertson, E. M. & Bencosme, S. A. *Am. J. Obstet. Gynecol.* **102**, 397–403 (1968).
24. Miller, C. J., McChesney, M. & Moore, P. F. *Lab. Invest.* **67**, 628–634 (1992).
25. Miller, C. J., McGhee, J. R. & Gardner, M. B. *Lab. Invest.* **68**, 129–145 (1993).
26. Spira, A. I., Marx, P. A., Patterson, B. K., et al. *J. Exp. Med.* **183**, 215–225 (1996).
27. Marx, P. A., Spira, A. I., Gettie, A., et al. *Nature Medicine* **In Press**, (1996).
28. Reimann, K. A., Tenner-Racz, K., Racz, P., et al. *J. Virol.* **68**, 2362–2370 (1994).
29. Pope, M., Elmore, D., Ho, D. & Marx, P. *Submitted* (1996).
30. Zhou, L.-J. & Tedder, T. F. *J. Immunol.* **154**, 3821–3835 (1995).
31. Mosialos, G., Birkenbach, M., Ayehunie, S., et al. *Am. J. Pathol.* **148**, 593–600 (1996).
32. Biberfeld, P., Porwit-Ksiazek, A., Bottiger, B., Morfeldt-Mansson, L. & Biberfeld, G. *Cancer Res.* **45**, 4665–4670 (1985).
33. Fox, C. H., Tenner-Racz, K., Racz, P., Firpo, A., Pizzo, P. A. & Fauci, A. S. *J. Infect. Dis.* **164**, 1051–1057 (1991).
34. Tenner Racz, K., Racz, P., Dietrich, M. & Kern, P. *Lancet* **1**, 105–106 (1985).
35. Tenner Racz, K., Racz, P., Schmidt, H., et al. *AIDS* **2**, 299–309 (1988).
36. Schuurman, H. J., Krone, W. J., Broekhuizen, R. & Goudsmit, J. *Am. J. Pathol.* **133**, 516–524 (1988).
37. Armstrong, J. A. & Horne, R. *Lancet* **II**, 370–372 (1984).
38. Tenner-Racz, K., Bofill, M., Schulz-Meyer, A., et al. *Am. J. Pathol.* **123**, 9–15 (1986).

65

DENDRITIC CELLS EXPRESS AND USE MULTIPLE HIV CORECEPTORS

Drew Weissman,[1] Andrea Rubbert,[1] Christophe Combadiere,[2] Philip M. Murphy,[2] and Anthony S. Fauci[1]

[1]Laboratory of Immunoregulation
[2]Laboratory of Host Defenses
NIAID, NIH
Bethesda, Maryland

INTRODUCTION

Dendritic cells (DC) are the first immunocompetent cells to encounter antigen at areas of inflammation in mucous membranes[1], which are the major sites where the initiation of HIV infection occurs. HIV enters a mucous membrane and interacts with Langerhans cells (LC)/DC resulting in binding of the virus to the cell with or without infection. The cell then migrates and delivers virus to the paracortical region of the draining lymphoid tissue; LC/DC also provide activation stimuli to CD4$^+$ T cells which become infected leading to replication and spread of virus[2,3]. Recently, in vivo data in the macaque model have clarified certain pathogenic events associated with primary simian immunodeficiency virus (SIV) infection. SIV was placed in the vaginal vault and infected cells were then identified and followed using in situ PCR technology. DC in the lamina propria of the cervicovaginal mucosa were found to contain SIV DNA 2 days after exposure to virus. Infected cells were observed in the sub-capsular and paracortical regions of the draining lymph nodes; this series of events mirrors the course that DCs take upon receiving a signal to migrate from the tissues to lymphoid organs[4]. Thus, in an animal model of HIV, DC appeared to be responsible for bringing virus from the site of inoculation to the paracortical T cell regions of the draining lymphoid organs leading to viral replication and systemic spread of infection.

Recently, a series of newly identified HIV coreceptors that are necessary for infection of CD4$^+$ cells has been identified. These coreceptors are members of a family of 7-transmembrane, G protein-linked receptors that bind chemokines[5–10]. Of the multiple HIV coreceptors that have been identified, two appear to play major roles in the infection of CD4$^+$ cells, but have different specificities for macrophage (M) and T cell line (T) tropic HIV[5–10]. The first is CCR5, whose natural ligands are the beta-chemokines macrophage inflammatory protein (MIP)-1α, MIP-1β, and RANTES and is used by M-tropic strains of HIV[5–9]. M-tropic viruses are observed after transmission and throughout most of the clini-

Dendritic Cells in Fundamental and Clinical Immunology
edited by Ricciardi-Castagnoli, Plenum Press, New York, 1997

cally latent period of infection[11]. The second coreceptor, CXCR4 (fusin), whose ligand is the alpha-chemokine stromal cell derived factor (SDF)-1, is used by T-tropic strains of HIV which are found principally after progression to AIDS[10]. Other less predominant HIV fusion coreceptors have been identified among the chemokine receptor family and include CCR2b and CCR3, the eotaxin receptor present on eosinophils.[5–9]

Monocyte derived DC (MDDC), obtained by GM-CSF and IL-4 treatment of monocytes[12], have been demonstrated to chemotax to multiple beta-chemokines, including MIP-1α, monocyte chemoattractant protein (MCP)-1, and RANTES[15]. Given the roles of DC in the initiation[4] and propagation[3] of HIV-infection and the observation that DC can be infected in vivo[4], we studied the expression of HIV coreceptors on different types of DC, including LC, peripheral blood DC, and MDDC.

MATERIAL AND METHODS

Isolation and Study of DC

MDDC were obtained by GM-CSF (50 ng/ml) and interleukin (IL)-4 (100 ng/ml) (R & D Systems, Minneapolis, MN) treatment of Percoll purified monocytes as described previously[12]. Peripheral blood DC were purified as described previously[3]. Northern blotting was performed as described[13]. CD40 cross-linking was performed by coating plates with anti-CD40 mAb (10 ug/ml) (PharMingen, San Diego, CA), followed by removal of unbound Ab, and incubation of cells on coated plates for 24 hours. LC were obtained by trypsin digestion (0.025% for 30 min) of epidermal sheets.

Chemotaxis Assays

96 well chemotaxis plates with 5 or 8 μm pores were used (Neuro Probe, Inc, Cabin John, MD). 200 ng/ml of chemokines MIP-1α, MIP-1β, RANTES, IL-8 (R & D), and eotaxin (Peprotech, Rocky Hill, NJ) were placed in the lower well or both wells (chemokinesis control), and 1000–2,000 cells were placed above the membrane. After 2 hours, the number of cells in the lower well was counted. Samples were run in triplicate and averaged. Epidermal cells containing LC (2–4%) were stained with fluorescein-labeled mAb specific for HLA-DR and CD1a (PharMingen) and the number of fluorescent cells in the lower well was counted.

HIV Infection of DC

Various populations of DC were pulsed with HIV (Ba-L or IIIB) for 2 hours, washed, cultured for 2 days, lysed, and analyzed for the presence of HIV gag DNA by PCR using the SK38/39 primer[14]. Chemokines (200 ng/ml), SDF-1 containing supernatant (50% V/V), or OKT4a mAb (20 μg/ml) (Ortho Diagnostic Systems, Raritan, NJ) were added to some cells at the time of viral pulsing until the end of the culture.

RESULTS

Initially, the ability of different types of DC to chemotax through porous membranes was investigated. We found that MDDC chemotax in response to RANTES, MIP-1α, MIP-1β, IL-8, MCP-1, MCP-3 and eotaxin (Figure 1A and data not shown); peripheral

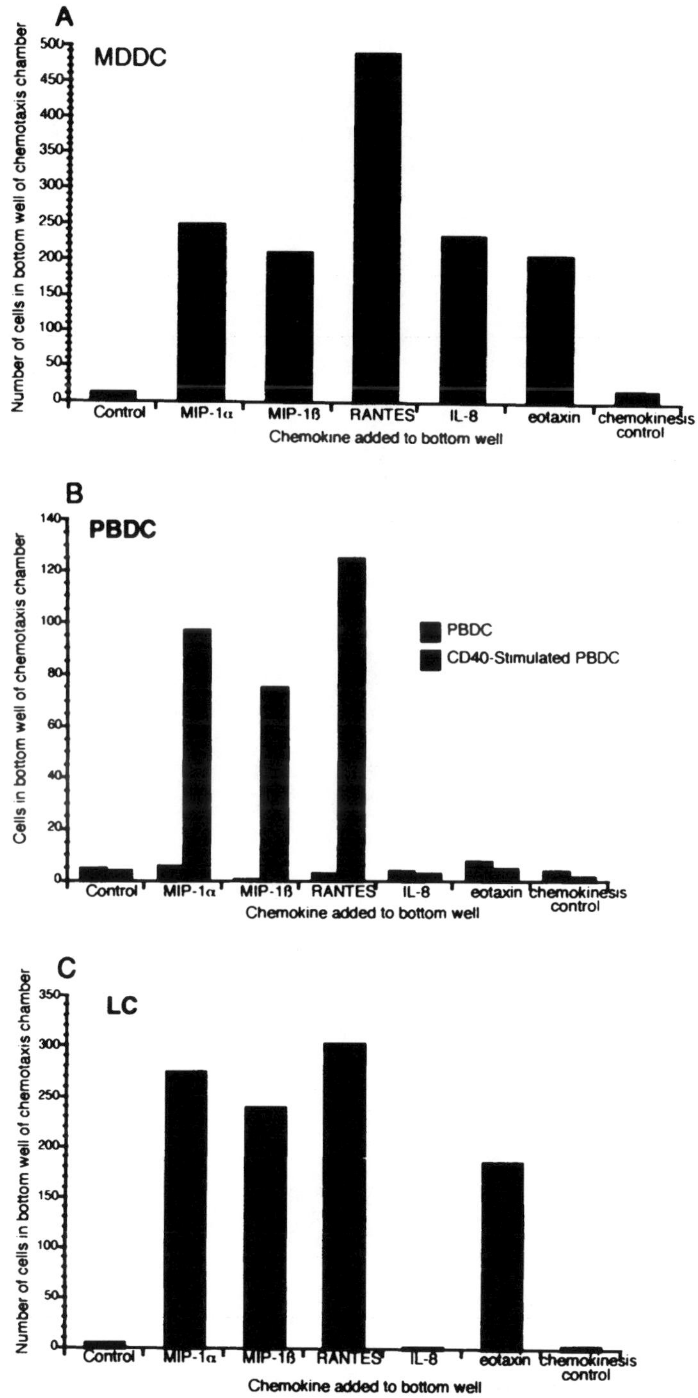

Figure 1. DC chemotax in response to multiple chemokines. A) MDDC, B) peripheral blood (PB) DC, either unstimulated or stimulated for 24 hours on anti-CD40 coated plates, and C) LC were placed in the upper well of 96-well chemotaxis chambers for 2 hours followed by counting the number of cells that migrated to the lower well. A single chemokinesis control, chemokine in top and bottom wells, for RANTES is shown; similar results were observed for the other chemokines tested.

blood DC chemotaxed in response to MIP-1α, MIP-1ß, and RANTES only if they were activated with CD40 cross-linking (Figure 1B), and LC responded to RANTES, MIP-1α, MIP-1β, and eotaxin, but not to IL-8 (Figure 1C). The response to eotaxin, which is known to activate CCR3, a receptor previously reported only on eosinophils[13], was surprising. To extend the chemotaxis results, northern blotting demonstrated the presence of specific mRNA for CCR3 in MDDC, as well as mRNA for CCR1, CCR2b, CCR5, and low levels of mRNA for CXCR-4 (Fusin) (data not shown).

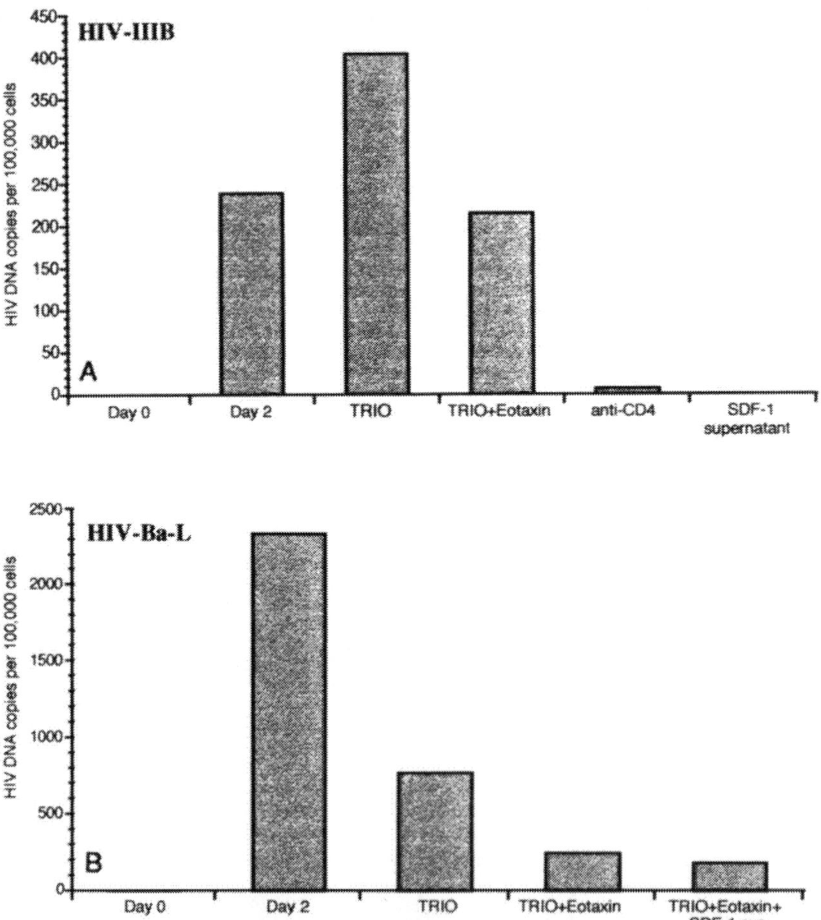

Figure 2. Infection of MDDC by T-tropic virus is inhibitable by SDF-1, while M-tropic virus infection is incompletely inhibited by a combination of chemokines. MDDC were infected with HIV-IIIB (A) or HIV-Ba-L (B). Day 0 are cells that were lysed immediately after addition of virus and are a control for gag DNA in the viral stock. HIV gag DNA (SK 38/39 primer) was measured 2 days after infection. ACH-2 controls were used for quantification on a phosphorimager. 200 ng/ml of each chemokine were added at the initiation of viral pulsing; TRIO represents the combination of MIP-1α, MIP-1ß, and RANTES at 200 ng/ml each. Similar results were observed with up to 1000 ng/ml (data not shown). Viral entry into T cell blasts was completely inhibited by similar concentrations of SDF-1 or the TRIO (data not shown).

The ability of T-tropic and M-tropic strains of HIV to infect DC was determined using HIV gag DNA PCR. MDDC were infectable by HIV-IIIB and this infection was inhibited by treatment of the DC with anti-CD4 mAb or SDF-1 containing supernatant. No inhibition was observed when RANTES, MIP-1α, MIP-1β, and eotaxin were added (Figure 2A). MDDC were also infected with the M-tropic strain HIV-Ba-'L. The addition of RANTES, MIP-1α, MIP-1β (TRIO) resulted in a moderate amount of inhibition (68%) while the addition of eotaxin and SDF-1 to the TRIO increased the inhibition to almost 95% (Figure 2B). SDF-1 supernatant alone did not inhibit HIV-BAL infection of MDDC (data not shown).

DISCUSSION

DC have been demonstrated to be the first cells infected in vaginal tissue after exposure to SIV, and their role in the initiation of HIV infections remains an important area of research. We have analyzed the expression and use of various HIV fusion coreceptors on different populations of DC. Peripheral blood DC did not express known functional ß-chemokine receptors as determined by a lack of chemotaxis in response to appropriate ligands. However, when they were activated through CD40 cross-linking, they were able to chemotax to MIP-1ß, suggesting upregulation of expression of CCR5, the only known chemokine receptor that binds MIP-1ß. LC and MDDC chemotaxed in response to multiple ß-chemokines including eotaxin, and by Northern blot analysis the MDDC expressed mRNA for the CCR3 receptor, which had previously only been found on eosinophils[13]. T-tropic strains of HIV infected MDDC, and this infection could be blocked by the addition of SDF-1, suggesting that a SDF-1 receptor was being used, preliminary results suggest that the SDF-1 receptor on DC is not CXCR4 (fusin). Infection with M-tropic strains could not be completely blocked by the addition of multiple chemokines (Figure 2B). Interestingly, RANTES, MIP-1α, and MIP-1β, could only moderately suppress infection with an M-tropic strain of HIV and the addition of eotaxin and SDF-1 increased this suppression. This suggested that other HIV coreceptors in addition to CCR5 are used by M-tropic strains of HIV in the infection of DC. Identification of CCR3 mRNA in DC and the observation that these cells can chemotax in response to its agonist, eotaxin, suggest an additional role for DC in allergic diseases. Moreover, the ability of eotaxin to suppress Ba-L infection in DC implicates CCR3 as an HIV coreceptor used by M-tropic viruses to infect DC. The delineation of HIV coreceptor usage by DC will lead to a better understanding of the pathogenesis of acute infection and may lead to the development of agents that can interfere with these early events.

REFERENCES

1. McWilliam, A. S., D. Nelson, J. A. Thomas, and P. G. Holt. 1994. Rapid dendritic cell recruitment is a hallmark of the acute inflammatory response at mucosal surfaces. J Exp Med 179:1331.
2. Cameron, P. U., P. S. Freudenthal, J. M. Barker, S. Gezelter, K. Inaba, and R. M. Steinman. 1992. Dendritic cells exposed to human immunodeficiency virus type-1 transmit a vigorous cytopathic infection to $CD4^+$ T cells. Science 257:383.
3. Weissman, D., Y. Li, J. M. Orenstein, and A. S. Fauci. 1995. Both a precursor and a mature population of dendritic cells can bind HIV. However, only the mature population that expresses CD80 can pass infection to unstimulated $CD4^+$ T cells. J Immunol 155:4111.
4. Spira, A. I., P. A. Marx, B. K. Patterson, J. Mahoney, R. A. Koup, S. M. Wolinsky, and D. D. Ho. 1996. Cellular targets of infection and route of viral dissemination after an intravaginal inoculation of simian immunodeficiency virus into rhesus macaques. J Exp Med 183:215.

5. Alkhatib, G., C. Combadiere, C. C. Broder, Y. Feng, P. E. Kennedy, P. M. Murphy, and E. A. Berger. 1996. CC CKR5: a RANTES, MIP-1alpha, MIP-1beta receptor as a fusion cofactor for macrophage-tropic HIV-1. Science 272:1955.
6. Dragic, T., V. Litwin, G. P. Allaway, S. R. Martin, Y. Huang, K. A. Nagashima, C. Cayanan, P. J. Maddon, R. A. Koup, J. P. Moore, and W. A. Paxton. 1996. HIV-1 entry into $CD4^+$ cells is mediated by the chemokine receptor CC-CKR-5. Nature 381:667.
7. Deng, H., R. Liu, W. Ellmeier, S. Choe, D. Unutmaz, M. Burkhart, P. Di Marzio, S. Marmon, R. E. Sutton, C. M. Hill, C. B. Davis, S. C. Peiper, T. J. Schall, D. R. Littman, and N. R. Landau. 1996. Identification of a major co-receptor for primary isolates of HIV-1. Nature 381:661.
8. Berson, J. F., D. Long, B. J. Doranz, J. Rucker, F. R. Jirik, and R. W. Doms. 1996. A seven-transmembrane domain receptor involved in fusion and entry of T-cell-tropic human immunodeficiency virus type 1 strains. J Virol 70:6288.
9. Choe, H., M. Farzan, Y. Sun, N. Sullivan, B. Rollins, P. D. Ponath, L. Wu, C. R. Mackay, G. LaRosa, W. Newman, N. Gerard, C. Gerard, and J. Sodroski. 1996. The beta-chemokine receptors CCR3 and CCR5 facilitate infection by primary HIV-1 isolates. Cell 85:1135.
10. Feng, Y., C. C. Broder, P. E. Kennedy, and E. A. Berger. 1996. HIV-1 entry cofactor: functional cDNA cloning of a seven-transmembrane, G protein-coupled receptor. Science 272:872.
11. Koot, M., I. P. Keet, A. H. Vos, R. E. de Goede, M. T. Roos, R. A. Coutinho, F. Miedema, P. T. Schellekens, and M. Tersmette. 1993. Prognostic value of HIV-1 syncytium-inducing phenotype for rate of $CD4^+$ cell depletion and progression to AIDS. Ann Intern Med 118:681.
12. Sallusto, F., and A. Lanzavecchia. 1994. Efficient presentation of soluble antigen by cultured human dendritic cells is maintained by granulocyte/macrophage colony-stimulating factor plus interleukin 4 and downregulated by tumor necrosis factor alpha. J Exp Med 179:1109.
13. Kitaura, M., T. Nakajima, T. Imai, S. Harada, C. Combadiere, H. L. Tiffany, P. M. Murphy, and O. Yoshie. 1996. Molecular cloning of human eotaxin, an eosinophil-selective CC chemokine, and identification of a specific eosinophil eotaxin receptor, CC chemokine receptor 3. J Biol Chem 271:7725.
14. Stanley, S., M. A. Ostrowski, J. S. Justement, K. Gantt, S. Hedayati, M. Mannix, K. Roche, D. J. Schwartzentruber, C. H. Fox, and A. S. Fauci. 1996. Effect of immunization with a common recall antigen on viral expression in patients infected with human immunodeficiency virus type 1. N Engl J Med 334:1222.
15. Sozzani, S., F. Sallusto, W. Luini, D. Zhou, L. Piemonti, P. Allavena, J. Van Damme, S. Valitutti, A. Lanzavecchia, and A. Mantovani. 1995. Migration of dendritic cells in response to formyl peptides, C5a, and a distinct set of chemokines. J Immunol 155:3292.

66

IN VITRO HIV INFECTION OF DENDRITIC CELL PRECURSORS

Effect on Dendritic Cell Differentiation and Function

Bruno Canque, Michelle Rosenzwajg, Sandrine Camus, Micael Yagello, Martine Guigon, and Jean Claude Gluckman

Laboratoire d'Immunologie Biologique and Laboratoire d'Immunologie Cellulaire de l'Ecole Pratique des Hautes Etudes
Faculté de médecine and hôpital de la Pitié-Salpêtrière
75651 Paris Cedex 13, France

1. INTRODUCTION

Cells from the dendritic lineage presumably play an important role in the pathogenesis of human immunodeficiency virus (HIV) infection[1,2]. Langerhans cells (LC), are probably among the first targets for HIV in genital mucosae[3,4]. Later, in T-cell dependent areas of lymphoid organs, dendritic cells (DC) may also be involved in both virus transmission to $CD4^+$ T cells and the generation of an effective anti-HIV immune response[2]. There is now agreement that mature DC/LC can take up the virus, reverse transcribe its RNA, cluster to and form syncytia with $CD4^+$ T cells, but the real degree of their permissivity to HIV still remains debated[1,5,6]. A precise knowledge of the relationships of cells of the DC lineage with HIV is important for understanding AIDS pathogenesis as well as for delineating possible new immunotherapy strategies. Studying their interactions with HIV in different models is therefore warranted.

DC are are known to be of hematopoietic origin[2,7–10]. They may be generated *in vitro* from $CD34^+$ hematopoietic progenitor cells by using combinations of granulocyte/macrophage-colony stimulating factor, tumor necrosis factor-α and stem cell factor[7–10]. Using such an *in vitro* differentiation system, we recently characterized a population of $CD13^{hiLin-}$ cells that appears after 5 days of culture and comprise precursors of the dendritic lineage ($CD1a^+$) but also of macrophages ($CD14^+$) and of granulocytes ($CD15^+$)[7,8]. $CD13^{hiLin-}$ cells express CD4 and they are therefore potential targets for HIV.

We examined here whether DC precursors are susceptible to HIV, and if DC differentiation and/or function are affected by HIV infection[10].

2. RESULTS AND DISCUSSION

2.1. HIV Infection of Day 5 Cultured Bulk Cells

Day 5 cells from $CD34^+$-derived cultures, which are a mixture of $CD13^{lo}$ and $CD13^{hiLin-}$ cells with a minority of already $CD1a^+$, $CD14^+$ and $CD15^+$ cells, were exposed for 3 hrs to 500 $TCID_{50}$ of macrophage-tropic (HIV-$1_{Ba\text{-}L}$, HIV-1_{ADA}, HIV-$1_{JR\text{-}FL}$) or lymphotropic (HIV-1_{LAI}) virus strains. The cells were further cultured for 8 days without medium change, but fresh cytokines were added every 3 days. Under these conditions, high viral production was noted only with the macrophage-tropic strains. On day 8 post-infection (PI), p24 levels in supernatants ranged from 50 to 1000 ng/ml in HIV-$1_{Ba\text{-}L}$ or HIV-1_{ADA}-infected cultures, and reached 50 ng/ml in HIV-$1_{JR\text{-}FL}$-infected cultures. Low virus production levels (p24 ^{2}4ng/ml) were noted in HIV-1_{LAI}-infected cultures. As reported[11], the kinetics of macrophage-tropic virus replication paralleled that of the differentiation of adherent macrophages in cultures, suggesting that the latter cells were mainly responsible for amplifying virus production. Under the conditions used, replication of macrophage-tropic strains was highly cytopathic for both the adherent and non-adherent (NA) cell populations: syncytia appeared on day 5 PI in cultures infected with HIV-1_{ADA} and on day 7–8 with HIV-$1_{Ba\text{-}L}$. NAC numbers were then reduced to 54±23% of controls ($p= 0.057$ by the paired Student t-test) with the latter viral strain. There was no cytopathicity in HIV-1_{LAI}- or in HIV-$1_{JR\text{-}FL}$-infected cultures at that time.

2.2. Characterization of HIV-Infected Cells in the Cultures

NAC from HIV-$1_{Ba\text{-}L}$-infected cultures were examined by immunocytochemistry on day 7 PI. Most $p24^+$ cells were involved in syncytia that were also stained by CD14, CD1a and LC-specific anti-S100 antibodies, indicating that they comprised cells of the MA lineage as well as DC. The latter point was confirmed by the finding of individual $CD1a^+$ or $S100^+$ DC clustering and fusing to syncytia. Individual $S100^+p24^+$ DC were rarely found, indicating that, if at least some DC were indeed productively infected by HIV, they were not major contributors to virus production. Altogether, these observations suggest two nonexclusive possibilities based on the natural tendency of DC to cluster: that DC were infected by HIV and rapidly formed syncytia with other DC as well as with other cells; that, even if or when not productively infected, DC readily clustered to and then fused with syncytia primarily formed by the major HIV-producing cells in this system, which were most probably of the MA lineage according to CD14 labeling and the viral strains that replicated best. Moreover, the presence of DC and their tendency to cluster may well be one reason for the unusually strong cytopathicity occurring with a reputedly non-syncytium-inducing strain such as HIV-$1_{Ba\text{-}L}$. These data also suggest in addition that the fusogenic capacity of DC is not restricted to T cells[2,5,12].

2.3. Susceptibility to HIV of $CD1a^+$ DC

DC in infected cultures harbored HIV DNA was then examined by nested pol PCR of single viable $CD1a^+CD15^-$ cells FACS sorted from day 7 PI HIV-$1_{Ba\text{-}L}$-infected cultures after being fixed in 4% paraformaldehyde (PFA). Viral DNA amounts were then about 5- to 10-fold greater in $CD1a^+CD15^-$ than in $CD1a^-CD15^+$ cells, though sometime lower than in unsorted bulk cells (Table 1).

Table 1. PCR detection of viral DNA in HIV-infected cultured cells

		Inverse of endpoint HIV DNA$^+$ dilution in cells			
Initially infected cells	Day PI	CD1a$^+$CD15$^-$	CD1a$^-$CD15$^+$	CD13^{loLin-}	CD13^{hiLin-}
(1) day 5 bulk	7#	5 - 10	1		
(2) sorted day-8 cultured cells	1	50 - 100	5		
(3) sorted day-5 cultured cells	1			10 - 50	1
(4) sorted day-5 CD13^{hiLin-} cells	2#	5	0 - 1		

The cells analysed were as follows: (1) day 7 PI paraformaldehyde (PFA)-fixed sorted CD1a$^+$CD15$^-$ and CD1a$^-$CD15$^+$ cells; (2) sorted CD1a$^+$CD15$^-$ and CD1a$^-$CD15$^+$ cells from day 8 noninfected cultures exposed to HIV-1$_{Ba-L}$; (3) sorted day 5 CD13^{loLin-} and CD13^{hiLin-} cells incubated for 24 hrs with HIV-1$_{Ba-L}$ prior to the PCR; (4) PFA-fixed sorted CD1a$^+$CD15$^-$and CD1a$^-$CD15$^+$ cells derived from day 5-sorted CD13^{hiLin-} cells that were then incubated for 24 hrs with HIV-1$_{Ba-L}$.
Relative HIV pol DNA contents of the different cell samples were determined by nested PCR performed in serial dilutions of lysates of the HIV-infected cells in HIV-negative A301 cell lysates (endpoint dilution assay). When already HIV-infected cells were assayed, they were first fixed with PFA before FACS-sorting (noted as #). Sensitivity of the assay was 1 copy per 10^4 unfixed cells and 10–100 copies per 10^4 4% PFA fixed cells

We next showed that HIV DNA could also be found in single CD1a$^+$ cells exposed to the virus after FACS-sorting from uninfected day 8 cultures, which indicated their susceptiblity to HIV infection *in vitro* (Table 1).

These findings indicate that, in both cases, at least one round of reverse transcription had occurred in these cells. Therefore, in addition to MA, DC appear to be infected by HIV in this system.

Using HIV-1$_{LAI}$, viral DNA was also noted in CD1a$^-$CD15$^+$ cells sorted from infected cultures, or shortly after exposure to the virus, suggesting that these cells or their CD4$^+$ precursors had been infected. No attempt was made to further investigate this point.

2.4. Susceptibility to HIV of the CD13^{hiLin-} Cell Population

The presence of viral DNA in DC of cultures that had been infected 7 days earlier did not allow to discriminate whether DC precursors were infected at the onset or if virus spread only later from the other infected cells to DC that subsequently differentiated in culture. To examine this point, HIV DNA was assessed in day 5 CD13^{hiLin-} cells that had been exposed to HIV after sorting. Indeed, although CD13^{hiLin-} cells were CD4$^+$ and represented 44±8% of the cells at that time, it is possible that productive infection of the bulk population could only involve the minor populations of CD1a$^+$ and/or CD14$^+$ cells already present in cultures. Viral DNA was preferentially detected in day-5 sorted CD13^{hiLin-} vs. CD13^{lo} cells 24 hrs after exposure to HIV-1$_{Ba-L}$, and it was also found in the short-term (48-hr) CD1a$^+$ progeny of such CD13^{hiLin-} cells (Table 1). These data indicate that CD13^{hiLin-} cells and, among these, DC precursors are susceptible to HIV.

2.5. Effect of HIV Infection on DC Differentiation and Function

Despite the strong cytopathicity occurring in HIV-1$_{Ba-L}$-infected bulk cell cultures, FACS analysis on day 7 PI failed to show significant modifications of the proportion of DC and, apart from CD4 downmodulation related to HIV production, overall expression of accessory and costimulatory molecules and of HLA-DR was similar to that of controls.

In order to assess DC antigen-presenting function, graded amounts of cells from day 7 HIV-infected or uninfected cultures, or of cells cultured with HI virus, were added to allogeneic normal PBL in the continuous presence of DDI used to prevent HIV propagation to responder cells. The ability of DC from HIV-infected cultures to stimulate the mixed leucocyte reaction was affected but in a limited manner(<10 fold) if any. Due to the use of DDI this effect, which could not be accounted for by phenotypic changes of the stimulating cells, was independent of the transmission of HIV from infected DC to primary $CD4^+$ T cells nor could it be induced by viral supernatants. Reduced T lymphocyte responsiveness to infected NA cells was therefore problably due to a functional defect of the stimulator cells, the basis of which remains unclear[1,13].

3. CONCLUSION

The observations presented here indicate that precursors of DC and their *in vitro* differentiated progeny are susceptible to HIV. Apart from the consequences of viral cytopathic effects, virus production in these cells and in adjacent cells of the MA lineage does not apparently affect DC differentiation and only in a limited manner DC antigen-presenting function. The *in vivo* relevance of these findings remains to be established.

ACKNOWLEDGMENTS

We are grateful to B. Asjö (Bergen, Norway) and to the NIH AIDS Research and Reference Reagent Program (USA) for the gift of virus strains used in this study. We thank C. Pioche for her invaluable help in cell sorting. Supported by the Agence Nationale de Recherche sur le SIDA, Université Paris 6, the Centre National de la Recherche Scientifique (ERS 107), and the Association pour la Recherche sur les Déficits Immunitaires Viro-Induits (Paris, France). BC has benefited from a fellowship of the Société Française d'Hématologie.

BIBLIOGRAPHY

1. Knight, S. (1994). A problem of antigen presentation ? Current Biol 4: 131.
2. Cameron, P., M. Pope, et al. (1996). Dendritic cells and the replication of HIV-1. J Leuk Biol 59: 158.
3. Zambruno, G., A. Giannetti, et al. (1995). Langerhans cells and HIV infection. Immunol Today 16: 520.
4. Spira, A., P. Marx, et al. (1996). Cellular targets of infection and route of viral dissemination after an intravaginal inoculation of simian immunodeficiency virus into Rhesus macaques. J Exp Med 183: 215.
5. Pope, M., M. Betjes, et al. (1994). Conjugates of dendritic cells and memory T lymphocytes from skin facilitate productive infection with HIV-1. Cell 78: 389.
6. Pope, M., S. Gezelter, et al. (1995). Low levels of HIV-1 infection in cutaneous dendritic cells promote extensive viral replication upon binding to memory $CD4^+$ T cells. J Exp Med 182: 2045.
7. Caux, C., C. Dezutter-Damboyant, et al. (1992). GM-CSF and TNF-α cooperate in the generation of dendritic langerhans cells. Nature 360: 258.
8. Rosenzwajg, M., B. Canque, et al. (1996). Human dendritic cell differentiation pathway from CD34+ hematopoietic precursor cells. Blood 87: 535.
9. Siena, S., M. Di Nicola et al. (1995). Massive ex vivo generation of funtional dendritic cells from mobilized $CD34^+$ blood progenitors for anticancer therapy. Exp Hematol 23: 1463
10. Canque, B., M. Rosenzwajg, et al. (1996). The effet of in vitro human immunodeficiency infection on dendritic cell differentiation and function. Blood, in press.
11. Kitano, K., C. Abboud, et al. (1991). Macrophage-active colony-stimulating factors enhance human immunodeficiency virus type 1 infection in bone marow stem cells. Blood 77: 1699.
12. Frankel, S., B. Wenig, et al. (1996). Replication of HIV-1 in dendritic cell-derived syncytia at the mucosal of the adenoid. Science 272: 115.
13. Ludewig, B., H. Gelderblom, et al. (1996). Transmission of HIV-1 from productively infected mature langerhans cells to primary $CD4^+$ T lymphocytes results in altered T cell response with enhanced production of IFN-γ and IL-10. Virology 215: 51.

67

THE ROLE OF DENDRITIC CELLS IN THE TRANSPORT OF HIV TO LYMPH NODES ANALYSED IN MOUSE

C. Masurier,[1] N. Guettari,[1] C. Pioche,[1] R. Lacave,[1] B. Salomon,[1] F. Lachapelle,[2] D. Klatzmann,[1] and M. Guigon[1]

[1]ERS 107 CNRS, UPMC and
[2]INSERM U134
CHU La Pitié-Salpêtrière
Paris, France

INTRODUCTION

The migration route of dendritic cells (DC) from peripheral tissues to lymph nodes (1) and their capacity to pass HIV infection to T cells (2–4) support a critical role for DC in the early events of HIV infection. It is assumed, but not yet demonstrated, that DC are the first HIV target cells in the mucosa and that their migration to draining lymph nodes will result in the transmission of HIV to T lymphocytes. Interestingly, this can be studied in a murine model, although murine cells are not susceptible to HIV infection. Indeed, (i) murine DC pre-incubated with HIV are able to transfer it to human T lymphocytes as efficiently as human DC (4); (ii) murine DC can be manipulated *in vitro*, re-injected, and their *in vivo* migration can be analysed.

The aim of our work was to develop a murine model to study HIV transport by DC, especially after mucosal inoculation, in order to test new molecules for blocking infection. In the present study, we have analysed the capacity of mouse bone marrow-derived DC injected subcutaneously to migrate and to carry infectious HIV into draining lymph nodes.

MATERIALS AND METHODS

DC were derived from the bone marrow of either DBA/2 mice or transgenic mice expressing the human CD4 (hCD4) under the control of MHC class I promoter (5). We used these mice with the idea that the expression of CD4 onto the surface of DC could favor the binding of HIV, although it is clear that we did not expect productive infection of these DC. In addition the expression of a transgene could be used as a marker for tracing DC migration.

Bone marrow cells were cultured for 7 days in the presence of GM-CSF as previously described (6, 7), which resulted in 40–70% of MHC-class II cells showing the morphological and cytochemical characteristics of DC. In some experiments, DC were further purified by sorting by flow cytometry.

In order to analyse their migration, DC were stained with a fluororescent dye (PKH2) and injected subcutaneously into the footpad of mice. The detection of fluorescent cells was performed on frozen sections of popliteal and inguinal lymph nodes 6 and 24 hours later.

In order to investigate their role in the transport of HIV-1, DC generated from normal and hCD4 transgenic mice bone marrow were incubated with either heat-inactivated (Hi) or infectious HIV-1_{LAI} supernatant for 2 hours, washed twice and injected into the footpad of syngeneic non transgenic mice. Popliteal and inguinal homo-and contralateral lymph nodes were harvested 24 hours later. HIV-1 gag and hCD4 sequences were detected by RT-PCR followed, when appropriate, by a specific hybridization of gag RT-PCR products. The persistence of the HIV virulence after transport to lymph nodes was assessed by cocultivation of popliteal and inguinal lymph node cells with the human lymphoid HUT-78 cell line. The kinetics of viral p24 production in culture supernatants was determined by ELISA.

RESULTS

Mouse Bone Marrow-Derived DC Fix and Transmit HIV-1 to T-Cells

After incubation of bone marrow-derived DC with either Hi-HIV-1 or infectious HIV-1, the virus was detectable by RT-PCR in DC immediately, 3 and 24 hours after the pulse. When HIV-1-pulsed-DC were cocultured with HUT-78 cells, a viral p24 production was observed after 7 days of coculture (data not shown). These findings demonstrate that murine bone marrow-derived DC even in absence of infection are able to fix HIV and to transmit HIV infection to CD4 lymphoid cells, as previously shown for splenic DC (4).

Mouse Bone Marrow-Derived DC Can Migrate to Draining Lymph Nodes

After the subcutaneous injection of PKH2-stained DC into the mouse footpad, stained cells were found on sections of the draining homolateral popliteal lymph node which was the first one draining the footpad injected. Fluorescent cells were already detected 6 hours after DC injection, and their numbers increased until 24 hours. This observation indicates that bone marrow-derived DC can migrate to the draining lymph node after subcutaneous injection.

Transport of HIV-1 to Lymph Nodes by DC

HIV RNA sequence was detected in the homolateral popliteal node 24 hours after injection of Hi-HIV-1-pulsed DC, but not in the homolateral inguinal nor in the contralateral popliteal and inguinal lymph nodes (figure 1). Similar results were obtained after injection of DC obtained from hCD4 mice: the viral RNA as well as hCD4 RNA were detected in the same homolateral popliteal node of mice injected with unsorted and sorted DC (table 1). This finding supports the participation of injected DC in the transport of HIV.

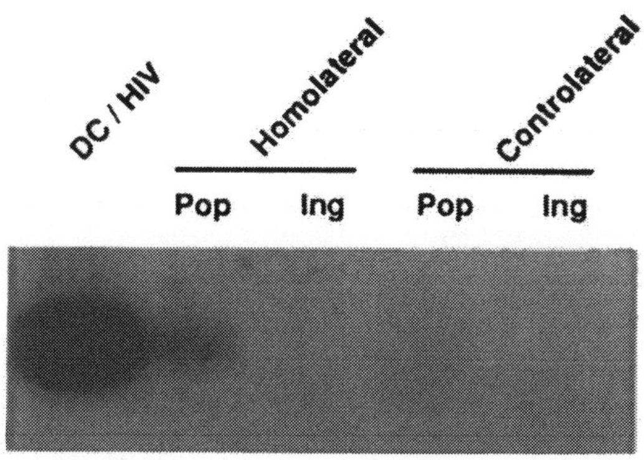

Figure 1. Dendritic cells (DC) from DBA/2 mice were incubated with Hi-HIV-1 supernatant for 2 hours, washed extensively, and injected into the footpad of syngenic mice. HIV-1 gag sequences were amplified by RT-PCR followed by specific hybridization of PCR products, 24 hours after injection, in homo-and contralateral popliteal (Pop) and inguinal (Ing) nodes. Positive control is DC incubated with HIV-1 before injection.

HIV Is Still Infectious after Transport to Lymph Nodes

After mice were injected with HIV-1 or with HIV-1 pulsed DC, lymph node cells were cultured with the human HUT-78 lymphoid cells. p24 production revealing the presence of infectious HIV was only observed with the draining popliteal node cells after 15 days of coculture (data not schown). These results demonstrate that HIV retains its infectivity after transport from the periphery to the lymph nodes in mice.

DISCUSSION AND PERSPECTIVES

Our data confirm the previous observations of Cameron et al.(4) on the capacity of murine DC to transmit HIV to human lymphoid cells. Furthermore, they indicate that the early events following HIV transmission can be studied in mice. We are now using this model to study the HIV migration from the vaginal mucosa to its draining lymph nodes.

Table 1. After a two-hour incubation with Hi-HIV-1 supernatant, DC derived from hCD4 transgenic mice were injected into the footpad of mice. 24 hours after, the detection of HIV in homo- and contralateral popliteal (Pop) and inguinal (Ing) nodes was performed by RT-PCR with gag primers. The detection of hCD4, that is only expressed by the injected DC, was similarly performed with appropriate primers

	RT-PCR detection							
	gag				hCD4			
	Homolateral		Contralateral		Homolateral		Contralateral	
	Pop	Ing	Pop	Ing	Pop	Ing	Pop	Ing
Unsorted DC / HIV	+	–	–	–	+	–	–	–
Sorted DC / HIV	+	–	–	–	+	–	–	–

REFERENCES

1. Austyn, J. M. 1996. New insights into the mobilization and phagocytic activity of dendritic cells. J. Exp. Med. 183: 1287–1292.
2. Pope, M. , S. Gezelter, N. Gallo, L. Hoffman, R. M. Steinman. 1995. Low levels of HIV-1 infection in cutaneous dendritic cells promote extensive viral replication upon binding to memory $CD4^+$ T cells. J. Exp. Med. 182: 2045–2056.
3. Weissman, D. , L. Yuexia, J. M. Orenstein, A. S. Fauci. 1995. Both a precursor and a mature population of dendritic cells can bind HIV. J. Immunol. 155: 4111–4117.
4. Cameron, P. U. , P. S. Freudenthal, J. M. Barker, S. Gezelter, K. Inaba, R. M. Steinman. 1992. Dendritic cells exposed to human immunodeficiency virus type-1 transmit a vigorous cytopathic infection to $CD4^+$ T cells. Science. 257: 383–387.
5. Lorès, P. , V. Boucher, C. Maclay, M. Pla, H. Von Boehmer. J. Jami, F. Barré-Sinoussi. J-C. Weill. 1992. Expression of human CD4 in transgenic mice does not confer sensitivity to human immunodeficiency virus infection. AIDS Research and Human Retroviruses. 812, 8: 12 : 2063–2071.
6. Inaba, K. , M. Inaba, N. Romani, H. Aya, M. Deguchi, S. Ikehara. S. Muramatsu, R. M. Steinman. 1992. Generation of large numbers of dendritic cells from mouse bone marrow cultures supplemented with granulocyte/macrophage colony-stimulating factor. J. Exp. Med. 176: 1693–1702.
7. Salomon, B. , P. Lorès, C. Pioche, P. Racz, J. Jami, D. Klatzmann. 1994. Conditional ablation of dendritic cells in transgenic mice. J. Immunol. 152: 537–548.

68

THE HIV-1 LIFE CYCLE IS BLOCKED AT TWO DIFFERENT POINTS IN MATURE DENDRITIC CELLS

Angela Granelli-Piperno,[1] Dongling Chen, Bernhard Moser,[2] and Ralph M. Steinman[1]

[1]Laboratory of Cellular Physiology and Immunology
The Rockefeller University
New York 10021
[2]Theodor Kocher Institute
Bern, Switzerland

INTRODUCTION

The proliferative status of the host cells influences the life cycle of retroviruses[1,2]. For oncogenic retroviruses cell division is considered a prerequisite for integration and viral replication. Lentiviruses differ from oncogenic retroviruses because they can complete their replicative cycle indipendently from cell division. The HIV-1 lentivirus uses a gag targeting cell to access the nucleus and integrate into macrophages. HIV-1 productively infects monocytes and terminally differentiated macrophages in specific organs like brain and cultured blood monocytes[3,4]. The predominant target for HIV-1 in the blood is a CD4$^+$ T lymphocyte, but activation of T cells is required for productive infection. Nonetheless, HIV-1 is capable of infecting and persisting in resting T cells without producing virions[5]. Once infected the resting T cell can synthesize an incomplete form of viral DNA[5] without complete reverse transcription. Following stimulation of the infected quiescent cells, productive infection occurs.[5,6] Dendritic cells represent a distinct lineage of white cells that derive from CD34$^+$ progenitors in the bone marrow. They are motile and widely distributed in most of the tissues and in all components of lymphoid system [for review see[7]]. Dendritic cells are specialized antigen-presenting cells for T cells *in situ*, both for self-antigens during T cell development and foreign antigens during immunity. Although relatively few in number, dendritic cells are effective antigen presenting cells because they express not only high levels of MHC class I and II but also several of the accessory molecules that are required for T cell binding and activation[8,9]. In many tissues dendritic cells express CD4 as in skin[10], tonsil[11], thymus[12], and several mucosae[13]. Recently, methods to generate large quantities of mature dendritic cells from blood precursors have been described[14].

Dendritic Cells in Fundamental and Clinical Immunology
edited by Ricciardi-Castagnoli, Plenum Press, New York, 1997

Whether HIV-1 can productively infect blood dendritic cells is still controversial. Some studies report that blood dendritic cells are susceptible to HIV-1 infection[15,16], while others report no infection in dendritic cells but transmission of cytopathic infection to stimutated T cells[17–20]. These discrepancies may be due to the state of maturation of the DCs.

Transcription of the HIV-1 genome depends on the intracellular environment into which the virus integrates. This transcription is regulated by a complex interplay between viral proteins [tat, rev] and cellular transcription factors interacting with the HIV-1 LTR, which contains two NF-kB and three Sp1 binding sites. Activated T cells, but not resting T cells, replicate virus efficiently owing to active nuclear NF-kB proteins. In contrast, mature DCs are infected with HIV-1 poorly or not at all.

In this study we have investigated HIV-1 infection in mature DCs generated in culture from blood progenitors to determine the level at which virus replication in DCs is controlled.

NF-kB BUT NOT Sp1 ARE DETECTED IN DCS GENERATED FROM BLOOD PRECURSORS

DCs with mature phenotype were generated by culturing ER⁻ cells for 7 days with GM-CSF and IL-4 and then for 4 more days with macrophage conditioned medium[14]. These differentiated DCs have a characteristic stellate shape, are not plastic adherents and express a panel of markers that are characteristic of DCs from other sources [i.e. CD83, p55, CD86, HLADR].

Previous work showed that dendritic cells isolated from skin and blood express all the members of the NF-κB/Rel family in high amount, but lack or have undetectable levels of Sp1 protein[21]. Cell lysates of 11 day cultured dendritic cells, depleted of T cells and B cells by cell sorting, were analyzed by western blot for the presence of the above mentioned transcription factors. As seen in fig. 1 mature and fully differentiated dendritic cells do not express Sp1 but do express the NF-κB/Rel proteins. Moreover, Sp1 binding activity was also undetectable in nuclear extracts of DCs [not shown].

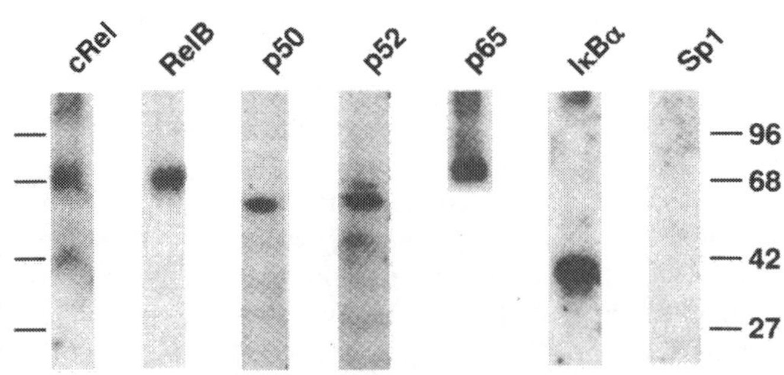

Figure 1. Expression of NF-kB/rel but not Sp1 proteins in mature DCs. 11 day cultured DCs were sorted [large FCS, fluorescent negative for CD3 and CD20]. Cell lysates [10μg] of sorted cells were immunoblotted and the membrane probed with antibodies to the indicated Rel and Sp1 proteins. Bound antibodies were visualized with peroxidase anti-immunoglobulin, followed by ECL.

HIV-1 ENTRY INTO DCs

Previous work showed that dendritic cells purified from blood are resistant to HIV-1 infection and are not productively infected[19]. To investigate whether the $CD4^{+}$ cytokine-derived dendritic cells are susceptible to HIV-1 infection, we have pulsed dendritic cells with HIV-1 and have used polymerase chain reaction to analyze and quantitate reverse transcripts in the infected cells. We focused on the detection of early reverse transcripts [R/U5] and of full length reverse transcripts [LTR/gag]. The signal for R/U5 DNA sequences was weak at time 0h, 90 min after adding virus to purified DCs. HIV-1 DNA then increased for 6–8h to reach a plateau up to 24h [fig 2A and 3]. Relative to a standard curve, we detected 10^3–10^4 copies of R/U5 sequences for 5×10^4 DCs. The R/U5 DNA was weaker in T blasts [fig. 2A]. Soluble CD4 as well as CD4 blocking antibodies inhibited virus entry into DCs [fig. 2B].

Some controls ruled out absorption of virions containing R/U5 DNA sequences. Few transcripts were found if virus was added on ice. The signal obtained by infecting the cells at 37^0 C was insensitive to trypsinization indicating internalization [fig. 2C]. Moreover, heat inactivation of the virus did not lead to detection of the R/U5 reverse transcripts [fig. 2D]. Pretreatment with AZT only partially inhibited the synthesis of R/U5, most probably a result of inefficient incorporation of nucleotide analogs into short reverse transcripts [fig. 2D].

Several HIV-1 isolates entered and began reverse transcription in DCs. The M-tropic strains Ba-L and SF-162 were particulary efficient followed by ADA and JR-FL [not shown]. The T-tropic, TCLA adapted IIIB, was also highly infectious [fig 2].

In contrast to the R/U5 sequences, few LTR/gag sequences were detected in infected DCs [fig 3]. As seen in fig. 3, the signal is detected as expected in infected T blasts because infection is productive in T blasts.

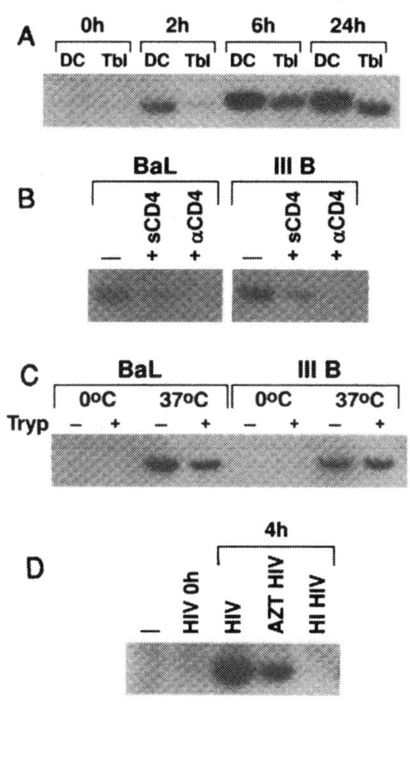

Figure 2. Early reverse transcripts [R/U5] are detected in HIV-1 infected DCs. A] Kinetic of synthesis of R/U5 sequences. DNA was prepared at different times from infected DCs and T blasts and then R/U5 sequences amplified by PCR. B] Inhibition of HIV-1 IIIB entry into DCs by soluble CD4 [20μg/ml] and anti CD4 [2μg/ml]. C] Reverse transcripts are not detected when DCs are infected on ice, and are trypsin resistant if infection occur at 37^0 C for 5h. D] Heat inactivation [HI, 56^0C for 30 min] of IIIB abolishes infection of DCs.

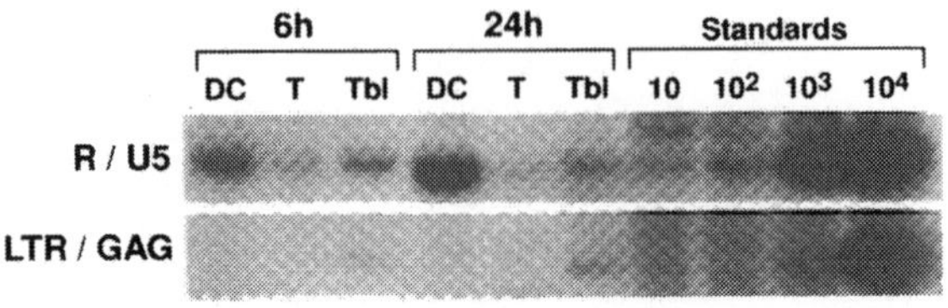

Figure 3. Stages of reverse transcription of the IIIB isolate in DCs, T cells, and T blasts at different time points. DNA extracted from 5×10^4 cells was amplified for R/U5, LTR/gag sequences.

CHEMOKINE RECEPTORS FOR HIV-1 INFECTION IN DCs

The identity of chemokine receptors that mediate HIV-1 entry into DCs was assessed with a panel of chemokines as blockers. The entry of Ba-L was almost completely blocked by RANTES and by MIP-1α, which are known ligands for the CCR5 receptor for M-tropic HIV-1 [fig. 4][22,23]. No block was found with SDF-1, the recently described ligand for the fusin/LESTR, CXCR4 receptor for TCLA isolates[24,25]. In contrast, the entry of IIIB into DCs was sensitive [50% or more inhibition] to SDF-1 [fig. 4]. Several other chemokines were without detectable effects when added at a dose of 100nM. These included the CC chemokines eotaxin, MCP-1, and MCP-3, and the CXC chemokines, IL-8, γ Ip-10, Mig, and PF-4.

CONCLUSIONS

These studies establish new features on the capacity of HIV-1 to infect cells of the immune system. First, HIV-1 does enter DCs and with greater efficacy than T cells. The infection is detected by the novo production of early reverse transcripts. Second, DCs are receptive to both M-tropic and TCLA strains, and consistent with prior findings, distinct chemokine receptors are used for each. For M-tropic viruses the CCR5 predominates since

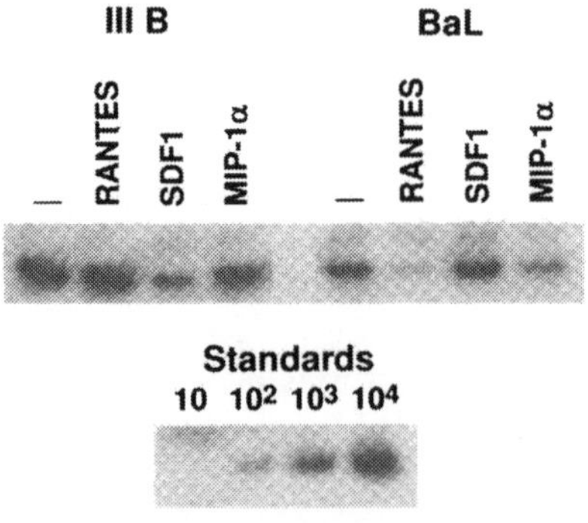

Figure 4. Blocking of Ba-L and IIIB entry into DCs with the selected ligands for the chemokine receptors. DCs were preincubated for 30 min with 100nM of the indicated chemokines and then infected with HIV-1 isolates. After 5h DNA was prepared and amplified for the R/U5 sequences.

patients with mutated CCR5 are resistant to entry of Ba-L[26]. Third, the chemokines RANTES, MIP-1α are potent inhibitors of M-tropic isolates in DCs. Fourth we believe that productive infection does not occur in mature DCs due to the lack or unsufficient level of the Sp1 transcription factor, a cellular factor required for the complete synthesis of the full length DNA provirus.

Our results indicate that the HIV-1 life cycle is blocked at two different points in mature DCs; one block is at the level of reverse transcription, probably before the synthesis of the plus-strand sequences. The lack of Sp1 may generate the second block.

BIBLIOGRAPHY

1. Varmus, H. *Science* **240**, 1427–1434 (1988).
2. Brown, P. O., Bowerman, B., Varmus, H. E. & Bishop, J. M. *Cell* **49**, 347–356 (1987).
3. Gartner, S., Markovits, P., Markovits, D. M., Kaplan, M. H., Gallo, R. C. & Popovic, M. *Science* **223**, 215–219 (1986).
4. Meltzer, M. S., Skilman, D. R., Gomatos, P. J., Kalter, D. C. & Gendelman, H. E. *Annu. Rev. Immunol.* **8**, 169(1990).
5. Zack, J. A., Arrigo, S. J., Weitsman, S. R., Go, A. S., Haislip, A. & Chen, I. S. Y. *Cell* **61**, 213–222 (1990).
6. Stevenson, M., Stanwick, T. L., Dempsey, M. P. & Lamonica, C. A. *EMBO* **9**, 1551–1560 (1990).
7. Steinman, R. M., Schuler, G., Romani, N. & Kaplan, G. in *Atlas of Blood Cells: Function and Pathology* (eds Zucker-Franklin, D., Greaves, M.F., Grossi, C.E. & Marmont, A.M.) Vol.2nd, 359–377 (Lea & Febiger, Philadelphia, PA, 1988).
8. Young, J. W., Koulova, L., Soergel, S. A., Clark, E. A., Steinman, R. M. & Dupont, B. *J. Clin. Invest.* **90**, 229–237 (1992).
9. O'Doherty, U., Steinman, R. M., Peng, M., et al. *J. Exp. Med.* **178**, 1067–1078 (1993).
10. Nestle, F. O., Zheng, X.-G., Thompson, C. B., Turka, L. A. & Nickoloff, B. J. *J. Immunol.* **151**, 6535–6545 (1993).
11. Hart, D. N. & McKenzie, J. L. *J. Exp. Med.* **168**, 157–170 (1988).
12. Sotzik, F., Rosenberg, Y., Boyd, A. W., et al. *J. Immunol.* **152**, 3370–3377 (1994).
13. Pavli, P., Hume, D. A., Van de Pol, E. & Doe, W. F. *Immunol.* **78**, 132–141 (1993).
14. Bender, A., Sapp, M., Schuler, G., Steinman, R. M. & Bhardwaj, N. *J. Immunol. Methods* **196**, 121–135 (1996).
15. Langhoff, E., Terwilliger, E. E., Bos, H. J., et al. *Proc. Natl. Acad. Sci. USA* **88**, 7998–8002 (1991).
16. Macatonia, S. E., Patterson, S. & Knight, S. C. *Immunol.* **67**, 285–289 (1989).
17. Cameron, P. U., Lowe, M. G., Crowe, S. M., et al. *J. Leuk. Biol.* **56**, 257–265 (1994).
18. Cameron, P. U., Forsum, U., Teppler, H., Granelli-Piperno, A. & Steinman, R. M. *Clin. Exp. Immunol.* **88**, 226–236 (1992).
19. Cameron, P. U., Freudenthal, P. S., Barker, J. M., Gezelter, S., Inaba, K. & Steinman, R. M. *Science* **257**, 383–387 (1992).
20. Pope, M., Betjes, M. G. H., Romani, N., et al. *Cell* **78**, 389–398 (1994).
21. Granelli-Piperno, A., Pope, M., Inaba, K. & Steinman, R. M. *Proc. Natl. Acad. Sci. USA* **92**, 10944–10948 (1995).
22. Alkhatib, G., Combadier, C., Broder, C. C., et al. *Science* **272**, 1955–1958 (1996).
23. Dragic, T., Litwin, V., Allaway, G. P., et al. *Nature* **381**, 667–673 (1996).
24. Feng, Y., Broder, C. C., Kennedy, P. E. & Berger, E. A. *Science* **272**, 872–877 (1996).
25. Oberlin, E., Amara, A., Bachelerie, F., et al. *Nature* **382**, 833–835 (1996).
26. Granelli-Piperno, A., Moser, B., Pope, M., et al. *Submitted* (1996).

69

INFECTION OF HUMAN DENDRITIC CELLS BY MEASLES VIRUS INDUCES IMMUNE SUPPRESSION

Dominique Kaiserlian,[1] Isabelle Grosjean,[1] and Christophe Caux[2]

[1]INSERM Unit 404
"Immunité et Vaccination"
Institut Pasteur de Lyon
Avenue Tony Garnier
69365 Lyon CX 07, France
[2]Schering-Plough
Dardilly, France

1. INTRODUCTION

Measles is characterized by lifelong immunity and a transient immunosuppression which, in developing countries, is responsible for a high morbidity and a high mortality consecutive to secondary infections. Strickingly, the immune suppression is coincident with MV-specific immunity and continues for several weeks after apparent recovery from measles. The immune suppression is characterized by the loss of delayed-type hypersensitivity skin test responses to recall antigens, such as tuberculin, and by the inhibition of antibody production and cellular immune responses to new antigens (1).

Measles virus (MV) infection is initiated during the first two days of the incubation period by virus replication in tracheal, bronchial and pulmonary epithelial cells (2). Virus is then transmitted from the respiratory tract to local lymphoid tissues where replication gives rise to syncitia of lymphoid or reticuloendothelial giant cells. Dendritic cells resident in mucosal tissues play an important role in local immune surveillance against invading pathogens. The DC of the respiratory tract are located within the upper layers of epithelial cells of the epithelium and extend their dendrites towards the luminal side of the tissue (3,4). DC of the respiratory mucosae are specialized in the uptake of antigen penetrating through surface epithelia, its transport and presentation to T cells in draining lymph nodes, where specific immune responses are then initiated (5). We have hypothesized that DC could be a target of MV and that MV-induced immune suppression could result from impaired antigen presentation by MV-infected DC.

2. HUMAN CULTURED DENDRITIC CELLS CAN BE INFECTED BY MEASLES VIRUS

Human dendritic cells were derived from cord blood CD34+ hematopoietic stem cells by culture in the presence of GM-CSF and TNF-α, as previously described (6). The cells recovered on day 9 of culture (mostly CD1a+ dendritic cells) were infected with a vaccine-like strain (Halle) or a wild type strain of MV (Lys-1) isolated from a patient with acute measles. Dendritic cell infection by MV was demonstrated by the appearance of syncitia of giant cells containing the viral nucleocapsid and by expression MV enveloppe glycoproteins on the cell surface. MV could be titrated from the cell cultures indicating that the infection was productive.Infection of DC with MV (Hallé), but not with MV (Lys-1) requires virus binding to CD46, the putative cell surface receptor for MV (7).

3. MEASLES VIRUS INFECTED DENDRITIC CELLS CAUSE IMMUNE SUPPRESSION

CD1a+ DC differentiated from CD34+ stem cells are able to process and present protein antigens in peptides associated with cell surface MHC class II to CD4+ T cells (8). Infection of DC with MV did not affect their ability to internalize an irrelevant antigen through receptor-mediated endocytosis. Alternatively, MV-infected DC could no longer stimulate T cells in allogeneic MLR. Furthermore, addition of as few as 1 MV-infected DC to cultures containing uninfected DC and allogeneic CD4+ T cells completely inhibited T cell proliferation. The inhibitory effect of DC required DC infection with live MV and is associated with transmission of the virus to T cells without affecting T cell viability. Studies are in progress to determine whether inhibition of T cell proliferation is due to transmission of the virus to T cells or could be also induced by the MV-infected DC independently of virus secretion; the effect of MV infection of DC on cytokine production and expression of accesory or costimulatory molecules is being studied.

4. CONCLUSION

Taken together our data show that DC can be infected by both a vaccine-strain and a wild type train of MV, and suggest that these cells may allow dissemination of MV to T cells after *in vivo* infection. Immunosuppression may result from MV transfer to T cells as well as from impaired antigen presentation of MV-infected DC. These observation further indicate that a subunit vaccine may be preferrable to a live attenuated vaccine, especially for vaccination of young children in developing countries.

REFERENCES

1. Starr S, Berkovitch S, Effects of Measles, gamma globulin-modified measles and vaccine measles on the tuberculin test . N. Engl. j. Med. 1964; 2701: 386.
2. Sakaguchi M, Yoshikawa, Yamanouchi K, Sata T, Nagashima K, Takeda K. Growth of measles virus in epithelial and lymphoid tissues of Cynomolgus monkeys. Microbiol. Immunol. 1986; 30:1067
3. Holt PG, Schon-Hegrad MA, Phillips MJ, McMenamin PG, Ia positive dendritic cells form a tightly meshed network within the human airway epithelium. Clin. Exp. Allergy. 1989; 19:597.

4. Schon-Hegrad MA, Oliver J, McMenamin PG and Holt PG. Studies on the density, distribution and surface phénotype of ijntraepithelial class II major histocompatibility complex antigen (Ia)-bearing dendritic cells in the conducting airways. J. Exp. Med. 1991; 173:1345.
5. Holt PG, Schon-Hegrad MA, and Oliver J. MHC class II-bearing dendritic cells in pulmonary tissues of the rat. Regulation of antigen presentation by endogenous macrophage populations.
6. Caux C, Dezutter-Dambuyant C, Schmitt D, and Banchereau J. GM-CSF and TNF-a cooperate in the generation of dendritic Langerhans cells. Nature; 360:258.
7. Naniche D, Varine-Krishnan G, Cervoni F, Wild F, Rossi B, Rabourdin-Combe C and Gerlier D. Human membrane cofactor protein (CD46) acts as a cellular receptor for measles virus. J. Virol. 1993; 67:6025.
8. Caux C, Massacrier C. Dezutter-Dambuyant C, Vanbervliet B, Jacquet C, Schmitt D and Banchereau J. Human dendritic Langerhans cells generated in vitro from CD34+ progenitors can prime naive CD4+ T cells and process soluble antigen. J. Immunol. 1995; 155:5427–5435.

IMMORTALIZED CELL LINES WITH DENDRITIC MORPHOLOGY DERIVED FROM MICE LACKING BOTH TYPE I AND TYPE II INTERFERON RECEPTORS PRESENT MHC II RESTRICTED ANTIGEN TO T CELLS AND INDUCE MIXED LEUKOCYTE REACTIONS

R. Nunez, M. Ackermann, and M. Suter

Institute for Virology
Veterinary Faculty
University of Zürich, Switzerland

INTRODUCTION

Interferons (IFN) are potent regulators of the immune response. They modulate host defense against infections and induce growth inhibition of tumors. Two types of IFN families have been described. The type I IFN family includes α, β and ω IFN, the type II IFN is represented by IFNγ as the only member. IFN molecules bind to specific receptors present on the cell surface. It is thought that the binding of IFN with its specific receptor induces the plethora of functions associated with IFN (6).

Mice lacking both type I and II interferon receptors (AG129) have been generated (1, 13, 17) and the role of interferons in anti viral response was analyzed. The animals displayed a high susceptibility to virus such as influenza, VSV and LCMV (13, 17). The susceptibility was explained by a lack of a specific anti viral CTL response in a later phase of the infection, probably because of an exhaustion of the CD8 subset of T cells. As expected, the humoral immune response in AG129 was biased towards an IgG1 response (7, 13, 17). The role of IFN in the function of dendritic cells (DC) is not yet fully understood.

Therefore, cells from lymphoid (thymus, spleen) and non lymphoid tissue (brain, eye, heart, kidney, lung and skin) of AG129 mice were isolated and immortalized by using the MIB2-N11 retrovirus shown to immortalize DC (8, 12).

The cell lines obtained have been analyzed morphologically and functionally by determining antigen uptake and stimulation of T cells.

Dendritic Cells in Fundamental and Clinical Immunology
edited by Ricciardi-Castagnoli, Plenum Press, New York, 1997

MATERIAL AND METHODS

Mice

Inbred AG129 (I-A^b) with deleted IFN receptors I and II (IFN r o/o) were kindly provided by Rolf Zinkernagel (1, 13, 17). DNA from nucleated cells in blood obtained from AG 129 mice and from selected cell lines (see below) was tested by PCR as described (17) in order to confirm the lack of the IFNrs.

Media and Antigens

Iscoves modified Dulbeccos culture medium (Gifco BRL, Life technologies, Basel, Switzerland) was used supplemented with 10 % FCS and a mixture of antibiotic-antimycotic (Gibco BRL) containing penicillin 100 U/ml, streptomycin 100 μg/ ml and amphotericin B 0.25 μg/ml. The antigens ovalbumin (OVA) and hen egg lysozyme (HEL) were from Sigma Chemical Co. (St Louis, MO).

Cell Lines

The cell lines used and their source are listed in Table 1.

Antibodies and FACS Staining

Cell surface expressed molecules were determined with monoclonal antibodies (mAb) and analyzed by FACS using standard procedures (8). For this, cultured cells were detached by a gentle trypsin-EDTA (Gibco) treatment, the cells counted and the viability determined by trypan blue exclusion. Only cells with more than 95% viability were used. In order to block the Fc receptors, cell suspensions (10^6 cells/ml) were incubated on ice with mAb 2.4G2 (8). After one wash in FACS buffer (PBS containing 5% FCS and 0.01 % sodium azide), cells were resuspended in equal volume of FACS buffer and aliquots of 100 μl of the cell suspensions (10^6 cells/ml) were seeded in individual wells of a 96 well plate. Appropriate concentrations of different mAb (given in Table 2) were added to the seeded cells and the mixture incubated on ice for 30 minutes. After three wash cycles with FACS buffer, cells were either fixed with FACS buffer plus 2% formaldehyde for direct

Table 1. Cell lines used

Cell line	Characteristics	Source *
VN11	Glia origin	PRC.
D2SC	Spleen origen	PRC
MT-2	Macrophage cell line	PRC
Hd-1AC5	HEL specifity, I-E^d	SIK
E3	Ovalbumin specifity, I-E^k	SIK
E8	Ovalbumin specifity, I-E^k	SIK
BO97.10.5	Ovalbumin specifity, I-A^b	PM, LA

*The following colleagues have established these cell lines and generously provide them for this study. PRC, Paola Ricciardi-Castagnoli, Pharmacology, University of Milan; SIK, Stephen I. Katz, Department of Dermatology, NIH, Bethesda, Md; PM, Philippa Marrack, The Jewish Medical Center, Denver, Co; LA, Luciano Adorini, Roche Milano Ricerche, Milano, Italy (4).

Table 2. Monoclonal antibodies

Antigen	Clone*	Antibody	Labeling
Mac-1/ CD11b	M1/70	Rat IgG2b	FITC-anti CD11b
CD90	53-2.1	Rat IgG2a	FITC anti CD90 (Thy 1.2)
I-Ak/I-Ab	10-3.6/AF6-120.1	MCWB IgG2a/ MBalb/c IgG2a	FITC anti I-Ak/ FITC anti I-Ab
CD4(L3T4)	h129.19	Rat IgG2a	FITC rat IgG2a
B-220	RA3-6B2	Rat IgG2a	PE anti CD45R
	R35-95	Rat IgG2a	FITC rat IgG2a
	R35-95	Rat IgG2a	PE rat IgG2a
	R35-38	Rat IgG2b	FITC rat IgG2b
	R35-38	Rat IgG2b	PE rat IgG2b
CD45	3OF11.1	Rat IgG2b	PE anti CD45
CD8(Ly-2)	53-6.7	Rat IgG2a	PE anti CD8

*PharMingen, San Diego, Ca.

immunofluorescence (IF) staining or incubated with biotin labelled phycoerythrin followed by streptavidin (PharMingen, SanDiego, CA). FACS analysis was performed with a FACS equipment (Coulter Electronics, Hialiah, FL). At least 5000 cells were analysed. Analysis of the data was performed with the PC lysis software (Becton Dickinson, San Jose, CA).

Immortalization of Cells

Newborn and adult mice were sacrificed under anesthesia and submerged in cold ethanol for 5 minutes. Tissues were removed and processed by gentle teasing until cell suspensions were obtained. The cells were counted and the viability was determined. Skin of newborn mice was teased off the body, cut in small pieces and incubated with 1% PBS buffered trypsin-EDTA for 30 min. at 37 °C. Then, epidermal cell suspensions were obtained by gently pipetting the cells through an 18 gauge needle several times. Cell suspensions seeded in 25 ml tissue culture flasks (Corning Glass Works, Corning, NY) were infected with 1:1 volume of 0.2 μM filtered cell free supernatants from the cell line containing the MIB2-N11 recombinant defective retroviral vector that carries v - myc MH2 gene (8, 12). After incubation for 1 h at 37 °C, half the volume was replaced by fresh medium and the culture was further incubated at 37 °C, in 5% CO_2. Twice a week half of the medium was changed with fresh medium. About 4–6 weeks after infections first cell clusters were observed. The clusters were transferred to new flasks (Corning) after a gentle trypsin-EDTA treatment.

Antigen Uptake by the Immortalized Cells as Determined by FACS and Confocal Microscopy

Determination of antigen uptake by the immortalized cells was assayed using 2×10^5 cells pulsed with OVA conjugated with fluorescein (OVA-FITC, Molecular Probes, Eugene, OR) as antigen. Graded doses (0.001 to 1mg/ml) of antigen was added to cells kept on ice for 10 min. Thereafter the cell antigen mixture was incubated at 37 °C for various amounts of time. Control cells incubated with the same amount of antigen were kept on ice for the same period of time. Antigen uptake was stopped by adding ice cold FACS buffer (PBS containing 5% FCS and 0.01 % sodium azide). Subsequently, the cells were washed three times with FACS buffer, fixed with FACS buffer containing 2% formaldehyde followed by FACS analysis. Data on antigen binding of control cells kept on ice troughout the entire experiment were used as control.

Determination of antigen uptake by cells analyzed by confocal microscopy was performed by seeding and culturing 10^4 cells/well in Lab-tek chamber slide systems (Nunc, Inc., Naperville, IL). After 24 hours in culture, the medium was removed from the adherent cells, and replaced at various intervals by fresh medium containing OVA-FITC. After three gentle washes with cold FACS buffer, the cells were fixed with 1% formaldehyde in FACS buffer and analyzed the same day using a Meridian confocal microscope (5, 11).

Antigen-Processing and Stimulation of T Cell Hybridomas

In order to evaluate the antigen-processing activity of the AG129 derived cell lines, T cell hybridomas with defined antigen specificity were used (4) (see Table 1). Graded numbers (100 to 2000 cells/well) of antigen presenting cells (APC) were cocultured in triplicates with 5×10^4 to 5×10^5 hybridoma cells in flat bottom 96 well plates (Corning) with or without varying concentrations of antigen. The activation of the hybridoma cells was determined by the production of IL-2, measured in the supernatants by using a commercial ELISA (R&D Systems, Minneapolis, MN). Supernatants were harvested after 48 hours of coculture. Data reported in pg, represent the mean value of triplicate wells. 5.3 pg of IL-2 are equivalent to 1 unit of standard NIBSC/WHO 93/566.

Mixed Leucocyte Reaction (MLR)

MLR was performed by cocultivating variable numbers of the immortalized cells from AG 129 ($H-2^b$) mice as stimulating cells for 5×10^4 to 2×10^5 responder cells from spleen of Balb/c mice ($H-2^d$) Spleen cells were depleted of adherent cells by incubation in polystyrene flasks at 37°C for two hours. Thereafter, the floating cells were harvested, counted and seeded in flat bottom 96 well plates and kept in culture for at least 7 days. Before harvesting, tritiated thymidine (185 MBq = 5 mCi, Dupont-New England Nuclear, Boston , MA) was added at a final concentration of 1 μCi /well for another 18 hours. The cells were harvested onto filter paper (Wallac, Turku, Finland) with a LKB Wallac cell harvester (Wallac) and thymidine incorporation was measured in a Betaplate-liquid. Scintillation counter (Wallac). Data are listed as counts per minute (cpm) and represent the means of triplicate wells.

RESULTS AND DISCUSSION

1. Immortalization of DC Like Cells from Neonatal AG129, Adult AG129 and Adult G129

Cells prepared from brain, lung, eye, thymus, kidney, heart, spleen, liver, lymph node and skin obtained from neonatal AG 129 mice were infected with the MIB2-N11 retrovirus. After one month in culture several cell lines with a variety of morphological features were obtained. The cells grew without externally added cytokines. In contrast and despite several attempts cell lines from brain and lung only were obtained from adult AG129 or from adult G129 mice using similar tissues as with the neonatal AG129. Some of the cell lines have been kept in culture for about one year.

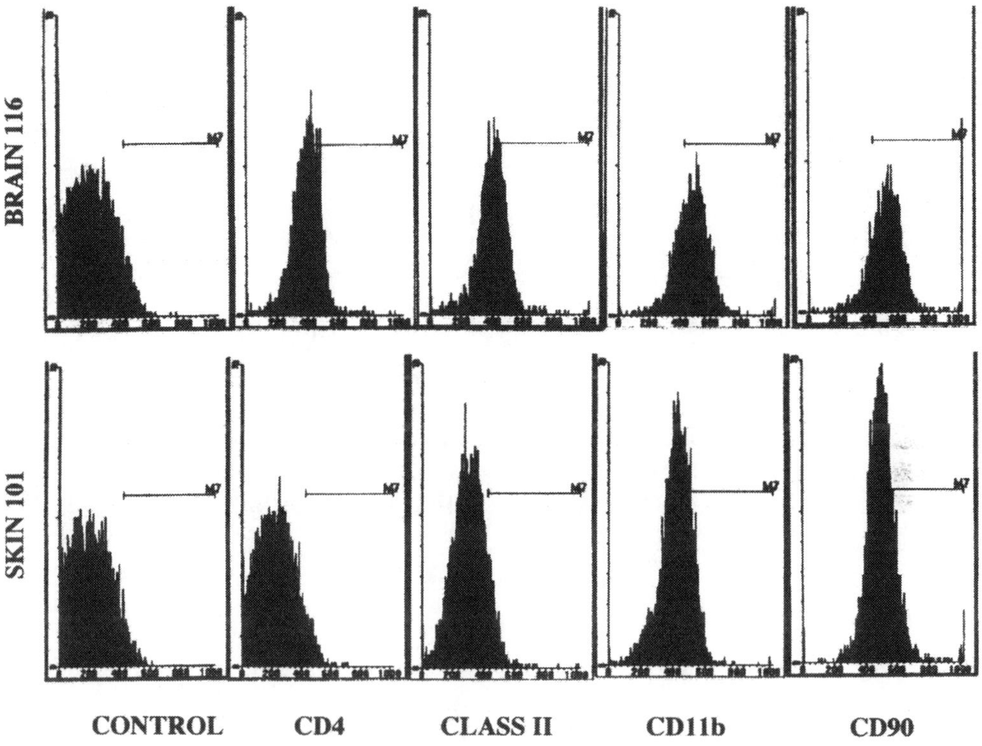

Figure 1. A representative FACS analysis of the cell lines 116 (top panel) and 101 (lower panel). The control FACS profile was obtained by staining of the cells with R35–95 (Rat IgG2a).

2. Surface Analysis of the DC Lines

To further characterize the morphological characteristics of the cell lines with dendritic morphology, cell surface markers were analyzed. The cells were Mac1+ and expressed low amounts of MHC class II molecules. The skin, brain and thymus derived cell lines expressed significant amounts of the Thy 1 marker, but expressed weak amounts or no CD4 molecules. A representative FACS profile of the cell lines, 101 and 116 which have been analyzed in more detail is shown in Figure 1.

3. Antigen Uptake and Processing by DC Lines

At certain juvenile stages in the differentiation, normal DC have the capacity to take up antigen and to process it. At a later stage in the development of such DC this function is lost (5, 11). We thus analyzed antigen uptake by both FACS and confocal microscopy. Fig. 3 clearly shows that FITC labeled OVA was taken up within 30 min. and was processed within 2 hr. Therefore, we consider the DC lines we have obtained as juvenile DC like cell lines. It was thus important to analyze the capacity of these cells to present antigen to T cells.

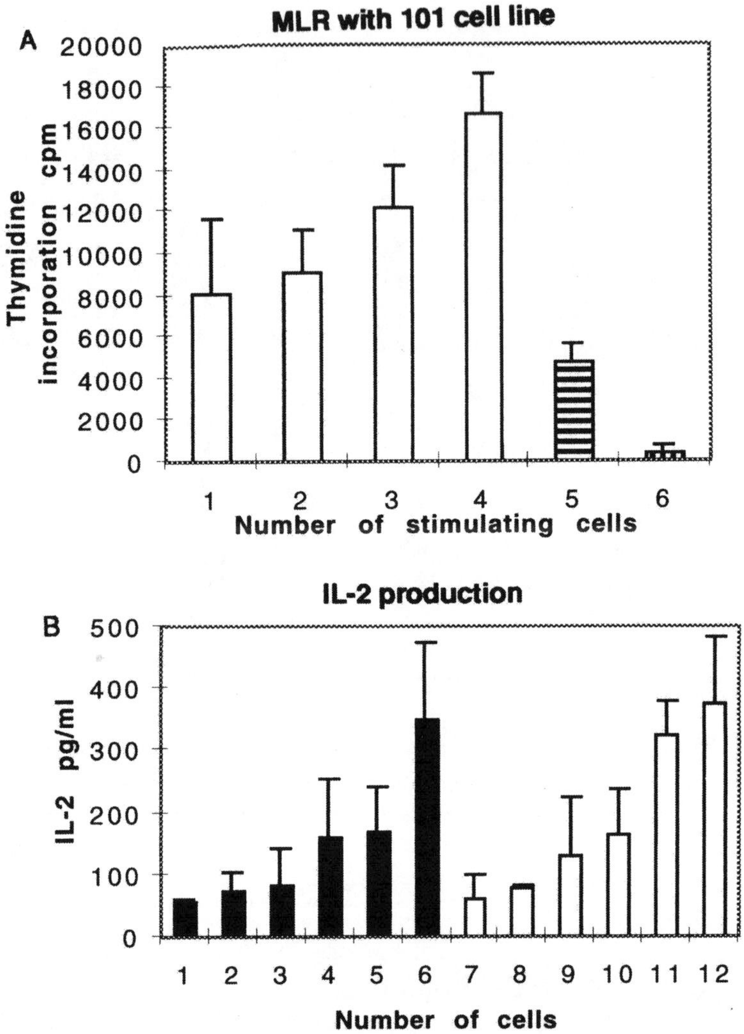

Figure 2. (A) shows the thymidine incorporation (cpm) of a representative MLR using the skin derived cell line 101 as stimulating cells (white bars). 500 cells (bar 1), 1000 cells (bar 2), 1500 cells (bar 3), and 2000 cells (bar 4) as the stimulater were used. The responder spleen cells without stimulating cells are displayed in bar 5. In bar 6 the proliferation of 2000 stimulating cells alone is shown. Standard deviation of triplicates are shown on top of the bars. (B) shows the IL-2 production by hybridoma T cell BO97.105 after antigen stimulation (OVO) using the cell lines 116 as APC (black bars) and 101 (white bars). 100 cells (bars 1 and 7), 150 cells (bars 2 and 8), 200 cells (bars 3 and 9), 500 cells (bars 4 and 10), 1000 cells (bars 5 and 11), 2000 cells (bars 6 and 12) were used as APC). The production of IL-2 by the hybridoma T cell BO97.105 without APC, in presence of antigen, is less than 15 pg/ml (data not shown).

4. MLR and Antigen Presentation

Antigen presentation and priming of naive T cells are some of the most important functions of dendritic cells *in vivo* and can be shown *in vitro* by an MLR (2, 3, 9, 14, 15, 16). Figure 2a, shows the results of a representative assay with the skin derived cell line 101 indicating that as few as 500 stimulating cells were able to induce a MLR in a dose

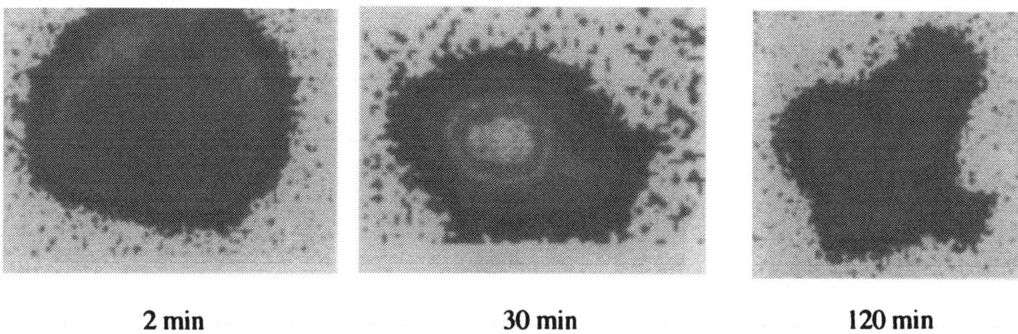

Figure 3. Confocal microscopy images of the antigen OVO-FITC taken up by the cell line 101 at 2 minutes (left panel), 30 minutes (center panel), and 120 minutes (right panel).

dependent manner. Similar results were also obtained with the brain derived cell line 116 (data not shown).

The capability of the cell line 101 and 116 to process and present the MHC II restricted soluble antigen OVA to the T cell hybridoma BO97.105 was also investigated. Proliferation of the T cell line was measured by the IL-2 production. As few as 100 antigen presenting cells (APC) triggered this T cell hybridoma to produce significant amounts of IL-2 (Figure 2b). Hence, the DC-like cells we have obtained are able to process antigen, induce MLR and present antigen to T cells.

The results extend previous observations that retrovirus immortalized immature dendritic cells could induce priming and proliferation of T cells (5). We also confirmed the observation that tissue from neonatal mice was more suitable to obtain cell lines than tissue from adult animals. Interestingly, immortalisation of cells from IFN o/o mice turned out to be easier than from wild type 129 animals (data not shown) indicating the role of the IFN in tumor biology (6).

On a functional basis the cells were capable of taking up antigen, process it and present peptides to T cells in an MHC restricted manner. Despite low or undetectable MHC molecules T cell stimulation does not appear to be altered when compared to similarly immortalized cells (8) or to data obtained from *in vivo* experiments (7, 17). It is possible that MHC up regulation by IFN—if needed at all for efficient T cell stimulation—may also be induced by DC-T cell interaction in the absence of IFN.

Encouraged by the data, the cell lines will be further analyzed in order to better understand the role of IFN type I and II in the process of antigen uptake and presentation by DC or the susceptibility of these cells for infectious agents.

ACKNOWLEDGMENTS

We thank Paola Ricciardi-Castagnoli for kindly providing access to her experience, cell lines, human resources (in particular Manfred Lutz and Claudia Winzler) and valuable reagents. Cell lines were provided by Stephen I. Katz, Department of Dermatology, NIH, Bethesda, MD, by Philippa Marrack, The Jewish Medical Center, Denver, CO, by Luciano Adorini, Roche Milano Ricerche, Milano, Italy, by Gerold Schuler, University of Erlangen, Erlangen, and by Nikolaus Romani, University of Innsbruck.

This investigation was funded by grants from the Kanton Zürich, Swiss National Science Foundation and BBW.

REFERENCES

1. Huang, S., Hendriks, W., Althage, A., Hemmi, S., Bluethmann, H., Kamijo, R., Vilcek, J., Zinkernagel, R. M. and Aguet, M., Science 1993. 259: 1742–5.
2. Katz, S. I., Tamaki, K. and Sachs, D. H., Nature 1979. 282: 324–6.
3. Katz, S. I., Cooper, K. D., Iijima, M. and Tsuchida, T., J Invest Dermatol 1985. 85: 96s-98s.
4. Koch, F., Trockenbacher, B., Kampgen, E., Grauer, O., Stossel, H., Livingstone, A. M., Schuler, G. and Romani, N., J Immunol 1995. 155: 93–100.
5. Lutz, M. B., Assmann, C. U., Girolomoni, G. and Ricciardi-Castagnoli, P., Eur J Immunol 1996. 26: 586–94.
6. Moritz, T., Kloke, O., Nagel-Hiemke, M., Kummer, G., Wandl, U. B., Opalka, B., Plappert, B., Kempeni, J., Seeber, S., Niederle, N., van den Broek, M. F., Muller, U., Huang, S., Aguet, M. and Zinkernagel, R. M., Cancer Immunol Immunother 1992. 35: 342–6.
7. Muller, U., Steinhoff, U., Reis, L. F., Hemmi, S., Pavlovic, J., Zinkernagel, R. M. and Aguet, M., Science 1994. 264: 1918–21.
8. Lutz, M. B., Granucci, F, Winzler C., Marconi G, Paglia P, Foti M, Assmann CU, Cairns L Rescigno M, Ricciardi-Castagnoli P. J. Immunol methods 1994, 174: 261- 279.
9. Romani, N., Gruner, S., Brang, D., Kampgen, E., Lenz, A., Trockenbacher, B., Konwalinka, G., Fritsch, P. O., Steinman, R. M. and Schuler, G., J Exp Med 1994. 180: 83–93.
10. Sallusto, F. and Lanzavecchia, A., J Exp Med 1994. 179: 1109–18.
11. Sallusto, F., Cella, M., Danieli, C. and Lanzavecchia, A., J Exp Med 1995. 182: 389–400.
12. Sassano, M., Granucci, F., Seveso, M., Marconi, G., Foti, M. and Ricciardi-Castagnoli, P., Oncogene 1994. 9: 1473–7.
13. Steinhoff, U., Muller, U., Schertler, A., Hengartner, H., Aguet, M. and Zinkernagel, R. M., J Virol 1995. 69: 2153–8.
14. Steinman, R. M., Annu Rev Immunol 1991. 9: 271–96.
15. Stingl, G., Tschachler, E., Groh, V., Wolff, K. and Hauser, C., Immunol Ser 1989. 46: 3–72.
16. Stingl, G. and Steiner, G., Curr Probl Dermatol 1989 1989. 18: 22–30.
17. van den Broek, M. F., Muller, U., Huang, S., Aguet, M. and Zinkernagel, R. M., J Virol 1995. 69: 4792–6.

INFECTION OF HUMAN THYMIC DENDRITIC CELLS WITH HIV-1 INDUCES THE RELEASE OF A CYTOTOXIC FACTOR(S)

Dominique Bergeron, Sylvie Beaulieu, Marielle Lafontaine, Isabelle Courchesne, and Eric A. Cohen

Département de microbiologie et immunologie
Université de Montréal
C.P. 6128, succursale Centre-ville
Montréal, Québec, Canada, H3C 3J7

1. INTRODUCTION

Lymphoid organs have been reported to represent the major site for the establishment and the propagation of human immunodeficiency virus (HIV-1) (1). HIV-1 infection is characterized by a permanent high-level of viral replication associated with a high-level turnover of $CD4^+$ T cells (2). The late stages of the disease are characterized by a progressive depletion of $CD4^+$ T cells that eventually lead to the development of the acquired immune deficiency syndrome (AIDS). A variety of direct and indirect mechanisms have been proposed to account for the decline of the $CD4^+$ T cells (3). Among them, inappropriate activation of apoptosis, a programmed cell death process, has been proposed to contribute to the $CD4^+$ T cell depletion (4). HIV-induced apoptotic cell death has been demonstrated not only *in vitro* but also in infected individuals and is now considered to be a major mechanism of HIV-mediated cell death (5–7).

Several observations have underscore a possible role of the thymus in the pathophysiology of AIDS. Principally, the presence of HIV infection has been detected in a variety of thymic cells and severe aberrations have been reported in thymuses of HIV-1-infected adults and fetuses (8–12). Apoptosis of uninfected cells, presumably mediated by HIV-1-infected bystander cells, was observed in lymph nodes obtained from infected individuals and in thymuses of HIV-infected severe combined immunodeficiency (SCID) mice reconstituted with human thymuses (5, 13–15).

Thymic dendritic cells (DCs) are recognized to play a major role in the intrathymic differentiation of T cells. In the present study, we investigated the ability of HIV-1 infected purified DCs to mediate cell killing of various uninfected hematopoietic primary cells.

2. MATERIALS AND METHODS

Cell Isolation and Culture

Normal thymic tissue were obtained from children undergoing cardiovascular surgery. Dendritic cells were isolated and characterized according to the DS purification method described previously (16). This method allows the recovery of >80% HLA-DR$^+$/CD2$^-$ cells which exhibit ultrastructural morphological features and functional activities described previously for thymic DC.

In certain experiments, HLA-DR$^+$ (DCs) and HLA-DR$^-$ (contaminating non DCs) cell subsets from DC population already isolated with the DS purification method were further purified by sorting anti-HLA-DR mAbs-reactive versus non-reactive cells (FACStar Instruments, Becton-Dickinson, Mountain View, CA). CD4$^+$ and CD8$^+$ cell subsets were obtained from unfractionated thymic cells by magnetic beads depletion of anti-CD8 or anti-CD4 mAbs reactive cells respectively.

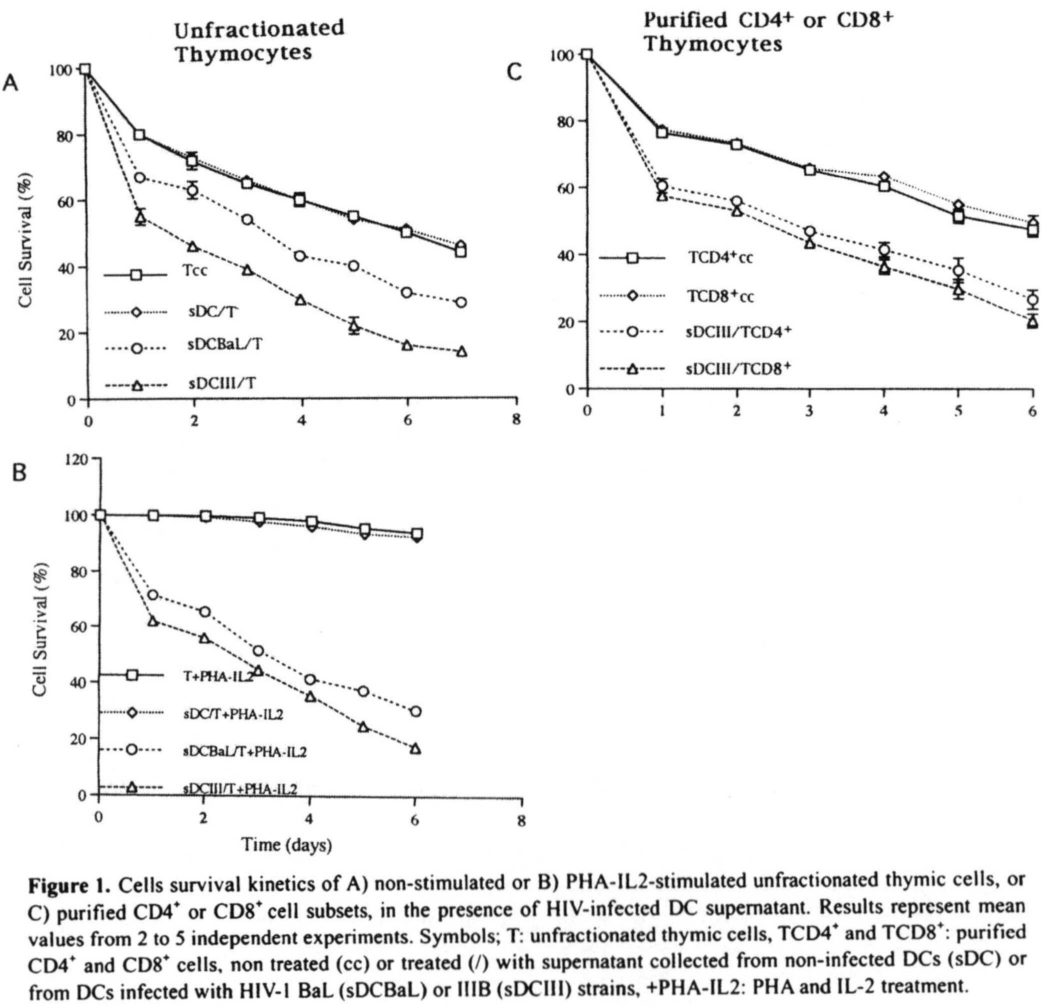

Figure 1. Cells survival kinetics of A) non-stimulated or B) PHA-IL2-stimulated unfractionated thymic cells, or C) purified CD4$^+$ or CD8$^+$ cell subsets, in the presence of HIV-infected DC supernatant. Results represent mean values from 2 to 5 independent experiments. Symbols; T: unfractionated thymic cells, TCD4$^+$ and TCD8$^+$: purified CD4$^+$ and CD8$^+$ cells, non treated (cc) or treated (/) with supernatant collected from non-infected DCs (sDC) or from DCs infected with HIV-1 BaL (sDCBaL) or IIIB (sDCIII) strains, +PHA-IL2: PHA and IL-2 treatment.

Virus Stock Preparation and HIV-1 Infection

Original DS-purified DCs, or their highly purified HLA-DR$^+$ or HLA-DR$^-$ subsets were mock-infected or infected at a m.o.i. of 0.1 $TCID_{50}$/cell, or on the basis of identical RT, with viral stocks of the T tropic HIV-1 IIIB or macrophage tropic HIV-1 BaL strains prepared and titrated as described in Beaulieu *et al.*, (1996) (8). To evaluate the involvement of viral infection of DC in the release of the cytotoxic factor(s), DCs were treated with AZT (1μM) prior and during HIV-1 infection. Supernatants from the different cultures were collected after 24 hours, filtered through a 0.22-mm syringe-loaded filter unit and used in the cell cytotoxicity assay.

DC Lysate

To evaluate the cytotoxicity of cell lysates, non-infected DCs were lysed by freeze and thaw (4 times) and lysis was verified by microscopic observation. Lysates were filtered through a 0.22-mm syringe-loaded filter unit and used in cell cytotoxicity assay.

Mitogenic Stimulation and Cycloheximide Treatment

As indicated, thymic cells were stimulated using 7.5 μg/ml of phytohemagglutinin (PHA) combined to 10 U/ml of interleukine 2 (IL-2), or treated with cycloheximide (50 μM) 2 hrs prior and during the cell cytotoxicity assays.

Cell Cytotoxicity Assay

Cells were incubated in the presence of either DC lysates or cell-free supernatant collected from the indicated DC cultures at a 1:10 dilution. Cell viability was monitored daily in the different cultures using the trypan blue exclusion assay previously described in Beaulieu *et al.*, (1996) (8).

3. RESULTS AND DISCUSSION

As presented in Fig.1A, a specific killing of purified unfractionated human thymic cells is induced upon treatment with cell-free supernatant collected from HIV-1-infected DCs. Kinetic of viable cells revealed that thymic CD4$^+$ and CD8$^+$ cell subsets are both susceptible to the cytotoxic effect of the HIV-infected DC supernatant (Fig. 1C). In contrast, purified thymic DCs are not themselves affected by the presence of the HIV-infected DC supernatant (8). We tested the susceptibility of activated thymic cells to the presence of HIV-infected DC supernatant. Viability of thymic cells is increased upon treatment with PHA combined to IL-2 (Fig. 1 B). In these cultures, the addition of HIV-infected DC supernatant has resulted in a dramatic decrease of cell viability (Fig. 1B). These results show that HIV-infected DC supernatant is not only cytotoxic for cells already committed in a spontaneous culture killing process but is also cytotoxic for activated thymic cells. HIV-infected DC supernatant was also added either to freshly purified resting or stimulated human PBMCs. Killing of stimulated PBMCs upon treatment with HIV-infected DC supernatant is observed although freshly purified resting PBMCs were resistant to this cytotoxic effect (not shown).

Purified thymic DCs do not represent an homogeneous cell population. Indeed, approximately 20% of the cells represent a HLA-DR$^-$ cell subset. The induction of thymocyte killing was observed when supernatants collected from either HIV-infected original DS-purified DCs or highly purified HLA-DR$^+$ cells were added to unfractionated thymocytes (Fig. 2A). In contrast, supernatants collected from the HIV-infected HLA-DR$^-$ contaminating non-DC subset do not induce thymocyte killing. These data clearly indicate that among purified DCs, the HLA-DR$^+$ cell subset is responsible for the release of a thymotoxic factor(s) upon HIV infection.

We compared viabilities of thymocytes untreated or treated with AZT prior and during the addition of the HIV-1-infected DC supernatant. The presence of AZT did not significantly affect the induction of thymocyte killing (Fig. 2B). These data clearly demonstrate that the cell killing induced by the HIV-1-infected DC supernatant does not result from a viral infection, but is due to the presence of one or several cytotoxic factor(s) in the supernatant.

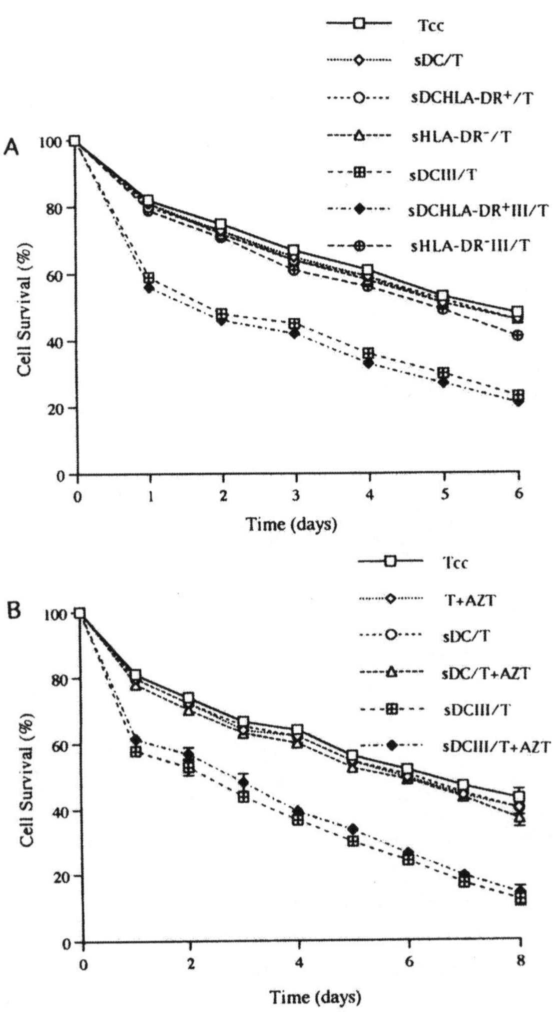

Figure 2. A) Cell survival kinetics of unfractionated thymocytes treated with supernatant collected from DC populations isolated using the DS purification method (sDC) or from their highly purified DC (sDCHLA-DR$^+$) or contaminating non-DC (sHLA-DR$^-$) cell subsets and either non-infected or infected with the HIV-1 IIIB strain (III). B) Effect of AZT treatment on thymocyte susceptibility to the HIV-infected DC supernatant. Results represent mean values from at least 2 independent experiments. Symbols are as described in the legend of Fig.1; +AZT: thymocytes were treated with AZT prior and during their treatment with the indicated supernatants.

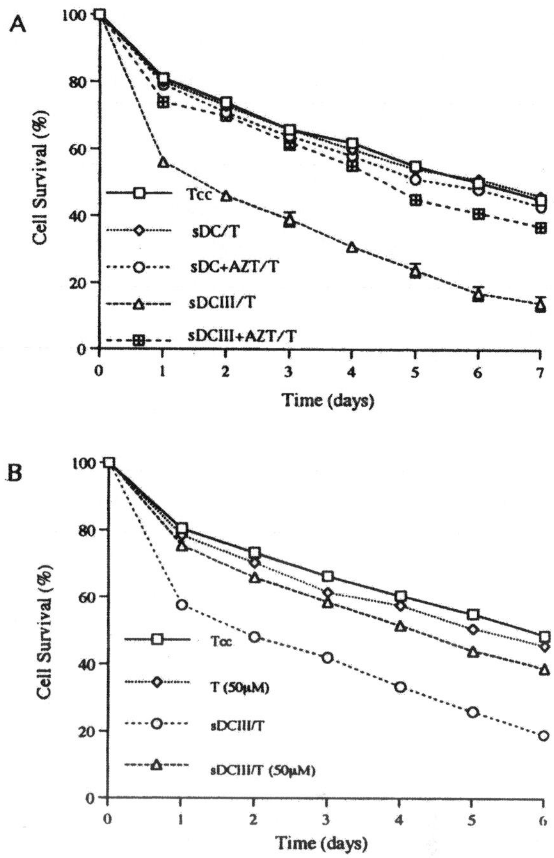

Figure 3. A) Effect of AZT on the release of the cytotoxic factor from HIV-infected DCs. B) Effect of cycloheximide treatment on thymocyte susceptibility to HIV-infected DC supernatant. Results represent mean values from at least 2 independent experiments. Symbols are as described in the legend of Fig.1 and 2; (50 μM):thymocytes were treated with cycloheximide.

The treatment of DCs themselves prior and during HIV-1 infection with AZT prevents the release of the cytotoxic factor(s) from DCs (Fig.3A). This suggests that viral infection is necessary to induce this DC response.

The cytotoxic factor(s) could be already active in DCs and passively release upon HIV-1 infection and/or cell death. However, our results revealed that high concentrations of non-infected DC lysate are unable to induce the thymocyte killing (not shown). This indicates that following infection, the cytotoxic factor(s) need to be either activated or expressed *de novo*.

We next determined whether *de novo* protein synthesis is necessary for the induction of thymocyte killing upon treatment with HIV-infected DC supernatant. For that purpose, thymocytes were treated with cycloheximide prior and during the addition of the cytotoxic supernatant. A rescue of viability was observed upon cotreatment with cycloheximide demonstrating that the cell killing induced upon treatment with the HIV-infected DC supernatant required *de novo* protein synthesis and thus is an active cellular process (Fig.3B).

4. CONCLUSION

In vitro HIV-1 infection of thymic DC is unique as its trigger killing *in trans* , namely in the absence of cellular contact. *In vitro* and *in vivo* observations make thymic

DCs as a potential intermediate in the induction of killing of bystander uninfected thymocytes in HIV-infected individual: 1) DCs participate in the thymic microenvironment, 2) these cells were shown to be a target for HIV-1 infection (8), 3) they produce a cytotoxic factor(s) upon infection and finally 4) this factor(s) induce killing of multiple population of thymocytes.

Taken together these findings suggest that *in vivo* infection of thymic DCs by HIV-1 could results in the release of a soluble cytotoxic factor(s) in the thymic microenvironment which could eventually participate in apoptosis of bystander non-infected thymic cells. We are currently investigating the role of cytokines, such as soluble form of the TNF family members (TNF-α, FasL, CD30L...), as candidate for such cytotoxic factor. The thymocyte killing, combined with HIV-associated peripheral destruction of $CD4^+$ T cells, may lead to their rapid depletion, resulting in early disease progression, particularly in pediatric AIDS given the dominant role of the thymus in children.

5. REFERENCES

1. Pantaleo, G., and A. S. Fauci. 1995. Annu. Rev. Immunol. 13:487–512.
2. Ho, D. D., A. U. Neumann, A. S. Perelson, W. Chen, J. M. Leonard, and M. Markowitz. 1995. Nature 373:123.
3. Levy, J. A. 1993. Microbiol. Rev. 57:183–289.
4. Ameison, J. C., and A. Capron. 1991. Immunol. Today 12:102.
5. Finkel, T. H., G. Tudor-Williams, N. K. Banda, M. F. Cotton, T. Curiel, C. Monks, T. W. Baba, R. M. Ruprecht, and A. Kupfer. 1995. Nature Medicine 1:129.
6. Muro-Cacho, C. A., G. Pantaleo, and A. S. Fauci. 1995. J. Immunol. 154:5555.
7. Oyaizu, N., and S. Pahwa. 1995. J. Clin. Immunol. 15:217.
8. Beaulieu, S., A. Kessous, D. Landry, S. Montplaisir, D. Bergeron, and E. A. Cohen. 1996. Virology 222:214.
9. Schnittman, S.M., S. M. Denning, J. J. Greenhouse, J. S. Justement, M. Baseler, J. Kurtzberg, B. F. Haynes, and A. S. Fauci. 1990. Proc. Natl. Acad. Sci. USA 87:7727.
10. Valentin, H., M.-T. Nugeyre, F. Vuillier, L. Boumsell, M. Schmid, F. Barré-Sinoussi, and R. A. Pereira. 1994. Virology 68:3041.
11. Joshi, V. V., J. M. Oleske, S. Saad, C. Gadol, E. Connor, R. Bobila, and A. B. Minnefor. 1986. Arch. Pathol. Lab. Med. 110:837.
12. Mano, H., and J.-C. Chermann. 1991. AIDS Res. Hum. Retroviruses 7:83.
13. Aldrovandi, G. M., G. Feuer, L. Gao, B. Jamieson, M. Kristeva, I. S. Y. Chen, and J. A. Zack. 1993. Nature 363:732–736.
14. Bonyhadi, M. L., L. Rabin, S. Salimi, D. A. Brown, J. Kosek, J. M. McCune, and H. Kaneshima. 1993. Nature 363:728.
15. Su, L., H. Kaneshima, M. Bonyhadi, S. Salimi, D. Kraft, L. Rabin, and J. M.McCune. 1995. Immunity 2:25.
16. Beaulieu, S., D. Landry, D. Bergeron, E. A. Cohen, and S. Montplaisir. 1995. J. Immunol. Methods 180:225.

DENDRITIC CELL SURFACE MOLECULES

A Proliferating Field

D. N. J. Hart, G. J. Clark, J. W. Dekker, D. B. Fearnley, M. Kato, B. D. Hock, A. D. McLellan, T. Neil, R. V. Sorg, U. Sorg, K. L. Summers, and S. Vuckovic

Haematology/Immunology/Transfusion Medicine Research Group
Christchurch Hospital
Christchurch, New Zealand

Dendritic cells (DC) are specialist antigen presenting cells (APC) derived in common with other leukocytes from bone marrow stem cells.[1,2] A myeloid derived precursor[3] gives rise to immature circulating blood DC which enter the tissues and after interacting with antigen migrate to the T lymphocyte dependent areas of lymph nodes, where they deliver stimulatory signals to responding T lymphocytes. Recent studies suggest the growth and differentiation of myeloid DC is heavily influenced by cytokines, notably flt-3 ligand, SCF, GM-CSF, IL-4 and TNFα[4-6]. A unique DC precursor in blood can be distinguished from freshly isolated blood monocytes.[7,8] However, considerable evidence suggests that monocytes exposed *in vitro* to certain cytokine combinations (notably including IL-4, which downregulates CD14) can acquire many, if not most DC characteristics.[9] The DC series also includes a lymphoid precursor derived cell[10] a subset of which, in the mouse, has an inhibiting effect on responding T lymphocytes. Lymphoid precursor cells have been described in man[11,12] but their function is unknown.

The definition of cell surface and intracellular molecules expressed selectively by DC is an essential prerequisite to understanding the growth and differentiation of both lymphoid and myeloid derived DC as well as understanding their unique stimulatory or inhibitory properties. The difficulties involved in isolating pure DC for immunization and screening has frustrated attempts to produce specific monoclonal antibodies (mAb) to DC. Nonetheless, a series of useful reagents 33D1,[13] NLDC-145,[14] and N418[15] have been used to study mouse DC. Reagents applicable to studies on human DC have until recently been more limited. Our laboratory has used both mAb and molecular techniques in an effort to define new molecules associated with DC differentiation, activation or function.

The CMRF-44 mAb reacts with an antigen expressed in low levels on a subset of circulating human blood DC and its expression is upregulated rapidly so that all DC become positive during a short period of *in vitro* culture.[16] We have found CMRF-44 extremely useful for identifying the DC population present in low density cultured T lymphocyte depleted peripheral blood mononuclear cells (PBMC).[17,18] The CMRF-44 mAb has also proved useful in defining the DC populations emerging in cytokine cultured PBMC preparations (Vuckovic et al, in preparation). Double labelling studies allow CMRF-44$^+$ DC to be distinguished from CD14$^+$ macrophages in the T lymphocyte areas of lymph node. The antigen recognized has some features suggesting it is a glycolipid but further attempts to isolate and characterize the target antigen are in progress.

Another mAb CMRF-56, which reacts with activated blood DC, has been produced recently using a similar immunization and screening process. This mAb is of particular interest as it reacts with an antigen on activated DC, which is absent from other resting and in vitro activated leucocyte populations including monocytes. The CMRF-56 mAb has a different specificity to both CMRF-44 and the CD83 mAb HB15[19,20]and these latter two mAb cap antigen independently of the CMRF-56 antigen on the Hodgkins disease derived cell line L428. Furthermore, CMRF-56 does not bind to either recombinant CD83 antigen or to CD83 antigen expressing COS cell transfectants. Analysis of the kinetics of antigen expression by freshly isolated blood DC following *in vitro* culture indicated that the CMRF-44 antigen is upregulated prior to the CD83 and CMRF-56 antigens. Interestingly, a smaller proportion of cytokine cultured PBMC express the CMRF-56 antigen compared to CMRF-44 and the functional properties of this population are under investigation.

The DC cell surface receptors which may be involved in antigen loading are also under investigation. The CMRF-35 mAb recognizes an antigen which has some features reminiscent of Fc receptors.[21,22] Attempts to clone the gene have led to the discovery of a family of related cDNAs and we are currently attempting to clarify their role, if any, as IgM binding proteins. Data on the human homologue of mouse DEC-205, a pattern recognition molecule with lectin-like characteristics,[23] has been sought. The human cDNA is very similar to the mouse sequence although its tissue distribution may be somewhat wider (Kato et al, in preparation).

Studies on the cytokines produced by DC have been difficult to date because of the problems involved in obtaining high purity DC preparations. We have shown previously that human tonsil DC did not produce IL-1,[24] nor did blood DC produce TNFα.[25] A recent RT-PCR analysis by Zhou and Tedder of CD83 purified human DC suggested a relatively restricted expression of cytokines.[26] However, as suggested previously[2] activation of DC may be critical to obtain DC cytokine production and release. We therefore studied DC production of IL-7 before and after a period of *in vitro* culture. DC produced significant IL-7 mRNA after activation and a series of experiments suggest that IL-7 produced by DC is important in DC activation of T lymphocytes (Sorg et al, in preparation).

The technique of differential display is being applied by a number of laboratories to the search for DC specific gene products. Our own studies have generated four cDNA clones which show restricted expression in a range of leucocyte populations including DC. One of these has been selected on the basis of its sequence homology for more extensive investigation and attempts to isolate a full length cDNA clone are in progress.

Whilst none of the molecules described above are expressed exclusively by DC, the ability to use mAb or RT-PCR/*in situ* hybridization to follow their expression is contributing significantly to studies on DC growth and differentiation, including DC infiltration/function in arthritis and malignancy. In time the isolation of these molecules and molecular characterization should allow their function to be determined.

REFERENCES

1. Steinman RM: The dendritic cell system and its role in immunogenicity. Annu Rev Immunol 9:271, 1991
2. Williams L, Egner W, Hart DNJ: Isolation and function of human dendritic cells. Int J Cytology 153:41, 1994
3. Egner W, McKenzie JL, Smith SM, Beard MEJ, Hart DNJ: Identification of potent mixed leucocyte reaction-stimulatory cells in human bone marrow. J Immunol 150:3043, 1993
4. Szabolcs P, Avigan D, Gezelter S, Ciocon DH, Moore MAS, Steinman RM, Young JW: Dendritic cells and macrophages can mature independently from a human bone marrow-derived, post-colony-forming unit intermediate. Blood 87:4520, 1996
5. Siena S, Di Nicola M, Bregni M, Mortarini R, Anichini A, Lombardi L, Ravagnani F, Parmiani G, Gianni AM: Massive ex vivo generation of functional dendritic cells from mobilized $CD34^+$ blood progenitors for anticancer therapy. Exp Hematol 23:1463, 1995
6. Saraya K, Reid CDL: Stem cell factor and the regulation of dendritic cell production from CD34(+) progenitors in bone marrow and cord blood. Br J Haematol 93:258, 1996
7. Egner W, Andreesen R, Hart DNJ: Allostimulatory cells in fresh human blood: heterogeneity in antigen presenting cell populations. Transplantation 56:945, 1993
8. Thomas R, Davis LS, Lipskey PE: Comparative accessory cell function of human peripheral blood dendritic cells and monocytes. J Immunol 151:6840, 1993
9. Peters JH, Ruppert J, Gieseler RKH, Najar HM, Xu H: Differentiation of human monocytes into CD14 negative accessory cells: do dendritic cells derive from the monocytic lineage? Pathobiol 59:122, 1991
10. Ardavin C, Shortman K: Thymic dendritic cells and T cells develop simultaneously in the thymus from a common precursor population. Nature 362:761, 1993
11. Galy A, Travis M, Cen D, Chen B: Human T, B, natural killer and dendritic cells arise from a common bone marrow progenitor cell subset. Immunity 3:459, 1995
12. Res P, Martinezcaceres E, Jaleco AC, Staal F, Noteboom E, Weijer K, Spits H: CD34(+) CD38 (dim) cells in the human thymus can differentiate into T, natural killer, and dendritic cells but are distinct from pluripotent stem cells. Blood 87:5196, 1996
13. Nussenzweig MC, Steinman RM, Witmer MD, Gutchinov B: A monoclonal antibody specific for mouse dendritic cells. Proc Natl Acad Sci USA 79:161, 1982
14. Kraal G, Breel M, Janse M, Bruin G: Langerhans cells, veiled cells and interdigitating cells in the mouse recognised by a monoclonal antibody. J Exp Med 163:981, 1986
15. Metlay JP, Witmer-Pack MD, Agger R, Crowley MT, Lawless D, Steinman RM: The distinct leucocyte integrins of mouse spleen DC as identified with new hamster mAb. J Exp Med 171:1753, 1990
16. Hock BD, Starling GC, Daniel PB, Hart DNJ: Characterisation of CMRF-44, a novel monoclonal antibody to an activation antigen expressed by the allostimulatory cells within peripheral blood, including dendritic cells. Immunol 83:573, 1994
17. McLellan AD, Starling GC, Hart DNJ: Isolation of human blood dendritic cells by Nycodenz discontinuous gradient centrifugation. J Immunol Methods 184:81, 1995
18. Fearnley DB, McLellan AD, Mannering SI, Hock BD, Hart DNJ: Isolation of human blood dendritic cells using the CMRF-44 monoclonal antibody: implications for studies on antigen presenting cell function and immunotherapy. Blood In press:1996
19. Zhou L, Schwarting R, Smith HM, Tedder TF: A novel cell-surface molecular expressed by human interdigitating reticulum cells, Langerhans cells, and activated lymphocytes is a new member of the Ig superfamily. J Immunol 149:735, 1992
20. Zhou L, Tedder TF: Human blood dendritic cells selectively express CD83, a member of the immunoglobulin superfamily. J Immunol 154:3821, 1995
21. Daish A, Starling GC, McKenzie JL, Nimmo JC, Jackson DG, Hart DNJ: Expression of the CMRF-35 antigen, a new member of the immunoglobulin gene superfamily, is differentially regulated on leucocytes. Immunol 79:55, 1993
22. Jackson DG, Hart DNJ, Starling GC, Bell JI: Molecular cloning of a novel member of the immunoglobulin gene superfamily homologous to the polymeric immunoglobulin receptor. Eur J Immunol 22:1157, 1992
23. Jiang W, Swiggard WJ, Heufler C, Peng M, Mirza A, Steinman RM, Nussenzweig MC: The receptor DEC-205 expressed by dendritic cells and thymic epithelial cells is involved in antigen processing. Nature 375:151, 1995
24. McKenzie JL, Prickett TCR, Hart DNJ: Human dendritic cells stimulate allogeneic T cells in the absence of IL-1. Immunol 67:290, 1989

25. McKenzie JL, Calder VC, Starling GC, Hart DNJ: Role of tumor necrosis factor α in dendritic cell mediated primary mixed leukocyte reactions. Bone Marrow Transplantation 15:163, 1995
26. Zhou L, Tedder TF: A distinct pattern of cytokine gene expression by human CD83+ blood dendritic cells. Blood 86:3295, 1995

ANALYSIS OF DENDRITIC CELLS AT THE GENETIC LEVEL

Gill Marland,[1] Franca C. Hartgers,[1] Remco Veltkamp,[1]
Monika F. C. Königswieser,[1] Dan Gorman,[2] Terrill McClanahan,[2]
Carl G. Figdor,[1] and Gosse J. Adema[1*]

[1]Department of Tumor Immunology
University Hospital Nijmegen
Philips van Leydenlaan 25
6525 EX Nijmegen, The Netherlands
[2]DNAX Research Institute
901 California Avenue, Palo Alto, California

1. SUMMARY

To increase our understanding of dendritic cell (DC) function we have used two approaches to search at the genetic level for molecules which are specifically expressed by these cells. First, we have performed random sequencing of cDNA libraries prepared from DC. Second, we have employed differential display PCR (DD-PCR). DD-PCR is a powerful technique for the identification at the RNA level of molecules which are expressed in a cell type-specific manner. In our study, we have compared RNA from DC with RNA from a panel of leukocyte cell lines. Here we present a summary of our findings using these two approaches, and show that both methods are complementary and can be used to identify molecules that are specific to DC.

2. INTRODUCTION

Among the professional antigen presenting cells of the immune system, dendritic cells (DC) stand out as the most adept. In particular they appear to be the only cells capable of priming naive T cells. Their potent antigen presenting ability can be explained in part by their unique life cycle, since DC exist in two maturation states. As immature cells, DC reside in the periphery and are efficient in the uptake and processing of antigen. They can take up large amounts of antigen by macropinocytosis, and also by mechanisms which involve the macrophage mannose receptor[1]. Maturation of DC occurs upon antigen uptake, following which the DC migrate to lymph nodes where they efficiently present antigen to T cells.

* To whom correspondence should be addressed.

Molecular analysis of DC function has been hampered by the relatively low amounts of DC present in blood mononuclear cells, and the absence of DC-specific markers. The recent description of methods for the generation of DC in vitro has broadened the possibilities for the exploitation of these cells[2,3]. In particular these cells could be used in a clinical setting, for example in the generation of anti-tumor responses in vivo. Increasing our understanding of DC function would allow for the optimization of such immunotherapeutic strategies. We have therefore employed two methods, random sequencing of cDNA libraries and differential display PCR (DD-PCR), for the identification of DC-specific molecules.

3. MATERIALS AND METHODS

3.1. Cells

DC were generated in vitro from human monocytes using a modification of described methods[2,3]. Monocytes were obtained by leukapheresis followed by centrifugal elutriation, resulting in a population of cells that were greater than 85% $CD14^+$. These cells were cultured in the presence of 800U/ml GM-CSF and 500U/ml IL-4 for 5–7 days.

3.2. Random Sequencing of DC cDNA Libraries

Complementary DNA libraries were prepared from the following populations; 1) DC cultured as in 3.1, 2) DC further stimulated with TNFα (15 ng/ml) and IL-1 (75 LAF U/ml) for 4 or 20 hrs, or 40% monocyte conditioned medium (MCM) for 4 or 20 hrs, and 3) DC further stimulated with LPS (2 μg/ml) for 4 or 20 hrs. Messenger RNA was isolated, and first and second strand cDNA was prepared using the SuperScript™ Plasmid System (Gibco BRL). The resulting double-stranded cDNA were then cloned unidirectionally into the pSport1 vector (Gibco BRL), and sequenced from the 5′ end. Nucleotide sequences were analyzed against the non-redundant GenBank and EMBL databases using the BLAST program[4]. Clones were considered identical to known sequences if there was greater than 90% sequence identity over at least 100 bp.

3.3. DD-PCR

DD-PCR was performed essentially as described[5]. The 3′ primers used were $T_{12}CN$, where N represents A, C, G or T. The 5′ primers were randomly-designed oligonucleotides of 10 bases. PCR was performed in the presence of ^{35}S-dATP to allow visualization of the products following separation by denaturing polyacrylamide gel electrophoresis. PCR products that were reproducibly cell specific were eluted from the gel, reamplified by PCR and cloned using the TA cloning kit (Invitrogen). Cellular specificity of the clones was determined by RT-PCR followed by Southern hybridization.

4. RESULTS AND DISCUSSION

4.1. Random Sequencing of DC cDNA Libraries

The three cDNA libraries used for random sequencing were selected in an attempt to cover all the known stages of DC maturation, as TNFα, IL-1, MCM and LPS have all

Table 1. Categories of cDNA sequences identified in DC libraries

Category	Number	Percentage	Examples
Unknowns	101	39%	
Repetitive sequences	20	8%	Alu sequences
			CA repeats
Housekeeping genes	45	17%	ribosomal proteins
			GAPDH
			cystatin B
Mitochondrial genes	21	8%	
Nuclear proteins	14	5%	X box binding protein
			RNA helicase
			histone H3.3
Cytoskeletal proteins	11	4%	actin
			vimentin
			tubulin
Signal transduction	14	5%	phospholipase C-α
			rho A GTP-binding protein
			protein tyrosine phophatase ε
Molecules involved in Antigen presentation/uptake	9	4%	class I, II, invariant chain
			CD1b
			mannose receptor
Ferritin	10	4%	
Cell surface receptors	7	3%	α-catenin
			CD23
			transferrin receptor
Secreted proteins	6	3%	IL-1RA
			prothymosin α
Oncogenes	2	< 1%	
Unclassified	2	< 1%	NMB (pMEL17 homolog)
			MAGE-11
Total sequenced	262	100%	

been shown to induce DC maturation[1,3,6]. This would therefore maximize the chances for the identification of important molecules from all stages of DC function. More than 250 clones from the three libraries were randomly sequenced. These were categorized as shown in table 1. Many of the clones encoded proteins that were expected to be expressed by DC. Some examples of these are also shown in table 1. The largest category represented the unknown sequences, which accounted for 39% of the clones. Of the unknown sequences, several clones (13%) showed high homology (around 75%) with known proteins, and therefore may encode new members of gene families. These include adhesion molecules, kinases and signaling molecules. We are now further analyzing a small panel of these clones which, based on their homology with known proteins, may shed light on DC function. One of these clones encodes a new member of the C-C chemokine family. Further investigation of this clone by Northern analysis has revealed that it is specifically expressed by DC. We are currently analyzing the functional properties of this new chemokine. The findings will be published elsewhere.

4.2. DD-PCR

The second approach we have adopted for the analysis of DC at the molecular level is DD-PCR. For this purpose, RNA from 3 independent DC donors was compared to RNA

Table 2. Categories of cDNA sequences identified by DD-PCR

Category	Number	Examples
Novel sequences	14	
Identity with EST	4	
Known sequences	5	hClC-K2 (chloride channel)
		oriP binding protein
		mitochondrial ubiquinone binding protein
		SS RP1 (high mobility group box protein)
		ribosomal protein L3
Total sequenced	23	

from 3 T cell lines (Jurkat, CEM and Peer), 3 EBV-transformed B cell lines (JY, BR and SR) and 3 monocytic cell lines (U937, THP-1 and MonoMac 6). Three DC donors were used in an attempt to eliminate the identification of donor-specific molecules. Seventeen different primer combinations were tested, and 8 bands which were DC-specific in two independent experiments were rescued from gels and cloned. Sequence analysis revealed that most of the bands contained multiple PCR products, resulting in the identification of 23 clones. None of these clones was identical to any of the clones identified by random sequencing. Database analyses of the nucleotide sequences of these 23 clones revealed that 5 were identical to known cDNA, and 18 were either novel or identical to expressed sequence tags (table 2). 17 of these 18 clones were analyzed further.

DD-PCR is known to result in the cloning of many mRNA of low abundance which fail to be detected by Northern blotting[7]. This hinders the verification of the cellular specificity of interesting clones. We opted, therefore, to do an initial screening by RT-PCR followed by Southern hybridization. The results showed that one clone, 17d620, was specific to DC. Screening of a DC cDNA library with an oligonucleotide corresponding to 17d620 yielded a cDNA of approximately 1.4 kb. Sequence analysis revealed that the 5′ 650 bp were identical to the mRNA encoding the macrophage mannose receptor. The remaining 750 bp, which contained the entire clone identified by DD-PCR, represented novel sequence.

The gene for the mannose receptor has 30 exons which code for a protein that contains a cysteine-rich domain at the N-terminus followed by a fibronectin type II repeat, 8 carbohydrate recognition domains (CRD), a transmembrane spanning region and a short cytoplasmic domain[8]. The point at which our cDNA changed from the published sequence was at the junction between exon 14 and intron 14 which occurs within CRD 4. This CRD has been shown to be essential for the efficient binding and endocytosis of carbohydrate ligand[9].

Further RT-PCR analysis on RNA from DC, using a 5′ primer specific for the mannose receptor and a 3′ primer specific to the novel sequence, revealed the presence of three more variants. Nucleotide sequence analysis of variant 1 revealed that it contained sequence identical to intron 14 of the mannose receptor. Subsequently the entire sequence of intron 14 was determined from a genomic clone which spanned exon 13 to intron 15 of the human mannose receptor gene. The intron was found to be approximately 1 kb long and contained all the novel sequence from the 4 variants, suggesting that the variants may result from alternative RNA splicing (figure 1). In support of this view, the putative 3′ cleavage sites used in all variants were preceded by the consensus sequence for a 3′ (acceptor) splice site which consists of a pyrimidine-rich region followed by the sequence NYAG, where N is any base and Y is a pyrimidine.

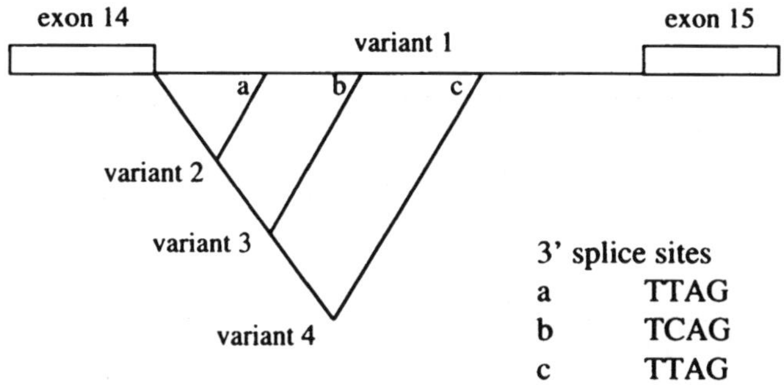

Figure 1. Scheme showing exon 14, intron 14 and exon 15 of the macrophage mannose receptor gene. Variants 2, 3 and 4 all share the 5′ donor splice site from exon 14, but use different 3′ acceptor splice sites at the points a (180 bp from the beginning of intron 14), b (355 bp) and c (710 bp) respectively. Variant 1 contains sequence from intron 14 without deletions, and most likely represents unspliced mRNA.

The macrophage mannose receptor, which has a restricted expression pattern, has recently been reported to be highly expressed by DC, and has been shown to be involved in antigen uptake for presentation to T cells[1]. If the variants that we have identified are translated in DC, they would give rise to truncated proteins due to the presence of stop codons within intron 14. These truncated proteins would lack CRD 4–8 which have been shown to be necessary for the binding of known ligands[9]. They would also lack the transmembrane region, suggesting that they could be secreted. However they would still contain the cysteine-rich domain at the N-terminus, followed by a fibronectin type II repeat and CRD 1–3, which could be involved in the binding of as yet unknown ligands. This would be suggestive of a role for these truncated molecules in the regulation of mannose receptor expression or function. However, several observations suggest that these variants are expressed in DC at very low levels. Firstly, screening of the cDNA library indicated an abundance of 1/100,000 mRNA. Secondly, PCR with two competing 3' primers, one of which bound within intron 14 (to detect the variants) and the other within exon 15 (to detect the mannose receptor), yielded only one product corresponding in size to the mannose receptor. It is, therefore, hard to imagine a functional role for the variants in DC, given their low expression in comparison with the mannose receptor. However, they may be expressed more abundantly at specific points in the DC life cycle which we have not yet analyzed. The significance of the production of these variants by DC therefore remains unclear.

5. CONCLUDING REMARKS

We have used two approaches for the analysis of DC at the genetic level. We have shown that both random sequencing of cDNA libraries from DC and DD-PCR can lead to the identification of DC-specific molecules, and that the two methods complement each other. Although only a limited number of clones have been analyzed so far, there was no overlap between the clones identified by random sequencing with those identified by DD-PCR. While random sequencing of cDNA libraries affords no selection for cell-specific molecules, it avoids the technical difficulties which can be encountered when using DD-PCR, for example the identification of high numbers of false positives. We are currently investigating clones identified using both techniques, and hope that the findings will help to provide a better understanding of DC function.

6. ACKNOWLEDGMENTS

We thank Dr Maureen Taylor for kindly providing the genomic clone of the macrophage mannose receptor, Karin Bacon for help in preparing the cDNA libraries, and Dr. G. Zurawski for advice.

7. REFERENCES

1. Sallusto F., Cella M., Danieli C. and Lanzavecchia A. (1995) Dendritic cells use macropinocytosis and the mannose receptor to concentrate macromolecules in the major histocompatability complex class II compartment: downregulation by cytokines and bacterial products. J. Exp. Med. 182:389–400
2. Sallusto F. and Lanzavecchia A. (1994) Efficient presentation of soluble antigen by cultured human dendritic cells is maintained by granulocyte/macrophage colony-stimulating factor plus interleukin 4 and downregulated by tumor necrosis factor α. J. Exp. Med. 179:1109–1118
3. Romani N., Gruner S., Brang D., Kämpgen E., Lenz A., Trockenbacher B., Konwalinka G., Fritsch P.O., Steinman R.M. and Schuler G. (1994) Proliferating dendritic cell progenitors in human blood. J. Exp. Med. 180:83–93
4. Altschul S.F., Gish W., Miller W., Myers E.W. and Lipman D.J. (1990) Basic local alignment search tool. J. Mol. Biol. 215:403–410
5. Liang P., Averboukh L. and Pardee A.B. (1993) Distribution and cloning of eukayotic mRNAs by means of differential display: refinements and optimization. Nucleic Acids Res. 21:3269–3275
6. O'Doherty U., Steinman R.M., Peng M., Cameron P.U., Gezelter S., Kopeloff I., Swiggard W.J., Pope M. and Bhardwaj N. (1993) Dendritic cells freshly isolated from human blood express CD4 and mature into typical immunostimulatory dendritic cells after culture in monocyte-conditioned medium. J. Exp. Med. 178:1067–1076
7. Guimarães M.J., Lee F., Zlotnik A. and McClanahan T. (1995) Differential display by PCR: novel findings and applications. Nucleic Acids Res. 23:1832–1833
8. Taylor M.E., Conary J.T., Lennartz M.R., Stahl P.D. and Drickamer K. (1990) Primary structure of the mannose receptor contains multiple motifs resembling carbohydrate-recognition domains. J. Biol. Chem. 265:12156–12162
9. Taylor M.E., Bezouska K. and Drickamer K. (1992) Contribution to ligand binding by multiple carbohydrate-recognition domains in the macrophage mannose receptor. J. Biol. Chem. 267:1719–1726

74

SUBTRACTIVE cDNA CLONING

A New Approach to Understanding Dendritic Cell Biology

Kiyoshi Ariizumi, Paul R. Bergstresser, and Akira Takashima

Department of Dermatology
University of Texas Southwestern Medical Center
Dallas, Texas

INTRODUCTION

Dendritic cells (DC) represent a specific subset of leukocytes that play a critical role in the initiation of T cell-mediated immune responses. Recently, we have established, from the epidermis of BALB/c mice, long-term DC lines, termed the "XS series", which resemble resident epidermal DC (i.e., Langerhans cells) in many respects, including their morphology, surface phenotype, antigen presenting capacity, and cytokine and cytokine receptor mRNA profiles (1–3). Taking advantage of these stable lines, we sought to study unique attributes of DC at a molecular level by identifying molecules that they express selectively. For this aim, we employed a subtractive cDNA cloning strategy in which cDNA prepared from the XS52 DC line was subtracted with mRNA isolated from the J774 macrophage line. In this brief communication, we outline our methods as well as the identities of cDNAs isolated in this manner.

MATERIALS AND METHODS

Cells

The XS52 DC line was established from the epidermis of newborn BALB/c mice; its phenotypic and functional characteristics have been described previously (1–3).

Construction of a Subtractive cDNA Library

A subtractive cDNA library was constructed using methods reported by Rubenstein et al. (4). Briefly, poly(A)+ RNAs (5 μg) isolated from XS52 cells were reverse-transcribed into cDNAs, ligated unidirectionally into the lambda ZAP II vector, and then converted into a single-stranded phagemid library (in the anti-sense orientation).

Subsequently, this cDNA library (1.2 μg) was hybridized with biotinylated, poly(A)+ RNA (50 μg) isolated from the J774 macrophage line. Unhybridized cDNAs were purified by the streptavidin-phenol extraction method and converted into a double-stranded form by Taq DNA polymerase as described previously (4).

Library Screening

The subtracted cDNA library was screened sequentially by colony hybridization, slot blotting, and Northern blotting. In colony hybridization, colonies were transferred onto nylon membranes and hybridized with total cDNA probes prepared from XS52 DC poly (A)+ RNA and with probes from J774 macrophage poly (A)+ RNA; we selected only those colonies that showed preferential hybridization with XS52 probes (but not with J774 probes). In slot blotting, the cDNA inserts were PCR-amplified, slot-blotted onto nylon membranes, and hybridized with XS52 probes and with J774 probes; once again, we selected the clones that were hybridized preferential with XS52 cDNA probes. In Northern blotting, cDNA inserts were excised by enzymatic digestion and then ^{32}P-labeled, and total RNAs (10 μg) isolated from XS52 DC line, as well as the J774 macrophage, 7–17 dendritic epidermal T cell (DETC), and Pam 212 keratinocyte lines were hybridized with these probes. We selected the cDNA clones that were expressed selectively by XS52 cells.

DNA Sequence Analyses

The selected cDNA clones were sequenced using an automated DNA sequencer and then analyzed for homologies the nucleotide sequences that were registered in the EMBL 46 and GenBank 95 nucleic acid database.

RESULTS

Summary of Construction and Screening of Subtracted cDNA Library

Through subtraction of XS52 DC-derived cDNAs with mRNA isolated from J774 macrophages, we sought to eliminate the abundant cDNA sequences (e.g., house keeping genes) that are expressed in common by both cell types. In fact, we did remove over 95% of the starting cDNA clones in this procedure, as estimated by counting colony forming units before and after subtraction.

Approximately 12,000 independent clones of this subtracted cDNA library were subjected to three rounds of screening. In the first screening with colony hybridization, about 50% of the clones were clearly detectable with total cDNA probes prepared from XS52 cells; these clones most likely represented genes that were expressed at relatively high levels. Most of these clones, however, were also detectable with J774 cDNA probes, indicating that our "subtracted" cDNA library still contained relatively large numbers of cDNAs that were expressed in common by DC and macrophages. On the other hand, 226 clones were detectable only with XS52 probes, but not with the J774 probes (Fig 1). We selected the clones showing this differential hybridization profile. When these clones were examined by slot blotting, more than 50% (140/226 clones) showed preferential hybridization with XS52 probes (Fig 2). These 140 clones were then tested in Northern blotting; we selected 50 clones that were expressed preferentially by XS52 DC and at detectable levels (Fig 3).

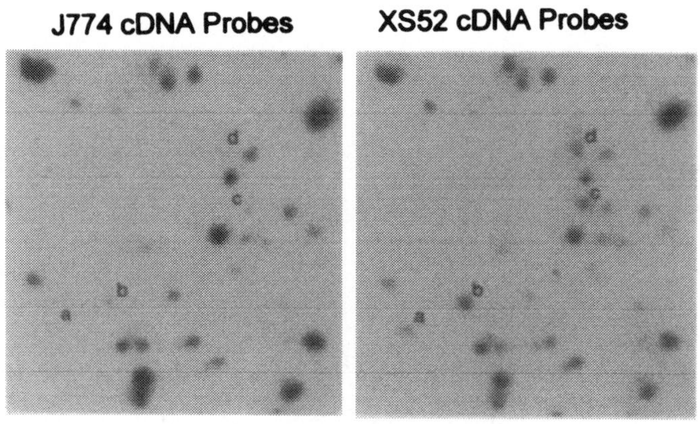

Figure 1. First screening (colony hybridization): Twelve thousand colonies in the subtracted cDNA library were hybridized with total cDNA probes prepared from J774 macrophages (left) or XS52 DC (right). We selected only those clones (indicated with *a* through *d*) that hybridized preferentially with XS52 probes.

Identities of cDNA Clones Isolated by Subtractive Cloning

Through DNA sequencing and cross-hybridization, we found that our cDNA pool (50 independent clones) selected from the three rounds of screening contained 11 distinct genes. Homology search of their sequences revealed 6 genes that encode currently recognized polypeptides in mice, including C10 (a β-chemokine) (5), IL-1β (6), cathepsin C (a cysteine protease) (7), spermidine/spermine N1-acetyltransferase (SSAT) (8), and A1 (a

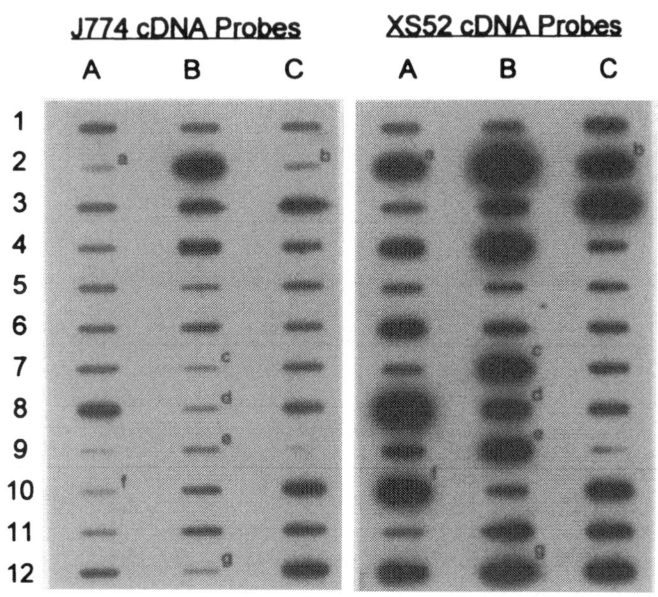

Figure 2. Second screening (slot-blotting): cDNA clones selected in the first screening were slot-blotted and hybridized with J774 macrophage probes (left) or XS52 DC probes (right). We selected 140 of the cDNA clones (indicated with *a* through *g*) that hybridized preferentially with XS52 probes.

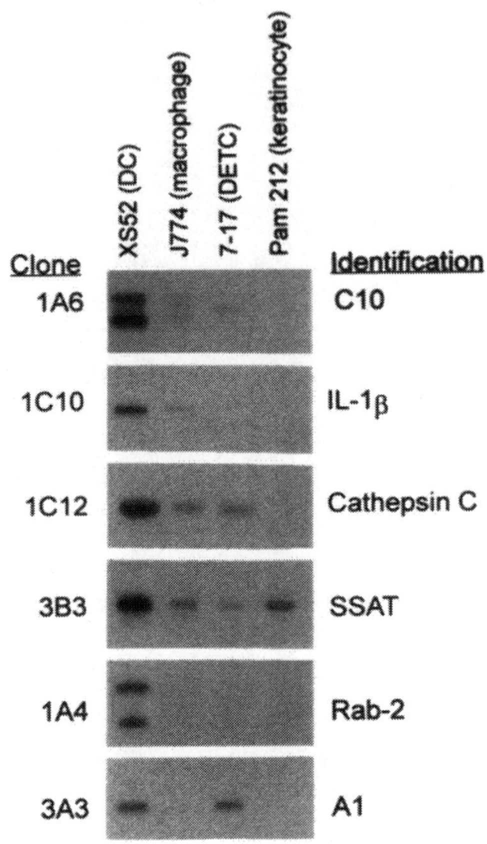

Figure 3. Third screening (Northern blotting): Data shown represent Northern blot analyses in the third screening. Note that all six genes are expressed most abundantly by the XS52 DC among the tested cell lines.

hemopoietic cell-specific early response gene) (9). We have also identified a gene that encodes a mouse equivalent of rat rab-2 (a ras-related protein) (10). Importantly, XS52 cells expressed all six genes constitutively and at relatively high levels. Moreover, these genes were expressed in substantially lower levels by the J774 macrophage line, the line used for subtraction, and they were minimally expressed or undetectable in Pam 212 keratinocytes and 7–17 DETC (Fig 3). The remaining 5 genes showed no significant identity to nucleotides currently registered in the EMBL or GenBank data base. These results validate that a subtractive cDNA cloning strategy is applicable to identify genes that are expressed preferentially by DC.

DISCUSSION

Recent establishment of short-term and long-term DC lines has made it feasible to apply modern technologies in molecular biology to study the biology of DC. In the present study, we demonstrate that a subtractive cDNA cloning strategy can be used to identify genes that are expressed preferentially by DC. In this regard, other methodologies have also been used to identify genes that are expressed in a tissue- or cell type-specific manner. These include differential hybridization (screening of a cDNA library with total or subtracted cDNA probes), PCR-based differential display, and PCR-based construction of subtractive cDNA libraries. Each methodology has advantages (and disadvantages). For example, differential hybridization is ideal for identifying genes that are expressed abundantly in a given tissue or cell type; differential display can be conducted without construction of cDNA libraries; and PCR-based subtractive cDNA cloning requires relative small amounts of mRNA. The subtractive cDNA cloning strategy, which we employed in this study, has been used successfully to identify several immunologically relevant molecules, including T cell receptors (11), CD4 (12), and CD8 (13). A major technical limitation of this method is that we can only identify cDNA clones that are expressed at relatively high levels. In fact, one half of our cDNA clones were discarded in the first screening without further analysis, simply because they were not detectable with total cDNA probes from XS52 cells. Such discarded clones may have contained one or more cDNAs that are expressed in a DC-specific manner, but at relatively low levels, and that encode polypeptides playing important roles in DC function. Nevertheless, the present study validates the feasibility of our approach in identifying DC-associated molecules by subtrative cDNA cloning. It also verifies that our XS52 cell line is clearly distinguishable from macrophages by the preferential expression of selected genes.

Our observations that XS52 DC express predominantly IL-1β, C10, and cathepsin C mRNA are consistent with previous reports that: a) Langerhans cells express IL-1β mRNA constitutively (14–16), b) Langerhans cells express mRNA for several chemokines (e.g., macrophage inflammatory protein-1α and 1γ) (14,16,17), and c) DC are capable of digesting complex protein antigens, a process that requires several proteases (18). On the other hand, the functional relevance of mRNA expression of SSAT, rab-2, and A1 remains unknown. It is tempting to speculate that the respective products of these genes play roles in the development, maturation, and/or function of DC. In addition to these six genes, we have identified five novel genes that are expressed preferentially by XS52 DC. We believe that further analysis of these genes will provide a new insight into the biology of DC.

ACKNOWLEDGMENT

This study was supported by National Institutes of Health grants RO1 AR44189, R01 AR35068, R01 AR40042, R01 AR43777, AND 1P30 AR41940.

REFERENCES

1. Xu S, Ariizumi K, Caceres-Dittmar G, Edelbaum D, Hashimoto K, Bergstresser PR, Takashima A: Successive generation of antigen-presenting, dendritic cell lines from murine epidermis. J Immunol 154:2697–2705, 1995.
2. Xu S, Ariizumi K, Edelbaum D, Bergstresser PR, Takashima A: Cytokine-dependent regulation of growth and maturation in murine epidermal dendritic cell lines. Eur J Immunol 25:1018–1024, 1995.
3. Takashima A, Edelbaum D, Kitajima T, Shadduck RK, Gilmore GL, Xu S, Taylor RS, Bergstresser PR, Ariizumi K: Colony-stimulating factor-1 secreted by fibroblasts promotes the growth of dendritic cell lines (XS series) derived from murine epidermis. J Immunol 154:5128–5135, 1995.
4. Rubenstein JLR, Brice AEJ, Ciaranello RD, Denney D, Porteus MH, Usdin TB: Subtractive hybridization system using single-stranded phagemids with directional inserts. Nucleic Acids Research 18:4833–4842, 1990.
5. Orlofsky, Berger MS, Prystowsky MB: Novel expression pattern of a new member of the MIP-1 family of cytokine-like genes. Cell Regulation 2:403–412, 1991.

6. Gray PW, Glaister D, Chen E, Goeddel DV, Pennica D: Two interleukin 1 genes in the mouse: Cloning and expression of the cDNA for murine interleukin 1beta. J Immunol 137:3644–3648, 1986.
7. Ishidoh K, Muno D, Sato N, Kominami E: Molecular cloning of cDNA for rat Cathepsin C. J Biol Chem 266:16312–16317, 1991.
8. Fogel-Petrovic M, Kramer DL, Ganis B, Casero RA, Jr., Porter CW: Cloning and sequence analysis of the gene and cDNA encoding mouse spermidine/spermine N^1-acetyltransferase - a gene uniquely regulated by polyamines and their analogs. Biochimica et Biophysica Acta 1216:255–264, 1993.
9. Lin EY, Orlofsky A, Berger MS, Prystowsky MB: Characterization of A1, a novel hemopoietic-specific early-response gene with sequence similarity to bcl-2[1]. J Immunol 151:1979–1988, 1993.
10. Touchot N, Chardin P, Tavitian A: Four additional members of the *ras* gene superfamily isolated by an oligonucleotide strategy: Molecular cloning of YPT-related cDNAs from a rat brain library. Proc Natl Acad Sci USA 84:8210–8214, 1987.
11. Hedrick SM, Nielsen EA, Kavaler J, Cohen DI, Davis MM: Sequence relationships between putative T-cell receptor polypeptides and immunoglobulins. Nature 308:153–158, 1984.
12. Maddon PJ, Littmen DR, Godfrey M, Maddon DE, Chess L, Axel R: The isolation and nucleotide sequence of a cDNA encoding the T cell surface protein T4: a new membrane of the immunoglobulin gene family. Cell 42:93–104, 1985.
13. Kavathas P, Sukhatme VP, Herzenberg LA, Parnes JR: Isolation of the gene encoding the human T-lymphocyte differentiation antigen Leu-2 (T8) by gene transfer and cDNA subtraction. Proc Natl Acad Sci U S A 81:7688–7692, 1984.
14. Matsue H, Cruz PD, Jr., Bergstresser PR, Takashima A: "Langerhans cells are the major source of mRNA for IL-1β and MIP-1α̧ among unstimulated mouse epidermal cells 192 3, 8 192." 1992. (UnPub)
15. Enk AH, Katz SI: Early molecular events in the induction phase of contact sensitivity. Proc Natl Acad Sci USA 89:1398–1402, 1992.
16. Heufler C, Topar G, Koch F, Trockenbacher B, Kampgen E, Romani N, Schuler G: Cytokine gene expression in murine epidermal cell suspensions: interleukin 1 beta and macrophage inflammatory protein 1 alpha are selectively expressed in Langerhans cells but are differentially regulated in culture. J Exp Med 176:1221–1226, 1992.
17. Mohamadzadeh M, Poltorak AN, Bergstresser PR, Beutler B, Takashima A: Dendritic cells produce macrophage inflammatory protein-1 gamma, a new member of the CC chemokine family. J Immunol 156:3102–3106, 1996.
18. Romani N, Koide S, Crowley M, Witmer-Pack M, Livingstone AM, Fathman CG, Inaba K, Steinman RM: Presentation of exogenous protein antigens by dendritic cells to T cell clones. J Exp Med 169:1169–1178, 1989.

75

INCREASED LEVEL OF PHOSPHOTYROSINE IN HUMAN DENDRITIC CELLS UNDER STIMULATION WITH CONTACT SENSITIZERS BUT NOT IRRITANTS

D. Becker, J. Willemsen, U. Kühn, R. van Brandwijk, and J. Knop

Department of Dermatology
Johannes Gutenberg-University
Mainz, Germany

1. INTRODUCTION

In the last years evidence was presented for the activation of dendritic cells (DC) under stimulation with contact sensitizers. Most data were obtained for murine Langerhans cells (LC) whereas in man blood-derived dendritic cells were found to be a more suitable model to study the mechanism of DC activation by haptens. The first observation was the upregulation of MHC class II molecules on murine Langerhans cells in vivo[1] followed by their migration from the epidermis into regional lymph nodes[2]. Very early events during this activation include the upregulation of IL-1ß in murine LC[3] as well as the endocytotic activation of this cell type[4]. Based on the last observation attempts were made to use this mechanism for predictive in vitro testing of contact sensitizers employing murine LC,[5,6] as well as human dendritic cells.[7]

To uncover the molecular details of this hapten-mediated activation we studied the phosphorylation of tyrosine residues, a central event of signal transduction pathways.[8,9] Up to now only few data on the expression and function of the regulating enzymes of tyrosine phosphorylation, protein tyrosine kinases (PTK) and protein tyrosine phosphatases (PTP) have been presented for DC[10,11,12], but nothing is known on the regulation of tyrosine phosphorylation under stimulation with contact sensitizers.

2. RESULTS AND DISCUSSION

2.1. Induction of Tyrosine Phosphorylation by Stimulation with Hapten

Mononuclear cells from buffy coats of healthy blood donors were stimulated with the strong contact sensitizer 5-chloro-2-methyl-4-isothiazolin-3-one (MI/MCI). The

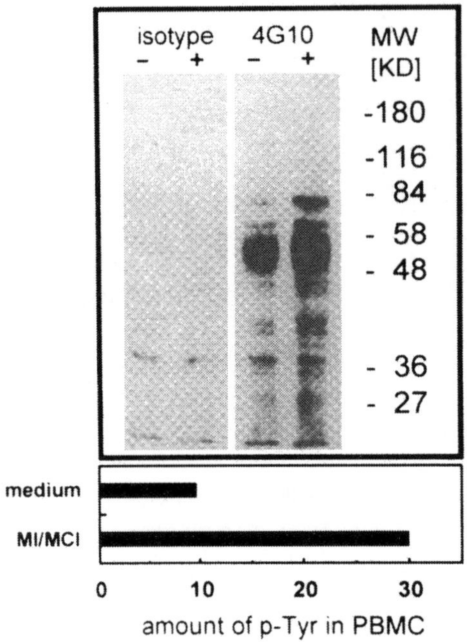

Figure 1. PBMC from healthy blood donors were stimulated with (+) or without (-) 0.4 µg/ml MI/MCI for 15 min. at 37°C, solubilised in the presence of Triton X 100 and several inhibitors for proteinases and phosphatases. Cell lysates obtained from each 10^7 PBMC were subjected to SDS-PAGE followed by detection of p-Tyr containing proteins using a Western blot technique and the p-Tyr specific antibody 4G10 or isotype antibody as well as ECL reagents. The amount of p-Tyr in PBMC as detected by flow cytometric analysis is shown in the lower panel as fluorescence intensity of p-Tyr-bound specific antibody.

amount of phosphorylated tyrosine residues (p-Tyr) was quantified by flow cytometry using a specific antibody after permeabilization of cell membranes with the detergent saponin. The capacity of this approach to detect changes in the phosphorylation of tyrosine-proteins was confirmed by use of a Western blot technique. Following stimulation with MI/MCI, p-Tyr was found to be upregulated using both techniques (Fig.1).

2.2. Analysis for the Regulation of Tyrosine Phosphorylation in DC under Stimulation with Hapten

Using a multi-color flow cytometric technique selective analysis of human DC in mononuclear cell preparations for their total content of p-Tyr was possible. The basal level of p-Tyr was quite low but a rapid and significant increase became apparent during the first 15 minutes of stimulation with MI/MCI (Fig.2A).

This hapten-mediated effect was not detectable by stimulation in the cold and required viable cells with intact cell membranes (Fig.2B). When DC were stimulated under optimal conditions followed by permeabilization and further incubation with ATP in a buffer used for tyrosine kinase assays[13] a more pronounced increase of p-Tyr was demonstrated for MI/MCI-stimulated cells in comparison with the medium control (Fig.2B). The potent PTK inhibitor tyrphostin B46 was able to block this difference and the ATP competitor genistein abolished any further phosphorylation at all (Fig. 2B). To exclude a direct inhibitory effect of MI/MCI on PTP, untreated cells were subjected to in vitro phosphorylation in the presence of this hapten or the potent PTP inhibitor sodium vanadate. Whereas the later one resulted in excessive formation of p-Tyr, MI/MCI had no effect (Fig. 2B). These data suggest an active role of PTK for the increased tyrosine phosphorylation under stimulation with a contact sensitizer rather than a passive inhibition of PTP.

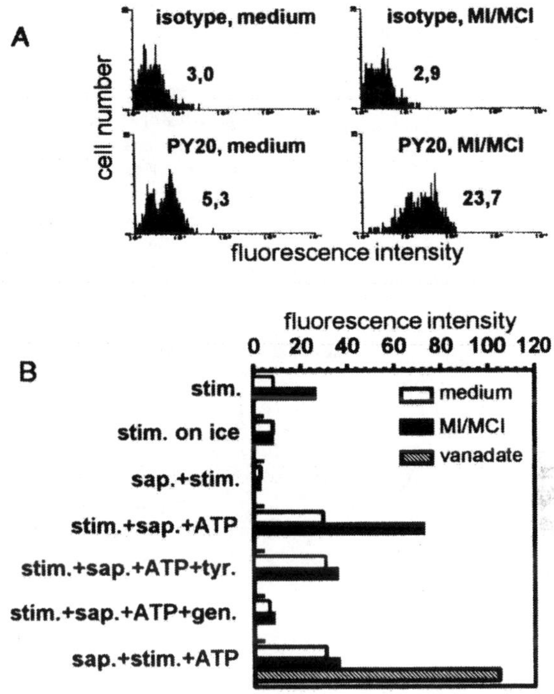

Figure 2. Selective flow cytometric analysis of DC for the intracellular amount of p-Tyr following stimulation with MI/MCI. Examples for DC populations under stimulation with or without MI/MCI are shown (A). In (B) the absolute fluorescence intensities of p-Tyr-bound antibodies are depicted. The following experiments were performed: stim.: stimulation of viable cells at 37°C; stim. on ice: stimulation of viable cells on ice; sap+stim.: permeabilisation of cells with saponin before stimulation; stim.+sap.+ATP: further incubation of stimulated cells after permeabilisation with saponin in the presence of ATP; tyr.: in vitro phosphorylation in the presence of 100 µM tyrphostin B46, gen.: in vitro phosphorylation in the presence of 100 µg/ml genistein; sap.+stim.+ATP: stimulation with MI/MCI or sodium vanadate in the presence of ATP after prior permeabilisation with saponin. The results of a typical experiment are shown.

2.3. Selective Induction of Tyrosin Phosphorylation by Haptens But Not Irritants

The suitability of the p-Tyr induction under stimulation with small chemicals as parameter for predictive testing of contact sensitizers was studied by testing subtoxic concentrations of water soluble conservatives, irritants and strong experimental contact sensitizers. Whereas compounds with significant risk to induce contact hypersensitivity in animal models and man (MI/MCI, dibromodicyanobutane, Bronopol, formaldehyde, thimerosal, 2,4,6-trinitroclorobenzene, 2,4-dinitrofluorobenzene) were able to induce vigorous tyrosine phosphorylation, neither the very weak sensitizers phenoxyethanol and methyl paraben nor the irritants benzoic acid, benzalkonium chloride and sodium lauryl sulphate were able to elicit a response (Fig.3).

In aggregate the activation of human DC by contact sensitizers includes augmented phosphorylation of tyrosine residues. It strongly depends on the activity of PTK although the details of this process need to be uncovered. A possible mechanism might be an increased activity of PTK on the molecular level by conformational changes and phosphorylation at regulatory sites as demonstrated for PTK of the Src-family[14]. Alternatively recruitment of PTK into complexes with phosphorylation sites for PTK, as discussed for certain signal transduction pathway in lymphocytes[15], might be responsible. Although a direct inhibition of PTP by haptens was not demonstrated in our experiments a translocation of regulatory PTP, induced by haptens in the viable cell only and leading to a local imbalance between PTK and PTP, could not be ruled out with certainty.

Irrespective of its further molecular characterization this hapten-specific mechanism should be very useful for predictive *in vitro* testing of new chemical compounds.

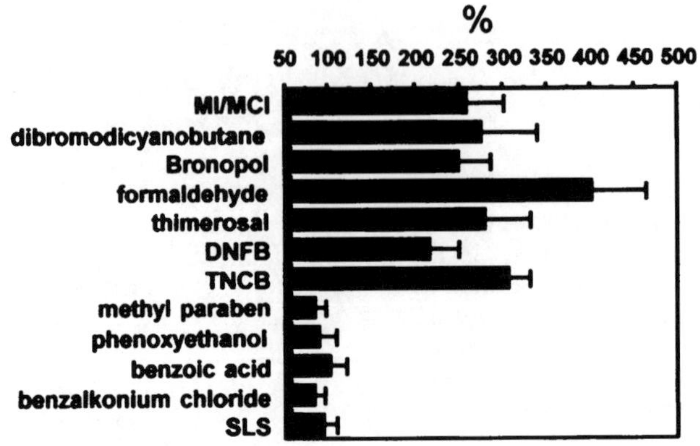

Figure 3. Reactivity of strong experimental contact sensitizers (8 μg/ml DNFB, 0.4 μg/ml TNCB), water soluble preservatives with known sensitizing capacity (0.4 μg/ml MI/MCI, 2 μg/ml dibromodicyanobutane, 5 μg/ml Bronopol, 50 μg/ml formaldehyde, 1μg/ml thimerosal) and weak sensitizers or irritants (50 μg/ml sodium lauryl sulphate, 2 mg/ml benzoic acid, 1 μg/ml benzalkonium chloride, 2 mg/ml phenoxyethanol, 1 mg/ml methyl paraben). All compounds were tested in the highest but still subtoxic concentration, the fluorescence intensity of p-Tyr-bound antibody is shown in percent of the medium control. The mean and SD of at least four experiments are shown.

3. ACKNOWLEDGMENTS

This work was supported by a grant from the Deutsche Forschungsgemeinschaft (Kn 120/6).

4. REFERENCES

1. S. Aiba, and S.I. Katz, Phenotypic and functional characteristics of in vivo-activated Langerhans cells, J. Immunol. 145:2791 (1990)
2. M.L. Kripke, C.G. Munn, A. Jeevan, J.M. Tang, and C. Bucana, Evidence that cutaneous antigen-presenting cells migrate to regional lymph nodes during contact sensitization, J. Immunol. 145:2833 (1990)
3. A.H. Enk, V.L. Angeloni, M.C. Udey, and S.I. Katz, An essential role for Langerhans cell-derived IL-1 beta in the initiation of primary immune responses in skin, J. Immunol. 150:3698 (1993)
4. D. Becker, U. Neiss, S. Neis, K. Reske, and J. Knop, Contact allergens modulate the expression of MHC class II molecules on murine epidermal Langerhans cells by endocytotic mechanisms, J. Invest. Dermatol. 98:700 (1992)
5. D. Becker, G. Kolde, K. Reske, and J. Knop, An in vitro test for endocytotic activation of murine epidermal Langerhans cells under the influence of contact allergens, J. Immunol. Methods. 169:195 (1994)
6. U. Lempertz, U. Kühn, J. Knop, and D. Becker, An approach to predictive testing of contact sensitizers *in vitro* by monitoring their influence on endocytotic mechanisms., Int. Arch. Allergy Immunol. 111:64 (1996)
7. D. Becker, U. Kühn, U. Lempertz, A. Enk, J. Saloga, and J. Knop, Flow-cytometric screening for the modulation of receptor-mediated endocytosis in human dendritic cells: Implications for the development on an *in vitro* technique for predictive testing of contact sensitizers., J. Immunol. Method in press
8. R.M. Perlmutter, S.D. Levin, M.W. Appleby, S.J. Anderson, and J. Alberola Ila, Regulation of lymphocyte function by protein phosphorylation, Annu. Rev. Immunol. 11:451 (1993)
9. H. Sun, and N.K. Tonks, The coordinated action of protein tyrosine phosphatases and kinases in cell signaling, Trends Biochem. Sci. 19:480 (1994)
10. T. Bieber, M. Jurgens, A. Wollenberg, E. Sander, D. Hanau, and H. de la Salle, Characterization of the protein tyrosine phosphatase CD45 on human epidermal Langerhans cells, Eur. J. Immunol. 25:317 (1995)
11. I.D.Haidl, D.H. Ng, S. Rothenberger, P. Johnson, and W.A. Jefferies, Detection of restricted isoform expression and tyrosine phosphatase activity of CD45 in murine dendritic cells, Eur. J. Immunol. 25:3370 (1995)

12. M. Jurgens, A. Wollenberg, D. Hanau, H. de la Salle, and T. Bieber, Activation of human epidermal Langerhans cells by engagement of the high affinity receptor for IgE, Fc epsilon RI, J. Immunol. 155:5184 (1995)
13. A.L Burkhardt, and J.B. Bolen, Immune-complex assays for tyrosine protein kinases, in: Current Protocols in Immunology, ed.: J.E. Coligan et al., John Wiley & Sons, Inc., Volume 3, 11.4.1–11.4.9 (1993)
14. J.A. Cooper, and B. Howell, The when and how of Src regulation, Cell 73:1051 (1993)
15. B.M. Sefton B.M., and J.A. Taddie, Role of tyrosine kinases in lymphocyte activation, Current Opinion in Immunology, 6:372 (1994)

INSULIN-LIKE GROWTH FACTOR TYPE I (IGF-I) SUPPORTS GROWTH OF V-relER DENDRITIC CELL PROGENITORS

Jaime Madruga, Nicolás Koritschoner, Gabi Stengl, Signe Knespel, Petr Bartunek, and Martin Zenke

Max-Delbrück-Center for Molecular Medicine
MDC
Robert-Rössle-Str. 10
D-13122 Berlin, Germany

INTRODUCTION

Over the past few years, several protocols have been established to differentiate antigen-presenting dendritic cells (DC) from DC progenitors grown in vitro (1–4, see also review of Peters et al., 5). While these procedures yield fully competent DC in large numbers, there is also some heterogenity in the cultures obtained since the cell populations are not clonal. Several DC lines were also established (6–10) which however in most instances recapitulate only part of the DC specific phenotype. Recently we have developed an in vitro differentiation system for DC which is based on the conditional hormone-inducible V-relER oncogene (11, 12). V-rel represents a retrovirus-transduced version of c-rel and is a member of the NF-kB/rel transcription factor family. In the chimeric V-relER gene, V-rel is fused to the hormone-binding domain of the human estrogen receptor, thus making the transforming capacity of V-rel hormone-inducible. Accordingly V-relER is capable of transforming chicken bone marrow cells in the presence of estrogen while it is inactive in the absence of hormone. Most importantly, V-relER transformed bone marrow cells can be grown as clonal homogenous cell populations to large cell numbers and induced to differentiate into DC after experimentally "switching off" V-relER activity by removing estrogen or by adding the estrogen antagonist ICI 164,384 (12, Fig. 1).

The gene expression repertoire of DC progenitor cells versus fully differentiated DC can now be assessed by employing differential cDNA cloning strategies. As a first step in this direction we have determined tyrosine kinase expresssion in V-relER cells using a set of degenerate primers and PCR amplification (modified from 13). We reasoned that identification of appropriate receptor tyrosine kinases would lead us to their cognate ligands which, when applied to DC, might provide some growth promoting activity. One of the cDNAs identified in the course of this study encodes the type-I insulin-like growth factor

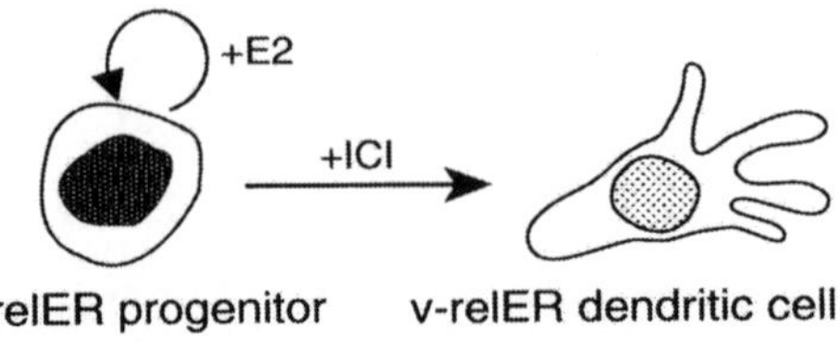

Figure 1. The V-relER differentiation system for DC. The conditional hormone-inducible V-relER oncoprotein transforms progenitor cells present in chicken bone marrow. V-relER cells proliferate in the presence of estrogen (+E2). Addition of an estrogen antagonist ICI 164,384 allows to experimentally "switch off" V-relER activity and induces progenitors to differentiate into fully competent antigen-presenting dendritic cells (+ ICI).

receptor IGFR. Additionally, we show that IGFR is functional in V-relER cells since the administration of IGF-I ligand supports sustained growth of V-relER progenitors in vitro.

MATERIALS AND METHODS

Cell Culture

Cell culture and differentiation conditions for V-relER cells are described (11, 12). In proliferation assays V-relER clones 22 and 25 were used. Recombinant human IGF-I was obtained from: IGF-I (A) Promega, Madison, WI, USA (#G5111), IGF-I (B) R&D Research, Minneapolis, MN, USA (#291-G1) and a Long R^3 IGF-I recombinant analog from Sigma, St. Louis, MO, USA (#I-1271); IGF-I was used at 10ng/ml. In ^{3}H-thymidine incorporation assays V-relER cells were seeded at 7×10^5 cells per well in microtiter plates and ^{3}H-thymidine incorporation was measured in triplicates one day after addition of IGF-I (14).

Cloning of Tyrosine Kinase-Specific cDNA

Total RNA from V-relER cells was prepared as previously described (15). First strand cDNA was synthesized from 2μg of total RNA employing MMLV reverse transcriptase and random hexamers using conditions recommended by the manufacturer (Promega, Madison, WI, USA). For PCR amplification a set of degenerate primers for the region encoding the catalytic domain of tyrosine kinases was designed according to 13. First strand cDNA were subjected to 35 cycles of PCR amplification using the following cycle profile: 94°C for 30 sec, 42°C for 60 sec and 63°C for 60 sec. PCR products were purified from agarose gels, reamplified under the same conditions as above and cloned into pCRII vector (Invitrogen, San Diego, CA, USA). Selected clones were sequenced by an ABI DNA sequencer.

RESULTS

To analyse tyrosine kinase gene expression, V-relER cells were grown in the presence of estrogen or induced to differentiate into V-relER DC by ICI 164,384 treatment (3 days); RNA was prepared and used for cDNA synthesis. Degenerate tyrosine kinase-specific primers (modified from 13) and PCR amplification were used to generate a pool of 203 bp PCR fragments which were cloned in pCRII plasmid vector. The identity of indi-

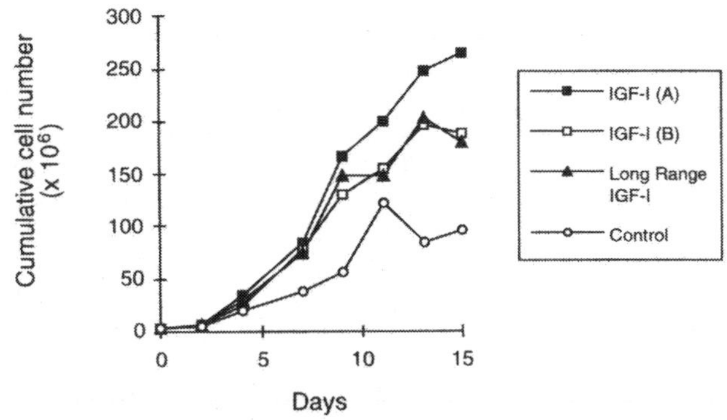

Figure 2. IGF-I stimulates proliferation of V-relER progenitors. V-relER cells were grown in standard culture medium in the presence of estrogen and IGF-I (A), IGF-I (B) or Long Range IGF-I as shown. Cells numbers were determined at the time points indicated in the graph. Cell density was readjusted to 3.5×10^6 cells/ml by adding fresh medium containing IGF-I (10ng/ml). Control: no factor added.

vidual tyrosine kinase-specific PCR products was determined by DNA sequencing. By this approach we were able to identify a number of receptor and non-receptor tyrosine kinases expressed in V-relER cells. One of the highly expressed genes was the type-I insulin-like growth factor (IGF-I) receptor IGFR, a member of the insulin receptor family. IGFR was expressed in both V-relER progenitors and differentiated progeny (DC) at similar levels (16).

To determine the impact of IGF-I on proliferation of V-relER progenitors, cells were grown in the presence and absence of IGF-I using standard growth conditions. Three different sources of recombinant human IGF-I were used: IGF-I (A), IGF-I (B) and a modified long range IGF-I (see Materials and Methods). Cumulative cell numbers were determined in regular time intervals. Fig. 2 shows that IGF-I efficiently enhanced V-relER cell proliferation with recombinant human IGF-I (A) being the most effective. IGF-I (B) and long-range IGF-I exhibited a comparable but lesser activity in this assay.

To extend these results we measured the effect of IGF-I on DNA synthesis in V-relER progenitors by ^{3}H-thymidine incorporation. Recombinant human IGF-I (A) and (B), and long range IGF-I were tested, now also employing different serum concentrations (0.5% and 2.5%, and no serum). Figure 3 shows that all of the IGF-I samples tested enhanced V-relER cell proliferation under serum-free conditions and at low serum concentrations (0.5%) as reflected by an increase in ^{3}H-thymidine incorporation, when compared to controls without addition of factor. In contrast, the effect of IGF-I was lost at higher serum concentrations (2.5%), presumably due to the presence of an IGF-I-like activity in serum. Thus, IGF-I effectively supported V-relER cell proliferation in two assays: first by measuring cumulative cell numbers and second by determining the rate of DNA synthesis in ^{3}H-thymidine incorporation experiments.

DISCUSSION

In the present study we attempted to identify cell surface receptors that, when activated by ligand, support growth of DC progenitors in vitro and, thus, might qualify for Dendritic Cell Growth Factor (DC-GF) receptor. We argued that ligands of receptor tyrosine kinases might represent a particular rich source for such growth factor activities. Therefore we have determined tyrosine kinase gene expression in DC using our V-relER differentiation system. Among various receptor tyrosine kinases identified, one of them, the type I insulin-like growth factor receptor IGFR, supported sustained growth of V-relER DC progenitors in culture when activated by IGF-I ligand. We consider this an im-

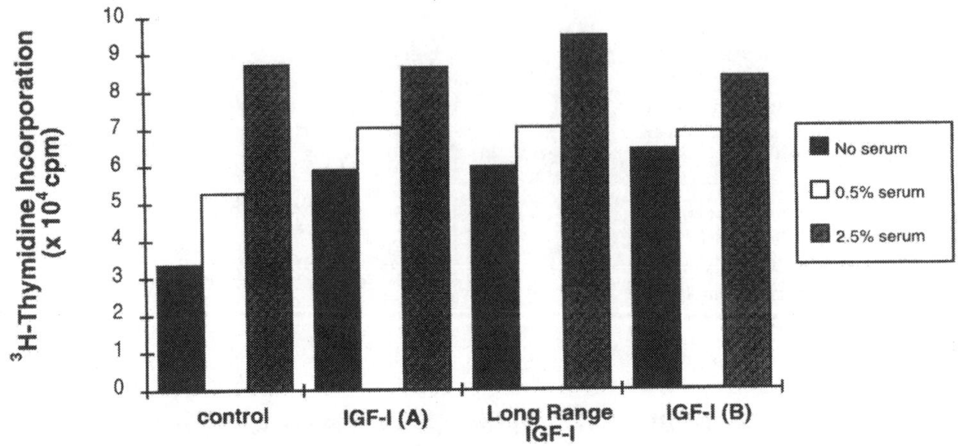

Figure 3. IGF-I enhances the rate of DNA synthesis in V-relER progenitor cells. ^{3}H-Thymidine incorporation was measured for V-relER cells grown under different serum concentrations: No Serum, 0.5% FCS and 2.5% serum (2% FCS + 0.5% chicken serum; see Materials and Methods). Data shown represent the average of triplicate values (7×10^5 cells per microtiter plate well). Control: no factor added.

portant finding, however emphasize that other growth factor receptors might exist, which might also be more restricted in their activity to specific DC progenitor subtypes. In line with this idea, activation of the c-kit/stem cell factor (SCF) receptor expands the DC progenitor pool present in CD34+ precursors from human bone marrow or cord blood (17–21). Additionally, different growth factor activities might have to act in concert to support optimal growth of DC precursors, as observed for SCF plus GM-CSF and TNFα.

It is also clear that there is a growing need to identify protein factors with DC-GF activity to support expansion of DC progenitor populations in vitro under synthetic, well-defined and serum-free conditions. This is because the potential of DC as antigen-presenting cells for tumour-specific epitopes in immunotherapy of cancers is currently being assessed. Such studies require growth and in vitro differentiation conditions in the absence of heterologous or animal sera (like eg fetal calf serum) and then autologous sera become a limiting factor. The identification of IGF-I as a protein factor that supports growth of DC precursors in vitro (at least in the V-relER system) should allow some of these limitations to be overcome.

IGFR expression was detected in both V-relER progenitors and differentiated V-relER DC containing an active and inactive V-relER oncoprotein, respectively. Thus, IGFR expression appears not to be subject to regulation by V-relER. Additionally, the presence of IGFR in V-relER DC suggests that IGFR might have a function also in fully differentiated DC, eg by enhancing their survival. Some initial experiments would be in line with such an idea.

ACKNOWLEDGMENTS

We thank Dr. A. E. Wakeling for ICI 164,384, Irene Gallagher for typing the manuscript and our colleagues for many stimulating discussions. This work was in part supported by the Institute of Molecular Pathology (IMP), Vienna, Austria when JM, NK, PB and MZ were located at IMP and in part by an Austrian Academic Exchange Service (ÖAD) fellowship granted to NK.

REFERENCES

1. Caux, C., C. Dezutter Dambuyant, D. Schmitt, and J. Banchereau. GM-CSF and TNF-alpha cooperate in the generation of dendritic Langerhans cells. *Nature* 360: 258–261 (1992).
2. Romani, N., S. Gruner, D. Brang, E. Kampgen, A. Lenz, B. Trockenbacher, G. Konwalinka, P.O. Fritsch, R.M. Steinman, and G. Schuler. Proliferating dendritic cell progenitors in human blood. *J. Exp. Med.* 180: 83–93 (1994)
3. Sallusto, F., and A. Lanzavecchia. Efficient presentation of soluble antigen by cultured human dendritic cells is maintained by granulocyte/macrophage colony-stimulating factor plus interleukin 4 and downregulated by tumor necrosis factor alpha. *J. Exp. Med.* 179: 1109–1118 (1994)
4. Strunk, D., K. Rappersberger, C. Egger, H. Strobl, E. Kromer, A. Elbe, D. Maurer, and G. Stingl. Generation of human dendritic cells/Langerhans cells from circulating CD34+ hematopoietic progenitor cells. *Blood* 87: 1292–302 (1996)
5. Peters, J.H., R. Gieseler, B. Thiele, and F. Steinbach. Dendritic cells: From ontogenetic orphans to myelomonocytic descendants. *Immun. Today* 17: 273–278 (1996)
6. Paglia, P., G. Girolomoni, F. Robbiati, F. Granucci, and P. Ricciardi Castagnoli. Immortalized dendritic cell line fully competent in antigen presentation initiates primary T cell responses in vivo. *J. Exp. Med.* 178: 1893–1901 (1993)
7. Elbe, A., S. Schleischitz, D. Strunk, and G. Stingl. Fetal skin-derived MHC class I+, MHC class II- dendritic cells stimulate MHC class I-restricted responses of unprimed CD8+ T cells. *J. Immunol.* 153: 2878–2889 (1994)
8. Girolomoni, G., M.B. Lutz, S. Pastore, C.U. Assmann, A. Cavani, and P. Ricciardi Castagnoli. Establishment of a cell line with features of early dendritic cell precursors from fetal mouse skin. *Eur. J. Immunol.* 25: 2163–2169 (1995)
9. Lutz, M.B., C.U. Assmann, G. Girolomoni, and P. Ricciardi Castagnoli. . Different cytokines regulate antigen uptake and presentation of a precursor dendritic cell line. *Eur. J. Immunol.* 26: 586–594 (1996)
10. Volkmann, A., J. Neefjes, and B. Stockinger. A conditional immortalized dendritic cell line which differentiates in contact with T cells or T cell-derived cytokines. In press.
11. Boehmelt, G., A. Walker, N. Kabrun, G. Mellitzer, H. Beug, M. Zenke, and P.J. Enrietto. Hormone-regulated v-rel estrogen receptor fusion protein: reversible induction of cell transformation and cellular gene expression. *EMBO J.* 11: 4641–52 (1992)
12. Boehmelt, G., J. Madruga, P. Dorfler, K. Briegel, H. Schwarz, P.J. Enrietto, and M. Zenke. Dendritic cell progenitor is transformed by a conditional v-Rel estrogen receptor fusion protein v-RelER. *Cell* 80: 341–352 (1995)
13. Wilks, A. F., R. R. Kurban, C. M. Hovens and S. J. Ralph. The application of the polymerase chain reaction to cloning members of the protein tyrosine kinase family. *Gene* 85: 67–74 (1989)
14. Beug, H., P. Steinlein, P. Bartunek, and M. J Hayman. Avian Hematopoietic Cell Culture: In vitro Model Systems to Study Oncogenic Transformation of Hematopoietic Cells. *Methods Enzymol.* 254: 41–76 (1995)
15. Chomczynski, P. and N. Sacchi. Single-step method of RNA Isolation by Acid Guanidinium Thiocyanate-Phenol-Chloroform Extraction. *Anal. Biochem.* 162: 167–159 (1987)
16. Koritschoner, N., P. Bartunek, S. Knespel, and M. Zenke. Assessing the gene expression by domain-directed differential display. In preparation.
17. Szabolcs, P., D. Avigan, S. Gezelter, D.H. Ciocon, M.A. Moore, R.M. Steinman, and J.W. Young. Dendritic cells and macrophages can mature independently from a human bone marrow-derived, post-colony-forming unit intermediate. *Blood* 87: 4520–4530 (1996)
18. Rosenzwajg, M., B. Canque, and J.C. Gluckman. Human dendritic cell differentiation pathway from CD34+ hematopoietic precursor cells. *Blood* 87: 535–544 (1996)
19. Saraya, K., and C.D. Reid. Stem cell factor and the regulation of dendritic cell production from CD34+ progenitors in bone marrow and cord blood. *Br. J. Haematol.* 93: 258–264 (1996)
20. Santiago Schwarz, F., K. Laky, and S.E. Carsons. Stem cell factor enhances dendritic cell development. *Adv. Exp. Med. Biol.* 378: 7–11 (1995)
21. Young, J.W., P. Szabolcs, and M.A. Moore. Identification of dendritic cell colony-forming units among normal human CD34+ bone marrow progenitors that are expanded by c-kit-ligand and yield pure dendritic cell colonies in the presence of granulocyte/macrophage colony-stimulating factor and tumor necrosis factor alpha. *J. Exp. Med.* 182: 1111–1119 (1995)

77

DIFFERENTIAL mRNA EXPRESSION IN UNTREATED AND TNF-α ELICITED MURINE DENDRITIC CELLS PRECURSORS

Patrizia Rovere,[1] Jeannine Trucy,[1] Valérie S. Zimmermann,[1] Samuel Granjeaud,[1] Dominique Rocha,[1] Catherine Nguyen,[1] Paola Ricciardi-Castagnoli,[2] Bertrand R. Jordan,[1] and Jean Davoust[1]

[1]Centre d'Immunologie INSERM-CNRS de Marseille-Luminy
Case 906, 13288 Marseille, France
Email davoust@ciml.univ-mrs.fr
[2]CNR Cellular and Molecular Pharmacology Center
20129, Milano, Italy

1. SUMMARY

We have compared the pattern of gene expression in long term cultured precursor dendritic cells (DC), either untreated (immature) or cultured for two days in the presence of recombinant murine (rm)-TNFα[1] (mature). The hybridization signature of complex cDNA probes prepared from total RNA extracted from immature and mature DC were analyzed using a mouse thymic cDNA library, gridded on high density filters. For each clone spotted on the filters, we have measured using an imaging plate device the hybridization signals of the complex probe obtained from immature or mature DC. Comparative analysis of these values allowed us to identify differentially expressed gene products. Our goal is to identify a new set of genes induced or repressed during DC maturation elicited by rmTNFα treatment.

2. INTRODUCTION

To perform their antigen presentation function, immature DC residing in non-lymphoid tissues need to be activated[2]. This process, referred to as DC maturation is characterized by profound changes in MHC class II distribution, antigen processing capacity, expression of costimulatory molecules and a marked rearrangement of adhesion molecules that is likely to allow DC migration to lymphoid organs[3–5]. Little is known on gene up- or down-regulation during this maturation process. To identify genes likely to be induced or repressed at different stages of DC maturation, we have generated large numbers of spleen-derived murine DC, that maintain characteristics of immature DC during the *in vitro* culture[1]. This DC population can be induce to mature using rm-TNFα. To identify differential gene expression, cDNA complex probes were generated from total RNA extracted from immature and mature DC. The relative amount of sequence-specific mRNA in the complex probe is related to the transcriptional activity of the corresponding gene. Clones from a mouse thymus cDNA library have been gridded in high-density colony filters[6–8] allowing the simultaneous screening of a large number of clones. A high throughput method was implemented to study gene expression in murine immune system, using an imaging plate device and specially developed software to quantify the hybridization signals[9].

3. MATERIALS AND METHODS

3.1. Mice

6–10 weeks of age C57BL6 mice were purchased from Charles River Laboratories (Calco, Como, Italy).

3.2. Cells

long term cultured murine splenic DC were propagated in Iscove's modified Dulbecco medium (Sigma) containing 10% heat-inactivated fetal bovine serum (Hyclone, Utah), 100 IU/ml penicillin, 100 μg/ml streptomycin, 2mM L-glutamine (all from Sigma, MO), 500 μM 2-βME supplemented with 30% NIH-3T3 supernatant containing 8 ng/ml rmGM-CSF, according to[1].

3.3. Complex cDNA Probes

cDNA complex probes were prepared as described in[8]. Briefly, total RNA was extracted from DC using Trizol (Gibco-BRL). Oligo dT and Reverse Transcriptase in the presence of ^{32}P-dCTP were used to ^{32}P-labelled cDNA production.

3.4. Hybridization Conditions

Hybridization conditions for filter hybridization the total probe (30 10^6 cpm) labelled as above was added to 50 ml of hybridization mixture (5x SSC, 5x Denhart, 0.5% SDS in DEPC water). Prehybridization was performed for 1 day at 65°C in the presence of the hybridization mixture alone. Hybridization was performed in the same mixture, adding the labelled probe and 150 μg/ml of ssDNA sonicated and denatured for 1 day at 65°C. The filters were then washed 5 times in 0.1xSSC, 0.1% SDC, for 1 hour at 65°C.

3.5. Signal Quantification

The hybridization filter was exposed to an imaging plate for 3 days and than scanned in a FUJIX BAS 1000 (Fuji) system. This result in a 2-megabyte image file representing the distribution of radioactivity on the filter. Hybridization signatures were determined by a modified version of the Bioimage software (Millipore, USA) running on a Unix workstation[9]. After hybridization, exposure to an "Imaging Plate Fuji" and quantification, comparative analysis of the signals was performed with Excel and Kaleidograph softwares[9] to identify differential clones.

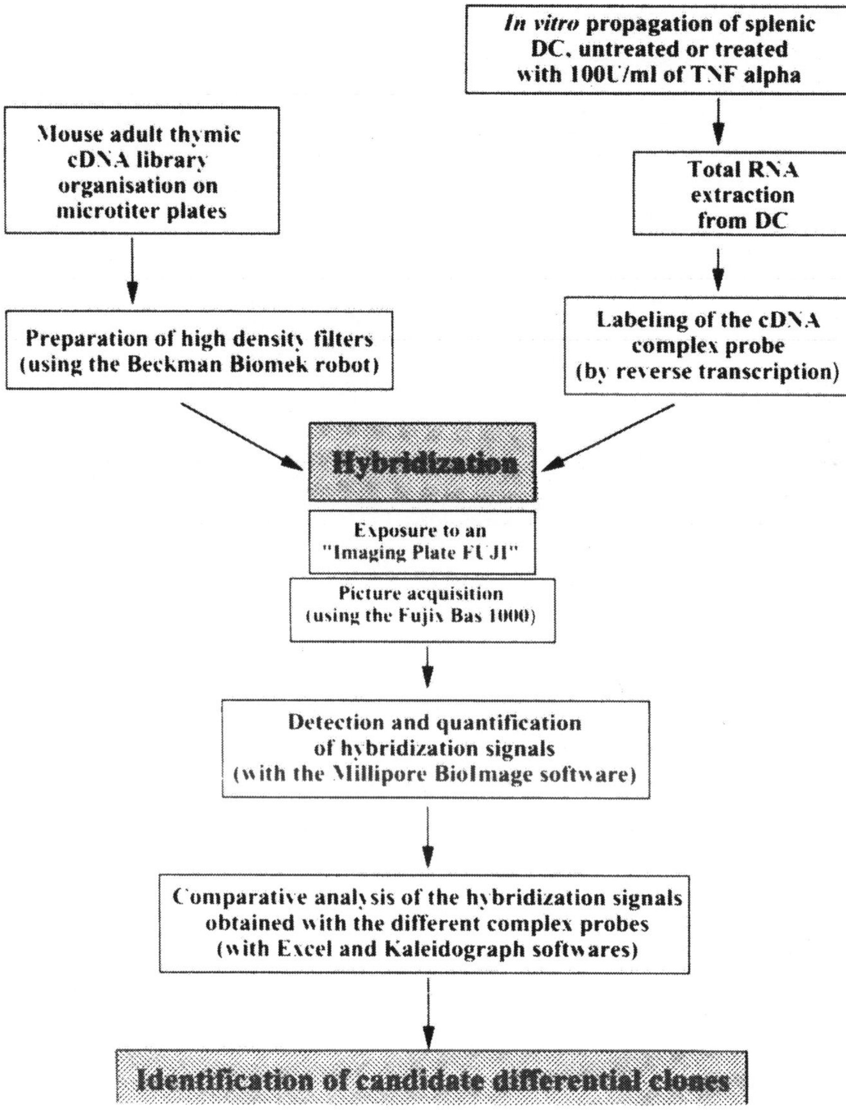

Figure 1. Flow chart of the differential screening.

4. RESULTS AND DISCUSSION

DC maturation was induced by culture for 2 days in the presence of rm-TNFα. Total RNA were extracted from cells in the two conditions and complex labelled cDNA probes were produced for hybridization on high density filters as described in[8]. Fig.1 shows a schematic representation of the method we used to quantify differential gene expression in immature and mature DC.

1536 cDNA clones derived from a thymic cDNA library were spotted in a 4×4 array on an 8×12 cm nylon filter. The filters were then hybridized with the complex probe. Fig.2 shows an example of a filter hybridized with a probe prepared from bone marrow derived

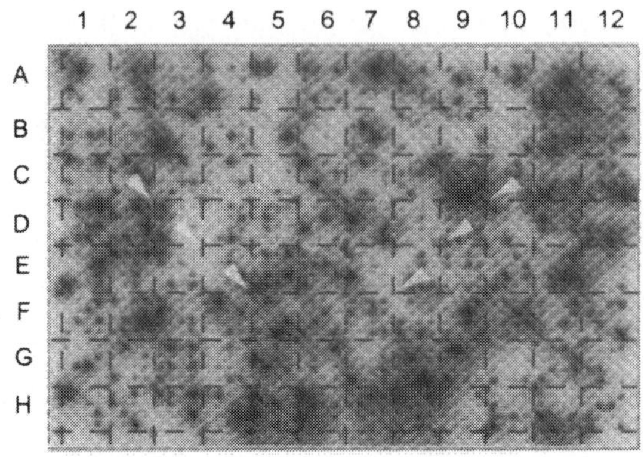

Figure 2. Localization of positive and negative controls in a high density filter containing mouse thymic cDNAs. High density filter containing 1536 colonies from a thymic cDNA library hybridized with a complex cDNA probes obtained from bone marrow derived DC total RNA. 90 positive controls are spotted at the low left corner of each 16 clones array and is represented by equal amounts of cytochrome C cDNA. 6 negative controls (poly A RNA) indicated by white arrows in the figure, are localized at positions C3, D4, E5, E8, D9 and E10 of the grid.

DC. At the lower left corner of each 4×4 squares, 90 positive controls, representing equal amounts of Arabidopsis Thaliana cytochrome C cDNA were spotted and revealed in each experiment (Arabidopsis Thaliana cytochrome C cDNA has been added during each hybridization at a known concentration). White arrows indicate at positions C3, D4, E5, E8, D9, C10, the six negative controls, represented by poly A (oligodT has been added in each hybridization to block possible recognition of poly A tails and non-specific binding).

Fig.3 shows an autoradiograph of the data obtained after imaging plate exposure. Panel A represents immature DC-derived cDNA hybridization; panel B TNF-α elicited DC-derived cDNA hybridization. Two examples of differentially expressed clones are shown by black arrows. Table 1 summarizes the results obtained by the quantification of the different hybridization images and the comparison of the different cDNA hybridized on the same high density filter. The columns report for each single clone the actual intensity of the hybridization signal (int.) and the intensity normalized on the basis of the vector hybridization alone (int./norm.), obtained using cDNA probes from mature and immature DC. The ratio between int./norm in mature and int./norm. in immature DC was also evaluated. The clones of interest were selected on the basis of different criteria reported in the table on the hatched cells reported above any list of clones. In particular the first selection, aimed to identify genes up-regulated by TNF-α treatment, excluded clones having low values of intensity (defined on the basis of the less intense of the 90 positive controls), and included clones whose signal upon TNF-α treatment increased more than 3 times (ratio between mature and immature DC > 3). The second selection was aimed to identify clones down-regulated during TNF-α treatment; the criteria were accordingly modified, as reported in Table 1. Finally in the third selection we have collected the clones poorly expressed (<0.015) in immature DC but highly stimulated after TNF-α treatment (>0.1). Most of the identified clones have not yet been sequenced (N.S.) or after sequencing have been classified as unknown genes (New) (see Table 1). Sequencing of the unknown clones and *in situ* hybridization will be used to identify new genes expressed during DC maturation and to investigate their *in vivo* expression. Three known gene products, namely thioredoxin, cytoplasmic β-actin (cyto. β-actin) and β-2 microglobulin (β-2 μ

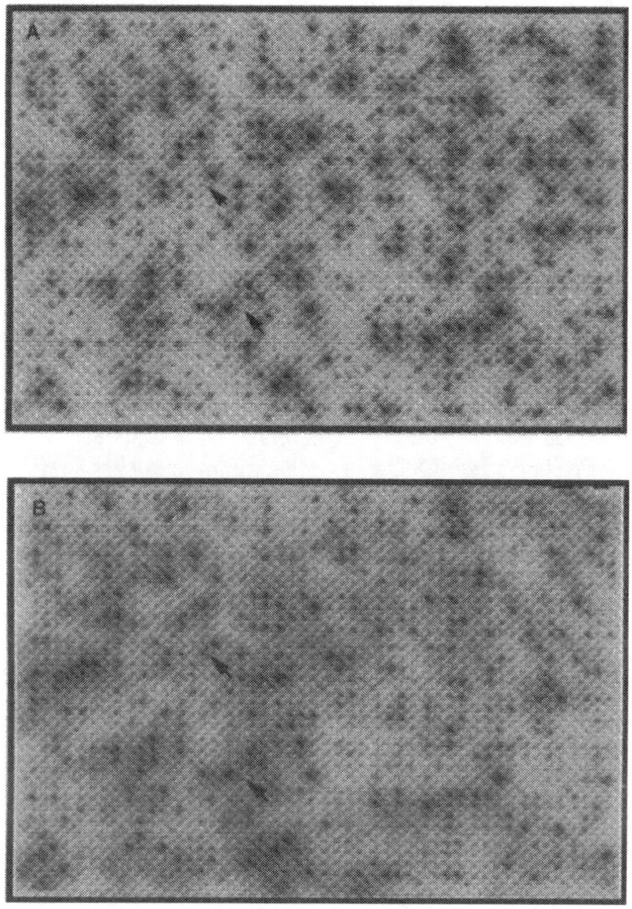

Figure 3. Identification of signals differentially expressed upon TNF-α treatment. Hybridization of high density filters with complex cDNA probes obtained from immature (panel A) and mature (panel B) DC total RNA. Two differentially expressed clones (MTA.D04.071, corresponding to β2-microglobulin and MTA.F05.078, corresponding to an unknown gene) are indicated by black arrows.

globulin) were up-regulated during DC maturation. The latter gene represents an internal control of our detection system, as TNF-α is a well known inducer of β-2 microglobulin expression in cells belonging to different lineages[10]. Cyto. β-actin is a cytoplasmic isoform of actin, associated to stress fiber bundles and to the peripheral actin-rich lamellipodia structures[11]. Under the control of highly conserved 3'UTR sequence in the RNA, cyto. β-actin mRNA is directed to the cell periphery and the proper regulation of this process has been implicated in the control of actin cytoskeleton and in the regulation of the morphology[12]. This is of particular interest, as a well characterized step of DC maturation is represented by the migration of DC from the periphery to lymphoid organs[2]. We have recently started the characterization of the molecular mechanisms underlying the profound changes in adhesion properties and cytoskeleton organization underlying this migration[1]. Thioredoxin, the third gene massively up-regulated by DC after TNF-α treatment, is a small ubiquitous redox regulatory molecule which through dithiol/disulfide exchange activity regulates the oxidation state of cell protein thiols[13]. Cytokine synthesis is tightly regulated

Table 1. Genes differentially expressed upon TNF-α treatment

Spot Name	Int. Immature DC	Int./norm. ImmatureDC	Int. Mature DC	Int./norm. MatureDC	Ratio	Gene
	>0.1		**>0.05**		**>3**	
MTA.A05.072	0,163	0,713	0,075	2,170	3,045	N.S.
MTA.A07.071	0,319	1,918	0,185	9,177	4,785	N.S.
MTA.A12.079	0,958	6,583	0,396	22,919	3,482	Mitochondrial
MTA.B09.075	0,138	1,006	0,155	3,166	3,148	Mitochondrial
MTA.C05.072	0,188	1,966	0,168	6,482	3,298	N.S.
MTA.C12.078	0,234	1,759	0,094	5,440	3,094	Mitochondrial
MTA.D01.079	0,122	1,087	0,560	4,137	3,808	New
MTA.D09.065	0,187	1,363	0,138	6,846	5,023	N.S.
MTA.D09.079	0,209	2,094	0,200	7,717	3,685	N.S.
MTA.F06.065	0,306	2,103	0,310	6,728	3,200	N.S.
MTA.F10.068	0,107	0,715	0,113	3,924	5,490	N.S.
MTA.F10.069	0,136	0,909	0,191	3,316	3,650	N.S.
MTA.G08.067	0,271	2,507	0,313	8,361	3,335	N.S.
MTA.G09.073	0,106	0,772	0,121	2,334	3,022	**Thioredoxin**
MTA.G09.075	0,183	1,158	0,324	4,167	3,598	Mitochondrial
MTA.G10.076	0,318	2,941	0,306	10,626	3,613	N.S.
MTA.H12.072	0,856	8,234	0,520	25,796	3,133	**Cyto.β actin**
	>0.015		**>0.02**		**<0.3**	
MTA.A07.067	0,098	1,683	0,023	0,444	0,264	N.S.
	<0.1		**>0.3**			
MTA.F05.075	0,090	0,601	0.366	4,539	7,550	N.S.
	<0.015		**>0.1**			
MTA.A12.065	0,000	0,000	0,158	10,973	–	Mouse EST
MTA.B03.079	0.000	0,000	0,177	2,049	–	N.S.
MTA.C01.076	0,000	0,000	0,172	1,457	–	N.S.
MTA.C02.067	0,000	0,000	0,153	3,125	–	N.S.
MTA.C02.078	0,012	0,289	0,208	2,889	10,011	New
MTA.C02.079	0,000	0,000	0,122	1,926	–	Human EST
MTA.D01.080	0,000	0,000	0,412	2,601	–	N.S.
MTA.D04.071	0,008	0,101	0,143	2,759	27,244	**β2 μ glob.**
MTA.D05.073	0,000	0,000	0,191	6,030	–	N.S.
MTA.D05.079	0,000	0,000	0,169	3,452	–	N.S.
MTA.D06.067	0,007	0,060	0,110	1,910	31,767	N.S.
MTA.D06.068	0,014	0,168	0,101	0,974	5,787	N.S.
MTA.D11.079	0,000	0,000	0,595	8,983	–	N.S.
MTA.F03.070	0,000	0,000	0,118	1,366	–	N.S.
MTA.F05.078	0,000	0,000	0,127	1,378	–	N.S.
MTA.H01.077	0,000	0,000	0,131	0,858	–	New
MTA.H05.076	0,000	0,000	0,173	2,146	–	N.S.
MTA.H06.070	0,000	0,000	0,182	1,975	–	N.S.

by redox-dependent reactions and thioredoxin is known to increase the expression of TNF-α in macrophage cell lines[13]. This is the first report concerning thioredoxin expression and regulation in DC. Upregulation of thioredoxin during DC maturation, *i.e.* of a potent costimulatory signal for the expression of different cytokines, including TNF-α itself, may provide a molecular explanation for the elusive requirements necessary for *in vitro* and *in vivo* maturation of DC.

5. ACKNOWLEDGMENTS

This work was partially supported by the Italian Ministry of Public Health (Instituto Superiore SANITA) AIDS grant 9408-92 and by the National Council for Research (CNR).

6. REFERENCES

1. Winzler et al., 1997, *J. Exp. Med.* 185:317.
2. Steinman RM, 1991. *Annu. Rev. Immunol.* 9:271.
3. Schuler and Steinman, 1985.*J. Exp. Med.* 161:526.
4. Inaba et al., 1994. *J. Exp. Med.* 180:1849.
5. Roake et al., 1995. *J. Exp. Med.* 181:2237.
6. Rouvier and Golstein, 1993. *In "Cytotoxic cells"* (M.Sitkovsky and P.Henkart, eds.) p.237, Birkhauser, Boston, MA.
7. Lennon and Lehrach, 1991. *Trends Genet.* 7:314.
8. Nguyen et al., 1995. *Genomics* 29:207.
9. Granjeaud et al., 1996. *Genetic Analysis* 12:151.
10. Drew et al., 1993. *J. Immunol.* 150:3300.
11. Gimona et al., 1994. *Cell Motil. Cytoskeleton* 27:108.
12. Kislauskis et al., 1994. *J. Cell Biol.* 127:441.
13. Schenk et al., 1996. *J. Immunol.* 156:765.

DIFFERENTIAL GENE EXPRESSION IN CULTURED HUMAN LANGERHANS CELLS IN RESPONSE TO PHAGOCYTIC STIMULATION

Josef Unger and William Phares

Genetics Department
Sandoz Forschungsinstitut
Brunnerstrasse 59
A-1235 Vienna, Austria

1. INTRODUCTION

In order to identify genes whose differential expression in Langerhans cells (LC) is increased by a potent, specific stimulus that is associated with the antigen-processing and -presenting activity of these cells, we have begun the analysis of phagocytic stimulation of freshly isolated human LC in culture (cLC) using zymosan particles. Previously it has been shown that cLC actively take up zymosan, and that phagocytosis of these particles is mediated by the mannose/ß-glucan receptor(s)[1]. Genes that are preferentially expressed in cLC following phagocytic stimulation were identified by mRNA Representational Difference Analysis (RDA)[2,3] and DNA sequence analysis of cloned RDA difference products (DP). In order to address the question of cell type-specific molecular activation patterns, a similar set of experiments with monocytes cultivated with GM-CSF and IL-4 (MoDC) was included in this analysis. In total, more than 150 recombinant plasmid clones containing DP were isolated, sequenced and compared with primate sequence databases and EST databases.

2. METHODS

LC were isolated from epidermal sheets obtained from fresh surgical specimens by modification of a previously described method[4]. These cell preparations were reproducibly 70–80% CD1a$^+$ and HLA-DR$^+$ as assessed by FACS analysis. For each experiment, LC were placed in duplicate cultures in RPMI 1640 medium supplemented with 10% FCS, 100 U/ml GM-CSF (Sandoz) and 10 U/ml IL-4 (R&D Systems); zymosan (Sigma) was prepared as described[1] and added to one of the cultures, and the second served as the unstimulated control. After 24 and 48 hr, cells were withdrawn from both cultures, total

Dendritic Cells in Fundamental and Clinical Immunology
edited by Ricciardi-Castagnoli, Plenum Press, New York, 1997

RNA was extracted (Trizol, BRL), and cDNA was prepared (MuMLV reverse transcriptase, BRL). For the RDA, cDNA prepared from LC obtained from three different donors was pooled in order to normalize variability between specimens. MoDC were prepared essentially as described[5], using monocytes from peripheral blood purified by centrifugal elutriation (kindly provided by W. Neruda and H. Loibner); for the latter, only a 48 hr time point was taken from stimulated and control cultures for the RDA. RDA was performed essentially as described[2], but using the thermostable DNA polymerase from *T. brockianus* (Finnzyme). Quantitative RT-PCR was performed with modifications of a described method[6], with β-actin as internal control; for these analyses, one of the primers for each sequence to be amplified was end-labelled with ^{32}P, and the PCR products were denatured and resolved on urea-polyacrylamide gels. Quantification was performed by phosphoimager analysis. Linearity of amplification with respect to the β-actin control was demonstrated by performing the amplifications over a 1000-fold range of template dilutions, for all templates employed.

3. RESULTS

3.1. RDA Difference Products from Phagocytically-Stimulated cLC and MoDC

RDA is a method that consists of three rounds of heteroduplex hybrid-competition coupled to PCR amplification, designed to selectively enrich for cDNA sequences uniquely expressed (or quantitatively overexpressed) in a target versus a driver cell popu-

Table 1. Summary of the database analysis of DP isolated from MoDC and cLC after phagocytic stimulation

	MODc	Lc 48hr	Lc 24hr
Number of analysed clones	48	87	23
Oxidative burst associated			
Cytochrome b X62996	10	22	
Cytochrome b U95000		3	
Metabolic enzymes or associated	1	1	
DNA binding proteins	1		
Protein synthesis	2	1	
Chaperone		1	
Proteases			
Cathepsin B	2		
Protein phosphotyrosyl phosphatases		1	
Membrane-associated			
Fc-epsilon receptor γ-chain	2		
CD53	1		
TNF-α RI	2		
Microtubular proteins	1		
Glucocorticoid-inducible			
Lipocortin	1		
Repetitive DNA / Alu sequences	1	12	
Unknown function, or not well characterized		1	
Unknown	18	42	23

lation. In the experimental design described here, sequences were enriched from the phagocytically stimulated dendritic cell cultures by competition with material from the parallel, unstimulated cultures. Therefore, the RDA difference products should represent gene sequences whose differential expression is increased by phagocytic stimulation. Sequence analysis of cloned RDA difference products from 24 hr- and 48 hr-stimulated cLC, and 48 hr-stimulated MoDC, is summarized in Table 1.

48 clones from phagocytically stimulated MoDC, 87 clones from cLC stimulated for 48 hr, and 23 clones from cLC stimulated for 24 hr have been analyzed. Of the sequences that could be identified by database analysis, most clones correspond to genes associated with the oxidative burst phenomenon, e.g. cytochrome b and cytochrome c oxidase, both encoded by the mitochondrial genome. Because all of the RDA difference products for cytochrome b were greater than 96% homologous with the corresponding database entry, this homology level was set as the criteria for a positive sequence identity. A number of other identified sequences whose expression may be relevant to immune activity correspond to Cathepsin B, FcεRI γ-chain, CD53, TNFα-RI and lipocortin. A very large number of clones were also isolated that contained DP whose identity could not be established by database sequence comparison. In the case of the 24 hr-stimulated cLC, all of the isolated DP clones correspond to only two unique, unknown sequences that have been termed Lc509 and Lc526.

3.2. Quantitative RT-PCR Analysis of Gene Expression in cLC

To assess by a second criteria the selection by RDA of some of the DP, gene expression was analyzed by quantitative RT-PCR relative to β-actin in unamplified cDNA from LC, and from cLC cultured for 24 and 48 hr with and without zymosan stimulation (Fig. 1).

With Lc509 and cytochrome b (Fig. 1, upper left and right panels, respectively) the differential expression in stimulated cultures could easily account for the selective enrichment by RDA. The very high levels of expression of cytochrome b in stimulated cultures, about 20 times that of β-actin, would also account for the large numbers of cytochrome b clones isolated as DP. The results with Lc526 (upper middle panel) at the 24 hr time point would indicate that also small differences in expression levels would contribute to RDA enrichment. However, the analysis of TNFα-RI and Cathepsin B expression (both of which were isolated by RDA as DP clones from stimulated MoDC, but not from stimulated cLC) revealed also significantly increased differential expression in stimulated versus unstimulated cLC (lower left and middle panels, respectively). And for CD11c, which was not isolated by RDA from either stimulated MoDC or cLC, considerable increased differential expression in cLC upon phagocytic stimulation is seen by RT-PCR analysis (lower right panel), and at the 48 hr stimulation time point an extremely high relative level of expression, nearly 100 times of that of β-actin, was obtained.

3.3. Southern Blot Analysis of DP Enrichment during Sequential RDA Cycles

One possible explanation for the discrepancy between RDA enrichment and RT-PCR analysis of (especially) CD11c expression in stimulated versus unstimulated cLC is that certain sequences may not in fact be progressively enriched by RDA. Therefore, Southern blot analysis of the DP cycles during RDA of cLC was performed, to follow the progressive enrichment of Lc509 and CD11c sequences (Figure 2). In the case of Lc509

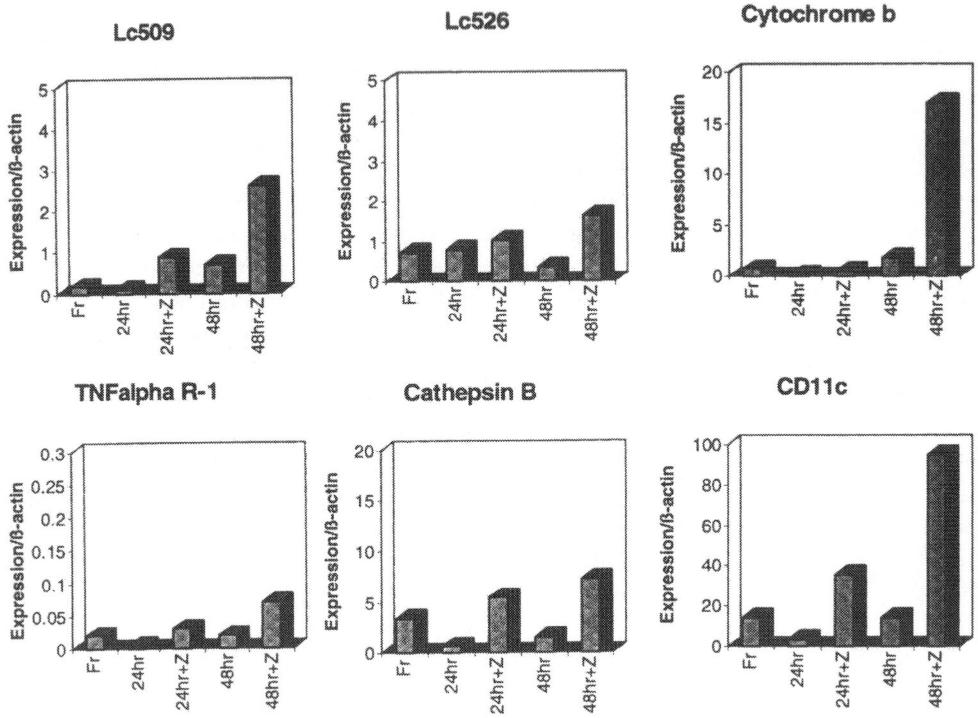

Figure 1. Quantitative RT-PCR of DP in freshly isolated LC (Fr), unstimulated cLC controls (24 hr, 48 hr), and 24 hr and 48 hr zymosan-stimulated cLC (24 hr +Z, 48 hr +Z).

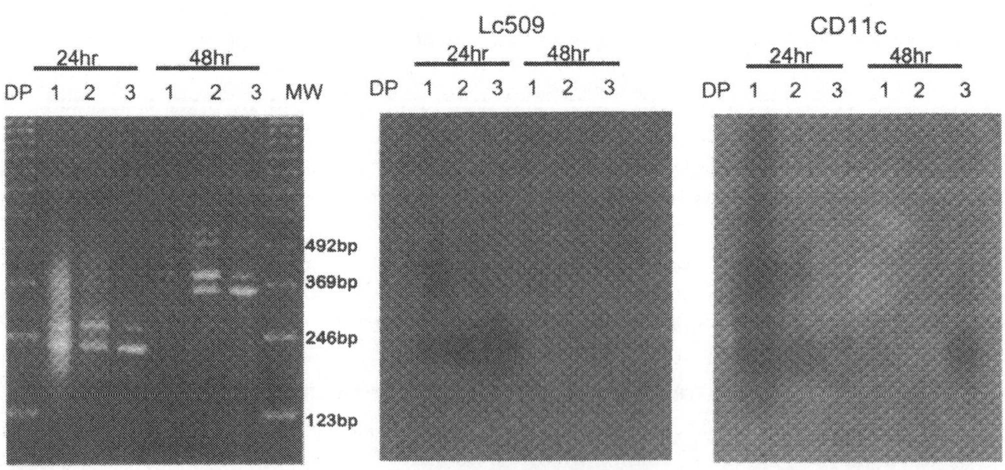

Figure 2. Electrophoretic and Southern blot analysis of DP resulting from differential subtractions (cDNA from stimulated cLC competed with that from unstimulated cLC). In the left panel is shown the discrete pattern of DP products obtained from the first (DP 1), second (DP 2) and third (DP 3) cycles of RDA, from 24 hr and 48 hr-stimulated cLC cultures, as revealed by ethidium bromide staining. Southern blot analyses of this gel with Lc509 and CD11c probes are shown in the middle and right panels, respectively.

sequences, a selective enrichment in the RDA cycles is clearly detectable, but this is not the case for CD11c. Possibly the high level of expression in unstimulated cells simply outcompetes the amplification of the RDA procedure; however, this effect should be more carefully examined. In general, it is important to bear in mind that any of the various techniques relying on cDNA synthesis and PCR amplification (e.g., RT-PCR, RDA, RDD, or related techniques) all have a degree of inherent variability.

4. CONCLUSIONS

Several of the DP obtained by RDA would be consistent with the acquisition of an enhanced professional immune function of cLC and MoDC following phagocytosis. The isolation of DP corresponding to cytochrome b and Cathepsin B sequences is consistent with elevated antimicrobial[7,8] and antigen-processing[9] activities, respectively. The isolation of FcεRI γ-chain, TNFα RI and CD53 may suggest signal transduction effects and a potentiation to respond to further stimulatory signals.

The experiments described here have examined differential gene expression in cLC only after 24 hr of phagocytic stimulation; an RDA enrichment was also attempted at earlier time points (2+4+6 hr stimulation, combined), however, no DP were obtained (data not shown). This would indicate that LC isolated from epidermis and immediately placed in culture with GM-CSF/IL4 may have a slow gene induction response following phagocytosis. This could reflect either that phagocytic stimuli *per se* are weak for generating transcriptional effects in cLC, or cLC have already begun a differentiation step associated with a weak response. In addition, or alternatively, a slow gene induction may be simply consistent with a relatively slow phagocytic activity, as previously observed in murine cLC[1]. How this activity may relate to that of LC in the skin is yet another question to be addressed.

The number of isolated sequences obtained by RDA that could not be identified by database comparisons is striking. One possibility is that DC may represent a particularly rich source of previously unknown genes. cDNA clones corresponding to several of the unknown difference products from Langerhans cells are currently being sought by hybridization of large, unsubtracted cDNA libraries.

5. ACKNOWLEDGMENTS

The authors wish to thank B. Volc-Platzer for assistance in obtaining skin specimens, and for suggestions and comments; C. Foster and A. Elbe for advice in early attempts to purify LC; W. Neruda and H. Loibner for MoDC; and K. Kleine for assistance with database analyses.

6. REFERENCES

1. Reis e Sousa, C., Stahl, P. D. and J. M. Austyn. 1993. Phagocytosis of antigens by Langerhans cells *in vivo*. J. Exp. Med. 178: 509–519
2. M. Hubank, D. G. Schatz. 1994. Identifying differences in mRNA expression by representational difference analysis of cDNA. Nuc. Acids Res. 22: 5640–5648
3. Lisitsyn, N., Lisitsyn, N. and M. Wigler. 1993. Cloning the difference between two complex genomes. Science 259: 946–951

4. Jürgens, M., Wollenberg, A., Hanaus, D., de la Salle, H. and T. Bieber. 1995. Activation of human epidermal Langerhans cells by engagement of the high affinity receptor for IgE, FcεRI. J. Immunol. 155: 5184–5189.
5. Sallusto, F. and A. Lanzavecchia. 1994. Efficient presentation of soluble antigen by cultured human dendritic cells is maintained by granulocyte/macrophage colony-stimulating factor plus interleukin 4 and downregulated by tumor necrosis factor α. J. Exp. Med. 179: 1109–1118.
6. Melby, P.C., Darnell, B.J. and V.V. Tryon. 1993. Quantitative measurement of human cytokine gene expression by polymerase chain reaction. J. Immunol. Meth. 159: 235–244.
7. Rabinovitch, M. 1995. Professional and non-professional phagocytes: an introduction. Trends Cell Biology 5: 85–87.
8. Jesaitis, A.J. 1995. Structure of human phagocyte cytochrome b and its relationship to microbicidal superoxide production. J. Immunol 155: 3286–3288.
9. Matsunaga Y., Saibara T., Kido H. and N. Katunuma. 1993. Participation of Cathepsin B in processing of antigen presentation to MHC class II. FEBS 324: 325–330.

ISOLATION OF DIFFERENTIALLY EXPRESSED GENES IN EPIDERMAL LANGERHANS CELLS

R. Ross, J. Schwing, K. Kumpf, A. Endlich, and A. B. Reske-Kunz

Clinical Research Unit
Department of Dermatology
Johannes Gutenberg-University
D-55101 Mainz, Germany

1. INTRODUCTION

Epidermal Langerhans cells (LC) represent immature dendritic cells (DC) resident in the skin, which are not yet able to prime naive T cells[1]. In vitro cultivation of LC in the presence of keratinocytes, supplying survival and differentiation signals, induces maturation events in LC[2]. These are highlighted by the downregulation of the biosynthesis of MHC class II molecules[3], by the upregulation of the surface expression of adhesion and costimulatory molecules like CD80, CD86, CD54 and CD58[4,5], and by the acquisition of a potent immunostimulatory capacity for T cells[6]. Mature LC are potent inducers of naive T cells. Thus LC represent an ideal model system to investigate the maturation of DC (reviewed in 7).

We studied freshly isolated LC (fLC) and cultivated LC separated from epidermal cell suspensions after 1–3 days of cultivation (cLC) with respect to differential gene expression using molecularbiological techniques, namely differential display and differential screening of a cDNA library prepared from cLC cultured for 3 days. With the two techniques we compared cDNA expression patterns of fLC and cLC and selected for differentially expressed cDNAs. As fLC and cLC are expected to share the housekeeping genes and typical LC-specific genes, it is likely that differentially expressed genes are involved in maturation events.

2. DIFFERENTIAL DISPLAY

The differential display protocol we used, which is based on the original protocol of Liang and Pardee[8], is outlined in Figure 1. Several modifications were introduced including additional verification steps. Briefly, cells were harvested and mRNA was isolated according to standard procedures, followed by reverse transcription of mRNA using $T_{11}MN$

Dendritic Cells in Fundamental and Clinical Immunology
edited by Ricciardi-Castagnoli, Plenum Press, New York, 1997

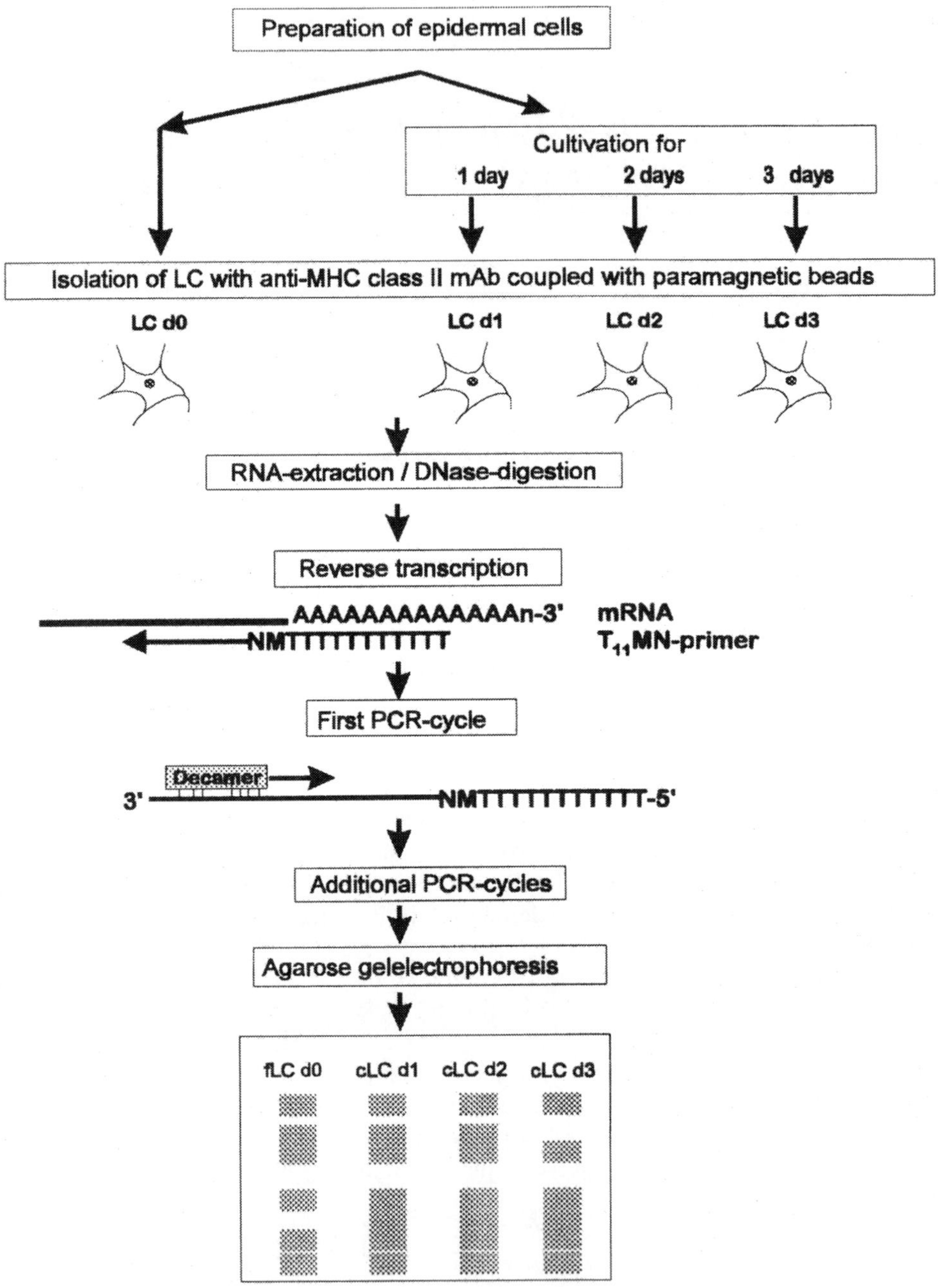

Preparation of epidermal cells
Cultivation for
1 day
2 days
3 days
Isolation of LC with anti-MHC class II mAb coupled with paramagnetic beads
LC d0
LC d1
LC d2
LC d3
RNA-extraction / DNase-digestion
Reverse transcription
AAAAAAAAAAAAAAn-3'
mRNA
NMTTTTTTTTTTT
T11MN-primer
First PCR-cycle
Decamer
3'
NMTTTTTTTTTTT-5'
Additional PCR-cycles
Agarose gelelectrophoresis
fLC d0
cLC d1
cLC d2
cLC d3

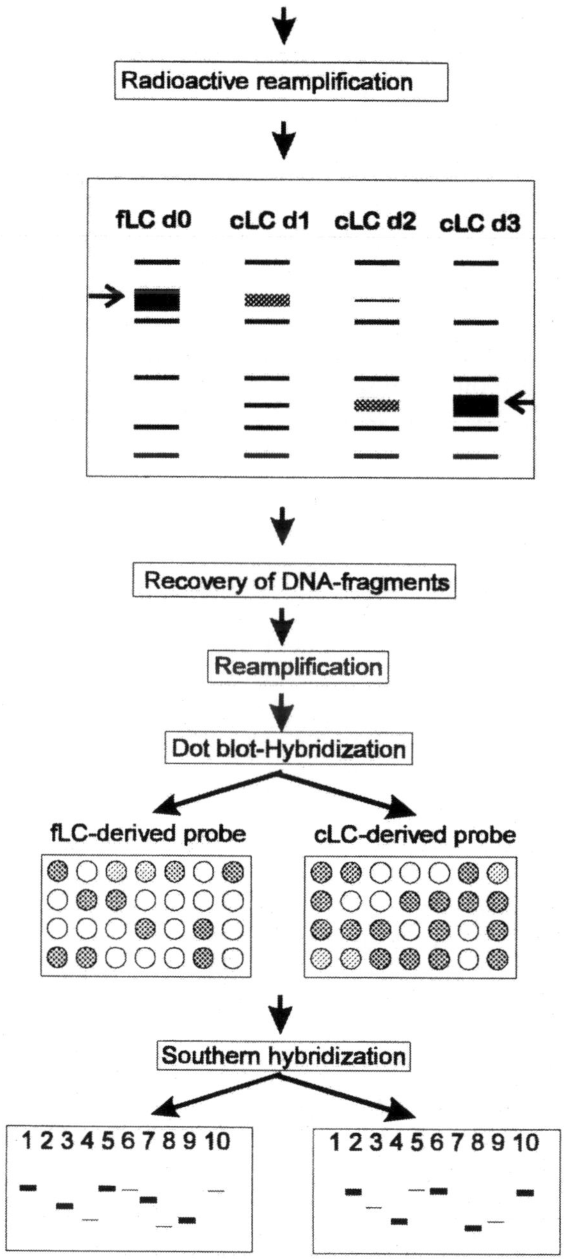

Figure 1. Differential display procedure. For further explanation see text.

primers (M = G, A or C; N = G, A, C or T) in order to reverse transcribe only part of the total mRNAs. Other parts of the total mRNAs were reverse transcribed with additional $T_{11}MN$ primers in parallel reactions. PCR was performed under low stringency using a randomly selected decamer as second primer after heat denaturation of the products. The cDNA-products were separated on agarose gels in order to monitor the success of the amplification. This additional step allows equalization of the overall product yield of the corresponding PCR-reactions from fLC and cLC before the samples were subjected to reamplification with radiolabelled nucleotides. The nonradioactive amplification avoids unequal amplification of the complementary DNA-strands because of unbalanced nucleotide composition. It also avoids premature stops of DNA-synthesis due to incorporation of partly destroyed radiolabelled nucleotides.

Figure 2 shows a typical result. Differential display reactions using fLC and 3-day cLC with the same primer combination were loaded next to each other onto the gel. For all reactions the same oligo-dT primer was used while the decamer primer was diffferent for the reactions shown in lanes 1/2 as compared with lanes 3/4, 5/6 and 7/8. Though the separation of the gel is not sufficient to distinguish all fragments, it clearly shows that similar patterns were obtained with fLC and cLC using the same pair of primers, while different decamer primers yielded different results. Similar product yields were obtained with fLC and cLC and therefore the same amounts were subjected to radioactive reamplification.

After reamplification in the presence of radiolabelled nucleotides the products were separated on high resolution polyacrylamide gels and visualized by autoradiography. Products obtained in parallel reactions starting from fLC and cLC cultivated for 1, 2 and 3 days were loaded onto the same gel. Differentially expressed cDNAs (arrows in Fig. 1) were cut out of the gel and recovered as described by Bauer et al.[9]

Two verification procedures were persued before the cDNA fragments were cloned and sequenced: The fragments were reamplified using the same pair of primers as was employed for differential display, and aliquots of the products were transferred to nitro-cellu-

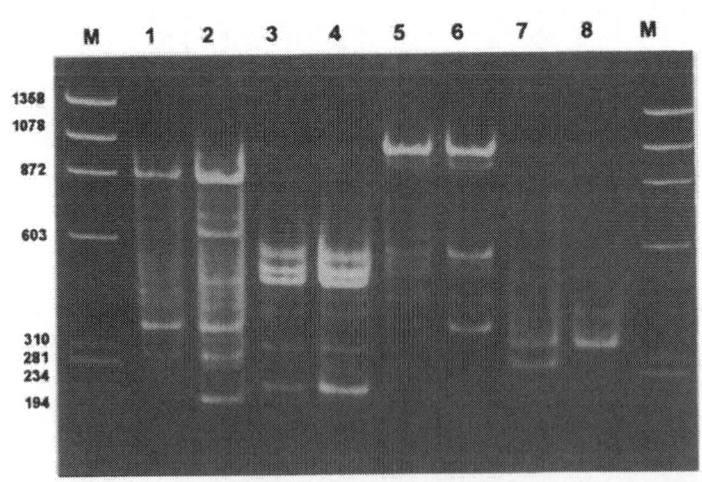

Figure 2. Agarose gel separation of nonradioactive differential display reactions.The same oligo-d T primer was used for all reactions. The decamer primer differed for the reactions shown in lanes 1/2, 3/4, 5/6 and 7/8. Even numbers: differential display reactions using flC; uneven numbers: differential display reactions using 3-day cLc. Marker (M) : ΦX 174 x Hae III. Molecular weight is denoted in basepairs.

lose filters by dot blotting. Two replica-filters were hybridized with radiolabelled total cDNA of fLC and cLC, respectively. In order to obtain sufficient amounts of total cDNA for radiolabelling we prepared double stranded cDNA, ligated double stranded adaptor-molecules to the ends and amplified the cDNAs with adaptor-specific primers. After verification of differential expression, the remaining aliquots were loaded on agarose gels. fLC- and cLC-specific fragments were loaded alternately onto the gel. Afterwards electrophoresis sandwich blots were performed. As a result replica-filters were obtained from the same gel. One filter was hybridized with fLC-specific probes while the other filter was hybridized with cLC-specific probes. The alternate loading of fLC- and cLC-specific fragments verified that both filters were of good quality and that both probes worked well. Differentially expressed cDNA-fragments were cloned and sequenced.

3. DIFFERENTIAL SCREENING

For differential screening we constructed a cDNA library of cLC in lambda Zap II containing $18 \cdot 10^6$ independent clones. Recombinant phages were plated, and phages from the plaques were transferred to nitro-cellulose filters twice independently in order to obtain replica-filters. One of the filters was screened with total radiolabelled cDNA from fLC while the corresponding filter was screened with radiolabelled cDNA from cLC. Recombinant phages solely hybridizing with probes derived from cLC, but not from fLC, were isolated and sequenced.

4. CONCLUSIONS

With the two methods outlined we isolated genes differentially expressed in fLC and cLC. Approximately 1–3% of the total cDNAs appear to be differentially regulated. Based on statistical estimates, the technique of cDNA-library screening is likely to result in the isolation of the most abundant differentially expressed cDNAs. In fact we isolated the most abundant differentially expressed genes repeatedly. In contrast, the amplification attained by differential display depends on primer annealing. Therefore maximum identity with the randomly selected primer sequences will result in strong signals even for weakly expressed proteins, typical of many cytokines and transcription factors. Indeed, the differentially expressed cDNAs isolated by differential display were different from the genes obtained by differential screening. Furthermore, it is possible to isolate upregulated as well as downregulated genes by differential display while only upregulated genes are detectable by differential screening of a cLC-cDNA library.

Computer database searches revealed that we isolated known as well as unknown cDNAs. The corresponding proteins of the known cDNAs are for example involved in gene regulation, signal transduction, structural maturation and antigen processing, and their differential expression appears reasonable in the light of the functional capacities of LC. The main focus of our research is directed towards the exploration of the function of the so far unknown genes, upregulated in cLC.

5. REFERENCES

1. K. Inaba, G. Schuler, M.D. Witmer, J. Valinsky, B. Atassi, and R.M. Steinman, Immunologic properties of purified epidermal Langerhans cells. J.Exp.Med. 164: 605–613 (1986).

2. C. Heufler, F. Koch, and G. Schuler, Granulocyte/macrophage colony-stimulating factor and interleukin-1 mediate the maturation of murine epidermal Langerhans cells into potent immunostimulatory dendritic cells. J.Exp.Med. 167: 700–705 (1988).
3. D. Becker, A.B. Reske-Kunz, J. Knop, and K. Reske, Biochemical properties of MHC class II molecules endogenously synthesized and expressed by mouse Langerhans cells. Eur.J.Immunol. 21: 1213–1220 (1991).
4. M.B.M. Teunissen, H.A.H. Rongen, and J.D. Bos, Function of adhesion molecules lymphocyte function-associated antigen-3 and intercellulár adhesion molecule-1 on human epidermal Langerhans cells in antigen-specific T cell activation. J.Immunol. 152: 3400–3409 (1994).
5. C.P. Larsen, S.C. Ritchie, R. Hendrix, P.S. Linsley, K.S. Hathcock, R.J. Hodes, R.P. Lowry, and T.C. Pearson, Regulation of immunostimulatory function and costimulatory molecule (B7–1 and B7–2) expression on murine dendritic cells. J. Immunol. 152: 5208–5219 (1994).
6. G. Schuler and R.M. Steinman, Murine epidermal Langerhans cells mature into potent immunostimulatory dendritic cells in vitro. J.Exp.Med. 161: 526–546 (1985).
7. R.M. Steinman, K. Inaba, and G. Schuler, Cutaneous dendritic cells: distinctive antigen-presenting cells for experimental models and disease states. In: The Immune Functions of Epidermal Langerhans Cells. (ed. H. Moll), R.G. Landes Company, Austin, USA p. 1–19 (1995).
8. P. Liang and A.B. Pardee, Differential display of eucaryotic messenger RNA by means of the polymerase chain reaction. Science 257: 967–997 (1992).
9. D. Bauer, P. Warthoe, M. Rohde, and M. Strauss, Detection and differential display of expressed genes by DDRT-PCR. PCR Methods Applic. 4:97–108 (1994).

80

ABSENCE OF FOLLICULAR DENDRITIC CELLS (FDC) IN MUTANT MICE WITH DEFICIENT TNF/LYMPHOTOXIN (LT) SIGNALING

Michel Le Hir and Hans-Pietro Eugster

Institute of Anatomy of the University of Zürich-Irchel
8057 Zürich, Switzerland
Institute of Toxicology of the University and ETH Zürich
8603 Schwerzenbach, Switzerland

1. INTRODUCTION

1.1. Morphology

FDC occur as clusters in the B cell areas (follicles) of lymph nodes, spleen, tonsil and Peyers patches. Anatomically these clusters appear as a locally specialized compartment of a cellular meshwork that spans the secondary lymphoid tissues. FDC are mostly considered to differentiate from the fibroblast-like cells forming that meshwork (1–6). However there is experimental support for the alternative view that FDC might originate from the bone marrow (2,7,8). It is possible to distinguish FDC by ultrastructural criteria from other cells of the reticular meshwork (4,6,9). Identification by light microscopy requires specific markers, e. g. antibodies against complement receptors (CR) (10). The monoclonal antibody FDC-M1 detects FDC in the mouse (11). Binding of immune complexes, e. g. peroxidase-anti-peroxidase, in vivo as well as in tissue sections constitutes a further criterion for identification of FDC (12).

1.2. Function

The putative functions of FDC in affinity maturation of antibodies and in the production of memory B cells are linked to their capacities to store and to deliver antigens on the one hand (13,14), and to deliver costimulatory signals to B cells on the other hand (1,15–17). Testing those hypothetical functions relied so far on cocultures of lymphocytes with enriched FDC preparations (15,16) or with a FDC cell line (1). The availability of mouse mutants lacking FDC offers new opportunities for assessing the functions of FDC in vivo.

2. FDC DEFICIENCY IN MICE

2.1. Severe Combined Immunodeficiency

The development of FDC is dependent on immunocompetent lymphocytes. Thus, FDC are lacking in SCID mice but they do differentiate after reconstitution with normal lymphocytes (18). Clearly, on account of the defective maturation of T and B cells, this strain is not appropriate for evaluating specifically the functional consequences of the lack of FDC.

2.2. LT and TNF/LT Deletion Mutants

These mice display profound alterations of the lymphoid system. Lymph nodes are missing and a structural organization of the spleen is lacking. In the spleen neither germinal centers nor FDC are detectable (19,20). Adoptive transfer of wild-type bone marrow cells leads to the differentiation of FDC (Ryffel et al., submitted) and germinal centers (21) in the mutant recipients. Inversely, irradiated wild-type mice lose their FDC and germinal centers after adoptive transfer of bone marrow from mutant mice.

2.3. TNF Receptor Type 1 Deletion Mutant (TNFR1-/-)

TNFR1-/- mice also lack FDC (19) (Fig. 1) and germinal centers (19,21) (Fig. 2) but in contrast with LT-/- mice they display normal development and architecture of lymphatic organs. Among the FDC-deficient mouse strains they seem to represent the only case with apparently normal B and Tcell compartments. Thus they represent so far the best model to study the effect of the lack of FDC.

3. FUNCTIONAL CONSEQUENCES OF FDC DEFICIENCY

3.1. Formation of Germinal Centers

Germinal centers are lacking in murine strains lacking FDC (19,20). This probably reflects a role of FDC in the organization of germinal centers.

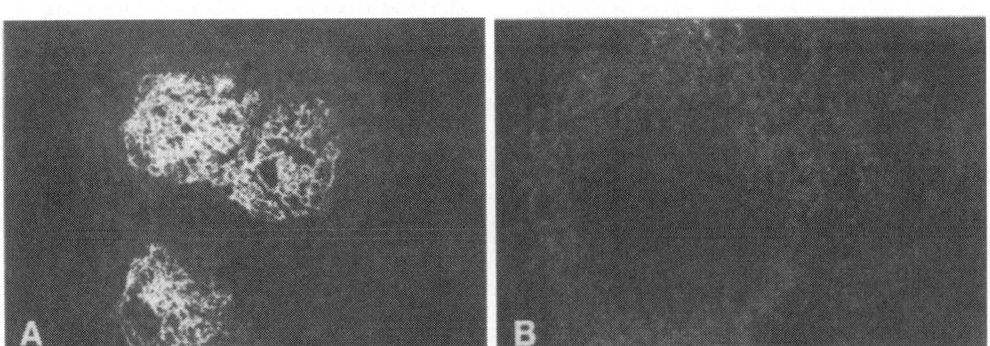

Figure 1. Complement receptor 1 immunoreactivity (clone 8C12) in the spleens of wild-type (A) and of TNFR1-/- (B) mice. FDC meshworks are absent in the TNFR1-/- mice. The weak diffuse immunoreactivity represents B cells. Magnification: x140.

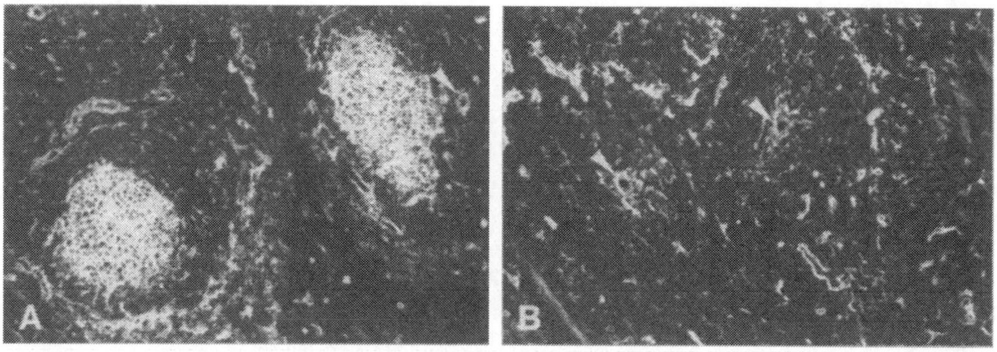

Figure 2. Peanut agglutinin (PNA) binding in the spleens of wild-type (A) and of TNFR1-/- (B) mice. Accumulations of PNA-positive germinal center B cells are visible in the wild type only. Arrowheads: central arterioles. Magnification: x140.

3.2. Isotype Switch

This takes place in the FDC-deficient strains LT-/- (20) and in TNFR1-/- (19). However FDC might be important for isotype switch at low doses of antigen (20).

3.3. Time Course of the IgG Response

Until now the most striking alteration of antibody response in TNFR1-/- mice is the early leveling-off of the IgG after a primary or a secondary challenge with SRBC injected intraperitoneally (19) (Fig. 3A). This does not represent an intrinsic inability to produce IgG for longer periods of time since the increase of antibody titers can be extended in time if SRBC are repeatedly injected (Fig. 3B). Thus, FDC seem to allow a sustained IgG production by constituting an antigen repository.

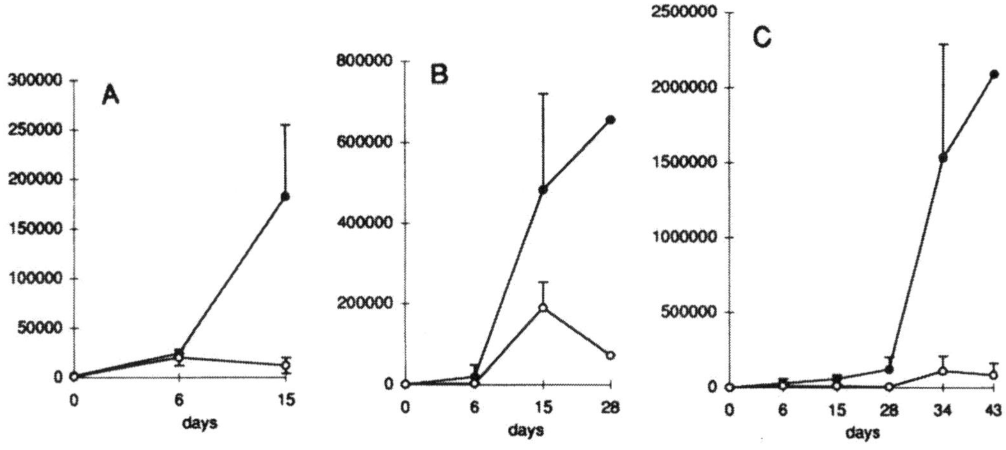

Figure 3. Specific anti-sheep erythrocytes (SRBC) IgG1 titer. SRBC were injected i. p. at day 0 (**A**), at days 0, 4, 8 and 12 (**B**) or at days 0 and 28 (**C**). Titers are expressed as the reciprocal value of the dilution showing an optical density of 0.1 above background. Similar patterns were obtained for IgG2a, IgG2b and IgG3. Mean values for 5 mice are given with the SD. Closed circles: wild type; open circles: TNFR1-/- mice.

3.4. B-Cell Memory

Memory has not been specifically investigated in LT-/- or in TNFR1-/- mice. However, the weak secondary response to SRBC in the latter strain (Fig. 3C) might well represent a failure to produce normal amounts of memory B cells.

3.5. Affinity Maturation

The importance of FDC for affinity maturation might depend on the dose of antigen since it occurs in LT-/- mice at a high dose, but not at low doses of antigen (20).

REFERENCES

1. Clark, E. A., K. H. Grabstein, A. M. Gown, M. Skelly, T. Kaisho, T. Hirano, and G. L. Shu. 1995. Activation of B lymphocyte maturation by a human follicular dendritic cell line, FDC-1. *J Immunol* 155:545–555.
2. Heinen, E., A. Bosseloir, and F. Bouzahzah. 1995. Follicular dendritic cells: Origin and function. In An antigen depository of the immune system: Folicular dendritic cells. M. H. Kosco-Vilbois, editor. Springer, Berlin. 15–47.
3. Yoshida, K., M. Kaji, T. Takahashi, T. K. van den Berg, and C. D. Dijkstra. 1995. Host origin of follicular dendritic cells induced in the spleen of SCID mice after transfer of allogeneic lymphocytes. *Immunology* 84:117–126.
4. Groscurth, P. 1980. Non-lymphatic cells in the lymph node cortex of the mouse. II. Postnatal development of the interdigitating cells and the dendritic reticular cell.*Pathol Res Pract* 169:235–254.
5. Heusermann, U., K. H. Zurborn, L. Schroeder, and H. J. Stutte. 1980. The origin of the dendritic reticulum cell. An experimental enzyme-histochemical and electron microscopic study on the rabbit spleen. *Cell Tissue Res* 209:279–294.
6. Villena, A., A. Zapata, J. M. Rivera-Pomar, M. G. Barrutia, and J. Fonfria. 1983. Structure of the non-lymphoid cells during the postnatal development of the rat lymph nodes. Fibroblastic reticulum cells and interdigitating cells. *Cell Tissue Res* 229:219–232.
7. Szakal, A. K., Z. F. Kapasi, S. T. Haley, and J. G. Tew. 1995. An antigen depository of the immune system: Follicular dendritic cells. Springer, Berlin. 1–13.
8. Kapasi, Z. F., M. H. Kosco-Vilbois, L. D. Shultz, J. G. Tew, and A. K. Szakal. 1994. Cellular origin of follicular dendritic cells. *Adv Exp Med Biol* 355:231–235.
9. Nossal, G. J., A. Abbot, J. Mitchell, and Z. Lummus. 1968. Antigens in immunity. XV. Ultrastructural features of antigen capture in primary and secondary lymphoid follicles. *J Exp Med* 127:277–290.
10. Kinoshita, T., J. Takeda, K. Hong, H. Kozono, H. Sakai, and K. Inoue. 1988. Monoclonal antibodies to mouse complement receptor type 1 (CR1). Their use in a distribution study showing that mouse erythrocytes and platelets are CR1-negative. *J Immunol* 140:3066–3072.
11. Kosco, M. H., E. Pflugfelder, and D. Gray. 1992. Follicular dendritic cell-dependent adhesion and proliferation of B cells in vitro. *J Immunol* 148:2331–2339.
12. Yoshida, K., T. K. van den Berg, and C. D. Dijkstra. 1993. Two different mechanisms of immune-complex trapping in the mouse spleen during immune responses. *Adv Exp Med Biol* 329:377–382.
13. Szakal, A. K., M. H. Kosco, and J. G. Tew. 1989. Microanatomy of lymphoid tissue during humoral immune responses: structure function relationships.*Annu Rev Immunol* 7:91–109.
14. Klaus, G. G., J. H. Humphrey, A. Kunkl, and D. W. Dongworth. 1980. The follicular dendritic cell: its role in antigen presentation in the generation of immunological memory. *Immunol Rev* 53:3–28.
15. Kosco-Vilbois, M. H., D. Gray, D. Scheidegger, and M. Julius. 1993. Follicular dendritic cells help resting B cells to become effective antigen-presenting cells: induction of B7/BB1 and upregulation of major histocompatibility complex class II molecules. *J Exp Med* 178:2055–2066.
16. Burton, G. F., D. H. Conrad, A. K. Szakal, and J. G. Tew. 1993. Follicular dendritic cells and B cell costimulation. *J Immunol* 150:31–38.
17. Tew, J. G., M. H. Kosco, G. F. Burton, and A. K. Szakal. 1990. Follicular dendritic cells as accessory cells. *Immunol Rev* 117:185–211.
18. Yoshida, K., T. K. van den Berg, and C. D. Dijkstra. 1994. The functional state of follicular dendritic cells in severe combined immunodeficient (SCID) mice: role of the lymphocytes. *Eur J Immunol* 24:464–468.

19. Lehir, M., H. Bluethmann, M. H. Koscovilbois, M. Muller, F. Dipadova, M. Moore, B. Ryffel, and H. P. Eugster. 1996. Differentiation of follicular dendritic cells and full antibody responses require tumor necrosis factor receptor-1 signaling. *J. Exp. Med.* 183:2367–2372.
20. Matsumoto, M., S. F. Lo, C. J. L. Carruthers, J. J. Min, S. Mariathasan, G. M. Huang, D. R. Plas, S. M. Martin, R. S. Geha, M. H. Nahm, and D. D. Chaplin. 1996. Affinity maturation without germinal centres in lymphotoxin-alpha-deficient mice. *Nature* 382:462–466.
21. Matsumoto, M., S. Mariathasan, M. H. Nahm, F. Baranyay, J. J. Peschon, and D. D. Chaplin. 1996. Role of lymphotoxin and the type i tnf receptor in the formation of germinal centers. *Science* 271:1289–1291.

81

A ROLE FOR CR2 IN FDC-B CELL INTERACTIONS

D. Qin,[1] J. Wu,[1] G. F. Burton,[1] A. K. Szakal,[2] and J. G. Tew[1]

[1]Department of Microbiology and Immunology
P.0. Box 980678, Medical College of Virginia
Virginia Commonwealth University
Richmond, Virginia 23298–0678
[2]Department of Anatomy, Division of Immunobiology
P.O. Box 980709, Medical College of Virginia
Virginia Commonwealth University
Richmond, Virginia 23298–0678

1. INTRODUCTION

If animals lack C3 or if C3 is destroyed by cobra venom factor, antibody responses are dramatically depressed[1–5]. Treatment of animals with mAb against CR2 or CD19, which are part of the CD19, CR2 TAPA-1 complex on B cells, also result in dramatically depressed antibody responses[6–8]. Furthermore, if animals are treated with a soluble construct of CR2, which will bind C3 fragments, the ability to mount a humoral response is markedly suppressed[8]. In addition, a recent study of CR2 knockout mice revealed that B cell expression of CR2 is required for immune responses to T-dependent antigens[9]. Furthermore, complement markedly lowers the threshold at which B cells respond to antigen and this effect may be attributable to the co-ligation of CR2 to BCR via the association with C3b-associated antigen[10,11]. It is also known that cross-linking of CR2 on the B cell by multiple C3b fragments on a carrier renders B cells more easily stimulated by mitogens including anti-μ[12,13]. These results suggest that CR2 is associated with an important signaling mechanism which is involved when B cells proliferate and differentiate into antibody forming cells (AFC).

Follicular dendritic cell (FDC) are known to provide potent costimulatory signals which promote B cell proliferation[14] but the nature of the molecules involved is not known. However, FDC are known to retain immune complexes on their surface for long periods of time[2,15]. Immune complexes can activate C3 and lead to generation of iC3b, C3dg and C3d and these molecules can serve as CR2 ligands (CR2L). Thus CR2L may be retained on FDC surfaces and CR2L-CR2 interactions might be a part of the mechanism of FDC costimulation. Fragments of C3 are known to be present on the FDC surface[16,17] and FDC-B cell interactions are intimate with FDC processes wrapping around B cells. We reason that the complement fragments in the immune complexes on the FDC may engage and co-ligate BCR and CR2 and this may relate to why B cells may be stimulated by FDC associated Ag-Ab complexes[18]. In this study we sought to determine if blocking the CR2-CR2L interactions using a mAb reactive with CR2 might interfere with the ability of FDC to provide costimulatory signals for B cell proliferation and antibody production. The data support the hypothesis that CR2-CR2L interactions are important in FDC-B cell interactions.

2. MATERIALS AND METHODS

2.1. Mice

Female BALB/cByJ, 6–8 weeks of age, were purchased from Jackson Laboratory, Bar, Harbor, ME. Food and water were supplied *ad libitum* and the mice were used between 8–20 weeks of age. Mice were immunized with OVA as explained by Wu, et al.[18]

2.2. FDC and B Cell Isolation and Cell Cultures

FDC and B cells were isolated from the draining lymph nodes of BALB/c mice as explained by Wu, et al.[18] In some experiments non-immunized mice were used as a source of FDC and B cells, and these FDC would bear only environmental antigens (EAg). Enriched FDC preparations (1×10^5 cells) were added to 3×10^5 B and T memory cells in 96 well tissue culture plates (Cat. No. 3595, CoStar; Cambridge, MA) containing 200 μl complete culture medium per well.

2.3. Monoclonal Antibodies Reactive with CR1 and CR2

The anti-CR1,2 (7G6) and anti-CR1 (8C12)[6,19] hybridomas were given by Dr. Kinoshita (Osaka University, Osaka, Japan).

2.4. Proliferation and Antibody Assays

The cultured cells were stimulated with anti-δ+ IL4 or LPS and ^{3}H-thymidine incorporation assay was used to monitor B cell proliferation[14]. Culture media was collected from the lymphocyte cultures each 6–7 days and anti-OVA specific IgG was measured by means of a solid phase ELISA as described by Helm, et al.[20] and Wu, et al.[18]

3. RESULTS

We first sought to determine if addition of anti-CR2 could inhibit FDC mediated costimulation. The data from a typical experiment are presented in Table 1. Note that addition of FDC to anti-δ and IL-4 stimulated B cell cultures resulted in a marked increase in the proliferative response and this is typical of FDC mediated costimulation (increased from about 20,000 CPM to 70,000 CPM). The addition of anti-CR2 to cultures containing the FDC significantly reduced the costimulatory response mediated by the FDC. In contrast, addition of anti-CR2 to cultures lacking FDC had no inhibitory effect. Adding more

Table 1. Effect of anti-CR2 on FDC mediated costimulation of mitogen induced B cell proliferation

Stimulator	Costimulator	Inhibitors	cpm±SE	p Value
α–δ + IL4	FDC	None	72033±3030	–
α–δ + IL4	FDC	α-CR1,2	54207±3780	< 0.02
α–δ + IL4	FDC	α-CR1	71621±2660	NS
α–δ + IL4	FDC	Isotype	73373±4970	NS
α–δ + IL4	None	None	21653±1250	<0.001
α–δ + IL4	None	α-CR1,2	20367±1380	< 0.001

Note: 1. The cpm±SE of other control groups were: B cell alone 1161±102, FDC alone 791±25, B+FDC (no mitogen) 2734±142.
2. Anti-CR1,2=7G6, anti-CR1=8C12, used at 1μg/well. NS=not significant.

anti-CR2 further inhibited FDC mediated costimulation but these higher doses also resulted in some inhibition of B cell proliferation in the absence of the FDC. The same results have also been obtained when LPS was used as the primary signal, i.e., about half of the FDC mediated costimulatory effect could be eliminated by addition of anti-CR2. In short, anti-CR2 appeared to be able to reduce FDC mediated costimulation under conditions where it has no direct effect on B cells in the absence of the FDC.

We next sought to determine if the ability of FDC to use immune complexes to induce memory responses *in vitro* could be inhibited by the addition of FDC and the data from a typical experiment are presented in Table 2. Note that in the absence of FDC the immune complexes were unable to elicit an *in vitro* secondary antibody response. However, the addition of FDC from normal animals results in a potent secondary response (over 10 μg of specific antibody/ml produced over a 6 day period). The addition of anti-CR2 reduced this response by about 50% and the addition of anti-CR1 resulted in a slight inhibitory effect. In short, anti-CR2 appeared to be able to reduce FDC's ability to present immune complexes to specific B cells and elicit *in vitro* secondary antibody responses.

4. DISCUSSION

Previous results prompted the hypothesis that the FDC enhanced B cell proliferation and antibody production may be mediated, at least in part, by the C3 fragments on the FDC engaging B cell-CR2. To begin testing this hypothesis, we sought to determine if anti-CR2 would block FDC mediated augmentation of anti-δ or LPS induced B cell prolif-

Table 2. Effect of anti-CR2 on FDC mediated costimulation of antibody response

Cell Cultures	Inhibitors	Anti-OVA IgG (ng/ml)	p Value
Ly^{OVA}+IC^{OVA}+FDC^{EAg}	none	11861±1372	–
Ly^{OVA}+IC^{OVA}+FDC^{EAg}	anti-CR1,2	4664±340	<0.005
Ly^{OVA}+IC^{OVA}+FDC^{EAg}	anti-CR1	7527±826	<0.01
Ly^{OVA}+IC^{OVA}+FDC^{EAg}	isotype	9883±1165	NS
Ly^{OVA} only	none	< 10	< 0.001

Note: 1.The culture supernatant was replaced with fresh culture medium at day 7 and the antibody production was measured at day 14.
2. Anti-CR1,2=7G6, anti-CR1=8C12, used at 10μg/well. IC^{OVA}= OVA-anti-OVA immune complex in antigen excess. NS=not significant.

eration. The data indicated that anti-CR2 reduced the FDC mediated augmentation of anti-δ or LPS induced B cell proliferation. In contrast, the addition of anti-CR2 had very little effect on mitogen induced B cell proliferation in the absence of FDC suggesting that the antibody blocked FDC-B cell interactions. Furthermore, potent secondary antibody responses were elicited *in vitro* when FDC were isolated and cultured with memory T & B cells in the presence of appropriate antigen-antibody complexes. Addition of anti-CR2 in such cultures inhibited this *in vitro* secondary response. These data support the concept that CR2L-CR2 interactions between FDC and B cells are important and provide support for a basic two signal model. Signal one would be from the FDC-Ag to B cell BCR and the second would be delivered via C3 fragments on the FDC engaging CR2 on the B cell. This two-signal model has parallels with other two signal models including MHC-Ag to TCR with B7-CD28 interactions providing the co-stimulatory signals.

A weakness in the data is that the inhibition of FDC mediated costimulation by the anti-CR2 was not complete. It is possible that there are other costimulatory molecules and that the CR2L-CR2 interactions are only a part of the FDC effect. We noted that high concentrations of anti-CR2 were able to interfere with B cells in the absence of FDC in our *in vitro* experiments. The dramatic inhibition mediated by anti-CR2 in the *in vivo* studies may relate to an ability of high concentrations of an antibody to directly interfere with the B cell. Alternatively, the lack of complete inhibition in this study may be attributable to the ability of the Fc portion of the anti-CR2 to bind FcR on FDC and thus providing a direct link between the B cell CR2 and FDC. The direct binding of FDC to B cell through anti-CR2 might provide a positive signal to the B cell. In future experiments we plan to use Fab where both coligation and cross-linking of receptors will be minimized. In one experiment with F(ab)2 we have seen better inhibition than with the intact antibody. In fact, our basic hypothesis would be strengthened if the intact anti-CR2 antibody can provide a costimulatory signal (even a weak signal) when in contact with the FDC while the Fab anti-CR2 cannot. We look forward to these extended experiments with interest.

5. REFERENCES

1. J.G. Tew, T.E. Mandel, and G.A. Miller, Immune retention: immunological requirements for maintaining an easily degradable antigen *in vivo*, *Aust. J. exp. Biol. med. Sci.* 57:401–414 (1979).
2. T.E. Mandel, R.P. Phipps, A. Abbot, and J.G. Tew, The follicular dendritic cell: long term antigen retention during immunity, *Immunol. Rev.* 53:29–59 (1980).
3. A. Berrebi, L. Bassous-Guedj, E. Vorst, S. Dagan, M. Shtalrid, and A. Freedman, Further characterization of prolymphocytic leukemia cells as a tumor of activated B cells, *Am. J. Hematol.* 34:181–185 (1990).
4. R. Burger, J. Gordon, G. Stevenson, G. Ramadori, B. Zanker, U. Hadding, and D. Bitter-Suermann, An inherited deficiency of the third component of complement, C3, in guinea pigs, *Eur. J. Immunol.* 16:7–11 (1986).
5. E.C. Bottger, S. Metzger, D. Bitter-Suermann, G. Stevenson, S. Kleindienst, and R. Burger, Impaired humoral immune response in complement C3-deficient guinea pigs: absence of secondary antibody response, *Eur. J. Immunol.* 16:1231–1235 (1986).
6. B. Heyman, E.J. Wiersma, and T. Kinoshita, In vivo inhibition of the antibody response by a complement receptor-specific monoclonal antibody, *J. Exp. Med.* 172:665–668 (1990).
7. S. Gustavsson, T. Kinoshita, and B. Heyman, Antibodies to murine complement receptor 1 and 2 can inhibit the antibody response in vivo without inhibiting T helper cell induction, *J. Immunol.* 154:6524–6528 (1995).
8. T. Hebell, J.M. Ahearn, and D.T. Fearon, Suppression of the immune response by a soluble complement receptor of B lymphocytes, *Science* 254:102–105 (1991).
9. D.A. Croix, J.M. Ahearn, A.M. Rosengard, S. Han, G. Kelsoe, M. Ma, and M.C. Carroll, Antibody Response to a T-dependent Antigen Requires B Cell Expression of Complement Receptors, *J. Exp. Med.* 183:1857–1864 (1996).

10. D.T. Fearon and R.H. Carter, The CD19/CR2/TAPA-1 complex of B lymphocytes: linking natural to acquired immunity, *Annu. Rev. Immunol.* 13:127–149 (1995).
11. D.T. Fearon, The CD19-CR2-TAPA-1 complex, CD45 and signaling by the antigen receptor of B lymphocytes, *Curr. Opin. Immunol.* 5:341–348 (1993).
12. G.C. Tsokos, J.D. Lambris, F.D. Finkelman, E.D. Anastassiou, and C.H. June, Monovalent ligands of complement receptor 2 inhibit whereas polyvalent ligands enhance anti-Ig-induced human B cell intracytoplasmic free calcium concentration, *J. Immunol.* 144:1640–1645 (1990).
13. R.H. Carter and D.T. Fearon, Polymeric C3dg primes human B lymphocytes for proliferation induced by anti-IgM, *J. Immunol.* 143:1755–1760 (1989).
14. G.F. Burton, D.H. Conrad, A.K. Szakal, and J.G. Tew, Follicular dendritic cells (FDC) and B cell co-stimulation, *J. Immunol.* 150:31–38 (1993).
15. J.G. Tew, T.E. Mandel, and P.L. Rice, Immune elimination and immune retention: the relationship between antigen retained in the foot and the elicitation of footpad swelling, *Immunology* 40:425–433 (1980).
16. J.G. Tew, M.H. Kosco, G.F. Burton, and A.K. Szakal, Follicular dendritic cells as accessory cells, *Immunol. Rev.* 117:185–211 (1990).
17. M. Yamakawa and Y. Imai, Complement activation in the follicular light zone of human lymphoid tissues, *Immunology* 76:378–384 (1992).
18. J. Wu, D. Qin, G.F. Burton, A. Szakal, and J.G. Tew, Follicular Dendritic Cell (FDC) Derived Ag and Accessory Activity in Initiation of Memory IgG Responses in Vitro, *J. Immunol.* 157:(1996).
19. T. Kinoshita, J. Takeda, K. Hong, H. Kozono, H. Sakai, and K. Inoue, Monoclonal antibodies to mouse complement receptor type 1 (CR1). Their use in a distribution study showing that mouse erythrocytes and platelets are CR-1 negative. *J. Immunol.* 140:3066(1988).
20. S.L.T. Helm, G.F. Burton, A.K. Szakal, and J.G. Tew, Follicular dendritic cells and the maintenance of IgE responses, *Eur. J. Immunol.* 25:2362–2369 (1995).

82

DISAPPEARANCE OF THE ANTIGEN EXCHANGING ABILITY OF FDC INDUCED BY REPEATED ANTIGEN INJECTIONS

Toshiko Yoshida and Kenichi Takaya

Second Department of Anatomy
Faculty of Medicine
Toyama Medical and Pharmaceutical University
Toyama, Japan

1. INTRODUCTION

Follicular dendritic cells (FDCs) retain the antigen on their membranes usually in the form of antigen-antibody complexes which disappear slowly with time (1). FDC is unable to retain the antigen for a long time, and replaces the old antigens with new ones, even if the antigen is the same (2). Recently FDC was reported to degenerate in individuals manifesting persistent generalized lymphadenopathy after infection with human immunodeficiency virus (HIV). It has been considered that the antigen retention is indispensable to form the FDC network. In this case FDC retained the antigens but the network was disrupted (3). It may be due to the incomplete antigen retaining ability of FDC for the practice of its role.

We thought that the FDC's ability of the binding, retention and replacement of antigens was related to the adjustment of the immune system. We make this clear by the repeated injection of the same antigens into the rat footpads and studying under light and electron microscopes.

2. MATERIALS AND METHODS

We examined the immunized and non-immunized 6 weeks female rats after repeated antigen injections. To clarify the localization of the antigen, horseradish peroxidase (HRP) binding colloidal gold (diameter, 5 and 20 nm) and HRP binding beads with fluorescence (diameter, 50 and 1500 nm) were used in addition to the HRP solution as antigens. The labeled HRP was injected at the beginning and /or end of the repeated injections.

The popliteal lymph node of the rats were dissected after perfusion of 1/2 Karnovsky's fixative solution. They were sectioned at 100 m thick with a vibratome (Lancer Series 1000)

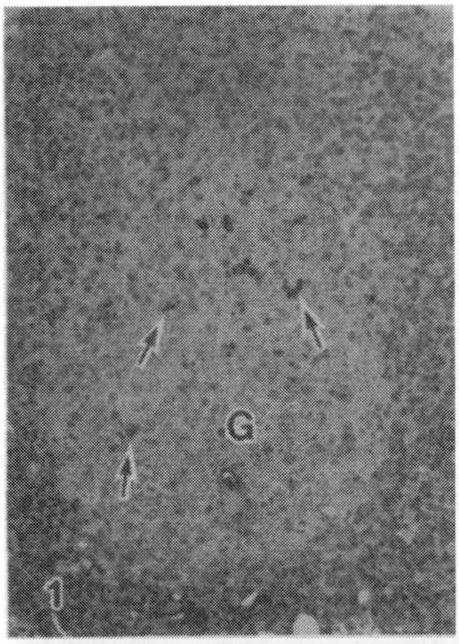
G
1

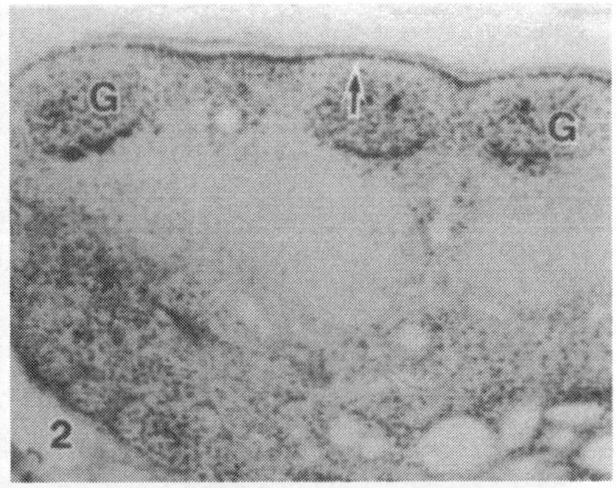
G
G
2

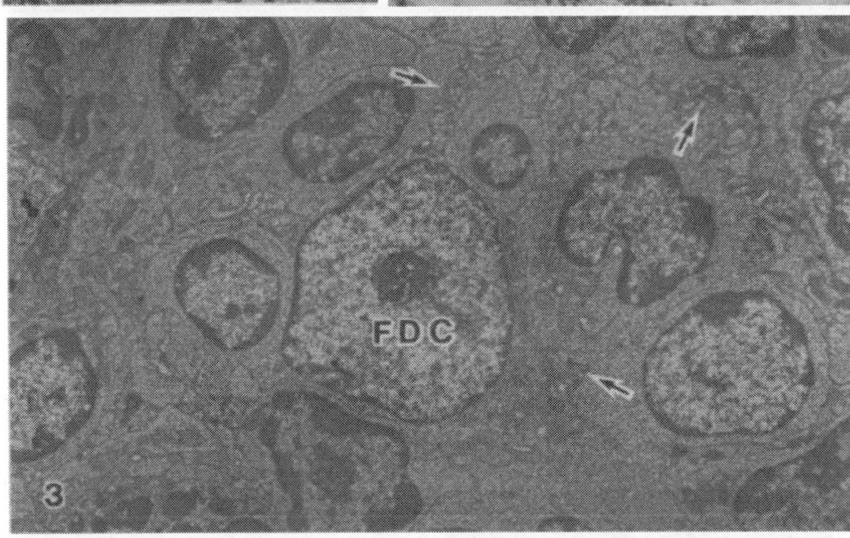
FDC
3

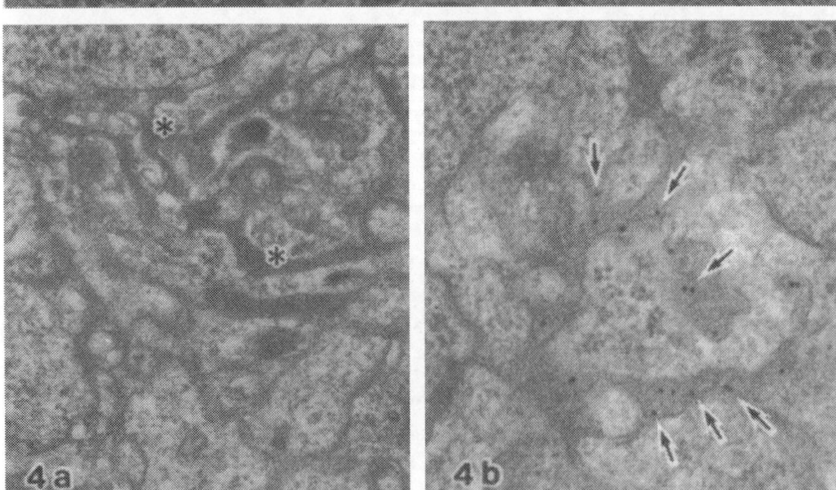
*
*
4a
4b

and reacted with 0.5% 3.3 diaminobenzidine (DAB; Sigma Co.) in 0.5% Tris HCl buffer (pH 7.6). The sections were observed by light and fluorescent microscopy and postfixed with 2% OsO4 and were treated by the conventional methods for electron microscopy.

The photographed slides of the vibratome sections were processed for image analysis and the areas were calculated about the number of pixels and the tone in the gray scale.

3. RESULTS

3.1. Non-Immunized Rats

After 25–30 times injection of HRP, the last injected HRP-G was observed, but after 60–65 time injections, no HRP-G was seen. The germinal center (GMC) was developed in the cortical area (Fig.1). By the image processing with the computer, the HRP reaction areas were almost the same. However, the tone of the GMC was darker in 25 times than 60 times injection rats. As the injection times increased, the intensity of the color in the GMC became weaker.

3.2. Immunized Rats

In all the repeated injections group, HRP-G which was last injected was not seen and the HRP reaction deposits were scattered in the follicle. The deposits were the material phagocytosed by the tingible body macrophages. In the border between the subcapsular sinus and cortical area there was the HRP reaction (Fig.2). In this area, the cells which elongated their processes into the subcapsular sinus contained HRP, HRP-G and HRP-b. These cells were considered to be antigen transporting cells (ATC). In the GMC typical FDCs elongated their processes among the lymphocytes and made up the complex labyrinth by the invaginating membrane (Fig.3) . The dense material, HRP deposits, existed in and between the cytoplasmic invagination. However, no HRP-G or HRP-b injected at the start and end of the repeated injections was observed (Fig. 4a). On the other hand, in the pre-immunized rats without repeated injections, the purple color of HRP-G was distinct in the GMC. FDCs retained many HRP-G on their cytoplasmic membrane (Fig.4b).

Figure 1. The popliteal lymph node of the non-immunized rats which were injected of the HRP solution 60 times in the footpads repeatedly. Germinal center (G) developed but no HRP reaction was observed in the follicle. Tingible body macrophages (arrows) contained scattered metachromatic materials. Stained with toluidine blue. x63.

Figure 2. Popliteal lymph nodes in the pre-immunized rats 5 days after the 20 times repeated injection. At the last injection HRP-G was used instead of HRP solution. HRP reaction was seen in the follicle and at the border (arrow)between the subcapsular sinus and the cortical area. The reaction was seen in the tingible body macrophage. No color of the HRP-G or no reticular reaction of HRP which indicated the antigens on FDC was observed. G: germinal center x 34.

Figure 3. Electron micrograph of the lymph node of the pre-immunized rat with repeated injection. The HRP-b was injected at the last injection into the footpads. Typical FDC (FDC) was seen in the germinal center. The processes of FDC made up the complex labyrinth in which homogenous materials were present. x5,300.

Figure 4. High magnification of the processes of FDC of the pre-immunized rat. a: With the repeated injection, the homogenous materials(*) in the intercellular matrix were located but no labeled antigen (HRP-b) injected last was observed. x 40,000. b: Many HRP-Gs (arrows) which were injected last were located in the intercellular matrix of FDC after a small number of injections. x 40,000.

4. DISCUSSION

FDC not only retains the same antigens for a long period of time and if the same antigen invades again into the lymph node, FDC is able to exchange the pre-existing antigens with new ones quickly (2). In our study after the repeated antigen injections, the last injected antigen, HRP-G was not found on the membrane of FDC. FDC may have exchanged the pre-existing antigens with newly arrived antigens, because the labeled antigen which was injected at an early time was not seen in the GMC. These phonemena certainly indicated that the repeated injection induced the reduction of the ability of antigen binding and replacement. Recently Masuda et al. (1994)(4) reported that the ability of FDC to trap new antigens disappeared around the second and third weeks of infection with murine acquired immune deficiency syndrome (MAIDS). They said that a loss of retained antigen would be expected to be followed by a loss of specific serum antibody.

ATC was reported by Szakal et al. (5) as a non-phagocytic cell which was related with antigen transport into the follicle and might be a precursor of FDC. We observed ATC played an important role in the antigen transport into the cortical area (6). After repeated injection the cell at the border between the subcapsular sinus and cortical area had HRP. ATCs may trap much antigens and suppress the immune reaction.

This investigation suggested that repeated injections induced the reduction of the ability of the binding and replacement of antigens on FDC. The negative action may be related with the adjustment of the lymph node to reduce the over-reaction of the immune response.

5. CONCLUSION

After repeated antigen injection, ATC existing in the border between the subcapsular sinus and cortical area had the HRP-G and HRP-b. On the FDC in the GMC the last injected labeled antigens, no HRP-G or HRP-b was not seen. Repeated injections of the same antigens induced the reduction of the ability of binding and replacement of the antigens of FDC. Reduction of the ability of FDC in binding and exchange of antigens may regulate the immune system to suppress the over reaction to the antigens.

6. REFERENCES

1. J.G. Tew, T. E. Mandel. Prolonged antigen half-life in the lymphoid follicles of specifically immunized mice. *Immunology* 37:69 (1979).
2. J.G. Tew, T. E. Mandel. The maintenance and regulation of serum antibody levels: evidence indicating a role for antigen retained in lymphoid follicles. *J. Immunol.* 120:1063 (1978).
3. G.S. Wood, C.F. Garacia, R.F. Dorfoman, R.A. Warnke. The immunohistology of follicle lysis in lymph node biopsies from homosexual man. *Blood* 66:1092 (1985).
4. A. Masuda, G.F. Burton, B.A. Fuchs, B.S. Bhogal, R. Rupper, A.K. Szakal, J.G. Tew. Follicular dendritic cell function and murine AIDS. *Immunology* 81:41 (1994).
5. A.K. Szakal, K.L. Holmess, J.G. Tew. Transport of immune complexes from the subcapsular sinus to lymph node follicles on the surface of non-phagocytic cells, including cells with dendritic morphology. *J. Immunol.* 131:1714(1983).
6. T. Yoshida, K. Takaya, A. Sakamaki. The relationship between follicular dendritic cells and antigen transporting cells in the rat lymph node during secondary response. *Dendritic cells* 4:159 (1994).

IN VITRO PRIMING TO TUMOR-ASSOCIATED PROTEINS

Susan L. Hand,[1] Helga Bernhard,[1,2] Eric S. Huseby,[1] William D. Rubin,[1] Mary L. Disis,[1] and Martin A. Cheever[1]

[1]University of Washington
Seattle, Washington
[2]University of Mainz
Mainz, Germany

Cancer can be cured in mice by adoptive transfer of T cells specific for the malignant cells or by vaccination to tumor-specific antigens. The application of immunotherapy to the treatment of human cancer hinges on the identification of human tumor antigens to which specific immunity can be elicited.

Recent efforts to identify human tumor antigens have revealed that some tumor-associated antigens recognized by T cells or serum antibodies are non-mutated self proteins expressed by normal as well as malignant cells.[1,2] These self antigens include proteins expressed during fetal development but only on rare adult tissues (e.g., MAGE-1[3]), proteins related to the differentiation state of the tissue of malignant origin (e.g., tyrosinase,[4] Melan-A/MART-1[5,6] and gp100[7]) as well as proteins that are expressed at higher levels by malignant cells than by normal cells (e.g., HER-2/neu[8–10]).

These findings indicate that the immune repertoire to some self proteins has not been deleted and that responses to self proteins can be generated in some situations, without clinical signs of autoimmunity. Despite this, the presence of the tumor alone does not seem to be sufficient to generate an adequate immune response, since tumors grow and progress in vivo. Reasons for inadequate or non-existent immune responses to potential tumor antigens could include: 1) functional tolerance, 2) a lack of the appropriate co-stimulation necessary for T cell priming, 3) tumor-induced immuno-suppressive factors, 4) dominant responses to other tumor-associated proteins, and 5) lack of immunogenicity, due to the protein not being processed in such a way that it can be presented to T cells. Given any one of the first four possibilities, a protein might still be an appropriate target for immunotherapy if measures were devised to elicit or augment immunity against it. Possible strategies include stimulation of an individual's T cells in vitro with tumor or purified tumor antigen and autologous antigen presenting cells (APC) in order to generate specific T cells for adoptive transfer, or immunizing patients with a vaccine containing the tumor antigen. The latter would be a straightforward method of eliciting immunity; however, the practical barriers to initiating human vaccine trials are great. The development of effective and reproducible methods for generating immune T cells in vitro would, therefore, be very advantageous.

In vitro, T cells may be primed under the appropriate conditions of co-stimulation by the appropriate APC, away from any negative influences of the tumor microenvironment. In addition, a system for priming T cells in vitro could be used to answer questions that are essential for the development of effective vaccines and adoptive immunotherapy: Which tumor-associated proteins are most immunogenic; which epitopes of these proteins are presented; what forms of an antigen are most effective for priming class I- versus class II-restricted responses; and which proteins elicit responses that are physiologically relevant, in other words, responses in which T cells can recognize autologous tumor or APC presenting antigens shed by the tumor. Moreover, in vitro it should be possible to bias the type of immune response generated (e.g. cellular or humoral) by using separated populations of CD4+ or CD8+ T cells and different combinations of cytokines (e.g. IL-4, IL-12, or interferon-γ) during the sensitization step[11–14].

To address these issues, we have established a system using autologous, bone marrow-derived dendritic cells (DC) as APC to rapidly generate antigen specific T cells from the PBL of normal individuals. DC are considered to be the most effective stimulators of T cell immunity.[15,16] The use of cultures enriched for DC as APC can improve the sensitivity of assays for detecting primed responses and has allowed the induction of priming in vitro.

We and others have shown that functional DC can be elicited and grown in vitro from CD34+ hematopoietic progenitor cells (HPC) derived from bone marrow or peripheral blood.[17–19] Culture of CD34+ HPC with GM-CSF and TNF-α yields a heterogeneous population containing cells with typical dendritic morphology. Phenotypic studies demonstrate a loss of the CD34 molecule over one week and an increase in cells expressing surface markers associated with DC: CD1a, CD80 (B7/BB1), CD4, CD14, HLA-DR, and CD64 (FcγRI).[17,18] While these DC have excellent ability to stimulate an allogeneic MLR, in our hands they had variable function when presenting specific protein antigens. Moreover, the proliferation of T cells in response to autologous DC in the absence of exogenous antigen (the autologous MLR) was quite high. There is much evidence demonstrating that DC can exist in two functionally distinct stages of maturation.[20–23] Immature DC process and present intact protein antigens, but have relatively low levels of surface class II and co-stimulatory molecules. Mature DC are specialized for activating naive T cells, having increased cell surface levels of class II and co-stimulatory molecules, but have down-regulated the ability to process and present intact antigens. Maturation of DC in vitro has been shown to be induced by cytokines,[22,23] CD40-ligand interactions,[24] or T cells[25] (and our unpublished data) and sometimes occurs 'spontaneously' after culture of DC in vitro. These observations suggested that the DC populations grown in GM-CSF and TNFα might have included cells of a relatively mature phenotype, i.e. cells that were highly effective at stimulating an MLR, but less efficient at presenting exogenous antigen.

As one of our goals was to establish a system allowing processing and presentation of intact protein antigens for priming T cell responses, we focused on optimizing conditions to generate APC with these capabilities. To this end, we tested a number of different cytokines for stimulating in vitro growth of DC. In these experiments, bone marrow derived CD34+ cells from healthy donors were used as the starting population. The CD34+ HPC were cultured with GM-CSF alone, GM-CSF and TNFα, GM-CSF and IL4, or GM-CSF and IL6. These cytokines were chosen based on their reported abilities to influence DC differentiation and/or growth.[17,23,26] Growth of DC appeared to be similar with TNF-α,

IL4, and IL6 in addition to GM-CSF; but, in the majority of experiments using tetanus toxoid as antigen, the background autologous MLR was lowest using IL6. This is an important point since a lower background proliferation can aid in detecting specific responses, whereas high backgrounds can obscure them. An additional factor contributing to the reduction of background proliferation was the use of serum-free medium. Based on these experiments, our current method of deriving DC capable of priming in vitro, but with the lowest relative autologous MLR, uses GM-CSF and IL6 in serum-free medium.

Using DC populations derived by this method, we have been able to prime naive (defined here as being positive for the CD45RA molecule) CD4+ T cells to the foreign antigens keyhole limpet hemocyanin (KLH) and ovalbumin (OVA). Of note, priming was detectable by the seventh day of culturing T cells with dendritic APC. Specificity was validated by showing that DC-stimulated CD4+ T cells responded to the priming antigen, but not to an irrelevant antigen, when tested three weeks following initial stimulation with antigen and DC. It should be noted that monocytes grown under the same conditions as the CD34+ HPC cultures did not present KLH to naive (CD45RA+) CD4+ T cells, whereas DC cultures did. This result validates that the CD34+ HPC-derived APC cultures have functional properties characteristic of DC.

Our current studies, using this priming system, focus on the elicitation of immunity to the self, tumor-associated protein HER-2/neu, as a prototype antigen. Our goal is to apply the methods and concepts developed in this system to investigations of the immunogenicity of other potential tumor antigens. HER-2/neu is a non-mutated, self protein that is overexpressed by certain human tumors. It is a cell surface, transmembrane protein of 185 kD, consisting of a cysteine-rich extracellular domain (ECD) that functions in ligand binding and an intracellular domain (ICD) with kinase activity.[27] It is a member of the epidermal growth factor receptor family and is presumed to function as a growth factor receptor.[27] Monoclonal antibodies to HER-2/neu are capable of inducing either stimulatory or inhibitory signals to the receptor.[28,29] The HER-2/neu protein is expressed during human fetal development and at low levels on the epithelial cells of many normal adult tissues.[30] In normal cells, the HER-2/neu gene is present as a single copy.[31] Amplification of the gene and/or protein overexpression has been identified in many human malignancies, including cancers of the breast, ovary, uterus, stomach, and lung.[32–36] In breast cancer, HER-2/neu is overexpressed on the malignant tissue of 20–40% of invasive ductal carcinoma patients, as well as in 50–60% of ductal carcinomas in situ, suggesting that overexpression of HER-2/neu may be related to cancer formation.[37] HER-2/neu protein overexpression correlates with a significantly worse clinical outcome in most subsets of breast cancer patients.[38]

Evidence to support the use of HER-2/neu as a target for immunotherapy has recently been accumulating. Studies have shown that antibodies directed against HER-2/neu protein can be detected in the sera of some patients with breast cancer.[8,39] The frequency of antibody to HER-2/neu is higher in patients with breast cancer than in normal individuals and higher in patients with HER-2/neu-positive cancers than in patients with HER-2/neu-negative cancers, suggesting that immunity developed as a consequence of the expression of HER-2/neu by the cancer cells.

In addition, preliminary studies have shown that some patients with breast cancer have existent CD4+ T cell responses to HER-2/neu protein and/or peptides[8,39] (and unpublished observations). When overexpressed, the ECD of HER-2/neu is often shed and present in soluble form in the extracellular environment, both in vitro, in the culture media of rapidly growing HER-2/neu-positive breast cancer cell lines[40,41] and in the sera of patients with breast cancer.[2,43] Since CD4+ T cells generally recognize soluble, exogenous antigens processed and presented by APC in the class II MHC pathway, ECD shed by tumor should be available for presentation to CD4+ T cells. In addition to ECD shedding, HER-2/neu may be present in the extracellular environment due to tumor necrosis. In animal models, CD4+ T cells are effective in tumor immunotherapy provided that the protein recognized is abundant and available for processing the class II MHC antigen processing pathway. Thus, HER-2/neu-specific CD4+ T cells have the potential to be useful in immunotherapy.

CD8+ cytotoxic T lymphocyte (CTL) immunity to HER-2/neu has been found in ovarian cancer patients.[9,10,44] Furthermore, HER-2/neu peptide-specific CTL have been generated in vitro from normal PBL, and some CTL could lyse HER-2/neu positive cancer cells.[39,45] In these studies, conducted prior to the optimization of our dendritic APC culture method, synthetic HER-2/neu peptides capable of binding HLA-A2 were used to repeatedly stimulate PBL from normal individuals homozygous for HLA-A2. CTL were generated that could lyse autologous B cells incubated with the immunizing peptide. In one of the two instances, the peptide specific CTL could also lyse HER-2/neu-positive, HLA A2-positive breast cancer cells.[39]

Taken together, the data suggest that some individuals can develop immunity to HER-2/neu by virtue of its overexpression on autologous cancer cells. Moreover, they support the idea that immune responses to an overexpressed self protein need not lead to destructive autoimmunity.

We are currently testing the use of cultured DC as APC to elicit CD4+ and CD8+ T cells specific for HER-2/neu from normal individuals. The expectation is that specific T helper cells and CTL will be more easily generated using DC as APC than by the method previously used to generate HER-2/neu-specific CTL, which required multiple restimulations. The use of DC as APC should aid in the identification of immunogenic forms of HER-2/neu which are suitable for eliciting both class I- and class II-restricted T cell responses.

In preliminary experiments using the DC system described above, CD4+ T cells from normal donors were primed to HER-2/neu peptides derived from both the ECD and the ICD. CD4+ T cell lines specific for the ECD-derived peptide p319–333 were established from two donors. A T cell line specific for the ICD-derived peptide p776–790 was also generated from one of the donors. The ability of these cell lines to recognize whole, recombinant HER-2/neu is the next step to determine whether peptide-specific T cells can respond to protein. In addition, studies will determine whether CD4+ T cells can be primed to intact HER-2/neu protein presented by autologous DC.

Experiments have also been done in attempt to elicit CD8+ CTL to HER-2/neu using autologous DC as APC. Bone marrow-derived DC from four normal donors were infected with recombinant vaccinia virus expressing either the ICD or the ECD of HER-2/neu, and these APC were used to stimulate autologous CD8+ T cells. Two subsequent stimulations were done at one week intervals, using autologous monocytes infected with the same recombinant virus. The T cells were tested for cytotoxic function following the third stimulation. Each T cell line tested recognized and lysed autologous Epstein-Barr Virus (EBV) immortalized B cells that were infected with either the HER-2/neu-recombinant vaccinia virus or the control wild type vaccinia (20–60% lysis at 20:1 effector to target ratio), but did not lyse uninfected, autologous EBV-immortalized B cells, demonstrating that the bulk of the response had been elicited to the virus. One cell line has been cloned by limiting dilution and screened for reactivity with the HER-2/neu-recombinant virus. One hundred and fifty six clones grew and were tested. The results of two rounds of screening showed that 41 clones (26%) were either non-lytic or non-specifically lytic, 105 (67%) were vaccinia-specific, and 10 clones appeared to preferentially recognize the recombinant virus over the wild type virus. Further testing is needed to validate specificity for HER-2/neu.

This system using bone marrow-derived DC for in vitro priming has promise to be a useful tool to aid in the determination of which tumor-associated proteins may be appropriate targets for T cell immunotherapy and which peptides, fragments, or forms of a protein are most immunogenic. When combined with T cell cloning, it can be used to determine which antigen-specific T cells are capable of responding to tumor cells expressing the protein.

REFERENCES

1. Houghton, A.N. Cancer antigens: immune recognition of self and altered self. J Exp Med 180, 1 (1994).
2. Pardoll, D.M. A new look for the 1990s. Nature 369, 357 (1994).
3. Van der Bruggen, P., Traversari, C., Chomez, P., Lurquin, C., De Plaen, E., Van den Eynde, B., Knuth, A. & Boon, T. A gene encoding an antigen recognized by cytolytic T lymphocytes on a human melanoma. Science 254, 1643 (1991).
4. Brichard, V., Van Pel, A., Wolfel, T., Wolfel, C., De Plaen, E., Lethe, B., Coulie, P. & Boon, T. The tyrosinase gene codes for an antigen recognized by autologous cytotoxic T lymphocytes on HLA-A2 melanomas. J Exp Med 178, 489 (1993).
5. Coulie, P.G., Brichard, V., Van Pel, A., Wolfel, T., Schneider, J., Traversari, C., De Plaen, E., Lurquin, C., Szikora, J.-P., Renauld, J. & Boon, T. A new gene coding for a differentiation antigen recognized by autologous cytologic T lymphocytes on HLA-A2 melanomas. J Exp Med 180, 35 (1994).
6. Kawakami, Y., Eliyanu, S., Delgado, C.H., Robbins, P.F., Rivoltini, L., Topalian, S.L., Miki, T. & Rosenberg, S.A. Cloning of the gene coding for a shared human melanoma antigen recognized by autologous T cells infiltrating into tumor. Proc. Natl. Acad. Sci. USA 91, 3515 (1994).
7. Bakker, A.B.H., Schreurs, M.W.J., de Boer, A.J., Kawakami, Y., Rosenberg, S.A., Adema, G.J. & Figdor, C.G. Melanocyte lineage-specific antigen gp100 is recognized by melanoma-derived tumor-infiltrating lymphocytes. J. Exp. Med. 179, 1005 (1994).
8. Disis, M.L., Calenoff, E., McLaughlin, G., Murphy, A.E., Chen, W., Groner, B., Jeschke, M., Lydon, N., McGlynn, E., Livingston, R.B., Moe, R. & Cheever, M.A. Existent T cell and antibody immunity to HER-2/*neu* protein in patients with breast cancer. Cancer Res 54, 16 (1994).
9. Fisk, B., Blevins, T.L., Wharton, J.T. & Ioannides, C.G. Identification of an immunodominant peptide of HER-2/neu protooncogene recognized by ovarian tumor-specific cytotoxic T lymphocyte lines. J Exp Med 181, 2109 (1995).
10. Peoples, G.E., Goedegebuure, P.S., Smith, R., Linehan, D.C., Yoshino, I. & Eberlein, T.J. Breast and ovarian cancer-specific cytotoxic T lymphocytes recognize the same HER2/neu-derived peptide. Proc Natl Acad Sci, USA 92, 432 (1995).
11. Le-Gros, G., Ben-Sasson, S.Z., Seder, R., Finkelman, F.D. & Paul, W.E. Generation of interleukin 4 (IL-4)-producing cells in vivo and in vitro: IL-2 and IL-4 are required for in vitro generation of IL-4-producing cells. J Exp Med 172, 921 (1990).
12. Swain, S.L., Weinberg, A.D., English, M. & Huston, G. IL-4 directs the development of Th2-like helper effectors. J. Immunol. 145, 3796 (1990).
13. Manetti, R., Parronchi, P., Giudizi, M.G., Piccinni, M.P., Maggi, E., Trinchieri, G. & Romagnani, S. Natural killer cell stimulatory factor (interleukin 12 [IL-12]) induces T helper type 1 (Th1)-specific immune responses and inhibits the development of IL-4-producing Th cells. J Exp Med 177, 1199 (1993).
14. Macatonia, S.E., Hosken, N.A., Litton, M., Vieira, P., Hsieh, C.S., Culpepper, J.A., Wysocka, M., Trinchieri, G., Murphy, K.M. & O'Garra, A. Dendritic cells produce IL-12 and direct the development of Th1 cells from naive CD4+ T cells. J Immunol 154, 5071 (1995).
15. Van-Voorhis, W.C.V., Valinsky, J., Hoffman, E., Luban, J., Hair, L.S. & Steinman, R.M. Relative efficacy of human monocytes and dendritic cells as accessory cells for T cell proliferation. J Exp Med 158, 174 (1983).
16. Steinman, R.M. The dendritic cell system and its role in immunogenicity. Ann Rev Immunol 9, 271 (1991).
17. Caux, C., Dezutter-Dambuyant, C., Schmitt, D. & Banchereau, J. GM-CSF and TNF-alpha cooperate in the generation of dendritic Langerhans cells. Nature 360, 258 (1992).

18. Bernhard, H., Disis, M.L., Heimfeld, S., Hand, S., Gralow, J.R. & Cheever, M.A. Generation of immunostimulatory dendritic cells from human CD34+ hematopoietic progenitor cells of the bone marrow and peripheral blood. Cancer Res 55, 1099 (1995).
19. Szabolcs, P., Moore, M.A.S. & Young, J.W. Expansion of immunostimulatory dendritic cells among the myeloid progeny of human CD34+ bone marrow precuraors cultured with c-kit ligand, GMCSF, and TNF-α. J Immunol 154, 5851 (1995).
20. Romani, N., Koide, S., Crowley, M., Witmer-Pack, M., Livingstone, A.M., Fathman, C.G., Inaba, K. & Steinman, R.M. Presentation of exogenous protein antigens by dendritic cells to T cell clones. Intact protein is presented best by immature, epidermal Langerhans cells. J Exp Med 169, 1169 (1989).
21. Streilein, J.W. & Grammer, S.F. In vitro evidence that Langerhans cells can adopt two functionally distinct forms capable of antigen presentation to T lymphocytes. J Immunol 143, 3925 (1989).
22. O'Doherty, U., Steinman, R.M., Peng, M., Cameron, P.U., Gezelter, S., Kopeloff, I., Swiggard, W.J., Pope, M. & Bhardwaj, N. Dendritic cells freshly isolated from human blood express CD4 and mature into typical immunostimulatory dendritic cells after culture in monocyte-conditioned medium. J Exp Med 178, 1067 (1993).
23. Sallusto, F. & Lanzavecchia, A. Efficient presentation of soluble antigen by cultured human dendritic cells is maintained by granulocyte/macrophage colony-stimulating factor plus interleukin 4 and downregulated by tumor necrosis factor a. J Exp Med 179, 1109 (1994).
24. Caux, C., Massacrier, C., Vanbervliet, B., Dubois, B., Van Kooten, C., Durand, I. & Banchereau, J. Activation of human dendritic cells through CD40 cross-linking. J Exp Med 180, 1263 (1994).
25. Kitajima, T., Caceres-Dittmar, G., Tapia, F.J., Jester, J., Bergstresser, P.R. & Takashima, A. T cell-mediated terminal maturation of dendritic cells. J Immunol 157, 2340 (1996).
26. Santiago-Schwarz, F., Tucci, J. & Carsons, S.E. Endogenously produced interleukin 6 is an accessory cytokine for dendritic cell hematopoiesis. Stem Cells 14, 225 (1996).
27. Coussens, L., Yang-Feng, T.L., Chen, Y.L.E., Gray, A., McGrath, J., Seeburg, P.H., Liberman, T.A., Schlessinger, J., Francke, U., Levinson, A. & Ullrich, A. Tyrosine kinase receptor with extensive homology to EGF receptor shares chromosomal location with *neu* oncogene. Science 230, 1132 (1985).
28. Harwerth, I.M., Wels, W., Marte, B. & Hynes, N. Monoclonal antibodies against the extracellular domain of the erbB-2 receptor as partial ligand agonists. J Biol Chem 269, 15160 (1992).
29. Lupu, R., Dickson, R. & Lippman, M. The role of erbB-2 and its ligands in growth control of malignant breast epithelium. J Steroid Biochem Mol Biol 43, 229 (1992).
30. Press, M., Cordon-Cardo, C. & Slamon, D. Expression of HER-2/neu proto-oncogene in normal human adult and fetal tissues. Oncogene 5, 953 (1990).
31. Kallioniemi, O., Kallioniemi, A., Kurisu, W., Thor, A., Chen, L.-C., Smith, H., Waldman, F., Pinkel, D. & Gray, J. *ERBB2* amplification in breast cancer analyzed by fluorescence in situ hybridization. Proc Natl Acad Sci USA 89, 5321 (1992).
32. Berchuck, A., Kamel, A., Whitaker, R., Kerns, B., Olt, G., Kinney, R., Soper, J., Dodge, R., Clarke-Pearson, D. & Marks, P. Overexpression of HER-2/neu is associated with poor survival in advanced epithelial ovarian cancer. Cancer Res 50, 4087 (1990).
33. Berchuck, A., Rodriguez, G., Kinney, R.B., Soper, J.T., Dodge, R.K., Clarke-Pearson, D.L. & Bast, R.C. Overexpression of HER-2/neu in endometrial cancer is associated with advanced stage disease. Am J Obstet Gynecol 15 (1991).
34. Kern, J.A., Schwartz, D.A., Nordberg, J.E., Weiner, D.B., Greene, M.I., Torney, L. & Robinson, R.A. $p185^{neu}$ expression in human lung adenocarcinomas predicts shortened survival. Cancer Res 50, 5184 (1990).
35. Slamon, D., Godolphin, W., Jones, L., Holt, J., Wong, S., Keith, D., Levine, W., Stuart, S., Udove, J., Ullrich, A. & Press, M. Studies of the HER-2/neu proto-oncogene in human breast and ovarian cancer. Science 244, 707 (1989).
36. Yonemura, Y., Ninomiya, I., Yamaguchi, A., Fushida, S., Kimura, H., Ohoyama, S., Miyazakil, I., Endou, Y., Tanaka, M. & Susaki, T. Evaluation of immunoreactivity for erbB-2 protein as a marker of poor short term prognisis in gastric cancer. Cancer Res 51, 1034 (1991).
37. Allred, D., Clark, G., Molina, R., Tandon, A., Schnitt, S., Gilchrist, K., Osborne, C., Tormey, D. & McGuire, W. Overexpression of HER-2/neu and its relationship with other prognostic factors change during the progression of *in situ* to invasive breast cancer. Hum Pathol 23, 974 (1992).
38. Slamon, D., Clark, G., Wong, S., Levin, W., Ullrich, A. & McGuire, W. Human breast cancer: correlation of relapse and survival with amplification of the HER-2/neu oncogene. Science 235, 177 (1987).
39. Cheever, M.A., Disis, M.L., Bernhard, H., Gralow, J.R., Hand, S.L., Huseby, E.S., Qin, H.L., Takahashi, M. & Chen, W. in *Immunol Rev* (eds. Moller, G.) (Munksgaard Intl Pub, Copenhagen, 1995).

40. Alper, Ö., Yamaguchi, K., Hitomi, J., Honda, S., Matsushima, T. & Abe, K. The presence of c-*erb*B-2 gene product-related protein in culture medium conditioned by breast cancer cell line SKBR3. Cell Growth and Differentiation 1, 591 (1990).
41. Zabrecky, J., Lam, T., McKenzie, S. & Carney, W. The extracellular domain of p185/*neu* is released from the surface of human breast carcinoma cells, SKBR3. J Biol Chem 266, 1716 (1991).
42. Mori, S., Mori, Y., Mukaiyama, T., Yamada, Y., Sonobe, Y., Matsushita, H., Sakamoto, G., Akiyama, T., Ogawa, M., Shiraishi, M., Toyoshima, K. & Yamamoto, T. In vitro and in vivo release of soluble erbB-2 protein from human carcinoma cells. Jpn J Cancer Res 81, 489 (1990).
43. Leitzel, K., Teramoto, Y., Sampson, E., Mauceri, J., Langton, B., Demers, L., Podczaski, E., Harvey, H., Shambaugh, S., Volas, G., Weaver, S. & Lipton, A. Elevated soluble c-erb-2 antigen levels in the serum and effusions of a proportion of breast cancer patients. J Clin Oncol 10, 1436 (1992).
44. Ioannides, C.G., Fisk, B., Fan, D., Biddison, W.E., Wharton, J.T. & O'Brian, C.A. Cytotoxic T cells isolated from ovarian malignant ascites recognize a peptide derived from the HER-2/neu proto-oncogene. Cell Immunol 151, 225 (1993).
45. Disis, M.L., Smith, J.W., Murphy, A.E., Chen, W. & Cheever, M.A. In vitro generation of human cytotoxic T cells specific for peptides derived from the HER-2/neu proto-oncogene protein. Cancer Res 54, 1071 (1994).

84

DEVELOPMENT OF DENDRITIC CELL-BASED GENETIC VACCINES FOR CANCER

T. Tüting, T. Zorina, D. I. Ma, C. C. Wilson, C. M. De Cesare, A. B. De Leo, M. T. Lotze, and W. J. Storkus

Department of Surgery and the Pittsburgh Cancer Institute
University of Pittsburgh
Pittsburgh, Pennsylvania, 15213

INTRODUCTION

Cytotoxic T lymphocytes (CTL) are an important component ot the host's immune response to cancer[1,2]. A number of genes encoding tumor-associated antigens (TAA) and their peptide products which are recognized by CTL in the context of major histocompatibility complex (MHC) class I molecules have recently been identified[3,4]. Our group has focused on the translation of these new insights into the development and application of novel immunotherapies.

Dendritic cells (DC) play a crucial role for the induction of primary, cell-mediated immune responses[5,6]. Advances in culture techniques allowing for the generation of large numbers of immunostimulatory DC from precursor populations in bone marrow and blood[7-11] have stirred considerable clinical interest to use DC as a biological adjuvant for the induction of CTL-mediated antitumor immunity. Cultured DC pulsed with peptides constituting relevant CTL epitopes can induce protective and therapeutic antitumor immune responses in mouse tumor models with defined CTL epitopes[12-18]. Of note, such responses can also be elicited with completely normal self-peptides derived from p53[17]. This resembles the expected clinical situation, since melanoma peptides recognized by CTL also derive from completely normal germline-encoded genes. Self-reactive CTL capable of lysing HLA-matched allogeneic melanoma cells can be generated in vitro using autologous cultured human DC pulsed with synthetic melanoma peptides[19-21]. A clinical trial with peptide-pulsed DC for the treatment of patients with malignant melanoma is about to start at the University of Pittsburgh.

As an alternative to peptides, DNA-based vaccines have recently been investigated for the immunotherapy of cancer[22-28]. We describe here the possibility to insert genes encoding TAA directly into immunostimulatory DC leading to endogenous production and processing of relevant antigenic peptides. DC were genetically modified using a particle-mediated gene transfer technology[29-32] and their ability to mediate protective antitumor immunity assessed in vivo in two well-characterized murine tumor models.

MATERIALS AND METHODS

Animals

Female C57BL/6 mice, 6–10 weeks old, were obtained from Taconic (Germantown, NY). Female BALB/cJ mice, 6–10 weeks old, were obtained from The Jackson Laboratory (Bar Harbor, ME). All animals were maintained in a specific pathogen-free facility (Central Animal Facility, University of Pittsburgh).

Cell Lines

C3 is a sarcoma obtained by transfecting mouse embryo cells of C57BL/6 origin with the entire genome of the human papilloma virus type 16[33] and was maintained in RPMI 1640 supplemented with 10% heat-inactivated FCS, 2mM l-glutamine, 50 μM HEPES, and 100 IU/ml penicillin, and 100μg/ml streptomycin (all reagents from GIBCO-BRL, Gaithersburg, MD). CMS4 is a chemically induced BALB/c sarcoma[17] which is continuously passaged in vivo.

Peptides

The H-2D^b-binding peptide HPV-$E_{7(aa49-57)}$ has the sequence RAHYNIVTF[33]. The H-2K^d-binding wild-type peptide $p53_{(aa232-240)}$ has the sequence KYMCNSSCM[17]. Peptides were synthesized by standard F-moc chemistry and purified by HPLC in the Peptide Synthesis Facility of the University of Pittsburgh Cancer Institute (shared resource).

Plasmid DNA

The plasmid pCMVluc contains a firefly luciferase gene driven by the CMV immediate-early gene enhancer/promoter and was kindly provided by Agracetus (Middleton, WI). The plasmid pCI-E_7 was constructed by subcloning the open reading frame of human papilloma virus (HPV) 16-E_7 into the expression plasmid pCI (Promega, Middleton, WI). Using PCR-techniques a SalI-NotI-fragment was generated from a plasmid containing the entire genome of HPV16 and ligated into pCI. The plasmid pCI-$p53_{(aa225-285wt)}$ was constructed by subcloning bp675 to 855 of the mouse p53 cDNA previously isolated from the CMS4 sarcoma[17] into the expression plasmid pCI. Using PCR-techniques a SalI-NotI-fragment was generated including artificial start and stop codons allowing for translation of the resulting minigene and ligated into pCI. All inserts were sequenced in both directions to exclude mutations introduced by PCR. Plasmids were grown in E. coli strain DH5α and purified using Qiagen Plasmid Maxi Kits (Qiagen, Chatsworth, CA).

Generation of Dendritic Cells

Dendritic cells were prepared from bone marrow as previously described[17] with minor modifications. Briefly, bone-marrow cells were harvested from femurs and tibias, immunodepleted of lymphocytes by treatment with anti-B220, -CD4, and -CD8 mAb followed by rabbit complement (Accurate, Westbury, NY), and cultured (37 °C, 5% CO_2) in 6-well plates at 4×10^5 cells/4ml/well in RPMI 1640 supplemented with 10% heat-inactivated FCS, 2mM l-glutamine, 1mM sodium pyruvate, 0.1 mM nonessential amino acids, 50 μM HEPES, 100 IU/ml penicillin, and 100μg/ml streptomycin (all reagents from GIBCO-BRL, Gaithersburg,

MD) as well as 1000 U each of murine rGM-CSF and rIL4 (Schering-Plough, Kenilworth, NJ). Loosely adherent cells are harvested on day 8 by gentle pipetting. 40–70% of DC routinely expressed the surface markers MHC Class II, B7.1, and B7.2.

Particle-Mediated Gene Transfer to Dendritic Cells

Plasmid DNA was precipitated onto 0.95 or 2.6μm gold particles at a density of 2 μg of DNA per mg of particles as previously described[32]. Briefly, gold particles and DNA were resuspended in 100μl of 0.05 M spermidine (Sigma Chemical Co., St. Louis, MO) and DNA precipitated by the addition of 100μl of 1 M $CaCl_2$. Particles were washed in dry ethanol to remove H_2O, resuspended in dry ethanol containing 0.075mg/ml of PVP (Sigma Chemical Co., St. Louis, MO), and coated onto the inner surface of Tefzel tubing using a tube loader. The tubing was cut into 0.5 inch segments resulting in the delivery of 0.5 mg gold coated with 1μg plasmid DNA per transfection with the Accell helium pulse gun. Gold particles, tubing, tube loader, and the Accell helium pulse gun were kindly provided by Agracetus (Middleton, WI). Bone marrow-derived DC were transfected in suspension in 6-well plates. DC were harvested, pelleted by centrifugation, 2×10^6 cells resuspended in 20μl of fresh medium, and spread evenly in the center of a well. Cells were bombarded at a pressure of 300 psi of helium and fresh medium added immediately.

Assays for Expression of Luciferase

Expression of luciferase was determined 8–72 hours after gene transfer to dendritic cells. Cells were washed with Hanks balanced salt solution (HBSS, GIBCO-BRL, Gaithersburg, MD), lysed in 100μl cell culture lysis reagent (Promega, Madison, WI), and stored at -80°C. Samples were thawed, cell debris pelleted, and 10μl of cell extract assayed using Luciferase Assay System (Promega, Madison, WI) and an Autolumat LB 953 (eG&G Berthold) set to integrate emission data over 10s. The level of sample luminescence was recorded as relative light units (RLU).

Immunization of Mice

After particle-mediated gene transfer DC were irradiated (3000 rad), washed twice with HBSS, and 2.5×10^5 cells in 0.2ml HBSS injected in the tail vein of syngeneic mice on days 0 and 7. On day 14 mice were injected with 2×10^6 C3 tumor cells or with 1.5×10^5 in vivo grown CMS4 tumor cells subcutaneously in the flank. Tumor growth was assessed every 3 to 4 days and recorded as tumor area (mm^2) by measuring the largest perpendicular diameters. Data are reports as the average tumor area +/- SEM. All experiments include 5 mice per group and were performed twice.

RESULTS

Transgene Expression Following Particle-Mediated Gene Transfer to DC

We used the firefly luciferase gene as a sensitive reporter gene in order to optimize the parameters for particle-mediated gene transfer to DC with the Accell helium pulse gun. Maximal levels of transgene were detected 16 hours after transfection (Figure 1). DC transfected with an irrelevant plasmid only produced background levels of biolumines-

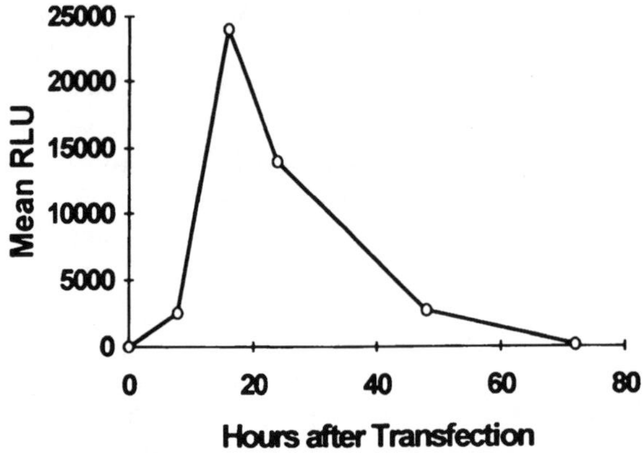

Figure 1. Luciferase expression in bone marrow-derived murine DC after particle-mediated gene transfer.

cence. Bombardment of $2x10^6$ DC with 1μg plasmid DNA coated onto 0.5 mg gold particles at a pressure of 300 psi of helium resulted in the highest transgene expression.

DC Genetically Modified to Express TAA Mediate Protective Antitumor Immunity in Vivo

The efficacy of a vaccine using DC genetically modified to express TAA was first assessed using a viral TAA. The C3 sarcoma is an HPV16-transformed cell line[32] latently expressing the known TAA HPV16-E_7 including the defined CTL-recognized H-$2D^b$-restricted peptide epitope derived from aa49–57[12,13,32]. C57BL/6 mice were immunized with DC transfected with pCI-E_7 or pulsed with the E_7(aa49–57) peptide. DC transfected with the empty expression plasmid backbone and naive mice served as controls. Immunization either with DC expressing E_7 or pulsed with the E_7 peptide resulted in protection against a subsequent challenge with C3, while tumors grew in naive animals or animals immunized with backbone-transfected DC (Fig. 2).

To model more closely the expected clinical situation, we examined whether DC genetically modified to express a tumor-associated self antigen can also mediate antitumor immunity. The CMS4 sarcoma expresses high levels of p53 molecules because of a missense mutation at codon 194. This leads to natural processing and presentation of the H-$2K^d$-binding 'self'-epitope derived from aa232–240[17]. BALB/c mice were immunized with DC transfected with pCI-$p53_{(aa225-285wt)}$ or pulsed with the p53wt(aa232–240) peptide. Again, DC transfected with the empty expression plasmid backbone and naive mice served as controls. Upon challenge with CMS4, a significant reduction of tumor growth could be observed in the groups immunized with peptide-loaded or gene-inserted DC. Interestingly, mice immunized with backbone-transfected DC also showed a considerable reduction of tumor growth compared to naive mice (Fig. 3).

DISCUSSION

The insertion of genes into primary cells such as dendritic cells is not an easy task. The use of viral vectors is probably more efficient for gene transfer to DC[34] than nonviral methods, but expression plasmids have important advantages: Several genes can readily

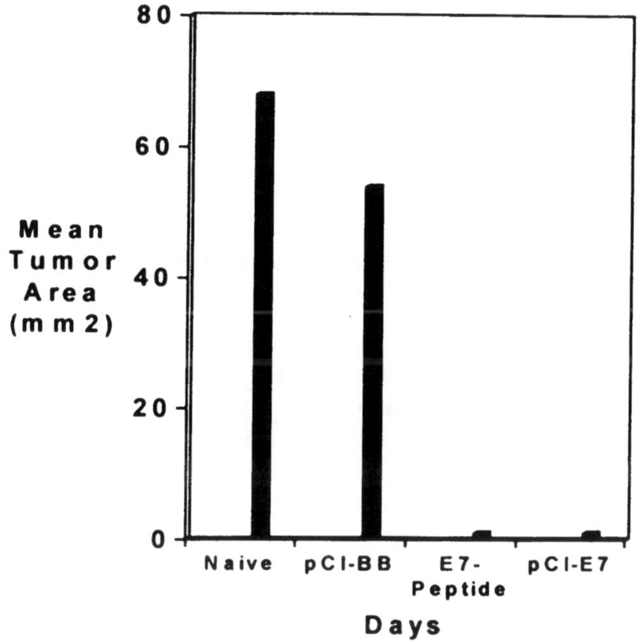

Figure 2. DC genetically modified to express HPV16-E_7 mediate protective antitumor immunity against challenge with an HPV16-transformed tumor cell line. Mean tumor size was measured 21 days after tumor challenge.

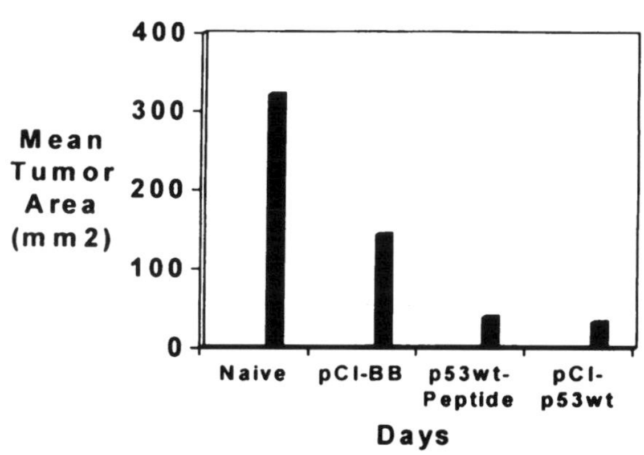

Figure 3. DC genetically modified to express a fragment of wild-type p53 mediate protective antitumor immunity against challenge with a tumor cell line overexpressing p53. Mean tumor size was measured 21 days after tumor challenge.

be transfected simultaneously, DNA of high purity can easily be produced in large quantities, is highly stable, and there is no risk of recombination. We show here that particle-mediated gene transfer is feasible, but nonviral gene delivery to DC clearly needs to be improved further.

Our results indicate that the genetic insertion of TAA directly into DC can induce protective antitumor immunity in vivo not only to viral but also to self-antigens. While peptide-based vaccines require prior knowledge of the patient HLA-haplotype as well as the relevant T cell epitopes, DNA-based vaccines lead to the expression of the entire TAA. This allows for the presentation of multiple and unknown MHC class I and potentially class II epitopes and mimics a viral infection.

Both tumor models investigated here are clinically relevant. More than 90% of human cervical carcinoma harbor human papillomavirus (HPV), predominantly HPV16. HPV16, as well as other "high risk" HPV, are considered to play an important role in the pathogenesis of human cervical cancer. CTL epitopes have been identified for several HLA haplotypes and active specific immunotherapy is pursued for prevention and treatment of cervical carcinoma[35]. Furthermore, a large number of human cancers overexpress p53 due to mutations. CTL epitopes have been identified for HLA-A2 and peptide-specific CTL can be generated in vitro[36]. Most recently, it was shown that wild-type p53 epitopes are naturally processed and presented by human tumor cells[37], suggesting the possibility to target p53 as a tumor antigen. Successes with gene therapy approaches directed to replace mutated p53 in cancer cells[38] might in part also be a result of the induction of immune responses to p53, a hypothesis favored by tumor immunologists which is currently under investigation. Potential safety considerations of DNA-based vaccines such as insertional mutagenesis, the induction of unwanted autoimmune responses or, conversely, of immunologic tolerance have to be carefully assessed.

Preliminary evidence suggests that antigen-specific $CD4^+$ T cells may also be induced using TAA-transduced DC. Cotransfection of cytokine genes such as IFN-α and IL-12, which are both produced by antigen presenting cells during viral infections, can modulate the nature of the resulting immune responses. Finally, human DC genetically modified to express MART-1 or gp100 can stimulate antigen-specific CTL in vitro supporting the translational potential of this approach.

ACKNOWLEDGMENTS

This work was supported in part by a fellowship from the Deutsche Forschungs Gemeinschaft to Thomas Tüting and NIH grants CA57840 and CA 67407 to Walter J. Storkus.

REFERENCES

1. Kast, W. M. et al. Eradication of adenovirus E1-induced tumors by E1A-specific cytotoxic T lymphocytes. Cell 59, 603–607 (1989)
2. Rosenberg, S.A. et al. Use of tumor-infiltrating lymphocytes and interleukin-2 in the immunotherapy of patients with metastatic melanoma : preliminary report. N. Engl. J. Med. 319, 1676–1680 (1988).
3. Van den Eynde, B., & Brichard, V.G. New tumor antigens recognized by T cells. Current Opinion in Immunology 7, 674–68 (1995).
4. Boon, T., & Van der Bruggen, P. Human tumor antigens recognized by T lymphocytes. J. Exp. Med. 183, 725–729 (1996).

5. Steinman, R.M. The dendritic cell system and its role in immunogenicity. Annu. Rev. Immunol. 9, 271–296 (1991).
6. Stingl, G., & Bergstresser, P. R. Dendritic cells: a major story unfolds. Immunol. Today 16, 330–333 (1995).
7. Inaba, K., Steinman, R.M., Witmer-Pack, M., Aya, H., Inaba, M., Sudo, T., Wolpe, S., & Schuler, G. Identification of proliferating dendritic cell precursors in mouse blood. J. Exp. Med. 175, 157–1167 (1992).
8. Inaba, K., Inaba, M., Romani. N., Aya, H., Deguchi, M., Ikehara, S., Muramatsu, S., & Steinman, R.M. Generation of large numbers of dendritic cells from mouse bone marrow cultures supplemented with granulocyte-macrophage colony-stimulating factor. J. Exp. Med. 176, 693–1702 (1992).
9. Caux, C., Dezutter-Damboyant, C., Schmitt, D., & Bancherau, J. GM-CSF and TNF-alpha cooperate in the generation of dendritic Langerhans cells. Nature 360, 258–261 (1992).
10. Romani, N., Gruner, S., Brang, D., Kämpgen, E., Lenz, A., Trockenbacher, B., Konwalinka, G., Gritsch, P.O., Steinman, R.M., & Schuler. Proliferating dendritic cell progenitors in human blood. J. Exp. Med. 180, 83–93 (1994).
11. Szabolcs, P., Moore, M.A. S., & Young, J.W. Expansion of immunostimulatory dendritic cells among the myeloid progeny of human CD34+ bone marrow precursors cultured with c-kit ligand, granulocyte-macrophage colony-stimulating factor, and TNF-α. J. Immunol. 154, 5851–5861 (1995).
12. Mayordomo, J.I., Zorina, T., Storkus, W.J., Celuzzi, C.M., Falo, L.D., Kast, W.M., Ildstad, S.T., DeLeo, A.B., & Lotze, M.T. Bone marrow-derived dendritic cells pulsed with tumor peptides elicit protective and therapeutic anti-tumor immunity. Nature Med. 1, 1297–1302 (1995).
13. Ossevoort, M.A, Feltkamp, M.C.W., van Veen, K.J.H., Melief, C.J.M., & Kast, W.M. Dendritic cells as carriers for a cytotoxic T lymphocyte epitope-based peptide vaccine in protection against a human papillomavirus type 16-induced tumor. J. Immunother. 18, 86–94 (1995).
14. Celluzi, C.M., Mayordomo, J.I., Storkus, W.J., Lotze, M.T., & Falo, L.D. Peptide-pulsed dendritic cells induce antigen-specific, CTL-mediated protective tumor immunity. J. Exp. Med. 183, 283–387 (1996).
15. Paglia, P., Chiodoni, C., Rodolfo, M., & Colombo, M.P. Murine dendritic cells loaded in vitro with soluble protein prime cytotoxic T lymphocytes against tumor antigen in vivo. J. Exp. Med. 183, 317–322 (1996).
16. Zitvogel, L., Mayordomo, J.I., Tjandrawan, T., DeLeo, A.B., Clarke, M.R., Lotze, M.T., & Storkus, W.J. Therapy of murine tumors with tumor peptide pulsed dendritic cells : dependence on T cells, B7 costimulation, and Th1-associated cytokines. J. Exp. Med. 183, 87–97 (1996).
17. Mayordomo, J.I., Loftus, D.J., Sakamoto, H., De Cesare, S.M., Appasamy, P. M., Lotze, M.T., Storkus, W.J., Appella, E., & DeLeo, A.B. Therapy of murine tumors with p53 wild-type and mutant sequence peptide-based vaccines. J. Exp. Med. 183, 1357–1365 (1996).
18. Porgador, A., Snyder, D., & Gilboa, E. Induction of antitumor immunity using bone marrow-generated dendritic cells. J. Immunol. 156, 2918–2926 (1996).
19. Bakker, A.B.H., Marland, G., de Boer, A.J., Huijbens, J.F., Danen, E.H.J., Adema, G.J., & Figdor, D.G. Generation of antimelanoma cytotoxic T lymphocytes from healthy donors after presentation of melanoma-associated antigen-derived epitopes by dendritic cells *in vitro*. Cancer Res. 55, 5330–5334 (1995).
20. van Elsas, A., van der Burg, S.H., van der Minne, C.E., Borghi, M., Mourer, J.S., Melief, C.J.M., & Schrier, P.I. Peptide-pulsed dendritic cells induce tumoricidal cytotoxic T lymphocytes from healthy donors against stably HLA-A*0201-binding peptides from the Melan-A/MART-1 self antigen. Eur. J. Immunol. 26, 1683–1689 (1996).
21. Tjandrawan, T., Mäurer, M..J., Castelli, C., Lotze, M.T., & Storkus, W.J. Autologous dendritic cells pulsed with synthetic melanoma peptides elicit specific CTL effector cells in vitro. Manuscript submitted.
22. Conry, R.M., LoBuglio, Loechel, F., Moore, S.E., Sumerel, L.A., Barlow, D.L., & Curiel, D.T. A carcinoembryonic antigen polynucleotide vaccine has in vivo antitumor activity. Gene Ther. 2, 59–66 (1995).
23. D.M., & Wu, T.-C. Treatment of established tumors with a novel vaccine that enhances major histocompatibility class II presentation of tumor antigen. Cancer Res. 56, 21–26 (1996).
24. Ciernik,I.F., Berzofsky, J.A., & Carbone, D. Induction of cytotoxic T lymphocytes and antitumor immunity with DNA vaccines expressing single T cell epitopes. J. Immunol. 156, 2369–2375 (1996).
25. Warnier, G., Duffour, M.-T., Uyttenhove, C., Gajewski, T., Lurquin, C., Haddada, H., Perricaudet, M., & Boon, T. Induction of a cytolytic T-cell response in mice with a recombinant adenovirus coding for tumor antigen P815A. Int. J. Cancer 67, 303–310 (1996).
26. Chen, P.W., Wang, M., Bronte, V., Zhai, Y., Rosenberg, S.A., & Restifo, N.P. Therapeutic antitumor response after immunization with a recombinant adenovirus encoding a model tumor-associated antigen. J. Immunol. 156, 224–231 (1996).
27. Irvine, K.R., Rao, R.B., Rosenberg, S.A., & Restifo, N.P. Cytokine enhancement of DNA immunization leads to effective treatment of established pulmonary metastases. J. Immunol. 156, 238–245 (1996).

28. Zhai, Y., Yang, J.C., Kawakami, Y., Spiess, P., Wadsworth, S.C., Cardoza, L.M., Couture, L.A., Smith, A.E., & Rosenberg, S.A. Antigen-specific tumor vaccines. Development and characterization of recombinant adenoviruses encoding MART1 or gp100 for cancer therapy. J. Immunol. 156, 700–710 (1996).
29. Yang, N.-S., Burkholder, J.K., Roberts, B., Martinell, B., and McCabe, D. In vivo and in vitro gene transfer to mammalian somatic cells by particle bombardment. Proc. Natl. Acad. Sci. USA 87, 9568–9572 (1990).
30. Burkholder, J.K., Decker, J., & Yang, N.-S. Rapid transgene expression in lymphocyte and macrophage primary cultures after particle bombardment-mediated gene transfer. J. Immunol. Methods 165, 149–156 (1993).
31. Yang, N.-S., & Sun, W.-H. Gene gun and other non-viral approaches for cancer gene therapy. Nature Med. 1, 481–483 (1995).
32. Pertmer, T.M., Roberts, T.R., and Haynes, J.R. Influenza virus nucleoprotein-specific immunoglobulin G subclass and cytokine responses elicited by DNA vaccination are dependent on the route of vector DNA delivery. J.Virol. 76:6119–6125 (1996).
33. Feltkamp, M.C.W., Smis, H.L., Vierboom, M.P.M., Minnaar, R.P., de Jongh, B.M., Drijfhout, J.W., ter Schegget, J., Melief, C.J.M., and Kast, W.M. Vaccination with cytotoxic T lymphocyte epitope-containing peptide protects against a tumor induced by human papillomavirus type 16-transformed cells. Eur. J. Immunol. 23, 2242–2249 (1993).
34. Henderson, R.A., Nimgaonkar, M.T., Watkins, S.C., Robbins, P.D., Ball, E.D., and Finn, O.J. Human dendritic cells genetically engineered to express high levels of the human epithelial tumor antigen mucin (MUC-1). Cancer Res. 56, 3763–3770 (1996).
35. Ressing, M.E., Sette, A., Brandt, R.M., Ruppert, J., Wentworth, P.A., Hartmann, M., Oseroff, C., Grey, H.M., Melief, C.J.M., and Kast, W.M. Human CTL epitopes encoded by human papillomavirus type 16 E6 and E7 identified through in vivo and in vitro immunogenicity studies of HLA-A*0201-binding peptides. J. Immunol. 154:5934–5943 (1995).
36. Houbiers, J.G.A., Nijman, H.W., van der Burg, S.H., Drijfhout, J.W., Kenemans, P., man de Velde, C.J.H., Brand, A., Momburg, F., Kast, W.M., and Melief, C.J.M. In vitro induction of human cytotoxic T lymphocyte responses against peptides of mutant and wild-type p53. Eur. J. Immunol. 23:2072–2077 (1993).
37. Theobald, M., Biggs, J., Dittmer, D., Levine, A., and Sherman, L. Targeting p53 as a general tumor antigen. Proc. Natl. Acad. Sci. USA 92:11993–11997 (1995).
38. Roth, J.A., Nguyen, D., Lawrence, D.D., Kemp, B.L., Carrasco, C.H., Ferson, D.Z,., Hong,W.K., Komari, R., Lee, J.J., Nesbitt, J.C., Pisters, K.M.W., Putnam, J.B., Schea, R., Shin, D.M., Walsh, D.L., Dolormente, M.M., Han, C.-I., Martin, F.D., Yen, N., Xu, K., Stephens, L.C., McDonnell, T.J., Mukhopadhyay, T., and Cai, T. Retrovirus-mediated wild-type p53 gene transfer to tumors of patients with lung cancer. Nature Med. 2:985–991 (1996).

85

IMMUNOTHERAPY OF CANCER

Generation of CEA Specific CTL Using CEA Peptide Pulsed Dendritic Cells

Susan E. Alters, Jose R. Gadea, and Ramila Philip

RPR GenCell
5301 Patrick Henry Dr.
Santa Clara, California 95054–1114

1. ABSTRACT

Antigen specific cytotoxic T lymphocytes (CTL) are being studied for their potential immunotherapeutic benefit in the treatment of cancer. Carcinoembryonic antigen (CEA) is an oncofetal protein best known for its overexpression in the majority of colorectal, gastric, pancreatic, non small cell lung, and breast carcinomas. We are using dendritic cells (DC) pulsed with the CEA CTL peptide epitope to generate CEA specific CTL. DC from HLA A2$^+$ donors were isolated by culturing plastic adherent PBMC in GMCSF and IL4 for 7 days. As expected these DC expressed the relevant cell surface molecules including HLA DR, CD58, CD80, and CD86. The DC were strippped of their endogenous peptides, pulsed with the A2 restricted CEA peptide, irradiated and used to stimulate autologous CD8$^+$ T cells in the presence of IL7. Using this approach we have been able to generate CEA specific CTL from the PBMC of breast and pancreatic carcinoma patients as well as normal donors. These CTL can lyse CEA peptide pulsed T2 targets as well as HLA A2$^+$ tumor cells expressing the CEA antigen. This data is being used to support a phase I active immunotherapy clinical protocol using DC pulsed with CEA peptide to treat patients with metastatic malignancies expressing CEA.

2. INTRODUCTION

Immunotherapy protocols for cancer have shown promise in experimental animal models and in the treatment of some human malignancies[1,2]. While approaches such as vaccination with non specific immunogens or systemic administration of cytokines have been disappointing[3], an emerging understanding of antigen specific cellular recognition supports the potential role of specific immunotherapy in human malignancy.

CEA is an oncofetal antigen which is expressed on most adenocarcinomas of the colon, rectum, pancreas, and stomach, as well as on 50% of breast cancers and 70% of non-small cell lung carcinomas[4]. It has also been found on normal adult colonic mucosa. Recently, Schlom and his colleagues have identified peptides corresponding to human MHC class I CTL epitopes within the CEA protein[5]. The immunodominant peptide in this study was a nine amino acid sequence, YLSGANLNL. This peptide was designated CAP-1 and is an HLA-A2 restricted CTL epitope. When CAP-1 was used to stimulate peripheral blood lymphocytes of patients previously immunized with recombinant human CEA vaccinia virus, CEA specific CTL lines could be generated, confirming the immunogenicity of the CEA antigen[5]. However, no CTL could be generated from unimmunized patients.

DC are potent antigen presenting cells (APC) which have been shown to stimulate both a naive and memory T cell response *in vitro*[6,7]. Recent studies have indicated that DC may be more potent in stimulating an anti-tumor immune response when compared to other aproaches, including viral vectors expressing tumor antigen genes, gene modified tumor cells, naked DNA, or peptide emulsified in adjuvant. In this study, we show that DC pulsed with the HLA-A2 restricted CEA peptide can generate a CEA specific CTL response *in vitro* from both unimmunized carcinoma patients and healthy donors.

3. MATERIALS AND METHODS

3.1. Dendritic Cell Isolation

DC were isolated using a modification of Romani et al.[8] Briefly, 1.5×10^8 PBMC from an HLA-A2$^+$ donor were allowed to adhere to T150 culture flasks for 2 hours at 37°C. The adherent cells were then cultured in 30 ml of RPMI-10% FCS medium containing 800 units/ml GMCSF and 500 units/ml IL-4 for 6–7 days.

3.2. Peptide Stripping and Loading

For peptide stripping, DC were washed once in cold 0.9% NaCl with 1% BSA solution, resuspended at 1×10^7 cells/ml in stripping buffer (0.13M L-ascorbic acid, 0.06M sodium phosphate monobasic [pH3], 1% BSA, 3 mg/ml b2microglobulin, 10mg/ml peptide) and incubated for 2 min on ice. The cells were then neutralized with 5 volumes of cold neutralizing buffer (0.15M sodium phosphate monobasic [pH 7.5], 1% BSA, 3 mg/ml b2M, 10mg/ml peptide) and spun at 1500 rpm for 5 min. Finally, the cells were resuspended in peptide solution (PBS-CMF, 1% BSA, 30 mg/ml DNAase, and 40 mg/ml peptide) and incubated for 4-hr at room temperature. After incubation, the cells were irradiated (3000 Rad) and washed prior to being used for stimulation.

3.3. CTL Generation

CD8$^+$ responders for peptide specific CTL were generated by adhering PBMC on AIS (Santa Clara, CA) CD8 T25 MicroCELLector^R flasks. CD8 captured cells were stimulated with irradiated DC loaded with CEA peptides (DC-CEA) at a stimulator to responder (S:R) ratio of 1:3. These cells were cultured in RPMI-10% FCS containing 10ng/ml IL-7. At day 10–12, the lymphocytes were restimulated with DC pulsed with CEA peptide (1:5 S:R ratio). Responders were restimulated weekly to a total of 3–4 restimulations at S:R ratios ranging between 1:5 to 1:15. As controls, CD8 captured cells were stimulated with IL-7 only (10ng/ml) or PHA/IL2 (2ug/ml PHA and 500 U/ml IL-2).

3.4. Cytotoxicity Analysis

HLA-A2 restricted CEA recognition by CTL was assessed by a standard 4-hr ^{51}Cr release cytotoxicity assay. Recognition of CEA peptide by CTL was assessed using T2 cells pre-incubated for 2–4 hr with peptide at 40mg/ml. In addition CEA specific cytotoxicity was evaluated on the CEA expressing cell lines SW403 (HLA-A2^{+}) and SW1417 (HLA-A2^{-}).

4. RESULTS

4.1. Generation of DC

DC were used as APC to present CEA peptide to CD8^{+} T cells. The DC were differentiated from PBMC adherent progenitors using GMCSF and IL4 as described[8]. The yield of DC (based on FACS analysis) generated in this study was approximately 6% of the starting PBMC population; the purity of these cells was about 60%. DC expressed high levels of HLA DR, the adhesion molecules CD54 (ICAM-1) and CD58 (LFA3), as well as the costimulatory molecules CD80 (B7.1) and CD86 (B7.2). DC did not express CD3, CD14, CD15, CD16, CD19 or CD20 as expected (data not shown).

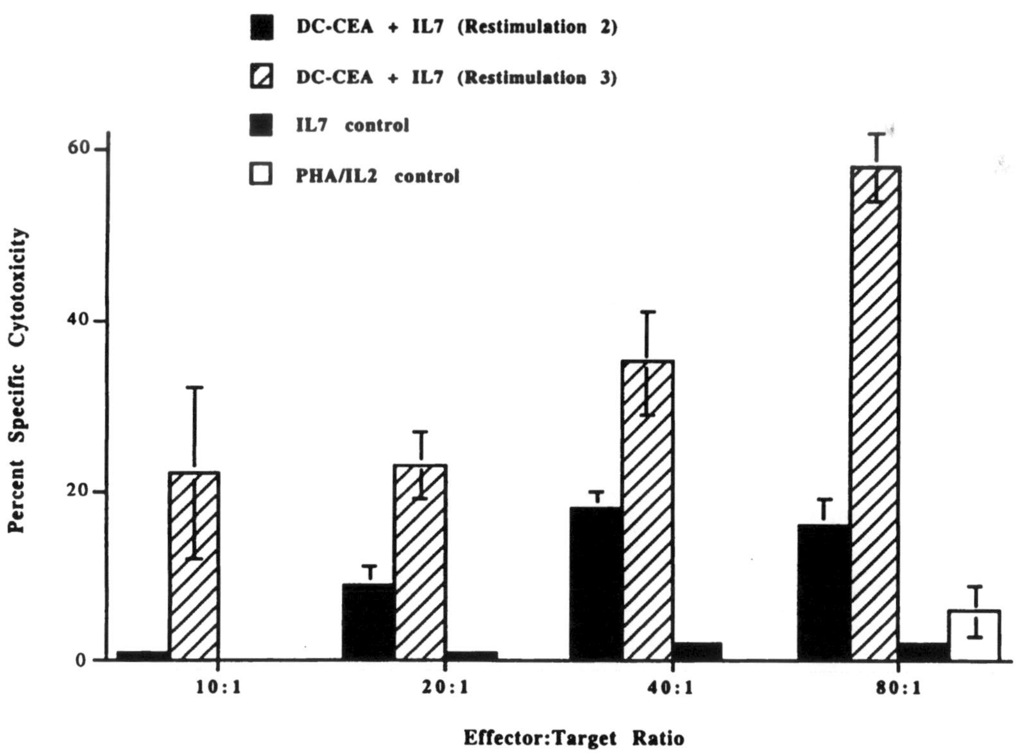

Figure 1. Generation of a CEA peptide response from an HLA-A2^{+} pancreatic carcinoma patient. CD8 captured responders were stimulated with CEA pulsed DC and cultured in 10ng/ml IL-7. Cytotoxicity was assayed after 2 (shaded) or 3 (slash lines) rounds of restimulation with CEA peptide pulsed DC + IL7 at effector:target ratios of 80:1, 40:1, 20:1, 10:1. Control cultures were cultured in 10ng/ml IL7 (■) or PHA/IL2 (□). The background response on empty T2 cells has been subtracted.

4.2. Generation of CEA Specific CTL from HLA-A2⁺ Carcinoma Patients

DC were used to generate CEA peptide specific CTL from an HLA-A2^{+} pancreatic carcinoma patient as described. After two and three rounds of restimulation with DC-CEA + IL7 the CTL effectors were tested for cytotoxicity using T2 cells pulsed with CEA peptide as targets. CEA peptide specific cytotoxicity ranged from 15–20% over background levels after two rounds of restimulation with DC-CEA + IL7 and increased to 60% after one additional round of restimulation (Figure 1). Effector cells that had been stimulated with IL7 only or with PHA/IL2 showed no CEA specific cytotoxicity.

Having shown that CTL generated using DC-CEA + IL7 were CEA peptide specific, we next assessed whether the effector CTL could recognize endogenously processed and presented antigen. DC were isolated as above, pulsed with CEA peptide, and used to stimulate autologous CD8 captured T cells from an HLA-A2^{+} breast carcinoma patient in the presence of IL7. As shown in Figure 2, CEA peptide induced CTL effectors lysed SW403, the CEA^{+} HLA-A2^{+} tumor target, at E:T ratios of 70:1 down to 9:1. Since previous studies have determined that the CEA peptide was HLA-A2 restricted, we assessed the ability of these CTL to kill SW1417, a CEA^{+} HLA-A2^{-} target[5]. Lysis of this tumor target was minimal. Again, effector cells that had been cultured with IL7 only showed no specific cytotoxicity on any target.

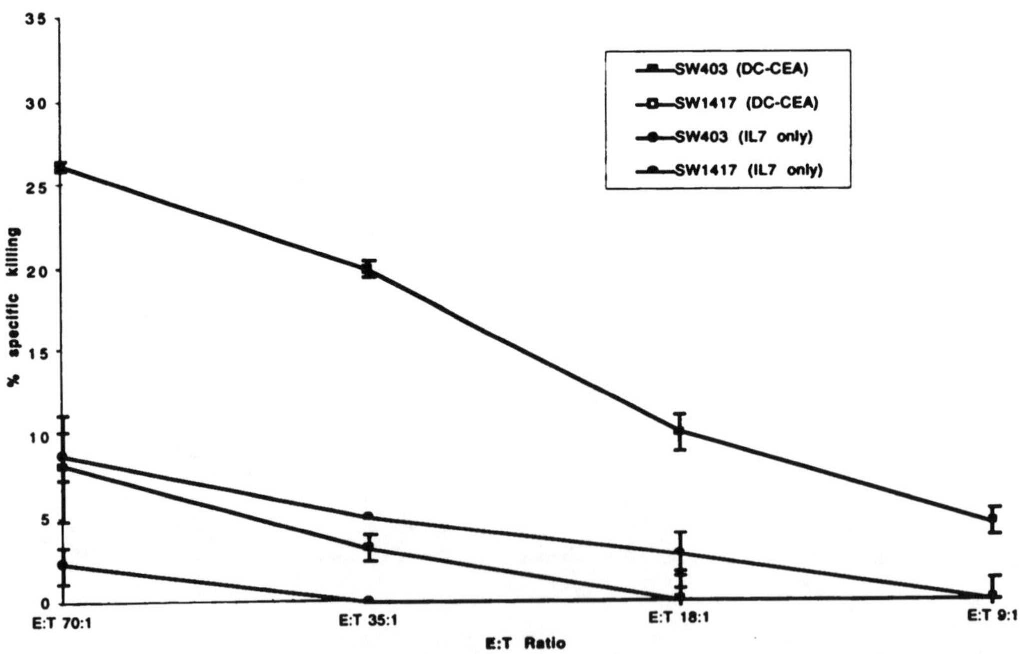

Figure 2. Generation of an HLA-A2 restricted, CEA specific response from an HLA-A2^{+} breast carcinoma patient. The cytolytic activity of CTL generated by stimulation with CEA pulsed DC + IL7 (10ng/ml) was assayed on an A2^{+} CEA^{+} (SW403; ■) and an A2^{-}CEA^{+} tumor cell line (SW1417; □) at effector:target ratios of 70:1, 35:1, 18:1, 9:1. Cytolytic activity of control cells generated with IL7 only was also tested (SW 403; ●) (SW1417; O). Cytotoxicity was assayed after four rounds of restimulation.

4.3. Generation of CTL from an HLA-A2⁺ Healthy Donor

We next looked at whether CTL could be generated from the blood of healthy HLA-A2⁺ donors. As seen for the carcinoma patients, CEA specific CTL could also be generated from a healthy donor (Figure 3). After 3 rounds of restimulation with DC-CEA in the presence of IL7, 30% killing was detected on T2-CEA as compared to T2 cells pulsed with an irrelevant peptide at E:T of 80:1, and 10% killing was detected at E:T of 40:1. Cells cultured in IL7 only showed no response as expected.

5. DISCUSSION

The CEA antigen which is expressed in many cancer types is a potential target for tumor immunotherapy. However, the generation of an immune response against the CEA antigen has historically been difficult to achieve. Recently, Schlom and colleagues were able to generate CTL lines by stimulating PBL with the CAP-1 peptide and IL2 from patients previously vaccinated with the CEA gene inserted into vaccinia virus[5]. However, no CTL could be generated from unimmunized patients.

In our study, we were successful in generating CEA peptide specific CTL from the blood of both unimmunized patients as well as healthy donors. The use of DC as APC in the CTL protocol may account for the favorable outcome. DC used in our studies express

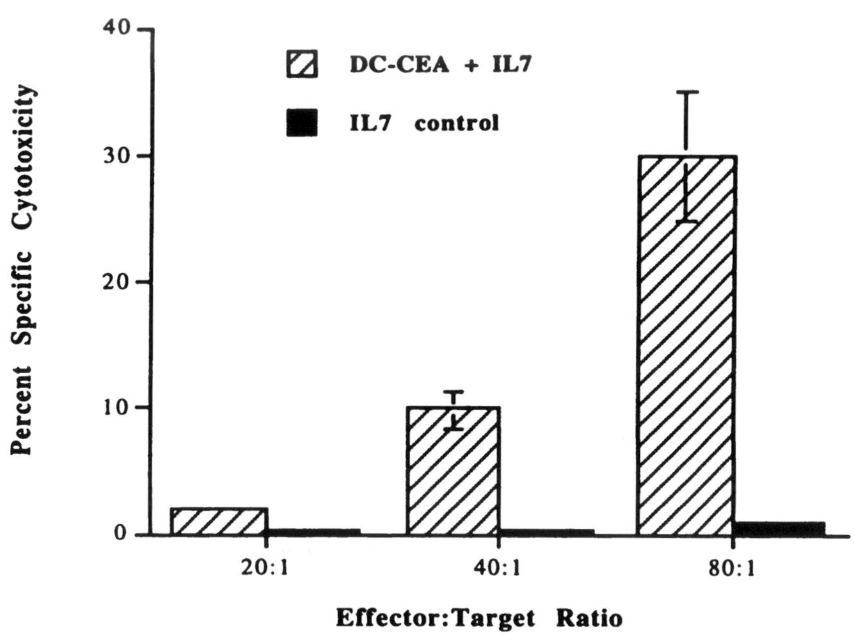

Figure 3. Generation of a CEA peptide response from an HLA-A2⁺ healthy donor. CD8 captured responders were stimulated with CEA pulsed DC and cultured in 10ng/ml IL-7. Cytotoxicity was assayed after 3 rounds of restimulation with CEA peptide pulsed DC + IL7 (slash lines) at effector:target ratios of 80:1, 40:1, 20:1, 10:1. Control cultures were cultured in 10ng/ml IL7 (■). The background response on T2 cells pulsed with an irrelevant peptide has been subtracted.

high levels of HLA-DR, costimulatory molecules such as CD80, CD86 and adhesion molecules such as CD54 and CD58. This may explain the ability of DC to play a critical role in the initiation of the cellular immune response. DC have been shown to stimulate naive T cells to recognize and respond to a variety of antigens *in vitro*[7,9] and can induce protective or therapeutic immunity when administered as a vaccine to animals *in vivo*[10–14]. Recently Bakker et. al.[15] have demonstrated CTL specific for tyrosinase, gp100, and MART-1 from healthy donors using peptide loaded DC as APC. In our current study we show that CEA specific CTL can be generated from unvaccinated carcinoma patients and healthy donors when the CEA peptide is presented by DC. The CTL generated using peptide-loaded DC were capable of lysing CEA peptide-loaded T2 cells as well as tumor cell lines expressing the CEA antigen. These studies indicate that using a specialized APC, like the DC, may allow for the induction of an effective cellular immune response without the need for prior vaccination. This data is being used to support a phase I active immunotherapy clinical protocol using DC pulsed with CEA peptide to treat patients with metastatic malignancies expressing CEA.

6. REFERENCES

1. Rosenberg, S. A et. al. A progress report on the treatment of 157 patients with advanced cancer using lymphokine activated killer cells and interleukin-2 or high-dose interleukin-2 alone. *N. Engl. J. Med.* **316**, 889–897 (1987).
2. Rosenberg, S.A. Immunotherapy of cancer using interleukin-2: Current status and future prospects. *Immunol. Today* **9**, 58–62 (1988).
3. Rosenberg, S.A., et al. Experience with the use of high-dose interleukin-2 in the treatment of 652 patients with cancer. *Ann. Surg.* **210**, 474–85 (1989).
4. Thompson, J.A., Gruner, F., and Zimmermann, W. Carcinoembryonic antigen gene family: molecular biology and clinical perspectives. *J. Clin. Lab. Anal.* **5**, 344–366 (1991).
5. Tsang, K. Y., et al. Generation of human cytotoxic T cells specific for human carcinoembryonic antigen epitopes from patients immunized with recombinant vaccinia-CEA vaccine. *J. Natl. Canc. Ins.* **87**, 982-990 (1995).
6. Inaba, K., Metly, P.J., Crowley, M.T., and Steinman, R.M. Dendritic cells pulsed with protein antigens *in vitro* can prime antigen-specific, MHC-restricted T cells in situ. *J. Exp. Med.* **172**, 631–640 (1990).
7. Mehta, D.A., Markowicz, S., and Engleman, E.G. Generation of antigen-specific $CD8^+$ CTLs from naive precursors. *J. Immunol.* **153**, 996–1003(1994).
8. Romani, N., et al. Proliferating dendritic cell progenitors in human blood. *J. Exp. Med.* **180**, 83–93 (1994).
9. Mehta, D.A., Markowicz, S., and Engleman, E.G. Generation of antigen-specific $CD4^+$ T cell lines from naive precursors. *Eur. J. Immunol.* **25**, 206–1211 (1995).
10. Mayordomo, J.I., et al. Bone marrow-derived dendritic cells pulsed with synthetic tumor peptides elicit protective and therapeutic antitumour immunity. *Nature Med.* **1** (12), 1297–1302 (1995).
11. Celluzzi, C.M., Mayordomo, J.I., Storkus, W.J., Lotze, M.T., and Falo, L.D. Peptide-pulsed dendritic cells induce antigen-specific CTL-mediated protective tumor immunity. *J. Exp. Med.* **183**, 283–287 (1996).
12. Zitvogel, L., et. al. Therapy of murine tumors with tumor peptide-pulsed dendritic cells: Dependence on T cell, B7 costimulation, and T helper cell 1-associated cytokines. *J. Exp. Med.* 183, 87–97 (1996).
13. Ossevoort, M.A., Feltamp, M.C.W., van Veen, K.J.H., Melief, C.J.M., and Kast, W.M. Dendritic cells as carriers for a cytotoxic T-lymphocyte epitope-based peptide vaccine in protection against a human papillomavirus type 16-induced tumor. *J. Immunother.* **18**, 86–94 (1995).
14. Porgador, A., and Gilboa, E. Bone-marrow-generated dendritic cells pulsed with a class 1-restricted peptide are potent inducers of cytotoxic T lymphocytes. *J. Exp. Med.* **182**, 255–260 (1995).
15. Bakker, A. B., et. al. Generation of antimelanoma cytotoxic T lymphocytes from healthy donors after presentation of melanoma-associated antigen-derived epitopes by dendritic cells *in vitro*. *Cancer Res.* **55**, 5330–5334 (1995).

TUMOR-INFILTRATING DENDRITIC CELLS ARE DEFECTIVE IN THEIR ANTIGEN-PRESENTING FUNCTION AND INDUCIBLE B7 EXPRESSION

A Role in the Immune Tolerance to Antigenic Tumors

Pascal Chaux, Nathalie Favre, Bernard Bonnotte, Monique Moutet, Monique Martin, and François Martin

CJF INSERM 94–8
Department of Biology and Therapy of Cancer
Faculty of Medicine, 7 Bd Jeanne d'Arc
F-21033 Dijon, France
Fax: (33) 80 67 17 88; e:mail: fmartin@u.bourgogne.fr

INTRODUCTION

Many human and experimental cancers express abnormal proteins, including the products of mutated oncogenes, tumor suppressor genes, or viral genes, fusion proteins resulting from translocations, or products of normal but silent genes (1). At least some of these abnormal proteins should be recognized as antigens by T lymphocytes and induce an immune response leading to tumor rejection. However, these antigens are tolerated by the immune system, and tumors, instead of being rejected, progress and ultimately kill their host. Among other mechanisms, a defect in the presentation of tumor antigens to the immune system could explain this tolerance.

IMMUNOGENIC AND TOLEROGENIC TUMOR CELL VARIANTS: THE PRO/REG TUMOR SYSTEM

Several clones of tumor cell variants have been obtained from a tumor cell line, derived from a chemically-induced rat colon carcinoma (2). Certain clones, named REG, give rise to tumors when sc injected into syngeneic hosts, but after 3 to 6 weeks, these tumors completely regress. REG tumor regression induces a strong but specific immune protection against the parental tumors and derived clones. Other clones, named PRO, are specifically rejected in syngeneic rats pre-immunized by a prior REG cell rejection, even when as many as ten million PRO cells were injected. This demonstrates that PRO cells express potent antigens, capable of being recognized by immune effector cells to induce tumor cell death. However, ten thousand PRO tumor cells, when injected into naive syngeneic rats, are sufficient to give rise to progressive, metastatic and lethal tumors. When rats bear a PRO tumor, they are specifically tolerant to the highly immunogenic REG tumor cell variants from the same cancer (3). Tumor resection abolishes this tolerance. These results show that PRO cells are both antigenic (since they are rejected by tumor-immune hosts), and tolerogenic (as they prevent the rejection of an immunogenic cell variant).

PRO TUMORS ARE INFILTRATED BY DENDRITIC CELLS

The dendritic cell is the paradigm of the antigen-presenting cell and its absence, rather than its abundance should be expected in a tolerogenic tumor. Paradoxically, PROb tumors are infiltrated by a dense network of dendritic-shaped cells that express major histocompatibility complex class I and class II molecules, but not the specific markers of rat T or B lymphocytes and macrophages (4). These tumor-infiltrating dendritic cells could have been expected to process and present to T cells the antigens that are released by the adjacent tumor cells. Tumor-infiltrating dendritic cells are located inside the tumor nests in direct contact with tumor cells, and are also found together with T cells and macrophages in the belt of inflammatory cell that surrounds the tumor.

Dendritic cell progenitors infiltrate PROb tumors as early as twenty-four hours after a subcutaneous injection of tumor cells. Many dendritic cells are seen to migrate from the blood vessels surrounding the injection site. This suggests that tumor cells produce one or several factors that attract circulating dendritic cell progenitors.

TUMOR-INFILTRATING DENDRITIC CELLS: FUNCTION AND PHENOTYPE

We isolated and purified these tumor-infiltrating dendritic cells to explore their function and phenotype. Indeed, tumor-infiltrating dendritic cells, when compared to splenic dendritic cells, were poor stimulators of primary allogeneic T cell response, as measured by T cell proliferation and cytokine production. [^{3}H]thymidine incorporation into allogeneic T cells was about 10-fold lower, and the concentrations of IL-2 and interferon-γ in the mixed culture supernatant were about 5- and 20-fold lower, respectively, when tumor-infiltrating dendritic cells were used instead of splenic dendritic cells for allogeneic T cells stimulation. The number of allogeneic T cells recovered after 7 days of mixed primary culture increased when T cells were mixed with splenic dendritic cells, whereas it decreased when they were mixed with tumor-infiltrating dendritic cells. This suggests that tumor-infiltrating dendritic cells cause T lymphocyte deletion.

These dendritic cells express major histocompatibility complex class I and class II molecules, as well as ICAM-1 and LFA-3 adhesion molecules. However, when stained with CTLA4-Ig fusion molecule, less than ten percent of tumor-infiltrating dendritic cells expressed B7, an essential costimulatory signal for T cells. The same result was obtained after a culture period allowing epidermal Langerhans cells to up-regulate B7 expression. A combination of GM-CSF and TNF-α, or cell-associated CD40-ligand, all known as potent stimulators of B7 expression on other dendritic cells (5), did not induce B7 expression by tumor-infiltrating dendritic cells. This suggests that the lack of B7 expression does not result from an immature status of tumor-infiltrating dendritic cells, but rather results from a long-lasting blockade due to the tumor microenvironment.

THE TUMOR-INFILTRATING DENDRITIC CELL: AN INDUCTOR OF T CELL TOLERANCE?

When an antigen is presented in the absence of the B7 costimulatory signal, anergy or apoptosis of antigen-specific T cell is expected, in the place of the normal, protective immune response (6–8). After a first exposure to tumor-infiltrating dendritic cells, allogeneic T cells responded poorly to a second challenge to splenic dendritic cells, while they responded normally to splenic dendritic cells from a third rat strain. In contrast, allogeneic T cells that had been exposed to splenic dendritic cells during the primary stimulation evoked a typical secondary response to a rechallenge with splenic dendritic cells. This suggests that B7-defective tumor-infiltrating dendritic cells decrease the capability of T cells to respond to the same antigen.

INFLAMMATORY CELLS INFILTRATING HUMAN COLORECTAL CANCER STROMA ARE ALSO DEFECTIVE IN B7 EXPRESSION

The defect in B7 expression of tumor-infiltrating dendritic cells was not restricted to the PROb tumor experimental model, but was also found for dendritic cells and macrophages infiltrating twenty-five individual samples of human colorectal carcinomas. These inflammatory cells were stained for CD68, a common marker for macrophages and dendritic cells, and most of them expressed strongly MHC class II molecules. However, antibodies to B7–1 and B7–2 stained either no cell or only less than 1% of the inflammatory cells infiltrating tumor stroma. Thus, like rat PRO cell tumors, human colorectal carcinomas are infiltrated by inflammatory cells that express MHC class II, but not B7 molecules.

REJECTION OF IMMUNOGENIC REG TUMOR CELL VARIANTS INVOLVES B7, BUT NOT B7-EXPRESSING TUMOR-INFILTRATING DENDRITIC CELLS

The rat colon cancer REG cell variants give rise to tumors that spontaneously regress in a few weeks. This regression was delayed or suppressed when rats received one or several injections of CTLA4-Ig, a fusion molecule that specifically inhibits B7–1 and B7–2 binding to their receptor on T cells (10). However, neither REG cells, nor freshly isolated MHC class II+ cells infiltrating REG tumors expressed B7 (11). This suggests that the target of CTLA4-Ig was not located inside the tumor. In contrast, MHC class II+ B7+ cells were found in the draining lymph nodes and in the spleen, suggesting that lymphoid tissue, rather than the tumor itself, was the site of tumor-antigen presentation to tumor-specific T cells. REG cells are more sensitive to apoptosis than PRO cells in vitro and many apoptotic figures were observed locally during the 48 first hours following REG cell subcutaneous injection. Accordingly, the efficient immune response generated by REG tumor cells could depend on their capacity to release tumor antigens upon apoptosis. These antigens could reach the secondary lymphoid tissues either as a soluble or particulate form, or after being endocytosed and processed by tumor-infiltrating dendritic cells that could migrate into spleen or lymph nodes.

CONCLUSION

Experimental but also human colorectal tumors are heavily infiltrated by dendritic cells that express MHC class II molecules but do not express B7, even after stimulation. The failure of tumor-infiltrating dendritic cells to express B7 may prevent them from presenting efficiently tumor antigens and could instead induce the anergy or the deletion of tumor-specific T cells. The tolerance to antigenic tumors could be explained by this abnormal presentation of tumor antigens inside the tumor.

REFERENCES

1. Boon, T., Cerottini, J.C., Van Den Eynde, B., Van Der Bruggen, P. and Van Pel, A., Tumor rejection antigens recognized by T lymphocytes. Annu. Rev. Immunol., 12, 337–365 (1994).
2. Caignard, A., Martin, M.S., Michel, M.F. and Martin, F., Interaction between two cellular subpopulations of a rat colonic carcinoma when inoculated to the syngeneic host. Int. J. Cancer, 36, 273–279 (1985).
3. Caignard, A., Pelletier, H. and Martin, F., Specificity of the immune response leading to protection or enhancement by regressive and progressive variants of a rat colon carcinoma. Int. J. Cancer, 42, 883–886(1988).
4. Chaux, P., Hammann, A., Martin, F. and Martin, M., Surface phenotype and functions of tumor-infiltrating dendritic cells: CD8 expression by a cell subpopulation. Eur. J. Immunol., 23, 2517–2525 (1993).
5. Caux C., Massacrier, C., Vanbervliet, B., Dubois, B., Van Kooten, C., Durand, I. and Banchereau, J., Activation of human dendritic cells through CD40 cross-linking. J. Exp. Med., 180, 1263–1272 (1994).
6. Schwartz, R.H., Costimulation of T lymphocytes: the role of CD28, CTLA4, and B7/BB1 in interleukin-2 production and immunotherapy. Cell, 71, 1065–1068 (1992).
7. Gimmi, C.D., Freeman, G.J., Gribben, J.G., Gray, G. and Nadler, L.M., Human T-cell clonal anergy is induced by antigen presentation in the absence of B7 costimulation. Proc. Natl. Acad. Sci. (USA), 90, 6586–6590 (1993).
8. Harding, F.A., McArthur, J.G., Gross, J.A., Raulet, D.H. and Allison, J.P., CD28-mediated signalling costimulates murine T cells and prevents induction of anergy in T cell clones. Nature, 356, 607–609 (1992).
9. Chaux, P., Moutet, M., Faivre, J., Martin, F. and Martin, M. Inflammatory cells infiltrating human colorectal carcinomas express HLA class II but not B7–1 and B7–2 costimulatory molecules of the T cell activation. Lab. Invest. 74, 975–983 (1996).
10. Linsley, P.S., Brady, W., Urnes, M., Grosmaire, L.S., Damle, N.K. and Ledbetter, J.A., CTLA4 is a second receptor for the B cell activation antigen B7. J. Exp. Med., 174, 561- 569 (1991).
11. Chaux, P., Martin, M.S. and Martin, F. T-cell costimulation by the CD28 ligand B7 is involved in the immune response leading to rejection of a spontaneously regressive tumor. Int. J. Cancer 66, 244–248 (1996).

87

ROLE OF B7 COSTIMULATION IN THE INDUCTION OF T AND B CELL RESPONSES BY DENDRITIC CELLS *IN VIVO*

Laurence Lespagnard, Pascal Mettens, Jacques Urbain, and Muriel Moser

Laboratoire de Physiologie Animale
Université Libre de Bruxelles
Rue des Chevaux, 67
B-1640 Rhode-Saint-Genèse, Belgium

INTRODUCTION

The adjuvant properties of dendritic cells have been amply demonstrated. A single injection of splenic DC, which have been pulsed extracorporeally with antigen, induces primary T (1) and B (2) cell responses *in vivo*. Recent reports have shown that tumor-bearing animals or patients can be immunized against their own tumors by injecting autologous DC pulsed with tumor associated antigens (3, 4). *In vivo* priming using elements of the immune system avoids the toxic effect of external adjuvants and leads to an antigen-specific, anamnestic immune response.

There is evidence that DC are the antigen-presenting-cells of the primary response, as they have the unique capacity to sensitize naive T cells. This property develops during a process of maturation that occurs spontaneously *in vitro* (5) and can be induced *in vivo* (6). During this process, DC upregulate the expression of class II MHC, B7–1 and B7–2 and lose their capacity to process antigens. Therefore, mature DC express very high levels of antigen/MHC complexes as well as of costimulatory molecules of the B7 family.

In this paper, we evaluated the role of B7 costimulatory molecules in the induction of primary responses. We measured the humoral and cellular responses of syngeneic mice injected with antigen-pulsed DC with or without blocking anti-B7 antibodies.

RESULTS

In a preliminary experiment, we injected 3×10^5 DC, pulsed with human gamma globulins (HGG), intravenously into DBA/2 mice that received one or two injections of 200 µg anti-B7–1 plus anti-B7–2 mAbs or were left untreated. The levels of HGG specific antibodies were measured in individual sera two weeks later. The data in Figure 1A show

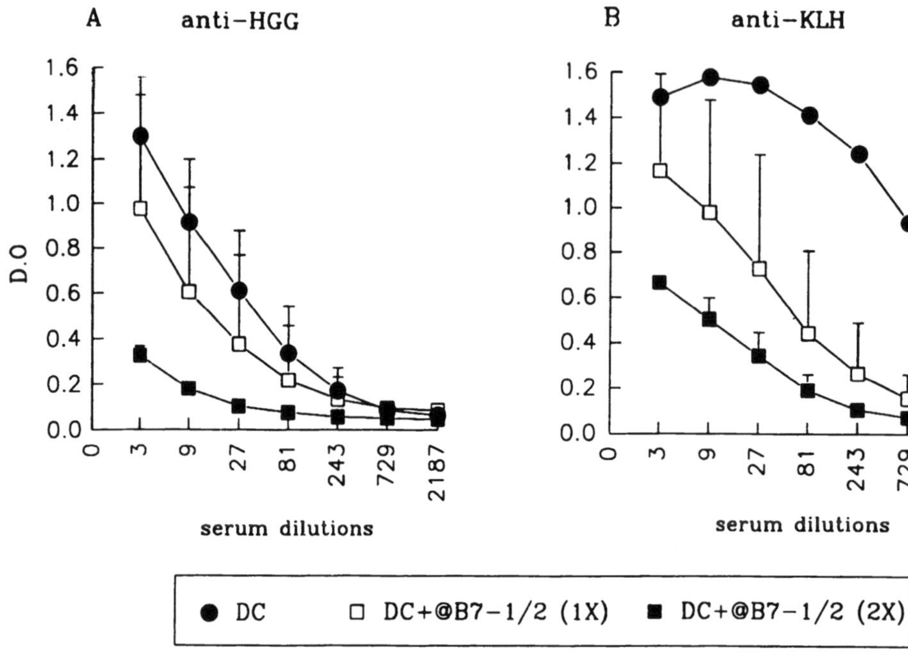

Figure 1. DBA/2 mice (5 per group) received an intravenous injection of 3×10^5 DC pulsed extracorporeally with HGG (A) or KLH (B) and one or two intraperitoneal injection(s) of 80 μg anti-B7–1 mAb (16–10A1) plus 80μg anti-B7–2 mAb (GL1). All mice were bled 13 days later and the levels of antibodies specific for HGG (A) or KLH (B) were measured in individual sera.

that two injections of antibodies abrogated the humoral response induced by DC, whereas one injection partially inhibited the synthesis of specific antibodies. Similar data were obtained with DC pulsed with keyhole limpet hemocyanin (KLH), as shown in Figure 1B. In subsequent experiments, all anti-B7-treated animals received two injections of both mAbs, one day before and several hours after the priming with DC.

We next compared the B7-dependency of the immune responses induced by the same antigen pulsed on DC or emulsified in Complete Freund's Adjuvant (CFA). The data in Figure 2 show that injection of anti-B7–1 and anti-B7–2 antibodies strongly inhibited the primary responses of both groups of mice. By contrast, the secondary response of mice primed with DC in the absence of functional B7 was impaired following injection of a small dose of KLH in saline, whereas animals immunized by injection of KLH/CFA in the presence of blocking antibodies mounted a anamnestic response near control levels. These observations could result from the persistence of the antigen when injected with CFA, or to a B7-independent priming of KLH-specific T and/or B lymphocytes. Furthermore, the treatment with either antibody alone only weakly affected the primary humoral response, suggesting that B7–1 and B7–2 molecules may have some redundant function.

A recent report showed that T cell-dependent antibody production was profoundly inhibited in transgenic mice that secrete a soluble form of CTLA4, but that T cells were primed normally in the same animals. In particular, mCTLA4-Hγ1 transgenic and control littermates responded equally well to immunization with KLH-pulsed dendritic cells, and the *in vivo* expansion of antigen-specific T cells was even enhanced in mCTLA4-Hγ1 transgenics (7, 8). Therefore, we injected 3×10^5 antigen-pulsed DC in the fore and hind

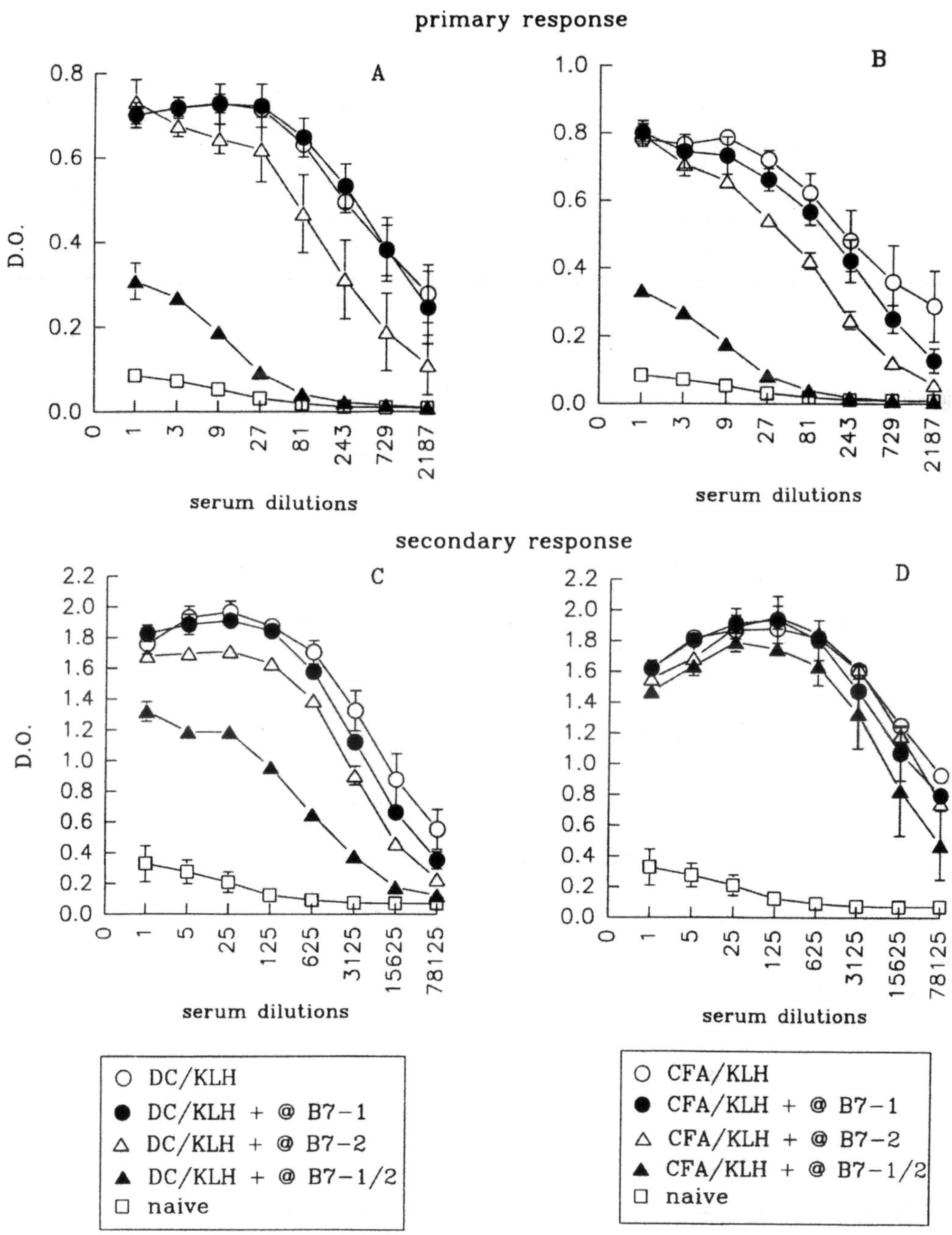

Figure 2. DBA/2 mice (5 mice per group) were primed by an intravenous injection of 3×10^5 KLH-pulsed DC (A, C), or an intraperitoneal injection of 10 µg KLH emulsified in CFA (B, D). Some groups of mice were further treated with 80 µg anti-B7–1 mAb, 80µg anti-B7–2, or both mAbs. One group of naive mice was included as control. The primary humoral response (A, B) was measured 13 days after immunization. For secondary humoral response (C, D), all mice received 10 µg KLH intravenously one month after priming, and were bled 10 days later.

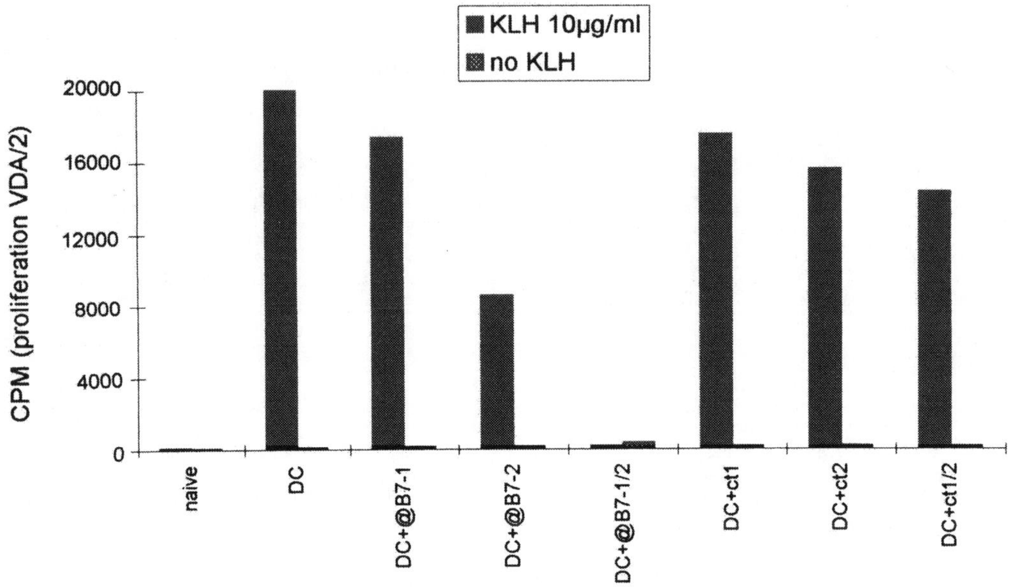

Figure 3. DBA/2 mice were immunized by injection of 3×10^5 KLH-pulsed DC into the footpads, and were treated with 80 µg anti-B7–1, anti-B7–2 or both mAbs one day before and several hours after DC injection. Control groups were injected with isotype-matched antibodies (ct1: parsi19, hamster IgG1 and/or ct2: rat IgG2a). 5 days later, the draining lymph nodes were harvested and the lymph node cells were cultured with or without KLH (10µg/ml). The IL-2 secretion was measured in the supernatant by a bioassay.

footpads and harvested the draining lymph nodes 5 days later, according to a protocol described by Inaba *et al.* (1). The results in Figure 3 show that lymph node cells from mice primed with KLH-pulsed DC proliferated and secreted IL-2 when restimulated with antigen in culture. Injection of the combination of antibodies, however, completely abrogated the proliferation and the secretion of IL-2. The blockade of B7–1 only did not affect the T cell response, whereas the inhibition of B7–2 partially suppressed the proliferation (not shown) and secretion of IL-2 (Figure 3) by lymph node cells.

DISCUSSION

The main observation of this paper is that recognition of B7–1 or B7–2 is required for the induction of T cell responses and T cell-dependent B cell responses. Dendritic cells are the cells primarily responsible for the priming of naive T cells *in vivo* and express some costimulatory activity in situ. The development in sequence of the capacity to process antigen and to sensitize T cells that correlate with upregulation of expression of B7–1 and B7–2 may contribute to the potent adjuvant property of dendritic cells. We show indeed that the capacity of DC to induce primary T and B responses is strictly dependent on B7 costimulation.

The administration of anti-B7–1 mAb has no significant effect on B and T cell responses, whereas injection of anti-B7–2 partially inhibits T cell proliferation and IL-2 secretion. The observation that blockade of B7–2 does not affect the humoral response but diminishes T cell function suggests that B7–2 is not required for antibody synthesis by B cells and that T cell help generated by B7–1 only is sufficient to induce antibody production. Treatment with both mAbs abrogates the immune response, indicating that B7–1 and B7–2 have some redundant properties. There is some evidence that B7–1 and B7–2 may have distinct function in determining the development of Th1 versus Th2 lymphocyte (9, 10). Experiments are in progress to measure the lymphokines produced by lymph node cells from mice primed with antigen pulsed DC with or without treatment with selected anti-B7 mAbs.

Our data do not support the hypothesis that B7 costimulation is required for the effector function rather than the proliferation and expansion of antigen-specific $CD4^+$ T cells, as reported by Lane *et al.* (8), but clearly show that priming of antigen-specific T cells is B7-dependent. The same authors showed that blocking T cell help to B cells resulted in impaired class switching, reduced somatic mutation and absence of germinal center formation.

The two-signal theory of activation predicts that antigen presentation in the absence of costimulation would induce a state of tolerance in T lymphocytes (11). Thus, the antigenic signal and the costimulatory signal are both required to induce optimal activation of T cells, and the outcome of T cell receptor occupancy (activation/tolerance) is determined by the presence or absence of the costimulatory signal. Since DC express very high levels of antigen/MHC complexes, it may be interesting to test whether priming with DC in the presence of anti-B7–1 and anti-B7–2 mAbs would induce antigen-specific unresponsiveness. Experiments are under way to test whether injection of antigen-pulsed DC in the presence of anti-B7–1/2 mAbs is a null event or induces a form of peripheral tolerance. We will use thymectomized animals to prevent emergence of newly educated T cells from the thymus.

REFERENCES

1. Inaba K, Metlay JP, Crowley MT and Steinman RM. *J Exp Med* 1990; 172:631.
2. Sornasse T, Flamand V, De Becker G, Bazin H, Tielemans F, Thielemans K, Urbain J, Leo O and Moser M. *J Exp Med* 1992; 175:15.

3a. Flamand V, Sornasse T, Thielemans K, Demanet C, Bakkus M, Bazin H, Tielemans F, Leo O, Urbain J and Moser M. *Eur J Immunol* 1994; 24:605.

3b. Zitvogel L, Mayordomo JI, Tjandrawan T, DeLeo AB, Clarke MR, Lotze MT and Storkus WJ. *J Exp Med* 1996; 183:87.

4. Hsu FJ, Benike C, Fagnoni F, Liles TM, Czerwinski D, Taidi B, Engelman EG and Levy R. *Nature Medicine* 1996; 2:52.
5. Larsen CP, Ritchie SC, Hendrix R, Linsley PS, Hathcock KS, Hodes RJ, Lowry RP and Pearson TC. *J Immunol* 1994; 152:5208.
6. De Smedt T, Pajak B, Muraille E, Lespagnard L, Heinen E, De Baetselier P, Urbain J, Leo O and Moser M. *J Exp Med* 1996; 184:1.
7. Ronchese F, Hausmann B, Hubele S and Lane P. *J Exp Med* 1994; 179:809.
8. Lane P, Burdet C, Hubele S, Scheidegger D, Müller U, Mc Connell F and Kosco-Vilbois M. *J Exp Med* 1994; 179:819.
9. Freeman GJ, Boussiotis VA, Anumanthan AJ, Bernstein GM, Ke XY, Rennert PD, Gray GS, Gribben JG and Nadler LM. *Immunity* 1995; 2.523.
10. Kuchroo VK, Das MP, Brown JA, Ranger AM, Zamvill SS, Sobel RA, Weiner HL, Nabavi N and Glimcher LH. *Cell* 1995; 80:707.
11. Quill H and Schwartz RH. *J Immunol* 1987; 138:3704.

POSITIVE AND NEGATIVE REGULATION OF DENDRITIC CELL FUNCTION BY LIPOPOLYSACCHARIDE *IN VIVO*

Thibaut De Smedt, Bernard Pajak, Eric Muraille, Jacques Urbain, Oberdan Leo, and Muriel Moser

Département de Biologie Moléculaire
Université Libre de Bruxelles
Rue des Chevaux, 67
B-1640 Rhode-Saint-Genèse, Belgium

1. INTRODUCTION

There is evidence that induction of primary immune responses *in vitro* and *in vivo* relies on the presentation of antigen by dendritic cells[1]. The unique property of DC to sensitize naive T lymphocytes may depend on a specialization of function over time that results in a reciprocal regulation of the capacity to process antigens and to activate T cells. Fresh DC process antigens and generate high levels of peptides/MHC complexes, whereas cultured DC have lost this capacity but have gained the property to prime T lymphocytes. This process of maturation occurs spontaneously in culture and correlates with increased expression of class II, B7–1 and B7–2 costimulatory molecules[2–6].

Little is known, however, on the maturation of DC *in vivo*. Recent studies have shown that Langerhans cells migrated in skin explants and concomitantly increased their immunostimulatory properties, suggesting that cells of the dendritic family may undergo maturation *in vivo*[7].

We have shown recently that systemic administration of LPS provoked a redistribution of DC populations in spleen[8]. R. Steinman and collaborators have indeed identified two subsets of splenic DC, both of which react strongly with N418 anti-CD11c[9]. One subset occupies the periarterial sheaths (the T cell area) and stains strongly with M342 and NLDC-145, whereas the other type occupies the marginal zone and lacks M342 and NLDC-145. The injection of 25μg LPS induced within 6 h the disappearance of N418$^+$ cells from the marginal zone and an increase in numbers of N418$^+$ cells in the T cell area. We interpreted these observations as the migration of most DC from the marginal zone to the T cell area, although we could not rule out the possibility that DC in T cell area were newly emerging DC. To clarify this point, we visualized DC in cryosections from spleens at different time after LPS injection.

2. RESULTS AND DISCUSSION

2.1. Injection of LPS Induces DC Migration and Maturation Within 6 Hours

LPS was injected to BALB/c mice and cryosections from spleens 2.5, 3, 3.5, 4, 6 hours later were stained with DC-specific N418 mAb. The data in Figure 1 show that the majority of DC were detected in the marginal zone between red and white pulp, and that a few cells were labeled in the T cell area in the white pulp. LPS administration resulted in a redistribution of DC, as all N418$^+$ cells were found 6 hours later in the T cell area (Figure 1f). As shown in panels b-e, DC were gradually detected in intermediate regions located between the marginal zone and the periarteriolar sheaths. Therefore, the simplest interpretation would be that DC migrate from the marginal zone to the T cell area, and that DC in the peripheral area give rise to the DC that localize in the center of the white pulp.

We next tested whether LPS induced the maturation of DC. Spleen cells from untreated mice and animals injected with LPS 2, 4, 6 or 8 hours previously were enriched for DC on a BSA gradient[10] and analyzed for B7–1, B7–2 and class II MHC expression by FACS. The data in Figure 2 show that LPS administration results in a time-dependent increased expression of B7 and class II molecules. We have recently reported that DC from LPS-injected animals displayed functional properties of mature DC, as assessed by a decreased capacity to process antigens and an increased capacity to sensitize naive T cells [8]. Therefore, we conclude that immature DC undergo maturation and migrate to the T cell area upon injection of LPS. A similar change in phenotype was induced by LPS injection in SCID mice (Figure 2, right panel), indicating that functional T and B lymphocytes were not involved.

2.2. LPS Administration Results in Loss of Splenic DC 48 Hours after Injection

The staining of spleen cryosections revealed that the migration of DC was followed by the loss of most N418$^+$ DC 48 hours after LPS injection[8]. We performed immunofluorescence studies on DC-enriched, low density spleen cells. Figure 3 shows that low-density spleen cells from mice injected 48 h previously lacked the DC population identified as N418$^+$. This observation as well as the defective APC function of spleen cells from the same animals[8] are in favour of a loss of cells rather than a change in phenotype. A similar loss of splenic DC was induced in SCID mice (Figure 3), excluding a role for T and B cells in DC depletion.

The disappearance of DC 48 hours following LPS administration may be due to death (by apoptosis?) or to migration of DC out of spleen. However, as most cellular trafficking in the spleen is thought to occur through the venous sinuses, loss of LPS-activated DC would imply their migration of DC from the T cell area back to the red pulp through the marginal zone. Analysis of spleen cryosections 16, 18, 20, 22, 24 hours after LPS injection failed to reveal N418$^+$ cells around the marginal zone, suggesting that DC do not leave the spleen but rather die in situ (data not shown). Experiments are in progress to detect apoptotic cells in the T cell area at different time after LPS administration.

3. CONCLUSION

Collectively, our data indicate that LPS has time-dependent opposite effects on DC function *in vivo*. A single injection of endotoxin induces the maturation and the migration

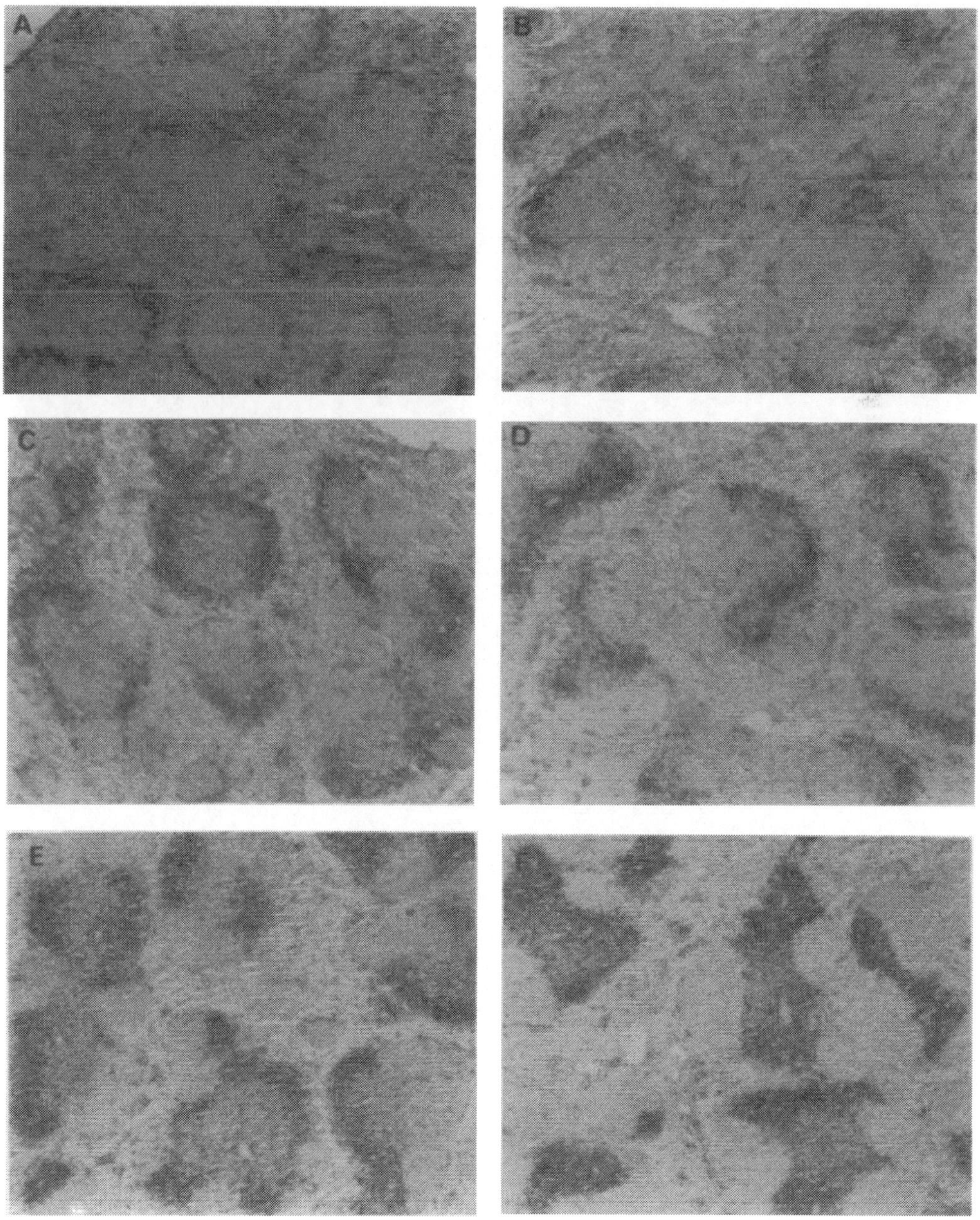

Figure 1. LPS induces migration of N418-positive cells from the marginal zone of the spleen to the T cell area. Immunoperoxydase labeling of cryostats sections from spleens of control mice (A) and mice treated 2.5 (B), 3 (C), 3.5 (D), 4 (E) or 6 (F) hours previously with 25 µg LPS intravenously. Sections were stained with anti-CD11c mAb and were counterstained with hematoxylin. The original magnification was 10.

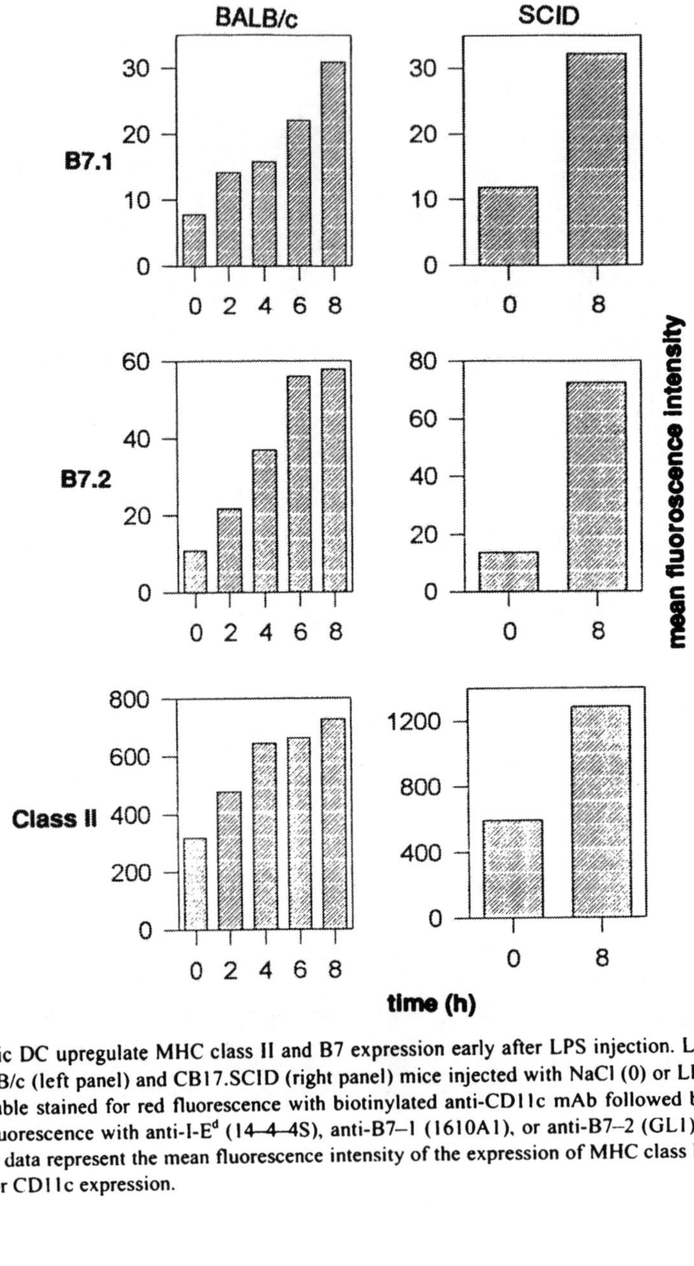

Figure 2. Splenic DC upregulate MHC class II and B7 expression early after LPS injection. Low density spleen cells from BALB/c (left panel) and CB17.SCID (right panel) mice injected with NaCl (0) or LPS 2, 4, 6, 8 hours earlier were double stained for red fluorescence with biotinylated anti-CD11c mAb followed by streptavidin-PE and for green fluorescence with anti-I-E^d (14–4–4S), anti-B7–1 (1610A1), or anti-B7–2 (GL1) mAbs coupled to fluorescein. The data represent the mean fluorescence intensity of the expression of MHC class II, B7–1 and B7–2 on cells gated for CD11c expression.

of DC, and subsequently provokes the disappearance of these cells from the spleen. This sequence of events may be beneficial to both the host and the pathogen, since it down-regulates the pathogen-specific immune response and prevents overstimulation of the immune system.

The colocalization of mature DC and T lymphocytes suggests that DC that have captured exogenous bacterial products, process and present the antigens and undergo maturation while migrating to T cell area where they activate T cells specific for bacterial antigens. However, the discovery of a third subset of splenic DC that express CD8 and Fas-ligand and induces apoptosis in T lymphocytes [11] may challenge this simple view.

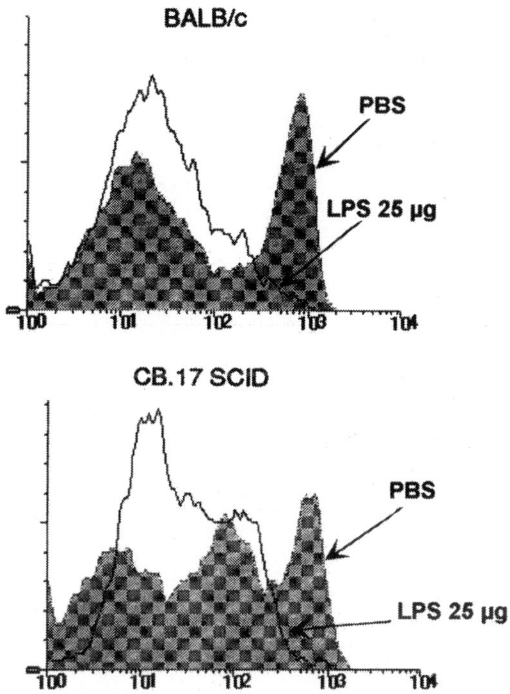

Figure 3. LPS injection results in the loss of N418-positive cells from mouse spleen after 2 days. Low-density spleen cells from BALB/c and CB17.SCID mice injected with PBS or 25 μg LPS 48 hours earlier were stained with anti-CD11c coupled to fluorescein.

4. REFERENCES

1. Steinman, R.M. 1991. The dendritic cell system and its role in immunogenicity. *Annu. Rev. Immunol.* 9:271–296.
2. Inaba, K., M Witmer-Pack, M. Inaba, K.S. Hathcock, H. Sakuta, M. Azuma, H. Yagita, K. Okumura, P.S. Linsley, S. Ikehara, S. Muramatsu, R.J. Hodes, and R.M. Steinman. 1994. The tissue distribution of the B7–2 costimulator in mice: abundant expression on dendritic cells in situ and during maturation in vitro. *J. Exp. Med. 180*:1849–1860.
3. Moser, M., T. De Smedt, T. Sornasse, F. Tielemans, A.A. Chentoufi, E. Muraille, M. Van Mechelen, J. Urbain, and O. Leo. 1995. Glucocorticoids down-regulate dendritic cell function in vitro and in vivo. *Eur. J. Immunol.* 25:2818–2824.
4. Romani, N., S. Koide, M. Crowley, M. Witmer-Pack, A.M. Livingstone, C.G. Fathman, K. Inaba, and R.M. Steinman. 1989. Presentation of exogenous protein antigens by dendritic cells to T cell clones. Intact protein is presented best by immature, epidermal Langerhans cells. *J. Exp. Med.* 169:1169–1178.
5. Crowley, M.T., K. Inaba, M.D. Witmer-Pack, S. Gezelter, and R.M. Steinman. 1990. Use of the fluorescence activated cell sorter to enrich dendritic cells from mouse spleen. *J. Immunol. Methods.* 133:55–66.
6. Koch , F., B. Trockenbacher, E. Kämpgen, O. Grauer, H. Stössel, A.M. Livingstone, G. Schuler, and N. Romani. 1995. Antigen processing in populations of mature murine dendritic cells is caused by subsets of incompletely matured cells. *J. Immunol.* 155:93–100.
7. Larsen, C.P., R.M. Steinman, M. Witmer-Pack, D.F. Hankins, P.J. Morris, and J.M. Austyn. 1990. Migration and maturation of Langerhans cells in skin tranplants and explants. *J. Exp. Med.* 172:483–493.
8. De Smedt, T., B. Pajak, E. Muraille, L. Lespagnard, E. Heinen, P. De Baetselier, J. Urbain, O. Leo, and M. Moser. 1996. Regulation of dendritic cell numbers and maturation by lipopolysaccharide in vivo. *J. Exp. Med.* 184:1413–1424.

9. Agger, R., M. Witmer-Pack, N. Romani, H. Stossel, W.J. Swiggard, J.P. Metlay, E. Storozynsky, P. Freimuth, and R.M. Steinman. 1992. Two populations of splenic dendritic cells detected with M342, a new monoclonal to an intracellular antigen of interdigitating dendritic cells and some B lymphocytes. *J. Leukoc. Biol.* 52:34–42.
10. Crowley, M., K. Inaba, M. Witmer-Pack, and R.M. Steinman. 1989. The cell surface of mouse DC: FACS analyses of DC from different tissues including thymus. *Cell. Immunol.* 118:108–125.
11. Süss, G. and K. Shortman. 1996. A subclass of dendritic cells kills CD4 T cells via Fas/Fas-ligand-induced apoptosis. *J. Exp. Med.* 183:1789–1796.

DENDRITIC CELLS INDUCE IMMUNITY TO CUTANEOUS LEISHMANIASIS IN MICE

Heidrun Moll and Stefanie Flohé

Research Center for Infectious Diseases
University of Würzburg
Röntgenring 11, D-97070 Würzburg, Germany

1. INTRODUCTION

In humans, the protozoan parasite *Leishmania major* produces a range of cutaneous disease manifestations. The skin lesions vary from small papules to non-ulcerated plaques to large ulcers with defined, raised edges. Satellite lesions are common. *L. major* infections are caused by introduction of parasites into the skin during a blood meal of an infected sandfly; they heal spontaneously after a few months to more than a year and are associated with severe scarring. In mammalian hosts, the parasites are obligatory intracellular and reside within macrophages and dendritic cells, which not only serve as a habitat for the parasite, but also fulfil antigen-presenting and antimicrobial effector functions.[1,2]

Human leishmaniasis can be imitated experimentally in mice and the outcome of the infection is genetically determined. Therefore, the model of murine infection with *L. major* has been used for studying the immunological and genetical parameters of disease resistance and susceptibility. Mice of resistant inbred strains, such as C57BL/6, are able to control the disease and develop only small, self-healing skin lesions, whereas susceptible BALB/c mice develop disseminating skin ulcers with fatal impact. These differences in the course of infection can be attributed to the lymphokine secretion pattern of *L. major*-reactive $CD4^+$ T cells.[3] Resistance-mediating T cells have the phenotype of T helper (Th) 1 cells and secrete interferon γ (IFN-γ), whereas disease-promoting T cells have the lymphokine production profile of Th2 cells releasing interleukin (IL) 4, IL-5 and IL-10. The bulk of evidence suggests that $CD8^+$ T cells do not exert a protective role during primary infections. Furthermore, antibodies are not involved in either the clearance of parasites or the establishment of immunity to subsequent infections.

2. VACCINATION AGAINST EXPERIMENTAL LEISHMANIASIS

Vaccination with live *Leishmania* parasites to produce self-healing lesions at an inconspicious site has been practiced for a long time in the Middle East. This method in-

Dendritic Cells in Fundamental and Clinical Immunology
edited by Ricciardi-Castagnoli, Plenum Press, New York, 1997

duces resistance in at least 70% of the individuals treated, but serious clinical complications associated with the live vaccine emphasize the need for an attenuated or defined vaccine against cutaneous leishmaniasis.[4]

In the mouse model, significant levels of protection against cutaneous leishmaniasis have been achieved by a variety of manipulations, including sublethal irradiation prior to infection,[5] injection of anti-IgM, anti-CD4 or anti-IL-4 antibodies,[6–8] treatment with IL-12,[9,10] and immunization with killed or disrupted parasites or defined parasite antigens.[11–13] The efficacy of vaccination depends on the way of immunization. Intravenous immunization with killed *L. major* induces protection in genetically susceptible BALB/c mice, whereas subcutaneous injection of the same antigen preparation is ineffective or even exacerbates the disease.[14] These findings suggest that the site of first antigen encounter is critical in determining the type of the resulting immune response. It is possible that efficient vaccination depends on the acquisition of antigen by certain antigen-presenting cells in the microenvironment that favor the induction of a protective T cell response. Therefore, a detailed knowledge of the specialized functions of different antigen-presenting cells in cutaneous leishmaniasis is important for the development of vaccination strategies.

3. THE ROLE OF DENDRITIC CELLS IN CUTANEOUS LEISHMANIASIS

Macrophages, the professional phagocytes, are the typical host cells for *Leishmania* and carry the major parasite load in the infected tissues. However, they are inefficient in presenting antigen to naive T cells for induction of the primary immune response. The principal sensitizing signal for initiation of the *Leishmania*-specific T cell immune response is presumably provided by Langerhans cells.

After cutaneous infection with *L. major*, the local inflammatory response induces Langerhans cells to cross the epidermal-dermal junction for subsequent phagocytosis of parasites in the dermis.[15] Parasite uptake by Langerhans cells is mediated by the attachment of a lipophosphoglycan on the surface of *L. major* to the host cells' complement receptor CR3. The mannose receptor, which has been suggested to be involved in antigen capture by dendritic cells,[16] does not contribute to endocytosis of *L. major* by Langerhans cells, as the uptake could not be inhibited by mannan[15] or antibodies against DEC-205 (Blank and Moll, unpublished), a homologue of the macrophage mannose receptor.[17] The parasite load of infected Langerhans cells remains consistently low, indicating that they are able to restrict the intracellular replication of the organisms. Unlike macrophages, however, Langerhans cells are unable to release nitric oxide (NO), the most important effector molecule required for the killing of *Leishmania*, and cannot be stimulated to express the cytokine-inducible NO synthase (iNOS).[18] The ability of Langerhans cells to control the parasite growth, on the one hand, and their failure to develop iNOS-mediated leishmanicidal activity, on the other hand, may be well adapted to the highly specialized accessory function of these cells: a distinguished feature of *L. major*-infected Langerhans cells is the transport of parasites from the infected skin to the draining lymph nodes, which can be detected as early as 24–48 hours after infection, and presentation of parasite antigen to naive T cells for triggering the primary immune response to *L. major*.[19] The factors regulating the translocation of Langerhans cells involve the cytokines tumor necrosis factor α, IL-1 and macrophage inflammatory protein 1α (Arnoldi and Moll, submitted).

Dendritic cells are critical for the regulation of the *Leishmania*-specific T cell immunity not only in the early phase after infection, but also at later stages after cure of skin lesions in genetically resistant mice. It has been shown that a small number of virulent parasites persist in these animals, predominantly in the lymph nodes draining the initial site of infection.[20] The persistence of *Leishmania* is not caused by dormant organisms but reflects a continuous process of parasite growth and killing.[21] Both macrophages and dendritic cells harbor persistent parasites. However, only dendritic cells but not macrophages are able to present residual parasite antigen to *L. major*-reactive T cells in vitro.[22] These findings raise the possibility that the survival of a few pathogens may be required for the continuous restimulation of memory T cells and the maintenance of protective immunity.

4. LANGERHANS CELLS AS VEHICLES FOR VACCINATION AGAINST *L. major*

Given their unique accessory functions both in the initiation of the *L. major*-specific T cell response and the long-term presentation of residual parasite antigen, Langerhans cells/dendritic cells are ideal candidates to serve as vehicles for vaccination against cutaneous leishmaniasis. Their potential as a natural adjuvant has already been demonstrated in cancer models.[23,24] Mice immunized with dendritic cells that had been pulsed with tumor-associated antigens were protected against a subsequent inoculation with cancer cells.

To determine whether dendritic cells are able to induce immunity to *Leishmania* infection in vivo, we immunized genetically suceptible mice with Langerhans cells that had been pulsed in vitro with *Leishmania* lysate. The results show that a single treatment mediates resistance against a subsequent challenge with virulent parasites (Figure 1). The protection was associated with a pronounced decrease in the parasite burden. Furthermore, it correlated with a change in the lymphokine production pattern upon in vitro restimulation of lymph node cells with *L. major* antigen.

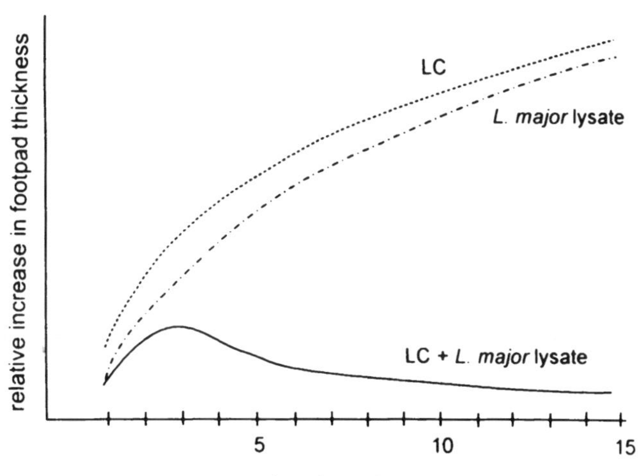

Figure 1. Schematic presentation of the course of lesion development in *L. major*-infected BALB/c mice that had been immunized with antigen-pulsed Langerhans cells (LC).

Analysis of the mechanisms underlying this adjuvant effect of Langerhans cells suggests that it involves (1) the ability of *L. major*-activated Langerhans cells to produce IL-12 which is known to promote the development of Th1 cells associated with resistance to *Leishmania* infection, (2) the presentation of the host-protective *Leishmania* lipophosphoglycan by Langerhans cells, which can not be achieved by macrophages, and (3) the modulation of major histocompatibility complex class II expression after exposure of Langerhans cells to *Leishmania* antigen (Flohé and Moll, submitted). Taken together, the data indicate that dendritic cells may serve as initiators of host-protective immune responses to intracellular microorganisms and, thus, as delivery systems for vaccination or immunotherapy. The recently developed methods for establishing dendritic cell lines and obtaining sizeable numbers of dendritic cells from human peripheral blood provide valuable tools for this approach.

5. REFERENCES

1. R.G. Titus, C.M. Theodos, A. Shankar, and L.R. Hall, Interactions between *Leishmania major* and macrophages, *in*: "Macrophage-Pathogen Interactions", T. Zwilling and T. Eisenstein, eds., Marcel Dekker, New York (1993), pp. 437.
2. H. Moll, S. Flohé, and C. Blank, Langerhans cells in cutaneous leishmaniasis, *in*: "The Immune Functions of Epidermal Langerhans Cells", H. Moll, ed., R.G. Lande, Austin (1995), pp. 159.
3. S.L. Reiner and R.M. Locksley, The regulation of immunity to *Leishmania major*, *Annu. Rev. Immunol.* 13:151 (1995).
4. F.Y. Liew and C.A. O'Donnell, Immunology of leishmaniasis, *Adv. Parasitol.* 32:161 (1993).
5. J.G. Howard, C. Hale, and F.Y. Liew, Immunological regulation of experimental cutaneous leishmaniasis. IV. Prophylactic effect of sublethal irradiation as a result of abrogation of suppressor T cell generation in mice genetically susceptible to *Leishmania tropica*, *J. Exp. Med.* 153:557 (1981).
6. D.L. Sacks, P.A. Scott, R. Asofsky, and F.A. Sher, Cutaneous leishmaniasis in anti-IgM-treated mice: enhanced resistance due to functional depletion of a B cell-dependent T cell involved in the suppressor pathway, *J. Immunol.* 132:2072 (1984).
7. R.G. Titus, R. Ceredig, J.-C. Cerottini, and J.A. Louis, Therapeutic effect of anti-L3T4 monoclonal antibody GK1.5 on cutaneous leishmaniasis in genetically susceptible BALB/c mice, *J. Immunol.* 135:2108 (1985).
8. M.D. Sadick, F.P. Heinzel, B.J. Holaday, R.T. Pu, R.S. Dawkins, and R.M. Locksley, Cure of murine leishmaniasis with anti-interleukin 4 monoclonal antibody. Evidence for a T cell-dependent, interferon gamma-independent mechanism, *J. Exp. Med.* 171:115 (1990).
9. F.P. Heinzel, D.S. Schoenhaut, R.M. Rerko, L.E. Rosser, and M.K. Gately, Recombinant IL-12 cures mice infected with *Leishmania major*, *J. Exp. Med.* 177:1505 (1993).
10. J.P. Sypek, C.L. Chung, S.E.H. Mayor, J.M. Subramanyam, S.J. Goldman, D.S. Sieburth, S.F. Wolf, and R.G. Schaub, Resolution of cutaneous leishmaniasis: interleukin 12 initiates a protective T helper type 1 immune response, *J. Exp. Med.* 177:1797 (1993).
11. M.J. McConville, A. Bacic, G.F. Mitchell, and E. Handman, Lipophosphoglycan of *Leishmania major* that vaccinates against cutaneous leishmaniasis contains an alkylglycerophosphoinositol lipid anchor, *Proc. Natl. Acad. Sci. USA* 84:8941 (1987).
12. D.M. Yang, N. Fairweather, L. Button, W.R. McMaster, L.P. Kahl, and F.Y. Liew, Oral *Salmonella typhimurium* (AroA⁻) vaccine expressing a major leishmanial surface protein (gp63) preferentially induced Th1 cells and protective immunity against leishmaniasis, *J. Immunol.* 145:2281 (1990).
13. D. McMahon-Pratt, D. Rodriguez, J.-R. Rodriguez, Y. Zhang, K. Manson, C. Bergman, L. Rivas, J.F. Rodriguez, K. Lohman, N.H. Ruddle, and M. Esteban, Recombinant vaccinia viruses expressing GP46/M-2 protect against *Leishmania* infection, *Infect. Immun.* 61:3351 (1993).
14. F.Y. Liew, C. Hale, and J.G. Howard, Prophylactic immunization against experimental leishmaniasis. IV. Subcutaneous immunization prevents the induction of protective immunity against fatal *Leishmania major* infection, *J. Immunol.* 135:2095 (1985).
15. C. Blank, H. Fuchs, K. Rappersberger, M. Röllinghoff, and H. Moll, Parasitism of epidermal Langerhans cells in experimental cutaneous leishmaniasis with *Leishmania major*, *J. Infect. Dis.* 167:418 (1993).

16. F. Sallusto, M. Cella, C. Danieli, and A. Lanzavecchia, Dendritic cells use macropinocytosis and the mannose receptor to concentrate macromolecules in the major histocompatibility complex class II compartment: downregulation by cytokines and microbial products, *J. Exp. Med.* 182:389 (1995).
17. W. Jiang, W.J. Swiggard, C. Heufler, M. Peng, A. Mirza, R.M. Steinman, and M.C. Nussenzweig, The receptor DEC-205 expressed by dendritic cells and thymis epithelial cells is involved in antigen processing, *Nature* 375:151 (1995).
18. C. Blank, C. Bogdan, C. Bauer, K. Erb, and H. Moll, Murine epidermal Langerhans cells do not express inducible nitric oxide synthase, *Eur. J. Immunol.* 26:792 (1996).
19. H. Moll, H. Fuchs, C. Blank, and M. Röllinghoff, Langerhans cells transport *Leishmania major* from the infected skin to the draining lymph node for presentation to antigen-specific T cells, *Eur. J. Immunol.* 23:1595 (1993).
20. T. Aebischer, Recurrent cutaneous leishmaniasis: a role for persistent parasites?, *Parasitol. Today* 10:25 (1994).
21. S. Stenger, N. Donhauser, H. Thüring, M. Röllinghoff, and C. Bogdan, Reactivation of latent leishmaniasis by inhibition of inducible nitric oxide synthase, *J. Exp. Med.* 183:1501 (1996).
22. H. Moll, S. Flohé, and M. Röllinghoff, Dendritic cells in *Leishmania major*-immune mice harbor persistent parasites and mediate an antigen-specific T cell immune response, *Eur. J. Immunol.* 25:693 (1995).
23. V. Flamand, T. Sornasse, K. Thielemans, C. Demanet, M. Bakkus, F. Tielemans, O. Leo, J. Urban, and M. Moser, Murine dendritic cells pulsed in vitro with tumor antigen induce tumor resistance in vivo, *Eur. J. Immunol.* 24:605 (1994).
24. S. Grabbe, S. Beissert, T. Schwarz, and R.D. Granstein, Dendritic cells as initiators of tumor immune responses: a possible strategy for tumor immunotherapy?, *Immunol. Today* 16:117 (1995).

IMMUNE RESPONSE IN DENDRITIC CELL DEPLETED MICE

Benoît Salomon,[1] Pieter J. M. Leenen,[2] and David Klatzmann[1]

[1]Laboratoire de Biologie et Thérapeutique des Pathologies Immunitaires
CNRS ERS 107, CERVI
Hôpital de la Pitié-Salpêtrière, Paris, France
[2]Department of Immunology
Erasmus University
Rotterdam, The Netherlands

1. INTRODUCTION

Dendritic cells (DC) are regarded as professional antigen presenting cells capable of efficiently stimulating T cells. In addition, *in vitro* stimulation of naive T cells can only be achieved efficiently by dendritic cells (DC) (1–3). After antigen administration, DC are the only cells bearing immunogenic fragments and capable of stimulating specific naive T cells *in vitro* (4, 5). Furthermore, B cells seem dispensable in T cell-dependent immune response *in vivo* (6, 7). However, as yet there is no direct evidence that DC are absolutely required for the generation of an *in vivo* primary immune response. A possible way to approach this question is to use an animal model in which DC are depleted. There are yet no natural animal model of a genetic deficit of DC. Recently, a generated *relB* knock out mice have a DC deficit in lymphoid tissues. However, secondary lymphoid tissues loss their normal structure, with a strong atrophy of splenic white pulp and lymph nodes, rendering complicated the use of this model to study the role of DC in immune response (8, 9).

We have generated an animal model aimed at generating a conditional ablation of dendritic cells in transgenic mice (10). The strategy consisted in the targeted expression of an *Herpes simplex* virus type I thymidine kinase (TK) in DC. TK allows the phosphorylation of the non toxic nucleoside analogue ganciclovir into a phosphorylated form that, after further phosphorylation by cellular enzymes, is converted to triphosphated ganciclovir. This latter molecule is incorporated into elongating DNA, blocking DNA replication and causing cell death (11). Thus, in transgenic mice, dividing cells expressing TK are the only ones to be killed upon ganciclovir administration. Therefore, the initiation and the duration of DC depletion can be fully controlled.

To target TK expression in DC, we have used regulatory sequences derived from the HIV long terminal repeat (LTR). This choice was dictated by a previous report showing

Dendritic Cells in Fundamental and Clinical Immunology
edited by Ricciardi-Castagnoli, Plenum Press, New York, 1997

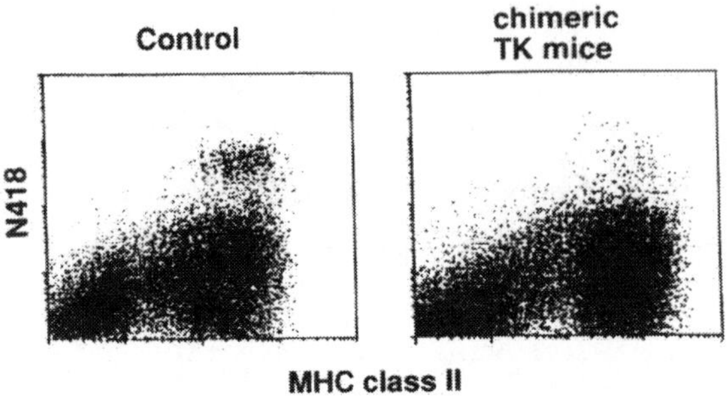

Figure 1. Splenic DC depletion in ganciclovir treated chimeric mice. Control and chimeric TK mice were treated with ganciclovir at 50 mg/Kg/d for 7 days as previously described (10). Spleen cells were then mechanically dissociated, double stained with the DC specific antibody N418 and the antibody M5/114 against MHC class II, and analysed by flow cytometry. Dendritic cells are identified as a discrete population by high level CD11c/N418 and MHC class II expression.

the preferential expression of a reporter gene under the control of HIV-LTR in Langerhans cells in transgenic mice (12). There are no clear molecular explanations for such a tissue specificity of the HIV-LTR. However, the presence of NFκB sites in these regulatory sequences could be an explanation since DC express constitutively high level of the NFκB/rel transcription factors (13, 14).

We previously reported that in LTR-TK transgenic mice, expressing TK under the control of HIV-LTR, the transgene was indeed preferentially expressed in DC. The treatment of these mice with ganciclovir resulted in a strong DC depletion notably in the spleen (10). This was also associated with a wasting syndrome and a thymic atrophy (10, 15). To determine whether these phenomena were directly related to DC depletion, we derived chimeric TK mice that were generated by transplantation of bone marrow cells from transgenic mice into lethally irradiated normal recipient of the same genetic background. As with transgenic mice, we reproduced in chimeric TK mice the DC depletion upon ganciclovir treatment (Figure 1). Importantly, this was not associated with a wasting syndrome and thymic atrophy, indicating that these two phenomenons were probably due to a leaky TK expression in non hematopoietic cells of LTR-TK transgenic mice (15). Therefore, we used these chimeric TK mice for studying the immune response in DC deficient mice.

2. RESULTS AND DISCUSSION

First, we analysed the extent of DC depletion in ganciclovir-treated chimeric TK mice in various tissues by flow cytometry and immunohistology analyses in order to delineate the best experimental system for studying immune response in the absence of DC. In the spleen, the major DC subpopulation localised in the marginal zone of the white pulp is completely depleted, while the other DC subpopulation localised in the T-cell area (Leenen et al. this volume) is only partially affected. Noticeably, the architecture of the spleen is not affected, and the proportions of lymphoid and myeloid cells are not significantly modified by the ganciclovir treatment. In lymph nodes, the ganciclovir treatment affects only the subpopulation of DC expressing lower levels of MHC class II (manuscript in preparation). The heterogeneity of the depletion of the different DC sub populations in these tissues must be related to their turn over, since ganciclovir kills only dividing cells expressing TK (manuscript in preparation). In any case, the strong and specific depletion of splenic DC observed in our model prompted us to analyse an immune response detectable in the spleen. We chose to use the classical antibody response in the spleen to sheep red blood cells (SRBC). After intraperitoneal injection of SRBC, the immune response is revealed by the detection of plaque forming units.

At least 6 weeks after bone marrow graft, chimeric TK mice and control mice (lethally irradiated mice reconstituted with bone marrow cells from non transgenic mice) were treated for 7 days with ganciclovir. 8 to 24 hours later, the mice were immunised with SRBC and 4 days later, spleen cells were recovered and the numbers of specific IgM and IgG secreting B cells were determined by plaque assay. Meanwhile, the extent of DC depletion was assessed at the time of immunisation on similarly treated mice (Figure 1).

These experiments showed that the primary immune response to SRBC is abolished in these DC deficient mice (Figure 2), indicating that DC are absolutely essential to initiate a T-dependent primary immune response. Similar results were obtained when the immune response was analysed 6 days, instead of the current 4 days, after the injection of SRBC (not shown).

Because of the conditional aspect of the DC depletion in our system, the role of DC in secondary immune response can also be analysed. In such experiments, mice were first primed with SRBC in the absence of ganciclovir treatment. Two months later, the mice were treated by ganciclovir for 7 days and 8 to 24 hours later they were re challenged with SRBC. This experiment showed that the secondary immune response was diminished but still persistent in DC deficient mice (Figure 2).

These studies indicate for the first time that, at least in the spleen, DC are absolutely required for generating primary immune responses to a T-dependent antigen. These cells seem less essential in secondary immune response probably because other antigen presenting cells can stimulate memory T cells (6, 16, 17). Further studies should now be done to generalise this observation to other antigens and other lymphoid tissues.

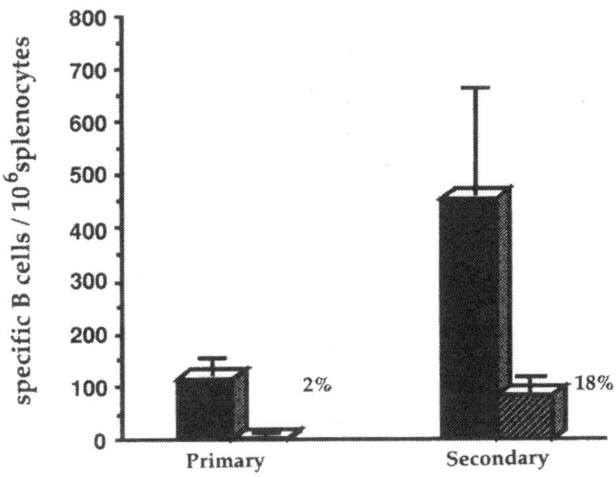

Figure 2. Primary and secondary immune response in ganciclovir treated chimeric mice. Primary immune response (primary). Control (black bar) and chimeric mice (hatched bar) were treated with ganciclovir at 50 mg/Kg/d for 7 days as previously described (10). 8–24 hours later, mice were immunised intraperitonally (10 mice per group) with 200 μl of 10% SRBC. Four days later, the number of SRBC-specific plasma cells secreting IgM and IgG were counted by the technique of plaque forming units. Secondary immune response (secondary). Control (black bar) and chimeric mice (hatched bar) were immunised with SRBC, 2 months later they were treated with ganciclovir at 50 mg/Kg/d for 7 days, 8–24 hours after cessation of ganciclovir, the mice were re challenge with the antigen, and the antibody response was analysed as for primary response. About 10 mice per group were used.

3. REFERENCES

1. Steinman, R.M., B. Gutchinov, M.D. Witmer, and M.C. Nussenzweig. 1983. Dendritic cells are the principal stimulators of the primary mixed leukocyte reaction in mice. *J Exp Med* 157:613–627.
2. Croft, M., D.D. Duncan, and S.L. Swain. 1992. Response of naive antigen-specific $CD4^+$ T cells in vitro: characteristics and antigen-presenting cell requirements. *J Exp Med* 176:1431–1437.
3. Young, J.W., and R.M. Steinman. 1990. Dendritic cells stimulate primary human cytolytic lymphocyte responses in the absence of CD4+ helper T cells. *J Exp Med* 171, no. 4:1315–1332.
4. Crowley, M., K. Inaba, and R.M. Steinman. 1990. Dendritic cells are the principal cells in mouse spleen bearing immunogenic fragments of foreign proteins. *J Exp Med* 172, no. 1:383–386.
5. Guéry, J.-C., F. Ria, and L. Adorini. 1996. Dendritic cells but not B cells present antigenic complexes to class II-restricted T cells after administration of protein in adjuvant. *J Exp Med* 183:751–757.
6. Ronchese, F., and B. Hausmann. 1993. B lymphocytes in vivo fail to prime naive T cells but can stimulate antigen- experienced T lymphocytes. *J Exp Med* 177, no. 3:679–690.
7. Epstein, M.M., F. Di Rosa, D. Jankovic, A. Sher, and P. Matzinger. 1995. Sccessful T cell priming in B cell-deficient mice. *J Exp Med* 182:915–922.
8. Burkly, L., C. Hession, L. Ogata, C. Reilly, L.A. Marconi, D. Olson, R. Tizard, R. Cate, and D. Lo. 1995. Expression of relB is required for the development of thymic medulla and dendritic cells. *Nature* 373, no. 6514:531–536.
9. Weih, F., D. Carrasco, S.K. Durham, D.S. Barton, C.A. Rizzo, R.P. Ryseck, S.A. Lira, and R. Bravo. 1995. Multiorgan inflammation and hematopoietic abnormalities in mice with a targeted disruption of relB, a member of the NF-kB/Rel family. *Cell* 80:331–340.
10. Salomon, B., P. Lorès, C. Pioche, P. Racz, J. Jami, and D. Klatzmann. 1994. Conditional ablation of dendritic cells in transgenic mice. *J Immunol* 152:537–548.
11. St. Clair, M.H., C.U. Lambe, and P.A. Furman. 1987. Inhibition by ganciclovir of cell growth and DNA synthesis of cells biochemically transformed with herpesvirus genetic information. *Antimicrobial Agents and Chemotherapy* 31:844–849.
12. Leonard, J., J.S. Khillan, H.E. Gendelman, A. Adachi, S. Lorenzo, H. Westphal, M.A. Martin, and M.S. Meltzer. 1989. The human immunodeficiency virus long terminal repeat is preferentially expressed in Langerhans cells in transgenic mice. *AIDS Res and Hum Retrovir* 5, no. 4:421–430.
13. Carrasco, D., R.P. Ryseck, and R. Bravo. 1993. Expression of relB transcripts during lymphoid organ development : specific expression in dendritic antigen-presenting cells. *Development* 118:1221–1231.
14. Granelli-Piperno, A., M. Pope, K. Inaba, and R.M. Steinman. 1995. Coexpression of NF-kB/Rel and Sp1 transcription factors in human immunodeficiency virus 1-induced, dendritic cell-T-cell syncitia. *Proc Natl Acad Sci USA* 92:10944–10948.
15. Salomon, B., C. Pioche, P. Lores, J. Jami, P. Racz, and D. Klatzmann. 1995. Conditional ablation of dendritic cells in mice: comparison of two animal models. *Adv Exp Med Biol* 378:485–7.
16. Cassell, D.J., and R.H. Schwartz. 1994. A quantitative analysis of antigen-presenting cell function: activated B cells stimulate naive CD4 T cells but are inferior to dendritic cells in providing costimulation. *J Exp Med* 180, no. 5:1829–1840.
17. Croft, M., L.M. Bradley, and S.L. Swain. 1994. Naive versus memory CD4 T cell response to antigen. Memory cells are less dependent on accessory cell costimulation and can respond to many antigen- presenting cell types including resting B cells. *J Immunol* 152, no. 6:2675–2685.

DENDRITIC CELL BASED THERAPY OF CANCER

Michael T. Lotze,* Michael Shurin, Ian Davis, Andrew Amoscato, and Walter J. Storkus

Departments of Surgery,and Molecular Genetics and Biochemistry
University of Pittsburgh School of Medicine
Biological Therapeutics Division, University of Pittsburgh Cancer Institute
Pittsburgh, Pennsylvania 15261

INTRODUCTION

The recent identification of techniques both in the mouse and in man to culture dendritic cells which could be adoptively trasnferred for therapy have opened new avenues of research. We have applied both synthetic and natural or acid eluted peptides from the tumor cell surface in order to elicit effective antitumor responses in vivo and in vitro in murine tumor models as well as in vivo in human preclinical trials. Based on these studies, we will soon initiate a human clinical trial utilizing human dendritic cells pulsed with synthetic melanoma derived peptides presented by HLA-A2. The delivery of genes into human dendritic cells to constitutively express cytokines, costimulatory molecules or tumor antigens represents another potential approach to the development of dendritic cell based therapies.

POTENTIAL TARGETS FOR DC LOADING: MELANOMA TUMOR ANTIGENS DEFINED BY T-LYMPHOCYTES

Several novel melanoma antigens have recently been identified with the help of cytotoxic T cells recognizing MHC class I presented epitopes on different HLA alleles. Of note, all melanoma antigens recognized by cytotoxic T cells have far thus been identified using individual patient's TIL or PBL. The tumor-associated proteins yielding antigenic peptides (T cell epitopes) are therefore capable of being recognized *in vivo* by the patients own immune system. The passive transfer of CD8+ T- lymphocytes recognizing such T

* Corresponding author: Department of Surgery, University of Pittsburgh, 300 Kaufmann Bldg., 3471 Fifth Ave., Pittsburgh, Pennsylvania 15213. Phone: 412–624–9375; FAX: 412–624–1172.

cell epitopes results in objective clinical responses of usually short duration in some patients (1,2).

The first human melanoma antigen identified was the MAGE-1 gene mapping to the X -chromosome(3). This antigen yields a peptide recognized by HLA-A1 (4), or HLA-Cw16 (5) restricted CTL. The HLA-A1 allele is present in about 26% of the Caucasian population , and the MAGE-gene is expressed in 50% of all melanoma lines tested; approximately 10–15 % of melanoma patients may therefore express this antigen. MAGE-1 is not expressed in normal human tissue, with the exception of the testes. Expression of the MAGE-antigen is not restricted to melanoma; MAGE-1 is also found in some Epstein-Barr virus transformed B-cell lines, in breast cancer, lung cancer, and in tumors derived from neuroectodermal origin, such as neuroblastomas and glioblastomas (6). Glioblastomas are also recognized by MAGE-1 specific T cells, suggesting that an anti-MAGE-1 directed cellular immune response may also be useful in the treatment of glioblastomas and generally, any MAGE-1 positive tumor (7). Expression of MAGE-1 in various tumors may explain the cross reactivity of anti- melanoma cytotoxic T-cells with tumors of different histology, such as Ewing sarcomas. Both of these cell types have a common embryonic origin (i.e., derived from the neural crest) (7).

MAGE-1 belongs to a broad family of closely related genes. Several MAGE family members have been detected and one of these (MAGE-3) also yields a peptide recognized by cytotoxic T - lymphocytes in the context of HLA-A1 (8) and HLA-A2 (9). In contrast to MAGE-1, MAGE-3 is expressed in 80% of melanoma cell lines tested. MAGE-3 could also serve as a potential target for immunotherapy in other tumors, unrelated to melanoma: 56 % of head and neck squamous cell carcinomas, 30% of non small cell lung carcinomas, 12% of breast tumors, and 16 % of colorectal carcinomas also express MAGE-3 (8).Different gene - families with a similar tissue distribution like the MAGE-family, still under investigation, have recently been identified and provide additional immunogenic peptides as potential targets for immunotherapy including the melanoma associated antigens encoded by BAGE or GAGE (10,11).

Another target for immunoreactive T cells in melanoma patients is the key enzyme in melanin production, tyrosinase (12). Tyrosinase provides at least two different peptides recognized by anti melanoma - directed CTL restricted by HLA-A2 (13). A tyrosinase related protein (gp75) has been identified as providing an HLA-A31 presented peptide to (CD8+) cytotoxic anti-melanoma specific T-cells (14). Of note, antibodies reacting with the gp75 protein have been identified in sera from patients with melanoma indicating that gp75 provides both, T and B cell epitopes for immune reactivity (14). Further characterization of the gp75 gene product(s) revealed that the immunogenic peptide epitope presented by HLA-A31 appears to be generated by an alternative open reading frame (ORF) detected in both melanocytes, and melanoma cells (15).

HLA-A2 is the most frequent MHC class I allele in Caucasians and appears to be the predominant restriction element for an anti-melanoma directed immune response. MART-1(Melanoma Antigen Recognized by T cells-1) or Melan-A encodes a previously unrecognized 13kD protein expressed in melanoma, normal dermal melanocytes, as well as the retina and provides three closely related peptides as targets for cytotoxic T cells (16–18). A different HLA-A2 presented target for antimelanoma directed T cells is encoded by the previously described gene product gp100 or Pmel 17, providing two T cell epitopes for immune recognition (2,19,20) . The gene product of gp100 is involved in melanosomal formation and is also recognized by the monoclonal antibody HMB45, employed routinely in the detection of cells of melanocytic origin. The expression of gp100 is therefore also restricted to cells of melanocytic lineage. Of note, passive transfer of anti - gp100 directed

T-lymphocytes, resulted in preferential localization of these immune effector cells in melanoma lesions, correlating with objective clinical responses (2).The shared expression of antigens presented by normal melanocytes and melanomas may provide one mechanism for the clinical observation of vitiligo, which in some cases is associated with regression of melanoma lesions. Anichini and coworkers reported that anti-melanoma specific T-lymphocytes are also capable of recognizing normal, HLA matched, melanocytes (21). Thus, several protein antigens provide peptide epitopes for an anti-melanoma directed T-cell response. Additionally, most of these proteins provide multiple peptide epitopes allowing for recognition by the immune system and presentation by dendritic cells.

DC IMMUNOBIOLOGY (ACTIVATION AND CYTOKINE PRODUCTION)

Dendritic cells (DC) clearly serve as antigen processing, presenting, and transporting cells ferrying potentially immunogenic epitopes to the draining lymphatics where clonotypic T cells (and B cells) may be activated, mainly in the paracortical regions of the lymphoid tissue (22,23). The process of cognate DC-T (or DC-B or DC-T-B) recognition results in an intercellular dialogue spoken in biochemical terminology involving both cell surface expressed receptor /ligand signals and secreted cytokines (24,25). In particular, costimulatory molecules such as CD40, CD80, and CD86 appear to play significant roles serving as "signal 2" elements required for T cell activation (26,27). Using a series of monoclonal antibodies directed against DC expressed molecules putatively involved in DC-T cell collaboration in crosslinking studies, we have recently determined that a subset of human peripheral blood derived DCs may be activated to synthesize and secrete IFN-α, IFN-γ, and IL-12 (Ma *et al.*, unpublished data). The levels of secreted cytokine were typically on the order of 50–100 pg/ml/10^6 DC/48h as determined by ELISA, with 35–80% of CD86+ DC producing the factor as validated by intracellular staining with cytokine-specific mAb. Zhou and Tedder (25) have similarly determined that human CD83+ DC constitutively synthesize, or may be induced with phorbol ester to synthesize, IL-1α, IL-1β, IL-6, IL-8, IL-9, IL-10, TNF-α, TGF-β, IFN-γ, GM-CSF, M-CSF, and G-CSF as determined by PCR analysis with cytokine-specific primer panels. The complex mixture of such cytokines produced by DCs in the DC-T cell microenvironment may profoundly bias the nature (i.e. Th0-like vs. Th1-like vs. Th2-like) of the resultant antigen-specific T cell response (25).

DC-BASED TUMOR THERAPIES: SYNTHETIC PEPTIDES, STRIPPED TUMOR PEPTIDES, AND TUMOR ANTIGEN cDNAs

The application of autologous dendritic cells in tumor therapies has received intensive interest in the past several years (reviewed in refs. 28–30). Animal models have clearly documented the ability of syngeneic bone marrow-derived dendritic cells to serve as effective immunogens of anti-tumor CTL *in vivo* when prepulsed with tumor-derived peptide epitopes (31–35). Thus dendritic cells generated over the course of 7–8 days of culture in the presence of GM-CSF and IL-4 or TNF-α, could be cocultured with microgram quantities of synthetic peptides derived from proteins such as ovalbumin, HPV-16 E7, ß-galactosidase, MUT1 (connexin 37), or p53, allowing for the effective loading of these peptides into DC expressed MHC class I complexes. The subsequent vaccination and boosting of mice with as few as 10^5 of these epitope-charged cells, allowed for the animal to reject a subsequent challenge with a normally lethal challenge dose of a tumor expressing the naturally processed and presented epitope used in the vaccine (31–33). Of significant interest, in some cases, animals rejecting such a tumor challenge were also able to then subsequently reject a challenge with an otherwise identical tumor that fails to lack the epitope used in the original vaccination (32). This concept of "epitope spreading" which may be of crucial interest in the induction of immunity against antigenically heterogenous human tumors. Recently Malcolm Brenner and associates have demonstrated that stripped peptides from the A20 B-cell lymphoma in the mouse, when applied onto GM-CSF and IL-4 expanded dendritic cells and adoptively transferred into a host bearing a 4d tumor led to the late (six to eight weeks) development of an autoimmune disorder that mimicked graft versus host disease. This may have been due to the rich display of Class II presented epitopes on this tumor. Certainly this should provide some sense of caution when considering DC therapies in human clinical trials.

OTHER POTENTIAL METHODS OF ANTIGEN TRANSFER: HEAT SHOCK PROTEINS

Recent evidence has suggested another pathway for the transfer of peptide antigenic determinants from infected or otherwise abnormal cells into the antigen processing pathways of antigen-presenting cells (APC). Heat shock proteins (hsp) are a class of proteins whose expression is upregulated by various stressors, including heat or glucose deprivation.(36). hsp have been shown to have important functions in the immune response. They act as molecular chaperones (37), playing critical roles in the movement of proteins and peptides from the cytosol to the endoplasmic reticulum, where they facilitate the assembly of MHC class I molecules with β2-microglobulin and peptide (38–41). hsp have also been shown to mediate the transfer of immunologic determinants from tumor cells, allowing for the vaccination of naive animals. It seems likely that hsp may be involved in antigen recognition by γδ T cells. For example, hsp60 is recognized by specific γδ T cells on lymphoma cells (42). gp96 is expressed in the endoplasmic reticulum (ER) (43,44). In the ER, gp96 accepts peptide from the TAP transporter protein (45,46) and passes it on to the MHC class I complex in an ATP-dependent fashion (47). hsp 70, found predominantly in the cytosol, may provide protection against reactive oxygen species (including nitric oxide) or against cytokines such as TNF-α or IL-1 (48). The critical role of hsp in class I peptide presentation was demonstrated several years ago. Cell lysates from chemically induced tumors could be used to vaccinate naive mice against that tumor type but not against other tumor types (46,49–51). Fractionation of the tumor lysates showed that the responsible components were hsp, in particular gp96, hsp90 and hsp70. More recent work has shown that naive mice can be vaccinated with hsp70, gp96, and to a lesser extent hsp90 from tumors (51). Immunization with gp96 leads to a CD8+ T cell response but not necessarily to a CD4+ response (52). This effect could be blocked by carageenan or silica, suggesting that macrophages (or perhaps DC's) were required.

Tumor cells may not express MHC class I molecules on the cell surface. However, hsp can be used to cross-prime across histocompatibility mismatched animals, even if the appropriate MHC molecule is missing from the original cell (53). These results suggest that hsp-based immunotherapy may be useful in priming precursor CTL even after cell lysis induced by infection, antibodies, or CTL (Srivastava et al., 1994). In addition to the

Table 1. Relationship between DC infiltration and prognosis in malignancy

Tumor	Reference	DC infiltration
Basal Cell	Brit Jl Derm 127:575, 1992	? Improved
Breast	J Pathol 163:25, 1991	?Improved Prognosis
Bronchoalveolar	Eur Jl Ca 28A:1365, 1992	No effect
Cervix	Am Jl Clin Path 99:200, 1993	Less in HPV+ tumors
Cervix	Cancer 70:2839, 1992	Improved prognosis*
Cervix, Stage III	In Vivo 7:257, 1993	Markedly improved prognosis*
Cervix/Penile	J Urol 147:1268, 1992	Less infiltrate with HPV infection
Cervix/HIV	Gynec Oncol 48:210, 1993	Less infiltrate in AIDS related
Esophageal	Virchows Arch 61:409, 1992	Markedly improved prognosis
Esophageal	In Vivo 7:239, 1993	Direct relationship to Grade
Gastric	Int Surg 77:238, 1992	Marked improved prognosis*
Gastric	Cancer 75:1478, 1995	More in tumor draining LNs
Gastric, Stage III	In Vivo 7:223, 1993	Marked improved prognosis*
Hodgkins Disease	Am J Cl Path 101:761, 1994	Follicular DC improves prognosis*
Lung	Pathology 25:338, 1993	Marked improved prognosis*
Lung	J Clin Invest 91:566, 1993	Related to GM-CSF production
Melanoma	J Invest Derm 100:269, 1993	Inverse tumor thickness/prognosis
Mycosis Fungoides/Sezary Syndrome	In Vivo 7:277, 1993	Markedly improved prognosis*
Prostate	Prostate 19:73, 1991	Improved prognosis

*Indicates statistically significant difference when compared to either normal or less DC's.
In contrast a comparatively poorer clinical prognosis is observed for patients with lesions that exhibit a sparse infiltration with DC and metastatic lesions are frequently deficient in DC infiltration.

well-documented roles of hsp in MHC class I complex assembly, there is now evidence that hsp may be involved in antigen presentation via the MHC class II pathway hsp may therefore provide a means of gaining access to the class II pathway for the manipulation of immune responses *in vivo* and *in vitro*.

DCs OBSERVED WITHIN HUMAN TUMORS

Of significant clinical interest, the histologic infiltration of DCs into primary tumor lesions has been associated with significantly prolonged patient survival and a reduced incidence of metastatic disease in patients with oral, head and neck tumors, nasopharyngeal tumors, bladder, lung, esophageal, and gastric carcinoma (55–103). Efforts to enhance DC delivery to tumors is thus of great interest.

ESTABLISHMENT OF MOUSE BONE MARROW AND SPLEEN DERIVED DC CULTURES

DC colonies develop from mouse CD34+ bone marrow progenitors during culture (7 days, 37°C, 5 % CO_2) in complete media (CM) (RPMI 1640, 2 mM L-glutamine, 50 µg/ml gentamicin sulfate, 10 mM Hepes, 10 % FBS (Gibco)) supplemented by mrGM-CSF (1000 U/ml) and mrIL-4 (1000 U/ml). The marrow single cell suspensions were treated with Abs and rabbit complement (Gibco) to deplete cells that express the lymphocyte antigens B220, CD4, and CD8. Treatment with Abs (partially purified supernatant of hybridoma cell cultures TIB-146, TIB-207, and TIB-105 for B220, CD4, and CD8, respectively) and complement removed detectable B and T lymphocytes from suspension cultures.

Splenocytes were obtained after digestion. These were similarly depleted of B and T lymphocytes and cultured for 7 days in GM-CSF and IL-4. The low level of GM-CSF supplementation ensures a low proportion of co-developing macrophages (Mϕ). Northern blot analysis and biosynthetic labeling revealed that under low GM-CSF conditions, class II synthesis by bone marrow Mϕ does not take place, so DC which are generated under these culture conditions are the only cells constitutively expressing class II molecules (104). IL-4 acts at least in part by suppressing monocyte differentiation (105). Moreover, using GM-CSF and IL-4 has been shown to be permissive for DC antigen capture and processing (106). PgE_2 down-regulates HLA-DR expression (107). In addition, PgE_2 significantly reduces the number of GM-CFU per femur and spleen in mice (108) and inhibits proliferation of GM-CFU in human (109). There is evidence that bone marrow cells are capable of producing NO (110) which, in turn, is associated with DC apoptosis (111). Addition of a cyclooxygenase inhibitor, indomethacin (1 μg/ml, Sigma) and a specific inhibitor of the L-arginine-dependent nitric oxide synthesizing pathway, N^G-nitro-L-arginine methyl ester (50 μM, L-NAME, Sigma), resulted in a significant elevation (4–6 folds) of DC number in 7-day cultures.

In routine 7 day cultures the yield of DC is about $50–60 \times 10^6$ DC per mouse (4 femori) and $8–12 \times 10^6$ DC per mouse spleen. This relatively low yield of splenic DC is one of the limiting factors in the use of these cells as an immunotherapeutic tool. GM-CSF has been shown to enhance the yield of DC generated *in vitro* from murine BM progenitors as well as peripheral blood (112). Other cytokines such as kit ligand (Szabolcs et al., 1995), TNF-α (114) and IL-4 (115) synergize with GM-CSF in *in vitro* generation of DC. Interestingly, FLT3 Ligand (FL) has been used as a component of multicytokine-supported culture (IL-1, IL-7, IL-3, IL-6, KIT ligand, TNF, GM-CSF and FL) medium for short-term DC cultures (116). However, until recently no growth factor was known to increase the amount of DC in lymphoid tissue. Maraskovsky et al. (117) have shown that daily administration of FL (10μg) subcutaneously for 9 days resulted in a significant increase in DC in spleen, lymph nodes, peripheral blood, bone marrow, Peyers patches, liver and skin. Experiments in our laboratory confirm these findings. Splenocytes from FL-treated animal were comprised of 15–20% DC in contrast to 0.5% DC found in the saline treated group. In FL treated mice there was a 17–20 fold increase in the yield of splenic DC and a 2–3 fold increase in BM-derived DC after 7d in GM-CSF/ IL-4 as compared with the yield from saline treated control mice. The DC were distinguished by their cytologic features, high levels of MHC class II products, B7.1, B7.2 markers, and ability to stimulate naive T-cells. Recently we have demonstrated synergy with IL-12 delivery in murine models and some evidence of antitumor effects in early tumor models when applied as a single agent.

FUNCTIONAL ASSAYS

The important functional role of DC is their initiation of primary responses such as the mixed lymphocyte reaction (MLR). DC, obtained from bone marrow and splenic culture were irradiated (2000 rads from ^{137}Cs) and added in graded doses to 3×10^5 allogeneic or syngeneic T cells in 96-well flat-bottom tissue culture plates (Falcon) in 200 μl of CM per well. T-cells were obtained from spleen cell suspension by passing them through a nylon wool column (Robbins Scientific) after lysing erythrocytes (155 mM NH_4Cl in 10 mM

Table 2. Synthetic peptides used in murine tumor models

Peptide	H-2 Restriction	Expressed by Tumor
OVA (257-264)	K^b	MO5 Melanoma
E7 (49-57)	D^b	C3 Sarcoma
MUT1 (Connexin 37 52-59)	K^b	3LL Lung Carcinoma
Mutant p53 (232-240)	K^d	MethA Sarcoma
Wild-type p53 (232-240)	K^d	CMS4 Sarcoma

Tris-HCl buffer, pH 7.5, 25°C). Proliferation was measured by the uptake of [^{3}H]thymidine (1 μCi per well, 5 Ci/mmol; NEN) pulsed at 80–96 h. Bulk peritoneal macrophages were 1/50th to 1/100th as active as cultivated BM-DC in allogeneic stimulation. However, only DC induced a syngeneic MLR. DC from FL treated animals were found to be more active in MLR than controls.

MURINE TUMOR PEPTIDE-PULSED DC ELICIT PROTECTIVE ANTI-TUMOR IMMUNITY *IN VIVO*

We directly assessed the ability of cultured DC to serve as biologic adjuvants promoting anti-tumor immune responses in murine models. DC precursors were harvested from murine bone-marrow, grown for 8 days in the presence of rmGM-CSF and recombinant murine interleukin-4 (rmIL-4), and shown to demonstrate the characteristic DC "veiled" morphology and a CD3-, CD14-, CD20-, CD40+, CD80+, CD86+, class II+ phenotype (31). Cells were then pulsed with a synthetic tumor-associated peptide (Table 2)(31), prior to extensive washing to remove unbound peptide, and injected into the tail-vein of B6 mice on days 0 and 7. Animals were then challenged on day 14 with relevant or irrelevant tumors injected s.c. in the right flank and tumor progression monitored (31). Animals vaccinated in this manner displayed complete and specific resistance to antigen-relevant tumor challenge in each of 4 distinct tumor model systems (M05 melanoma (H-2^b), C3 sarcoma (H-2^b), 3LL lung carcinoma (H-2^b), and Meth A sarcoma (H-2^d)). In comparative experiments, we demonstrated that DCs grown in GM-CSF + IL-4 were significantly better *in vivo* stimulators of anti-tumor immunity than DCs grown in GM-CSF + tumor necrosis factor-alpha (TNF-α), DCs grown in GM-CSF alone, or cultured, skin-derived langerhans cells (LC) (31,32). The observed efficacy of DC/peptide vaccines was directly correlated to the ability of the vaccine to prime *in vivo* splenic anti-tumor CTL and was not dependent upon the route of vaccine administration since i.v., i.p., and s.c. routes all yielded animals resistant to tumor challenge (data not shown).These results, as well as those of others (33,118–120), supporting the general effectiveness in DC/peptide-based vaccines.

Animals were inoculated in the hind flank with twice the minimum tumorigenic dose of a given tumor 5–14 days prior to treatment.Tumor peptides were either synthetic (Table 2)(31, 120) or were eluted from explanted tumor tissue by a mild acid elution technique previously demonstrated to preferentially recover soluble MHC class I eluted peptides previously expressed on the treated cell surface (121). Peptides were pulsed onto d8 bone marrow derived DC (GM-CSF + IL-4) for 2h, prior to extensive washing and inoculation of peptide-loaded cells i.v. in the tail vein of the tumor bearing-animals every 4–7 days. Therapies consisting of synthetic tumor peptide pulsed DC cured 80–100% of C57BL/6 animals bearing days 14 C3 tumors or days 7 3LL lung carcinomas, typically within 7–10 days (31). 1–2 DC/peptide vaccines are sufficient for successful therapy and long-term survival. All disease-free animals were subsequently shown to specifically reject rechallenge with higher doses of relevant tumor but not irrelevant tumors (data not shown). In the C3 model, treatment was also assessed in days 21 and days 28 models, with DC/E7 peptide vaccines curing 60% and 20% of animals, respectively (31).

These results observed for DC/synthetic tumor peptide vaccines led us to evaluate the therapeutic benefit of vaccines consisting of DCs pulsed with unfractionated peptides extracted from tumor cell MHC molecules. This strategy was designed to serve as a mechanism whereby vaccines might be constructed for tumor histologies where TAA and TAA-derived peptide epitopes have not yet been identified, i.e. most non-melanoma histologies. Peptides were extracted from freshly explanted tumors grown s.c. in syngeneic donor animals. After enzymatic digestion was performed to obtain a single cell tumor suspension, cells were briefly incubated in mild (pH 3.3) acid to denature their cell surface MHC class I complexes. Peptides were isolated, concentrated, and desalted from the cell-free supernatant, with aliquots of this peptide-containing solution pulsed onto d8 bone marrow-derived DC (GM-CSF + IL-4). Peptide pulsed, syngeneic DC were inoculated i.v. into C57BL/6 (H-2^b) mice bearing either the CL8–1 melanoma, MCA205 sarcoma, or C3 sarcoma, or into BALB/c (H-2^d) mice bearing the TS/A mammary adenocarcinoma, every 4–7 days. In most cases, these treatments reproducibly resulted in transient stasis (up to 30–40 days) of disease progression, but in no cures (121). The exception to this finding was the C3 model (d14) in which all mice were cured within 2 weeks of initial treatment. In all cases, therapeutic efficacy of DC/tumor peptide vaccines was tumor antigen-specific since irrelevant tumor peptides were ineffective vaccine components, and most of the therapeutic effect could be attributed to the first 2 vaccines applied (data not shown). Furthermore, it is important to note that despite the virtual certainty of acid-extracting and vaccinating animals with diverse "self" peptide epitopes, no overt signs of autoimmune pathology were observed in any treated animal.

TRAFFICKING PATTERN OF I.V. INOCULATED DC IN TUMOR-BEARING ANIMALS

Most of these adoptively transferred APC rapidly localize to the spleen, liver, and inguinal lymph nodes of these animals (data not shown). The spleens, inguinal lymph nodes, and tumors of treated mice were evaluated for the immunostimulatory impact of DC/unfractionated tumor peptide vaccines in the MCA205 sarcoma model. Briefly, animals were inoculated with tumor on day 0, received i.v. DC/tumor peptide therapy on days 8 and 12, and were sacrificed, with tissues removed for examination on day 16. Single cell suspensions of spleen, lymph nodes, and tumor were cultured overnight with irradiated MCA205 tumor, with culture supernatants collected at 24h for analysis of cytokine production. Only those animals receiving DC/MCA205 peptide vaccines demonstrated therapeutic benefit and this was correlated with significantly elevated production of IFN-γ (10–50X control DC alone values) in the lymph node and spleen, but not tumors, of treated animals [data not shown, ref.120]. No elevation of cytokine production was observed for mice receiving DC alone or DC pulsed with unfractionated splenic-derived peptides vs. PBS controls, suggesting that the response was MCA205 tumor-specific. In separate experiments, neutralizing anti-cytokine antibodies were inoculated i.p. in tumor-bearing mice, prior to and during, the DC-peptide vaccination protocol. These studies support the critical requirement of IFN-γ, TNF-α, and IL-12 in the generation of a vaccine-induced anti-tumor immune response since each of these reagents virtually negated the therapeutic impact of the vaccine (121). The *in vivo* inhibitory effect anti-IL-12 only occurred if it was applied before or during the first DC-peptide vaccination, suggesting the importance of IL-12 during the initial "priming" of antigen-specific immunity. Anti-IL-4 reagents demonstrated a complementary pattern of inhibitory effects, with statistically significant diminishment of vaccine benefit occurring only after the second vaccination. This finding is consistent with the involvement of IL-4 in a later (i.e. possibly humoral) immune response to tumor.

REQUIREMENT FOR "HELPER" AND "KILLER" T CELLS AND FOR T CELL COSTIMULATION

This was determined by administration of anti-CD4 or anti-CD8 monoclonal antibodies and CTLA4-Ig fusion protein, a potent inhibitor of B7 family-associated costimulation. Each of these single reagents effectively neutralized the therapeutic benefit of DC/tumor peptide vaccines (120). Examination of an MCA205 tumor 8 days after receiving DC-based vaccines revealed a significant elevation in tumor infiltration by CD4+ and CD8+ T cells, as well as by, S100+ cells only in those tumor-bearing animals receiving DC-MCA205 peptide vaccines (Table 3). We believe that these latter cells are host APCs, possibly DCs, based on the previously reported cell type reactivity pattern established for the S100 antibody (121). Our finding that increased S100+ cellular infiltration correlates with enhanced survival of the tumor-bearing animal is consistent with numerous clinical reports noting a prognostic association betweentumor infiltrate with S100+ cells and enhanced patient survival (see below) (122–125).

HUMAN DC/TUMOR PEPTIDE VACCINES ELICIT POTENT ANTI-TUMOR CTL *IN VITRO*

We (126) and others (127–130) have recently documented the ability to also generate potent immunostimulatory DC from human peripheral blood in cultures containing

Table 3. DC/Tumor peptide vaccines are associated with infiltration by S100+ cells, CD4+ T cells, and CD8+ T cells. Animals bearing d8 MCA205 sarcomas were treated on days 8 and 12 with syngeneic DCs pulsed with either MCA205-derived peptides, splenic (normal) peptides, or no peptides. On day 16, tumors were harvested and examined for histologic infiltration by APC (S100+) or by T cells

	Tumor infiltration by		
Treatment group	CD4+	CD8+	S100+
Untreated	–	–	–
DC-No Peptide	+/-	+/-	–
DC-Splenic Peptide	+	+	–
DC-MCA205 Peptide	++++	++++	+++

Infiltration is quantitated on a relative - to ++++ scale.

GM-CSF + IL-4. Similar to the murine results reported above, the resulting cells exhibit the characteristic dendrites and "veiled" morphology associated with dendritic cells and express high levels of MHC (class I and class II) and costimulatory (CD40, CD80, CD86) molecules. These DC are efficient stimulators of mixed lymphocyte reactions using allogeneic responder lymphocytes. Further these cells may be induced to secrete IL-12, IFN-α, and IFN-γ, factors important to the induction and maintenence of a Th1-associated cellular immune response, when activated by T cell ligands (data not shown). Using serum-free AIM-V media (GIBCO-BRL) containing 1000 U rhGM-CSF/ml + 1000 U rhIL-4/ml, we were able to generate DC from the peripheral blood of 50 (50/50) normal donors and from 12 of 12 patients with melanoma. After 7 days of *in vitro* culture, yields were typically 8%-12% of the starting PBMC obtained from the donor, regardless of disease status (126). Thus, DC were readily generated from both normal donors and from cancer patients.

We selected 10 normal donors and 10 melanoma patients that were HLA-A2+, derived DC from their peripheral blood, and evaluated the ability of these DC when pulsed with melanoma TAA epitopes (Table 3) to elicit anti-melanoma CTL *in vitro* using autologous PBL responders (126). After 2–3 *in vitro* weekly restimulations with autologous DC pulsed with tumor peptide, anti-melanoma CTL responders were demonstrated from both normal donor and patient cultures stimulated with some, but not all, of the previously reported HLA-A2 presented melanoma epitopes (126,130,131). In particular, MART-1 27–35, MART-1 32–40, gp100 280–288, and tyrosinase 368–376 epitopes appeared to be immunogenic for most HLA-A2 responders, while the gp100 457–466, tyrosinase 1–9, and MAGE-3 epitopes were less or non-immunogenic (Table 4).

Based on these pre-clinical results, we have constructed an IRB/FDA-approved protocol using vaccines consisting of autologous DC pulsed *ex vivo* with the immunogenic peptides identified in Table 3 for the treatment of HLA-A2+ patients with AJCC stage III or IV melanoma. More recently we have been able to generate antitumor reactive T cells from an HLA-A2+ patient with melanoma who had her tumor metastases resected from the lung. Approximately 10^9 cells were obtained which were acid eluted, passed through a Sep-pak and lyophilized. 1/4 of this material was pulsed onto autologous 7day DC. PBMC were coincubated at a 30:1 ratio with DC's and restimulated on day 14 with autologous DC's pulsed with peptide and supplemented with 60 IU/ml IL-2. Peptide reactivity using

Table 4. DC's pulsed with melanoma peptides elicit HLA-A2 restricted, anti-melanoma CTL *in vitro*. PBL responses to 3 weekly autologous DC/tumor peptide stimulations was assessed in 4h cytotoxicity assays against HLA-A2+ melanoma 526

		CTL response induced in	
Melanoma peptide	Sequence	normal Donor PBL	melanoma Patient PBL
MART-1 27-35	AAGIGILTV	9/10	9/10
MART-1 32-40	ILTVILGVL	9/10	9/10
gp100 280-288	YLEPGPVTA	8/10	8/10
gp100 457-466	LLDGTATLRL	0/10	0/10
Tyrosinase 1-9	MLLAAVLYCL	0/10	0/10
Tyrosinase 368-376	YMDGTMSQV	7/10	6/10
MAGE-3 271-279	FLWGPRALV	1/3	1/4

peptide fractions obtained from the HLA-A2+ 526 melanoma and cytokine release or cytokine release and cytotoxicity against 526 but not the A2+ CIR cell line.

DCs GENETICALLY ENGINEERED TO SECRETE CYTOKINES OR TO EXPRESS TUMOR-ASSOCIATED ANTIGENS SERVE AS EFFECTIVE IMMUNOGENS FOR THE INDUCTION OF TUMOR-REACTIVE CTL *IN VITRO* AND *IN VIVO*

DC induction of a Th0 pattern of cytokine production (i.e. IL-4 and IFN-γ) within lymphoid tissue in vaccinated animals suggested that the efficacy of DC-based therapies might be further enhanced by biasing the immune response towards a more Th1-like response (i.e. preferential T cell production of IFN-γ but not IL-4). This suggested that we implement IL-12, a cytokine known to drive principally Th1-associated immunity, to augment a cellular anti-tumor response in our established tumor model. While DC can naturally produce IL-12 when appropriately stimulated (i.e. by CD40 ligation or T cell cluster formation, ref. 128), it was theorized that the provision of systemic rmIL-12 or the engineering of DC with an expression plasmid encoding mIL-12 under a strong viral promoter might increase the efficacy of DC/tumor peptide therapies, by providing physiologic doses of IL-12 at the outset of vaccination (132). Murine BM-DC (day 8) generated as outlined above were transfected using the DFG-mIL-12 retroviral vector (132), with transfected cells producing approximately 2ng mIL-12/ml/10^6 DC/24h. These IL-12 producing DC (DC-IL-12) were then pulsed with MCA205-derived peptides, prior to extensive washing and injection i.v. in the tail vein of B6 mice bearing d8 sub-cutaneous MCA205 tumors. Tumors in animals treated with DC-IL-12 pulsed with tumor peptides failed to progress for longer periods of time than tumors in animals treated with control DC pulsed with MCA205 peptides. This enhanced maintenance of tumor stasis correlated with increases in tumor-specific CTL in the spleens of DC-IL-12 + MCA205 peptide treated mice.

We have also evaluated the ability of DC engineered to express the melanoma-associated antigens MART-1 and/or gp100 to promote anti-melanoma CTL *in vitro* and *in vivo*. DCs transfected with these gene products are conceptually able to process and simultaneously present tumor epitopes on multiple APC-expressed MHC class I *and* class II molecules. This circumvents the requirement of patient HLA-typing prior to accrual in synthetic melanoma peptide-based vaccine protocols. Murine BM-derived (d8) or human PBMC-derived (d7) DC grown in IL-4 + GM-CSF containing medium were transfected with cDNAs encoding MART-1 and/or gp100 using the Accell gene gun. Expression of the transgenes (22–45%) was documented by flow cytometry after permeabilizing a portion of the DC with saponin and intracellular staining with MART-1-specific (kindly provided by Dr. S.A. Rosenberg, NCI) or gp100-specific (HMB-45) monoclonal antibodies and FITC-conjugated goat anti-mouse Ig reagents. Transfected DC were irradiated (2500 rad) and then used to stimulate autologous T cells *in vivo* in B6 mice (10^6 DC engineered to express both MART-1 and gp100, injected i.v. on days 0 and 7) or *in vitro* using using PBMC responder T cells derived from a normal HLA-A2+,-A3+ donor (10^6 engineered DC + 10^8 responder T cells, with boost of 10^6 engineered, autologous DC on day 7). Spleens were removed from the vaccinated mice on day 14 and restimulated for 7 days *in vitro* with irradiated MART-1+, gp100+, MHC I+, MHC II+ MELI-1 melanoma (10,000 rad) prior to assessment of anti-MELI-1 CTL reactivity in a standard 4h ^{51}Cr-release assay. Splenic CTL exhibited MELI-1 specific reactivity that was not observed if animals were instead vaccinated with DC transfected with the backbone plasmid.The cytolytic response appeared to contain both MHC Class I- and Class II-restricted T cell effector populations since anti-H-2D^b and anti-H-2IA^b reactive MoAb can each significantly inhibit the anti-MELI-1 killing observed.

Human CTL primed *in vitro* with autologous DC expressing either the MART-1 or gp100 antigens were evaluated in cytolytic assays at day 17 against a series of targets that included C1R.A2 (an EBV-B cell line expressing only HLA-A2), C1R.A3, gp100 or MART-1 transfectants of C1R.A2 or C1R.A3, or the allogeneic melanoma cell line Mel 526. Of note, HLA-A2+,-A3+ DC engineered with MART-1 cDNA promoted CTL responses against both C1R.A2.MART-1 and C1R.A3.MART-1, as well as Mel 526, but not against targets transfected to express gp100. The reciprocal result was true for CTL induced by DC.gp100. These results suggest that tumor antigens applied as cDNA in DC-based vaccines may simultaneously prime the expansion of bulk, tumor-specific CTL restricted by multiple host class I (and presumably class II) alleles.

USE OF UNPURIFIED ACID-ELUTED PEPTIDES FROM HUMAN TUMORS

We have performed several preliminary experiments using fresh human tumor and cell lines, in order to answer the following questions: 1. Is it possible to alter the quantity and nature of peptides expressed on the surface of malignant cells by *in vitro* manipulation? 2. Is it possible to use unpurified acid-eluted peptide preparations and APC to stimulate memory cytotoxic T cells? 3. Is it possible to use unpurified acid-eluted peptide preparations from freshly resected human tumors, and autologous peripheral-blood derived DCs, to activate naive PBL against that tumor target?; 4. Can other methods be used to transfer immunogenic tumor peptide determinants to DCs in order to stimulate naive T cells?

a. The human melanoma cell line 526 expresses the HLA-A2-restricted epitopes of the gp100 and MART-1/MELAN-A proteins. We have established conditions to maximize MHC class I expression of 526 cells. 526 cells were grown for 1–4 days in RPMI medium/10% FBS in the presence or absence of IFN-γ, and at 37°C or 40°C. At 24 hour intervals, the cells were harvested by trypsinization, stained with the W6/32 antibody for flow cytometric determination of MHC class I expression. IFN-γ causes a 1.6X increase in the level of expression of class I molecules on 526 cells at 24 hours (Table 5). A similar increase is seen at 48 hours but not at 72 hours.
b. 526 cells were also examined for alterations in the peptide profile under conditions of heat or IFN-γ. 526 cells were grown in RPMI medium/10% FBS for several days and then split into 32 T-225 tissue culture flasks. Prior to reaching confluency, the flasks were divided into four groups of 8 and treated either with or without IFN-γ 1000 U/ml, and at either 37°C or 40°C (approximately 2–4 x 10^8 viable cells/condition at the end of incubation). After 24 hours the cells were harvested by trypsinization, washed several times in PBS, and MHC-associated peptides were stripped from the cells by mild acid elution (citrate-phosphate buffer, pH 3.2, 4 ml for 1 minute). The acid-eluted peptides were concentrated on a C18 Sep-Pak, washed to remove salt, and eluted with 60% acetonitrile. The eluted peptides were then subjected to filtration to remove species of molecular

Table 5. Changes in MHC class I expression by 526 melanoma cells with heat or IFN-γ

Condition	Mean channel fluorescence	Fold increase over baseline
24 hr, 37°C	143	(Baseline)
24 hr, 37°C, IFN-γ	216	1.5
48 hr, 37°C	257	1.8
48 hr, 37°C, IFN-γ	467	3.3
72 hr, 37°C	108	0.8
72 hr, 37°C, IFN-γ	149	1.0
24 hr, 40°C	136	—
24 hr, 40°C, IFN-γ	202	1.5
48 hr, 40°C	195	1.4
48 hr, 40°C, IFN-γ	344	2.5
72 hr, 40°C	135	1.0
72 hr, 40°C, IFN-γ	275	2.0
96 hr, 40°C	74	0.5
96 hr, 40°C, IFN-γ	165	1.2

mass greater than 3000. The filtrate was concentrated and evaporated to near dryness for removal of organic solvents. The resulting solution was examined by electrospray ionization mass spectrometry. Treatment of 526 cells with IFN-γ results in a alteration of the mass ion profile. An approximate 1.5 to 2.0-fold increase in total mass ions was detected. This is confirmed in the total ion current plots of the non-treated and IFNγ-treated cells. The increase in complexity is particularly seen in the region between Da/e ratios of 800 and 1000. Specific alterations were seen at particular Da/e ratios: ions with ratios of 621.3, 665.2, 709.3, 753.3, 828.1 and 988.1 showed increases, while ions 791.0 and 982.9 were decreased with IFN-γ treatment. Finally, treatment of 526 cells at 40∞C was shown to induce increases in specific mass ions (Da/e ratios of 663.9, 678.9, 706.0, 709.7, 823.4, 828, 950.2, 987.9 and 993.3), although the overall complexity of ion species generated was reduced . The addition of 1000 U/ml of IFN-γ caused a further decrease in the intensities of the ions with Da/e ratios indicated above. These results indicate that IFN-γ not only upregulates class I expression on 526 cells, but that the presentation profile of particular peptides is altered.

c. Initial experiments have been performed to address the question of whether treatment of 526 cells with IFNγ or heat causes differences in the expression of biologically relevant class I-associated peptides. Peptides were eluted from 526 cells as described previously and were reconstituted in water such that equal volumes of reconstituted peptides represented equal numbers of starting cells. 2 ml of each peptide was diluted in 96-well U-bottom tissue culture plates at the following concentrations: neat (undiluted), 1/10, 1/100 and 1/1000. The C1R.A2 B cell line (which expresses HLA-A2) was used as stimulators in this assay, and MG12 or G280 cell lines were used as responders. These T cell lines were derived from patients with melanoma and recognize the melanoma epitopes gp100 and MART-1/MELAN-A, and gp100, respectively, in the context of HLA-A2. 10^4 ^{51}Cr-labelled C1R.A2 cells/well were plated in the wells containing peptide and were incubated for 1 hour. Following this, the MG12 or G280 responder cells were added and the plates were incubated for four hours, after which time supernatants were collected for estimation of ^{51}Cr release. CTL reactivity was seen even at a 1:1000 dilution of peptide. However, in this experiment no clear differences between the preparative conditions (± IFN-γ, ± heat) were seen. Further experiments are planned to attempt to quantitate the differences in peptide expression that were inferred from the mass spectrometry experiments described in (a) above.

REFERENCES

1. Robbins PF, El - Gamil M, Kawakami, Y, Rosenberg SA. Recognition of tyrosinase by tumor - infiltrating lymphocytes from a patient responding to immunotherapy. 1994; Cancer Res54:3124–3126.
2. Kawakami Y, Eliyahu S, Delgado CH, Robbins PF, Sakkaguchi K, Appella E, Yanelli JR, Adema GJ, Kimi T, Rosenberg SA. Identification of melanoma antigens recognized by tumor infiltrating lymphocytes associated with in vivo tumor rejection . 1994; Proc Natl Acad Sci 91:6458–6462.
3. Van der Bruggen P, Traversari C, Chomez P, Lurquin C, De Plaen E, Van Den Eynde B, Knuth A, Boon T. A gene encoding an antigen recognized by cytolytic T lymphocytes on a human melanoma. 1991; Science 254:1643–1047.
4. Traversari C, van der Bruggen P, Luescher IF, et al.: A nonapeptide encoded by human gene MAGE-1 is recognized on HLA-A1 by cytolytic T - lymphocytes directed against tumour antigen MZ2-E. J Exp Med 1992: 176:1453–1458
5. Van der Bruggen P, Szikora J-P, Boel, P, Wildmann C, Somville, M, Sensi M, Boon T. Autologous cytolytic T lymphocytes recognize a MAGE-1 nonapeptide on melanomas expressing HLA-Cw*16018. 1994. Eur J Immunol 24:2134–2140.
6. Zakut R, Topalian SL, Kawakami Y, Mancini M, Eliyahu S, Rosenberg SA. Differential expression of MAGE-1, -2, -3 messenger RNA in transformed and normal cell lines. 1993; Cancer Res 53:54–47.
7. Shamamian P, Mancini M, Kawakami Y, et al.: Recognition of neuroectodermal tumors by melanoma -specific cytotoxic T lymphocytes: evidence for antigen sharing by tumors derived from the neural crest. Cancer Immunol Immunother 1994; 39:73–79
8. Gaugler B, Van den Eynde B, van der Bruggen P, Romerao P, Gaforio J J , De Plaen E, Lethe B, Brasseur F, Boon T . Human gene MAGE-3 codes for an antigen recognized on a melanoma by autologous cytolytic T lymphocytes. 1994; J Exp Med 179:921–930.
9. Van der Bruggen P, Bastin J, Gajewski T, Coulie PG, Boel P, De Smet C, Traversari C, Townsend A, Boon T. A peptide encoded by human gene MAGE-3 and presented by HLA-A2 induces cytolytic T lymphocytes that recognize tumor cells expressing MAGE-3. 1994. Eur J Immunol 24:3038–3043.
10. Boel-P; Wildmann-C; Sensi-ML; Brasseur-R; Renauld-JC; Coulie-P; Boon-T; van-der-Bruggen-P BAGE: a new gene encoding an antigen recognized on human melanomas by cytolytic T lymphocytes. 1995. Immunity. 2: 167–75
11. Van den Eynde B, Peeters O, De Backer O, Gaugler B, Lucas S, Boon T. A new family of genes coding for an antigen recognized by cytolytic T lymphocytes on a human melanoma. J Exp Med 182:689–698, 1995.
12. Brichard V, Van Pel A, Wolfel T, Wolfel C, De Plaen, E. Lethe B, Coulie P, Boon T. The tyrosinase gene codes for an antigen recognized by autologous cytolytic T lymphocytes on HLA- A2 melanomas . 1993; J Exp Med 178:489–495.
13. Wolfel T. Van Pel, A, Brichard V, Schneider J, Seliger B, Meyer zum Buschenfelde K-H, Boon T. Two tyrosinase nonapeptides recognized on HLA-A2 melanomas by autologous cytolytic T lymphocytes. 1994; Eur J Immunol 24:759–764.
14. Wang R-F, Robbins PF, Kawakami Y, Kang X-Q, Rosenberg SA. Identification of a gene encoding a melanoma tumor antigen recognized by HLA-A31 - restricted tumor - infiltrating lymphocytes. 1995; J Exp Med 181:799–804.
15. Wang R-F, Robbins PF, Kawakami Y, Rosenberg SA. A novel human cancer antigen resulting from translation of an alternative open reading frame of TRP-1. Science, in press.
16. Kawakami Y, Eliyahu S, Delgado CH, Robbins PF, Rivoltini L, Topalian SL, Miki T, Rosenberg SA. Cloning of the gene coding for a shared human melanoma antigen recognized by autologous T cells infiltrating into tumor. 1994; Proc Natl Acad Sci 91:3515–3519.

17. Coulie PG, Brichard V, VanPel A, Wolfel T, Schneider J, Traversari C, Mattei S, De Plaen E, Lurquin C, Szikora J-P, Renauld J-C, Boon T. A new gene coding for a differentiation antigen recognized by autologous cytolytic T lymphocytes on HLA-A2 melanomas. 1994; J Exp Med 180:35–42.
18. Castelli C, Storkus WJ, Maeurer MJ, Martin DM, Huang EC, Pramanik BN, Nagabhushan TL, Parmiani G, Lotze MT. Mass spectrometric identification of a naturally - processed melanoma peptide recognized by CD8+ cytotoxic T lymphocytes. 1995; J Exp Med 181:363–368.
19. Bakker ABH, Schreurs MWJ, de Boer AJ, Kawakami Y, Rosenberg SA, Adema GJ, Figdor CG. Melanocyte lineage - specific antigen gp100 is recognized by melanoma derived tumor infiltrating lymphocytes. 1994; J Exp Med 179:1005–1009.
20. Cox AL, Skipper J , Chen Y, Henderson RA, Darrow TL, Shabanowitz J, Engelhard VH, Hunt DF, Slingluff CL. Identification of a peptide recognized by five melanoma - specific human cytotoxic T cell lines. 1994; Science 264:716–719.
21. Anichini A, Maccalli C, Mortarini R, Salvi S, Mazzocchi A, Squarcina P, Herlyn M, Parmiani G. Melanoma cells and normal melanocytes share antigens recognized by HLA-A2 restricted cytotoxic T cell clones from melanoma patients. 1993. J Exp Med 177:265–272.
22. Egner W, McKenzie JL, Smith SM, Beard ME, Hart DN. Identification of potent mixed leukocyte reaction-stimulatory cells in human bone marrow. Putative differentiation stage of human blood dendritic cells. Journal of Immunology 1993;150:3043–3053.
23. Inaba K., Witmer-Pack M., Inaba M. et al. The tissue distribution of B7–2 costimulator in mice:abundant expression of dendritic cells in situ and during maturation in vitro. J. Exp. Med. 180:1849–1860, 1994.
24. Kämpgen, Koch F, Heufler C, Eggert A, Gill LL, Gillis S, Dower SK, Romani N, Schuler G. Understanding the dendritic cell lineage through a study of cytokine receptors. Journal of Experimental Medicine 1994;179:1767–1776.
25. Zhou L-J, Tedder TF. A distinct pattern of cytokine gene expression by human CD83+ blood dendritic cells. Blood 1995;86:3295–3301.
26. Turcovski-Corrales SM, Fenton RG, Peltz G, Taub DD. CD28:B7 interactions promote T cell adhesion. European Journal of Immunology 1995;25:3087–3093.
27. Péguet-Navarro J, Dalbiez-Gauthier C, Rattis FM, Van Kooten C, Banchereau J. Schmitt D. Functional expression of CD40 antigen on human epidermal Langerhans cells. Journal of Immunology 1995;155:4241–4247.
28. Young JW, Inaba K. Dendritic cells as adjuvants for class I major histocompatibility complex-restricted antitumor immunity. Journal of Experimental Medicine 1996;183:7–11.
29. Paglia P, Chiodoni C, Rodolfo M, Colombo MP. Murine dendritic cells loaded in vitro iwth soluble protein prime cytotoxic T lymphocytes against tumor antigen in vivo. Journal of Experimental Medicine 1996;183:317–322.
30. Knight SC, Hunt R, Dore C, Medawar PB. Influence of dendritic cells on tumor growth. Proceedings of the National Academy of Science 1985;82:4495–4497.
31. Mayordomo, J.I., Zorina, T., Storkus, W.J., Zitvogel, L., Celluzzi, C., Falo, L.D., Melief, C.J., Ildstad, S.T., Kast, W.M., DeLeo, A., and Lotze, M.T. Bone marrow derived dendritic cells pulsed with synthetic tumour peptides elicit protective and therapeutic anti-tumor immunity. *Nature Med.* **1**: 1297–1302, 1995.
32. Celluzzi, C.M., Mayordomo, J.I., Strokus, W.J., Lotze, M.T., and Falo, L.D. Peptide-pulsed dendritic cells induce antigen-specific, CTL-mediated protective tumor immunity. *J. Exp. Med.* **183**: 283–288, 1996.
33. Flamand, V. Murine Dendritic Cells Pulsed *In Vitro* with Tumor Antigen Induce Tumor Resistance *In Vivo*, *Eur. J. Immunol.* **23**: 605–610, 1994.
34. Zitvogel, L., J. I. Mayordomo, T.Tjandrawan, A. B. DeLeo, M. R. Clarke, M. T. Lotze, and W. J. Storkus. 1996. Therapy of murine tumors with tumor peptide-pulsed dendritic cells: dependence on T cells, B7 costimulation, and T helper cell-associated cytokines .J. Exp. Med. 183:87–97.
35. Zou J-P, Shimizu J, Ikegame K, Takiuchi H, Fujiwara H, Hamaoka T. Tumor-immuotherapy with the use of tumor-antigen-pulsed antigen-presenting cells. Cancer Immunology Immunotherapy 1992;35:1–6.
36. Morimoto RI. Cells in stress: transcriptional activation of heat shock genes. Science 259: 1409–1410, 1993.
37. Georgopoulos C and Welch WJ. Role of the major heat shock proteins as molecular chaperones. Annu Rev Cell Biol 9: 601–634, 1993.
38. Mariethoz E, Tacchini-Cottier F, Jacquier-Sarlin M, Sinclair F and Polla BS. Exposure of monocytes to heat shock does not increase class II expression but modulates antigen-dependent T cell responses. Int Immunol 6: 925–930, 1994.
39. Polla BS and Mariethoz E. More evidence for a case for chaperones in antigen processing. Immunol Today 13: 421–422, 1992.

40. Suto R and Srivastava PK. A mechanism for the specific immunogenicity of heat shock protein-chaperoned peptides. Science 269: 1585–1588, 1995.
41. Pierce SK. Molecular chaperones in the processing and presentation of antigen to helper T cells. Experientia 50: 1026–1030, 1994.
42. Kaur I, Voss SD, Gupta RS, Schell K, Fisch P and Sondel PM.. Human peripheral gamma delta T cells recognize hsp60 molecules on Daudi Burkitt's lymphoma cells.J Immunol : 2046–2055, 1993.
43. Mazzarella RA and Green M. ERp99, an abundant, conserved glycoprotein of the endoplasmic reticulum, is homologous to the 90-kDa heat shock protein (hsp90) and the 94-kDa glucose regulated protein (GRP94). J Biol Chem 262: 8875–8883, 1987.
44. Booth C and Koch GLE. Perturbation of cellular calcium induces secretion of luminal ER proteins. Cell 59: 729–737, 1989.
45. Van Kaer L, Ashton-Rickardt PG, Ploegh HL and Tonegawa S. TAP1 mutant mice are deficient in antigen presentation, surface class I molecules, and CD4–8+ T cells. Cell 71: 1205–1214, 1991.
46. Li Z and Srivastava PK. Tumor rejection antigen gp96/grp94 is an ATPase: implications for protein folding and antigen presentation. EMBO J 12: 3143–3151, 1993.
47. Levy F, Gabathuler R, Larsson R and Kvist S. Cell 67: 265–274, 1991.
48. Jacquier-Sarlin MR, Fuller K, Dinh-Xuan AT, Richard M-J and Polla BS. Protective effects of hsp70 in inflammation. Experientia 50: 1031–1038, 1994.
49. Udono H and Srivastava PK. Heat shock protein 70-associated peptides elicit specific cancer immunity. J Exp Med 178: 1391–1396, 1993.
50. Udono H, Levey DL and Srivastava PK. Cellular requirements for tumor-specific immunity elicited by heat shock proteins: tumor rejection antigen gp96 primes CD8+ T cells in vivo. Proc Natl Acad Sci USA 91: 3077–3081, 1994.
51. Vanbuskirk A, Crump BL, Margoliash E and Pierce SK. A peptide binding protein having a role in antigen presentation is a member of the hsp70 heat shock family. J Exp Med 170: 1799–1809, 1989.
52. Udono H and Srivastava PK. Comparison of tumor-specific immunogenicities of stress-induced proteins gp96, hsp90, and hsp70. J Immunol 152: 5398–5403, 1994.
53. Arnold D, Faath S, Rammensee H-G and Schild H. Cross-priming of minor histocompatibility antigen-specific cytotoxic T cells upon immunization with the heat shock protein gp96. J Exp Med 182: 885–889, 1995.
54. Srivastava PK, Udono H, Blachere NE and Li Z. Heat shock proteins transfer peptides during antigen processing and CTL priming. Immunogenetics 39: 93–98, 1994.
55. Tsujitani S, Oka A, Kondo A, Ikeguchi M, Maeta M, Kaibara N. Infiltration of dendritic cells into regional lymph nodes in gastric cancer. Cancer 1995;75 (6 Suppl):1478–1483.
56. van Heerden WF, Raubenheimer EJ, van Rensburg EJ, le Roux R. Lack of correlation between DNA ploidy. Langerhans cell population and grading in oral squamous cell carcinoma. Journal of Oral Pathology and Medicine 1995;24(2):61–65.
57. Schreiner TU, Lischka G, Schaumburg-Lever G. Langerhans' cells in skin tumors. Archives of Dermatology 1995;131(2):187–190.
58. Zeid NA, Muller HK. Tobacco smoke condensate cutaneous carcinogenesis: changes in Langerhans' cells and tumor regression. International Journal of Experimental Pathology 1995;76(1)75:83.
59. Viac J, Schmitt D, Claudy A. TNF alpha expressing cells in epidermal tumors: correlation with ICAM-1 induction on keratinocytes and Langerhans cell distribution. In Vivo 1994;8(2):221–225.
60. Alavaikko MJ, Blanco G, Aine R, Lehtinen T, Felbaum C, Taskinen PJ, Sarpola A, Hansmann ML. Follicular dendritic cells have prognostic relevance in Hodgkin's disease. American Journal of Clinical Pathology 1994;101(6):761–767.
61. Girod SC, Kuhnast T, Ulrich S, Krueger GR. Langerhans cells in epithelial tumors and benign lesions of the oropharynx. In Vivo 1994;8(4):543–547.
62. Srivastava BI, Srivastava A, Srivastava MD. Phenotype, genotype and cytokine production in acute leukemia involving progenitors of dendritic Langerhans' cells. Leukemia Research 1994;18(7):499–511.
63. Kerrebijn JD, Balm AJ, Knegt PP, Meeuwis CA, Drexhage HA. Macrophage and dendritic cell infiltration in head and neck squamous-cell carcinoma: an immunohistochemical study. Cancer Immunology, Immunotherapy 1994;38(1):31–37.
64. Chan JK, Tsang WY, Ng CS, Tang SK, Yu HC, Lee AW. Follicular dendritic cell tumors of the oral cavity. American Journal of Surgical Pathology. 1994;18(2):148–157.
65. Zeid NA, Muller HK. S100 positive dendritic cells in human lung tumors associated with cell differentiation and enhanced survival. Pathology 1993;25(4):338–343.
66. Anonymous. Anticancer activity of dendritic cells. Symposium proceedings of the 4th International Conference of Anticancer Research. Crete. In Vivo 1993;7(3):185–313.

67. Becker Y. Dendritic cell activity against primary tumors: an overview. In Vivo 1993;7(3):187–191.
68. Imai Y, Yamakawa M. Dendritic cells in esophageal cancer and lymph node tissues. In Vivo 1993;7(3):239–248.
69. Yamakawa M, Kato H, Takagi S, Karube Y, Seki K, Imai Y. Dendritic cells in various human thyroid diseases. In Vivo 1993;7(3):249–256.
70. Tsujitani S, Kakeji Y, Maehara Y, Sugimachi K, Kaibara N. Dendritic cells prevent lymph node metastasis in patients with gastric cancer. In Vivo 1993;7(3):233–237.
71. Kessler II. Epidemiological considerations in the role of dendritic/Langerhans cells in human cancer. In Vivo 1993;7(3):305–312.
72. Meissner K, Loning T, Rehpenning W. Epidermal Langerhans cells and prognosis of patients with mycosis fungoides and Sezary syndrome. In Vivo 1993;7(3):277–280.
73. Muller HK, Dandie GW, Ragg SI, Woods GM. Langerhans cell alterations in cutaneous carcinogenesis. In Vivo 1993;7(3):293–296.
74. van Rensburg EJ, van Heerden WF, Raubenheimer EJ. Langerhans cells and human papillomaviruses in oesophageal and laryngeal carcinomas. In Vivo 1993;7(3):229–232.
75. Fink-Puches R, Smolle J. Langerhans cells in epithelial skin tumors. A quantitative immunohistological and morphometric investigation. In Vivo 1993;7(3):213–216.
76. Austyn JM. The dendritic cell system and anti-tumor immunity. In Vivo 1993;7(3):193–201.
77. Toriyama K, Wen Dr, Paul E, Cochran AJ. Variations in the distribution, frequency, and phenotype of Langerhans cells during the evolution of malignant melanoma of the skin. Journal of Investigative Dermatology 1993;100(3):269S-273S.
78. Tazi A, Bouchonnet F, Granssaigne M, Boumsell L, Hance AJ, Soler P. Evidence that granulocyte macrophage-colony-stimulating factor regulates the distribution and differentiated state of dendritic cells/Langerhans cells in human lung and lung cancers. Journal of Clinical Investigation 1993;91(2):566–576.
79. Spinollo A, Tenti P, Zappatore R, DeSeta F, Silini E, Guaschino S. Langerhans' cell counts and cervical intraepithelial neoplasia in women with human immunodeficiency virus infection. Gynecologic Oncology 1993;48(2):210–213.
80. Morelli AE, Sananes C, DiPaola G, Paredes A, Fainboim L. Relationship between types of human papillomavirus and Langerhans' cells in cervical condyloma and intraepithelial neoplasia. American Journal of Clinical Pathology. 1993;99(2):200–206.
81. Drexhage HA, Mooy R, Jansen A, Kerregijn J, Allaerts W, Tas MP. Dendritic cells in tumor growth and endocrine diseases. Advances in Experimental Medicine and Biology. 1993;329:643–650.
82. Tas MP, Simons PJ, Balm FJ, Drexhage HA. Depressed monocyte polarization and clustering of dendritic cells in patients with head and neck cancer: in vitro restoration of this immunosuppression by thymic hormones. Cancer immunology, Immunotherapy 1993;36(2):108–114.
83. Koretz K, Moller P, Schwartz-Albiez R. Plasminogen activators and plasminogen activator inhibitors in human colorectal carcinoma tissues are not expressed by the tumor cells. European Journal of Cancer 1993;29A(8):1184–1189.
84. Tsujitani S, Kakeji Y, Watanabe A, Kohnoe S, Maehara Y, Sugimaghi K. Infiltration of S-100 protein positive dendritic cells and peritoneal recurrence in advanced gastric cancer. International Surgery 1992;77(4):238–241.
85. Fivenson DP, Beck ER, Dunstan RW, Nickoloff BJ, Moore PF. Dermal dendrocytes and T-cells in canine mycosis fungoides. Support for an animal model of human cutaneous T-cell lymphoma. Cancer 1992;70(8):2091–2098.
86. Regezi JA, Nickoloff BJ, Headington JT. Oral submucosal dendrocytes: factor XIIIa+ and CD34+ dendritic cell population in normal tissue and fibrovascular lesions. Journal of Cutaneous Pathology 1992;19(5):398–406.
87. Tsujitani S, Kakeji Y, Orita H, Watanabe A, Kohnoe S, Baba H, Anai H, Maehara Y, Sugimachi K. Postoperative adjuvant immunochemotherapy and infiltration of dendritic cells for patients with advanced gastric cancer. Anticancer Research 1992;12(3):645–648.
88. Morelli AE, Ronchetti RD, Secchi AD, Cufre MA, Paredes A, Fainboim L. Assessment by planimetry of Langerhans' cell density in penile epithelium with human papillomavirus infection: changes observed after topical treatment. Journal of Urology 1992;147(5):1268–1273.
89. Stene MA, Holthoj-Asnong C, Cochran AJ. Effect of epidermal Langerhans cells from melanoma patients on lymphoproliferative responses. Melanoma Research 1992;2(1):57–62.
90. Daniels TE, Chou L, Greenspan JS, Grady DG, Hauch WW, Greene JC, Ernster VL. Reduction of Langerhans cells in smokeless tobacco-associated oral mucosal lesions. Journal of Oral Pathology and Medicine 1992;21(3):100–104.

91. Nakano T, Oka K, Takahashi T, Moria S, Arai T. Roles of Langerhans' cells and T-lymphocytes infiltrating cancer tissues in patients treated by radiation therapy for cervical cancer. Cancer 1992;70(12):2839–2844.
92. Bergfelt L, Larko O, Lindberg M. Density and morphology of Langerhans cells in basal cell carcinomas of the face and trunk. British Journal of Dermatology 1992;127(6):575–579.
93. Papadimitriou CS, Datseris G, Costopoulos JS, Bai MK, Ioachim-Velogianni E, Katsouyannopoulos V. Langerhans cells and lymphocyte subsets in human gastrointestinal carcinomas. An immunohistological study on frozen sections. Pathology, Research and Practice. 1992;188(8):989–994.
94. Kumar D, Sanchez RL, Kumar S. Dendrocyte population in cutaneous and extracutaneous Kaposi's sarcoma. American Journal of Dermatopathology 1992;14(4):298–303.
95. Tosi P, Sforza V, Santopietro R, Lio R, Gotti G, Paladini P, Cevenini G, Barbini P. Bronchiolo-alveolar carcinoma: an analysis of survival predictors. European Journal of Cancer 1992;28A(8–9):1365–1370.
96. Morelli AE, diPaola G, Fainboam L. Density and distribution of Langerhans cells in the human uterine cervix. Archives of Gynecology and Obstetrics 1992;252(2):65–71.
97. Furihata M, Ohsuki Y, Ido E, Iwata J, Sonobe H, Araki K, Ogoshi S, Ohmori K. HLA-DR antigen- and S-100 protein-positive dendritic cells in esophageal squamous cell carcinoma - their distribution in relation to prognosis. Virchows Archiv. B. Cell Pathology Including Molecular Pathology 1992;61(6):409–414.
98. Gallo O, Bianchi S, Giannini A, Gallina E, Libonati GA, Fini-Storchi O. Correlations between histopathological and biological findings in nasopharyngeal carcinoma and its prognostic significance. Laryngoscope 1991;101(5):487–493.
99. Gomez-Morales M, Alvaro T, Munoz M, Garcia del Moral R, Aguilar D, Caballero T, Aneiros J. Diffuse sclerosing papillary carcinoma of the thyroid gland: immunohistochemical analysis of the local host immune response. Histopathology 1991;18(5):427–433.
100. Wilson AJ, Maddoz PH, Jenkins D. CD1a and S100 antigen expression in skin Langerhans cells in patients with breast cancer. Journal of Pathology 1991;163(1):25–30.
101. Tefany FJ, Barnetson RS, Halliday GM, McCarthy SW, McCarthy WH. Immunocytochemical analysis of the cellular infiltrate in primary regressing and non-regressing malignant melanoma. Journal of Investigative Dermatology 1991;97(2):197–202.
102. Bigotti G, Coli A, Castagnola D. Distribution of Langerhans cells and HLA class II molecules in prostatic carcinomas of different histopathological grade. Prostate 1991;19(1):73–87.
103. Rucci L, Bani D, Gallo O, Arbi Riccardi R, Borghi Cirri MB, Fini-Storchi O. Interdigitating cells in the peritumoral infiltrate of laryngeal carcinomas: an immunocytochemical and ultrastructural study. Orl: Journal of Oto-Rhino-Laryngology and its Related Specialities. 1991;53(6):349356.
104. Scheicher C, Mehlig M, Dienes H-P, Reske K. Uptake of microparticle-absorbed protein antigen by bone marrow-derived dendritic cells results in up-regulation of interleukin-1α and interleukin-12 p40/p35 and triggers prolonged, efficient antigen presentation. Eur. J. Immunol. 25:1566–1572, 1995.
105. Jansen JH, Wientjens G-JHM, Fibbe WE, Willemze R, Kluin-Nelemans HC. Inhibition of human macrophage colony formation by interleukin 4. J. Exp. Med. 170:577–582, 1989.
106. Sallusto F, Cella M, Danieli C, Lanzavecchia A. Dendritic cells use macropinocytosis and the mannose receptor to concentrate macromolecules in the major histocompatibility complex class II compartment: Downregulation by cytokines and bacterial products. J. Exp. Med. 182:389–400, 1995.
107. Snyder DS, Beller DI, Unanue ER. Prostaglandins modulate macrophage Ia expression. Nature 299:163–165, 1982.
108. Gentile PS, Pelus LM. In vivo modulation of myelopoiesis by prostaglandin E2. II.Inhibition of granulocyte-macrophage progenitor cell (CFU-GM) cycle rate. Exp. Hematol. 15:119–126, 1987.
109. Pelus LM, Ottmann OG, Nocka KH. Synergistic inhibition of human marrow granulocyte-macrophage progenitor cells by PgE and recombinant interferon-α, -β, and -γ and an effect mediated by tumor necrosis factor. J. Immunol. 140:479–484, 1988.
110. Nussler AK, Billiar TR. Inflamation, immunoregulation, and inducible nitric oxide synthase. J. Leukoc. Biol. 54:171–178, 1993.
111. Lu L, Bonham, Chambers F, Watkins S, Hoffman R, Simmons D, Thomson A. Induction of nitric oxide production by mouse dendritic cells in response to interferon-γ, endotoxin and interaction with allogeneic T cells: Induction of nitric oxide synthase is associated with dendritic cell apoptosis. Submitted.
112. Inaba K, Steinmann RM, Pack MW, Aya H, Inaba M, Sudo T, Wolpe S, Schuler G. Identification of proliferating dendritic cell precursor in mouse blood. J. Exp. Med. 175:1157–1167, 1992(b).
113. Szabolcs P, Moore MAS, Young JW. Expansion of immunostimulatory dendritic cells among the myeloid progeny of human CD34+ bone marrow precursors cultured with c-*kit* ligand, GM-CSF, and TNF-α. J. Immunol. 154:5851–5861, 1995.
114. Romani N., Gruner S., Brang D. et al. Proliferating dendritic cell progenitors in human blood. J. Exp. Med. 180:83–93.

115. Sallusto F, Cella M, Danieli C, Lanzavecchia A. Dendritic cells use macropinocytosis and the mannose receptor to concentrate macromolecules in the major histocompatibility complex class II compartment: Downregulation by cytokines and bacterial products. J. Exp. Med. 182:389–400, 1995.
116. Galy A., Travis M., Cen D., Chen B. Human T, B, Natural Killer, and Dendritic Cells arise from a common bone marrow progenitor cell subset. *Immunity* 3:459–471, 1995.
117. Maraskovsky E., Brasel K., Teepe M., Roux E., Shortman K., Lyman S.D., McKenna H.J. In vivo administration of flt3 ligand results in the generation of large numbers of dendritic cells in the lymphoid tissue of mice. In press.
118. Porgador A. and Gilboa, E. Bone-Marrow-Generated Dendritic Cells Pulsed with a Class I-Restricted Peptide are Potent Inducers of Cytotoxic T Lymphocytes, *J. Exp. Med.* 182: 255–260, 1995.
119. Grabbe, S. Tumor Antigen Presentation by Murine Epidermal Cells, *J. Immunol.* 146: 3656–3661, 1991.
120. Mayordomo, J.I., Loftus, D.J., Sakamoto, H., Lotze, M.T., Strokus, W.J., Appella, E., and DeLeo, A.B. THerapy of murine tumors with p53 wild-type and mutant sequence peptide-based vaccines. *J. Exp. Med.* 183: 1357–1365, 1996.
121. Zitvogel, L., Mayordomo, J.I., Tjandrawan, T., DeLeo, A.B., Clarke, M.R., Lotze, M.T., and Storkus, W.J. Therapy of Murine Tumors with Tumor Peptide Pulsed Dependence on T-cells, B7 Costimulation, and Th1-Associated Cytokines, *J. Exp. Med.* 183: 87–98, 1996.
122. Kerrebijn, J.D., *et al.* Macrophage and Dendritic Cell Infiltration in Head and Neck Squamous-Cell Carcinoma; An Immunohistochemical Study, *Cancer Immunol. Immunother.* 38: 31–37, 1994.
123. Zeid, N.A. and Muller, H.K. S100 Positive Dendritic Cells in Human Lung Tumors Associated with Cell Differentiation and Enhanced Survival, *Pathol.* 25: 338–343, 1993.
124. Gallo, O., *et al.* Langerhans Cells Related to Prognosis in Patients with Larygeal Carcinoma, *Arch. Otolaryngol.* 117: 1007–1010, 1991.
125. Giannini, A., *et al.* Prognostic Significance of Accessory Cells and Lymphocytes in Nasopharygeal Carcinoma. *Pathol. Res. Pract.* 187: 486–502, 1991.
126. Tjandrawan, T., Maeurer, M.J., Castelli, C., Lotze, M.T., and Storkus, W.J. Autologous Dendritic Cells Pulsed with MART-1/Melan-A, gp100, tyrosinase, or MAGE-3 Peptides Elicit Anti-Melanoma CTL From both Normal Donor and Cancer Patient Peripheral Blood Lymphocytes *In Vitro*. Submitted for publication, 1996.
127. Hsu, F.J., Benike, C., Franjoli, F., Lilies, T.M., Tzerwinsky, T., *et al.* Vaccination of patients with B cell lymphoma using autologous antigen-pulsed dendritic cells. *Nature Med.* 2: 52–55, 1996.
128. F. Sallusto and A. Lanzavecchia, Efficient Presentation of Soluble Antigen by Cultured Human Dendritic Cells is Maintained by Granulocyte/Macrophage Colony-Stimulating Factor Plus Interleukin 4 and Downregulated by Tumor Necrosis Factor Alpha, *J. Exp. Med.* 179: 1109–1118, 1994.
129. Romani, N., Gruner, S., Brang, D., Kampgen, E., Lenz, A., *et al.* Proliferating dendritic cell progenitors in human blood. *J. Exp. Med.* 180: 83–93, 1994.
130. Bakker, A.B., Marland, G., De Boer, A.J., Danen, H., Adema, G.J., and Figdor, C.G. Generation of antimelanoma cytotoxic T lymphocytes from healthy donors after presentation of melanoma-associated antigen-derived epitopes by dendritic cells *in vitro*. *Cancer Res.* 55: 5330–5339, 1995.
131. Storkus, W.J., and Lotze, M.T. Biology of Tumor Antigens: Tumor Antigens Recognized By Immune Cells. In: V. DeVita, S. Hellman, and S. Rosenberg (Eds.), Biologic Therapy of Cancer, 2nd. ed. J.B. Lippincott Company, Philadelphia, 1995. pp. 64–77.
132. Zitvogel, L., Couderc, B., Mayordomo, J.I., Robbins, P.D., Lotze, M.T., and Storkus, W.J. IL-12 engineered dendritic cells serve as effective tumor vaccine adjuvants *in vivo*. *New York. Acad. Sci.*, in press, 1996.

IDENTIFICATION AND ISOLATION OF CD1a POSITIVE PUTATIVE TUMOUR INFILTRATING DENDRITIC CELLS IN HUMAN BREAST CANCER

Brendon J. Coventry,[1] Jonathan M. Austyn,[2] Steve Chryssidis,[1] Debra Hankins,[2] and Adrian Harris[3]

[1]Department of Surgery
University of Adelaide
Royal Adelaide Hospital
North Terrace, Adelaide, South Australia, 5000
[2]Nuffield Department of Surgery and
[3]Department of Clinical Oncology
University of Oxford
Headington, Oxford, United Kingdom, OX39DU

1. SUMMARY

Identification of dendritic cells (DC) in human tissues has been technically problematic due to the lack of truly specific immunohistochemical markers for DC's . Human dendritic cells express CD1a glycoprotein at certain points in their life cycle. CD1a positive cells are present in many human tumours and have been associated with improved survival. However, little information exists concerning the separation of DC from human tumours. The current study reports that human breast carcinomas have low densities of CD1a positive cells with dendritic morphology, and details are shown of a technique for successful separation of these cells from tumour tissues.

2. INTRODUCTION

T-cell recognition of antigens expressed on tumour cells, leading to MHC restricted tumour cell killing has been demonstrated both in-vitro and in-vivo[1,2]. Dendritic cells are potent accessory cells, important in initiating primary immune responses through presentation of antigens and co-stimulatory signals culminating in T cell activation. Failure of co-stimulation may arise from low DC numbers within tumours or reduction in the capacity of DC to effectively present antigen to T-cells. Increased intra-tumoural dendritic cell infiltrates have been reported to be associated with improved survival for thyroid carcinomas, colorectal cancers, lung carcinomas and gastric carcinomas[3–6]. The presence of DC in breast carcinomas appears not to have been documented. The relationship between DC, immunological response and tumour growth is currently unclear as are the possible mechanisms behind these observations. Progressive tumour growth is associated with loss of DC, and tumour regression is associated with increased DC density[7–9].

The CD1a molecule, a member of the CD1 antigen family, is a non-polymorphic MHC class I-related cell surface glycoprotein expressed in association with Beta 2-microglobulin[10]. It is expressed in cortical thymocytes, DCs/LCs, and interdigitating dendritic (reticulum) cells as well as some activated B-cells and macrophages[10]. CD1a has the limitation of not being specific for DC's, although it is one of the only markers for DC identification and has been extensively used. DC's appear to express CD1a at different stages of development, but not during other stages[9,11].

The aims of the studies were to (1) investigate the density and distribution of the dendritic cell population in invasive ductal breast carcinomas using CD1a antibody and (2) separate CD1a positive cells from human breast carcinomas. To our knowledge, the separation of DC from human tumours has not been reported in the literature. Separation of CD1a cells from the tumour microenvironment may provide a source of DC which may be able to be cultured and investigated for their antigen presenting capacity.

3. METHODS

3.1. Primary Antibodies

See Table 1.

3.2. Part I: CD1a Expression in Breast Tumours

21 infiltrating ductal breast carcinomas (IDC) were frozen then sectioned and immunostained using monoclonal antibodies and high sensitivity di-amino-benzidine tech-

Table 1. The primary antibodies used for Parts I and II

CD	Antibody	Dilution	Origin
Part I			
CD1a	M721	1/1000 (IgG2a)	Dako Ltd., Copenhagen
CD3	Leu4	1/60 (IgG1)	Becton Dickinson
CD14a	FMC32	neat supernatant	H. Zola, Adelaide
negative control	X63	1:100 (IgG1, IgG2a)	ATCC
Part II			
CD1a	NA1/34	neat	McMichael, Oxford
CD68	Y-1/82A	neat	Mason, Oxford
CD45	F10894	1/200	Fabre, Oxford
Class II	L203	1/100	Lampson et al.

Controls - without primary antibody; without secondary antibody (Bu22 IgG1 irrelevant antibody as primary); and without Avidin/Biotin complex. Human Tonsil was used for positive control tissue.

niques[12,13]. Stained cells were counted and averaged per 50 high power fields (x400). 3 normal breast specimens were also examined in the same way.

3.3. Part II: CD1a Expression and MNC Separation in Breast Tumours

3.3.1. Tissues. 12 IDC were collected in the operating theatre and cut under sterile conditions. Tissue was: (i) frozen for sectioning and immunostaining, and (ii) disaggregated using sterile mechanical and enzymatic methods.

3.3.2. Immunostaining. Blocking using human AB serum and 3% H_2O_2 in water, were performed. Avidin-biotin complex (Vector, Elite) AEC/H_2O_2 was used.

3.3.3. Cell Separation Technique. Tumour infiltrating mononuclear cells were separated using a method developed from previously described techniques[14–16].

1. Tumour was incubated for 45mins at 37°C with: Collagenase, trypsin, DNAse mixture (10mg Collagenase dissolved in 4ml RPMI [Sigma C-2139, 405 units/mg], 1ml DNAse solution [1mg Sigma D-0751, 2000Kunits /mg DNAse in 10 mls RPMI]; and 5ml Trypsin solution (2mg/ml) and dilute with 11.5ml RPMI = 2mg/ml], then washed and filtered through a sterile stainless steel strainer with cold Ca^{++}/Mg^{++} free RPMI.
2. Incubation with EDTA solution (10ml) for 10mins at 37°C and washed.
3. Incubation with Collagenase solution (20mg in 10ml) for 20mins at 37°C.
4. The tumour residuum was retained in the filter then pushed through the filter using a 5ml syringe plunger, until no tumour tissue remained.
5. R10 was used to wash enzyme solution, then centrifugation for 7mins @ 1200RPM, followed by resuspension in 1 ml R10. Viability was assessed using trypan blue. Dilutions to 1–2 x 105 cells /ml were made for cytospins. Tissue section and cytospin immunostaining of was assessed by two investigators (JMA, BJC).

4. RESULTS

4.1 Part I: CD1a Expression in Breast Tumours

4.1.1 CD1a Staining. Background levels of staining were minimal or absent, and crisp positive staining of DCs was observed. CD1a expression in normal breast tissue was either absent or only a few cells were seen in the entire section.

4.1.2. FMC 32, CD68, and Leu4. FMC32 (and CD68) labelled both monocytes and macrophages, and spatial correlation with other markers was used to determine the nature of cells labelled. In tissue sections CD1a and FMC32/CD68 labelled different cell populations. However, a spatial association was noted between CD1a and the CD3 (Leu4) positive T-cell population within the breast tumours.

4.2. Part II: CD1a Expression in MNC Separated from Breast Tumours

The separated mononuclear TIL assessed in each case with trypan blue dye exclusion demonstrated viabilities of greater than 95% in all cases.

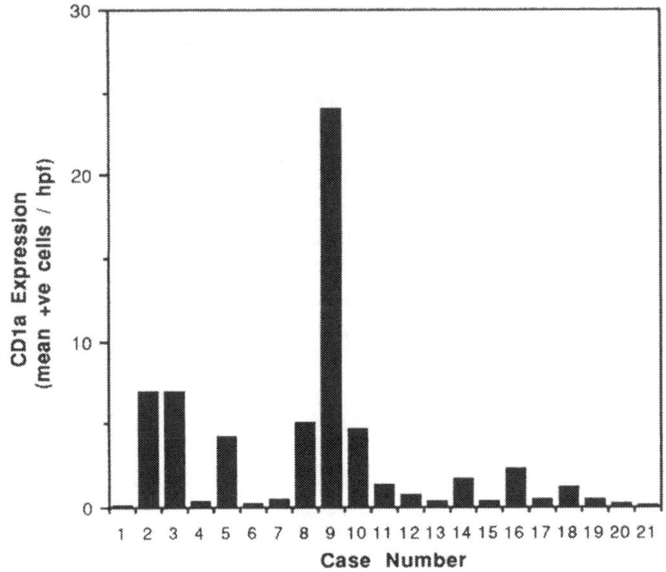

Graph 1. Comparison of DC grid counts for each case (cell bodies averaged per 50 high power fields x400 magnification).

4.2.1. Tissue Section and Separated Cell Immunostaining.
See Table 3.

5. DISCUSSION

Dendritic cells are potent accessory cells, important in initiating primary immune responses through presentation of antigens and co-stimulatory signals culminating in T cell activation. Co-stimulation of T-lymphocytes occurs through the binding of the B7 molecules

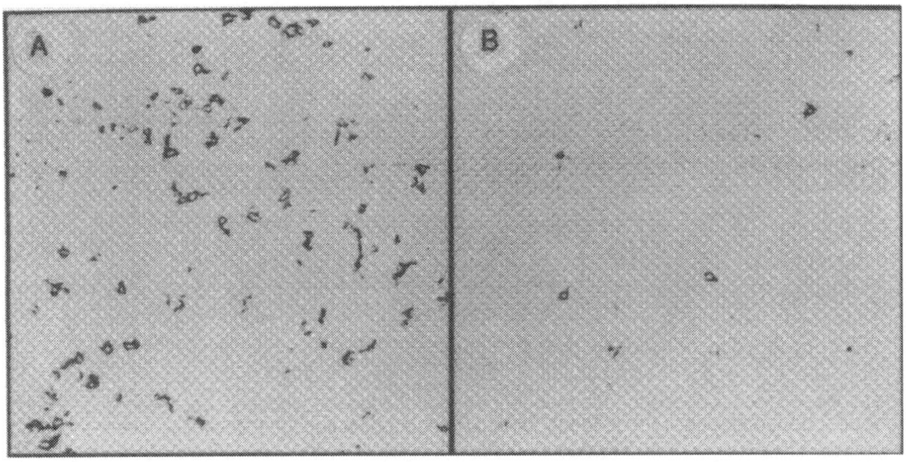

Figure 1. CD1a cell density in breast cancers. A: maximum density; B: typical density (X400).

Table 2. Mononuclear cell extraction efficiencies

Tumour	Total MN cells (cells/ml)	Sample weight (g)	Cells/g
CA1	45×10^4	0.5g	90×10^4
CA2	179×10^4	0.41g	436×10^4
CA3	792×10^4	0.71g	1115×10^4
CA4	44×10^4	N/A	
CA5	376×10^4	N/A	
CA6	1039×10^4	N/A	
CA7	3608×10^4	N/A	
CA8	302×10^4	N/A	
CA9	600×10^4	1.68g	357×10^4
CA10	316×10^4	0.9g	351×10^4
CA11	272×10^4	1.5g	181×10^4
CA12	142×10^4	0.39g	364×10^4

[N/A = Not Available]
Mean Cell Extraction/Gram of Tumour = 413×10^4
Range: 90 - 1115 $\times10^4$ Cells/Gram of Tumour

present on the DC surface to CD28 or CTLA4 molecules on the T cell surface at the same time as TCR binding to Class II molecules on the DC surface[17]. Recent information suggests that IL-2 production and IL-2R expression by tumour infiltrating T cells is defective in human breast cancers, which may be related to poor antigen presentation and incomplete co-stimulation from DC[18]. Deficient antigen processing and presentation has been recently proposed as a mechanism by which tumour antigens are not recognised by T-lymphocytes within the tumour microenvironment[19,20]. Failure of co-stimulation may arise from low DC numbers within tumours, reduction in the capacity of DC to effectively present antigen to T cells or down-regulation of DC function. Clinical evidence for this is provided by observations that induction of infiltration of DC into gastric cancers is associated with improved survival[21], and studies in

Table 3. Comparison of CD1a positive cells in (intact) tumour sections (TS) and in cytospins (CS) of extracted (disaggregated) cells, respectively, from the same breast tumour

Tumour	TS	CS
CA1	+	+
CA2	++	+/-
CA3	+	+
CA4	+	+
CA5	++	+
CA6	N-A	+
CA7	+	+
CA8	+/-	+/-
CA9	N-A	+
CA10	+	+
CA11	+	+/-
CA12	+	+

(N-A not assessed)
[Note: This represents a qualitative assessment of CD1a expression and is only semi-quantitative].

tumours (other than breast cancers) have shown a direct relationship between CD1a (DC) number within tumours and improved survival[3–6].

These studies have shown that CD1a positive putative dendritic cells are present in most human breast cancer tissues but in small numbers, and that CD1a positive cells can be isolated from breast carcinomas using a combination of mechanical and enzymatic techniques. However, in the present study it is not possible to define a specific density of DC critical for generation of an effective anti-tumour immune response. This aspect may be possible to evaluate when clinical survival and recurrence data are available at 5 and 10 years post-surgery. TIL viabilities indicate that culture of CD1a positive DC from the tumour microenvironment of breast cancers should be possible unless these cells are terminally differentiated and unable to divide. Culture of DC would then allow more detailed investigation of the functional role of DC in tumour antigen presentation and co-stimulation of T-lymphocytes in the anti-tumour immune response.

6. ACKNOWLEDGMENTS

These studies were supported by the Royal Australasian College of Surgeons (Marjorie Hooper Fellowship), and the Anti-Cancer Foundation of the Universities of South Australia. The authors acknowledge Angela Neville, Evy Hillenbrand (technical support), Prof Peter Morris and Prof David Mason (antibodies and advice) for their help.

7. REFERENCES

1. Cox, AL, Skipper J, Chen Y, Henderson RA, Darrow TL, Shabonowitz J, Engelhard VH, Hunt DF, Slingluff CL (1994) Identification of a peptide recognised by five melanoma specific human cytotoxic T-cell lines. Science 264: 716–719; .
2. Baxevanis, C.N., G.V. Dedoussis, N.G. Papodopoulos , and I. Missitzis (1994). Tumour specific cytolysis by tumour infiltrating lymphocytes in breast cancer. Cancer 74(4):1275–1282.
3. Schroder S, Schwartz W, Rehpenning W, Loning T, Bocker W (1988) Dendritic/Langerhans cells and prognosis in patients with papillary thyroid cancers. Am J Clin Pathol. 89: 295–300.
4. Ambe K., Mori M., Enjoji M. (1989) S100 protein-positive dendritic cells in colorectal adenocarcinomas. Cancer, 63: 496–503.
5. Zeid, NA; Muller, HK (1993) S100 positive dendritic cells in human lung tumours associated with cell differentiation and enhanced survival. Pathology 25: 338–343.
6. Tsujitani S., Furukawa T., Tamada R., Okamura T., Yasumoto K., Sugimachi K. (1987) Langerhans cells and prognosis in patients with gastric carcinoma. Cancer, 59: 501–505.
7. Knight S.C., Stagg A., Hill S., Fryer P., Griffiths S. (1992) Development and function of dendritic cells in health and disease. J. Invest. Dermatol., 99: 33S-38S.
8. Tazi A., Bouchonnet F.. Grandsaigne M., Boumsell L., Hance A.J., Soler P. (1993) Evidence that granulocyte macrophage-colony-stimulating factor regulates the distribution and differentiated state of dendritic cells/langerhans cells in human lung and lung cancers. J. Clin. Invest. 91: 566–576.
9. Austyn J.M. (1993) The Dendritic cell system and anti-tumour immunity. In Vivo, 7: 193–202.
10. Krenacs L., Tiszalvicz L., Krenacs T., Boumsell L. (1993) Immunohistochemical detection of CD1a antigen in formalin-fixed and paraffin-embedded tissue sections with monoclonal antibody 010. J. Pathol., 171: 99–104.
11. Toriyama K., Wen D.R., Paul E., Cochran A.J. (1993) Variations in the distribution, frequency, and phenotype of Langerhans cells during the evolution of malignant melanoma of the skin. J. Invest. Dermatol., 100: 269S-273S.
12. Coventry, BJ; Neoh, S; Mantzioris, B; Skinner, JM; Bradley, J; Zola, H (1994) Comparison of the sensitivity of immunoperoxidase staining methods with high-sensitivity fluorescence flow cytometry - antibody quantitation on the cell surface. J. Histochem. Cytochem. 42 (8): 1143–1149.

13. Coventry, BJ; Bradley, J; Skinner, JM (1995) Differences between standard and high-sensitivity immunoperoxidase staining methods in tissue sections - Comparison of immunoperoxidase staining methods using computerised video image analysis. Pathology 27: 221–224..
14. Inaba, K; Schuler, G; Witmer, MD; Valinski, J; Atassi, B; Steinman, RM (1986) Immunologic properties of purified epidermal Langerhans cells. J. Exp Med. 164: 605–613.
15. Lenz, A; Heine, M; Schuler, G; Romani, N (1993) Human and murine dermis contain dendritic cells. J. Clin. Investig. 92: 2587–2596.
16. Austyn, J; Hankins, DF; Larsen, CP; Morris, PJ; Rao, AS; Roake, JA (1994) Isolation and characterisation of dendritic cells from mouse heart and kidney. J Immunol. 152: 2401–2410.
17. Schwartz, RH (1992) Costimulation of T lymphoctes: The role of CD28, CTLA-4, and B7/BB1 in interleukin-2 production and immunotherapy. Cell 71: 1065–1068.
18. Coventry, BJ; Weeks, S;Heckford S;Sykes P; Skinner, J; Bradley, J (1996) Lack of interleukin-2 (IL-2) cytokine expression despite IL-2mRNA transcription in tumour infiltrating lymphocytes in primary human breast carcinoma : selective expression of early activation markers. J Immunol. 156: 3486–3492.
19. Chen, L; Linsley, PS; Hellstrom, KE (1993) Costimulation of T cells for tumour immunity. Immunol. Today 14(10): 483–486.
20. Restifo, NP; Esquivel, F; Kawakami, Y; Yewdell, JW; Mule, JJ; Rosenberg, SA; Bennink, JR (1993) Identification of human cancers deficient in antigen presentation. J. Exp. Med. 177: 265–272.
21. Tsujitani S., Okamura T., Baba H., Korenaga D., Haraguchi M., Sugimachi K. (1988) Endoscopic intramural injection of OK-432 and Langerhans' cells in patients with gastric carcinoma. Cancer, 61: 1749–1753.

DENDRITIC CELLS AND INTERLEUKIN 12 AS ADJUVANTS FOR TUMOR-SPECIFIC VACCINES

Ursula Grohmann,[1] Roberta Bianchi,[1] Maria Laura Belladonna,[1] Silvia Silla,[1] Daniela Surace,[1] Maria Cristina Fioretti,[1] and Paolo Puccetti[2]

[1]Department of Experimental Medicine
University of Perugia
Perugia, Italy
[2]University of Tor Vergata
Rome, Italy

1. DENDRITIC CELLS AS A MAJOR TOOL IN TUMOR IMMUNOTHERAPY

The demonstration of autologous CD8 cytotoxic T lymphocytes (CTL) with specificity for class I MHC-restricted tumor antigens is a cornerstone in the field of tumor immunology and immunotherapy, raising the critical issue of developing effective means of increasing the frequency of tumor-specific CTL and translating a trace repertoire into effective immune responses[1]. Dendritic cells (DC) are unmatched in their ability to present antigen to naive T cells in a way that activates the immune response[2]. Recent work in our laboratory has begun to address this challenge by using DC to elicit antigen-specific CD8 cell responses in vivo in mice[3–5]. In particular, our model system makes use of DC pulsed in vitro with a well-characterized peptide related to the tumor rejection antigen P815AB[6] of mastocytoma P815, and the tumor peptide-loaded DC are transferred into recipient mice, that are monitored for development of CD8 cell reactivity by a skin test assay.

2. USE OF A SKIN TEST ASSAY TO MONITOR CLASS I-RESTRICTED REACTIVITY IN VIVO TO TUMOR PEPTIDES AFTER INJECTION OF PEPTIDE-PULSED DC

Analogous to the occurrence of class II-restricted (classical delayed-type hypersensitivity, DTH) responses, skin test reactivity can be elicited in the footpads of immunized mice by local challenge with synthetic, class I-restricted peptides administered in saline[3]. Regardless of whether or not the eliciting peptide contains class II-restricted epitopes, its local biological half-life, which is limited in time, permits presentation only in association with MHC class I

Dendritic Cells in Fundamental and Clinical Immunology
edited by Ricciardi-Castagnoli, Plenum Press, New York, 1997

molecules, thus resulting in the selective activation of CD8 but not CD4 T lymphocytes. However, for effective priming to occur in the afferent induction of reactivity, the class I-restricted tumor peptide requires that DC co-present class II-restricted epitopes, that can be provided by unrelated (helper) peptides[4]. Thus, priming of DC with P815AB in a physical mixture with a tetanus toxin peptide (tt), but not with P815AB alone, will initiate P815AB-specific skin test reactivity mediated by CD8 cells. These findings, in line with recent data on DC as adjuvants for class I-restricted antitumor activity[7], make tumor immunotherapy with peptide-pulsed DC an attractive possibility, yet point to the possible requirements for MHC class II helper epitopes to implement class I MHC antigen-pulsed DC immunization strategies. In a more general context, several questions still remain to be answered regarding the possible use of DC as adjuvants in tumor immunotherapy. These include: a) what is the requirement in terms of costimulatory molecules and cytokines to obtain peptide-specific CD8 cell-mediated reactivity *in vivo* by DC priming? b) is an effective priming by peptide-pulsed DC accounted for by the overcoming of an otherwise insufficient immune response or does it involve the overcoming of tumor-specific anergy and/or suppressor activity? c) can peptide-pulsed DC induce specific T cell anergy rather than activation under inappropriate priming conditions? d) what is the longevity of the response, whether functional or tolerogenic, induced by peptide-pulsed DC?

2.1. Requirements for Costimulatory Molecules and Cytokines: The Role of IL-12

While our early studies using P815AB-pulsed DC to confer class I-restricted DTH on recipient mice had demonstrated the need for the co-presence of helper epitopes such as those

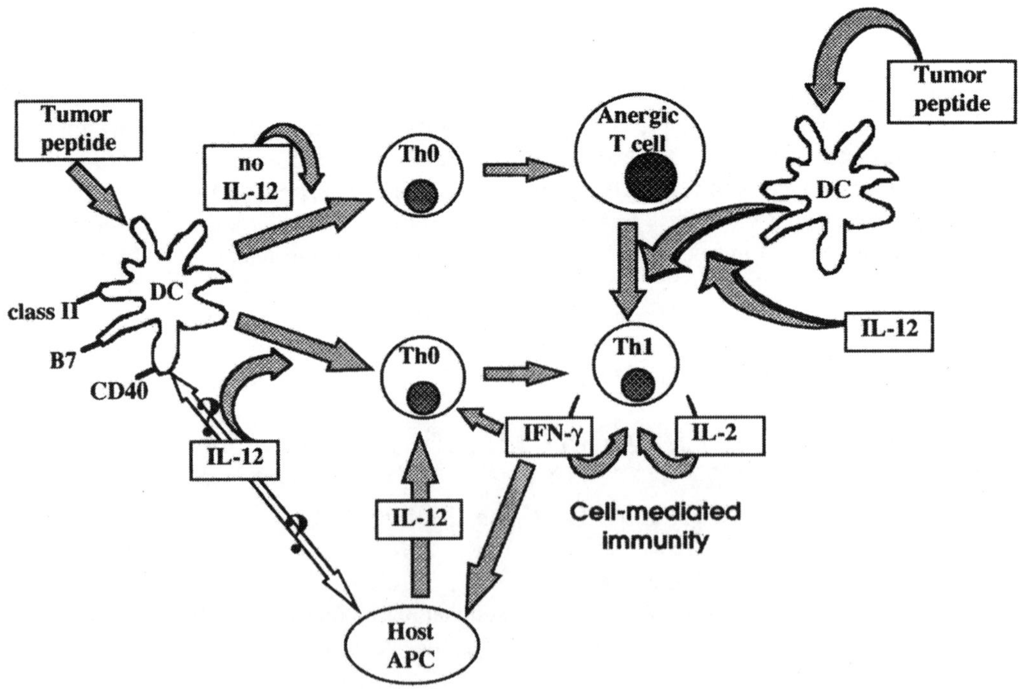

Figure 1. IL-12 effects on host priming with tumor peptide-pulsed DC.

provided by the tt peptide[4], recent data have shown that the tt peptide can be replaced by recombinant IL-12. Exposure of DC in vitro to IL-12 or administration of the cytokine in vivo, or a combination of both, can confer strong DTH on mice transferred with DC pulsed with P815AB alone[5]. A series of experiments aimed at providing a mechanistic explanation for these effects led us to the following conclusions: a) the exogenous IL-12 requires both CD4 and CD8 cells for activity; b) the immune response initiated by IL-12 relies on later production of IL-12 by the host (Fig. 1); c) the early adjuvanticity of exogenous IL-12 involves increased presentation and/or recognition of class II-restricted epitopes of P815AB. In particular, relative to this last point, recent experiments seem to indicate a dual role of IL-12 in promoting T cell reactivity: a portion of its effect, mostly affecting peptide presentation by DC, may be mediated by IFN-γ, while other effects are induced directly by IL-12 on the responding T cells, leading to increased recognition by P815AB-specific receptor bearing CD4 and CD8 cells[5]. The IFN-γ-dependent effects of IL-12 on antigen presentation may include increased expression of MHC class II and B7 molecules, whereas the effect on T cells (mostly CD4) involves triggering of Th1-biased differentiation.

2.2. Induction of Th1 Differentiation vs Anergy in the Responding CD4 Cells

In principle, the adjuvant activity of DC and IL-12 used in combination could involve the potentiation of an intrinsically insufficient immune response to P815AB. However, recent experiments have provided evidence that the signals received by the T cells on first encountering P815AB may determine whether receptor ligation results in cell activation or induction of P815AB-specific unresponsiveness[8]. This appears to be of particular interest because P815AB is also a self peptide expressed in murine testis and placenta[1,6], and there is a strong similarity between autoimmune and anti-tumor responses from a regulatory perspective. In fact, in the absence of added adjuvanticity such as that provided by a helper peptide or recombinant IL-12, P815AB presented by DC results in a state of specific unresponsiveness (anergy[9]), characterized by a profound inability of CD4 cells to produce IL-2 in vitro[8] (Fig. 1). P815AB-specific anergy is also observed after neutralization of endogenous IL-12 at the time of optimal priming with P815AB plus helper peptide[8]. Therefore, these data on induction of functional CD8 cell responses by P815AB-pulsed DC indicate that the necessity and sufficiency of IL-12 in our system relates not only to an intrinsic deficiency of the developing Th1 response, but also to regulation of a balance between activation and tolerogenic signals acting on T cells. Of note, recombinant IL-12 is capable of reverting the anergic state induced by a tolerogenic priming with P815AB-pulsed DC[8] (Fig. 1).

2.3. Longevity of the Response Induced by DC Priming

Thus, transfer of DC pulsed in vitro with a tumor peptide may either initiate a functional CD8 cell response or induce a state of specific unresponsiveness depending on the presence of costimulatory molecules and cytokines at the time of priming. The durability of either response is of great interest. Early studies in our laboratory using DC and peptide combination (P815AB + tt) showed that skin test reactivity mediated by CD8 cells can be observed for at least two months after DC transfer[3,4]. On assaying, in contrast, T cell unresponsiveness induced by DC plus P815AB alone, we have recently found that the unresponsive state has peculiar kinetics of detection, being maximal at about two weeks but no longer detectable at about six to seven weeks after priming[8]. This is consistent with the notion that the unresponsiveness induced by P815AB alone is a reversible state, as typically occurs in anergy[9]. These data also indicate that anergy induction may contribute to the "poor immunogenicity" of P815AB in vivo. Although representing an effective target for rejection responses in preimmunized mice, neither P815AB expressed on tumor cells nor the synthetic peptide will activate unprimed CD8 cells for in vivo reactivity[4,5]. As this condition (poor immunogenicity) is shared by most tumors, this again emphasizes the need for developing effective immunotherapy strategies, including the use of DC.

3. CONCLUSIONS

In conclusion, our data demonstrate that a synthetic peptide, related to a murine self protein and tumor rejection antigen, P815AB, can result in a reversible state of antigen-specific T cell anergy under priming conditions that are unsuitable for initiating functional responses in vivo, such as host transfer with peptide-pulsed DC without added adjuvanticity. This demonstrates that DC can present antigen in vivo not only in an immunogenic fashion, but also in such a way to induce specific tolerance. The anergic state induced by DC and peptide under the latter conditions involves unresponsiveness in CD8 cells, as detected by skin test assay in vivo, and suppression of IL-2 production by CD4 cells in vitro. Endogenous IL-12 has a key role in determining the outcome and in vivo priming with peptide-pulsed DC, acting both as adjuvant and inhibitor of anergy induction. Exogenous IL-12 not only prevents the onset of tolerance to P815AB, but is also able to revert unresponsiveness to this tumor peptide. On the one hand, our data emphasize the role of DC and IL-12 as adjuvants for class I-restricted antitumor immunity. On the other, they suggest that several precautions are warranted in designing possible DC-based antitumor strategies in humans. First, optimal CD8 cell priming may require the addition of class II helper epitopes. Second, DC can present tumor antigen peptides in a tolerogenic fashion, an occurrence which seems to be obviated by the presence of endogenous or exogenous IL-12.

4. ACKNOWLEDGMENTS

The authors are grateful to Prof. T. Boon for continued support in their studies with tumor-specific peptides and Genetics Institute, Inc., (Cambridge, MA) for the generous gift of recombinant IL-12. This work was supported by ACRO, CNR-Italy and AIRC.

5. REFERENCES

1. Boon T. and van der Bruggen P., J. Exp. Med. 183: 725 (1996).
2. Steinman R. M., Annu. Rev. Immunol. 9: 271 (1991).
3. Puccetti P., Bianchi R., Fioretti M. C., Ayroldi E., Uyttenhove C., Van Pel A., Boon T., and Grohmann U., Eur. J. Immunol. 24: 1446 (1994).
4. Grohmann U., Bianchi R., Fioretti M. C., Fallarino F., Binaglia L., Uyttenhove C., Van Pel A., Boon T., and Puccetti P., Eur. J. Immunol. 25: 2797 (1995).
5. Bianchi R., Grohmann U., Belladonna M. L., Silla S., Fallarino F., Ayroldi E., Fioretti M. C., and Puccetti P., J. Immunol. 157: 1589 (1996).
6. Van den Eynde B., Lethé B., Van Pel A., De Plaen E., and Boon T., J. Exp. Med. 173: 1373 (1991).
7. Young J. W. and Inaba K., J. Exp. Med. 183: 7 (1996).
8. Grohmann U., Bianchi R., Ayroldi E., Belladonna M. L., Surace D., Fioretti M. C., and Puccetti P., submitted for publication.
9. Quill H., J. Immunol. 156: 1325 (1996).

GENERATION OF IN VITRO AUTOLOGOUS HUMAN CYTOTOXIC T-CELL RESPONSE TO E7 AND HER-2/Neu ONCOGENE PRODUCTS USING EX-VIVO PEPTIDE LOADED DENDRITIC CELLS

H. Navabi,[1] B. Jasani,[2] M. Adams,[1] A. S. Evans,[4] M. Mason,[1] T. Crosby,[1] and L. Borysiewicz[3]

[1]Department of Clinical Oncology
Velindre NHS Trust Whitchurch, Cardiff
[2]Department of Pathology
[3]Department of Medicine
[4]Department of Gynaecology
UHW Healthcare NHS Trust, Cardiff

1. INTRODUCTION

Cytotoxic T-cells (CTL) have been shown to be capable of causing tumour-specific cell lysis when primed with tumour-specific antigenic peptides presented in conjunction with haplotype matched cell surface MHC class I molecules. For this approach to be successful for immunotherapy of tumours in vivo, it is essential that it is shown to be capable of generating adequate numbers of tumour-specific CTL either in vivo for active immunisation, or ex vivo for adoptive transfer therapy. A powerful means for ex vivo expansion of CTL has been recently shown to be antigen loaded autologous dendritic cells (DC)[1]. It has also become possible to generate large numbers of these cells from bone marrow or blood derived precursor cells cultured in the presence of GM-CSF and IL4[2]. Furthermore, murine studies have confirmed the efficacy of these cells to evoke significant ex vivo and in vivo responses against tumour specific antigenic peptides, in particular, HPV 16 E7 and HER-2/neu oncogene products[3], thereby creating the possibility of using these antigenic targets[4] for effective immunotherapy of gynaecological cancers such as cervical and ovarian carcinomas in which these tumour associated antigens are expressed with 93%[5] and 30%[6] frequency, respectively.

Dendritic Cells in Fundamental and Clinical Immunology
edited by Ricciardi-Castagnoli, Plenum Press, New York, 1997

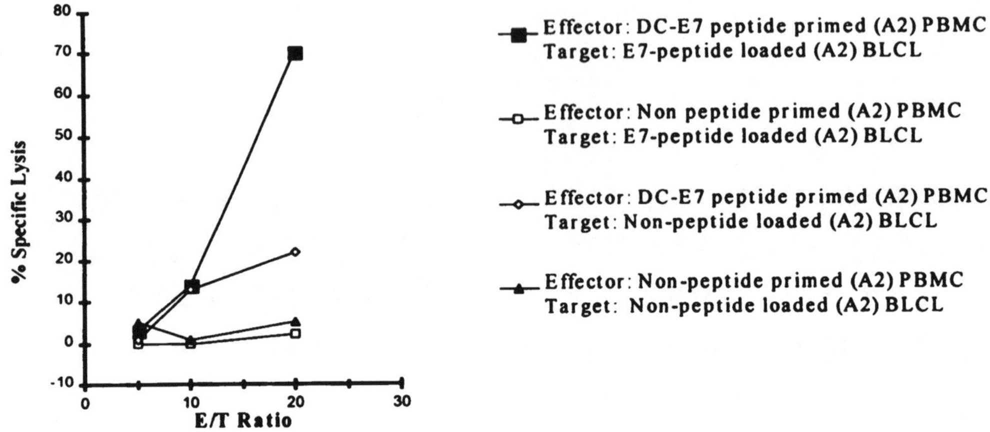

Figure 1. CTL responses using E7 loaded DC.

2. AIM

To assess the ability of ex vivo E7 or HER-2/neu peptide loaded DC to elicit adequate in vitro peptide specific CTL responses against autologous haplotype matched cell targets, as a feasibility study prior to clinical trials.

3. MATERIALS AND METHODS

The autologous dendritic cells were prepared from individual patients using the in vitro culture method of Romani et al (1994) as outlined in table 1. Their overall yield, morphology, immunophenotypology and FACS analysis based cell surface phenotype after 8 days in culture were found to be as described in in tables 2 and 3 and figures 4, 5 and 6 respectively.

The 8 day DC were loaded with 3 different varieties of peptides the sequences of which are shown in table 4. Individual peptide loaded DC were used to prime CTL using

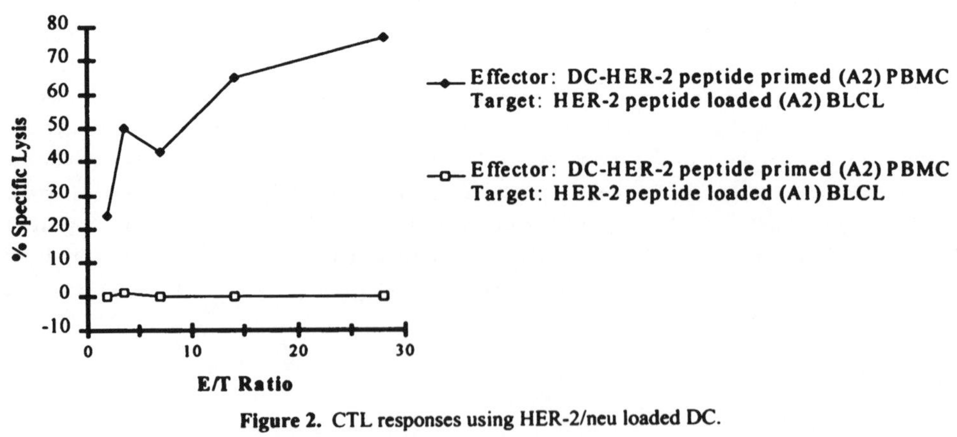

Figure 2. CTL responses using HER-2/neu loaded DC.

Table 1. Preparation of autologous dendritic cells

1. Isolation of peripheral blood monocytes using nycoprep 1.068 gradient centrifugation.
2. Lightly adherent fraction cultured with GMCSF (800 Iu/ml) + IL4 (500 Iu/ml) for 7 - 9 days.
3. Harvest poorly adherent fraction.
4. Identify DC using morphological (figure 1), immunocytochemical (figure 2) and FACS (figure 3) analysis, and determine yield.
5. Evaluate antigen presenting and T-cell priming ability of DC.

Table 2. Dendritic cell yields

Case	% Yield	Initial cell no./ 20 ml blood	Dendritic cell no. (day 9).
1	56%	1.5×10^6	8.3×10^5
2	42%	2.5×10^6	10.8×10^5
3	20%	2.8×10^6	5.6×10^5
4	25%	1.8×10^6	4.6×10^5
5	26%	1.9×10^6	5.1×10^5
6	16%	3.1×10^6	5.0×10^5
7	17%	2.3×10^6	4.0×10^5

1 - 2: Patient receiving radiotherapy for cervical cancer.
3 - 7: Healthy Volunteers.

the method outlined in table 5. The peptide primed CTL were used in CTL cytotoxicity assay (see table 6) which made use of EBV transformed B-cells (see table 7) as the target.

4. RESULTS

The strength and specificity CTL responses obtained using E7 HER-2/neu and flu peptides are graphically represented in Figures 1, 2 and 3. All these peptides were able to cause significant specific CTL responses at 20:1 effector : target ratios.

Table 3. Cell surface phenotype

	Case						
Marker	1	2	3	4	5	6	7
CD80 (B7.1)	+	+	++	+	+	+	+
CD86	+++	++	++	+++	+++	++	+++
HLA-DR (Class II)	++++	+++	+++	+++	+++	+++	+++
CD1a (Class I-like molecules)	+	±	+	±	+	±	+
CD54 (ICAM-1)	+++	++	++	+++	++	++	++
CD68 (monocytes, marophages)	±	+	±	+	+	±	+
CD20 (B-cells)	+2%	+4%	+5%	+7%	+7%	+2%	+2%

1 - 2: Patients receiving radiotherapy for cervical cancer
3 - 7: Healthy Volunteers

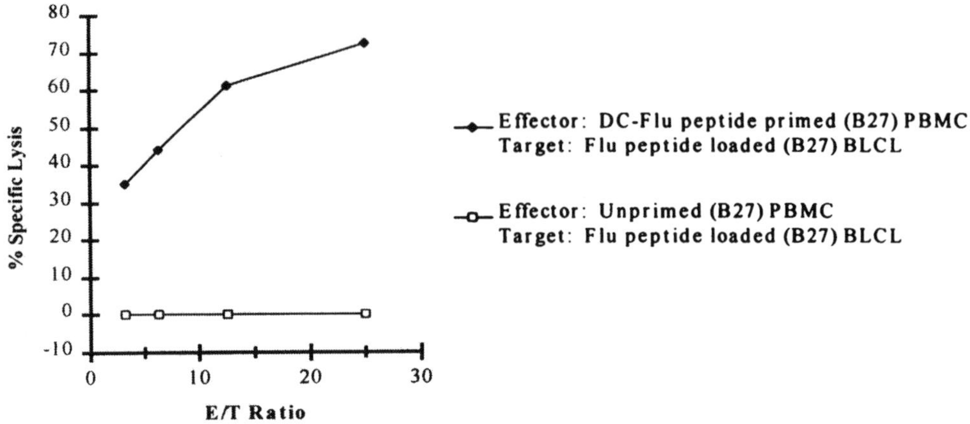

Figure 3. CTL responses using Flu loaded DC.

5. DISCUSSION

It is likely that for peptide-based immunotherapies to be successful that sufficient numbers of tumour specific CTL need to be recruited. Tumour specific CTL may be expanded in vivo in the human by immunisation of the tumour antigen. Borysiewicz and colleagues detected HPV specific CTL[7] in one of three evaluable patients with advanced cervical cancer after immunisation with a live recombinant vaccinia virus expressing the E6 and E7 proteins of the HPV16 and 18 virus. An alternative means of indirectly expanding tumour specific CTL in vivo which may prove to be suited to the clinical situation is to infuse ex vivo tumour antigen peptide pulsed autologous DC as anti-cancer therapy[8]. This approach is supported by recent murine studies[3,9]. In one tumour modal the murine C3 sarcoma which presents HPV16 E7 antigen treatment of animals bearing established

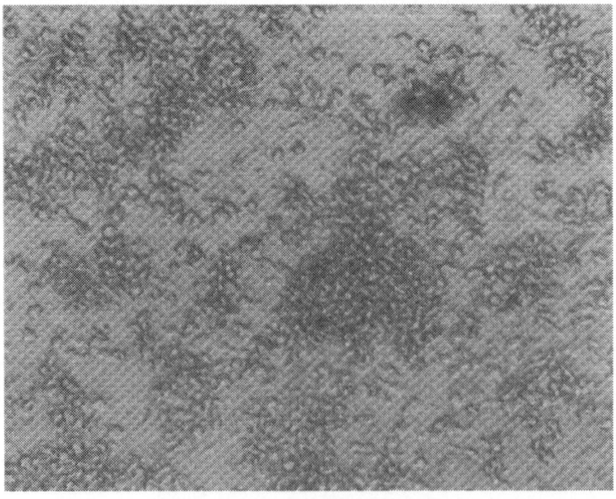

Figure 4. Morphological features of DC: day 8 – case 2.

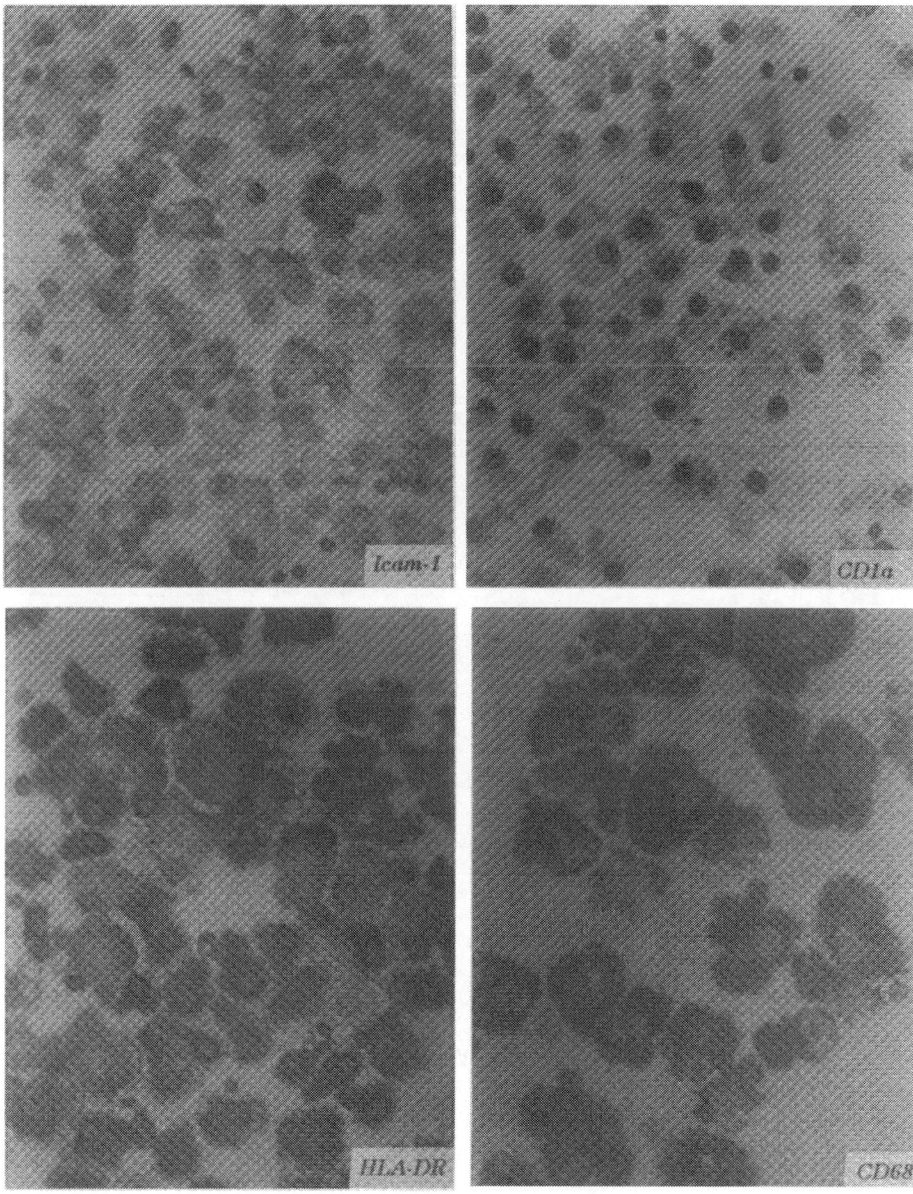

Figure 5. Immunophenotypic features of DC: day 8 – case 4.

macroscopic tumour (up to $1cm^2$) with DC loaded with HPV16 E7 peptide resulted in sustained complete eradication of tumour masses in 80% of mice[3].

The present study further confirms that it is possible to induce Class 1 restricted CTL responses using ex vivo peptide loaded dendritic cells obtained from human peripheral blood progenitors cultured with lymphokines IL4 and GMCSF[2]. Significant numbers of DC have been produced from peripheral blood of normal individual cancer patients. Specific CTL responses MHC Class 1 have been induced using a variety of peptides including flu, HER-2/neu and HPV16 E7. The latter two antigens may represent feasible tar-

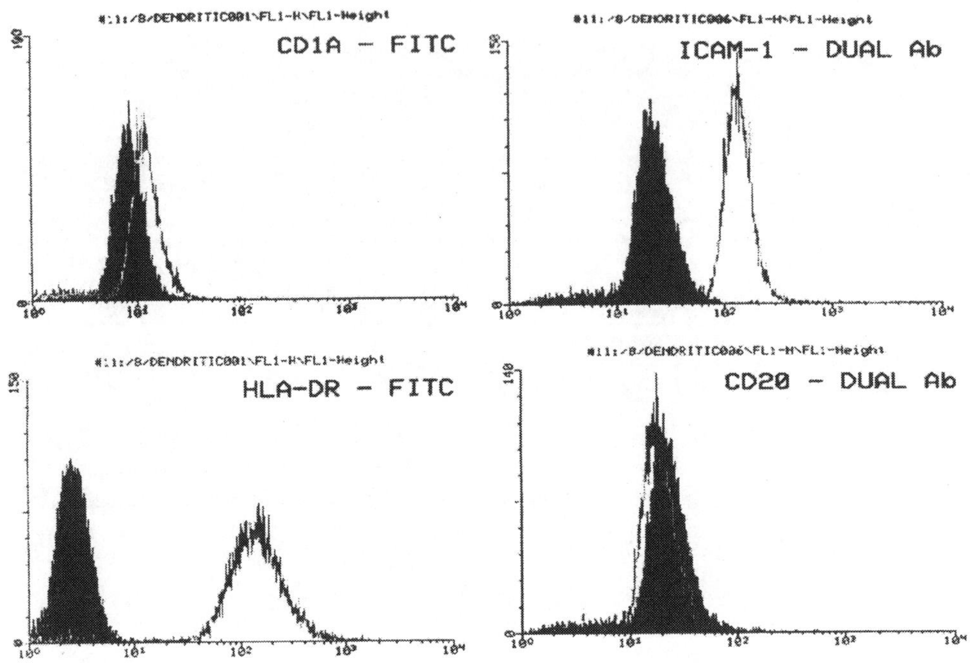

Figure 6. FACS characteristics of DC: day 8 – case 2.

Table 4. Target peptides sequences

Code	Class I	aa	Position
HER-2/neu	A2 specific	KIFGSLAFL	369 - 377
HPV16 E7	A2 specific	YMLDLQPETT	11 - 20
Flu	B27 specific	SRYWAIRTR	383 - 391

Table 5. Antigen specific priming of CTL

1. Load DC with peptide + B2 - microglobulin.
2. Irradiate peptide loaded DC.
3. Co-culture peptide loaded DC with autologous PBMC.
4. Culture in human serum + IL7 for 12 days and restimulate PBMC with peptide loaded DC after 7 days.
5. Add IL2 to culture days 12 - 14.
6. Harvest for CTL day 15.

Table 6. CTL cytotoxicity assay

1. Label + pulse target cells (autologous EBV transformed B cells) with 51Cr and synthetic peptide.
2. Mix target cells with autologous effector cells.
3. Incubate 4 - 6 hours.
4. Count radioactivity.

Table 7. Preparation of EBV transformed B cells as autologous cell targets

1.	Take 10ml of blood into tubes containing preservative free heparin
2.	Separate PBMCs using Histopaque as a density-gradient
3.	Pellet cells and add 1.0ml of EBV (B95.8 cell line) supernatant
4.	Incubate for 1 hour at 37°C
5.	Wash cells with RPMI
6.	Pellet cells resuspend in RPMI + 20% FCS. PHA (1% v/v) and 1% penicillin/streptomycin
7.	Aliquot into wells of a 24 well plate at 1 - 2 × 10^6 cells/well
8.	Culture at 37°C
9.	After two weeks foci of B cells should be visible
10.	Split cells 1:2 as necessary and feed twice weekly by removing half the supernatant and replacing it with fresh medium without PHA

gets for immunotherapy of cervical and ovarian cancer respectively. These data further support the rationale for testing autologous tumour antigen peptide primed DC as a potential means of immunotherapy in gynaecological cancer in clinical study.

REFERENCES

1. Saullusto F, Lanzavecchia A, (1994). Efficient presentation of soluble antigen by cultured human dendritic cells is maintained by granulocyte/macrophage colony stimulating factor plus interleukin-4 and down regulated by tumour necrosis factor alpha. *J Exp Med* 179: 1109–1118.
2. Romani N, Gruner S, Brang D, Kampgen E, Lenz A, Trockenbacker B, Konwalinka G, Fritsch PO, Steinman and Schuler G, (1994). Proliferating dendritic cell progenitors in human blood. *J Exp Med* 180, 83–93
3. Mayordomo JI, Zorina T, Storkus WJ, Zitvogel L, Celluzzi C, Falo LD, Melief CJ, Ildstad ST, Kast WM, Deleo AB, (1995). Bone marrow derived dendritic cells pulsed with synthetic tumour peptides elicit protective and therapeutic anti-tumour immunity. *Nature Medicine* 1(12):1297–1302
4. Dsis ML, Calenoff E, McLaughlin G, Murphy AE, Chen W, Lydon N, McGlynn E, Moe R, Cheever MA, (1994). Existent T cell and antibody immunity to HER-2/neu protein in patients with breast cancer. *Cancer Res* 54:16–22
5. Bosch FX, Mano MM, Munoz N, Sherman M, (1995). Prevalence of human papilloma virus in cervical cancer: a worldwide perspective. *Journal of the National Cancer Institute* 87:796–802
6. Slamon DJ, Godolphin W, Jones LA, Holt JA, Wong SG, Keith DE, (1989). Studies of the HER-2/neu proto-oncogene in human breast and ovarian cancer. *Science* 244:707–12
7. Borysiewicz LK,Fiander A,Nimako, Man S,Wilkinson GWG, Westmoreland D, Evans AS, Adams M,Stacey SN, Boursnell MEG, Rutherford E, Hickling JK, Inglis SC. (1996) A recombinant vaccinia virus encoding human papillomavirus types 16 and 18,E6 and E7 proteins as immunotherapy for cervical cancer. *Lancet* 347:1523–27 .
8. Grabbe S,Beissert,Schwartz and Granstein RD (1995) Dendritic cells as initiators of tumor immune responses: a possible strategy for tumor immunotherapy? *Immunology Today* 16(3)117–121.
9. Gabrilovich DI Nadaf S,Corak J,Berzofsky JA and Carbonne DP (1996).Dendritic cells in anti-tumour immune responses 11 Dendritic cells grown from bone marrow precursors but not mature DC from tumour bearing mice are effective antigen carriers in the therapy of established tumours . *Cellular Immunology* 170, 111–119.

INDEX

SPRINGER NATURE

GPSR Compliance

The European Union's (EU) General Product Safety Regulation (GPSR) is a set of rules that requires consumer products to be safe and our obligations to ensure this.

If you have any concerns about our products, you can contact us on ProductSafety@springernature.com

In case Publisher is established outside the EU, the EU authorized representative is:

Springer Nature Customer Service Center GmbH
Europaplatz 3
69115 Heidelberg, Germany

FSC
www.fsc.org
MIX
Papier | Fördert
gute Waldnutzung
FSC® C083411

Kolibri 360 GmbH
Ferdinand-Jühlke-Straße 7
99095 Erfurt, Deutschland
produktsicherheit@kolibri360.de